Proceedings of the International Conference on Optimization Design

(ICOD 2010)

Wuhan, China

March 18-20, 2010

ASME PRESS

ISBN: 978-0-7918-5958-2

Message from the General Chairs

With the development of science, optimization design has become more and more popular. During the past decades, many works have been done in optimization design and related many different fields. The 2010 International Conference on optimization design (ICOD 2010) will provide a high level international platform for worldwide researchers and engineers to exchange their latest research results in the emergent field of optimization design. We believe this great event will greatly enlighten the academia as well as the industry, accompanied with many new ideas and creative resolutions to optimization design and related topics.

On behalf of the organizing committee, it is our greatest pleasure to welcome you to this great event - the 2010 International Conference on Optimization Design (ICOD 2010) to be held on March 18-20, 2010 in Wuhan, China. This year, ICOD 2010 attracts broad interest throughout the world, with more than 369 submissions through email. After strictly peer reviewing, only 99 papers were selected as final papers. The desire for high quality is also shown by the distinguished invited speaker who will share their insights and recent research results with the attendants. As final note, we would like to take the opportunity to thank all of the individuals who have contributed to this conference, including authors, the reviewers and all the special sessions' organizers.

We would like to thank the program chairs, organization staff, and the members of the committees for their work. Thanks also go to American Society of Mechanical Engineers Press (ASME) for its great publishing service. And, we especially wish to deliver our special thank to Tara Smith for her wonderful service to this proceeding. It is our great pleasure to have you with us at the conference.

ICOD 2010 Conference Chairs
Xingkuan Wu, Wuhan University of Technology
Li Yu, Cardiff University

Message from the Program Chairs

Welcome to the 2010 International Conference on Optimization Design (ICOD 2010). Optimization Design and related research fields play a vital role in today's global economy, and certainly in China. ICOD 2010 bring together Chinese and international researchers to showcase recent research advanced in these focused topics.

We received 369 submissions, among which 99 submissions were accepted papers. Papers reflect research activities at forefront, which topics ranging the full spectrum of optimization design. The program team did an excellent job in soliciting participation, coordinating the review process, and promoting the technical program. We would like to thank every one of them for taking leadership roles in organizing the various aspects of program.

Another important component of the program is the reviewers who volunteered their time and professional expertise. We also would like to take this opportunity to thank every one of them. Last but not least, we would like to thank all the authors for contributing their quality works.

Finally, we welcome you to ICOD 2010 and we hope you will enjoy the program and get a chance to meet new friends and colleagues.

ICOD 2010 Program Chairs
Dongwen Xiang, China University of Geoscience, China
Jing Wu, Huazhong University of Science and Technology

Organizing Committees

Conference Chairs
Xingkuan Wu, Wuhan University of Technology, China
Li Yu, Cardiff University, United Kingdom

Program Chairs
Dongwen Xiang, China University of Geosciences, China
Jing Wu, Huazhong University of Science and Technology, China

Conference Steering Chairs
Hao Xie, Huazhong University of Science and Technology, China
Haoyan Zhao, Lund University, Sweden

Reviewers

Xiaozhou Tang, Wuhan University of Science and Technology, China
Yang Zhang, University of Leeds, United Kingdom
Li Song, Shandong University of Technology, China
Shixue Zhao, Anyang Normal College, China
Yi Han, Inner Mongolia Agricultural University, China
Jiaqi Lv, Cardiff University, United Kingdom
Liqun Guo, Wuhan Engineering University, China
Ruo Zhang, University of Sheffield, United Kingdom
Kang J, University of Sheffield, United Kingdom
Jianian Liang, Wuhan University of Technology, China
Lei zhang, Tongji University, China
Bing Zhu, Stockholm University, Sweden

2010 International Conference on Optimization Design

ICOD 2010

Table of Contents

"Oasis" Eco-park Landscape Design and Rainwater Utilization

Zhaoyang Song
College of Art and Design
Shandong Institute of Commerce and Technology
Jinan, China
Email: szy001@hotmail.com

Yuexia Mao
Planning Department
Real Estate Co., Ltd., Jinan Greenland
Jinan, China
Email: myx18@hotmail.com

***Abstract*—A modern competitive city must be of a good ecological environment and landscape, and. The city's development must be compatible with the natural environment, and the virtuous circle can provide sustainable impetus for the society. The construction of the ecological landscape must conserve enough green space in the future. In order to achieve a good result we integrate the ecological landscapes, rainwater utilization and the natural water cycle. It is believed that numbers of the eco-garden "point" will come together into a green "face", then our city can be turned into a green forest and attractive home.**

Keywords-*eco-landscape; rainwater utilization; landscape design, ecology; City*

I. INTRODUCTION

As China's urbanization development and economic take-off, urban construction and urban landscape has rapidly developed, wide flat streets, large and small squares, uniform green, so that Chinese cities look more beautiful, but the changes of the city's appearance does not bring a corresponding improvement of the urban climate. In recent years, some of China's Cities suffer from the dusty storm and the heavy rain. For instance, on July 18, 2007 a sudden heavy rain hit Jinan City, Shandong Province, the poor city changed from a "Spring city" into a "water city". The streets turned into rivers, underground shopping malls become a big pond. The heavy rain resulted in the great economic losses and heavy casualties. How are our cities? Why do they become so irritable and heartless? With the beautiful face the city is indeed so vulnerable.

The rainwater gardens, ecological landscape design concept is very prevalent in the United States in 1980s and 1990s. Residential areas in some cities are no longer blindly hardened the ground, but the attention to rainwater infiltration and ecological balance. The most famous is the "ArE Siskiyou" reconstruction project in Portland. After the transformation, there is much less pressure on municipal drainage, because runoff water, pond water and groundwater form a good natural water cycle and also a pleasing urban landscape. For another example, professional design 2008 ASLA Honor Award - bordered by a lake channel design (Passage to the Lake) is Michael Van Valkenburgh, a landscape architecture designed by landscape class residential project. The project site in the United States, Maine Stoneham City, is a residential lake access road and landscape design.

The houses take advantages of the lake's natural landscapes, clever design of the road, the greatest degree of hardening and the road avoid the damage to the environment. Lake of the wooden tables in the design simple and beautiful scenery, a variety of landscape design elements are actually used well.

At present, China's construction of a conservation-based Urban Green Space is also strongly advocated, and the so-called saving green space is "the least amount of land, the least water, the least the financial allocations, the least disturbance to the surrounding ecological environment of the green model.

II. "OASIS" ECOLOGICAL GARDEN LANDSCAPE DESIGN

Our case locates at the northern suburbs of Jinan with a total area of 300 acres and slowly tilts from southwest to northeast. Throughout are warm and humid monsoon climate. After the completion of the project set of dining, spa vacation, fishing, cultivating and selling flowers and plants, especially in shallow buried geothermal, mining, low cost, water temperature 52-62℃, the water is clear, rich in lithium, strontium, bromine, iron, etc. minerals and trace elements, are very seductive.

The project includes catering, spa vacation, fishing, nursery places for an integrated operation, including fishing ponds, regional location has been set, position in the land of the south-west, the entire project on the ground gently from northeast to southwest tilt the terrain started to design exactly in line with our idea: the rain from the northeast to southwest flow of such big terrain without the rectification, earthwork greatly reduced. local shape terrain, the basic use of the original terrain. Moreover, in addition to the main road outside the alternative routes are arranged to maximize the use of the permeable ground.

Item terrain is very simple, vertical change little, the highest point and lowest point of elevation difference 1.5 meters, the final decision to the southwest to the northwest side of the pond soil accumulation, in this orogeny. Hot Springs, accommodation, catering peripheral set to form a mountain, water, forest, housing patchwork landscape. Room quiet elegance, hot springs gurgling sounds make us feel pleasure. The main entrance of the venue and the municipal road between a 4-meter wide river, the two sides covered with reeds, weeds. We make use of them finishing in the river side of the construction of a permeable pavement for leisure, and to river water and the park connecting the water to keep part of the original style and form of the original venue for the continuation of a memory.

III. WATER, LANDSCAPE, RAIN WATER UTILIZATION AND DESIGN

On the surface rainwater collection and use of rain water to form the core of the landscape is the case. Our design is three-tiered design of the main road. Rainwater from the surface into the cave shrub planting areas and trees, after the tree hole, planting areas, after filtering sediment flow gullies, gullies flow through the ladder-shaped pond, and the gullies design "Water Curb Stone," the bottom gullies cobblestone BEDDING, beautiful and practical. The rain water is collected and flows into the ditch and the pond as the following flow chart:

Tree Point----→
Rain-→ Road-→ Ditch→Pond→ Discharge
Growing areas→

Fig. 1 Graphic Design road mop

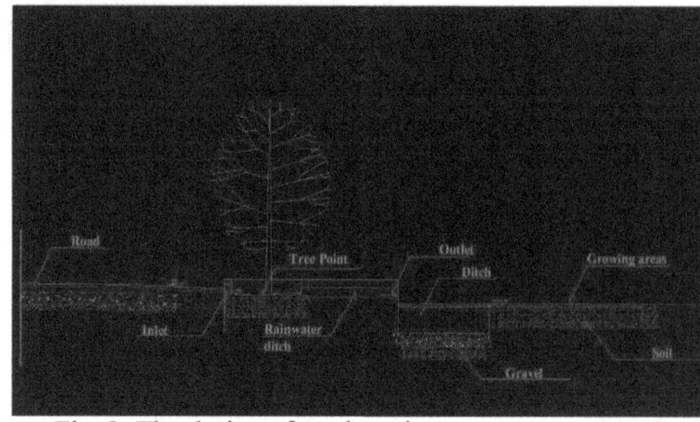

Fig. 2 The design of road sections

Main roads are divided into three levels: a main street planting area to flowers, evergreen shrubs and ornamental trees, mainly; the trees in order to position the main beautiful Acer mono Maxim; gullies, to aquatic plants and ornamental fish-based. Hardening the road, building square, planting, the permeable ground, gullies, and ponds staggered horizon, gullies with Chinese-style stone arch bridge connecting the two sides, ponds surrounded by a wooden plank road, river shuttle, static and dynamic binding, creating the most poetic scenery.

In the cities, the presence of lakes will be greatly improved micro-climate, Hangzhou West Lake, Dian-chi Lake in Yunnan Province are examples of the sea in Qingdao as well. Jinan, the high terrain south of the northern low-vary greatly, we can consider the design of expansion "of Da-ming Lake," and to increase water storage lakes. This can accumulate up surface water, so the rain can improve the situation, affect the micro-climate in Jinan, and form a virtuous circle. The airscape of "Oasis" Ecological Garden is shown in Fig.3

Fig.3 Airscape of "Oasis" Ecological Garden

IV. CONCLUSION

The development of the city needs a good climate and natural cycles, human and the natural environment will require a balance between the need of ecological landscape and improvement of the natural water cycle. People, nature, need to coordinate the development of the land is the urban landscape of the context, there is continued momentum, as people's yearning for a place.

Of course, each city has different features, advanced ideas and geographical characteristics of a combination is an effective approach. Plants, landscape I have with the local culture and philosophy of a unified, simple design on the outside, somewhere they might not receive the public recognition. A rain garden might not completely change the urban landscape, but the number of green "point" will come together into a green "face" and turn our city transformed into a green city, attractive homes.

[1] [China] Yu Kong-jian, LI Di-hua. Sustainable Landscapes [J]: Urban Environmental Design, 2007 (1) :7-12

[2] [United States] Frederick • Steiner. Life Landscape - Landscape Planning of the ecological approach (second edition) [M]. China Building Industry Press, 2004

[3] [China] Yu Kong-jian. Landscape: cultural, ecological and perception [M]. Beijing Science Press, 2005

“Plastic Limit” Triggered Rethinking the Design of Packaging

Haojun Cheng
College of Art and Design
Wuhan University of Technology
Wuhan, China
Email: chenghj@whut.edu.cn

Abstract----As people's environmental awareness, the limits of plastic bags use caused all thinking. The idea of finding alternative products can only be overcome temporary inconvenience, the positive long-term and effective approach should be deeper to explore. As a packaging design industry should be thinking and innovation from people, markets, technologies, system and management mechanisms. In packaging design and innovation as the core, matching the scientific system of rules. With the design concept and to influence consumer behaviour,rather than to have to cater to consumer groups is also very mature, I believe that the long-standing philosophy of environmental protection calls can be effectively implemented.

Keywords-Plastic limit, the design of packaging,thinking

I. *“PLASTIC LIMIT”EFFECT AND STIMULATE THINKING*

The time when widespread use of plastic bags is the 20th century, 80 years, from the Guangdong provided free of charge to customers of retail plastic shopping bags since then spread rapidly to the river north and south. Plastic shopping bags is not only easy for people to integrate the purchased items, but also for its use, such as: People have shopping bags to take home after not immediately thrown away, but the best use, the plastic bags to trap the litter baskets, to be filled with refuse, then joint and several plastic bags with rubbish thrown away. This not only saves work, while not dirty their own garbage basket. Others trivia in the life of the use of plastic shopping bags are everywhere at convenience. For a long time,Members of the public in life naturally cultivated everywhere rely on the use of free plastic shopping bags habits. When the free plastic shopping bags to people bringing lots of convenience, there was the media that: "This opens a new chapter in the history of the consumer." Can be as time goes by, this was the British "Guardian" as "the 20th century, the worst invention of mankind," more and more of a headache: non-degradable plastic bags, buried, will occupy the land, affecting crops to absorb nutrients, pollution, groundwater. Burned the harmful gases, damage to human health. Since then, plastic bags, and will have a "white pollution" of "reputation."

Faced with increasingly serious environmental problems, our government and the public has long recognized the serious harm of plastic bags to facilitate 1:00. Countries around the world reached a consensus, there are some countries had led the start restricting the use of plastic bags. November 20, 2007, in the U.S. west coast city of San Francisco provides a supermarket plastic bag ban entry into force of the law, which makes San Francisco became the first U.S. city in a plastic bag ban. The new environmental initiatives the city has also been in other U.S. cities and even states to emulate, many local councils are in the pipeline relevant legislation. In addition, Australia has banned plastic bags issued a law,The British Prime Minister Gordon Brown in November 2007 said it would ban the use of plastic bags.

With China's "plastic limit" the promulgation and implementation, people accustomed to shopping approach and many real-life problems follow. Such as: the temporary shop, grocery shopping away from home, the family's garbage, storage and so these past activities, rely on plastic shopping bags it appears that some "difficult" the. A ripple, a time "plastic limit" has become a hot topic of discussion. Although some suited, but can understand its purpose. China's large population base, greater use of plastic bags. To raise public awareness of environmental protection, paid use of plastic bags have a positive meaning. In order to save resources, in order to really improve our living environment,Removal of this artificial "white pollution", is also imperative. The face of temporary difficulties, and some people began to return to the use of the last bag, mesh bag or a rattan basket, bamboo baskets, there is the shopping cart with modern, "green bags", etc. to replace have been accustomed to shopping habits. The results or perceived: either not convenient enough, can not be widely; or not mature enough, find it difficult to solve practical problems. If there will be proper alternatives, the question naturally they will not exist. We believe that with the advancement of technology in the near future will be achieved.

But the real point of view to think about the environmental importance, should adopt a more proactive, more long-term, more effective approach. If from the product itself, packaging design and design management up to find solution also not a bad idea then?

As we all know, the fundamental role of the plastic bag storage, integration, protection of goods and easy to carry and so on, is a basic way of packaging, its main role is to integrate features of objects and easy to carry. Before assuming there is no plastic bags, how will the integration of the purchased thing? Packaging design what will be the trend? Therefore, we need to purchase goods from the market on the general characteristics and shopping behavior and needs of groups used to analyze and seek to not only easy to integrate, to carry, in keeping with green design, while not affecting the way business and the interests of consumers.

II. CHARACTERISTICS OF TODAY'S COMMODITY AND PACKAGING

Wherever Times is specified, Times Roman or Times New Roman may be used. If neither is available on your word processor, please use the font closest in appearance to Times. Avoid using bit-mapped fonts if possible. True-Type 1 or Open Type fonts are preferred. Please embed symbol fonts, as well, for math, etc.

2.1 Product Features

1. Diversity: shopping malls, supermarkets can be said that the product has rich, varied, everything, covering required in our lives, "clothing, food, housing, transportation, use."
2. Scale and size of different sizes: as small as a needle, large bottle of wine, and even furniture, home electrical appliances.
3. Patterns varied: There are rules of industrial products, agricultural products such as irregular. There are liquid, powder, granules different.
4. Material properties different: Some easily broken, and some easy to dirty, some should be waterproof and so on.
5. The product attributes are not the same: plenty of pharmaceuticals, some food, some items in different places.

In short, the rich product features, determine their physical and chemical characteristics of diversity. But also affect their packaging design commonalities and differences. Only based on specific product characteristics, there is an indisputable right of system design in order to make reasonable appropriate packaging.

2.2 Common characteristics of packaging

Common features of packaging is mainly reflected in their functions and roles, such as: protection of goods, to facilitate circulation and easy to use, promote sales, and even spread of cultural influence in many people's habit.

2.3 Packaging differences

Packaging design pay attention to is the different style of creativity and individuality, so the differences among the different interior design, there are materials and structures. The key is different businesses of different products at different times, for different consumers whose purpose is not the same.

III. CHARACTERISTICS OF THE FORMATION OF CONSUMERS SHOPPING HABITS

Free use of plastic bags in a wide range of environments, not just plastic caused environmental damage, even more terrifying is that people environmentally conscious ambiguity.

Influence consumer buying behavior of consumers themselves are the main factors factors, social factors, business and product factors. Vulnerable populations due to some things and the environment, long-term effects, but to form their own kind of thinking and behavior patterns known as "customary." Such habits tend to develop some sort of shopping behavior. The conditions in the free use of plastic bags, because plastic bags do on people's shopping has brought convenience constitute a market impact can be imagined, but also affect people's shopping habits and awareness of the general characteristics. Such as: random shopping (emotional shopping), but also the nature of shopping and so on.

IV. AFFECT PEOPLE'S IDEAS AND BEHAVIOR OF THE PACKAGING SYSTEM OF INNOVATION

System innovations include: design innovation; technological innovation; package management innovation; packaging evaluation system innovation; industry innovation system of laws and regulations.

4.1 Design Innovation

Design Innovation is the core element of the whole system solution is the most direct route. The key point of design innovation is the design philosophy, design methods and design process innovation. Many businesses are the needs of the user convenience, packaging designers began the famous "universal design" principles shown interest. Including packaging design specialization, standardization, universal and systematic. Involved in packaging graphics, color, shape and structure, and even processing technology, production and transportation all-round co-ordination. Such as: stacked-type package: Large supermarket shelves, the sales staff will be on the full use of Showcase space,merchandise stacked up as far as possible to conduct exhibitions, so that they can put in more provincial areas. Good packaging design has a bright graphic design and color scheme. So that the entire space visual impact will steeply increase, also help to promote sales. There are such as metal boxed biscuits, canned drinks, etc., at the bottom, and cover both the design of embossing groove, stacked up and just put on, take place is very convenient and safe. Many chocolate packaging carton packaging structure using a triangle, very strong, secure, convenience for customers and sales staff to take place. There are sets of packaging: that is the same type of goods sold and packaged as a set of combinations, such as tea, beverages, hair towels, pens, soap, etc. to use more complete package. Both increased sales to consumers and easy to carry.

The integration of packaging the module, use the change in the way, the system of uniform standards to make it split up. Conducive to a breakthrough in packaging design and development, and conducive to the integration of business sales, which will help the establishment of environmental awareness.

4.2 Technical Innovation

Technological innovation is promoting green, environmental protection and the protection of sustainable design, packaging design requirements of technological innovation. Technological innovation is the formation of an idea, and can be used to produce products to meet the needs of the market process. The implementation of technical innovation not only related to technological progress, but also the organizational culture and systems of innovation. Green packaging technology system should be set up raw materials, processing, handling and other aspects of innovation, to promote the development of packaging design.

1.Innovative materials

Habitat destruction of the source of raw materials, but also the carrier of packaging products and selected raw materials to control the root causes of its pollution, so the choice of raw materials and innovation are important. Packaging materials green technology innovation should be considered from two perspectives: First, natural materials; two eco-materials, or to the ecological transformation of traditional materials.

Emphasis on the direct use of natural resources, vigorously develop the natural material of green packaging materials packaging is an important way of innovation. Packaging materials made of natural advantages: First, shorter product life cycle chain, only after raw materials, manufacture, use and disposal stages of the cycle chain is short, on the environmental damage is small; second is renewable, abandoned after natural degradation of the environment No harm. Such as bamboo, grass and so on. Re-development of a conducive eco-materials, packaging design, development and creating more of their great environmental significance and market potential.At home and abroad are studying the development and trial of degradable plastics, and some high-tech materials, innovative packaging design and development of all have far-reaching significance.

2.Processing and manufacturing technology innovation

Processing and manufacturing technology is designed to protect the good packaging, processing and manufacturing is followed by raw materials, product life-cycle stages. First, the raw materials processing and manufacturing innovation is inseparable from technological innovation. Any kind of raw material will be subject to processing and manufacturing in order to become packaging products. Second, the original processing technology in the design and construction of environmental issues when considering incomplete or not suited to the existing manufacturing technology, packaging design, development needs,:, but also needs to be improved to meet new environmental requirements.

Processing and manufacturing technology innovation can be divided into two categories: development of manufacturing technology and equipment, and decontamination capabilities; it is to develop in order to reduce pollutants and waste has been generated for the purpose of the end of the treatment technology. Both are to achieve cleaner production.

3.Recycling processing technology innovation

In the packaging industry, recycling and reuse, waste disposal technologies is particularly important. Plastic bags can be taken abroad, the trade-in; detergent, shampoo, etc., may use an old plastic bottles filling directly in the supermarket, the price of the finished product than to buy a new bottle cheaper than half. These measures in order to warm and considerate way, imperceptibly lead the public to adopt healthy living habits. Therefore, part of packaging design can be taken into account the possibility of reuse.

In short, in the innovation process, we should focus on combination of innovation, and to get the combination of benefits. Packaging design innovation needs the support of advanced technologies, advanced technology is only through rational design can embody a certain value. Synchronization and combination of several innovative, innovation and efficiency can be maximized.

For some new green packaging technology for the promotion and application must go through a means to be guaranteed, that is to have green packaging application system to ensure that the main process of green packaging technology, equipment and materials in place and implemented in accordance with certain procedures to achieve the promotion of green packaging standards. Green packaging applications system, including the green channel to determine the supply of packaging materials, quality standards identification, the use of before and after treatment and recovery methods guidance. What kind of packaging suitable for what kind of products that can, through green packaging applications. Some people believe that solve the "white pollution" is the key to recovery and proper disposal. Countries around the world to adopt various measures to reuse and recycling of products was the responsibility of waste disposal in place, some foreign enterprises will be product packaging recycling and the products themselves see as important. Not only should pay attention to waste recycling, but also pay attention to the recovered material for commercial use. Ample evidence that, from the trash renewable resources, is a visionary. Abroad, the paper called the "urban forest resources" is a very vivid metaphor. Of course, the key is to reuse the design phase of the awareness-building to deal with technological innovation of energy is conducive to update and improve the quality of packaging products.

Modern packaging line at the service of convenience to people to meet others, guide others to be really "people-oriented." Packaging design must be further refined. Including the establishment of packaging management, evaluation system and the laws and regulations, constraints, and design applications, innovation and development.

4.3 Standard Management

People understand the past, the management of packaging, mainly in the production and recovery, often neglected the management of packaging pre-production, for example, packaging waste recovery organizations on the need for good management, the implementation of separate collection of waste in order to facilitate the recovery of packaging is indeed an indispensable way, but China's current need to establish a well-set design, production, recovery in one management system.

4.4 The establishment of reasonable and feasible evaluation system for packaging design

In business establishment and operation of green packaging management system, we need a system management tool for enterprise environmental impact and resource recycling and to evaluate the data, be guided by ISO. But the key point lies in the establishment of evaluation system should focus on prevention, not a "belated" like evaluation. Instead, they should establish a good source of design management and design of evaluation systems. Such as: 14000 series of standards life-cycle assessment method, using quantitative data - environmental behavior index indicated that the assessment results. Packaging life cycle of products from raw material extraction, processing, manufacturing, distribution, recovery reuse to the whole process of waste disposal.

4.5 A sound legal system, packaging design and line regulation

Application and promotion of green packaging, need to be protected by law. This is the world's major industrial nations are generally used in an effective measure. Of course, using a variety of policy measures conducive to the green packaging production and circulation, so that the implementation of the plastic limit not only control the expansion of the use of plastic bags, but also conducive to establishing a public environmental awareness.

In the packaging industry to take a mandatory administrative and legal constraints, after all, is a negative approach, indispensable, but positive prevention and to raise people's environmental awareness and to improve the result of the environmental impact of packaging, packaging design application of innovative and development is the essence.

It should be said of modern packaging design innovation is still in the stage, in many respects still far from perfect, as long as the various efforts will achieve greater progress. Take packaging for moon cakes. In 2006, China's first "moon cakes" for the first time the implementation of mandatory national standard for moon cake market has a significant impact. The new standard in view of the current market, the negative phenomenon of moon cakes, developed, and therefore its impact on the business of the moon cakes are very obvious, reflected in the moon cake packaging, mainly in the following areas:

1. Mooncake packaging size smaller decline in value of Mooncake Packaging;
2. More enterprises to choose the brand and product positioning, personalized packaging, the public version (Universal OEM) packaging will be gradually reduced;
3. Packaging design and packaging production and gradually separated from design companies do not have the strength of its market share shrink and career change;
4. Mooncake packaging is more simple, elegant, innovation will be mainly reflected in the packaging material and craft;
5. Mooncake packaging design style will be more close to the brand image, back to nature, and make things simple has become mainstream

V. SUMMARY

Of course, the face of worsening environmental problems, which need to be coordinated and efforts to take preventive measures. Is easy to produce environmental pollution, packaging industry should reflect a timely manner. Compared to Europe and the United States in foreign countries and Japan, China, packaging still exist, and as some of the differences. First, pay more attention to environmental protection and foreign packaging simplicity. Second, box-shaped with the screen looking for a more succinct. 3 is designed to more systematically. Fourth, the management of more standardized. And many of our domestic packages are only interested in images of luxury, but more importantly is our knowledge of packaging design and grasp to be improved. The spirit of "people-oriented" concept,"Plastic limit" for the product packaging design is just an opportunity, can not afford to limit the breeding of plastic and other negative measures, but should be a positive face, from the roots to find a breakthrough, with regard to all aspects of packaging design but also REN Zhong Road far.

REFERENCES

[1] Packaging Design Packaging Innovation and application of new technologies and technical standards for the quality of packaging inspection, Practical Handbook: Jilin Publishing House Audio (2005)

[2] Cao Xue, Li Wei : Product packaging and promotional design:China Light Industry Press (2002)

[3] Zhu Guoqin,Wu Feifei:Packaging:Shanghai People's Fine Arts Publishing House(2004)

A New Rural Look Created by Perseverant Pursuit of Tradition in Reconstruction of Ancient Dwellings

Wu Changyou
College of Urban Construction
Wuhan University of Science and Technology
Wuhan, China
Email: wucy1168@sina.com

Shen Jie
College of Urban Construction
Wuhan University of Science and Technology
Wuhan, China
Email: wabrian302@163.com

Abstract—**Exemplified by optimally planning and designing new dwellings and old residential extensions in reconstruction of ancient dwelling house clusters, this essay has singled out some valuable creative ideas in helping create China's new countryside by pursuing its tradition and culture. It begins with background information on the ancient dwelling house clusters. Next, an original planning concept of broken-line design in accordance with the local geography is recommended instead of traditional axis-symmetric idea so as to better preserve local tradition and culture. Then, a new design method is suggested for the layout of landscape in a specific building lot according to local individual conditions, and also for the spatial layout of new dwellings and extensions by replacing traditional quadrangles with novel three-sided courtyards for a better open living environment. Finally, helpful suggestions are offered for reference in planning, designing and reconstructing ancient dwelling houses clusters for an enjoyable modern life.**

Keywords- rural reconstruction; axis; cultural inheritance; environment; courtyards; suggestions

I. Introduction

A new subject of modern times as to how the living conditions can be improved for farmers and how a unique-looking rural village can be built in China is now drawing continuous attention from more and more architectural scholars and designers, who are trying hard to work out this problem : how can they contribute to the reconstruction of vast rural areas outside modern towns and cities. Owing to lack of experiences and also as latecomers in this field, three major misconnects have dominated their planning and design in the process of building new countryside: ideologically, they tend to consider pulling down all the ancient buildings to reconstruct new ones; conception ally, they are inclined to reject all the regional or traditional cultural elements to imitate the same western-style models in design; humanly, they are reluctant to consider ecological and environmental connection with human living needs except for the modernization of houses.

Recently, a creative idea of planning and design, however, has broken through the above misconceptions in building new houses complete with ancient dwellings in Wang Mingfan Village of Tongshan County in Hubei province, thus presenting numerous exquisite conceptions of planning and designing new rural villages in practice in relation to cultural and traditional inheritance, blending of traditional style and modern style, surroundings considerations and many other minor details.

Know as the "No.1 top ancient magnificent aristocrat residence in Hubei Province", the old Yuyuan Mansion House, also the official residence for Wan Mingfan, one of the high-ranking ministers in Qin Dynasty, is located in Wutian Village of Dalu Township in Tongshan Country, 5 kilometers south of the country town; the village is inhabited by 443 dwellers, constituting 91 households, of which 14 households live in the ancient "Mansion House". Despite its initial elegance struck off by time and weather to some extent, the "Mansion House" is still depicting its grandeur and magnificence, showing off its Chinese typical traditional architectural uniqueness, especially its features against the setting sun. The villagers have lived through all the hardships, uncertainties or whatever, for hundreds of years which, yet, have witnessed their native lifestyle and plain living morale that have embraced local cultural associations. In order to reproduce its historical attractiveness, new buildings around the ancient dwelling house clusters complete with Wang Mingfan Mansion House are all based on history and tradition, living function and spatial environment in an effort for persistent pursuit of ancient residential features connected with the hidden but ubiquitous cultural and traditional sediment. Some valuable creative ideas have been shared as follows in the process of reconstructing ancient dwelling house clusters in Wang Mingfan Village with reference to planning concept, design method and spatial layout of its whole building lot.

II. PLANNING CONCEPT: SUCCESSFUL PPLICATION OF BROKEN-LINE AXIS

The planning conception of dwelling house clusters in Wang Mingfan Village is based on the Yuyuan Mansion House, so all the new buildings are scattered around in a natural layout with traditional-style design elements so as to match up with its historical and cultural features, which shows a continuation of tradition and culture represented in the process of building new countryside, hence having formed a new creative planning idea which has been put into practice in planning new dwelling houses around the ancient dwelling house clusters in Wang Mingfan Village by way of resourceful design of axis.

The axis-symmetric idea has always been emphasized in the planning of Chinese traditional architectural clusters, and actually the planning of modern buildings also tends to be focused on the use of one or more simple geometric lines as axes in the layout of a building lot for the proper spatial layout. The axis of reference used in ancient dwelling house clusters of Wang Mingfan Village in Tongshan County is based on a bending curve , i.e. a broken-line of the outline of the building lot, the house clusters starting at the northeast main entrance of the building lot, extending alongside the exterior outline locus in North, West and South, shaping a bending curve round the ancient Yuyuan Mansion House and linking the village plaza. Alongside the axis progressively inwards, a main entrance is formed inside the building lot, along with a pedestrian street, central landscape, ancient Yuyuan Mansion House, and village plaza, as shown below in the bird's-eye view of Wang Mingfan Village.

Figure 1. A Bird's-eye View of Ancient Dwelling House Clusters in Wang Mingfan Village in Tongshan County.

A. Design of Axial Form

Three advantages are characterized by the choice of axial form for the dwelling house clusters of WangMingfan Village in Tongshan Country.

1) Certain implications of ancient streets

Affected by the metaphysics in Chinese Taoism, the layout of Chinese traditional dwelling houses is a set of irregular designers, which is typically reflected in the design of Chinese landscapes and gardens, whose designers usually seek for a naturalism implied in the form of a quiet and peaceful environment with twisting paths or winding roads, thus pushing forward the philosophy of pursuing nature-like landscapes in the design at its best. Such philosophical ideas are best demonstrated in the use of broken lines for the axial design of ancient dwelling house clusters in Wang Mingfan Village in which all the spatial layouts of building lots are organically related to each other, with all the spatial arrangements intersecting one after another, lanes and streets crisscrossed longitudinally and latitudinal. Visitors may be delighted by the attractive of the village when strolling leisurely in the lanes or streets, slowly onwards first to Yuyuan Mansion House, center of the village, and then to the open square next to the waters of a large pond. The choice of axial design in a broken line form gives a sensation of pleasantry for the whole building lot planned spatially in proper order , good proportion and splendid combination. "Fig. 2"

Figure 2. A Street look of planned ancient dwelling house clusters in Wang Mingfan Village.

2) Streamlined extension for business valnes.

The choice of axial design in a broken line form makes the streamline longer than all the other geometrical lines, thus increasing the total volume of all the frontage stores and businesses, all sorts of traditional handcraft workshops are set up inside for such activities as processing bamboo products, bean curd and foodstuff. This may not only increase farmers employment opportunities, hence improving their economy, but also help to preserve traditional culture and native lifestyle, which in turn increases the value of either farming land or building lot. as well as the value of business by extending the business lot properly.

3) Maximumized utilization of a building lot: improvement on living quality, tourist amusement and facilitation.

The axis design of ancient dwelling house clusters in Wang Mingfan Village is a typical design based on

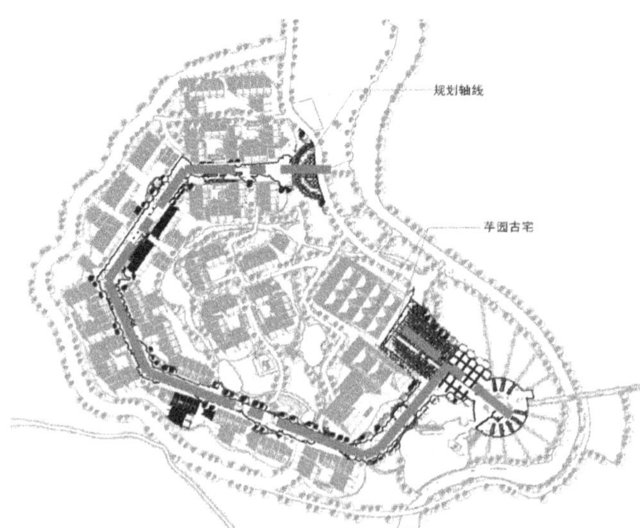

Figure 3. A typical view of old street in Wang Mingfan Village.

individual local conditions, of which the layout is formulated not on the commonly-used straight line or arc form as axis design, but on the detailed survey of building lots, especially the survey directed at their particular outline whose locus is used for axial design , thus mapping out the inner transporting network of the whole building lot, all parts of which can be reached conveniently, whether for transporting or sightseeing. In the design of village extension, all the new buildings are designed to be two or three storied with eaves height of 7.5 to 11 meters, lanes width of 2.5 to 4.5 meters, and street width of 12 to 15 meters. Such designs have made the whole building lot fully utilized with a good proportional space for lanes and streets, thus creating a comfortable and convenient environment for both living and working, as shown in the given figure. "Fig. 3"

B. *Frontage Design Alongside Axis*

1) *Varied forms of longitudinal frontages alongside axis*

The axial frontages vary with different spatial layouts and all the frontage buildings on both side a of pedestrian streets consist primarily of individual dwelling houses whose frontal designs are based on such traditional elements as railings, wood columns, square-grid-pattern windows, lanterns, signboards, miniature landscapes and so on. All the buildings are chiefly painted with three colors of white, red brown and greenish grey , forming a typical view of ancient streets of a small town. The central landscape inside the building lot is mainly composed of a soft and tranquil environment with pavilions, bridges, trees and plants, and all building structures are constructed with Chinese traditional style both in form and colors to achieve a unification of the whole building lot in every way , the highlights of which is focused on the formation of a unique environment of Ancient Yuyuan Mansion House on the north side and the waters of a vast pond on the south side of the axis end. In the reconstruction and preservation of ancient building environment, the combination of modern design with Chinese traditional architectural style has created a wonderful living environment where both local dwellers and sightseeing-goers may enjoy an agreeable mood and strong sensation of mixture of its past and modern history, culture, life style or changes in the beautiful and natural scenery. "Fig. 4"

Figure 4. Broken-line axial design of ancient dwelling house clusters in Wang Mingfan Village.

2) *Utilization of Cloud-Dragon-Pattern-Outlined Wall and Traditional Paint Color Tone.*

By refining and improving the symbolic elements of Ancient Yuyuan Mansion House, triangular-section walls outlined by cloud-dragon-patterns are utilized as one of the traditional features in planning and designing new dwelling houses in WangMingfan Village so as to match up with traditional style of ancient dwelling house clusters in the village, forming an organic connection with its tradition, customs and culture. The major difference between the old and new cloud-dragon-pattern-outlined walls lies in the forms and shapes of house eaves; large-angle-projecting eaves are used at both triangular-section end-walls of new buildings which may better give a sense of grandeur and elegance, plain background of ancient architectural features mixed with a modern building stylistic taste. "Fig. 5" "Fig. 6"

Figure 5. A new look of planned streets between ancient dwelling house chisters in WangMingfan Village.

Figure 6. Configuration of cloud-dragon-pattern wall of ancient Yuyuan Mansion House.

In order to keep in line with the major colors of paint used for Ancient Yuyuan Mansion House, three main colors of greenish grey, white and red brown are employed for all the new buildings, hence maintaining a unity of color tone for both, and also bringing on a new look of rural villages. "Fig. 7"

Figure 7. Frontage of ancient dwelling houses in Wang Mingfan Village.

III. DESIGN CONCEPTION:MODERN DESIGN COMBINED WITH TRADITIONAL LANDSCAPE DESIGN METHOD

A. Building layout based on Taoist Naturalistm

Generally, two types of planning concepts dominate the layout of China's traditional dwelling houses: the urban planning is affected by the philosophy of "spherical sky and square ground" arranging the houses in square-shape layout weaved with traffic system in a longitudinal, and horizontal intersected form like the Chinese character "井" (meaning a water well), whereas the rural dwelling house planning is mostly based on individual local conditions of a building lot , houses often scattered among hills, on river banks or over other in cultivable plain land ,thus constituting villages in a natural layout in rural areas. The layout of ancient dwelling house clusters in Wang Mingfan Village is just based on a modern planning concept combined with traditional Taoist naturalism in practice in which all the dwelling house clusters are built in a definite function division concept but without a distinct unified division form. All the ancient dwelling house clusters, facing all directions, appear to be scattered not only individually in the formation but also inner-connected logically on the whole, thus presenting thoroughly a plain primitive feature, along with a lively style in the reconstruction of new countryside.

B. Water scenic formations based on individual local conditions

Chinas traditional landscape design concepts are centered on the theory of "A beautiful view appearing no vivid without a body of water against it,." So in the design and planning of ancient dwelling house clusters in WangMingfan Village, the existing water resources are fully utilized to constitute a vivid view of water scenery either by preserving the natural water bank lines of the streams as much as possible or by reconstructing the bank sides in part of the streams. The water scenery in Wang Mingfan Village consists of three branches of a river, two streams in the Northwest, flowing zigzag and crisscross with both sides planted trees and flowers plus multi-directional frontages of dwelling houses, thus all of these compositions forming a beautiful view of a mountain village, whereas the waters of a huge pond connected with another branch of the river in the south have been reconstructed with modern planning and designing concepts combined with traditional landscape design method by which the natural bank lines are maintained to the maximum degree . In addition, some pavilions and other geometrical-shape platforms are built on the watersides, hence constituting a wide variety of beautiful water views against other natural surroundings to match up with the basic traditional style of ancient house clusters of the village.

The utilization of water resources for scenery planning of Wang Mingfan Village is based not only on the development of landscape values but also on the consideration of local living custom by keeping the original water ditches flowing within the reach of every household, and by doing so it can offer much convenience for the dwellers' daily life as well as enrich their modern life with primitive scenes and sights of cultural interest. "Fig. 8"

Figure 8. A flowing water channel in the village

C. Particular scenic spots with particular features.

According to Zen Buddhism, "one flower represents one world", the theory of which can be interpreted by modern design science as the principle of landscape design, and in other words, the whole concept of planning and design should be embodied in all its details, just as shown in a particular wood column or even in a pot of plants and flowers which may set off the whole layout wonderfully against all its details. The dwelling surroundings are closely associated with dwellers' everyday lives that designers should take into full consideration, that is to say ,the design of ancient dwelling house clusters should be based on the principle of taking humans as the basic fundamental element which is used as a starting point to center all the designs on the maximum comfort of dwellings and thus to form a harmonious basic tone of buildings on all the details. This basic design principle has been best used in the planning and design of ancient house clusters in Wan Mingfan Village. For instance, such landscape elements as stages, lanterns and potted plants are largely employed in the garden design of courtyards. All these elements not only can best meet dwellers' daily needs but also set off the natural and cultural features of ancient house clusters. And what's more, the choices of paint colors as well as all the artificial building components for ancient house clusters in Wan Mingfan Village particularly reflect the planners' and designers' perseverant pursuit of the conception of symbolizing local tradition and culture as required by the locals.

IV. CREATION OF UNIQUE DWELLING FORM WITH MODERN RURAL LIFESTYLE

A. Housing types

China's traditional handicraft economy has dominated typical housing types of villages, towns and cities. A particular dwelling house usually consists of three functions: daily life spaces, frontage stores and handicraft workshops. This typical housing type has also created China's particular rural and urban lifestyle and culture that are reflected in the unique features of marketplaces, lanes and streets. Such traditional and cultural practices are best pursued in the planning and design of every individual dwelling house in Wan Mingfan Village, like the forms of all spatial arrangements in the pedestrian streets: first floor frontage stores with handicraft workshops, and second floor spaces for dwelling and everyday life. Such designs in practice have helped to realize the idea of continuing local tradition and culture in organic combination with the desire of enjoying modern life.

B. Combinations of Courtyards

All the courtyards are well–organized in planning and designing ancient dwelling house clusters in Wang Mingfan Village. In the north of clusters on both sides of a street, every three individual houses form a household, thus enclosing a three-sided courtyard, and between streets, dwelling houses clusters are laid out with a plaza, stage, pavilions, greenbelts, potted plants, flows and woods, etc. Such combination, which may offer much more convenient contact between households inside and outside a courtyard, differs somewhat from China's traditional quadrangle which may probably isolate itself form the outside in contact with others. Obviously, the households with three-sided courtyards have succeeded to the advantages of traditional quadrangles, through some improvement on the layout, and therefore have more flexible and convenient living spaces and daily contacts for dwellers inside and outside. "Fig. 9" "Fig. 10"

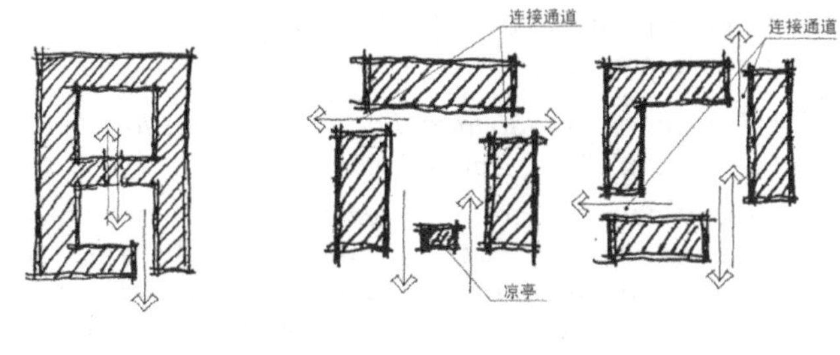

Figure 9. Comparison of traditional and modern space.

Figure 10. A Presenting View of planned Ancient Dwelling House Clusters in Wang Mingfan Village.

V. SOME SUGGESTIONS FOR RECONSTRUCTION OF NEW COUNTRYSIDE

A. Utilization of Local Cultural Resources in Presenting Rural Attraction

Recently, China's excessive urban construction has led to urban population explosion which has become one of the causes for traffic jams problem to which a fundamental solution can be found by reconstructing rural villages into optimized dwellings so as to decentralize urban population, just as what has been done by taking advantages of rural historical, cultural and tourist resources in the reconstruction of Wang Mingfan Village in Tongshan Country of Hubei Province.

B. Improvement on Typical Rural Business for Better Rural Life Quality

A big gap has been caused between urban and rural life quality because of China's unbalanced economic growth in both urban and rural areas, the gap between which can not be properly bridged just by implementing some advantageous economic policies for rural areas. To solve this social problem, it is essential to further develop rural economy to as much satisfactory degree as the urban economy, either by reconstructing rural villages into better-equipped communities for dwellers or by improving typical rural businesses such as handicraft industry, and rural tourism so as to raise rural life quality which may help to share urban population pressure.

C. *Utilization of Individual Local Conditions for Optimized Living Environment*

Social development often leads to a way for humans to enjoy their lives both materially and Spiritually, which seems to be affected strongly by the living environment in relation with their everyday moods and spirits. Generally, rural residents may enjoy plentiful land resources and excellent natural resources which , however, have not been well planned for their proper uses, thus giving a sense of "mess", "disorder" or irregularity in many ways of rural life. So, a unified thorough planning should be taken into account in the reconstruction of new countryside, where a comfortable, agreeable, tidy and orderly living environment can be well-created for a nice modern rural life for all dwellers.

D. *Correct Service Concept in Efforts for Avoidance of Mimicry and Copying*

The natives in the countryside are the end clients in the services of planning and design for the reconstruction of new countryside. Unlike the urban dwellers, the rural dwellers are often characterized by their regional ways of living and working. For example, the plateau residents' lifestyle and working practices differ sharply from those of mountainous villagers. Therefore, these factors should be carefully examined in the services of planning and design so that the most optimized dwelling houses can be constructed for the locals to satisfy their routine life and work.

VI. Conclusion

As what has been respectively examined above, full considerations should be given to these aspects in planning, designing and reconstructing villages, towns and cities: tradition, culture, economy, geography and daily life. In specific, planning work should be based on geographical environment, natural condition, local tradition, regional culture and native lifestyle; designing conception should be focused on the traditional architectural elements combined with modern life necessity; spatial layout should be centered on practical living function, working facilitation, traditional heritance and cultural continuation so that a more practical, agreeable and convenient living environment can be created for the local dwellers.

References

[1] Xiao Mo, "Idea of Architecture Volume 5", Qinghua University Press, Beijing, 2006

[2] Wang Xiaobai, "A Collection of Zhao Guanqian's Essays on Humans", Shelters and Environment, Sichuan University Press, Chengdu, 1998

[3] Dong Jianhong , "History and Theoretical Research of Urban Planning", Tongji University Press, Shanghai, 1999.

[4] Chen Congzhou, "On Chinese Gardens", Tongji University Press, Shanghai, 1986.

A Study on the Protection and Remolding of the Spatial Characteristics of a City in China

Zhu Jing
Architecture Department
Xi'an University of Architecture and Technology
Xi'an, China
Email: zhu_jing_nwu@126.com

Feng Jixin
The Humanities Department
Xi'an Physical Education University
Xi'an China
Email:578686876@qq.com

***Abstract*—In order to soundly and sequentially guide the city construction of the spatial feature in China, the protection and remolding of the city's spatial characteristics, based on the related data, is analyzed in this study. Moreover, the concepts of the spatial characteristics and special landscapes of a city are defined. And the procedure of formation, protection and deterioration of the city's historical spatial characteristics is discussed. Also the relationship between spatial characteristics and culture legacies of a city is developed. Meanwhile, some regulations and methods as to city constructions in China on the spatial feature should be conscientiously followed. At last, some suggested policies and strategies are supposed to protect and reform the City's Spatial Characteristics.**

Keywords-Space; Characteristics; City culture

I. INTRODUCTION

The characteristic of the city is a complex issue rather than simple because cities all over the world always differ from natural environment, culture and society. So the studying point of penetration on city's spatial characteristics is various. At the present, most researches are focused on microcosmic analysis, such as squares, commercial blocks, and historical streets. With the development of city construction, macroscopical and systematic analysis on city's spatial characteristics and its remolding is ignored and needed. In pace with the rapid urbanization in our country, mass constructions in cites are being carried on everywhere. And some cities' constructions merely chase after the modernization at the cost of spatial space and historical textures of their own city. Facing this issue, the author believes that a city's development and expansion should not take the city's history and specialty as the price. Both the molding and inheritance of city's characteristics extend and back up a city's culture and life, and show the city's nature as a life individual. In this thesis, through the study on the concept of the spatial characteristics of the city and its evolution, the author tries to propose the countermeasures on protecting and developing a city's spatial characteristics.

II. CITY SPATIAL CHARACTERISTICS AND SPECIAL SPACE

The City's spatial characteristics refers to a kind of space feature which is different from other cities' and contains many factors as follows: natural geographical environment, environment, history, culture, economy, and society. Among them the natural geographical environment is the construction basis of a city. Other factors are based on it and closely connected with each other to shape the city spatial characteristics.

A. Special Space

There are lots of studies on the concept of special space, which, until now, still not get a clear definition. Here, we proposed our definition. Special space refers to a kind of material space or what is called locales in City Landscape Science, that shows nature spirit, humanistic spirit, special history and architecture culture. This space grows on the nature harmoniously, which has a definitely edge, and it is a small part of the city space. Generally, it is microcosmic.

B. Spacial Characteristics

Cities which are congregated by the tertiary-industry and the secondary industry are always the center of an area. City scale enlarged continuously with its development. Many blocks that productions and activities occurs to grow up by the offcenter power and incenter power. By those power, blocks shaped various space on the proceed of differentiation, gather, scatter, algamation, extension, and endogenesis. Blocks grew up with sequence or crisscross distribution. Special space generated by those issue. This space is different from other cities in color, material scene and space rhythm. Scent, language, folk-custom are related with these factors[1]. The entire city environment space is not only a material space, it is also the carrier of human spirit space[2]. Furthermore, spatial characteristics means the space feature that is clustered by space, culture, and material factor in a certain area(including city and village), macroscopically.

C. City Characteristic is the Congregation of Lots of Special Spaces

City characteristics always connected with region. In the development of the city, special culture grows and shows by material and non-material way in many city spaces. These special spaces shaped the whole space characteristics of the city. People studied the constitution of city space and made some choices, arrangements and plans. Then special space stands up in the city[3]. Nowadays, in China, many cities made the same construction, and shows similar appearance.

Why? Firstly, the build of special space which can show the city spirit is ignored. Secondly, space constitution shows disorder. Therefore, if there are more special spaces which can show the city history, culture, nature and cultural heritage, and balanced with other spaces, the city space characteristics will be shown apparently. In short, various special spaces and its order congregation composed the city spatial characteristics.

III. SPACIAL CHARACTERISTICS AND TASTE OF CITY'S CULTURE

Places where surrounded by beautiful natural sceneries had always been chosen by our ancestors to build cities. Although some people say Chinese architectures pay great attention to fengshui, actualy, that is the pursuit of the harmony between nature and the people. The building of urban spatial characteristics includes the utilizing of natural space and building of human space.

A. The Natural Space Background is the Basisi of City's Spatial Characteristics

The physical space in the city special space is composed of natural space, buildings and structures. The natural landscape in city is very basical and primary, which shows zoology, value and ornamental[4]. And it influences city construction everywhere such as the city morphology, function, arrangement, and landscape. Every city is different from others according to the nature basis, which leads to diverse spatial characteristics. It guides the pattern of city space and the city general scene. Generally, diverse natural space shapes diverse city characteristics, such as water city, hill city, flower city and spring city[5].

B. The Accumulation of Human Space Forms as One of City Spatial Characteristics

Physical environment, which reflexes city characteristics and scene, is the carrier of culture. Therefore, human space is the heritage of the city, and it grows up on the physical environment. Diverse city culture tastes lead to diverse human spaces that are shaped by history, celebrities and customs.

From the origin to the growth of the city, modern life, which is the value orientation and value basis in social life, is influenced by human spirit everywhere[6]. In a certain way, the city culture characteristics is reflected by different period, managers, designers and planners. Moreover, the city space image is impacted far and deeply by the level of architecture designers and their works[7]. For instance, in Beijing, there are buildings such as the Site of Dadu of the Yuan Dynasty, the Great Wall, the Imperial Palace, Temple of Heaven, Summer Palace and Tian An Men Square etc. And in Xi'an, Shaan'xi province, there are historic architectures too, such as Han Chang'an Heritage, City Wall,Drum-tower, Bell Tower、Big Wild Goose Pagoda、Small Wild Goose Pagoda、the Forest of Steles and Shuyuanmen Ancient Cultural Street etc.

C. Summary

Above all, the relevant factors of city spatial characteristics can be analyzed as the following Fig. 1.

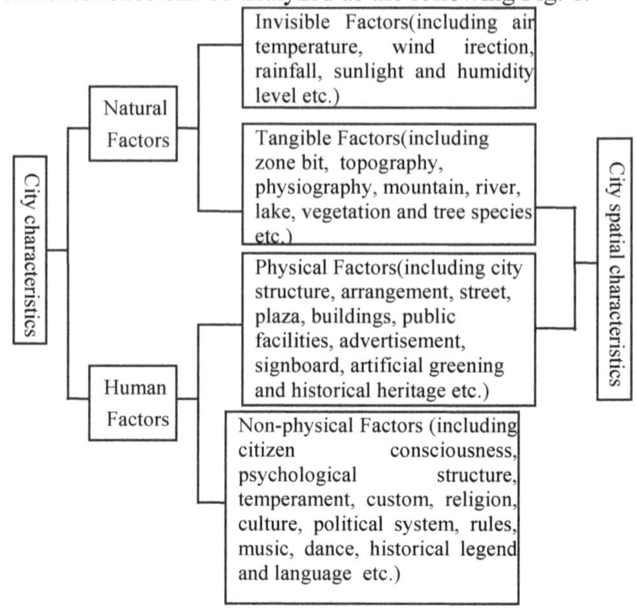

Figure 1. The analysis on the relevant factors of the city's spatial characteristics[7]

IV. THE FORMATION AND ALTERNATION OF CITY SPECIAL SPACE

City spatial characteristics are the outcome of city construction during thousands of yeas accumulation and choosing. Also it is resulted from the comparison and siftings between human and nature.

A. The Formation of City Characteristics in Ancient Agricultural Society

In the agricultural society, the productive forces were so underdeveloped that human being had to rely on the nature. Therefore, human beings and the nature must coexist harmoniously; otherwise, human beings would be punished. That's why the city constructions in the agricultural society must entirely suit measures to local conditions. In ancient times, under the conditions of underdeveloped transportation and technology, it was difficult to carry goods everywhere; and some isolated cities' constructions had to use local materials, respect the local history, culture and custom. Thus, the specialty of every region and nationality is showed through the applicability and representation of buildings and structures (Fig.2 is the relationship of these factors). And then every city has its own spatial characteristics and scene. Ancient cities always have their unique spatial characteristics, such as ancient Chang-an city, Luoyang city, Suzhou city, Lijiang city and Pingyao city etc.

B. The Reason for the Formation of no Characteristics and Lack of Characteristics in Modern Cities

1) Cities are treated as standard manufactured products by modern industry.

The city constructions have partly lost their original nature basis with the increasing urbanization, developed technology and productive forces. In some cities, the natural environment has been freely curved and reformed and not harmonious with its nature any more. Both standard module of construction and standard supply of commodity are manufactured according to the request of modern industry, which shape cities with similar buildings and building-styles just like produced by one factory. The box-like architectures with steel structures, concrete, and glass walls can be seen everywhere. Unified brand and style of McDonald's and KFC, similar neon lamps, dress and even songs form the similar street scenes and also can be seen everywhere.

City constructions are convenient, efficient, and even omnipotent by the support of developed technology and industrial manufacturing. With the promotion of the industrialization, standardization, the same mode and frontier technology, it has become possible that the same products are manufactured at the same time but in different places. Moreover, goods are amazingly identical in different areas. The formation of the city space, from architectures to dustbins, should have its own special features in different areas but now is losing its specialty and becoming similar.

2) Cities are treated as intensified commerce goods only for chasing the economic benefit

Since the industrial revolution, urbanization speeds up constantly. With the fast gathering of human population, companies and capitals in cities, the cities shaped in the agriculture period expand so rapidly that the original space frames cannot bear on. Considering the politics, religion and culture, relatively speaking, both economy and technology has become the leading power in the society and force the city's space to serve them[6]. Driven by the economic interest, the space constructions of the city are asked to intensively develop for getting the maximum benefits. So the standard city's construction pattern is an easy way to get this goal. The economic interest has become the key factor in city space construction. Other factors, such as environment, city morphology, arrangement, architecture style and skyline, are all ignored in city construction.

3) The culture dispersion under the economic globalization leads to the gradual disappearance of city characteristics by strong culture.

Culture is born in economic activities and in turn stimulates economic activities. It means that custom, knowledge, materials and behavior mode are spread by the society from one group or one society to another [8]. For instance, one of the consequences of the culture dispersion is the culture of fast food restaurant born in USA, which has flooded over most of cities and created similar streetscape in different cities. Economy and culture firmly rely on each other because economy feeds culture and conversely culture backs up economy. Under the opening-up policy and globalization, the rapidly developing Chinese economy has brought both benefit and foreign culture to China. As a result, the local culture is shocked by the powerful foreign culture. In this thesis, foreign culture is defined as strong culture. It is obvious that strong culture is always brought by the powerful foreign economy.

Space is shown by culture or vice versa. City space without culture is empty and lack of specialty. And city construction is always based on its own culture. On one hand, some city constructions of our country crave for things big and foreign and are keen on international bid and designing from foreigners. However, those foreign architects who are not familiar with traditional local culture, definitely will weaken and twist the originally traditional material culture in the local city-space. On the other hand, undeveloped regions easily set the strong culture that based on powerful economy as their models, imitate it and copy it for coming up with the developed regions [9]. All these reasons gradually speed up the disappearance and deterioration of special cities shaped in the ancient agricultural period.

4) Inappropriate administrative intervention and over-intervention

Due to the restriction of current management system, the administrative intervention still influences our constructions greatly. Then the administrative over-intervention will surely cause unreasonable problems. If city construction is guided by administrative intervention instead of scientific regulation, it must negatively effect on the city space, culture, and taste. Especially the ancient space characteristics and cultural environment in the historic heritage and historical cities must create the externally negative effect caused by the strong intervention. The New City Planning Law of the People's Republic of China has granted the seriousness of regulations to the unprecedented status, and made some rules on protecting a city's characteristics. It extensively includes that a permitted city-town regulation is the basis of city-town construction and regulating management, and should not be altered except legal proceeding; making and carrying out a city-town regulation serve for protecting the arable land, natural resource, historical culture heritage, local specialty, national characters and traditional scene. Unfortunately, on the other hand, the new law regulates that the relevant chief department and direct chief violated the law only could be asked to correct, or circulated a dispatch, or punished by the corresponding government, higher regulating department of the government or supervisory department[1]. However, from the view of a city's hundred-year-planning construction, these subtle punishments are too far to totally stop the administrative intervention. Currently, in the process of urbanization, due to the administrative intervention, more and more special cities marked by the historic towns, blocks and ancient architectures are being swallowed up by the daily developed and assimilated cities. The gradual shrinks of city's special spaces are weakening and eliminating the city's characteristics day by day.

There are many special spaces destroyed by the administrative intervention in China. For instance, The Zhouzhuang Ancient Town, aged 900 years, the symbol of the water township in southern Yangtze River, wholly

[1] The New City-town Planning Law of the PRC, 2008.1

remains the building patterns of the typical water town in southern China. However, a modern road across the town is planed to build by the local government. Although the road-building is stopped by many experts' appeals, the influence by the administrative intervention also can be seen. Another example, The Three Lanes & Seven Paths (well-known historical blocks in Fuzhou formed in Tang and Song Dynasty) is rebuilt as a commercial street. Some ancient buildings are luckily remained but scatter in the modern commercial buildings. And the originally historical patterns and atmosphere are totally destroyed. Three Lanes & Seven Paths exits in name only now. In Zhejiang Province, the government of Zhoushan city offended the national laws about heritage and historical city protection because of remaking the historical blocks in historical city Dinghai. As the following the ancient scene of the historic city has totally gone away. In Sichuan province, Tri-Star piles the ruins are well-known as the Chinese ancient cultural culture remains all over the world. And there are still lots of unresolved puzzles in it. However, a real estate company is preparing to build a hotel on it. In Hangzhou city, the united meeting of Party and government replaced the planning bureau to decide the location of the Wang Jiangmen Thermoelectric Plant. Unfortunately the factory was just located in the city center that was planed by the new general planning of Hangzhou city. At last, the construction of the plant was stopped with the cost of RMB 0.4 billion paid by the government [11].

5) Summary

Above all, the Fig. 3 shows the reason why modern cities are always lack of characteristics.

V. Conclusion: The Protection and Remoding of the City's Spatial Characteristics

A. The Protection of The City's Spatial Characteristics

Since the quality of the city's spatial characteristics is an important criterion to measure the suitability and attractiveness of a city, the protection and development of the city's spatial characteristics is becoming a significant issue in the study field. Concerning that the following ways are suggested to protect and develop the city's spatial characteristics.

- Improve relevant laws and regulations; strengthen the enforcement on the law of city planning. The seriousness and science should be reflected in the law. Then the regulations can be operated after avoiding administrative over-intervention.
- City planning and design should insist on the dialectical materialism, and firmly protect the historic buildings and natural special characteristics so that culture and nature characteristics can be shown in cities.
- People should regulate their view on city planning. City planning should serve for the long-term benefit and special space construction of a city instead of the overwhelming short-term benefit.

B. The Aufheben and Remolding of The City's Spatial Characteristics

The construction of the city's spatial characteristics in modern society should be paid more attention as the following points for building a special city which has good space arrangement, clear characteristics, coherence and orderliness.

- Improve and formulate laws and regulations that relate to historic heritage; reserve the valuable historic heritages and buildings; protect the integrity of historic blocks.
- The planning concept should be strategically and generally designed in advance. The whole logical process of a city's space how to come into being should be fully proved and repeatedly compared with others. This is equal to deepen and expand the original concepts. And a city's spatial characteristics should be put on an important status. Meanwhile, the showing-way of a city's spatial characteristics must be considered in order to find a unique way among the armored concrete buildings, histories and cultures to construct a city's feature. Afterwards the construction of a city's feature is successful if people can easily recognize where they are by the surroundings and architectures,.
- Emphasize on details. The spatial constructions of a city should be paid more attention to as well as the city's details, including the green land, architecture pieces, sculptures, cultural buildings and heritage inheritance etc.
- Suit measures to local conditions and seek truth from facts. It is necessary to take actions to protect the locally special nature and geography. The humanity spirits in traditional culture should be inherited, promoted, remolded, and innovated. The nature-respected and human-oriented ideas can create a way to seek the balance between human and nature. The planning will reflect locally special culture, find a best link for history and modern society, protect the link with the construction, and harmoniously unite the development and innovation.

Consider comprehensively, plan systemically, and coordinate all kinds of conflicts and contradictions. Regarding to strong culture, we should absorb its essence and discard the dross. It is both wrong to deny it completely and absorb it in a whole. At the time of developing a city, we should think over the system, comprehension and foresight, deal with the conflicts between micro-parts and macro-whole as well as the short-term construction and long-term development. Furthermore, the link between city management and special space construction, and the coordination between new space order and traditional culture are both very important.

References

[1] Shang'hai Construction Network Station, The Theory Study and Applying Strategy Analysis on Optimizing City Space Image in Shang'hai, www.shucm.sh.cn/home/ztyanjiu/lesson/article/09.asp.

[2] Pu Su, Urban Space, www.wenxinshe.org/040sp3/ReadNews.asp?NewsID=4928.

[3] Northeast News Network Station, Five New Nouns: Expressing the New Thinkings and Ideas to the Development of Shenyang, Capital of Liaoning Province in Northeast China, www.nen.com.cn.

[4] Kevin Lynch, The Image of the City, The MIT Press, pp15-100, 1960.

[5] Qikang. Context and Specilty——the Culture Characteristics of Urban Morphology, Urban Development Research, pp20-24, January 1997.

[6] The Development and Reformation Planning Bureau in Hangzhou, The Space Planning, Architecture and Human Spirit of Hangzhou, the Capital of Zhejiang Province , www.hznet.gov.cn/hzdpc/0401/5001.htm.

[7] Duanjin, A Study on the Perceiving Laws and Investigation of City Space Characteristics, Modern City Research, pp59-62, January 2002.

[8] Richard T. Schaefer. Sociology and Life (Illustration revised, the 9th printing). World Books Pressing Company, pp50-90, July 2008.

[9] Zheng Shiling, New Challenge : What should be done When Chineses Weak Culture are facing to Strong Construction , Architecture Times, 2004.

[10] People.com.cn, How to Protect Historical Culture Heritage in the New Century, www.people.com.cn/GB/guandian/183/2301/3328/20010117/381073.html.

[11] Qiu Baoxing, Challenges and Oppurtunities of the Chinese Urbanization, Chinese Architecture Industry Press, Beijing, pp. 150-330, May 2004.

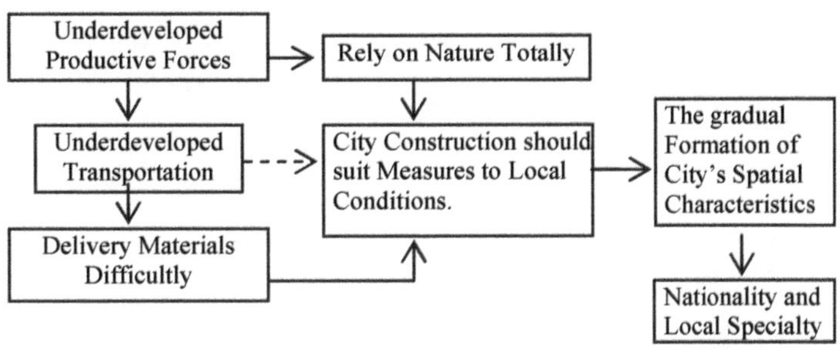

Figure 2. The formation on spatial characteristics of ancient cities

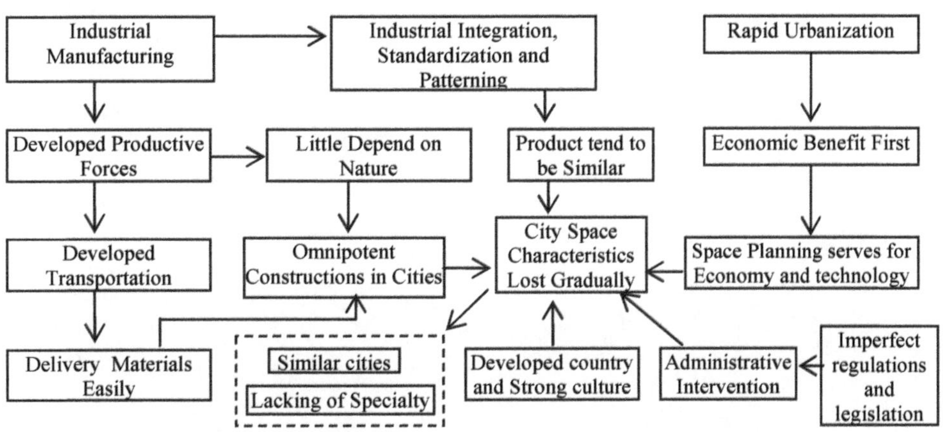

Figure 3. The formation on the lack of spatial characteristics of modern cities in China

A Theoretical Discussion on Methodological Study of Modern Design

Zhiyong Wang
School of Art and Communication
Southwest Jiaotong University
Chengdu, China
Email: matinee2005@yahoo.com.cn

***Abstract*—a reliable and efficient method should be confirmed firstly if any researcher does want to acquire knowledge. Without this helpful assistant however, there must be a fatal obstacle between the researchers and the object on which their study were focusing, and the result would be no better than this kind of contradictory situation: the knowledge was missing but the study is still in process. The same fact has always been found in theoretical study of Modern Design, even with more complicated conditions. Once any researcher has been involved in such a study, he (or she) will finally become aware of this reality, that, for instance, whatever principle the researchers had depend on when they were in study, they had to have tolerance with opinions, although such subjective views were opposed to their principles. It only because that no absolute foundation of methodology has ever been given on the study of Modern Design. Such a complication caused the emergence of scepticism on the research, while the very arbitrary views could be authorised as well. A caution for such problems is the main purpose of the presentation of this essay, although it has no ambition to offer any total solution for them. Since no dogmatism of academic study gave help for the research, a gradually process of correction for problems but makes progress which do benefit the study.**

Keywords-Theoretical Study; Modern Design; Methodology; Natural Science; Humanities Study

I. INTRODUCTION

When people were focusing on some abstract ideas instead of concrete objects, rock or water, for instance, their talking would be much more likely involved in arguments. We eagerly endeavoured to find its reason, but the reason itself will hardly be clarified by us.

Let us firstly look into a case which was very common in the researches on forensic science. Although such an example seemed far away from the issue we will discuss on, an analysis and comparison between their methodological studies could have the cared questions more easily uncovered. The specialists in forensic medicine often encounter such troubles. That is, when they were focusing on a research object (a corpse, for instance), it is very possible that some disagreements will be emerged as they tried to give a suitable conclusion which was led by some causes, for example, a specific reason for the death of a man. In such a seemed delicacy situation, however, the researchers will not stop to attempt to discover more evidences with a sustainable and reliable way until their new discovery can be presented to support their former conclusion, or conversely, to negate it. Suppose, a new analysis on the specimen of inside from the body tell the correct reason of its death was chemical poisoning, but the early wrong deduction was mechanical injury.

II. A PRESENTATION OF THE CHARACTERISTICS OF METHODOLODY FOR THE STUDY OF MODERN DESIGN

The brief explanation in the introduction revealed a key point. It is, as in those study as forensic science, although many disagreements could troubled the researchers when they were in study process, they at least had a reliable method which they could trust and depend on. The conclusions may not be consistent, but usually the method will not be doubted. The theory from Auguste Comte gave a good explanation for it: science should care of those entities which could be only understood by our experience. Based on particularly empirical observation, we could conclude the rules which would interpret the relationship between the observed phenomena. The scientists will know what is going to be happened by understanding the cause and the result of things [1]. It is to say, a positive method whose effectiveness was testified empirically played a key role in this process. The universality of such methodology was reflected in natural science. The problem is, however, should we view the study of modern design as the study of natural science? But any affirmative answer is very suspicious here. There is no doubt that, with the development of modern design, its many factors were involved in the natural science's researching. Such as the examination of safety in Modern Design, it could be related closely with the material science or mechanics, in which the subjective opinions can not be tolerable at any rate. But the issue is never as easy as we usually assume. If we only review the history of Modern Design or the history of modern Architecture, we could probably find that, in term not only of the ideas, but also of the forms, the historical interpretation have covered the issues massively with Sociology, Economics, Aesthetics, or even Political Ideology, etc. If compared with natural science research, the theoretical interpretation of modern design with such issues has had the traits which were included under the category of humanities study, where the very subjective attitudes or opinions could find their own places. From the view point of H. Windelband, the tolerance of subjectivity was a kind of remains which was common in humanities study and was only caused by the uncopiable historical process. So, the opinions or perspectives can not be as easily testified in an empirical way as done in natural science. Moreover, it is still has variety within such complicated studies as Modern Design. Assumed we were about to inspect the interactive relationship between the political situation before the Second World War with the development of Modern Design Movements then, say, the impact of Stalin government on Russian Constructivism or the Nazi power's influence on Bauhaus: only if the events happened in history exactly, we could acquire the significance of the historical knowledge whose availability was just from the related events which were viewed as the empirical evidences, although they happened only once and although we can not go back into the past in order to change the events (as the variables) to exam its validity, just like go back to keep Nazi group from get power. This knowledge could be doubted seriously by us for its deficiency as evidence, but it does not allow any more arbitrary. So, we improved our confidence on researching with such inductive knowledge, while it indeed enlarged the foundation of the theoretical study of Modern Design.

III. THE ISSUE OF AESTHETICS STUDY ON MODERN DESIGN

But we are not so as good fortune as ever. There are still many problems in this study that can never be resolved easily only by the logical deduction or by the collection of objective empirical evidences (whatever the evidences was from the repeated tests in nature or from the uncopiable history). The issue of aesthetics is one of these problems in the theoretical study of Modern Design. Within this particular field, we all have tried to establish the general criterion (as an efficient way) to guide the criticism, whereas anyone's subjective view can hardly be looked as the prejudice, for another prejudice could be lightly put on any existed view. This paradox could probably puzzle us. Supposing that we were talking about the architectural style of Garnier's Industrial City, for instance, upon its architectural forms with the character of cubic [2], the warmest approve could be brought as much as the fierce critics if we just care of its aesthetical forms in stead of the relationship with sociology or the meanings of political economics. Further more, each side can give no overwhelming testimony. If a forensic specialist was queried by his (or her) colleagues when they tried to clarify the identity of a unnamed body, he (or she) could show the fingerprint or DNA data as the proof, but anyone can not give the exact evidence like fingerprint when they remarked on the aesthetical issues of Garnier's Industrial City, no matter how their commentary is approve or against. Two strategic solutions would be presented to deal with this antinomy: (1) People will likely choose this standpoint that if we just concern the issue of aesthetics, the criteria which were from the existed aesthetical principles are the only permitted and available standards. But here the trouble is, comparing with the principles of logic theory, the aesthetical principles are only the regulated norms, or there were little absurdity when such regulations were violated. We could not imagine anyone can stay in two rooms at the same time, unless it is full absurdity logically. But when Marcel Duchamp showed his works "fountain", if someone sneer at it, such a sneer would be probably mocked again [3]. Or (2) people stand in line with scepticism, in which they began to doubt whether we can explain the aesthetical issues on Modern Design only based on the ideas from pure aesthetics. They go further out of the aesthetical field to look for the trace of the suitable answer. So, if anyone put a severe critics on the style of Bauhaus architecture because of its inhospitality or excessive functionalism, the sceptics could argued that such style will be the best, if we noticed that just the more in public begin to share the benefit from new architectural forms which is from the new architectural construct methods, technologies and materials, and again, they judged that it is very human and hospitable. Then, their criteria could have been revised under this thought and gave an affirmative review on its visible forms.

IV. THE ESTIMATION & REMINDING

With the first strategy, the supporters construct a believing system which was built upon their own religious sensibility, or they believe the self-sufficiency of traditional aesthetics to interpret the issues of aesthetics on Modern Design, and meantime they contempt the sceptics, seemed such suspicion as the Secondary rate in aesthetical appreciation or even being lack of it. On the other hand, because of the obstinacy and the spirit of suspicion, the sceptics will suffer the risk that they have to extend their endeavour to find answer of Modern Design's aesthetics issue in those fields which are out of Aesthetics study. So, in such a complicated process, the study of aesthetical appreciation on Modern Design will scatter itself into all kinds of other researching sphere, such as technology, political economics or material revolutions, etc.

The variety of chose strategies revealed the difference of ideas between supporters from each side. The former wish the study of aesthetical appreciation on Modern Design should be caressed only by aesthetical criteria tradition which was applicable in the study of Fine Art appropriately for so long ages. But the latter insisted that the study of Modern Design is a study which should be viewed as a true complex working based on the diversified scientific researching. Since this study based on science, the interpretation of its aesthetical appreciation should be acknowledged under the interactive conditions where the other regarded elements (historical, social or scientific elements, etc.) were bounded in. The latter's idea was reflected in the proposition of Tomas Maldonado (the second principal, Ulm Institute of Design) [4]: he supported that the essence of Modern Design should be rational, scientific and technological, he developed a extremely rational education system of design in this Institute, where he thought the education of design should based totally on the science, its educational purpose is only to train the designers with the character of scientific and have them serve the development of German industrialisation [5].

This essay is till now just being viewed as a notice to advise why a reliable method should be created in the theoretical study of Modern Design and what kind of problems it will confront with in this special research process, because the establishment of correct methodology is primary in the tasks, unless we could probably blindly advance in a wrong and confused direction but have little possibility to gain the considerable harvest in researching.

[1] Anthony Giddens, "Sociology", 4th ed., translated by Zhao Xudong, Beijing: Peking University Press, 2003, p7.

[2] Nikolaus Pevsner mentioned the new personalities of Garnier's Industrial City in his work "The Sources of Modern Architecture and Design" : the reinforced concrete were employed massively, but ornaments stayed in the less important. Beijing: SDX Joint Publishing Company, 2001, p151.

[3] It is reason why the subjective opinion can hardly be regarded as the prejudice.

[4] Wang Shouzhi, "A History of Modern Design", Beijing: China Youth Press, 2002, p290.

[5] However, such a position's disadvantage was to be the equal of its glamour, in the same book ("A History of Modern Design") the writer immediately mentioned when this position concerned more on the scientific process or on the rational and technological elements in industrialisation, but much less on the various psychological demands of human-being, it eventually led to the indifferent style with the lack of individuality and diversity.

Advanced DFT Features for an Industrial Application SoC

Zhikuang Cai, Kai Huang & Jun Yang
National ASIC Systems Engineering Technology Research Center
Southeast University
Nanjing, China
Email: whczk@seu.edu.cn

***Abstract*—With the development of process technologies and increased complexity of chips, SoC (System-on-Chip) testing becomes more and more difficult. Traditional testing methods cannot meet the current design requirements anymore. In this paper, we present a DFT (design-for-testability) scheme for an industrial application SoC chip basing on SMIC 130nm CMOS technology, it includes boundary scan test, memory BIST (built-in self-test), at-speed scan testing and parameter testing. Experimental results demonstrate that the proposed technique can gain high fault coverage and test compression ratio, which are 97.39% and 30%, satisfying the project demands.**

Keywords-DFT; BIST; SoC; at-speed testing

I. INTRODUCTION

At present, the semiconductor industry is introducing new manufacture techniques to meet the area, power and timing demands. As a result, SoC (System-on-Chip) testing becomes more and more difficult, traditional testing methods, such as boundary scan testing and full scan testing based on stuck-at fault model, cannot meet the current design requirements anymore [1][2]. Test engineers should try to optimize the previous DFT (design-for-testability) framework and test scheme.

There are two main challenges that must be addressed for at-speed test [3]. The first problem is that the number of patterns creating for at-speed test is several times larger than the corresponding stuck-at pattern amount. The second challenge is that we must provide the at-speed clock signal for testing. In addition, as a result of the use of multiple IP-cores for the chip, test scheduling and switching between the various models are particularly important.

In this paper, a complete DFT scheme for an industrial application SoC chip basing on SMIC 130nm CMOS technology is presented. It contains boundary scan test, memory built-in self-test, at-speed scan testing and parameter testing. As a supplement, delay test, test compression, and at-speed on-chip test clock generation are added to obtain high-quality test results [5][7].

The rest of the paper is organized as follows: section II explains the structure of the target chip and test goals; section III introduces various testing methods, and discusses some test difficulties; a complete test flow of the design is described in section IV, and also the experimental results are provided; the paper is concluded in section V.

II. BACKGROUND

The chip is designed in National ASIC System Engineering Technology Research Center, Southeast University. It integrates ARM core, PLL, MAC, ESRAM (64K), USB and so on. The design is targeted to operate at 300MHz and has four main clock domains, which are SYS_CLK, MAC_CLK, USB_CLK, CLK_5M, and there are other clock domains, such as TCK for JTAG and CLK_32K for RTC PAD. The overall structure is shown in Fig. 1.

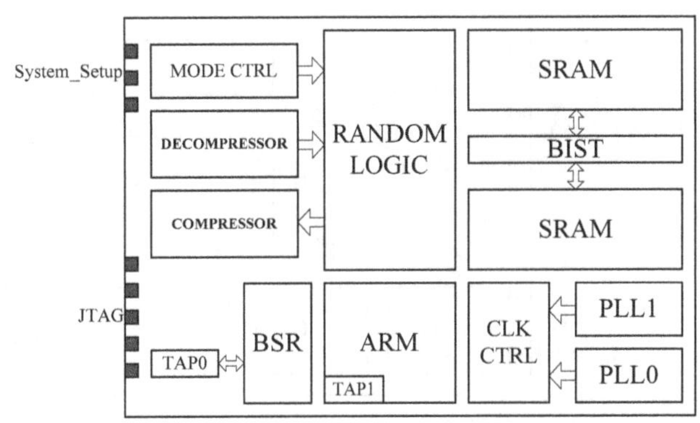

Figure 1. Schematic of the system structure.

The DFT scheme contains boundary scan test, memory built-in self-test, at-speed scan testing and parameter testing. Here, a mode control module is designed to generate each test control signal for every test mode, it has three input signals, system_setup [2:0]; and six output signals, which are func_mode, bsd_mode, mbist_mode, arm_mode, pll_mode and scan_mode. These signals are used to control the switching between the various modes in order to achieve good test results.

III. DFT FEATURES

A. Boundary Scan Test

Boundary scan test is achieved by adding boundary scan unit and control logic to each port. Using EDA tools, test circuit can easily be added on the chip, this step is usually completed before the scan chain insertion. There is a TAP module in ARM core itself, so if inserting a new TAP controller, there will be two on-chip TAP controllers, but only one set of JTAG ports. Since these two modules are based on the IEEE 1149.1 standard, there are two solutions: 1) maintains the two TAPs, using the mode control signal to select; 2) inserts ARM2BSR module, sharing a TAP controller and saving chip area.

B. Memory BIST

We adopt built-in self-test method to test embedded SRAM [4]. Here, two 32K SRAM can be carried out parallel tests. MBIST structure is shown in Fig. 2. MBIST module is composed of five parts: JTAG expansion module, control module, pattern generation module, address generation module and comparison module. In order to save I/O ports, we can reuse free scan chains in the core, exchanging test signals through the JTAG port. We also use dual-clock operation, a clock for the JTAG interface clock for the completion of the initial MBIST configuration and the final test results output; while the other synchronization clock for the system clock for the process of testing, so memory can perform at-speed tests to detect faults associated with the timing. The MBIST algorithm is March C-, fault models which can be measured include: AF, SAF, TF, SOF, CFin, CFid and so on.

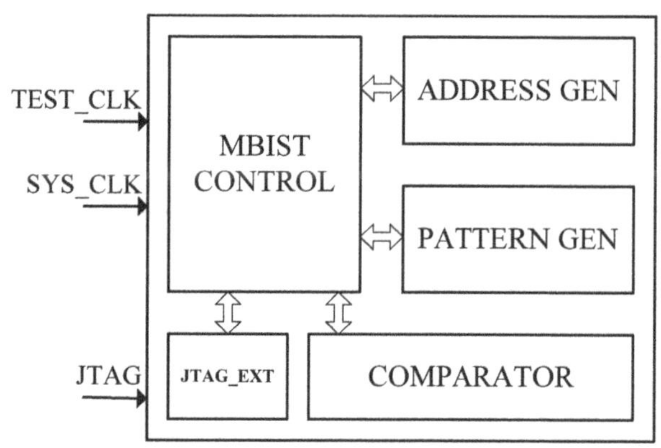

Figure 2. Schematic of memory BIST circuit.

C. At-Speed Test Clock Generation

At-speed testing for DSM (deep-submicron) defects requires the accurate application of high-speed clock pulses to apply the tests for these timing-related fault models. Since the frequency of ATE is limited, using on-chip clock to apply at-speed test for a high performance design is an economic solution. What is more, it is better to get a high quality at-speed test results, without additional skews from the test equipment or test fixtures [5][6].

This design includes an on-chip PLL for generating the at-speed testing clocks. It contains three parts: 1) CLK GEN module; 2) CLK CTRL module; 3) CLK CHAIN module. Fig. 3 shows the schematic of the at-speed testing clock generation circuit.

The CLK GEN module contains three parts: PLL, PMU and GCM. PLL provides high frequency clock signal, and PMU works as a divider which creates different clocks for multi clock domains. GCM is an important part for clock generation; it eliminates any possible glitch existing in clock signals.

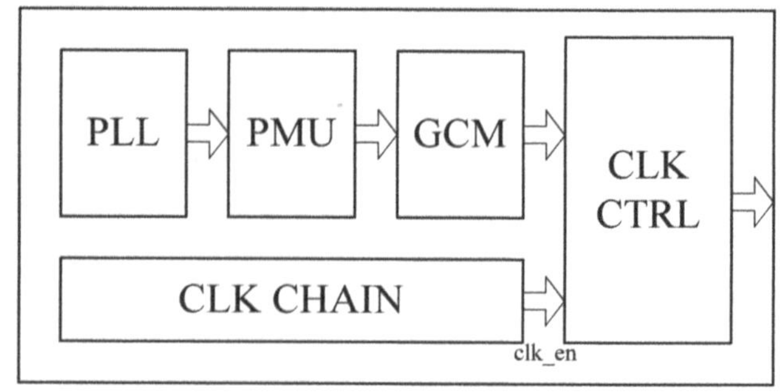

Figure 3. Schematic of at-speed test clock generation circuit.

The CLK CTRL module serves as an interface between the CLK GEN and system logic, and controls internal clocks to switch from a slow clock during shift to a fast clock during launch and capture. It is consisted of clock multiplexing logic, clock gating and a GCM module. The input signal clk_en is from CLK CHAIN, it determines whether switching clocks between slow clock and fast clock. The CLK CHAIN module is a common scan chain consisted of one or more scan cells SDFF. This chain determines a clock selection mechanism by generating output signals clk_en. Clock selection values are loaded into the clock chain as part of the regular scan load process. The SDFFs of Clock chain are connected to dedicated scan inputs and scan outputs and are outside the decompressor to compressor path, if the circuit is inserted with scan compression logic.

D. Parameter Testing

Parameter testing contains IDDQ testing and phase-locked loop output parameter testing. As a standard feature during testing or an important complement to scan test, IDDQ testing can detect many types of failures of the CMOS circuit, and other methods is almost impossible to detect these failures. Since the I/O PAD has an independent power supply rail, I/O PAD and the core logic can be tested separately; to avoid the use of internal pull-up or pull-down, if possible, drive or bypass buses to prevent them floating.

IV. Experimental Results

Based on 130nm standard CMOS process, we present a complete DFT scheme for the test chip. During synthesis stage, we use 1-pass scan synthesis, D flip-flops are replaced by scan flip-flops after integration, which is conducive to the timing optimization and reduce the hold timing problem. We must consider the design of the test compression circuit and at-speed test clock generation, this can greatly reduce the test costs. For scan ATPG, each fault model (Stuck-at fault, IDDQ fault, Delay fault, Transition fault) must be used to generate the corresponding test patterns. What is more, these vectors are required to run simulation for verification. After the completion of pre-simulation, STA (static timing analysis) is in process, and then standard anti-SDF simulation (post-simulation) is used to verify the vector problems in considering the correlative timing information. The ATE testing includes test procedure conversion and debugging, the format of general-generated test pattern is WGL, needed to be translated into the binary codes files. The process of debugging is to check the connectivity of the chip, and then, includes SCAN, BIST, BSD, FUNC and some parameters test.

This article only gives experimental analysis based on stuck-at fault model, carried out to compare the difference between before and after compression using of DFT MAX, and other test results (such as BSD, MBIST, IDDQ, etc.) are not discussed. Table I gives the chip test information which is not designed with scan compression, the chip has 10 scan chains, scan depth is 2330, based on the SA fault model, the test coverage is 97.71%, generating 2068 test vectors, and the chip area is 23011504 after integration. Calculated with the test frequency of 10MHZ, the time required to complete the test is 482ms in theory.

TABLE I. BASIC TEST INFORMATION OF THE CHIP

Type	Number
Fault Model	SA
Test Frequency	10MHZ
Scan DFFs	23295
Scan Chains	10
Scan Depths	2330
Test Patterns	2068
Fault Coverage	97.71%
Area	23011504.000000

DFT MAX compression uses adaptive scanning strategy, that is, its input signal decompression circuit is formed by the combination of MUX circuit, and output signal is compressed by the XOR network. During design, according to the formula M*N*1.2 to determine the structure of the circuit after the compression, in which M refers to the existence of the original number of scan chains, N is the smallest value of compression [7][8].

TABLE II. TEST INFORMATION OF THE CHIP AFTER COMPRESSION

Type	Number
Scan Chains	360
Scan Depths	66
Test Patterns	2421
Fault Coverage	97.39%
Area	24963267.000000

Table II gives the chip compression data, M and N are 10 and 30, respectively, when fault coverage and the chip area do not change, the basic testing compression ratio is up to 30 times, the effect was very satisfactory to the project required and test cost is reduced greatly.

V. CONCLUSION

In this paper, a DFT scheme containing boundary scan test, memory built-in self-test, at-speed scan testing and parameter testing is presented for an industrial application SoC chip. Experimental results show that compression-based test results are highly satisfactory and significantly reduce the test cost.

REFERENCES

[1] NIKIL K, PAREKHJI RA, "DFT for test optimisations in a complex mixed-signal SOC - Case study on TI's TNETD7300 ADSL modem device," 35th International Test Conference, 2004, pp. 773-782.

[2] DATE H, HOSOKAWA T, Muraoka M, "A SoC test strategy based on a non-scan DFT method," 11th Asian Test Symposium, 2002, pp. 305-310.

[3] LIN XJ, PRESS R, RAJSKI J, et al, "High-frequency, at-speed scan testing," IEEE DESIGN & TEST OF COMPUTERS, vol. 20, 2003, pp. 17-25.

[4] BS WANG, Q XU, "Test/Repair Area Overhead Reduction for Small Embedded SRAMs," 15th Asian Test Symposium, 2006, pp. 37-44.

[5] X. Fan, Y. Hu, L.-T. Wang, "An On-Chip Test Clock Control Scheme for Multi-Clock at-speed Testing," Proceedings of the Asia Test Symposium, 2007, pp. 341-346.

[6] Da Wang, Rui Li, Yu Hu, Huawei Li, and Xiaowei Li, "A Case Study on At-Speed Testing of a Gigahertz Microprocessor," IEEE International Symposium on Electronic Design, Test & Applications, 2008, pp. 326-331.

[7] Selina Sha, Bruce Swanson, "A Real Case of Significant Scan Test Cost Reduction," IEEE Computer Society Annual Symposium on VLSI, 2008, pp. 239-244.

[8] Synopsys, DFT MAX User Guide: Adaptive Scan, http://www.synopsys.com/, 2007.

An Analysis on the Influence of Lifestyle Studies on the Development of Industrial Design

Long Zhoujun
College of Fine Arts
Hubei Normal University
Huang Shi, China
Email: brokenangel@sohu.com

***Abstract*—Lifestyle symbolizes the way people live. It influences the forming of human culture and keep records of people's actions and patterns of living. Nowadays, people's lifestyles are getting more diversified. Changes are taking place in fashion trend; the demands segmentation of the consumption levels are urging us to make adjustments to the industrial design in accordance with the pace of globalization and the regional and cultural development in order to meet the demands of the times. Industrial design is playing one of the leading roles in forming and changing lifestyles. At the mean time, the introduction of the Lifestyle Concept is making the product development more people-oriented or culture-oriented in order to be more attractive to consumers and meet their real needs. For these reasons, this paper demonstrates that the studies on lifestyle design not only make industrial design more people-friendly but also improve the quality of lives.**

Keywords-Lifestyle, industrial design, development trend, consumer group, influence.

I. Definition and Origin of Lifestyle

A. *Definition*

Literally, lifestyle means life and attitude. In an ancient Chinese book the word "life" is defined as "all sorts of daily routine ". While the word "attitude" is explained in the book *Social Psychology* as "the continuous orientation of conception of people, things and objects in terms of recognition, emotions and behaviors". Generally speaking, a lifestyle is a certain way people live and the attitudes they hold toward life. The ways and attitudes result in different external tendencies of people's behaviors. These tendencies make up people various cognitions and feelings toward things and objects, affecting the designer's comprehension and understanding of industrial products design. Lifestyle symbolizes the way people live. This function of symbolizing not only affects the development direction of industrial design, but also keeps a record of the behaviors people have in daily life and the development of the ways of living.

B. *Origin*

The study on lifestyle was originated from psychology and sociology. It was applied to marketing management in the 1960s and introduced to the designing industry in 1980s. The basic conceptions and theories of lifestyle were first coined by Adler, a psychologist in 1927. He believed that a lifestyle referred to the goals set by an individual for himself and the means to achieve these goals. That is to say, a lifestyle is the means people use guided by a central goal. There have been many scholars who conduct discussions on lifestyle from different aspects after Adler, which laid a solid theoretical foundation for future studies[1].

II. The Applications of the Lifestyle Concept in Industrial Design

A. *Shifts in the Traditional Design Concept*

In traditional industrial society, the concept of industrial design was based on the large-scale mechanized production and the cost and value of production were integrated as a whole. The design only put emphasis on the functions of the products and the forms of products were subject to functions. Product competency mainly lay in quantities and performance rather than forms and types of the products. However, in the post-industrial society the value of products was affected by cost and other non-material factors. In the knowledge-based economy, the resource of production is knowledge rather than materials. For example, information of large amount created value by using non-materials such as knowledge. People's views on consumption also took a great shift, which has led the product design more form-oriented rather than function-oriented. Beauty in product design was enhanced and product brands also became more diversified. Any corporation, no matter how large it is, can't only produce one product as they did before or else they will lost market competency. Therefore, the relationship between functions and forms of products is not as closely connected as before. For many products, forms and functions are not so clearly divided so that even if there is a form, the form itself turns into something non-material, untouchable and unobservable and doesn't have shapes, colors and lines or other symbols any more. "In the end, the final products can't be put in front of us to be explained or interpreted. They become a function or something super-functional. The non-materialism of forms and the super-fictionalization have gradually separated design from the material level and turned it more spiritual." [2]. The shift in the design concept calls for products to be more people-friendly, fun and diversified than just being practical. Therefore, the industrial design in the knowledge-based economy is not purely a technical job. Instead, it is a kind of activity which combines technological production with human touch. Products produced in this activity not only have the practical functions but also show a humanistic spirit. People all want to live a happy life. They want their material desire to be satisfied, which is human nature and understandable. However, people need spiritual life other than material life. Material life is the basis and guarantee of human survival while spiritual life makes us feel alive and has made humans paramount among all species. [3]. That is to say due to the shift in the traditional design concept; industrial design is beginning to cater for people's needs for spiritual life other than that for material life.

B. Core of the Views on Lifestyle Design

Lifestyle research is just one of the marketing methods. In the process of product designing, however, practical market positioning can be conducted through analysis and planning of lifestyles in order to cater for consumers' tastes and meet the demands of the mainstay. In this way a bridge can be built between products and consumers for better communication, which are the core of the views on lifestyle design and the aim of applying the lifestyle theory to practical industrial product design.

From the aspect of designing, a lifestyle offers a whole new interpretation of the forms and applications of industrial product design and development. Except from introducing the relevant factors of lifestyles and design development, a theoretical view and the practical application of lifestyle design is stated. As a result the idea of lifestyle has been widely applied in areas concerning consumer behavior, marketing and designing in order to help with the segmentation and positioning in the market. If designers and salespersons can understand consumers' lifestyles, it will greatly facilitate their product planning, positioning, advertising and sales promoting. What's more, they can make further improvements in the products to better meet the demands of customers.

C. Ways to Apply the Lifestyle Design Concept to Industrial Design

Studies and research on lifestyles are targeted at a certain person or group of people about the features of their daily life. These features include the social setting, regional culture, geographic landscape, personal values, consumption view and worldview, personalities, interests and hobbies, interactions between them and the surroundings and the products they use. In other words, the lifestyle and the trend of social development can be reflected in the study and understanding of consumers' behavior and their attitudes.

Therefore the lifestyle design concept is mainly about knowing and getting hold of the possible tendencies of the consumer groups and then creates new products according to their needs, seek market opportunities, obtain initiative and the say in the market, and emerge victorious in competitions. For industrial product design, research on lifestyles is to study the great impact brought by the times, further observe the demands and tendency of designing so as to meet the needs of consumers of different levels. For the development of design for modern daily articles and products, research on lifestyles is not just for knowing about the market share of different consumer groups and the gaining knowledge of modern lifestyles. More importantly, we hope to obtain correct and effective information from the research and use it as key reference and data support for future product design and development.

III. Importance and Impact of Lifestyle Design in Industrial Design

A. Importance of Industrial Design in Social Life

Industrial design provides major support for the manufacturing industry and is the main component in modern service industry. Looking back at the history of the manufacturing industry in western developed countries, industrial design played the leading role in each stage of its development[4]. Traditionally speaking, manufacturing means to produce people's daily use by machines on large scales. As long as the articles have a certain function, they're treated as product no matter how they appear in form. However, with the coming of the knowledge -based economy, industrial design is needed not only in material life but also spiritual life. Anything that has a direct bearing on human survival needs industrial design. As a western scholar put it, "To some extent, every human action is designed as long as it is aimed at changing the status quo in order to make it better" [5].With the development of economic globalization and the coming of the era of service industry, industrial design is becoming increasingly important in building brands, forming corporate competency, making structural adjustment, enhancing the national strength and improving people's livelihood. Countries and corporations are attaching more importance to industrial design and it has become an essential component in social life. Furthermore, as economic globalization is gaining momentum, industries in different countries are becoming unprecedented interdependent. The specialization of international division if labor has been intensified and there had been great changes in the competitive landscape. As competition in the global market is getting fierce, the comparative advantage of products is being taken place by competitive advantage. So only by giving priority to design, R&D, development and innovation can we succeed in designing environmentally- friendly and energy-saving products with advanced functions. Successful design of products which meet the needs of the global market can improve the international product competency of China. In this sense, industrial design is key to China's manufacturing industry for our country to prosper in the global competition [6].

B. *Significance of Lifestyle Design*

With the evolvement of the times and the social, economic and cultural development, a corporation should consider consumers' habits and needs before developing a new product. This is also in line with the advancement of information technology and the trend of a more diversified consumer market. If a corporation can meet the multi-level needs of consumers, it will be able to get the dominating role and earn profits in this buyer-oriented market which is filled with fierce competition. From the perspective of designing and developing industrial products, the trend at present is to be consumer-oriented. Consumers are the final ones to test and use the products , therefore, their feelings and views on the product is of great importance and conducting research on the lifestyle design is necessary. The research and studies are actually a systematic and comprehensive analysis of consumers' daily activities, interests, opinions and hobbies. In this way, the understanding of the consumer groups' attitudes toward life can be obtained to help form the concept of the product and develop the design [7]. Such basic analysis can be of reference to consumer demands in designing industrial products, and provide key data support in the design and development process that follow.

Lifestyles can pre-determine the direction of developing product design based on consumer needs, expectations or complaints. This is also one of the major reasons for a differentiated consumer market as it meets the needs of consumers and starts developing a new lifestyle. People's needs and interests on various products are affected by their lifestyles. As a matter of fact, the impact brought by lifestyles can be observed in consumer behaviors and attitudes.

Discussions on lifestyle design applied to industrial design can act as a channel for communication between the product and its users. Knowing about and understanding the influence of tendencies in life and the times helps conduct further studies on the directions and priorities in product development. Only in this way can the real needs of consumers with different lifestyles be met, the quality of life improved and new trends or fashions of life be encouraged. [8].

C. *The Role of Lifestyle Studies in Industrial Design*

A lifestyle acts as the bridge between people and products and is the key factor in their interactions. Therefore the design of industrial products is closely connected to lifestyles. The diversity in levels and types of lifestyles include: the difference in age, marriage concept, environmental protection concept, social form, economic mode, the idea of consumption and personal values. Effectively and correctly comprehend the constructing members and the positioning of the market's targeted group of consumers is key to successful design of advanced industrial products as well as an effective way for the corporation to survive, develop and take the lead in the market.

As lifestyles symbolize the way people live, a new way of life has come to us, together with the information age, the changes in the fashion trends, priority given to consumers' needs and the development of culture and civilization. All this has made the ways of living suit people's wishes and live up to their expectations. There's no doubt that designing plays the key role in forming and changing lifestyles design of industrial products will become the component of life and lifestyle will also leave a great impact on industrial products. Lifestyle will continuously develop in a more human-friendly, standardized and dynamic direction. It is not only important to designers, but also essential to improving people's livelihood and the environment they live in.

Industrial design has changed the way people live and think in less than 100 years. It has raised people's desire for social and economic development and accelerated the improvement of society. The essence of industrial design is to create a new way of life. Product design, being the core in industrial design, has a direct bearing on people's life styles. Humans have lived on this planet for millions of years. The use of tools as means of living symbolizes the leap in people's way of living. The evolution of lifestyles reflected the history of both human evolution and development of industrial design. It created human civilization and the mode of living, and has laid the foundation for new lifestyles and trends in the future.

D. *Influence of Lifestyle on Future development of Industrial Design*

The living modes discussed in lifestyle design reflect the special quality of different groups of people. However these qualities may change since people live in different cultures, regions, have various tastes of life and different economic conditions. Thanks to the influence of the lifestyle design concept, now we can obtain the latest information of living modes , apply the information in industrial design and produce daily products that better meet future standards and with market potential.

The industrial design in China is just beginning to take shape and now is at the most important stage of development. At present the Chinese economy is growing at a fast speed. Therefore, the fast-growing economy can serve as an engine in encouraging industrial design. Design acts as a strategic tool for stimulating development of products and services, promoting innovation and competition among corporations. According to Mr. Michael Thomson, Executive Board Member of the International Council of Societies of Industrial Design, "In China, the necessary environment for industrial design companies to grow and prosper must be formed under strict legislation and affected by market forces both home and abroad. It will also be greatly influenced by external competition, demands at the exporting market as well as the educational system, because education encourage individuality in creativity and strategic mind. It needs the designers, together with the corporation, to strive to satisfy the needs of users from manufacturing and service industries, because consumers' experience of using the product is as important as the quality of the product itself". [9].We can see from his words that the development of China's industrial design calls on us to put customer's needs first and always have customer-oriented minds. We need to get a better understanding of people's needs based on innovation in design, seek sustainable development rather than focusing on low production cost.

Through the analysis and studies on lifestyles, people can better understand the trends of and fashions of the times .As time passes by , people's minds, living modes and needs are gradually affected by the advancement of science and technology. In the knowledge-based society with high level of consumption, the consumer market is showing energy, vitality and passion. Many new products and new ways to use products are changing people's lifestyles and living habits and new science and technology is also leaving its impact on our lives. As a result, industrial design cannot only be designated for the daily routine. We should take the lifestyle design concept into consideration, conduct deep and long-lasting discussions on designs in order to succeed in designing products which better meet the demands of consumers of different consuming level and improve their life quality. Therefore, knowing about the current lifestyles can help us not only get hold of the pace of the times, but also make inference of the direction and priority area in future lifestyle design, eventually making the new lifestyle better satisfy our needs and provides hope for our future.

REFERENCES

[1] Dr.Jui-che Tu. Life Style Design., [M].Asiapac Books,2004. p.15

[2] Teng Shouyao, Non-material Society. Preface, Sichuan People's Publishing House,p.7

[3] Teng Shouyao, Nie Zhenbin, Aesthetics and Designing in the Age of Knowledge Economy, Nanjing Publishing House,2006, p.52

[4] Wang Xiaohong, Ma Qianmai tr. Designing creates wealth, China Light Industry Publishing House,2006, p.1

[5] M. Diani Marco, Non-material Society, Teng Shouyao tr. Sichuan People's Publishing House,1998, p.106

[6] Wang Xiaohong, Ma Qianmai tr. Designing Creates Fortune, China Light Industry Publishing House,2006, p.2

[7] Dr.Jui-che Tu. Life Style Design., .Asiapac Books,2004, p.49

[8] Dr.Jui-che Tu. Life Style Design., .Asiapac Books,2004, p.54

[9] Wang Xiaohong, Ma Qianmai tr. Designing Creates Fortune, China Light Industry Publishing House,2006, p.41

An Exploration of the Residence Unit Design in Changsha

Wang Wei
School of Architecture
Hunan University
Changsha, China
Email: wangwei_19821121@163.com

Miao Lin
The Research Center for Internet Information Services
Hunan University
Changsha, China
Email: linlinei@yahoo.com.cn

Abstract—**To achieve the development goal of actively building a resource-conserving and environmental friendly city cluster advocated in the Two-Oriented Society development process in Changsha, Zhuzhou and Xiangtan, the researcher intends to design a residence unit compatible with the developments of two-oriented city cluster through researching and studying various elements such as local geography, climate, social economy, families and population, etc. The researcher will conduct a pilot experiment to analyze and study the optimized design of small and middle-sized residence units so as to propose a reference for improving residence units in Changsha.**

Keywords-small-sized;apartment spatia scale analysis; improved design strategies

I. Introduction

As the acceleration of urbanization and the booming economic development after reforms and opening up, there are tremendous changes happening in people's life. But housing is still one of the key issues for citizens and families in China. It provides inevitable material foundation for people and is the significant measurement criteria for the human happiness. So how to inherit the city's culture heritages, which means how to make the architecture more compatible with ecological environment "learn from the nature" [1], to create the most optimized residence unit is quite important.

II. Improved Design Strategies

A. Coming to the point

The way of living to some extent can reflect a time; it is one of the expressions of different regional culture. [2] Through a survey analysis to the ways of living in Changsha families, the researcher decides to proceed with four aspects. First, conduct a survey on the already established living environment, especially on the comprehensive environment of basic living units. This is to understand correctly the relationship between the Changsha local people's way of living and the activity behavior, and to elaborate positively the effects of decoration and furniture [3].Second, study the cognitive orientation of the residents to the living space, and infer the potential requirements of the residents. An analysis will be made about the comfort of the residence unit and the functional flexibility, and this will lay a basis for the improvements of the following residence unit design. Third, study the constitute of family members and design a more rational residence unit design according to the different development stages of people and the changes in the constitute of family members. Fourth, make a conclusion about the characteristics based on the Changsha people's understanding of living environment, their living awareness and fondness. The information collected through these four channels should be multi-dimensional and can cover all influencing factors of majority of the living ways. These factors are demonstrated in Diagram 1. These factors are rectified as the development of times, and possess a clear developmental venation.

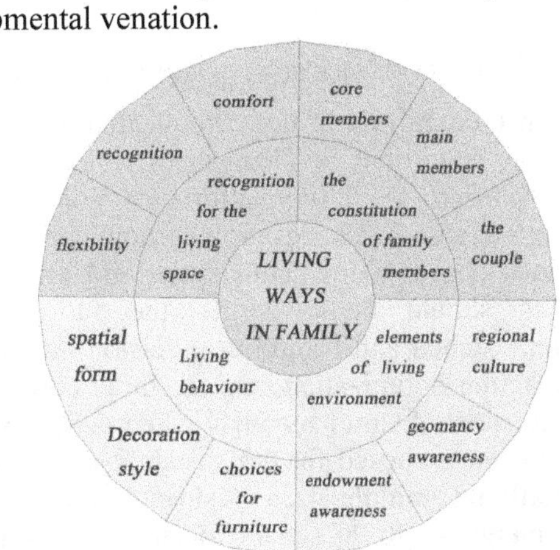

Diagram 1: The analysis chart for accommodations and elements of life

B. Space measures analysis of residence units

Small-sized residence unit is the fundamental environment of individual's surroundings. Space measures are very important to human, only when suitable space measures demonstrated in design can the space measure be essentially meaningful [4]. There should be certain space measures around a people. The visible measures includes outside measurement, height, weight, and other attributes; invisible measures contains auditory sense, visual sense, olfactory sense and other extended attributes. According to the interpersonal communication in family, there are five types of absolute measures in family,(Diagram 2)they are standing dialogue measures, sitting dialogue measures, public measures, interactive measures, and recognition measures. For these five measures, 1.5 meters and 2 meters are the turning points of the relationship among family members. With the space distance of 1.5m, people can conduct sitting or standing communication. With the space distance of 2m or 2.5m, the family members can conduct effectively public communication such as interaction or entertainments.

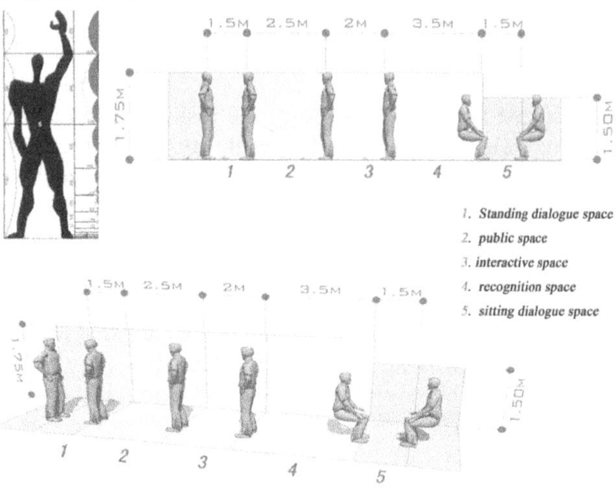

Diagram 2: The analysis chart for human space measures

III. THE STUDY ON CHOOSING RESIDENCE UNIT DESIGN

Based on the analysis of the complicated local climate and geography conditions in Changsha, five kinds of space have been chosen as the breakthrough points for the residence unit design in Changsha: bathroom space, storage space, kitchen space, living room space and lobby space. Some focus should be given to the rational plane optimization, kitchen and bathroom precision optimization, spatial interaction optimization, and the three optimizations should be considered simultaneously when designing. A, B and C have been proposed for the sustainable development of the small-and middle-sized residence units in two-oriented society: A. 70-90 square meters, B. less than 50 square meters, C. 50-70 square meters. (Diagram 3) These three kinds of residence units will benefit the building space resources research and study for the resources conserving and environmental friendly two-oriented society.

A. Rational Plane Optimization

As the wild fluctuation of spring temperature, the heavy summer raining, the autumn high temperature, and the infrequent winter chilliness in Changsha, the A, B and C residence units always keep upright and foursquare layout. About the functionalities, the design will position the overall structure of the residence unit according to the activity space required by the family members in Changsha, such as living room, dining room and other public space. The chosen controllable floor height is 2.8m to 3.0m, the clear heights in living room and bedroom are 2.6m to 2.8m, controllable clear heights in kitchen and bathroom are 2.4m to 2.6m, the ideal depth length in living room is 3.6m to 3.9m, and the depth length in bedroom is 2.2m to 4.2m. Considering the depth of residence unit is visually amplified, it is possible to adopt the design of longitudinal binary space to remit the visual impacts. The entry garden in the vestibule divides clearly the shoe cabinet and garden area. This can protect the privacy while remitting the entry space, and is also a representation of architectural etiquette.

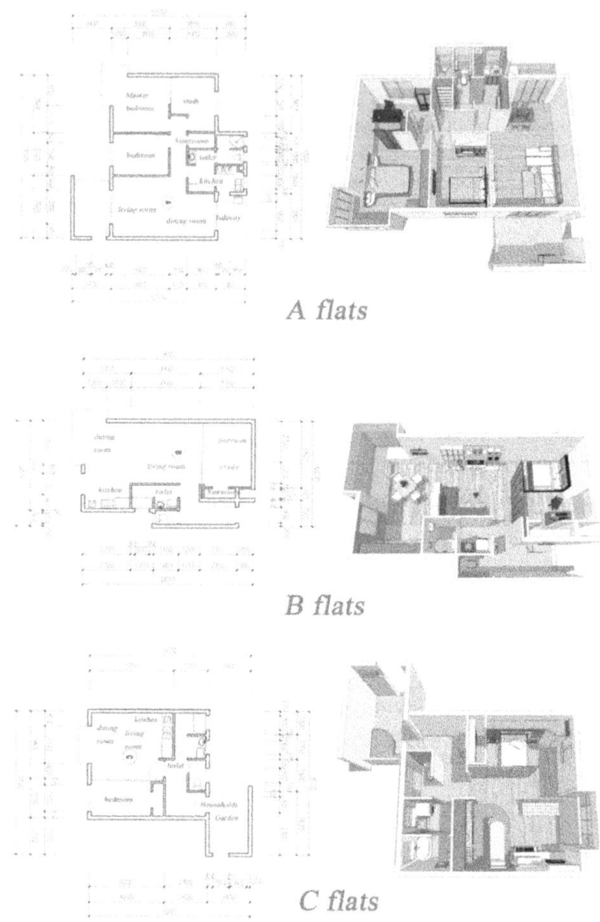

Diagram 3: The drawing of the ABC three kinds of suitable flats

B. Unit Optimal Combination

It is possible to create a more optimized combination to use land resources more efficiently based on the A, B and C residence units. One staircase of two households plan designs are broadly considered to be the best unit combination [5]. No matter in which inspective process of different elements such as sun exposure and ventilation, the one staircase of two households cannot save land resources effectively; only a unit combination building composed of one staircase of three households or four households or even more can realize the saving of public resources (such as elevators, stairs, shared pool area and occupied area). To solve these problems, the A, B and C three kinds of residence units are designed. The three small-and middle-sized unit combinations are with appropriate area scale, compact plane layout, clear unit subarea, and appropriate arrangements for kitchen ventilation and washing machine drainage. The appearance after the combination is simply and profusely and with low shape coefficient. This has provided better conditions for house ventilation, natural lighting and storage space (Diagram 4).

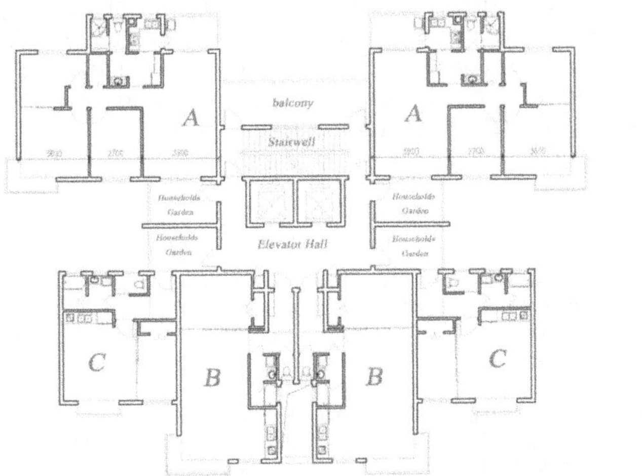

Diagram 4: The area analysis of the combination of ABC flats

IV. The Experiment Analysis of Ventilation and Natural Lighting after Unit Combination

A. Computer-based Simulation Experiments Analysis of Natural Lighting and Ventilation after Unit Combination

Comfort Index is used broadly in meteorology [6]. After the units' mutual combination, the ECOTECT test software was used to conduct natural lighting analysis after unit combination. According to the buildings lighting design standards (GB/T50033-2001) for the natural lighting experiments, it is required that the outdoor illumination should be 5000Lux, the calculation should be carried out based on the window sill section (the sill is 0.9m away from the floor), and the indoor surface reflection coefficient is 0.5 (pale color). As the calculation shows, the indoor average illumination intensity is 219.75Lux, the corresponding daylight factor is 4.4%. Compared with the Standard Visual Working Level III (the lowest lighting 100Lux), it is clear that majority of the unit meet the requirements.

Usually light-operated t illumination analysis concept is based on the research achievements of Building Research Establishment of England. It demonstrates that the energy conserving effects can be assessed by the ratio of the closing hours of the illumination device and the time of natural lighting. And the concept can be used to initially analyze and assess the energy-conserving effects of a natural lighting house equipped with light-operated illumination device. Hypothesis: sensor regulator is designed as switch mode, the minimum illumination intensity is 300Lux. The calculation finds out that in 77% of the natural lighting time, it is possible to switch off the manual illumination device (Diagram 5).

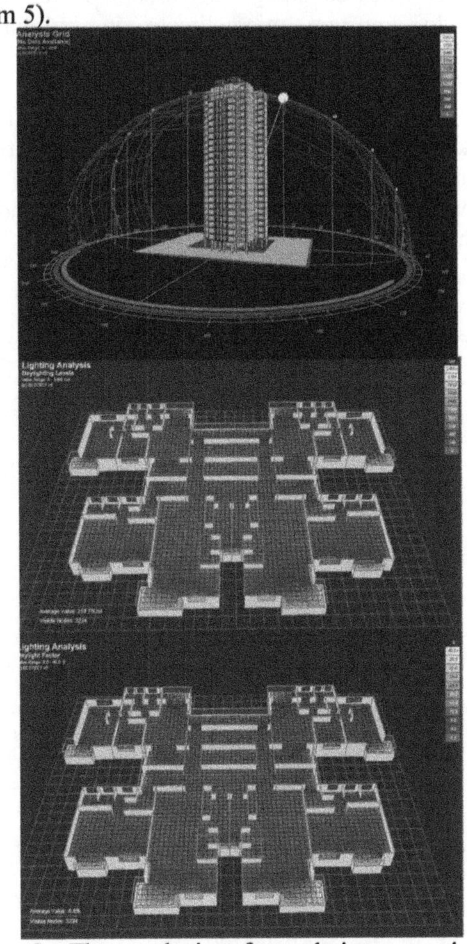

Diagram 5: The analysis of emulation experiments on lighting

B. The Experiments Analysis of Ventilation after Unit Combination

The analysis of ventilation is based on the κ-ε Equation, According to the ECOTECT climate data in Changsha, wind speed relaxation factor was set as 0.2, energy relaxation factor was assumed to be 0.5. According to The Special Atmosphere Dataset of Buildings Thermal Environment Analysis in China, the experiment was conducted to imitate the ventilation effects in summer nights. The outdoor wind speed was set as 2.4m/s, wind direction was south-wind. The wind vector plots of the sections whose distance to the floor was 0.9m (sitting) and 1.5m (standing) and the outdoor wind speed nephogram have been obtained (section spacing is 0.3m).

After the analysis of ventilation experiments and the sundial diagram of the building, the buildings with layout of southwest to east about 12 degree in Changsha can be perpendicular to the summer prevailing wind direction, and will get smooth and abundant lighting and ventilation. The form of empty space can make the natural wind smoothly goes through the buildings or communities. This can make it possible to take advantage of urban prevailing wind direction to improve microclimate in buildings and communities and to reduce the happening of thermal conduction effects. Considered Changsha locates at 28 north degree, the best sunlight direction is 10 degree south to east, while taking the summer prevailing wind direction into consideration, 10-15 degree south to east will be the best orientation to absorb abundant sunlight and extend winter sunlight hours (Diagram 6).

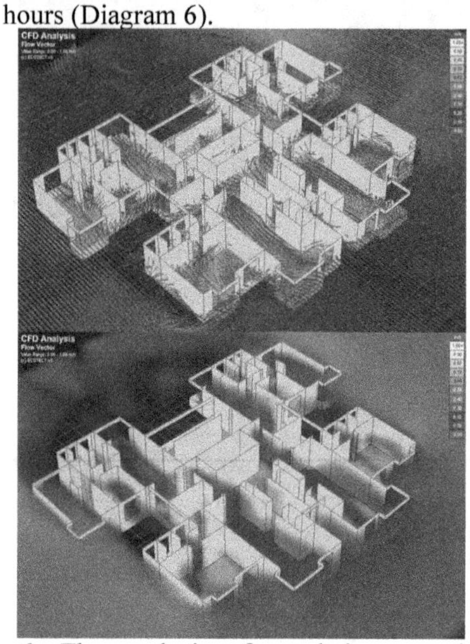

Diagram 6: The analysis of emulation experiments on ventilation

V. CONCLUSIONS

In the process of constructing the environmental friendly and resources conserving two oriented society, the small- and large-sized residence units are in a significant position. After a series of optimization strategies mentioned above, the A, B and C are the three units best suitable in terms of human spatial scales. After the optimal combination of the three residence units and the simulation experiments of lighting and ventilation based on the atmosphere data in Changsha, an ideal result has been achieved. In pilot area for the tow-oriented society construction, the optimal design strategies, the appropriate technology and the green building tools can be adopted to reduce resources consumption while ensuring the quality of the design so as to realize the important tactic of sustainable developments.

REFERENCES

[1] Lmcharge Ian, Design with Nature, Doubleday/Natural History press Doubleday & company. Inc. Garden city, New York, 1994.

[2] Song Chunhua, A Long Way for China's Residence Construction: A Thinking about the Residence Construction around the Turn of the Century, Construction Institute News Paper, vol.10, pp 4-8, 1999.

[3] E. Laumannn and J.S.House, Living Room Styles and Social Attributes: The Patterning of Material Artefacts in a Modern Urban Community [J], Sociology and Social Research, pp 85, 1970.

[4] Mc Harg, Design with Nature, National History Press, Garden city, New York, pp 33, 1971.

[5] Zhao Guanqian, Explaning the Concept and Planning Design for Small and Middle Sized Units: An Analysis of 90 Excellent Small and Middle Units Schemes, Construction Institute News Paper, vol.4, 2007.

[6] Dong Huiqing, Huang Haihong and Huang Xiangxing, Prediction System of Human Comfort Index, Atmosphere in Guangxi, vol.20 (3), pp 37-40, 1999.

Analysis Development of Animation with National Culture

Zhou Yan
Arts and Design College
Wuhan University of Technology
Wuhan, China
Email: zhouyan706@sina.com

Zhang Wei
Arts and Design College
Wuhan University of Technology
Wuhan, China
Email: power900@sina.com

***Abstract*—In Animation creation, we don't have to be obscured by the traditional material and formality, we still could glamour our people's traditional from the topics, characters and etiquette and so on. Chinese Animation must make progress in two aspect, one is their mentality be broaden not stiffen; second is the real realization of their traditional civilization so that we could find a path of our own development of Animation.**

Keywords- Animation, National Culture, development

I. Preliminary Remarks

Animation the magical fashionable and popular art form could be traced up to 20~30 thousands years ago when cave man complete the art piece, but as an independent art formality, Animation have a hundred year history. As the most ancient and modern popular phantasm, Animation has found its way to different aspect of our community.

Chinese Animation has had his glamour and also desolation. But when we notice the glamorous vision of western world, the mimicry and learn is natural and necessary. And there is one trend, that to give up our traditional cultural and spirit, and blindly mimic popular formality of the western world. Not to notice their conception overwhelming just to mimic the techniques and visional assignment. This is against the civilization trend of the world.

As a matter of fact, the western Animation had never given up their civilization heritage and nowadays the global civilization circumstance is multiple. In Animation creation this multiple atmosphere expresses itself in the aspect of the multiple style and forms. By means of their traditional art forms stress on regional character, visualize ethnological, daily lives, historical and cultural backgrounds. So as to stand their stance in the world of Animations, by this mean Chinese Animations must notice their heritage of traditional civilization. In aspect of mimic the foreigners, we must reflect our own.

II. Traditional civilization cast an influence over the Animations art form

As a form of civilization, Animation realizes the heritage of civilization. This heritage differs in aspect of Culture, we utilize as a language to show our own, philosophic, sentimental with the point of we want to attain this, we must deeply understand our own traditional civilization, inherit their merits, and create better Animations.

The traditions of art is the accumulation of historical, with the ebb of time we have divert in our aspect of the beauty conception, but the rural art aspect still influence our conception and value judgment even the creativity and style. Japanese and U.S. inheritance truce an external and internal aspect. The early Japan Animations inherited their old styles. "Ukiyo-e" a Kaleidoscope of world, "scroll painting" a Chinese style imitation and Carving and paint with colors we call "wood engraving" in those kind of old cultural, we trace the originality of Japan Animations. In Japan Animations, the "Light flower", the "Rice Box", the "Noodle Pile", the "Clutch Fan", all the daily utility in highly emerged in their cultural. The famous Japanese Animator Hayao Miyazaki his early masterpiece is fundamentally based on old Japan, "My Neighbor TOTORO", "POM POKO", "Memories Of Teardrops", we could find Japan style landscape, temple, wooden frame, straw mattress, their face are oriental, with these formidable styles, they have their style and recognized by their audience.

As respect to the U.S. the difference of culture make a different aspect. Broadway style dance is typical Hollywood. So they usually use this chorus and dance to express their sensations for example "Snow White and the Seven Dwarfs", "Cinderella", "Pinocchio", "Pocahontas", and the U.S. styled comics is highly entertaining, for example "Mickey and Donald", "Tom and Jerry".

The European style is highly artistic. Their painting puppet shoe tradition makes them show a profound beauty.

III. Animation Nowadays How they express the traditional cultural profoundness

In our times, every nation has a glamorous expression, and transportation and nets develop. We have a multicultural, this is necessary, nations have influence over nations, how we hold our tradition, this makes the problem and so is what we face in animation atmosphere say Oriental or Western, they never stop to influence one another, but their intrinsic are deep, the European civilization could trace to Egyptian, Greece and Rome, in their development many ethnics join their trace, but they never change the "Trend" even to make it more spectacular, "Goth" the Gothic they are barbarians.

The Spanish the Moors adds intrinsic of Arabic. The Renaissance adds the Oriental intrinsic.

The Buddhism is introduced from India and has induced his influence over Chinese culture mainly Confucianism and Taoism. But this enriches Chinese culture intrinsic. The Tang Dynasty has absorbed most cultural contents from outskirts, this tolerate attitude toward Ural people attain the Great Tang and exhibit the real Super Nature, the absorption of foreign culture and comparison will not make a people to lose his own culture.

As a form of civilization, the animation also has this absorption and comparison in different stages; we know this conception and dealings.

In animations the primitive style is to utilize the traditional material, the art form, scenery, character, costume backstage etc. But since the traditional boundary, for example Hayao Miyazaki his animations utilize large scale of European scenery and character, such as "Kiki's Delivery Service", "Howl's Moving Castle", and "Legend of Earthsea". To describe future time space the Scientific Phantasm material animations has been popular in Japan, for example "AKIRA" of Katsuhiro Otomo, "GHOST IN THE SHELL" of Mamoru Oshii.

The Hollywood has began to utilize large scale of different people and civilization animations, for example "Aladdin", "Pocahontas", "Prince of Egypt", "Mulan " and "The Hunchback of Notre Dame", and another trend leaded by another style had been largely by appear like "Monsters, Inc.", "Toy Story", "Cars", "The Incredibles". From these trends in the matter of materials and appearance they are no more ethnic. But their internal contents are richly engaged in their own civilization.

In the above mentioned Japanese Animation and their master pieces. They maybe use European act Scenery or alien space act, but their Soul Figure and the figures mentality is Japan. Their manuscript philosophy and judge of value and beauty is Japanese and the value of Bushido Ninja and the mystery etc. We could say its typical Japanese and traditionally Japanese.

And the extra, the U.S. animations, no matter the story derives from what nationality they rearrange the piece in their cultural and mentality, so that they match the normal ways, more comprehended and accepted by the audience. This is obvious in the "Mulan". In the Chinese conception "Mulan" is a traditional obedient girl, but in Hollywood, she is a girl who seeks self-fulfillment and in styling and details, they inherited U.S. cultural tradition, value of judge and humanism as the point.

As for the outlook of the people is a traditional Chinese girl, but in the internal and stylish conduct display the original creation animation reveals the American cultural mentality and the ways of their livelihood. In different aspect, the multi-nation art piece displays the style-rooted in their different culture.

IV. CHINESE ANIMATION'S INHERITANCE TO THEIR TRADITION CULTURE.

Chinese animation has a fertilized soil unique to the inheritance, Chinese 5000 years of civilization develops his own Artistic and culture system. No matter the civil artist or palatial, no matter the artist or the religious by means of the painter, artisan, skillful craftsman and their creations, accumulate abundance artistic showcase formality, from bronze, porcelain, carving brick, sculpture, calligraphy painting, lacquered pottery to the formidable paper-cutting. Engraved wooden-cut painting, leather-cutting opera figure, and the different regional dramas costume, architecture, all these assume an abundance so as to express in animations, and becomes the originality of Chinese animation.

The world famous Chinese style animations in one hand have traditional Chinese story materials and profoundness inner contents as a creation fundamental, in another hand they borrowed and utilize traditional art form for example the late 1950s "Little Carp Jump the Dragon Door", "The Monkey King's Rebel in Heaven" "Conquers the Dragon King" etc. They had borrowed Wooden Carving art piece, drama opera and the ancient Chinese painting style. These animations is close to Chinese tradition, add up to Chinese intrinsic, and they had once stand a stance in the past old good times.

With respect to the unceasingly development of social life, the western civilization technology and their living style emerge, the modern audience's mentality and judge of beauty develop, as well. Chinese animation must also follow the trend, must absorb new ingredient from foreign essences. In fact, the elderly generation of animators in China has already use this approach, for example "The Monkey's Rebel in Heaven" has borrowed Mexican drawing skill, in space treatment they use three dimension space perspective to manage the depth of space, many skills reveals the use of foreign skill.

In recent years Chinese animations artist make a few probes, trying to match a balance between traditional culture and borrowed ones. In the 2003, TV series animation "Neza" attains high level popularity and richly rewarded financially. In this animation they keep the intrinsic of the traditional, in order to satisfy the modern audience's favor, they use an inter-nation style of figure and show down. The boy "Neza" has an absolutely new brand, their artistic' style has betrayed the older one sketch. Basically, the style for example, has abandoned the little China boy's style, lovelier and more spirituous. The boys cloth is develop from the bellyband, the background architectural structure is of Dynasty Han-artistic design has changed the oriental two dimension perspective, and adopted three dimensional and focus perspective. The color is light shadowed. The perspective is more adapted for modern approach. In the boy's character construction, we maintained the traditional virtue and emerge modern conception. That is to say, we keep the traditional intrinsic and modern and realistic is the story.

In 2007 the series cartoon drama "The Adventure of Little Carp" is a similar trial. Its material derives from the traditional story "A Carp Over Dragon's Door" and we add up many contents modern, stage property, language, mentality, behavior are more accustomed for little audiences.

Compare to U.S. and Japan animation, Chinese animation is still seeking balance between inheritance of traditional and modern nowadays consideration, there is a long way to go. To this point, modern Hollywood and Japan experience is worthy our imitation. We don't have to be obscured by the traditional topics and formality; our civilization could still emerge from the inner part, character and judge.

V. THE PATH TO DEVELOPMENT OF CHINESE ANIMATION

Chinese animation development could neither proudly self-conceived, neither because the rapidly development of western techniques and be self-defiant, Chinese cartoonist want to seek his own animation development path must break off in two. One is to broaden his consideration not to stiffen his mind. Secondly is to really understand the inner part of traditional civilization.

And we must mention, creativity is very important in animation creation, to develop this creation power, we must broaden our mentality, abandon stiffen mental condition, compared to the U.S. and European animation Chinese animation usually appeared lack new idea and imaginative, to much instructive purposely maneuver can't touch our audience's heart. To change this condition, we could do the follow. For example, to break topics and formality development, so to make the audience conceive the true implication, add up to amusement of the animation, adding the imagination and phantasm. As for the personnel enrich the character make it believable and persuasive.

In the other hand, Chinese cultural root is not to be abandoned,. When Chinese animation is first created, they tried to follow this path, but now this road is obscured many artist is perverted against it. They think it a hindrance, they want be rid of it, but this is wrong. The traditional is the universal, Chinese animation must follow this instruction. We could be obvious when we look at the U.S. and Japan animation achievement. The strait-jacket of traditional Chinese animation is not because the traditional is not good, it is because we the generation in our development, no matter education or environment, is confined to reach the real Chinese. So we are not the real new Chinese generation; did not knowing our own civilization. This decides that we don't find a heritage to our own art, so as we see we must learn our own culture.

VI. CONCLUSION

In today globalization backgrounds, we need absorb and utilize foreign culture in the foundation of civil culture. Consecutively compose and de-compose the civil culture. But however the change, the development of Chinese animations should merge, rooted in the profoundness of civil culture, combine the new technology and foreign culture to really inherit and to recreate and Renaissance.

"The Civil is Universal". Only to root deep in Chinese culture Chinese animation could acquire his status in world animation scope. In the scope of animation, the large scales of international product have given us entrance to many a splendid art pieces and this is a great pressure to local cartoonist. To face this obscure, man must contrive consecutively, we learn the excellent compositional technique and creativity of the foreigner. To combine with the fashion of etiquette, broaden our mentality. Be rid of stiffen thought, learn and understand tradition culture implication, so could we march our civil animation future path.

REFERENCES

[1] Wu Guanying, "The Fine Arts Design of Animation ", Higher Education Press, Beijing, 2001.7, P28

[2] Wu Zhengchen, "Script Analysis of Ecology Animation Film", "Chinese Animation", 2005.11, P18

Campus Pastoral Landscape Ideas

Shen Jie
College of urban construction
Wuhan university science and technology
Wuhan, China
Email: sabrian302@163.com

Wu Changyou
College of urban construction
Wuhan university science and technology
Wuhan, China
Email: wucy1168@sina.com

Abstract—**Proposing garden-style scenic ideas for campus by analysis and thorough study of common design in current scenic features on domestic university campuses, related with China historical philosophy, the article explores opinions, composition, construction, senses and prospective forecast of campus landscape, intend to annotate ideas of campus landscape from the viewpoint of ecology, productivity, and anthrop-culture.**

Keywords-campus-landscape;garden-style;ecology; productivity ; humanity

I. ESTABLISHMENT OF GARDEN-STYLE

A. The impact of China traditional philosophy on campus scenic considerations

China has been focusing on advocating tradition, loving nature for thousand of years, not only Confucius claiming “universe is blended with space” （*MengZi*） but also Taoism believing “universe lives with me， and space and me are the one” （*zhuangZi*） . All these ideas connect daily lives tightly with human and space. Chinese have gradually formatted the unique oriental-taste opinion from this kind of philosophy system, and also created quantity of invaluable art products; the theory of “Taoist Naturalism” is more clearly cast by this kind of philosophy in garden construction.

There were Royal school yards, Governmental school yards， and other private schools as China colleges in the past. The former served Royal family, its school garden marking Royal Garden style, with neat symmetrically layout courtyards, and the difference between private school yards governmental schools lies in the choice of location, the former preferring being close to “mountain and forest”, “beautiful scenic spots”, “litterateur and artist center”, like most famous Tianfu School yard (in today’s Shangqiu, Henan Province), Bai Lu Dong School yard (Lushan, Jiangxi). The characteristic of school yard is as close to Nature as possible, in the form of wild life in mountains with forests, peaceful life isolated from realistic society, far away from towns and cities.

B. The network and features of modern campus scenic development

Four periods in campus construction in China

1) Late in Qing Dynasty

Main stream at that time lean to foreign, many ambitious scholars wished to by built more colleges which emphasized on educating for new country, so campus scale and design were not well researched.

2) Early in Twenty Century

Influenced by America and European culture, more American and European designers were invited in campus plan and construction, western scenic theory idea imported, so beautiful environment campus featuring combination of oriental and western were generated, i.e. Lingnan University, Dongwu Universtity, Wuhan University, Yanjing University, and others. The construction of those campuses was highly on research of environment and design, scenic elements of square, garden and courtyard were remarkably established.

3) After 1949

Under international and political impact, universities construction mainly leant Russian construction experiences, many campuses’ planning followed the lead of Moscow University, resulted in simply layout, lack of unique. At that time, school yard embodying big main building, huge square, and symmetric design, like China Geological University. (Wuhan)

4) From 80s of the last century to now

Economic reforming brought economy totem causing construction wave in China, especially in recent decade, along with developing of advanced education, university town construction, college merger and reform, and new school are leading a new wave of construction. These constructions have expressing new era in scenic design, planning tool by scientific efficiency, with center symmetric system, functional areas apart each other, so special environment of current campuses are coming up.

Due to short construction time, the speed gives you “a big mouth”, it results in some school yards have lack of “human & culture”, no coordinating between old and new buildings, deviating the style existed, disordering the whole campus layout. Not only hurt campus atmosphere in general, desultorily in sight as a plus. Some design had axial lead, but human and traffic direction were missed in construction, so that you won’t be able to realize it while you walk inside, no way to relate with culture.

C. Reflection of Farming Scenic thoughts in modern campus

Initially the forming of current school campus scenic is from social demand, at present, domestic university yards planning is impact by foreign culture, not thoroughly heritage with classical, not moving forward.

China classic art and garden focus on farming scenic, its form unique in art pattern, with supreme value, acknowledged by many in the world. The melting point between human and nature emphasized by China Classical Scenic Philosophy is the same as major content of ecology in our new era. What the farming style scenic idea of university campus on theoretic scenic thoughts is demanding following the philosophy of China Classic Garden, and refer to common regulation of modern scenic construction; In forming, the farming, forest and fishery products are mainly elements in scene; In thinking content by ecology, productivity, human and culture, in mental education by human kind, culture and nature harmonized, sharing life cycle, and compensating current campus construction in some degree.

II. Content of Farming Style Scenic View

A. Agricultural scenic view

To meet education goal, some agricultural universities layout test field for planting in campus, but most only focus on teaching function, not exploring the sense of scenic view. Using Huazhong Agricultural University as an example, its general layout is almost same as others in campus scenic view, greening doesn't feature in agriculture; and test field is separated with classroom and living section, as a teaching field siding in campus, so that campus scenic view and education are completely divided.

Growing inexpensive plants and flowers as landscape elements should be considered according to local individual conditions for the utilization of their vast scenic view values. For specific selection of agricultural plants, field size, seeding plan, attention must be given to the appropriate spatial arrangements, based on location, and proper layout, neither over-planned nor under-planned, so the value of scenic view on agricultural plants could be realized. This is not enough if only staying on theoretic research, and plentiful practices and summing-up must be done so as to cultivate planners' and designers' aesthetic sentiments

Wood has to be the key element of scenic view in campus. Neat row trees along both road sides, dotted, and along the square, are forming elegant school scenic view, among the phoenix tree, camphor tree, osier, and palm tree are most. Beside of green, the forestry value has to be considered, then land value could be explored maximum. But dotted planting tree must intensively consider the unique enjoyment with a thought of vision result and internal space, so fruit tree and mulberry tree are the option.

B. Fishery scenic view

Garden theory of China Mountain and water garden emphasizes "garden lives on water", how to plan the water scenic is the important means in beautifying campus scenic view. There are three kind of watering view based on source:

1) there are huge natural lake or near lake inside of campus.

2) there is a river or canal going through campus.

3) there are small human-made lake, walk-through pond and fountain which are key elements.

Primary thinking point on fishery scenic view are ultimately with productive value and unique scene which may include plants and animals like lotus, reed, met sequoia and fish, duck and goose.

C. Scenic of building and construction

Some scholars classified scenic view into three categories: Virtual Scenic, Material Scenic and humankind. Main buildings and general constructions are functional devices, and portion of scenic view, its special feature is the shaping of human-made and format, but also this is the hard one to control, how to harmonized with overall scenic view and surroundings, and how to express its unique characteristic and scenic contents, all need designers' heart and soul on it. Garden Scenic View doesn't encourage totally opposite with current, solely seeking classic construction, but emphasizing tiny section outside, internal space, constructional materials, must be considering the forming with ecology, unique feature and cultural content, and other constructive buildings, its meaning of scenic view is higher than its functional purposes, must keep its village principles, like straw house, bamboo bridge, and so on.

D. Scenic view of human and animals

Another characteristic of garden style scenic view in campus is human and others, in the poem of garden view in China ancient, it usually described the scene of busy farming to express human and natural harmonizing highly. The scenic elements are agriculture, forestry and fishery in garden style scenic view, thus another picture to express human and natural harmonizing by teachers and students living in the campus, dealing with seeding, picking and fishing.

III. Construction of Garden Style Scenic View

A. Contracture of garden style scenic view

The construction of garden style scenic view is divided into three aspects: one is intuitionist scenic view, Agricultural technology managerial system, and maintaining of scenic view.

1) Intuitionist scenic view

There is no big difference between ultimately intuitionists brought by Garden Style Campus Scenic Construction and classical, the former is also consisting of views in plant, construction, watering, square, road, human and culture. The fundamentally difference compared with classical is at substitution of scenic materials, and emphasizing on unique characteristic, ecological, considering human and culture.

2) Agricultural technology managerial system

Since garden campus scenic view uses agricultural plants, forestry products and hydrophytes as main scenic materials, its scenic system possesses productivities, so the construction of garden style scenic view must have professional agricultural management system in which including agriculture, garden, plants layout and other professional theoretic instruction, and irrigating, seeding and harvesting and so on practical experience. The point much be highly attention is how to plant in whole campus, surviving rate on combination seeding variety plants and aesthetics effecting, and how to match irrigating and road system in whole campus.

3) Maintaining of scenic view

In order to maintain scenic views in ordering and format, the regular mowing and trimming is request. The maintain workload on garden style scenic view in campus is greater than traditional campus because longer cycle period on agricultural plants, big changes on view degree, administration needs daily agricultural management, one is activities on agricultural plants, and the other is to maintain the beauty vision result of neat sorted on view of garden style. At the same time, the men activity of farming realizes another sense of scenic view.

B. Specific executing principles of garden style scenic view construction

To select seed in a specific area must follow the local condition in the garden scenic view construction, such as fruit in flat and open area, tea in hard soil on hillside, terrace is suitable on wetland. On combination seeding plants, vision effective is the key such as excessive arbor in rice area will absorb fertilizer; corn is not suitable with camellia, in order not to create chaos.

The construction of scenic view also needs seasonal rules in order to lead appropriate seasonal in vivid scenic through proper combination of plants. Road net designing firstly meet the need of teachers and students, and then match natural garden view in some degree. Beside the most important is that human must be positioned on top of all, comfort is based on viewing space, green materials in garden style campus scenic view decides all natural tastes, this requires certain agricultural knowledge and designer's layout vision.

C. Scenic management in garden style campus

1) Open campus in proper control

Often proposed by some scholars when summarizing university campus scenic, from closed to open will be primary developing network in campus model, partially focus planning idea on some western famous schools' "similar living in town when in school" which maximum sharing sources, convenient to education exchanging. But domestic much consider local environment, under current education and economical condition, domestic schools are unable to raise entire city civilization, so "only focus on openness" will only bring chaos in management and security, for sure now colleges could open campus inside town possible, and also share the same sources. This is not to insist with campus close management in universities, campus scenic is an important part of city view, and campus planning should consider scenic publicity like forestry view, water productivity in vast water area. Because garden scenic must ensure its productivities, most scenic view must be centralized management in order to manage production.

2) Proper staff agricultural technicians, both teachers and students are participating in farming

To have both ornament properties and productivities on campus scenic viewings, the maintenance and technicians will work together. Workshop needs to be setup more for technicians' experiments to avoid many visitors' presentation. And also teachers and students should practice properly in farming, re-education to realize viewing of human and cultural caring.

IV. MEANING OF REALIZING GARDEN STYLE SCENIC VIEW

A. Maximum utilizing land voiding constructive earthwork.

At present time, college campus construction is on hot waving, campus scale and geology is different in some degree, so the scenic system couldn't be applied in all campuses, garden style scenic system emphasizes garden view which includes terrace, hillside camellia, and other forms viewings, this will avoid re-design whole campus by teardown present campus, reducing constructive earthwork and saving on budget.

B. Duplicate meaning of vision and production in scenic view.

Rice field in Shenyang Architecture University is from "ideal re-construction was brought by one time simply exchanging", inhere viewing could bring productivity. Now campus scale is huge, if the productivity thought could be applied on scenic system except viewing, the harvest could be rather quantity, it is a exploring on land and scenic view, also inevitable developing trend on city scenic under farming field is reducing significantly.

C. Ecology

Environment has been becoming a global issue caused by industrial social developing post industrial revolution recent years. One after the other in many scholars proposed "ecology" idea in different industries, emphasizes its importance. Ecology is the actual carry-out principle of constancy developing, and also brand-new aesthetics standard on subject of land, environment protection and culture. Garden style scenic view is based on nature of agricultural activities, its possesses the function of metabolism, and also created a picture of back to nature, with variety interesting wild life as campus view, reflecting current era's needs on scenic idea.

D. Human and cultural caring

Garden style campus scenic view is to realize human and culture caring of scenic view by scenic forms and scenic activities. In form, campus scenic construction is based on agricultural element, its plants especially features local area, melting agricultural activities into campus view construction, in which the win-win situation expressing scenic unique and maintain local classical culture. Second from scenic activities, the unique scenic view forms in garden style campus possesses others scenic theories which have no human and view mutual involvement. College teachers and students participant in agricultural work, which is direct heritage on China classical culture, this kind of courses design is meaningful.

V. Prospects on tendency of Garden Style Campus Scenic View

College campus scenic feature is vivid carrier of campus culture, some of scholars believe students' knowledge is not only from books, but also half from agreeable living space created by campus environment. The planning of current campus scenic view must be focused on human needs, feature higher education, let students enjoy a leisurely and tranquil mood while walking in comfortable and quiet surroundings psychologically and physically. Garden style campus view represents the concept ecology, productivity, and human culture, the cultural heritage and the principle of agricultural activities on which the basic scenic system and basic design rule are based. This concept of planning and designing is dirtied at current construction of campus scenic view, keeping in steps with social development and time change. Hopefully, this proposal may serve as starting point for the designers and scholars to take into deep consideration in proposing scenic planning of campus for colleges in China, as well as a solution to improving current campus scenic construction in order to build a more and more beautiful campus view, and train more high quality genius students.

References

[1] Yu Kongjian,Positioning Contemporary Landscape Design The Art of Survival,China Building Industry Press, Beijing,2006

[2] WangYuncai,Landscape ecological planning principles, China Building Industry Press ,Beijing,2007

[3] SunQingzhu,An Introduction to campus culture,Shandong University Press .Beijing,2008.

Analysis on Spatial Form of Traditional Towns in Hubei Province

Wan Yanhua
School of Architecture and Urban Planning
Huazhong University of Science and Technology
Wuhan, China
Email: yan65919@tom.com

Chen Tao
School of Architecture and Urban Planning
Huazhong University of Science and Technology
Wuhan, China
Email: 279082272@qq.com

Song Yang
Wenhua College
Huazhong University of Science and Technology
Wuhan, China
Email: 279082272@qq.com

***Abstract*—From the analysis on the problem of research on China's traditional town, this paper points out the significance of analyzing spatial patterns of traditional town in Hubei Province. On this basis, the paper analyzes the influencing factors and elements of spatial patterns of traditional town in Hubei Province, and approaches the types and characteristics on linear pattern, multi-linear pattern and topo network pattern of the traditional town in Hubei Province.**

Keywords-traditional town; spatial form; spatial form of traditional town; Hubei Province

I. ORIGIN

A. The existing problems of research on traditional towns in China

- The limitations on the study of traditional towns.

In recent years, the tourism of China's traditional towns is persistent overheating, and introductory books on tourism, sightseeing of traditional towns have been coming in thick and fast, such as "China's Traditional Towns Tour", "Book Of Traditional Towns" and so on. Based on the purpose of tourism promotion, these books often focus on introducing the history and customs, heritage buildings of traditional towns, rather than taking traditional towns as an academic to study, therefore, there be the lack of academic sense of the deeply research on traditional towns. Of course, there are some academic research articles about traditional towns, but most of them discuss around towns of siting, spatial elements, protection and development and so on, and the depth and scope of their content are still narrow, especially the lack of systematic integrating, multi-level in-depth study on the traditional towns' spatial form.

The so-called traditional towns' spatial form "is a complex economic, social phenomena and processes, is a result from the interaction combined by human activities and natural elements in a particular geographical environment and a certain stage of historical development of society, is overall intention understood, perceived and reflected the whole city through a variety of ways by people "(Wujin, 1990); Therefore, studying traditional towns' spatial form not only help us understand the formation and development, evolution process of them, but also be conducive to continuation of historic look of town and excellent traditions truly, and even inspire us to learn the way of dealing with the harmony spatial form between human and nature.

- Imbalance of area of study on traditional town.

China stretches across a vast area, there are large number of characteristics traditional towns in all parts, but because of economic and social development of imbalances, study on China's traditional towns appears the situation of "attaching weight to east, light to central and western regions", "more on the lower reaches of Yangtze River, less on the upper and middle reaches of Yangtze River". This research pattern of traditional town is not conducive to grasp the overall features of China's traditional towns systematically for us, so at this stage we should aggressively to reverse the downstream of traditional towns' study, and try to reduce this imbalance on the study area.

B. The meaning of analysison spatial form of traditional towns in Hubei Province

Hubei Province is located in central China, the Yangtze River middle reaches, "thoroughfare of nine provinces" of the geography makes towns with charming of many advantages, downing from the intersection of historical and cultural information between southern and northern. Analysis of spatial form of traditional towns in Hubei not only help us understand the impact of factor, the constituent elements and composition methods, types and characteristics of spatial form of traditional towns in Hubei, but also protect fine tradition of them shooting the arrow at the target, break the current situation of regional imbalance on study, deepen and expand us study the depth and breadth of China's traditional town, and understand globalization and regional architectural creation and design correctly.

II. THE IMPACT FACTORS AND THE CONSTITUENT ELEMENTS OF SPATIAL FORM OF TRADITIONAL TOWNS IN HUBEI PROVINCE

Traditional towns in Hubei Province studied in this paper refer to traditional small towns in Hubei which emerged before the Ming Danasty, Qing Danasty and continued developing; such small towns still maintain more completely traditional spatial pattern, building groups and street space and environment atmosphere. The spatial form of traditional towns studied in the paper is not microscopic one between traditional towns' architecture and the inner yards, nor is macroscopical one between town sites and environment, but rather something present a whole spatial pattern, form and meaning combined with topography, agricultural land, small climate and social and human elements (such as economic, military, transportation) between the two, under the synergy by a certain structural relationship; it is the composite result that human activities and natural elements interact under the particular geographical environment and certain historical development of phase, and also people's mental imagery for the town of space cognition and reflection. Concretely speaking, the external macro-space form of traditional towns constitute of factors of spatial form of traditional forms the paper studied, while the inneral micro-space constitute of elements.

A. The impact factors of spatial form of traditional towns in Hubei

- The natural impact factors of spatial form of traditional towns' in Hubei.

1) Topogrephy and geomorphology

Topography and geomorphology plays a decisive role to the overall pattern of spatial patterns of traditional towns. In traditional times, people were lack of capacity of large-scale digging water, therefore, their settlements of traditional towns were all conform to topography and site selection and development. Topography, geomorphology and town settlements are nature unity to form so unique spatial form.

In Hubei Province, the spatial form of traditional towns has two ways to adapt to mountainous terrain: One is that mountains play a role of the formation of space-qualified outside the traditional town(Figure 1); Another is that traditional towns are built on the mountain, and mountain is the bottom interface of spatial form of traditional towns, the town streets either are parallel to the mountain contour lines, showing the level of linear development, or perpendicular to the contour line of the mountain showing platform-style development.

Figure 1. Mountains outside traditional towns in Hubei.

The network of rivers clouds in Jianghan Plain of Hubei, traditional towns built over river, born of river, so formate waterfront spatial form. Street walked with the river, housing built over the river, producing a linear response to river motive force.

2) Farmland Topogrephy and geomorphology

Self-sufficient small scale peasant economy try to make location of traditional towns in Hubei close to the agricultural labors place of residents —— fields(Figure 2); At the same time, it is in order to not or less account for fields, spatial form of traditional towns in Hubei present a variety of irregular shape. Diaojiaolou is a kind of residential architectural forms in southwest Hubei owing to not accounted for fields but adapt the slopes of the site conditions; The residential architectural form makes traditional towns spatial form vary as the chang of agricultural land in southwest H ubei.

Figure 2. Farmland outside traditional towns in Hubei.

3)Climatic factors Topogrephy and geomorphology

Hubei is located in east longitude 108°21'-116°27 ', latitude 29°05'-33°20'. The whole province's climatic conditions is broadly similar, showing a typical feature of the continental climate, hot in summer and cold winter, monsoon climate, rain heat in the same quarter, which is characterized as "volatile", and "great change"; Consequently, we should not only consider the spatial form of the town in summer cooling, but also the winter warmth. To this end, traditional towns in Hubei should be located in the spots where there are mountains in north, east and west surrounded and a placket by the South East (which is conducive to the summer monsoon through the placket into the townships) of the basin; At the same time, the dense layout should be adopted to block the strong summer sun with each other in town building complex combination.

- Social and human factors in spatial form of traditional towns in Hubei.

(1) Economic factor

Economic factor could be the most important factor affecting the spatial form of traditional towns in Hubei. The spatial division of morphological types of traditional towns reflects, in essence, its economic level: for example, spatial form of the economic under developed traditional towns in Hubei is a simple single-line type, while the economically developed regions, there is a complex network and double-track type.

(2) Military factor

In ancient times, wars happened frequently in Hubei Province, then there were some towns built up for military defense. Such as Luotian country, Shengli town, they have three-side mountains surrounded, one side ring of water, so they are easily defensible; within the heavily fortified, there is no lack of spatial forms processing to enhance the defensive means, such as "T" shape treatment between street and alley transfer node, the design of entrance barriers, and so defensive steps. Apart from some town built up for military purposes, spatial patterns of other towns in Hubei are more or less affected by the impact of military factors, being with a certain amount of character for defensive spatial morphology.

(3) Traffic Factor

The land transport in ancient Hubei Province is extremely inconvenient, then the water transport plays the most important role in the rise or fall of the traditional towns in Hubei, thus the spatial form of the traditional towns in Hubei, is subject to its water traffic in a certain extent. Because of this, most of the traditional towns are constructed relying on rivers in Hubei, river is also the basis for their development of spatial form. Usually traditional towns' space begins from a pier, spreading all around; while, around the pier there is a commercial transaction sales stall space or meeting space for folk, street and alley radiate from this space, then directly into every corner of town。

B. The constituent elements of spatial form in traditional towns of Hubei

Spatial form of traditional towns is ultimately constituted by a series of artificial elements, which contains the main construction groups (such as single-block-style building, courtyard-style building, courtyard-style building), street (such as streets, roadway, etc.), space-node (such as street intersection, the well bay, water ports or terminals, etc.).

- The first constituent element of spatial form in traditional towns of Hubei : architecture groups.

Spatial form of traditional towns is ultimately constituted by a series of artificial elements, which contains the main construction groups (such as single-block-style building, courtyard-style building, courtyard-style building), street (such as streets, roadway, etc.), space-node (such as street intersection, the well bay, water ports or terminals, etc.).

Architecture is the initial element in traditional towns' spatial form. Among traditional towns in Hubei Province, the number of single block building is not so large; due to the great hotness of summer and the coldness of winter in Hubei Province, thus, the residential shape form hence the characteristics of northern courtyard houses, forming the main house style "three-side surrounded courtyard dwellings "(Figure 3) beside with patio-style construction (Figure 4). These buildings have large differences in the bay, facewidth, number of storeys, height, form and change on the actual situation. It is precisely these differences, making spatial patterns of these towns in Hubei colorful and various.

Figure 3. Courtyard-style Architecture in Xiaohe Town of Xiaochang, Hubei.

Figure 4. The entrance node of Democratic Street in Xindian Town of Chibi, HuBei.

- The second constituent element of spatial form in traditional towns of Hubei: streets and Lanes.

Streets as the backbone in spatial form throughout towns in Hubei, link outside roads and spots throughout the towns. Meanwhile, Street is also a spot for commercial and social activities of traditional towns in Hubei, its space is relatively wide, often with tall buildings on both sides. So streets usually become the bustling areas and landmarks in the towns.

Lanes as branches perpendicular to the street access, contact with streets and t block, while also contact with streets and rivers. Lane with a smaller scale, can generally offer only single-pass, but cannot be the function of goods transport. So lane sit in the backseat in the alley system and spatial form of towns.

- The third constituent element of spatial form in traditional towns of Hubei: the node of space.

The main spatial nodes of traditional towns site in the start and ending, bindingsite, turning point, the part of elevation or depression of streets and lanes, showing a certain concessions for the construction interface, enlarging the space of street; therein, spatial nodes between the intersection of Street and Street often locate in the geometric center of town, which is an important punctuation between geographical and emotional space of traditional towns(Figure 5).

Figure 5. The entrance node of Democratic Street in Xindian Town of Chibi, HuBei.

Most traditional towns of Hubei draw water from the wells; the well bay is often located in the streets and lanes and gives spacious space for drawing water easily without traffic trouble. As a result it forms many well bay space caving in the lanes or the corners, or combining with the linear streets, and composing such a case just as punctuation of the paragraphs, and generating self-assured sense of rhythm (Figure 6). Therefore, well bay has become one of the constituent elements of spatial form of traditional towns in Hubei Province.

Figure 6. The well bay node of traditional towns in Hubei.

Water ports or terminals in the waterfront towns of Hubei is important symbol of spatial morphology. It refers to the facilities which are located in the revetment from the land down to the shore. For protecting the flood, revetment is often higher than the water, but the ports or terminals must be made close to the water. Terminal or water port is not only a place for towners drawing water, washing, parking boats, transporting goods, but also the end of streets to the rivers which combines the rivers and streets. Due to the presence of water ports or terminals, river and street are no longer located in two water level above the disparate elements of the space, but varied occlusal plane, space on the upper and lower links of the three-dimensional continuum.

III. THE TYPE AND CHARACTERISTIC OF SPATIAL FORM IN TRADITIONAL TOWNS OF HUBEI

With the interaction of specific external factors and internal elements of traditional towns in Hubei, traditional towns show several main spatial form types, such as single type, double-track-type, free network-type, and reveal regional characteristics of Hubei.

A. The spatial form types of traditional towns in Hubei

- The first spatial form type of traditional towns in Hubei: single type.

The traditional towns often developed by stage or market town, organized every function space of traditional towns by the main road as axes. Then lanes, which the main street branched, entered each residential groups. The whole streets and lanes of traditional towns were herringbone. The original function of traditional towns was single, but the function grown richly following the development of the market economy. The spatial form also transformed the complex form— double-track-type or network-type.

From the analysis of formation pattern, the spatial form of the single type is relatively simple, and basically is the combination of construction group and streets. From the environmental, the type appeared not only in both the mountainous of southwestern, northwestern, southeastern, northeastern of Hubei, but also the river network of Jianghan plains region. There are two subtypes in hilly zone. The first type is that the main street is parallel to the mountain high linear. The curvature of the main street is substantially in line with the contour lines and the elevation of the main street become indistinct. Yet the branch lane is perpendicular to the mountain of the contour lines, combined with the drain, and the elevation changes obviously. The second type is that the main street is perpendicular to the mountain high linear. The main street extends up and downs with step by ridge or valley. Yet the branch lane is parallel with the mountain contour lines. The construction groups layout are almost fronting water and with hill on the back.

There are also two subtypes on the river network area. The first type is that the main street is back on rivers, the main street intersperse the architectures such as shops as the axis. Yet the riverway is on side of the traditional towns. On the waterway water ports or terminals which load and unload of goods only play the role of transport. The back river-type pattern is common in the hilly areas in Hubei which have large volume of water flooding more frequently. For example, in the Yanxin longgang town, because of abundant of water in Longgang River, the main street is far away from it under the considerations of flood protection—the farthest place is 110 meters. Meanwhile, the main street with enough space might apply to in-depth development (Figure 7). The second type is pattern of the main street adjacent river. The side of the main street is on the river, and is parallel to the river way. The side of the river changes street. The main street not only meet the production, life, fire protection of the town, but also made the space landscape of the main street extremely rich. The river adjacent the town is little water and relatively narrow, so it is the possibility of a rare flood.

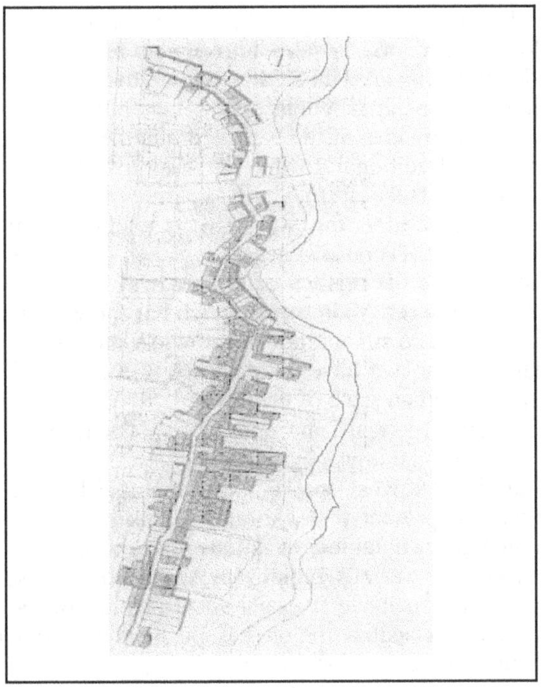

Figure 7. The spatial form back the river of Longgang Town in Yangxin, Hubei.

- The second spatial form type of traditional towns in Hubei: double-track-type.

Double-track-type spatial form is in developed on the basis of the single type spatial form. When the single type couldn't meet the development of the towns, it would be from a single model to both sides of the depth direction, forming the main street of "cross-type", "T-shaped type" or " parallel type".

"the cross type " spatial form is called "Phoenix Spreading the Wings" in fengshui. For example, Shipai town in Zhongxiang, with the "hill-street—center street—Market Street" as the phoenix body, the northest side gate as the phoenix head, the western Han River as phoenix tail, south and north respectively "west laifeng off" and "north gate street-the north gate key" for the two wings, consitutes of a cross-shaped main street framework, and locally known as " Phoenix Spreading the Wings ". Therefore, stone steles is also called Fengtai.(Figure 8)

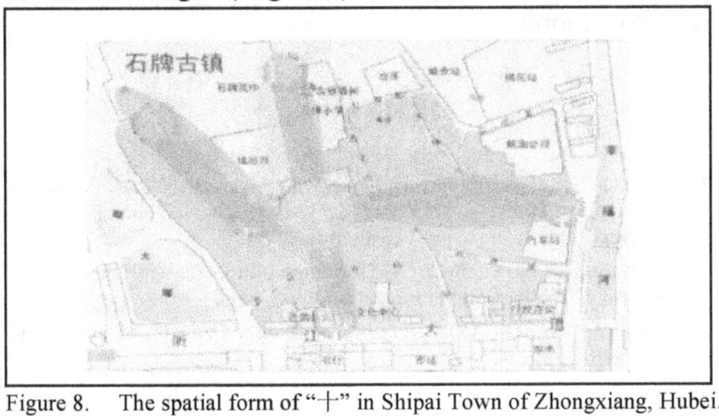

Figure 8. The spatial form of "十" in Shipai Town of Zhongxiang, Hubei.

"T-shaped type" spatial form is formed by two main streets intersection. Such as Xindian town of Chibi, in the influence of New River, over all architecture of the main streets is "T-shaped". The horizontal of "T" is the street along the river now. Ming and Qing Dynasties there were six terminals and temples arranged along the New River, which is true reflection of water and tea culture in ancient times. The vertical of "T" was Democracy Street and Construction Street, and was primary road of a land-based commodity distribution. Hence, with the development of economic and population, streets and lanes increases more and more. The other new streets such as Victory Street broke the original "T-shaped type" pattern, developing as intertwined space system of streets and alleys today.(Figure 9)

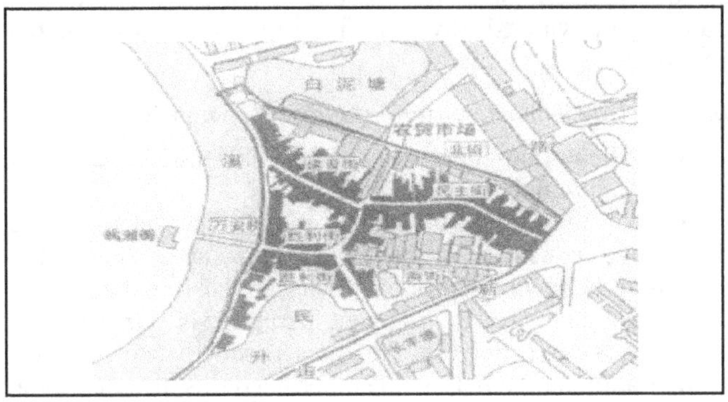

Figure 9. The spatial form of "T" in Xindian Town of Chibi, Hubei.

"Parallel Type" spatial form is formed by two main concatention streets or lanes. For example, Zhoulaozui Town in Jianli, the main street-Old Main Street stretche from west to east winds. South of the main street is tied for sub-street along the river—Along river Street. Lane communicated the main street and sub-street, taking on "parallel type" spatial form(Figure 10).

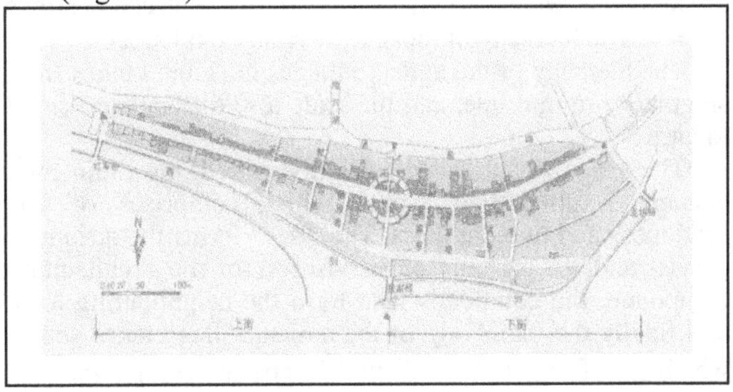

Figure 10. The spatial form of "二" of Zhoulaozui Town in Jianli, Hubei.

- The third type spatial form of traditional: free network-type.

When the linear spatial cannot meet traditional towns' development, with the the continuing expansion of towns' area, its internal street, alley develop into a free network of criss-crossing type under the influence of the elements in the terrain. This spatial form is more complicated than single-type and double-track-type spatial form, and its towns' district is more pronounced; generally, there is a large space for the node in traditional towns, which is multi-functional space for people doing business, meetings, entertainment, interaction .

There is an example of Baini traditional town in Chongyang. Since the completion of baini-bridge as well as the strength of Baini harbor, the area of old town experienced a development process of "point-shaped mesh to banding connection to the growth of settlement ribbon": initially only a few built settlements along the river; following as the Baini-bridge built during Ming Xuande period, it formed taking baini-bridge as centric market, and developed to a big city until the pier, then developed along the river; the completion of Yu Geng-bridge in Qing made the development of town crossing to the north shore, and traditional town basically form free of network-based spatial patterns.

B. *Spatial Characteristics of Ancient Town in Hubei*

- The terrain-compliying and flexible spatial pattern of traditional towns in Hubei

Unlike the diging-mountain and filling-water of the destructive construction in the modern age, the spatial pattern of the Hubei traditional towns emphasizes the terrain-adjustment and the environment paragenesis. On the one hand this is determined by the lower productive forces level in the traditional Hubei, on the other hand this also has some connection with the thought of Nature fUnity in traditional China.

The varied spatial pattern of the traditional towns in Hubei is result from their varied topographical phenomenas such as the the plain, the hill and the mountainous region, they are the optimal spatial patterns the traditional Hubei residents could obtain with the least construction investment in the specific geographic and geomorphic conditions. In other words, these spatial patterns and their construction process inquires the least nature transformations and destroys.

- The hierachy of the antient villages in Hubei

The hierachy of the antient villages in Hubei ranges from singularity to multiple, and the scale also differs from small to large.

The space in antient villages diversified, the rudimental element is the "Jian" space which comprises of one brightness and two darkness. The "Jian" forms the atrium or parvis, and the atriums or parvises form the architectural compound, the compouds then form the neighbouring area, and finally the mainbody of the antient village takes shape, which is formed by "Jian → atrium or parvis → architectural compound → street and lane → block → traditional towns". Thus the series of spaces defines the traditional villages.

- Human-scale spatial form of traditional towns in Hubei

The scale of traditional towns' spatial form is an important factor affecting the quality of traditional towns space form. People's awareness of traditional towns' spatial form depends on a series of basic standards. Such as the windows and doors, facades into almost as street sides of traditional towns in Hubei all follow the following ratio relationship: openings width and height ratio of 1:2, door frames on both sides of the wall width is the door wide 1/2, so that the traditional towns are easy-to-measure of the relationship in Hubei from whole to part.

At the same time, the Main Streets width (D) and street building height (H) ratio is generally 1:1 in traditional towns of Hubei, while the privacy of the roadway in a strong D/H in 1:0.5-1, between Main Street which has more than a sense of enclosure. The street, alley space width and elevation has a clear ratio, so we have easy access to a sense of spatial patterns for human scale of the town .

- Strongly significant nodes of spatial form in Hubei traditional towns

Traditional towns' spatial form in Hubei have a strong identity, and present that various regions of space for the town intersperse a number of unique nodes in space, they not only make the various regions of space create a sense of rhythm, but also make it easier identification of its spatial nature, so that spatial region has personality and identity, thus will become a regional landmark.

The constituent elements of the spatial nodes for spatial form are mainly water, port, street intersection, the small square, well bay, the head of bridge and so on in traditional towns , Hubei; they have a more prominent feature of the visual characteristics and functions as a visible symbol of space, producing space experiencc different from the surrounding environment, and to some extent, they dominated atmosphere of the region space.

IV. Conclusion

To sum up, the impact of factors and the constituent elements of traditional towns' spatial form in Hubei not only have common ground, but also peculiarities with the rest of districts in China; three types of spatial patterns of Hubei traditional towns is so as them. Meanwhile, the response to topography, grade mix, scale humane, the characteristics of strong identity for Hubei traditional towns are significant to protect the inherent qualities of Hubei traditional towns actively. Hope the paper can lead to insight attention for the wide awake people, and spread a much more comprehensive and in-depth research for protecting excellence tradition of Hubei traditional towns better, especially the protection of their spatial form.

References

[1] SONG Yang, "Hubei town space shape analysis and integrated protection study," Master's degree thesis. Wuhan,Huazhong University of Science and Technology, 2007

[2] Li Baihao, Wan Yanhua, "Architectural Culture Towns and Villages in China," Wuhan, Hubei Education Press, 2008.

[3] Wujin, "China's Urban Form,"Nanjing, Jiangsu Science and Technology Press, 1990.

Analysis on the Employment Characteristics and Strategies of Art and Design Graduates

Exemplified by School of Art and Design in Wuhan University of Technology

Guanghua Hu
School of Art and Design
Wuhan University of Technology
Wuhan, China
Email: hughua@163.com

***Abstract*-As Fine Art and Design graduates are characterized by the geographical selectivity, concentration of industry, occupational freedom, and career mobility and so on, there grows a demand-and-supply conflict between the graduates and the employers. Therefore, the concerned staff in the employment department should pay much attention to exploring innovative mechanisms for employment, providing career guidance for the graduates, and improving the employment rates. Specifically speaking, the measures include innovating teaching to support the employment, making assessment and incentive policies, strengthening graduates' correct attitudes toward employment, and making full use of the family's positive influence.**

Key Words-College; Art and Design; undergraduate; employment

I. INTRODUCTION

The enhancement of China's economic restructuring and development of college education result in the rapid expansion of college and university admissions, which in turn, makes the conflicts between the enlarged number of graduates and the social demand more prominent. The graduates face a serious employment pressure, not to say graduates of Art and Design majors. Therefore, the employment of college graduates, especially those majoring in art and design, is of great importance, involving a wide range, with a tight schedule, and with a full load. This issue is not only related to national long-term interests, but also concerned with thousands of families in the society. Thus, to effectively solve this thorny question and improve the quality of college education has a deep impact on the sustainable development of Art and Design School and imposes a far-reaching influence on the college's social prestige.

II. THE UNIQUENESS OF ART AND DESIGN GRADUATES' EMPLOYMENT

Art and Design graduates compose an important part in the college students in China. In terms of employment, they share something in common with both science and literature graduates. Nevertheless, due to their major's characteristics, personality traits and the current socio-economic situation and the impact of other factors, they get a distinctive artistic characteristic when coming to employment. Thus summarizing the employment characteristics of Art and Design graduates will help carry out employment education and guidance and improve the effectiveness.

A. The occupational freedom and changeability

Instead of signing Employment Agreement with employers, some graduates choose to put their registered residence and archives on trustee in talent exchange center, just to be a freelancer. Some even take initiatives to get design business on their own, as long as the work pays. The two kinds of no-employment of self-employment show art graduates' occupational freedom and changeability. Although the contract-signing rate is low, as a result of the various forms of employment, the employment rate is actually rather high. Thus, given this characteristic, the Art and Design School leaders should on one hand, encourage students to sign contracts, and on the other hand, provide the essential documents and statistics concerning those who choose to get employed in other ways. Last but not least, they should provide detailed information with patience for graduates in the process of trusteeship of registered residence and archives.

B. Graduates' attention on the company's location and the economic returns

The company's location grows to be an important factor for graduates. The graduates prefer those developed areas, coastal regions and capital cities. Take our School of Art and Design for example. Most of those majoring in Design of Industry seek southward for development, and environmental art and design, to eastern cities. Postgraduates tend to work as a professional teacher in college or school. Due to the higher cost (total number of tuitions, accommodation fees, meal expenses and etc in the four years borders on RMB10, 000) in education than other majors, graduates naturally pay great attention to the economic returns. Given the information mentioned here, the Art School should develop a scientific employment attitude among graduates and guide them to lower the expectancy and to grasp every opportunity to get employed.

C. *Company's attention on innovation ability and application ability*

There are more companies who stress applicants' innovation ability and application ability. It is estimated that a rather large number of applicants will pass in the first round of interview but they won't get the contracts immediately. They are to have a period of training and internship, after which the company will weed out some and keep those with innovation and application ability. In the light of this reality, the Art School also shifts its focus. First, cultivate graduates' ability to put what they have learned to a better use. Second, combine the graduation design, graduation internship and vocational internship so that graduates could turn the company's need as their topic of graduation design. Thirdly, strengthen the relationship between the school and the company to enable the company to better understand students' ability.

D. *Graduates' entrepreneurship*

Through entrepreneurship, the Art graduate can not only provide a job for himself, but also for others, which will boost the employment in the whole society. The entrepreneurship can bring along employment, exerting a far-reaching influence on alleviating unemployment. Our School of Art and Design has already accumulated some experience in this field, and students' entrepreneurship has developed and covered a certain size, with many students wining various prizes in different kinds of entrepreneurship competition. Liu Bin and other three postgraduates of 2009 got RMB 80,000 after participating in the first Wuhan Undergraduates Technology Innovation Competition and wining the third prize for their Comic and Animation CG Training Program, which is now under work. Li Hui, Xun Yong and other two, postgraduates of 2007, who were senior high schoolmates in a country in Enshi Hefeng County, opened a Fine Art Training Center in the summer vacation and gained RMB 1,000,000, which they later utilized to start an Art Restaurant in Wuhan, spreading its branches to other cities… The four once poverty-stricken undergraduates realized their dream and became Band of Brothers through entrepreneurship.

E. *New phenomenon emerging from the company*

The art and design companies, limited by the industry's unique features, are in little need of employees, employing no more than 2 applicants. What's more, owing to the Economic Crisis, this year, many companies in the art and design industry rescheduled their enrollment plan. Some removed the human resource reservation and thus cancelled their recruitment to lower the cost. Others even took the initiative and ended the contracts with signed graduates.

III. Exploring Innovation Mechanism for Employment and Facilitating the Job-Hunting

Based on our university's science and technology background, as well as the demand of applicable art design talents, out art school have realized that the issue of employment is an essential point in cultivating talents, a transitional process for students to develop to their full potential, and also a touchstone of the education. Hence, our art school will aim at improving students' comprehensive quality and competence in employment, and do our best to promote employment.

A. *Establishing operational system of employment, deploying the job, protecting the employment*

In order to do the job of the employment of graduates well who graduate in the year of 2009, the school sets up a leading group of employment of the school of art and design in which the rector and the president work as the headman, including all leaders of the school, every director, every department chairman and all counselors. The establishment of the leading group which has the system of double-headmen reflects the high regard for the employment of graduates fully. The employment of graduates carries out the responsibility system. Every leader of the school keeps in touch with a department. So it offers a full guarantee in staff and system. The job duties are that headman takes charge to set up the leading group of employment of the school, to manage the discussion and to supervise and urge the formulating and working of the employment plan and the construction of the employment base; vice-headman is mainly in charge of the specific implement of the employment plan; counselors of the senior classes are in charge the reception of the employment units, calculating and releasing the employment information, carrying out the policy and directing the gradate family; all directors ,department chairmen and tutors totally involve in contacting the employment units positively through their social relationship to recommend graduates to be employed. The leading group holds meetings on a term's basis, reports the progress of the work, researches and analyses the work task of next stage.

Strengthen the ideological and political education and build the correct concept of employment and job-seeking; all the members of the school pursue the policy of "inviting employment units into the college and walking out of the college", digging up the employment information and broaden the way of employment widely. Be guided by the market need to carry out the great discussion of discipline and major, to reform teaching system, to raise teaching quality and to raise graduates' competitive ability of employment; the tutors of graduates direct students to finish the graduation design well meanwhile to direct them to write the recommendation, including academic record, quality of knowledge capability and occupational skills for recommend them to work; searchers who have the need of requiring the research assistants can employ graduates to do the job of researching to solve the part of graduates' employment problem; all counselors should strengthen the service consciousness and improve the service levels to ensure the stable running of the employment. With high attention of the leading of the school, all departments give full cooperation to coordinating and it forms a graduates' employment pattern which is all-personnel, overall-procedural and omni bearing.

B. *Aiming at the need for the talents in economy society, innovating the teaching, supporting the employment*

School emphasizes the cultivation characteristic of applied art talent, strengthens the combination between professional education and market demand, pushes forward the reform of education and teaching unswervingly which is oriented by the employment and society demand and raises the implementing ability and the competitive power of graduates. According to the situation of this major, the school acts positively and strengthens the research and inspection with employment units, art colleges, professional practice bases and employment bases. It clears up the knowledgeable structure demand for the art talents, on which basis, the school carries out the great discussions about discipline and major, takes the cultivation of the talents as the goal, economic society demand as the guiding, the employment as the foothold. It refines the major connotation, adjusts the teaching plan and the practice teaching pattern which combines production, study and research and cultivation pattern of innovation talents are formed to make students have a profound theoretical attainment, handle the skillful practice operation ability and comprehensive application ability as well as they get more major knowledge. It raises the practice ability, employment ability, innovation ability and ability of starting a business and the competitive power of graduates to broaden the way of employment. Such as the student in industrial design major of this school Yang Bing, won gold metal of Germanic Red Point which is accepted by the whole world and regarded the top award of industrial design. He was employed by the Net Ease and the annual salary was 100 thousand. It is infrequent in bachelors.

C. *Making the incentive policy of employment examination, creating the atmosphere and incensing the employment*

The school specially makes " the measures for the implementation of employment for graduates in art colleges ", ruling that the school sets up contribution award of employment and commends and advertises the collectivity and individual who works distinctly and excellently; rewards the collectivity and individual who contacts the employment units positively, offers the employment units and successes in getting a job; rewards the collectivity and individual who sets up the employment units positively and makes it play a great role; rewards the tutor who takes graduates in the research project to solve the employment problem; the school encourages the teachers to direct graduates to do the pioneering work. At the same time, the school examines the rate of Graduate-study admission exam, rate of signing contracts and rate of employment in September of every year. If the record is excellent, it will get the reward from the office of employment management. Trough the incentive policy of employment is put into practice, the conditions and the capital source are ensured and the positivism of directing and servicing for employment of many teachers is protected. It makes all members take part in the job of employment.

D. *Strengthening the ideological education of employment of graduates, changing the idea, starting the business and employing*

The characteristics of art design major make the possibility and success rate of starting a business for a graduate in a high rate. Dong Yansong, who was entered the college in 2002 is animation major. When he was in the college, he likes shooting with a DV. After his graduation, he set up a company named HuBei Heaven Cinema Culture and Art Company with his schoolmates. They work for HuBei TV to make the program and help Publicity Department; Department of Culture of Province, the Group Rrovincial Party Committee and Wuhan Jin Newspaper to launched the First School Firm Festival successfully and offer the opportunities to the lower grade schoolmates for practice and part-time jobs. According to this new characteristic, the school strengthens the publicity and education of starting businesses, collects and clears up the successful models to propagandize in the website of college and the blog of the counselors extensively to wake up the consciousness of starting businesses of graduates; invites the people who started a business successfully to hold the activity of telling their stories to raise the skills of starting a business; use the directive lecture of starting a business to set up the propaganda and education and reduces the risk of starting a business; builds the pioneer park and the employment base for graduates and gives some support in policy.

E. *Removing the positive factors of society and family, forming resultant force, helping graduates be employed*

In 2008, the school uses the May Holiday to invite a parent of a student who entered the college in 2005 to talk with the students. The school makes a research about the career development after their graduation. On the issue of the plan after graduation, 73.3 percent of the parents choose to go to study, 13.3 percent choose to go aboard, 20 percent choose to work and there are also the parents who said they respect their children's choice. However, none chooses to start a business. Students also launch a research in Xiaonei Web about the plan after graduation. Under the condition that one can choose two options, going to study is 24 percent and starting a business is 20 percent. Contrasting to the result of parents, the plan about how to do in the future is just opposite. Under the circumstances, in order to unify thought, make a common view, the school communicates with parents positively by the letter in the holiday and telephone connection, develops the main part role of the graduates and the important role of the parents and society in the job of employment and builds a solid network of employment in which college, society, family and students work together.

F. *Collecting the employment information of enterprises widely, broadening the way, promoting employment*

Abundant information of the demand of the employment units is the basic premise of the employment for graduates. The school strengthens the connection with the employment units and receipts the employers warmly. The reception is judgmatical. The school uses every kinds of way to collect information; builds information data base of the employment units and graduates, clears up the specific condition and major; uses Bulletin board, website, QQ group and such carriers to release the information and make sure pass these information on every graduate. Through putting related work into practice, the mission is really disintegrated, target is really fulfilled. It pursues the actual effects and promotes the successful process of the job of employment.

G. *Strengthening the summary research of the employment, gaining experience, directing the employment*

The school must pay a great attention to the summary and the scientific research of the leaders who do political works, encourage counselors to take part in the training of employment, write the paper about the employment, declare the issue on employment area and explore the characteristic and the way of employment of graduates in art design major. The school raises the service level of employment by organizing counselors to take part in the enterprise training, in the progress of the school job fair and by holding the training of directing professional work. A series of trainings which are in different level and topic, one side, raise the staffs' understanding of policy, system and the law of employment in our country and raise the conscious of the role and importance of the employment; and on the other hand, improve and perfect the knowledgeable structure of demand of graduates underemployed to make sure to give these information to employment units in time and broaden the breadth and width of the graduates market of art design counselors about career development and education, raise the skills of the counselors for directing the employment and so raise the quality and level of employment service, strengthen the construction of working staff of employment in this school, strengthen counselors' abilities of directing the employment and prepare for the job of graduates employment in the side of theory and technique.

REFERENCES

[1] J. Zhou, "In-depth Study and Practice of the Scientific Outlook on Development , Mmake An All-out Effort for the Employment of Graduates in 2009" Video Conference on the Employment Work of graduates in 2009, 2008,12.

[2] W. Z. Xiao, "Employment-oriented Safeguard Mechanism for Employment of Art and Design Department," Journal of Changsha Social Work College, 2008, vol. 15, pp. 77-79.

[3] P. Wang, "Analysis on the Graduates Employment Characteristics of Shanghai Art College," Business Culture, 2008, 2, pp. 104.

[4] H.X. Wang and Y.Y. Ge, "Analysis on the Employment of Graduates in Domestic Art Colleges," Art Research, 2006, 4, pp.29-30.

[5] H.Y. Shen and W.J. Chen, "A Survey and Analysis on the Job-Hunting Intention and Action Among The Art Specialty," Higher Education Forum, 2006,2, pp.150-153.

Application of Artistic Wallpaper in Soft Interior Decoration Design

Yang Yi
College of Design Guangzhou Academy of Fine Arts
Guangzhou, China
Email: yy-alice@126.com

Abstract—**Wallpaper is an important factor in soft home decoration. This essay introduces basic types of wallpaper as well as some wallpaper with special decorative effects, and elaborates wallpaper's functions of visually changing the size of the room, lightening the users' mood and manifesting the interior artistic style.**

Keywords-wallpaper; interior; decoration; soft decoration design; home style

I. Introduction of Wallpaper

Soft decoration means using changeable and mobile furniture and various decorations for the overall design of interior decoration. Among various factors of soft decoration, wallpaper takes up the largest visual space of home environment, and reflects best the overall style of home decoration. In recent years, as the home design concept of "simplify fitments, stress decoration" prevails, soft decoration design has become a new highlight in interior decoration design. How to reasonably use wallpaper to build a comfortable home environment and create a personalised cultural and artistic atmosphere? Below we will carry out a preliminary exploration of the design and application of wallpaper.

A. Development of Wallpaper in China

Originated from Europe, wallpaper is hugely popular and acts as a necessity of life in Europe, America, Japan and other countries and regions. However, in China, current wallpaper is rarely used in households, and the usage rate in amusement places, restaurants and hotels is less than 20%.Wallpaper was introduced into China in the early 1970s. From early 1980s to early 1990s, wallpaper became popular as home decoration market thrived after the housing system reform of China. However, the wallpaper technology was immature at that time, and problems created, like colour fading, poor stain resistance, deformation, difficulty to clean or renovate, and short life. Besides, the introduction of large-scale import of paints of international brands into China, wallpaper gradually faded out of China's home decoration market.

Nowadays, China's wallpaper industry has already overcome the above mentioned problems. China's wallpaper is now totally able to compare with its overseas counterpart in terms of quality, pattern, variety, grade, function and other aspects. With the development of China's economy, the annually increasing living space and constant improvement of people's living standard, wallpaper will have plenty of room to grow in China.

B. Types of Wallpaper

There are different types of wallpaper: paper wallpaper, PVC wallpaper, fabric-backed wallpaper, textile wallpaper, glass-fibre wallpaper, etc. Each type of wallpaper has their own features due to the different substrate materials and manufacture technologies. For example, PVC wallpaper is water-resistant, durable, easy to clean, delicately printed and strongly textured; non-woven wallpaper is moisture-absorbing, breathable and flexible, and has a cashmere touch; Since it is made of processed minerals, glass-fibre wallpaper prevents the multiplying of microorganism and bacteria, and is resistant to bacteria and moisture.

According to decorative effect, wallpaper can be divided into ordinary PVC embossed wallpaper, metallic wallpaper, plant-fibre wallpaper, glass-fibre wallpaper, foam wallpaper, sand-textured wallpaper, embroidered wallpaper, etc. Below is the introduction of some wallpaper with special decorative effects:

1) Metallic wallpaper : Metallic wallpaper is made of PVC laminated to vacuum aluminium plating and coated with a protective surface. Metallic wallpaper is highly reflective and is able to decorate with various prints and embossments, creates the decorative effects of magnificent and glorious.

Figure.1 Metallic wallpaper

*2) Plant-fibre wallpaper :*Plant-fibre wallpaper has a paper substrate laminated to a layer of dried grass, bamboo, rattan, wood, leaves or other natural plants. Directly made of natural plant fibres, it is two to three times thicker than ordinary wallpaper and is natural and elegant with classic simplicity, it gives a feeling of returning to nature.

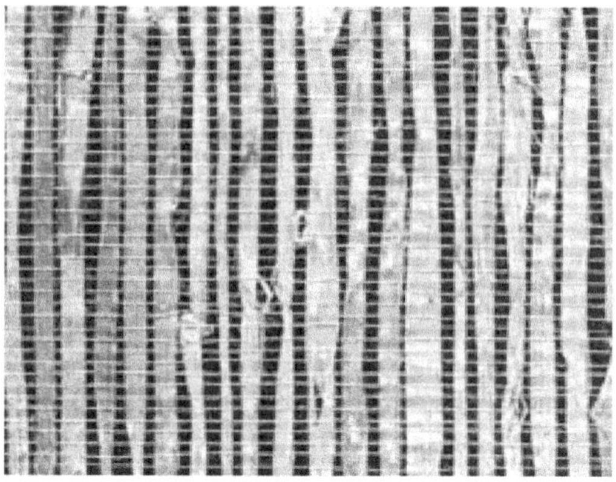

Figure.2 Plant-fibre wallpaper

3) Foam wallpaper :Foam wallpaper is made of paper substrate coated with PVC paste resin mixed with foamer, which is heated up and foams after printing. This type of wallpaper looks thicker and softer than ordinary wallpaper. With elastic and uneven surface, it can be printed with patterns resembling wood grain, mosaic or ceramic tile, which is vivid and gives a stereoscopic impression, thus creats a good decorative effect.

Figure.3 Foam wallpaper

4) Sand-textured wallpaper :Sand-textured wallpaper uses special technology to evenly spray tiny plastic sand particles on the surface of ordinary wallpaper according to the designed patter. It creates a natural sand texture. If it adds special process of glittery flakes and crystal silver, the wallpaper will present clear and glittery patterns in the lamplight, creats a strong decorative effect and displays a mysterious and splendid style.

Figure.4 Sand-textured wallpaper

5) Embroidered wallpaper :Embroidered wallpaper is a new type of wallpaper. It has a substrate of printed non-woven or vinyl wallpaper, on which embroidery effect is added through computer embroidery technology, forming embossed and stereoscopic patterns, which are colourful, elegant and delicate. It perfectly integrates traditional embroidery and wallpaper material, and thus has a high artistic value.

Figure.5 Embroidered wallpaper

II. Application of Artistic Wallpaper in Soft Interior Decoration Design

As mentioned above, among various factors of soft interior decoration, wallpaper is essential. When compared with other decorative factors, wallpaper has a great influence on decorative effect as well as psychological effect. It is also the most expressive part in the overall artistic style.

A. Wallpaper Visually Changes the Size and Height of Interior Space

Artistic wallpaper usually has certain colours and patterns on the surface. When used in a large area, these colours and patterns change people's visual sensation of the size of interior space.

For example, wallpaper splashed with bright colours in sharp contrast has a strong visual impact. When used in a large area, it makes the area appear to be rich and full. However, when used in a small room, it gives a confining, chaotic and depressing feeling. Therefore, when decorating a spacious room, using wallpaper splashed with colours or wallpaper with wide stripes helps to reduce the hollowness and manifest the artistic style of the design theme. For a small-size compact room, it is appropriate to use wallpaper patterned with small and compact figures in plain colour. Different walls can use the same hue with slightly different brightness, which helps to enlarge the spaciousness. If wallpaper with big complicated patterns is needed in order to highlight the theme, careful consideration should be given to the choice of area and colour. Under such circumstances, the area of wallpaper should be limited. It can be used as a decorative background wall, on which patterns with low saturation and low colour contrast should be used.

In addition, designers can use the function of wallpaper with stripes to visually change the height of a room. Overall, wallpapering with stripes can achieve two different visual effects. Vertical stripe pattern gives a visual illusion of upward extension. Wallpaper with vertical stripes extended from the floor to the ceiling allows the room to assume a greater height. If the room has a low ceiling, wallpaper with vertical stripes can be used to visually heighten the room and thus eliminate the depressing feeling. On the contrary, horizontal stripe pattern makes the room or one of its walls appear spacious. For the decoration of a room with high ceiling and narrow space, wallpaper with horizontal stripes can be chosen.

Moreover, designers can use wallpaper borders to adjust a high and narrow room. Wallpaper border can create a feeling of visual segmentation. The ceiling will naturally descend by placing the border thirty centimetres below the top of the wall and painting the upper part of the wall same colour as the ceiling. A border can also be pasted about ninety centimetres above the skirting, wallpaper with different patterns and colours can be used to decorate the two segmented parts of the wall. In this way, the high and narrow wall will be segmented with different effects and creates a horizontal extending feeling.

B. Wallpaper Lighten the Users' Mood

Artistic wallpaper changes people's psychological feelings of their living space. Users of rooms have different design requirements for the function of a space, and designers can choose and match decorative wallpaper according to users' psychological needs.

For example, an elder's room needs to be plain and dignified, thus it is appropriate to choose calm and composed colours, such as camel, brown, milky white, cream, pale purple and other simple but elegant colours. Patterns should be delicate and elegant with harmonious matches of colours and relatively lower purity of hue;

The design of children's room should meet children's psychological needs of being lively, cheerful and imaginative, thus the tone of design should be simple and brisk with bright colours, and the wallpaper can be patterned with blue sky and white clouds, animals and plants, as well as children's favourite cartoon figures, etc.

Red, yellow and other warm colours are usually suitable for dining-room wallpaper. According to colour psychology, warm colour keeps up people's spirit and lightens their mood, which helps to give them a good appetite. At present, in most designs of home space in China, dining-room is connected to the living-room, which means the dining-room's colour match usually follows the predominant colour of the living-room. Besides the selection of colour, wallpaper of dining-room can also adjust the diners' mood through patterns and matches of colours. For example, using lively and bright patterns with similar colours to food can make people happy and have a good appetite, such as patterns of fruit, bread, vegetables, animals, etc.

III. Wallpaper Manifests the Artistic Style of Interior Decoration

Artistic style of interior decoration is constituted of all sorts of hardware and shape, material, colour, pattern and other factors of soft decoration. Choosing and matching wallpaper which is one of the above factors usually follows this procedure: determining the overall artistic style determining the area of application– matching colours choosing and matching patterns.

Design of home space should reflect individuation and features. Designers should first carry out a detailed analysis of type of modern home decorative style and decorative features, and then choose and match wallpaper surrounding the predetermined theme style. Below are several common home styles in China and an analysis of choosing and matching wallpaper's types, colours and patterns.

1) European classical style: In the decorative style of traditional European royal court, colours of gold and silver are largely used to represent the room's magnificence and classic style. Heavy decorative colours are often used in wallpaper, such as gorgeous gold matched with blue and red, various glittery viridis, coral, peacock blue, etc. As for patterns, there are pomegranate patterns which were common during the Renaissance period, baroque henbane leaf patterns, rococo floral patterns and Neo-classical period patterns (of Greek/Roman theme), etc. With regard to the type of wallpaper, puffy faux-suede foam wallpaper and splendid metallic wallpaper are often adopted.

2) European country style: European country style is one of the main trends of modern home decoration. Its dominant life philosophy is comfort, relaxation, romance and elegance. Country style decoration features bright colours and simple but elegant patterns. In this style, PVC printed wallpaper, cloth-backed wallpaper and non-woven wallpaper are usually used. As for patterns, realistic floral clusters, small floral prints, branches and leaves, etc. are often used to form scatter patterns, matching stripes and checks or flowers, fruit and plant themes. Light colours are often used, such as white, pale yellow, pink, light blue, pale green and other cosy and elegant hues.

3) Western neo-classical style : Western neo-classical style is modern designers' re-interpretation of traditional European classical style. Low-key luxury is a byword for neo-classical style. In this style, factors are drawn from decorative details of classical style, lines are simplified and new technologies and materials are integrated. It presents simplicity and elegance with obvious characteristics of the times. In neo-classical style, colours are often used to reflect low-key luxury. Main colours used are gold, silver, black, grey, white, low-purity colours, dull red, purple grey, etc. Classical patterns in the baroque period are often deliberately redesigned and made modern through simplification, geometrization, partitioning or adding fashion elements.

4) Chinese neo-classical style: Chinese neo-classical style is the re-interpretation of traditional Chinese decorative style in modern times. It emphasizes on simplifying and being similar to classical home decoration in appearance, while having modern functions. There are different decorative schools of Chinese neo-classical style: magnificent school pursuing dignity and splendid, elegant school pursuing Zen atmosphere and scholarly air, and mix and match school featuring bright colours.

Magnificent school uses heavy and saturated colours, including mainly Chinese red, black, gold and ochre, followed by purple, peacock blue and sapphire, supplemented by glittery materials including gold and silver. These colours give an impression of magnificence, luxury and glory. Wallpaper with golden powder, metallic touch and hand-painting effect is used. Traditional Chinese patterns symbolizing good luck are often used, such as figures of lotus, peony, magpie, etc.

On the basis of frequently-used steel-grey, black and white, and brown, elegant school adds more natural hues, such as beige, celadon, greyish blue, black and white, grey, dark orange, etc. Wallpaper materials are often decorated by bamboo mat, organza, thick cotton-flax and light silk. Plant-fibre wallpaper with natural textures is also used. Making use of the contrast between different textures, this design style represents the "Zen" conception in Chinese culture;

Mixed and matched Chinese neo-classical style is close to Western neo-classical style. Traditional Chinese elements are still used for patterns, but the effects are more modern and decorative, and the matching of colours is more flexible. Some bright or European-style colours are also added to the supplementing colours of wallpaper, such as salmon, rose, peach, pink, beige, bright blue, lotus green, etc.

5) Natural style: Natural and casual style is the mainstream of modern home decoration. It holds dear natural and harmony, while being simple and environment friendly. Natural home style can be manifested in various forms. Elements of natural style can be drawn from natural materials including wood, rattan and sand, textures of stone, water and tree, as well as visual and simple images of plants and animals. Plant-fibre wallpaper can be used in natural style. It is also fashionable to use fabric patterns of supernatural style catering young people's taste for beauty, i.e. patterns of plants designed with aid of computer.

IV. FORECAST AND ANALYSIS OF THE TREND OF WALLPAPER

With the changes of people's interior decoration concepts and the detailed division of home decoration styles, wallpaper has an increasing important role in soft home decoration and becomes one of the major decorative materials. The future of wallpaper application is as below:

A. Trendy designs and diversified styles of wallpaper

Nowadays, there are many different kinds of wallpaper design to satisfy the different market needs. With various styles, wallpaper meets diversified cultural demands. Meanwhile, wallpaper design is closely following the trend of international home decoration, and with its strong visual expression, it has gradually become a major element in creating home decoration style.

B. Environment-friendliness of wallpaper

Today's, people are more aware of environmental issue, therefore, environmental friendly home decoration becomes an major trend, especially the "breathable" non-woven wallpaper and plant fibre wallpaper. Wallpaper also has to pass through "State Administration of Quality Supervision, Inspection and Quarantine" to ensure the wallpaper's quality meets the environmental standard.

C. Fun and diversified functions of wallpaper

Creative wall painting is another trend in modern interior design among young adults This clearly reflects fun and multi-function wallpaper has an expanding market; which includes heat sensitive wallpaper (pattern changes with temperature), creative labyrinth wallpaper (can paint with colour pencil) and luminous wallpaper with fluorescence effect etc.

V. CONCLUSION

Among soft home decoration factors, wallpaper takes up the largest space. Skilful application of artistic wallpaper can improve the living environment, mould the artistic style of interior decoration, and create an ideal living space for users. While the concept of soft interior decoration has been widely known with the modern home decoration trend, wallpaper is still left in the most obscure corner of home improvement market. Its various styles remain unknown to most people.

As the trends develops, I believe there will be huge increase in the use of wallpaper. It will have growing influence on soft interior decoration in the next few years, and make future home decoration more colourful and beautiful with its fashionable changes.

REFERENCES

[1] *History of Western Dyeing and Weaving Patterns*, Cheng Yifu (Japanese), Sun Jiliang trans., Textile Industry Press, March 2002

[2] *Collection of Classical Styles of Interior Decoration—Soft Decoration*, Lin Geng ed., Tianjin Science & Technology Translation & Publishing Corporation, January 2006

[3] *Wall Decoration Art*, Du Bingxu ed., Yan Get trans., Liaoning Science and Technology Press, May 2009

[4] [4] Styles and Schools of Interior Decoration, Zhang Qiman, China Architecture & Building Press, September 2006

[5] Basic Home Furnishing (Rayli BOOK), Friend of Housewives (Japan), trans. by Beijing Rayli Magazine, Light Industry Press, September 2005

[6] Basic Home Decoration (Rayli BOOK), Beijing Rayli Magazine, Light Industry Press, September 2005.

Application of Parametric Design in Contemporary Architecture Design

ChengBin
Civil engineering and Architecture institute
Southwest University of Science and Technology
MianYang, China
Email: chengbin1971@126.com

WangYu
Civil engineering and Architecture institute
Southwest University of Science and Technology
MianYang, China
Email: wangyu@swust.edu.cn

***Abstract*—Parametric design is a new design strategy based on digital technologies, now has become a vanguard of contemporary Avant-garde architecture as practice. This paper introduce the general meaning and characteristics of parametric design in the architectural design, and focus on how they interact and thus improve the efficiency of architecture design, and through the application in the pilot program, described the concrete manifestation of high-efficiency embodied in various stages of architectural design at length, thereby supporting the parametric design as a new wave of architectural design ideas in practice.**

Keywords- Parametric design ,Parametric modelling; Form-finding; Parametric figuration

I. INTRODUCTION

For some time now, digital technologies have had a substantial impact on architectural design. From the use of standard drafting packages to the more experimental use of generative design tools, they have come to play a major role in architectural production. and the digital technology in the field of architectural design applications have evolved from the earliest stage characterized by computer-aided design to the new stage which characterized by the simulation of artificial intelligence ,and algorithm-based parametric design. This is another revolution of technology-based update in the design following the post-modern movement. Currently in the Western academia, one of the more popular view is that we are in an era of multi-non-standardized, that is, the era of personalized design, parametric design has bred a technological base for this era.

Throughout the history of the evolution of human technology, technical updates are always able to improve production efficiency as the ultimate goal, that is, relatively little time, material and human inputs, to create more results. Every technological innovation can bring about a leap in efficiency after the collision resulting Between the technical and social needs. Parametric design is no exception, it would not only solve many previously unattainable design ideas, but also to improve the efficiency of the existing architectural design system. The core of parametric design is the adoption of relevant digital design software, to establish the parametric relationship between the design form and output shape, under certain design restrictions ,and generate a digital model which can flexibility to control . As for the argument, it can be a variety of design-related information, such as cost, density, climate and so on, which can be changed into the language which the computer may distinguish through the code extension. In this encoding process, using the records similar to the image-oriented (e.g.: Mapping, refers to objective and faithful record certain types information based on specific change of parameters, and to manifest themselves in the image-oriented) and the diagram and other techniques, as well as software such as Eco-Tech, we can convert these non-form information into readable data. This change brought about a qualitative leap in the architectural design, and promoted improvement in the efficiency of the industry.

II. CONCEPT AND CHARACTERISTICS OF PARAMETRIC DESIGN

Parametric design has been used in industry design for decades because of the availability of appropriate software, which are much less complicated to develop than those dealing with architectural design. Whereas in the beginning of this century, parametric design is becoming the potential main stream in architecture arena since the technology was able to tackle architectural design problems.

In the field of architectural design, Parametric design can be defined as building a mathematical or geometrical relationship flow by specific scripting software, which is controlled by multi-parameters according the design purpose. This new technology and design method brings considerable advantages for architecture design. It makes the design progress fully logically described. By varying the parameter at any section of the relationship flow, thousands of alternative options are generated for filtering the most optimized [1] .

Contemporary avant-garde architecture seek to address this societal demand via a rich panoply of parametric design techniques. However, what confronts us is a new style rather than merely a new set of techniques. The techniques in question-the employment of animation, simulation and form-finding tools, as well as parametric modeling and scripting–have inspired a new collective movement with radically new ambitions and values. In turn, this development has led to many new, systematically connected design problems that are being worked on competitively by a global network of design researchers. Over and above aesthetic recognisability, it is this pervasive, long-term consistency of shared design ambitions/problems that justifies the enunciation of a new style in the sense of an epochal phenomenon. Parametric design is a mature style, There has been talk of "continuous differentiation", versioning, iteration and mass customisation among other things for quite some time now within architectural avant-garde discourse [2].

The parametric design aimed for maximum emphasis on conspicuous differentiation and the visual amplification differentiating logics. Aesthetically, it is the elegance of ordered complexity and the sense of seamless fluidity, akin to natural systems that constitute the hallmark of parametric design. Parametric software enables the design of complex geometry, but this is not an inevitable outcome. It is just another way of facilitating design intent. It does have to impact on aesthetics.

III. How Parametric Design Demonstrates Its Efficiency In Contemporary Architectural Design

Parametric design not only allows the architect control previously unattainable complex forms, but also can enhance the efficiency of existing architectural design. The following five aspects, detailed parametric design concrete embodiment in improving the design efficiency .

A. Achieve the Complex and Organic Form with Cost-effective Way

The complex forms often stem from the plan complex context, extends in all directions the streamline which reaches, the environment energy organic response, and so on, for designers, also is a new design challenges. To the complex forms ,the most important aspect is how to precisely control its form , so that it economically feasible.

Parameter design can help designers to achieve a number of complex organic forms with the programming and script design. Architects all over the world are fascinated by algorithmic architecture which uses computer scripting to generate design solutions. Scripts are made out of series of written commands which can be programmed to solve complex geometric problems. Scripts can also be used to generate new, unexpected outcomes as they have the ability to evaluate, process and build on the given commands. In architecture scripting can be used in a variety of ways for different purposes ranging from automation of simple tasks to the management of entire design processes.

B. Efficiently Create a Variety of Programs

Option is inevitable that each project is to optimize the program needs. In the mainstream classical design methods, the architect will always subjective in accordance with the premise of separate treatment of different options for each option. , There will still spend a lot of energy to do screening. However, in the parametric design, the number of options does not depend on the design time, as long as the establishment of a parametric model, to make 10 options of time and make 100 or less the same time. On the other hand, the architect can completely do not waste time in the form of concerns about whether the current wishful, in the modeling process, only concerned about the parametric relationship between the t input and output .The completion of a model, it means the completion of a series. During the design process, where appropriate, the form of this aspect has been omitted.

The main advantage of parametric design is that the designer have an automated system for generating alternatives. That is really important. Normally, the designer would literally need to design hundreds of alternatives for one building. In order to do it realistically the designer need an automated process using all the data that allows the designer to evaluate those alternatives. At the moment the designers are dealing with hundreds, or maybe thousands of alternatives. When are looking to optimize everything, like return capital for the client, efficiency of the building, general aesthetics of the building, efficiency of the circulation, construction technologies, etc., it generates a huge number of alternatives, all of which go right from the concept design, all the way through to the construction completion. Because changes of design keep happening, the designer need to have an efficient way of dealing with global changes. In the old days, the designer may have been allowed to do one or two alternatives within a couple of months. But today they may have a week, or two days, or two hours, or even two minutes to make alternatives. That's why parametric design is able to allow the designer to iterate through the process.

C. Efficient Adjust Programs in response to Various Changes in the project

the classical way that the designer design a building is that the designer do a concept design, and then do the setting out of the geometry. Once the designer have done the setting out geometry, it will never change, because that determines the building. When the designer have design problems further down the design process, the designer cannot go back to change the fundamental design assumptions that were made at the very beginning. But parametric design allows the designer to do that. In parametric design, all the design details are still valid when the designer march back to the design assumptions, which will propagate themselves and take into account all the design details the designer have developed. That is really fundamental evolution. We used to take for granted that we had frozen the form of the building once setting out of geometry had been done which meant you could not re-visit the source in classical way. When the designer re-think the design process, the milestone is much less rigid in parametric design. When the client comes up with more requirements in later design stages, the response to the client can be to change the overall design rather than only changing the details. Overall, as long as the internal logic between the relationships of the form has not changed, parametric design can make you capability to round-trip "in the past" and "now"[3].

D. Optimum Design and Respond to the Design Constraints with the most Efficient Form

Here the parametric relationship refers to the algorithm, a form of link between the rules. Algorithm-based pattern generation is an objective, because it depends on the algorithm in the form of the output and input parameters for the formulation, rather than the subjective personal aesthetic preferences. But it is not exclusive initiative, because the algorithm development as well as the results from the generated choices still need to initiative participation. Algorithm-based parametric modeling to enable accurate quantitative restrictions to follow specific design conditions (including many of the principles of sustainable development) is possible.

E. Exchange the Design Information highly Effective

As the engineering division of labor become increasingly complex, very often, The operating efficiency of a project depends largely on its ability to efficiently exchanged design data between the different collaboration units. For example, structural engineers may only need to the axis model rather than the surface model, project budget division may require only the size of all components, equipment engineers need only a location of the cross-section or profile plane, and so on.

In the classical design process, each time with the new requirement from the other collaboration units, it means that architects need to start a new chart production. Moreover, if the construct physique is very complicated, the architect is very difficult to guarantee whether can tally accurately between each chart. Under the parametric design technique's support, through parametric modeling , the architect may export all of the technical data, generate all the technical drawings, and even big-like detail node. In practical engineering, the most powerful feature of parametric design is to keep the project design and production flow lines smoothly. In this regard, the most perfect development is the DP (Dirichlet Process). In the DP's parametric modeling environment, three-dimensional environment and two-dimensional environment and the physical properties of the list of components are connected together, as long as any changing of part ,the three-dimensional model, two-dimensional technical drawings and the physical properties of the list of components will be updated automatically. On the contrary, you can change the form of three-dimensional model by changing the data in Excel which contact with the three-dimensional mode. This parameterized model can always be used from the beginning of the conceptual design to construction plans, until the construction phase of construction, which design information maintains the coherence and consistency.

IV. Parametric Design's role at different stages of Architectural design

As we have mentioned above, parametric design allows the architect to improve the design efficiency, but in different phases of architectural design, its focus is radically different, specifically described below:

A. The Conceptual Design Phase

At this stage, the task of parametric design is based on the conditions and requirements for architectural design, set design issues need to be addressed and develop the technique that the problem of design can be resolved, and to determine the most appropriate software technology platform. Rely on this platform, to experiment with different options, and determine the final technique, at the same time, by adjusting the parameters ,to prove that the model can be operational, and to generate a number of listed options.

B. The Preliminary Design Phase

In the preliminary design stage, designers need to get feedback from the owners and all other relevant types of work, review the parameters set to confirm whether they can respond to possible future changes, at the same time, according to the feedback to adjust the input parameters of parametric model, analyze and evaluate the output form.

With in-depth and optimization program, to find a rational design basis is necessary, this process is known as form-finding. In the application of parametric design technology, designers need to set an operational 3d model According to the Context and concepts for their design. This is to encourage architects to rethink that the building is a self-organizing complex systems, In this system, Form-finding is based on a variety of design constraints (such as budget, the environment, function, scale, etc.) under the joint action of self-generated, rather than relying on subjective and intuitive command-style top-down design approach. Therefore, the parametric design still needed designer's creativity, its creativity more displays in considered how to construct the inspirational three dimensional parameter control model through the tracing during correlation factor's relations, but not subjectively to design the architectural forms.

C. The Expansion Conceptual Phase

At this stage, the design's main task is based on the the last phase of the evaluation program, to determine the final form of bulk sample ,and began exporting two-dimensional drawings for the other building Collaboration units to use, as well as material data for use in project budget. At roughly the shape is determined, the model can be developed on the detail design, maintenance and related technical personnel interaction. Another important task is to discover and correct the problem which data does not tally in the model, at the same time, update the two-dimensional drawings.

D. The Construction Documents Phase

The task of construction design stage is to export all technical drawings from the determined parametric model, in collaboration with other units to complete the relevant professional design. The mainly task of parametric model designer is to ensure that their information and data and parameter information derived model coherence for contractors and manufacturers.

In the construction design and facade tendering stage, the contractors will make their requirements based on their own manufacture difficulty and economy. In the traditional design process, due to technical limitations, the architect was not always able to control completely, but now we ourselves can control and verify the architectural effect in both detailed and overall sight with the help of parametric method. It is worth mentioning the interface with the constructors, in order to ensure geometry conversed precisely, none Mesh but Solid or Nurbs 3D-model should be import to their software platform, and of course, the tolerance should be limited[4].

V. Conclusion

Parametric design is the inevitable result that the digital technology continues intellectualization, Its emergence is no doubt a revolutionary push to the existing architectural design system, and it better suited to the existing, complex and fast changing design environment.

Parameter design can not only improve the design efficiency also allows the architect through two ways to achieve the innovation of strategies and methods: one is experimental application to the existing computer-aided design techniques which would breakthrough the existing software's modeling potential, such as access to software programs language, direct programming commands, such as Rhino, Max, and Maya's script; The second is to introduce an interdisciplinary technology to architectural design field and create a new architectural design techniques, such as Digital Project was a result of aircraft and motor vehicle design platform Ctatia.

The application and development of parametric design in the architectural design reflects the needs of our times and architectural design industry, new features, it has further optimized the social resources, improve the design and production efficiency.

[1] Gao Yan, "Cause Of Parametric Design:Interview with Labs Hesselgren, The Senior Associate Partner At KPF", Journal of World Architecture , June 2008,pp.22-27.

[2] Patrik Schumacher, "Parametricism: A New Global Style for Architecture and Urban Design", Journal of Architectural Design,Volume 79 Issue 4,2009,P.14 - 23..

[3] R. Aish and R. Woodbury, "Multi-level Interaction in Parametric Design", SmartGraphics, 5th International Symposium, August 2005,pp.151-162.

[4] Xu Weiguo, "Fold Theory,Nomadic Space:Interview On Parametric Architecture Design", Journal of World Architecture , August 2009 ,pp.16-17.

Application Scope and Study of Animation Virtualization

Liu Chang
Wuhan University of Science and Technology
Wuhan, China
Email: 43706219@qq.com

Abstract—**Animation is an interdisciplinary study. It can be extensively adopted by a number of other branches of learning. With prominent feature of virtualization, animation virtualization technology is beginning to serve a wide range of industries and aspects of life, where its optimization and interaction characters are best demonstrated. This paper studies the application scope of animation virtualization, manifests the optimization and interaction of animation virtualization in real world, and explores and develops the animation virtualization technology for it to play a more important role in today's fast-changing world.**

Keywords-optimization; real world; function virtualization; human performance; interaction

I. INTRODUCTION

With the fast development of science and technology, animation is no longer limited to bringing fun and joy to children, but beginning to serve a wide range of industries and aspects of life with its prominent features of virtualization.

Animation is an interdisciplinary subject. For instance, production of animation TV/films involves a considerable amount of TV/film knowledge, requiring both extraordinary skills in arts and mastering of cameras. Stop-motion animation has even higher requirements on the professionals such as modeling abilities, hand-made craftwork and camera shooting skills. It is no exaggeration to say that animation is an interdisciplinary subject that combines TV, film, drawing and computer science. As a disciplinary subject, it is widely applied by other branches of learning. For instance, TV commercials made by animation is ordinary today. So is multimedia presentation. However, animation is an interdisciplinary subject not only because of the above mentioned simple applications, but more importantly because of its virtualization feature.

II. APPLICATION OF ANIMATION VIRTUALIZATION IN ART DESIGN

A. Application of Animation Virtualization in Environmental Art Design

As is known to all environmental art designers, environmental art design involves all ground environments on the surface of earth and all the design categories related with decoration. They are illustrated in effect pictures no matter they are garden design, landscape design, facade or small environment design, with the goal of demonstrating the effect clearly and directly. With the constant development of 3D technology and the prevalence of 3D animation, environment art designers are now trying with 3D animation technology to simulate the real world and reflect the effects of design. This will release the burden of the effect picture and present the design more clearly and directly. By giving a camera to the effect picture, adjusting the position of the camera from picture to picture with the animation tools provided by 3DMax, and demonstrate the effect picture with multi-positions and multi-angles, it is as if a real camera is shooting in the real world. This multi-positional simulated demonstration, plus voice and sound effect, is both direct and full of fun. The pre-set camera will play video of the building under design or the garden, the landscape and the room decoration from different angles, accompanied by beautiful melodies. The pre-recorded illustrator voice will explain the design plans as the camera changes pictures. Such a method of animation virtualization of effect pictures is increasingly more prevailing today.

3D animation technology enables environment art design an excellent platform of demonstration, ushering the illustration of design effect into a new era.

B. Application of Animation Virtualization in Industrial Design

Industrial design is endowment of new quality and qualification to material, structure, form, color, surface processing and decoration with training, expertise, experiences and visual perception, as defined by International Council of Societies of Industrial Design (ICSID) on the Paris Conference in 1980. The core of traditional industrial design is product design, i.e., response to tools and equipments generated by the relationship between man and nature, including design of material equipments essential to life and existence such as tools, devices and products. However, no matter what material equipment it is, the design will finally have to be presented by models or effect pictures. Designers can draw effect pictures of their idea of design by hand or with 3D software on computer to show the exterior outlook and interior structure of the product, make models to test the outlook and the structure, and finally produce and present moulds to the public. Once the moulds are made, the shape, size, form and color of the product is directly visible, but the functions and operation method of the product are not as easy to see. Animation virtualization solves this problem by clearly showing the functionality and operation method of the product. When the moulds are made, the designer can display the product with animation, which can clearly show both the interior structure and the functionality. Moreover, narration that explains the idea of design frees the designer from long and tedious introduction words.

Product animation virtualization can also be applied by marketing, pre-sales and promotion of the products. It is an "electronic manual" that tells consumers how to use the product, which is as direct, clear, and interesting as a multimedia coursework of tool software.

With development of animation virtualization technology, designers today can conduct idea development, technique design and functionality simulation without actual models or prototypes. They can also present the outlook features, performance tests and safety tests and even start mass production. The application of animation virtualization enables industrial design with a new starting point in both terms of form and scientific analysis.

C. Application of Animation Virtualization in Industrial Design

TV and film are a most prominent cultural phenomenon in the 20th century. From a certain aspect, film-video art is an integrated subject that shares commonality with many other forms of arts. Many art elements of different art forms are absorbed by file-video arts and turned into an intrinsic element of itself, which thus becomes an integrated vide-audio art form. The introduction of digital technology, in means of animation virtualization mimicking of live performance, makes unprecedented breakthroughs of the video-audio effects of film-video arts, winning applauses from the audience with magical picture effects.

Animation virtualization hit the TV-movie industry in as early as 1933 with the launch of movie King Kong, despite that all the stunts in movies back then were produced in the stop-motion method. Computer technology enables wide application of digital technology in film-video arts, and stop-motion method was gradually replaced by computer. Human actors play against a blue background. Then via computer software, image of the performance is matted into a virtualized scenario. Movie 10,000 BC was played by human actors and actresses, but living environment of man in 10,000 BC was virtualized. This movie also virtualized the performance of mammoth by matting a virtual character into a real world scenario and synthesizing them together. For another example, in the new edition of King Kong, professionals virtualized the character King Kong in animation and synthesized it into real scenario with digital technology. Take Yangtze River VII for another example. The cute ET pet was totally virtualized, from the character to all its motions. For some movies, not only the scenario is virtualized, but the characters are virtualization of humans as well. Japanese TV drama Dinosaur Taskforce Koseidon that was broadcast in 1990s was a good example. Soldiers fighting with ETs are played by human actors, but the scenarios with dinosaurs were virtualized. According to sources, MTV film awards set an additional virtualized performance award for human-made movie characters.

Import of animation virtualization technology enables the means and ends to upgrade profoundly to meet the changing demands of technology, society and entertainment. With animation virtualization and animated stunts, the movies and TV dramas are more impressive visually and are more attractive to audiences.

The extensive application of animation virtualization in art design field boosts film-video art into a higher level, and has a significant influence on the display pattern as well as on industrial design method. It contributes to the art design industry and many other industries with the distinctive optimization features.

III. Optimization of Animation Virtualization

A. Reflection of the optimization of animation virtualization

Three-dimensional animation is able to turn designs on paper into virtual scenario, which saves labor, material and time cost and thus raises efficiency.

Urban plan of a metropolitan, for example, is first conceived in the mind of designers. Then it is roughly presented by drawings and shown to the public. Finally the draft drawings are more clearly reflected by a large amount of effect pictures. Because urban plans usually cover a very large space with a considerably amount of details and because each location and direction of the city will need to be shown in details in the effect pictures, a significantly large amount of effect pictures are needed. Hand-drawn effect pictures prevailed before computer came into being. When computer was used on a large scale, 3D software is used to produce effect pictures, whose effects are greatly improved compared with hand-drawn ones. However, pictures on paper are limited in the ability of reflecting the idea of the designer. Once the plan is implemented, improper designs will result in dismantlement and re-construction, wasting labor, material and time. What is worse, sometimes the plan has to be re-designed.

When virtualized animation came into birth, 3D software is able to generate virtualized animation from the planning and produce each section of the space under design, including streets, buildings, grassland, public utility, and even location of fire hydrants and high-voltage wires, linked by a camera in movement to realize 360 degree all-round display. The audiences will feel immersed in the environment, with strong sensual impacts as if they are in the real environment. With digital interface, the audience can also obtain data information of the project any time in the virtualized world. This greatly facilitates the planning, design, bidding, application and management of large-scale complicated projects, and helps the designer and management to conduct assistant design and assessment of different plans. Because virtualized animation is based on real-world data, the 3D scenario strictly following engineering standards and requirements of the project can represent the project under design very vividly, and the design flaws that are invisible in the past can now be easily detected. Moreover, we can easily amend data to adjust building height, change material and color of the building facade, and change the parameter of intensity of forestation. As losses caused by improper design, which cannot be rescued in the past, can now be avoided, assessment quality of the project is greatly enhanced. Plus virtualized automobiles and passengers, image of the urban planning is directly visible. The video can be made into multimedia files to show the public for them to take part in the assessment and in the meantime the plan is known by the public widely.

B. Optimization Functionality of Virtualized Animation

On traditional passenger aircraft, before the aircraft takes off, cabin crew will stand in front of the aisle to demonstrate to passengers about what to do before the take-off, where to get and how to wear oxygen mask and life vest, how to open the parachute, and where the emergency exits are.

On new aircraft, human demonstration is replaced by virtualized animation video. Before the airplane takes off, in front of each row of seat there will be a screen showing animation that simulates a passenger sitting on the seat and doing the above actions. It tells passengers what to be cautious about on the seat and what to do when the airplane is in emergency, such as where to get the oxygen mask and life vest, which part of the aircraft passengers should go to in an emergency situation, how the emergency exits are opened, and how passengers should escape from the exits, together with voice narration, as if it is real.

Functionality of virtual animation is far more than that on airplanes. In real life, in many situations virtual animation can replace human demonstration. For example, it can demonstrate when a patient is in emergency, what people around him or her should do and how to use fire hydrant in case of fire. In modern military exercise, plan and route of the operation can be made into a virtualized animation to simulate the real world war-game, which can be shown to officers and soldiers that will participate in the operation and for them to discuss. Anything improper in the plan can be corrected in a timely way.

With the development of computer technology, virtualized animation technology becomes more powerful. Its optimization functionality enables it to be extensively adopted by different industries with wider range of usage.

By leveraging animation virtualization technology, in the process of simulating the real world, the player can experience the total process of the event as if it is all real. They an also review data in the process of production, implementation, product/service supply, and logistics to make sure the plan is being executed correctly and appropriately. In this way, flaws and problems can be quickly detected, and the task can be fulfilled with less uncertain factors. The total virtualized animation can be operated in some cases to realize an interactive communication, so that the player participates in the virtual situation.

IV. Interaction of Animation Virtualization

Learning through games is no new concept. Today, knowledge such as actions to take in fire or earthquake emergencies is taught via animations.

The animation simulates the situation of a fire emergency. As the situation evolves in the games, a window pops up at every stage to alert the player of what objects to use, how to use them and what actions to take. The game designs different possibilities and options, as if the player is encountered with a real fire. The player will be presented with different scenario according to what choice he or she makes. Finally, the player will be able to escape from the fire with help from firemen and the game is over. The virtualized fire game props up an alert of what actions should be taken every step, so the players can play the game and at the same time learn how to deal with fires from this interactive animation game. Thus knowledge is learnt through game-play.

Similarly, an interactive game that virtualizes an earthquake can set a number of scenarios such as at home, on the way to office, on road, or in a car. When the earthquake breaks out, a piece of alert will appear on the computer screen to tell where the player should move to and how to protect him or herself. When the player does so, the scenario will move on, and multiple possibilities are designed for each scenario. For instance, if the player is stuck when he or she is in the middle of running away, the game will tell him or her how to make use of objects around to sustain life and get rid of the dangerous environment with assistance from relief people. For another instance, if, in the virtual environment, the player is injured by big stone and bloods severely, the game will tell him or her how to make use of objects available to handle the wound, rescue oneself, and wait for relief people. The player will also be taught about how to conduct first aid on people around who are hit by stones or are stuck in a confined space. If the player follows the instruction and makes every step correct, the system will praise him or her with "Great!"

Such interactive animation is already widely adopted by pre-schooling education because it is interesting and mysterious. When the kids finish a movement, they look forward to enter the next scenario, and different choice will lead to different results. As the kids learn from games, they do not feel boring, which is more effective than pressing knowledge into their minds.

It is popular with education websites in western countries to design courses into interactive animations. Take cooking course for example. The player click button on "how to make Italian noodles". For every step the player goes, the system will notice him or her about what gadget to use and how to operate. When the noodles are made ready, the system will evaluate the dish based on the operation of the player. In the occasion of science experiment, the player can click on "how to do experiment of laser". Similarly, a pop-up instruction appears at every step. When the player correctly finishes all the steps by following the instruction, the system will show "Experiment successful!" This method is applicable even in interpersonal communication courses, such as "how to do job interview". The scenario will be designed exactly the same as that in the real world. The player is asked to answer questions generated by the system in simulation of a real-world job interview. At the end of the interview, the system will evaluate the performance of the player and give comments and suggestions. The player can get familiar with the process of job interview with the help of this software and improve him or herself. This is especially helpful to graduate students who have never experienced job interviews.

The interactive emulation environment enabled by animation virtualization can make people immersed in it and feel as if in a real word situation. Moreover, human can interactively communicate with the machine, shifting from a mere recipient to an active participator.

SUMMARY

Animation virtualization technology is a new technology that vividly imitates visual, audio and motion actions of human in a specific environment by producing animation of real world scenario with the help of computers. It serves a wide range of industries by leveraging the immersion of animation virtualization technology and interaction and real-time reflection of computers.

In consideration of the current status of technological development, animation virtualization is still in an experimental stage. However, with the constant development of digital technology and evolvement of society, this technology will become increasingly more prevalent and more widely used to better serve the people.

REFERENCES

[1] Jing Xiong, and Dingkang Zhou, "Virtual Technology in an Internet Environment -- VRML", Computer and Modern Education, Feb 2004.

[2] Haiyan Cui, and Zuozhang Duan, "Application of Virtual Technology in Modern Education", China Modern Education Equipment, March 2005.

[3] Virtualization – www.baidu.com

Applied Analysis of Light and Shadow in Virtual Space

Xinfan Ma
College of Fine Arts & Design
Guangzhou University
Guangzhou, China
Email: 33153838@163.com

Abstract—**This essay illustrates how visual effect of light and shadow transcend the limitations of sight and lead people to experience another kind of psychic space. This essay explains that light and shadow connects people's mutual physiological properties and mental characteristics without limitations and enhances the sense of reality and interaction of the scene.**

Keywords- Virtual Space; Design of Light and Shadow; Psychology of vision; Psychological Effect

I. INTRODUCTION

Light and shadow are important ways for building the atmosphere of virtual space which means the psychological emotion of audience after watching certain scenes. For example, after Laura entered the tomb in Tomb Raider, the light behind her produced a shadow in front of her which demonstrated her moving direction like an arrow, formed a trust force from behind and showed her agility. Intensive effect of light and shadow strengthened the contrast and reached the climax.s

According to the Bible, God made light and shadow on the first day of creating the universe and all mankind. He changed the area which was formless and empty, and the darkness was over the surface of the deep, to become a beautiful earth which constitutes both light and darkness where human beings can live. As such, we can see that, light is the soul of the universe, it is also the soul of animation. By means of light, we can see colorful pictures, shapes of matters, rich colors and sense the quality feeling of physical images. Light is the foundation of visual information in screen and visual modeling. Through shadow and light, an unified artistic visual effect is demonstrated in the visual space. Therefore, light is the most basic and element in human experience It is the visual counterpoint which prolongs the energy and heat of life.

Among all the factors of visual forms, light, shadow and color are the format languages which have the most powerful stimulus to the sense organs. Light and shadow are more useful if speaking from the aspect of their influence on human spirit. Light and shadow is more suitable for demonstrating strong subjective emotions. This effect also endows a deeper meaning to the light and shadow in real life and virtual space. Pixar said: "A sense of reality must provide the audience with the experience of realness. We don't expect audience to think that these animations are shot by real photographers, but we need to make it close to reality as much as possible." In virtual space, light and shadow are indispensable elements in producing the sense of reality.

II. PERCEPTUAL EXTENSION OF LIGHT

Virtual space is the simulation of real condition; it is an actual new form as well as a perceptual extension of human toward simulation space which makes the perceptive environment wider and deeper. Luxo Jr, the Oscar nominated animation short film, is an interesting example in demonstrating light and shadow by means of simple scenes. It showed the dynamic of light as well as the extension of life. It becomes such a success not only because of its thought-provoking plot and personalized characters but also the perfect and real light and shadow in scene. Therefore, a real and credible virtual space may inspire audience's psychological and physical feelings. Proper application of certain visual function may break the limitations of language, enrich emotions, and enhance the artistic appeal as well as sentimental effect.

III. PSYCHO-VISUAL OF LIGHT AND SHADOW

Vision,the most important visual perception of human capabilities, can be perceived by the shape, size, color, brightness, mechanism and movement ect.of the outside world. Through the ratio of modality and the amount of space, color, light and shade, the processing of the relationship between deficiency and excess ect., the visual arts is vividly expressed with the ideas, concepts and emotions and so on.

The atmosphere of light and shade in the virtual scene is a visual information that the designer imparts to the audience. For instance, the light portrayed as divine power in The Prince of Egypt creates an impression that it's the holy light brought out with the advent of God, granting more superior status to the characters. By employing backlighting and oblique illumination, it can generate the light and shade chequered with realistic and virtual effect, rendering the scene distinct gradation and extensity. Shooting toward light and finding views against the light may dissociate the character from the background by drawing the outline of matters with the setting-off of light. It will provide a reasonable imaginative space as well as the optimum visual effect to satisfy audience's minds.

In the light and shadow study of visual art psychology, the physical properties of light are related with the psychological perception,the relationship of which is not merely linear.The hue and brightness of light associate closely with its frequency and intensity, as well as the background. Therefore, the use of light and shadow constantly creating suspense plot is one of the effective ways to attract audiences in animation. For instance,in Little Red

Riding Hood, the frequent use of light and shadow creates an atmosphere of tension at the crime scene. And in the dim blue light, the warning lights, searchlights, and flashlights flash high frequently.The indoor light contrasts sharply with the darkness around. That the large area of dark side and the small size of high brightness differ each other greatly with the inexplicable beam from bottom and overhead view, cause fear, and escalates the sense of crisis

Light and shadow has a subtle and unobservable effect on shaping character's psychology. Generally speaking, warm-toned sunlight makes leading role look like a performer or leader when shot from the angle of elevation. In The Lion King, audience can experience the different management of light and shadow on positive and negative roles: Simba always stands in sunshine while Scar in darkness. Furthermore, setting sun exerts dim light on a long shadow of the character, which demonstrates a depressing emotion, is one of the most common-seen scenes in romantic love stories. Shooting towards light and finding views against the light may dissociate the character from the background by drawing the outline of matters with the setting-off of light. It will provide a reasonable imaginative space as well as the optimum visual effect to bring about psychological satisfaction to the audience.

IV. From Psychological Outcomes to Perception

Sizes, shapes, lights, shadows and colors of the same object may be influenced by environment if seen from different angles, distances and intensities, but its structure remains unchanged. Therefore, light and shadow of space satisfy not only audience's psychological demands, but they also make people feel the dynamic of light and shadow ,remind them the music and strong emotion with the combinations of quality, direction and brightness, as well as having the tastes of sweet, sour, bitter and spicy in their mouths. Direction and intensity of light and shadow changes silent shapes of objects into vocal waves which is also a common-seen phenomenon of auditory sense. For example, sunlight penetrate thick forest which gives people tranquility of mind and a peaceful feeling. Although it is only a picture, the audience seem able to hear the noises of cicada. When Simba was born in the cartoon film of The Lion King, the old monkey climbed onto the mountain and raised him high in hands; golden sunlight shone on his body which demonstrates the holiness and greatness of the future king, but also enhances the auditory feeling by means of the rousing symphony. Figure 1. Scar is a crafty character, so scenes with him adopt dark yellow and shadow as contrast. Audience's emotion can be unconsciously influenced by the director. Building the positive and negative roles, themes of poetical justice and success of light over darkness can also be demonstrated in usage of light and shadow.

In Tour of a Dreamer, the director uses dark purple, deep blue and black to show that the world is a total wasteland in the beginning; but then with the emergence of bright colors, the director demonstrates the development of human history; and at last the world is swallowed by the bright golden flames. Changes of colors enlighten the progress of human civilization and life. Color contrast on vision lead audience to participate at imaginative and emotional angles The audiences seem listening to a rhythmic and magnificent symphony. With the on-going of the entire human civilization as well as the development of the plot, the hue, brightness and purity of colors in pictures penetrated each other and was changed into a dynamic and powerful scene and even surpassed the limitations of music. Audience can feel the changes of temperature from light and shadow; experience the rhythm of music as well as the sympathetic response of soul. Figure 2

"Light is the most important element in visual image. Besides its basic function of illumination, light decides the difference of hue, shape, outline, color, quality and depth of hue. It forms composition relations, balance, coordination and contrast as well as demonstrates emotions and creates an atmosphere and visual continuity." [3]. Light and shadow in virtual space may not only provide audience with auditory experience, but also a touching experience under certain lighting conditions. For example, in Ratatouille, the director highlighted the fur by means of backlight which means not only a visual enjoyment but also a touching experience. See picture three. Warm-toned pink light and shadow made the mouse look lovely. On the contrary, if the director used cold-toned green light and shadow, the audience will have the impression of cold, scary and wet.

Food may look more delicious if being shot directly or indirectly by color rendering light. A comfortable environment may make the audience feel that the food is very tasty and sense the visual pleasure. These all belong to physiological level, but only combined with pleasure and experiences of other organs can it create more senses of beauty. Just like enjoying a candlelight dinner in a western-style restaurant with lyric music and harmonious atmosphere, it is totally different if doing the same thing in a shabby tin house which circumstance is wet, noisy and dark.

Colors of characters and light in scenes are created on the basis of our lives for making matters more real. For example, the dark forest and kinky night in Tim Burton's Corpse Bride make the audience feel the depressing and scared; bright sunshine in In the Heat of the Sun make the audience feel the sun in winter as well as the scent of earth; glowing bamboo in In The Woods was like a cup of green tea which is full of pekoe, green, scented, dense, sweet and refreshing. Moonlight in Moonlight Over Lotus Pool of Zhu Ziqing is an example of perfect combination of light and shadow with synaesthesia. Moonlight and shadow as well as the lotus in breeze create a visual and scenting experience. Light and shadow of the moonlight in the pool as well as its color make the readers feel the scent of lotus, experience the coolness and relaxation under the moonlight and sense the artistic enjoyment. Visual images' influence on scenting experience will not only satisfy the audience's visual demands, but also their olfaction demands. These two aesthetic experiences interact with each other and inspire audience's emotional impulse which is demonstrated by light and shadow.

V. Conclusion

Design of light and shadow in space inspires visual enjoyment, surpass visual limitations and make people experience a different psychological condition. With the super perceptual experience of audiences toward light and shadow, rationally collocated light and shadow space may strengthen the reality of scene and serve as a special expression of space language. They will effectively inspire audience's emotions and make them have a deeper understanding of art language. Therefore, creating images by means of light and shadow which may inspire people's emotional and visual pleasure is the basic requirement for designing virtual space in order to satisfy audience's psychological and physical experience.

REFERENCES

[1] Pixar（American） Pixar Animation Studios, always devotes to making excellent computer animation products, the products have won the Academy Award for Animated Short Film, the best animation full-length film, and other technical awards for many times.

[2] Luxo Jr was produced by Luxo Jr in 1987, and won the nomination of Academy Award for Animated Short Film, and won the First Jury Award-Golden Gate Award, of Computer Image Film, San Francisco International Film Festival.

[3] Page 75 of Final Cut Pro User's Manual by Peterward, translated by Fan Zhongli, Huang Zhimin, Huaxia Press.

[4] Ma Xinfan, Animation Scene Preparation ［M］, Hunan Normal University Press, 2008.

[5] Likai, Jing jiniao, Zhang Shenshen, Miracle of Three-dimensional Creation ［M. Beijing: Engineering Industry Press, 2004.

[6] Deng Yongjian, The Arts of Light and Materials ［M］.Beijing: Posts & telecom press, 2006.

[7] Wu Yinwei., Light for Photography ［M］.Xi'an: Shanxi People's Fine Arts Publishing House, 1983.

Figure 1. Picture One The Lion King

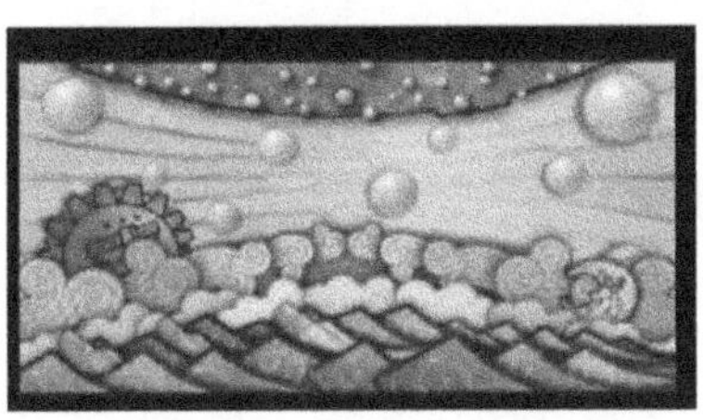

Figure 2. Picture Two Tour of the Dreamer

Figure 3. Scene in Ratatouille

Architectural Design Optimization of Transformation Under the Digital Age

Xie Ou
College of Fine Arts
Hubei Normal University
Huangshi, China
Email: sunyiqu0561@163.com

Abstract—**Architecture is a collection of culture, art, materials, structure and space. Throughout the thousands years of architectural history, we see the vast majority of buildings were present in the world in the forms of cuboids, cylinder, cone and other geometric shapes. Within a long period of time, architect has been using the two-dimensional drawing tools to express three-dimensional design ideas. Habit of long-term ruler use Impact thinking of the architect's design limits the creation of program design, but also limits innovation in terms of the construction forms. The emergence of digital technology and virtual reality technology help us solve this problem. By the help of new tools, architect can come up with better creative thinking in the early stage.**

Keywords-digital means; virtual technology; optimization; transformation

I. INTRODUCTION

Architecture is a collection of culture, art, materials, structure and space. Throughout the thousands years of architectural history, we see the vast majority of buildings were present in the world in the forms of cuboids, cylinder, cone and other geometric shapes. Within a long period of time, architect has been using the two-dimensional drawing tools to express three-dimensional design ideas. Habit of long-term ruler use Impact thinking of the architect's design limits the creation of program design, but also limits innovation in terms of the construction forms. The emergence of digital technology and virtual reality technology help us solve this problem.

II. TECHNOLOGY FACILITATE TRANSFORMATION

Technicization boosts development under digital age. Since the Industrial Revolution, The biggest change for human life is that technology replaced handicraft. In the last decade, the emergence of computer technology and network triggered series thinking in human society. Computer technology as a social component factors affecting the different areas. In terms of Architectural design approach, a new digital design system switched from habitual drawing tools, integration of architect, representative of three-dimensional design, computer technology, artificial intelligence, three-dimensional animation software, and digitalized virtual reality systems become an effective selecting means for designers.

Since 1960s, when computer has drawing function, computer graphics began to be applied to architectural design. When the computer-aided design help us accomplish the point, line, plane tasks, it has become a medium for thinking, design concepts presence and ways of creation. Today, with the constant update on hardware and software, computer has become an integral part of the architectural design process, more and more architect dependent on the computer-generated architectural forms to present their design intent. With the help of CAD, 3D, and other graphics software, building digital models can create all kinds of free surface, this modeling approach has helped the architect retain creative thinking component to the greatest extent during the early process of thinking. CATIA (Computer Aided Tri-Dimensional Interface Application) system as an advanced hybrid modeling techniques was soon accepted by architect. CATIA System program help architects inspire new ideas, its powerful curve and surface modeling features give free and full play to the designer's originality in the design process. However, the main reason for CATIA system surpasses other similar ones is that it replaces the simple space fixed-point technology of other systems. The system can define arbitrary surface, take any point as the definition point on the construction surface, you can quickly calculate the relevant values. As the system on architectural digital processing facilitate the Precision of Data, budgets, bidding and manufacture to promote contemporary complex architectural design and application. Into the 21st century, with the popularization of computer technology, virtual reality systems are no longer present in the large-scale professional equipment and platforms, which have become more accessible. The emerging of a large number of virtual realities authoring software allows the continuous optimization on the approach of architectural design. In the Process of transformation and development in technology, designers apply the free curve focused Organic form pattern into new tools usage instead of linear space pattern.

III. DIGITAL TECHNOLOGY'S APPLICATION IN THE FIELD OF ARCHITECTURAL DESIGN

The significance of architectural design was specifically brought up into discussion academic conference called architecture and computer in Boston Architecture center in 1964.After that, SOM architecture firm applied CAD into

Figure 1. Seattle Experience Music Project Museum

architecture and planning design services, and did a great job. Digital technology allows us to experience design concept in a way of intuitive visual perception, while the application of digital technology allows us to break the inherent icon way of thinking. Since CAD was invented in 1982, designers were free from burdensome hand drawings, to a more accurate and faster digital drawing. When architect Frank Gehry construct an "ideal arc" by French aviation manufacture software- CATIA in 1992, a new architecture form come into being. For instance, Guggenheim Museum in Spain, Seattle Experience Music Project Museum and Los Angeles Music Center in the U.S. are all adopt digital technology.

Without computer, these architectures with irregular shape are difficult to drawn precisely and post-construction as well. Probably just like Antoni Gaudi's Sagrada Familia Church, which cannot be completed by hand drawing in the pre-design stage, so people have to go to the site to construct the building with a great number of models. Despite more than 40 years construction, Sagrada Familia Church design was not being done until Gaudi's car accident. Today, digital technology has been adopted into this church. Architect use CAD to analyze the church structure and send data to the CNC device to produce related components. This digital technology accelerates the progress of the construction project. Because of the efficiency of this technology, this method was also used in the repayment of ancient buildings.

Netherlands firm UN Studio lead the way in attempting the 3D modeling, and worked out three-dimensional drawing. Designers can analyze, combine, modify their own idea by physical creation, and eventually form the best shape of the structure. Mercedes-Benz Museum designed by them was regarded as one of the most beautiful architecture in the world. Deconstruction master builder Hadid design the Guangzhou Opera House by adopting computer digital calculation, which covers an area of 46,000 square meters and cost about 1 billion Yuan. It is difficult to complete the task accurately by using the original method of hand drawing, not mention the perfect visual effect. It is the emergence of digital technology that we can see those amazing architecture and enjoy the glamour of design. From the above examples, it is demonstrated that digital and virtual technology are not only applied to building design but also now gradually penetrate into urban planning and residential design. This technology can help designers to come up with a better and more beautiful form of design work.

Figure 2. Guangzhou Opera House

IV. THE DEVELOPMENT AND INNOVATION ON DIGITAL AND VITURTAL TECHNOLOGY IN ARCHITECTURE DESIGN

Prior to the popularization of computers, many architectural design concepts are very difficult to be effectively and fully presented. As the well-known digital architect Liu Yudong in Taiwan said: While the designer wants to have endless imagination, and meet the specific space experience need, but it turn out to be in vain. In this way, we can only rely on so-called 'paper architecture to fill the little void. Although Paper architecture embodies nearly unlimited imagination and specific materials, still the inside feeling is very scarce. The new digital media computer undoubtedly brought about an unprecedented liberation throughout the history of human civilization. The three-dimensional modeling is a complicated and time-consuming work, but once it was done, the figures of façade, plane, and profile can be accurately calculated by computers, which lay a solid foundation for the later construction. Despite the conceptual design stage, many ideas are put into draft by hand drawing, which can only provide a prototype for concept thinking. They can not specify the drawing in 3 dimensional ways, not mention to measure the load-bearing strength of structure.

The application of design software not only efficiently supervise digital information, but also can share resources, cooperate and exchange during the design process, most importantly it can perfectly visualize designer's thinking. Application on Virtual reality technology are more consummate than the original three-dimensional model, it demonstrate a ready-made buildings in front of the owners, and through the role immersion in the virtual space so that the owners are able to feel the afterwards scenario.

The digital imaging technology has the draw on the original rendering in terms of represent ability. This interaction between designer, audience and investors can be conducted by man-machine dialogue.

Digital and virtual reality technology have already had wide application in architecture design, the majority of people have fully endorsed the active role played by computer and virtual reality system in the field of architectural design, with the help of these technologies, designers are given a larger space of thinking, the study of architectural forms and structures break the limitations of traditional design approach for thousands of years, more fancy ideas with the new design means toward a more reasonable direction in terms of structure and space. Today, we have to re-examine the classic formula of construction made by architect Vitruvius - applicable, solid and beautiful. This architect formula has been remembered by countless generations of architect, in which beautiful shape are placed in the final step to consider. In the past even a good planning can be given up due to the "structure is not feasible" reason.

As for the issue of whether function and structure top the priority list or modeling in architecture design has not been stop to discuss until today. Still many architect insist that architecture technology outweighs it's artistic. Digital technology has changed this opinion to some extent. This allows designers to better retain the original inspiration from visual point of view for the physical construction since start of the design. It may easily be conceived that China's 2008 Olympic Bird's Nest main venue, Water Cube swimming pool, the Guangzhou Opera House have become China's unique architectural landscape.

Irregular architecture, a architecture modeling design adopt by digital technology is known as physical construction generated by the use of mathematical tools to perform Irregular analysis and spontaneous calculation. Irregular architecture exhibition and Irregular experiment activities organized by Pompidou Center in Paris and Massachusetts Institute of Technology from 2003 to 2004 make a number of digital generated and shape based architecture plans stand out from the crowd, which pleasant a lot of architecture fans. And many designs in the exhibition have been put to actual construction. We can see that digital design based digital and virtual reality technology, has become the future trends, which lead the people to update the thinking on function of architecture, which broke the traditional concept of building exterior. Architecture is no longer just a shelter in traditional sense, but is possessed of new building with various shapes and sizes. Architect conducts in-depth study in terms of building structure and external shell.

Digital technology is not only applied to building design but also to the ecological architecture field. Through digital computing, architects can easily collect indicators like energy consumption, lighting, thermal comfort and energy efficiency cost analysis. People can take the initiative in architectural energy efficiency planning and efficiently adopt ecological architecture concept, which can be optimized in the early stage of design.

V. Conclusion

The impact of Digital and virtual reality technology for the architectural design is a long-term gradual process. During the past decades from 1960s, the initial computer-aided design gradually became the main approach for modern designers. Although the development of architecture under the digital age is still in the preliminary stage, but the digital means on the drawing, modeling, animation, free shape technology, virtual technology and classical hand drawing undergo a breakthrough transformation in terms of building performance, design form, architectural mass, and space and design idea.

Application of digital and virtual technology and overall improvement of design thinking, design methodology, space imagination, bring out the new digital architectures. In another word, we can say we've already in the process of digital architecture. Some people may question whether digitalization is really the most effective design methods in architectural design, Currently we can only see the advantages, and the con side will be gradually eliminated in the development of computer technology, possibly we will see more architectures like birds nestle, Guangzhou Opera House. Optimization and transformation of Technology bring about digital architectures, which lead to a new era after Modernism, and will have great impact on designer's ideas and thinking.

References

[1] Liu Yudong, Digit Architecture and Orential Experiment [M] ,Taipei: World Vision Publishing Co. Ltd, 2005

[2] Grigore C. Burdca, Virtual Reality Technology [M], Publishing House of Electronics Industry,2005

Article in a journal:

[3] Shen Xutao, Digital Architecture, Immoral city--- On Application of Virtual Reality Technology in Architectural Design [J] , Sichuan Architecture, 2006

[4] Wang Wei, Application of Computer Virtual Reality Technology in Architectural Design and Construction [J] Software Guidance, 2007

[5] Lu Aimin, Xiang Binren, Digital Forming of Ecological Architecture [J]. New Architecture, 2005

[6] Liang Zhiyao, Yan Jun, Application Cases of Ecological Architectural Technology in Architectural Desin, [J] Architectural Design and Technology, 2008

[7] Liu Zheng, Li Liping, Zhao Huimin, Technological Intergration and Creative Design [J], United Architectural Science Study, 2008

Architecture Optimization Design under Extreme Natural Environment

A Case Study of the Office and Living Quarter for SDIC Potash Project in Lop Nur, Xinjiang

Gao Bo
College of Architecture
Xi'an University of Architecture and Technology
Xi' an, China
Email: GB1949@163.com

Huang Lei
College of Architecture
Xi'an University of Architecture and Technology
Xi' an, China
Email: 76228460@qq.com

Zhu Jing
College of Architecture
Xi'an University of Architecture and Technology
Xi' an, China
Email: sherryzhujing@163.com

***Abstract*—This article comprehensively analyzes the plan about how to achieve the architecture optimization design under the extreme natural environment from the general layout, architecture form, relation between architecture and environment, and spirit of the construction site. The article discusses the adaptability of the architecture design and the measures to assort with the special factors concerning the site, and emphasizes the practice and significance of architecture optimization design too.**

Keywords-Lop Nur; extreme natural environment; optimization design; adaptability; harmony

I. ARCHITECTURE OPTIMIZATION DESIGN

The term - "Optimization Design" is a best research and solution plan for a problem, which adopts the systemic theories and measures to treat the relation between special systems to be finished and the relation between the whole and the part correctly with optimization of the whole as the target[1]. The word- "Best" can be regarded as the most satisfying result that can be achieved under the specific conditions.

The practice and significance of architecture design change with the development of the society. Although the current design is developed from the design in the past, the evaluation standard and the expected target have been boosted. The former target shall be modified and a new target shall be created along with the actualization of the plan, the renewal of design concept and the environmental development. Therefore, it is necessary to introduce the idea of adaptability in the field of architecture design, i.e. to use the idea of general harmonization and dynamic development to turn the architecture design to an adjustable concept system that is suitable for the environment and can be fed back[2]. Taking the office and living quarter for SDIC Potash Project in Lop Nur, Xinjiang as an example, this article explains the practice and significance of architecture optimization design further.

II. PROJECT INTRODUCTION

"Fig.1"As supplementary works of the key national construction project – Potash Project with an annual output of 1,200,000t in Xinjiang Lop Nur Potash Base, the office and living quarter of SDIC Xinjiang Lop Nur Potash Project Co., Ltd is located in the Lop Nur Saline in Bayinguoleng, Xinjiang on the east of Tarim Basin, and 390km away from Hami City. The construction site for the office and living quarter is 300-500m away from Potash Factory, and it is a rectangle site that is about 485m long from east to west and 251m wide from south to north, occupying a land of 121,651m2; additionally, the ground occupied by the construction site is flat and even with the surface layer composed of chemical sedimentary rocks (crystal of sodium chloride, glauberite and gypsum, etc) and the ground surface being hard pellicular salt.

"Fig.2"Three parts, namely office building, hotel and culture activity center, are planned under this project. The designs for the above three parts shall be assort with the two existing projects in construction (mess hall and lodging house) completely in order to form integrated facilities for working, living and supporting service.

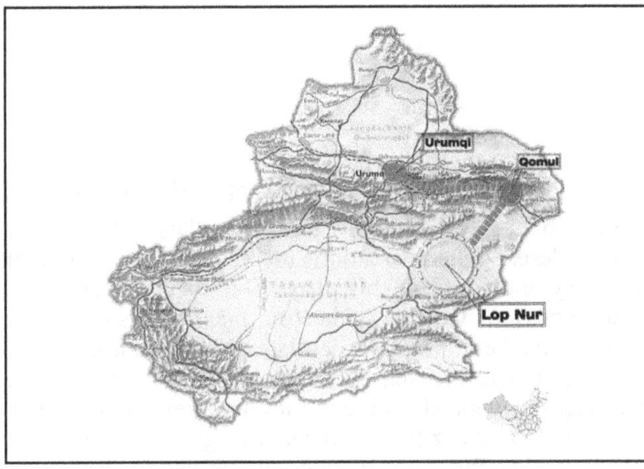

Figure 1. Lop Nur in xinjiang regin.

Figure 2. Relation between construction land and supporting facilities.

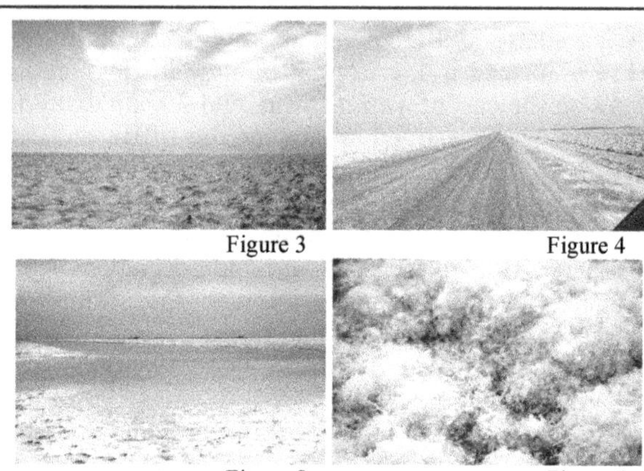

Figure 3. Heaven and earth.

Figure 4. Traffic road.

Figure 5. Potash lake.

Figure 6. Solid potassium salt.

III. Introduction on Extreme Environment

"Fig.3-6"Lop Nur is neither a snowy polar wasteland nor a lonely desert. There are only the most primary natural states- Sky and Earth in this place. Due to effect of saline soil and atrocious weather, almost no plant grows here, and no people live here. This is a desolate place; because of lack of any reference, the senses of people almost can not be controlled or balanced. The earth around melts and the sky blend entirely and hazily. For human beings, it is a forbidden zone of life; for construction practitioners, it is an inaccessible project base.

IV. Concept System

The base is a void and simplex space, a forbidden zone of human and architecture and a place lacking life so that it needs the green plant and a wonder that the life can survive from the hopeless straits. The remarkable architecture masterpiece, wonderful shape, special structure, new materials or high-efficiency surface is unnecessary for this place. The architecture constructed here shall have vigorous vitality filtering into the nature deeply and spreading the living space to any corner, and can produce a pyramid-type spirit with centricity, orientation, distinct boundary and powerful influence, which, namely the adaptability and organique of architecture is what we pursue[3].

As a complex artificial physical system, the architecture is endowed with all features of a system. The architecture adaptability design is an system behavior by adjusting the compositions of architecture to adapt to the objective external conditions on the basis of the overall view, i.e. through the adaptability design, an architecture meeting the sustainable development concept is created by adapting the architecture to the objective conditions and the internal relation under the given target and method.

The site occupied by this project has atrocious environmental conditions and no plant so that this site is too big to be measured or defined. Therefore, we shall create an artificial spatial dimension, boundary and green environment suitable for such extreme environment to make up its disadvantage, namely lack of vitality so as to turn the actual sizzzte to the best optimum site that we expect and look for.

V. Optimization Design System

A. *Optimization design for general layout*

The system thinking requires considering all factors as a whole when treating them so that we shall comprehensively consider the influence of relations between nature and social environment, between function and form, between design and evaluation, between construction and management on the architecture form, and take the negative effect of these factors after they are executed into account.

The routine standards for urban architecture are invalid in Lop Nur. Any complex shape and bitty space are negligible in such environment because there is no other architecture that can be compared with it; however, the simple geometry shape and uniform architecture mass seem more important. The uniformity can create a sense of harmony, and the equilibrium can create a sense of stability so uniformity and equilibrium which can separate the architecture from others and also unify the architecture and the environment are the most fundamental principle for beauty in form.

Figure 7. General plan.

"Fig.7"The new office building, hotel and culture activity center stated in this plan are unified with the lodging house and mess hall in construction by balanced axle distribution, i.e. with the south-north centerline of the mess hall in construction as the common axle of existing works and new works in the whole office and living quarter, on the south of the mess hall located on the central axle is the core office building and the entrance square of the whole quarter, on the west of the central axle is the lodging house in construction, on the east of the central axle is the cultural service zone composed of the hotel and the culture activity center, in which way, the whole architecture is optimized.

B. Optimization design for architecture form

Chinese people think that the "sky" can be divided into positive and negative parts with the positive strong and powerful and the negative gentle and mild, so that the aesthetic standard and artistic theory integrating rigidity and gentleness are formed for the relation between compositions of the form. In aesthetic standard and artistic theory, we shall pay attention to one basic concept that the square is created by circle and the circle is made by square, i.e. the straight line shall be assort with the curve and the curve shall be regarded as the basis.

Therefore, we find a specific spatial carrier- circle for the aforementioned centrality, orientation, distinct boundary, powerful influence and other ideology that we pursue in this design, and execute this spatial form through the office building that is at the front end of the central axle of the office and living quarter, and is also the highest building in the quarter. The function and mass of this office building are also suitable for this option. The lodging house, culture activity center and mess hall respectively on the west, east and north of the central axle are combined uniformly with the office building as core and with the big controlling the small or the main governing the secondary. This uniform combination not only shows the dominant status of the office building among different spatial forms and structures inside the base, but also gathers the infinite environment at this location, representing its pyramid-type spirit and making this field more wonderful.

"Fig.8"On both sides of the main office building are the switching tower and the sightseeing tower corresponding to both sides of main entrance of the whole quarter. The two high towers constructed of bricks and concrete and equipped with the most typical decorations and symbols of Islamic architecture in Xinjiang- Bangke building are also the scenic structures representing the enterprise sign. In the spatial form, the two towers enhance the geometrical center status and endocentric function of the round office building. The distinct roundness topic represented by them also makes contributions to the uniformity.

According to the spatial form of lodging house in construction, both the hotel and the culture activity center are constructed to be the form of X patched up by two triangles, and they are endowed with the spatial features that can be extended infinitely and increased gradually. Such design conforms to the primary purpose of the plan, namely to endow the architecture with vigorous vitality filtering into the nature deeply and spreading the living space to every corner.

Figure 8. Air scape.

Figure 9. Function division.

"Fig.9"In order to make the spatial forms on both sides of central axle similar and uniform, the lineal corridors are added on the diagonals, patching up a form of X under the condition that some modification space is left for the lodging house on the other side of the central axle in light of its structure, function and space so that the spatial similitude on both sides of central axle and the balanced and uniform integrity sense are achieved.

C. Optimization design for relation between architecture and environment

The harmonious coexistence of human beings and natural environment shall be achieved in the harmonious coexistence of human beings and other creatures, for which the architecture is indispensable because it provides a site for their harmonious coexistence. The ideal state is that all buildings in the city can enable people to enjoy the convenience of modern life, breathe in fresh air, have flowers and birds as partners and animals as neighbors just like in the forest, and become the carrier of the natural environment. The architects can create conditions for the harmony between human beings and nature through their designs. It is almost impossible to realize this ideal state in the extreme environment stated in this plan, but it is the duty

TABLE I. METEOROLOGICAL INFORMATION OF LOP NUR

S/N	Items	Quantity	Unit
1	Annual mean temperature	13.4	°C
2	Daily maximum temperature	48.0	°C
3	Daily minimum temperature	-17.5	°C
4	Annual mean rainfall	38.5	mm
5	Monthly maximum rainfall	15.0	mm
6	Annual maximum evaporation	5065	mm
7	Annual mean relative humidity	46.5	%
8	Maximum wind speed	26.0	m/s
9	Prevailing wind direction	NE、NNE	

Source: design data supplied by Party A

of optimization design to achieve such ideal state and make people have better living and working conditions.

As the main compensation approach for satisfying the demands in comfort, the architecture turns the natural environment to the state most suitable for people to live in. As a part of external environment, the relation between weather and architecture is very close. "Table.1"In Lop Nur, anything not only is bond by the nature but also must receive the test on mentality, physiology and spirit. The dry weather, difference in daytime and night temperature, lack of direction sense and loneliness of this place are unforgettable. The natural environment is too extreme to be described with any word. How is the design for the architecture in such environmental conditions optimized?

"Fig.10-14"The architecture under the extreme environment and weather shall be capable to resist the atrocious natural environment. Due to the climatic conditions of this place, namely torridity, dryness, great difference in daytime and night temperature and prevailing northeaster, we suggest the measures against wind and coldness and the daytime awning measures for the environment and climate here.

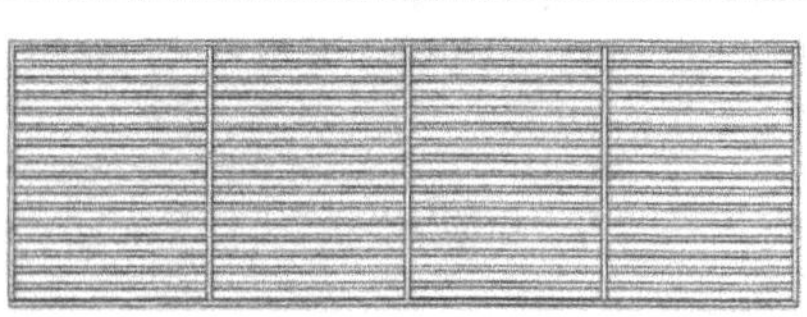

Figure10

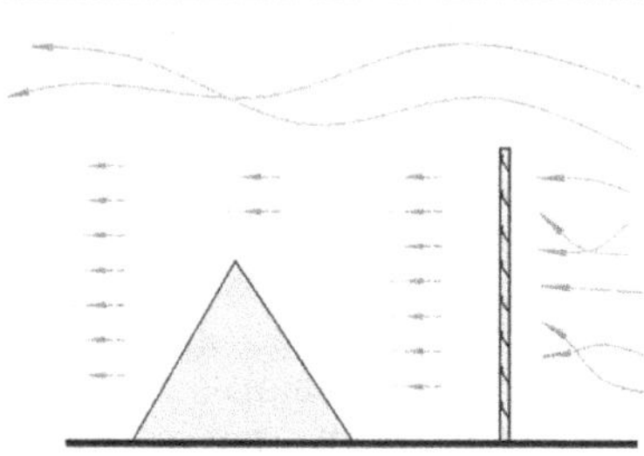

Figure 11

Figure 10. Wind fences.

Figure 11. Windbreak effect.

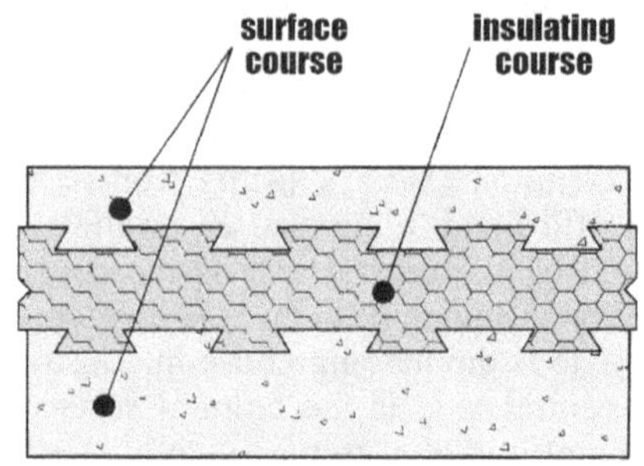

Figure 12

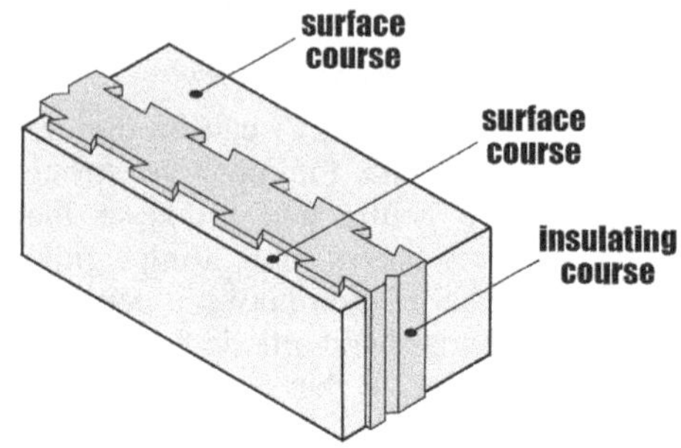

Figure 13

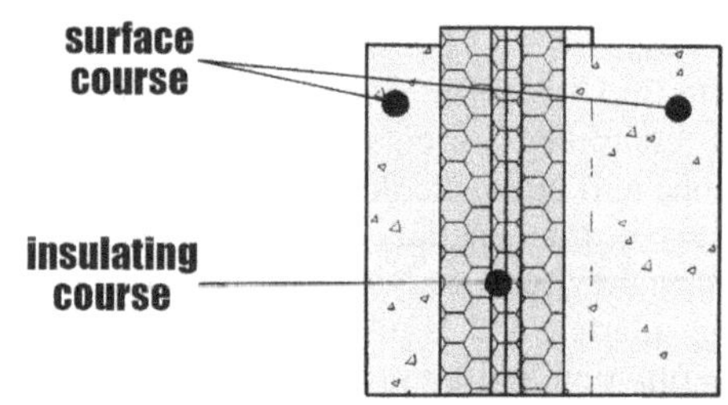

Figure 14

Figure 12. The plan of wall structure.

Figure 13. The axial side view of wall structure.

Figure 14. The section of wall structure.

To design the spatial form of architecture, the adaptability of architecture is mainly improved by spatial combination of architecture, change in architecture plane, setting of awning and special structures, and so on: the plane space is combined to a enclosed yard, and the external walls of architecture are quite closed with smaller outward window or without outward window so that the high fencing walls around can slow down the wind speed; all windows that are hopper windows are equipped with double-layer insulating glass to enhance their tightness and thermal insulation; the external walls of the architecture are constructed of dry-fasten stones, which minimize the area of windows on the eastern wall facing the prevailing wind, and minimize the effect of special wind and temperature change. "Fig.15-16"In the external spatial yard combined by the lodging house and culture service zone on both sides of central axle in the form of X, there is a piece of drawplate tent that can block the sunshine and offer people an appropriate outdoor activity site[4].

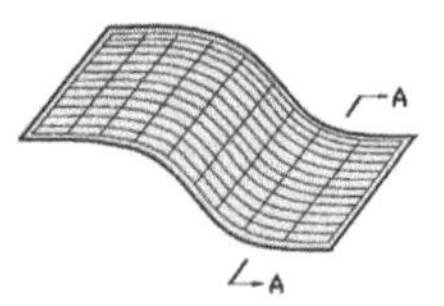

Figure15

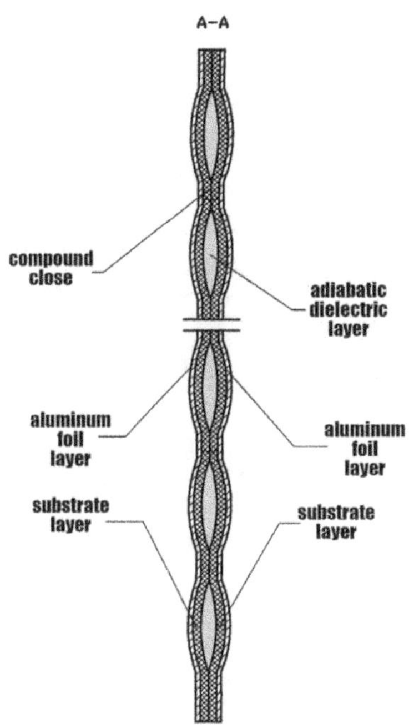

Figure 16

Figure 15. Sunshade membrane.

Figure 16. The section of sunshade membrane.

Figure17

Figure 18

Figure 17. Interior microenvironment.

Figure 18. The plan of interior microenvironment.

As we all known, it is a natural rule that the green plant can adjust a small environment through promoting the evaporation and ventilation of air[5]. The complementary existence of human beings and green plant endow the architecture with the features of green plant. The green plant together with architecture and human beings forms a complete architecture environment, realizes the recirculation of substance and energy in the nature so that without green plant, human beings can not live harmoniously. "Fig.17-18"However, due to the atrocious natural environment, up to now no plant that can grow outdoors in Lop Nur has been fostered. In order to create a complete natural environment suitable for human beings to live in, we can only grow potted plant indoors. Such potted plant is not only to beautify and decorate this place but also to melt into the extreme natural environment and the lifeless artificial environment so as to comfort, balance, entertain and ease human beings, optimize and improve the living environment.

VI. Conclusion

"As along as the natural condition in the base has its own features, the architecture shall be assort with the ambient environment as if it is grown naturally from the base." On the atrocious saline and alkaline land in Lop Nur, we shall discover its value and emphasize its value with architecture cognition: it has no valid routine standard so the simple geometry shape and the uniform architecture mass are preferred; it has no such ideology as centrality, orientation, distinct boundary and powerful influence so a specific spatial carrier- circle is determined; it is unable to resist the bad natural environment so the measures against wind and coldness and the daytime awning method are suggested; there is no plant capable to grow outside in this place so a lot of potted plant are grown indoors.

That anything is based on the nature but designed so skillfully is the charm of Lop Nur.

Acknowledgment

The successful completion of this project has benefited from the guidance and suggestion from Professor Zhimin Li of Xi'an University of Architecture and Technology, and heartful thanks for him.

References

[1] Shen Zhongying and Xiao Zijian, "New Essays on Dialectics of Nature." Shaanxi People's Press. Sep.2005, pp292.

[2] Wang Yang and Ye Weihua, "Integral Optimization & Dynamic Adaptability: Analysis on Significance of Architecture Adaptability Design," in World Architecture. Nov.2002, pp71-72.

[3] [USA] Kenneth Frampton, "Modern Architecture: A Critical History", May.2004, pp.366, 389

[4] Lu Donghong and Yan Minghui, "Analysis on Optimization of Energy-saving Design for Architecture in Regions Hot in Summer But Cold in Winter," in Shanxi Architecture. vol.34/8, Mar.2008, pp253-254.

[5] Chen Ying, Cao Qiang, Luo Shu-hua and Wu Huan, "Architecture and Environment," in Shaanxi Architecture, vol.156, Jun.2008, pp3-4.

Based on the Context of Urban-Rural Integration Optimal Design of Agricultural Landscapes

Yujing Zhu
Art Institute
Hubei Normal University
Huangshi, China
Email: yushachuan@163.com

Abstract: **This article from the urban-rural integration, are analyzed in modern Chinese cities and rural landscapes separate existence of the dual structure of the issue, the direction of development of agricultural landscapes to develop seven optimal design principles, and to submit its optimum design of the specific methods of operation, the final up to co-ordinate the development of urban and rural agricultural landscapes five strategic model, trying to explore a line with China's national conditions, the agricultural landscape development.**

Keywordst: urban-rural integration; agricultural landscape; agricultural landscape

I. Interpretation of the Integration of Urban and Rural Background

The Tenth NPC Standing Committee 30th meeting, a vote by the Town and Country Planning Act since January 1, 2008 shall come into force, past the Town Planning Act be annulled simultaneously. From the urban planning law to Town and Country Planning Act, the Chinese are breaking down based on urban-rural dual structure of the planning and management system, integrated into the urban and rural planning era.

Urbanization, urban-rural integration is the development of a new stage in the development of the productive forces to promote urban and rural residents mode of production, lifestyles and living the process of change in the way it is urban and rural population, technology, capital, resources and other elements of mutual integration, inter - resources, inter-market, and mutual service, and gradually reached between urban and rural areas in the economic, social, cultural, eco-coordinated development process.

Urban and rural areas are two areas with a particular connotation in the economic, social, political, cultural, lifestyle, and so many ways, completely different, but interdependent, mutual influence; there is also friction and antagonism, but also with the coordinated development of the inherent requirement. Socio-economic development trends in urban and rural areas step by step integration. Looking at the history of the world, both developed and developing countries, there is the trend towards urban-rural integration. Changes in the relationship between urban and rural areas, can be divided into: Rural education city, urban and rural isolation, urban and rural opposition, rural and urban integration of the urban-rural integration, and several other stages, which is the universal law of development between urban and rural areas.

"Suburbanization is a rural development model in a region or a country spread, that is, the spread of urbanization in the form of a region or a country casual." [1]The performance of the city's promotion of rural villages to cities to provide cheap industrial raw materials and food, transportation needs of industrial and commercial expansion of labor, as well as through external or implied in the form of industry and commerce for the city to provide capital accumulation.

Urban development on rural development and social progress of productive forces play a role in promoting expressed as: City to provide for rural markets for agricultural products, thus stimulating the development of agricultural commodity production; urban industrial development, labor productivity growth in agriculture and provided a strong the material basis and technological conditions; urban industrial expansion of the agricultural labor force has also provided more job opportunities to absorb surplus agricultural labor force; urban culture and way of life necessarily involves and affects the rural culture and way of life, the promotion of rural culture way of life to the scientific, modern, civilized development.

II. Optimal Design of the Framework of Agricultural Landscape

A. Analysis of Existing Situations

Domestic urban fringe status quo still has a significant urbanization and rural character of the binary mixed characteristics of various urban development activities conducted in the presence of a large number of abandoned land and natural villages of the construction is lagging behind, making the status quo is often a partial visual system, which is under construction the overall lack of coordination. Main landscape elements: the river valley, beaches, wetlands, forest, abandoned grassland, farmland, village settlements, water conservancy and hydropower projects, bridges, roads, factories, municipal public facilities, historic and cultural sites, heritage and landscape in the city's rapid development and spreading of this Under the lack of integrated planning and control system. Urban landscape was almost entirely outside the agricultural element of isolation, reinforced concrete is difficult to feel the fruit and fragrance, but also do not see the golden wheat, urban comedy rigid array, the lack of rural rustic charm.

In this context, the agricultural landscape and turned it into the only solution to the current development path, its meaning refers to the urban landscape and crop combination, by the aesthetics of urban space in the design, organic agriculture and landscape combined, while the creation of green ecological environment of the community, while consumption of agricultural products to meet the urban and rural residents, creating a new urban landscape model of urban-rural integration.

B. "7 Optimized" Design Principles:

1) Optimization of environmental and ecological nature: man-made environment and the natural environment should be integrated with each other, both the activities to meet the needs of the residents, but also to enhance the development of eco-efficiency of agricultural landscapes and sustainability.

2) Optimization of land use structure: to encourage the land market, improve the legal institutionalization of the land, saving arable land resources, thereby achieving large-scale, intensive management, to further improve the land utilization, land-output ratio.

3) Optimize the economic structure of agriculture: not only means the product structure, regional planting structure adjustment is more important is the industrial structure and the corresponding adjustment of the structure of employment.

4) Optimization of landscape science and technology content: should actively explore new technologies and new materials in modern agriculture, the use of landscape design, landscape opened up new possibilities.

5) Optimization of human culture: not just refer to the local cultural customs, but to tap the traditional farming culture and modern connotation of urban agriculture to give the agricultural landscape demonstration area unique cultural character. "It's the culture of the soil and the water and the cycles of nature; the culture of the land and the working of the land and food production."[2]

6) Optimizing the overall diversity: to ensure that model the surrounding environment of landscape elements and the coherence and integrity, simultaneously pleasing landscape demonstration area infected people, pleasant people.

7) Optimization of environmental protection and development: focus on the protection of environmental diversity and landscape diversity development.

C. Optimal Design Method of Concrete Operation

Optimization of urban agriculture in the landscape design of the concrete practice of mainly the following aspects: (Please see last page of this document for AN EXAMPLE of Figure 1.)

- In urban planning to retain the kinds of cultivated crops, and vigorously promote urban agriculture in the demonstration area implementation.
- The use of public space development planning area crops, a positive demonstration area will be developed into a healthy, sunny green Metro, the urban agriculture into the urban landscape.
- The use of the various communities in the development of public spaces, crops, use of community development in residential roof crops, to promote the implementation of green roofs.
- The establishment of a new agricultural technology demonstration sites, and vigorously promotes the new crop varieties and new agricultural technology, knowledge and popularity of the establishment of agricultural science and technology stations counseling community residents to develop pollution-free environment-friendly health food.

III. AGRICULTURAL LANDSCAPE OF URBAN-RURAL INTEGRATED DEVELOPMENT MODEL

The attached strategic model:

A. Strategy of Agricultural Industrialization

Integration of urban and rural development objectives of the agricultural landscape is to enhance the urban and rural areas in the landscape design of competitiveness and transformation of agricultural landscape and beautify people's living environment, enhancing environmental sustainability, development, and increase their income. Agricultural landscape of the idea of industrialization can promote agricultural landscape intensification and industrialization of agriculture around the city to do the article. Vigorously develop the processing industry, leading enterprises as the core operations, to leading enterprises to promote the strategic adjustment of agricultural structure to drive standardization of production, and optimizing the regional distribution of agriculture to extend the industrial chain, developing the brand of agricultural products, processing, packaging, preservation and storage of such follow-up efficiency, increase agricultural production and farmers, the degree of organization.

B. Green Product Strategy

Implementation of the strategy of green products, it is necessary to establish a good basis of agro-ecological environment and strive to build a virtuous cycle of agro-ecosystems, increase prevention and control water pollution, atmospheric pollution and agricultural non-point source pollution intensity. The development of green industry-led, closely rely on scientific and technological progress to promote green farming techniques and strive to foster a number of large scale, brand sound pollution-free agricultural products, green food and organic food from a "Plow-to-table" the whole process of agricultural products and processed products and intermediate inputs for quality and safety control.

C. Landscape Production Strategy

The urban landscape and crop combination, through the production of green space, agricultural experiment sites, residential areas surrounding the production of landscape areas, and so setting the agricultural production and urban landscape design organically combine to create the green side of the ecological environment of the community , while consumption of agricultural products to meet the urban and rural residents, creating a new urban landscape model of urban-rural integration. Consumption can also be extended to the general agricultural landscape tourism benefits, the green marketing of agricultural products, and the supporting product development and so on agricultural landscapes.

Effectiveness of the development of the landscape, the growing market cultivation, urban and rural areas and accelerating the integration process and the increasing urban-rural economic and social integration of agricultural landscapes for the rapid development of radiation vector, and accumulated a good basic condition.

D. Conservation modernization strategy

The latter part of the agricultural landscape maintenance and repair work should seize the pace of development of the times, to high-tech armed forces, at the same time pay attention to the symbiosis with the overall living environment, sustainable development. For example, in the design of "Hubei Qianjiang Li Rong new rural demonstration bases", the designer has to "make full use of the venue the status quo, in the pond grass, grass pigs rabbit, pig raising earthworms rabbit droppings, and then earthworms fed fish, to rely on recycling and three-dimensional planting and breeding techniques, to avoid the use of pesticides, chemical fertilizers, hormones and other harmful items, the establishment of Montrose, aquaculture, rice-duck farming and other ecological agriculture base. " [3]

E. Environment, the People First strategy

People First is the environment from human behavior, psychological needs, spiritual pursuits start to work with people to spend time creating a modern environment, landscape, in the implementation of the strategy process can be divided into specific: Viewable- revetment adjacent to reed marshes, lakes and the pond, the landscape for people to overlooking, appreciation, bringing a better visual enjoyment; Experience- people are able to participate, can be fun, be open, can be harvested, can be shared, so that the landscape can be joyful mood, enrich people's spiritual development.

IV. CONCLUSION

To sum up, the landscape of China's urban and rural integration development road has just begun; we need to further improve the understanding of modern agriculture, innovation and agricultural development ideas, a way of emancipating the mind and plan the agricultural landscape of development ideas. To fully mobilize the scientific research strength, with good planning and intelligence experts, to develop workable implementation plans. Necessary to strengthen leadership and coordination, the development of a mature one, step by step to start agricultural landscape transformation project. To do a good job of propaganda and mobilization and farmer education and training, improve social attention, and farmers participation, mobilize all positive factors for agricultural landscape services, to promote the formation of a joint force for the Development of the agricultural landscape.

[1] Oliver Gillam, B. YE Qimao, NI Xiaohui, "The limitless city," Beijing: China Building Industry Press, November 2007, p. 9,

[2] Jimena Martignoni, "Productive landscape-Tierra Atacama Hotel," Landscape Design, November 2008, No.6, pp. 42–47.

[3] ZHU Yujing, "Planning study of Hubei Qianjiang LiRong new rural demonstration bases," Anhui Agricultural Science, 2008 No.32, p.14060

Figure 1. Example of agricultural landscapes in the modern urban space

Based on the Optimal Design of Product Form Modeling of Cognitive Semantics of Symbolic Elements

Yanfang Peng
School of Mechanical Engineering
Southwest Petroleum University
Chengdu, China
Email: swpu-id6@swpu.edu.cn

***Abstract*—the advent of the information age demands products easy to identify, the accelerated pace of people's lives, but also require product easy to operate. Products to meet these requirements will need an accurate form of semantics, which is one of the optimal designs of products. Product semantics refers to the appearance of the product itself, through the Industrial Modeling Language, to represent the use of product-related content. In this paper, the product form of semantics from the user in the form of symbolic elements of cognitive psychology, analyzes the research on the product form the user requirements and preferences of semantics to arrive at product design modeling symbolic elements of the semantic use of the general laws, and through real cases analysis of the derived rules for authentication, reached its feasibility, so as to product form of cognitive semantics to contribute.**

Keywords Optimal design; product semantics; cognition; modeling symbols; Law

I. Introduction

Products are not only consumer goods, while acts of human life and life processes also play a normative role, and reflects the people's outlook on life and values, so the product itself in many functions other than his mouth, the information that people pay more attention to "exchange and communication ", there has been material and consumption and consumption changes to the information, products form an increasing proportion of the information conveyed by high .In addition, the product semantics of the design process is now still exist many problems, such as: in the product design process, designers rely too heavily on the experience of their own choice of the semantic factors, with a strong personal touch, their adaptability is bound to be affected; Moreover, a complete and accurate standardization brings expressive, but the lack of care for people's spiritual needs; completely customized to fully meet the spiritual needs of individuals, but a holistic view of things can cause communication difficulties, cause great inconvenience. In addition, the modern pace of life faster and faster, the pressure of living is also growing; customers need products to ease the psychological pressure, the semantics of the design can be a good way to solve this problem very well.

II. Product Form Semantic Cognition

People's awareness of the product refers to the existence and use of the product process, the human spirit and products that may arise between the interactions. People's perception of the product is a complex process, mainly through the product itself shape, color and other forms of factors play a role in transmission of information. Cognitive refers to the product users of the product form of semantic understanding and grasp of the core are interpreted as a form of deciphering the process. In the cognitive process, by product form of symbols on the users to stimulate and inspire the living with their own previous experience of related experience or conduct a link, so that products are identified and make the appropriate response. The form of symbol for products is constituted by the product shape, structure, color and material factors such as composition, so analysis of these patterns of cognitive characteristics of symbols to convey the product form of semantics is very important. In this paper, the symbol from the product of cognitive modeling to study the departure of its semantic regularity, so as to a product form of cognitive semantics to contribute.

III. Cognitive Modeling Symbols

A. Cognitive characteristics of modeling symbols

1) Sensory characteristics of modeling symbols.

a) *Point. In product design, some relatively small parts, such as products, knobs, switches, screws, light, trademarks and text elements such as the points dealt with in different forms in space arrangements, played embellishment, echo and balance of visual role. The following points of the different layouts generated by the feeling of properties listed here*, as in Table I as shown.

TABLE I. DIFFERENT POINTS OF THE ROW OF SENSORY CHARACTERISTICS OF FABRICS

Arranged form	***Sensory characteristics***
Single point	Loneliness, silence
Two points	There is a tension between sense and attractiveness of the points
Three points at any position	Sense of movement
Regular arrangement of points	Safe, tidy
Points of different sizes	Near and far, rhythm, light and dark, rhythm changes in spatial leve

The appearance of the product form, it can be perceived as a point of very large numbers, such as buttons, cooling holes, sound hole and so on, following the point of use in the product form to list specific as shown in table II.

TABLE II. POINTS IN THE APPLICATION OF PRODUCT MODELING IN

Name	Semantic Analysis
Computer Case cooling holes	Multiple points was scattering like row cloth, can convey the semantics of heat
Handset	Multiple points of the law-arranged to convey the semantics of holes out aloud
Stool	Closely arranged multiple points, forming a surface with a flat feel properties that can convey the semantics of sitting or lying

b) Line. Geometrical sense of line is an invisible entity, which is the point of moving the track; there is no thickness, only the length and direction. In semiotics, based on-line extension of the objective not only exists within the meaning of, but also has a corresponding content of referring to. Line extension function is to split screen function, the main function of the line determined by points, meaning the movement of factors, namely, direction, speed and speed stability. Point movement in a different direction was the formation of different forms of straight line, polyline, curve; points, while the velocity of the formation of the different lines in different images. A straight line, including horizontal, vertical, oblique; curves include the function curve and arbitrary curves. The connotation of the various lines have different feelings characteristics, the specific content, such as in Table III as shown.

TABLE III. CHARACTERISTICS OF DIFFERENT LINEAR SENSE

Linear	Sensory characteristics
Thin line	Slender, sharp, smooth
Chunky straight	Thick, simple, vigorous, non-smooth
Horizontal	Stable, calm, solemn
Perpendicular	Correct, tall and straight, warm
Slash	Scattering, is not stable, the combination of warmth and solemn
Rectangular polyline	Stable equilibrium
Obtuse angle Polyline	Slow, stiff, and sometimes a more elegant appearance
Acute Polyline	Active, sensitive, strong
Curve	Energy, momentum, freedom, fullness

Product lines include structural modeling of line and the product itself linear. Different types of lines have different feelings features, the following some examples of the use of linear in the product case, the specific content as shown in Table IV.

TABLE IV. LINE IN THE APPLICATION OF PRODUCT MODELING

Name	Semantic Analysis
Seat	Repeated with the linear arrangement of materials, create a sense of rhythm to convey a fluid, free of semantics
Computer Case	Chassis surface of the parabolic form of parting line has brought dynamism to the overall shape
Kettle	Arc form of sub-type line appear in an orderly sense of the edge of the curve to convey the direction of flow, but also reflects a sense of the flow of water

c) Surface. Geometry on the surface is a line of trajectory of moving to a different line to move a particular way constitute a different two-dimensional space, namely surface. Surface in the form of many factors, including geometric shape, combined-shaped, free-shaped, occasionally shape, each shape is also shown by the feeling of different characteristics, specific features such as the feeling in Table V as shown.

TABLE V. CHARACTERISTICS OF DIFFERENT FEELINGS SURFACE

Surface	Sensory characteristics
Geometric surface	Gives clear, correct, order, simple semantics, but it also may give a rigid, monotonous feeling
Combination type	Consists of two identical or dissimilar in shape to connect, overlap, or separate from the composition of rich semantics
Free-form	Gives the semantics to convey a more casual
Accidental shape	Occasional access to graphics, based on different audiences will have a different psychological awareness varied

d) Body. Length, width and depth together constitute the third element of the three-dimensional space, namely body. From the composition of characteristics of point of view, can be divided into two half-body three-dimensional, three-dimensional point, line three-dimensional, surface three-dimensional, three-dimensional pieces and so on. Body can also be divided into two kinds of flat three-dimensional and three-dimensional surface, which constitute the body of personality characteristics and personality characteristics of the surface, flat three-dimensional from two-dimensional composition, with a clear outline of the characteristics, gives a bold, strong, strong, bright semantic; surface-dimensional plane by a surface or surface with the combination of composition, gives a smooth, soft, full, smooth semantics.

2) *symbolic significance of modeling symbols.*

In addition to its inherent modeling symbol sense features, but also has a certain symbolic significance, that is implicit in the semantics, this semantics to shape borrowed from a degree of consensus representations to express, and it borrowed the objects hidden nature of meaning, the following will form the symbol of symbols intended to set out in Table Ⅵ .

TABLE VI. FORM THE SYMBOLIC MEANING OF SYMBOLS

Modeling Symbols	Symbolism
Streamlined	Speed
There are models	Life
Spiral	Life curve
Tall and straight, rigid body	Male
Gentle curved	Female

In the present product design, the extensive use of symbolic modeling notation in order to meet the user's psychological needs. The following symbols for the product modeling analysis of the symbolic significance. The specific content, such as in Table Ⅶ as shown.

TABLE VII. STYLE SYMBOL SYMBOLIC SIGNIFICANCE

Name	Symbolism
Armchair	Using a large arc of the curve, captured the women's ost representative features
Armchair	Organic surface, to convey the natural, vitality and life of the semantics
Lamps	Using spiral, and has strong vitality, to convey an extension of the semantics of the growth

B. Existing products, the semantics of modeling symbols to convey

1) Case Analysis.

Modeling is the product form of the most important factor is the product basis for their existence, so modeling symbols to convey the process of semantics plays a very important role. The following symbols to convey product for some form of success stories semantic analysis of the specific content as shown in Table Ⅷ.

TABLE VIII. SYMBOLS TO CONVEY PRODUCT SEMANTIC MODELING SUCCESS STORIES

Name	Product Semantic Analysis
Scissors	The handle of the scissors is designed negative-shaped fingers and palm, for hand holding a pair of scissors methods are indicative of a strong, very clear reminder of the scissors is used
Chair	Base with sharp curves, reflecting energy, soft, and conveyed the women's characteristics, and completely symmetric shape and shows that the sense of stability
Wall-mounted CD Player	The CD Player just pull the power cord of light can be started or stopped, in order to pass the old fan shape, building materials and the relationships between people are familiar with, arise out of the use of intuitive way
Hammoc k	Bed using a linear row of fine cloth, to convey the semantics of lightweight, sturdy the bottom of the form used to convey a stable, solid semantics
Sofa	Adopt the form of manpower, giving kind of visual experience, allowing users to get the feelings of being care
Red wine	The picture can be clearly seen from the cup mouth of the gap is to prevent users from using the cup when the nose and mouth contact, resulting in inconvenience. Users will feel the designers for their concern, naturally there will be a happy mood

2) Failure Case Study.

Of course, a lot of today's product modeling, there was also a considerable number of aspects of the product to convey the semantics of failure, resulting in the failure on the one hand because the designers mentioned earlier according to their own wishes and experience to convey semantics, on the other hand there is no semantic reason for the designer to shape consciousness. The following analysis of failed cases for some specific content, such as shown in table Ⅸ.

TABLE IX. MODELING SYMBOLS COMMUNICATE PRODUCT FAILURES SEMANTICS

Name	Product Semantic Analysis
Faucet	The novel shape of the tap, but tap the existing forms vary significantly, almost no link, users use the absence of any experience, may be only discovered after several failed attempts to use manner, and even drop out operation
Doorknob	To install the same on both sides of the door handles, reached a formal unity, but easily lead to consumer confusion: in the end is deliberately alienate or to the lira? Usually users will randomly push or sliding doors, at the first randomized trial after failure to determine the second approach is correct. However, if an emergency escape, so that the door will delay people's escape time
Tissue Holder	Designers use the image of the mouse and cheese design to add fun, but if there is no picture or text description, few people will recognize this is a paper towel rack

C. *Product modeling symbols to convey the general rules of product semantics*

Symbolic form through the inherent characteristics of the cognitive sense, as well as existing products, modeling the case of symbols to convey the semantic analysis, combined with the older generation of the relevant conclusions, summed up the product modeling product semantics of symbols to convey the general laws, mainly including the following:

a) A number of points, closely-arranged to form a surface, with surface features can be used in furniture design to convey the semantics of sitting or lying, for use in other specific products, it can convey heat, sound and other semantics.
b) Line can form multiple permutations of a strong sense of rhythm to convey a dynamic, life-semantics.
c) Thick, symmetrical shape can convey stability, and strong semantics.
d) The slender shape to convey the semantics of light.
e) The vertical patterns for conveying upright and strong semantics.
f) Curve shape for conveying the soft, dynamic semantics.
g) Bionic patterns give people a general feeling of ease, be able to convey a cordial and nice feeling of properties.
h) Can convey the organic form of life, the natural semantics. The fast pace of life will give the community to create some of Aura, so people have some breathing space.
i) Product modeling should be directional prompts, operation methods. For example, shape of the tap should be a clear direction in which the instructions are open switches; door handles need to note the shape or look inside pull deliberately alienate so.
j) New shape need to contact the user already has the experience, otherwise it will create new types of users do not understand or misinterpreted, leading to inconvenience the operation.
k) Product modeling should be consistent with the human body structure. For example, handles can be designed into the palm of your hand the negative shape, key negative surface designed finger-shaped so as to enhance the product indicative.

IV. CONCLUSION

In this article the research process, fully aware of the importance of product semantics, and now the semantic design of products that still exist some problems. Through user surveys, to a certain extent, understand the semantics of user preferences and product form requirements and summarized a series of conclusions, in future designs, these conclusions will play a guiding role.

Of course, in this study, there are some problems. The main problem is due to time constraint, the user only for a few investigations as part of the user to start such an investigation concluded that there is not necessarily the accuracy and one-sidedness, if the time to fully, it should be relatively large groups of people to investigate , and even a certain classification of the survey carried out to further understand the different groups of products form of semantic preferences and requirements, so that the conclusions more in line with the needs of users. In addition, in the conclusion of the actual application, the same, since it is only for a product design, and only use a small part of the conclusion, if sufficient time, it should be a lot of product design, to the conclusions drawn from use to be verified.

V. REFERENCES

[1] Chen Jie. Cognitive psychology and consumer product semantics [D]. Donghua University, 2004

[2] GU-ching. Product morphological analysis [M]. Beijing: Beijing University of Technology Press, 2007.9

[3] in the sails, Chen Yan. Image Design [M]. Wuhan: Huazhong University of Science and Technology Press, 2007.4

[4] Chen Shen Ren, Ma Haibo. Product design example of the semantic form of [M]. Beijing: Mechanical Industry Press, 2005

[5] Zhu Ying. Product shape design and consumer psychology research [D]. Donghua University, 2004

[6] Donald • A • Norman book Meiqiong translation. Design Psychology [M]. Beijing: CITIC Publishing House, 2003.10

[7] Hu Fei. Design symbols and product semantics [M]. Beijing: China Building Industry Press, 2005

[8] Xian-Rong Zhang. Designed to Semiotics [M]. Beijing: Chemical Industry Press, 2004

[9] Chen Shen Ren. Designed to shape semantics [M]. Beijing: Chemical Industry Press, 2006

Biotrickling Filter Optimization Measures and it Impact on the Removal Efficiency for H_2S

Ge Jie & Zhang Chengzhong
School of Environmental and Municipal Engineering
Xi'an University of Architecture and Technology
Xi'an, China
Email: xyzge@sina.com

***Abstract*—Biotreatment of Hydrogen sulfide (H_2S) is extensively studied for it's economically viable. In order to optimization the biotrickling filter capacity, microbial selection, support media options and optimization, tower design and appropriate operating conditions are usually taken into account. Get results from the simulation experiment, bacteria isolate and using as dominant species in biofilm can achieve stable industrial technique with loss cost. Modified ceramic has advantage of H_2S removal as filler with good microbial attachment and growth. Cross-flowing horizontal biotrickling filter has faster removal of pollutants in the filter to maintain the smooth flow suitable for handling the low H_2S concentrations and in no area limit place with suitable length. Vertical adverse biotrickling filter is more suitable for area limited place, high concentration and the large flowing with higher cost. When the biofilm thickness exceeds range, the filter should immediately stop running and washing.**

Keywords-component;*Biotrickling Filter, Bacteria Isolate, H_2S, Media Modified, Biofilm Washing*

I. INTRODUCTION

Hydrogen sulfide (H_2S) is extremely toxic odor to living organisms produced naturally during the natural circulation and human activities belongs to emissions control GB14554-93 in China[1]. It is produced in the oil industry, paper industry, wastewater processing, and food processing, animal husbandry and gas industry. Biotreatment compared to other available the physical and chemical technologies process has been extensively studied for its economically viable, less secondary pollution [2, 3].

The biotrickling filter had no height limitation, and hence could be designed as a tower, with a reasonable diameter. Biotrickling filter compared to other bioprocess equipment can better control of circulating water-phase reaction conditions (such as pH maintenance, temperature, nutrition, removal of reaction products) take advantage of hydrogen sulfide and other acid gases treatment significantly. In biotrickling filter, microorganisms are working under low nutrient conditions and less excess sludge with simple device. As long as proper control of the load conditions and gas-liquid contact conditions, can be achieved high efficiency of deodorization. In order to increase the biotrickling filter capacity, support media options, microbial selection, tower design and appropriate operating conditions are usually taken into account [4].

II. MATERIAL AND METHODS

A. *Experimental Equipment*

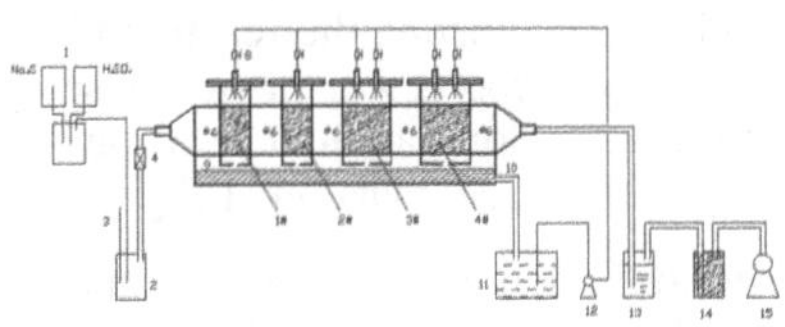

1 H_2S generating device, 2 buffer bottles, 3 air intake manifold, 4 gas flow meter, 5 filler layer 1 #, 2 #, 3 #, 4 #, 6 sampling hole, 7 nozzle, 8 valve, 9 liquid collection trough, 10 fluid holes 11 set liquid tank, 12 pump, 13 tail gas absorption bottle, 14 dryers, 15 vacuum pump

Figure 1. Schematic diagram of crossed flowing biotrickling filter system

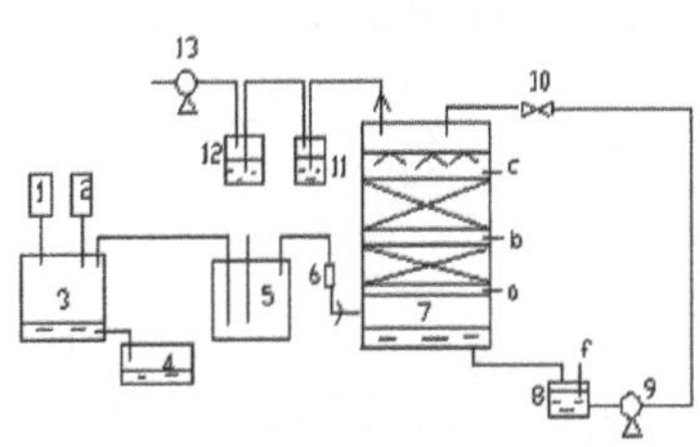

1-Na_2S solution; 2-H_2SO_4 solution; 3-H_2S generator; 4-waste water tank; 5-Gas mixing bottle; 6-gas flow meter; 7-bio-trickling reactor; 8-circulating liquid pool; 9-pump; 10-valve; 11-tail gas absorption bottle (Noah solution); 12-exhaust dryer; 13-exhaust fan; a, b, c-gas sampling point; f-fluid sampling point and the temperature, pH measurement Point

Figure 2. Schematic diagram of adverse flowing biotrickling filter system

The plexiglass biotrickling filter was developed for biodegrade H_2S. Experimental apparatus and process show in Figure1 and 2. H_2S odor is from the Na_2S solution and H_2SO_4 solution reaction in H_2S reactor. H_2S odor gets into the bio-trickling reactor by the vacuum pump. Experiments use gas-liquid crossed and adverse flowing operation, liquid pool adding to the cycle of nutrients to meet the needs of microbial growth.

B. *Deodorization Bacteria Source*

Sedimentation tank sludge; Heterotrophic bacteria have been screened by Laboratory.

C. *Determination and Methods*

H_2S concentration of Methylene blue spectrophotometric method; gas flow, gas flow meter; fluid flow, Flow meter; biomass, Weight measurement; void fraction , the void measurement.

III. RESULTS

A. Microorganism Selection

Microorganism is the main site of decomposition and transformation of pollutants have become the focus of researches. Most previous studies domesticated sewage sludge by target pollutants for waste gas purification directly [5-7]. The mixed bacteria have their own advantages, such as species rich are conducive to the formation a stable ecosystem. Microorganisms in sludge effectively degradation H_2S are Xanthomonas (*Xanthomonas* sp.), Anaerobic sulfur bacteria (T.*denitrificans*), Naples sulfur bacteria (*Tthiobacillus neaplitanus*), Thiobacillus thiooxidans (*Tthiobacillus thiooxidans*) [8], nitrogen and sulfur bacteria (*Thiobacillus denitrificans*), Thiobacillus ferrooxidans (*Thiobacillus ferrooxicans*), etc. [9-10]. At present, the sulfur-oxidizing bacteria has been using in many works, while it is has some defect such as: low pH, treatment systems, and the need to control the sulfur by adjusting the reaction conditions [11]. In the small number of reports *Xanthomonas* sp. [12] has better growing environment for partial neutral, and the product is sulfur bio-polymers for degradation H_2S [13].

In order to find out the removal capability in sewage domesticated directly and selection dominant species, simulate deodorizations of hydrogen sulfide (H_2S) experiment has been done. In the inlet flow 1L/min, residence time of 72s sprinkling water, 1.25L/h, circulating fluid pH value is neutral. The experimental results show in Figure 3.

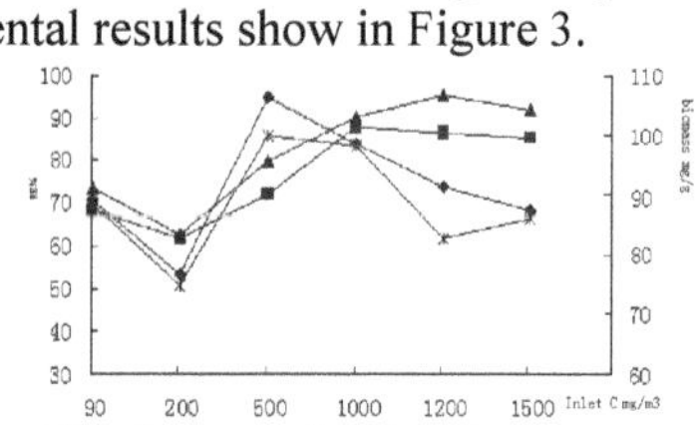

-■-RE of sewage -▲-RE of Xanthomonas sp.
-◆-Biomass of sewage -✱-biomass of Xanthomonas sp.

Figure 3. Removal Efficiency (RE) and Biomass in Sewage and *Xanthomonas* sp. Dominant Biotrickling Filter

It is found that the *Xanthomonas* sp. dominant biotrickling filter has high removal efficiency with less biomass. Artificial selection of the relating bacteria before biofilm has advantage of high activated can be used directly and it does not require a long period of domestication. In addition, the domestication process of sewage domesticated directly is not easily repeatability. So that in order to achieve stable industrial technique, microorganisms isolate first is a better way.

B. Support Media Options and Optimization

The desired features of a good support media should have high void fraction, high surface area per unit volume, low gas phase pressure drop, hydrophilic surface, low cost. Support media of the biotrickling filter is the principal place for biological fixation and main venues of pollutant treatments. Good packing can improve the concentration of microorganisms per unit volume; can affect the transmission capacity, buffer capacity and system load. The modification of the fillers can improve the biotrickling reactor removal efficiency of pollutants. Filler modification can be divided into two ways, one is to add ingredients in the preparation process such as: nanomodified ceramic [14], modified polypropylene biofiller [15]. The other is to change the surface properties such as: Foldesova [16] and others found through the thermochemical and physical methods to change the surface properties of ceramic increases the metal adsorption. At present the relevant surface modifications of filler used to remove contaminants have less reporting.

Ceramic as a carrier, $FeCl_3$ as a modifying agent to keep calcination temperature 400 ℃, calcination time of 3 hours, in 1.5mol/L concentration of the modifier prepared under conditions of optimization ceramic. The main features of modification ceramic in Table 1.

Table 1. Characteristic data of ceramic changes with modified

Feature	Shape	Size (mm)	Density (g/cm^3)	Void (%)	Equipotential pH	Biomass (mg/g)
unmodified	spheric	2—6	2.041	48	0.7—2.2	1.2
modified	spheric	2—6	2.381	55	8.5	1.6

In order to find out the removal capability in unmodified and modified ceramic, simulate deodorizations of hydrogen sulfide (H_2S) experiment has been done. In the inlet flow 1L/min, residence time of 195s sprinkling water, 5L/h, circulating fluid pH value is 6.65. The experimental results show in Figure 4.

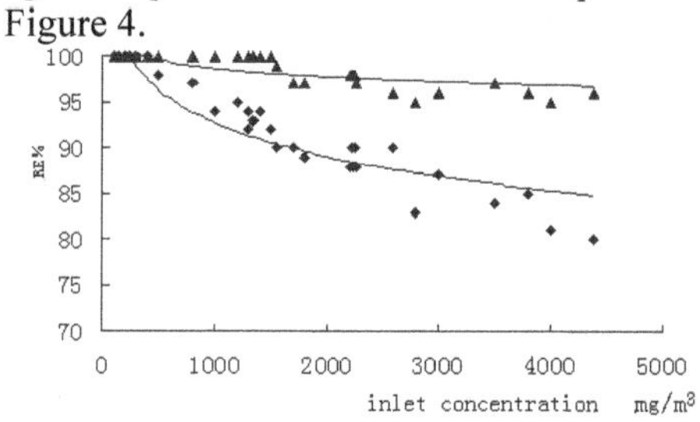

◆ RE of unmodified ceramic ▲ RE of modified ceramic

Figure 4. The removal efficiency of pottery particle and Orifice pottery particle

Removal efficiency of the modified ceramic was higher than unmodified ceramic. Modified ceramic maintain 100% when the inlet concentration is less than 1566 mg/m^3. With the inlet concentration increased, the modified ceramic has a high removal efficiency .when inlet concentration increased from 1566 to 4388mg/m^3, removal efficiency stabilized at between 99% and 95.52%. H_2S removal capacity of modified ceramic as filler in filter advantage becomes more apparent in the inlet concentration of 1182 ~ 4396mg/m^3. At this point the largest elimination capacity is 80.49 g / L • h.

C. Biotrickling Filter design and appropriate operating conditions

In order to find out the removal capability in flowing type and length of the tower, simulate deodorizations of hydrogen sulfide (H_2S) experiment has been done. Changing inlet flow, keeping residence time, sprinkling water 1.25L/h, circulating fluid pH value is neutral. The experimental results show as follow.

1) Cross-flowing Type Biotrickling Filter

Self-designed rectangular cross-flowing type of horizontal biotrickling filter as Figure 1, which is characterized by liquid is distributed more evenly. Due to horizontal biotrickling filter, each filling layer of the spray, respectively. Compared to Vertical trickling, it faster removal of pollutants in the reactor to maintain the smooth flow of the reactor. Practical applications in engineering, because liquid is a cross-flow, the pressure drop smaller, so power consumption is small; and the horizontal height of the trickling filter low, ease of maintenance.

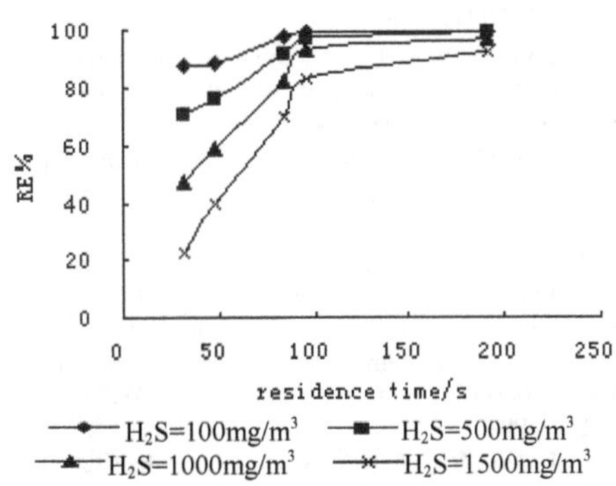

Figure 5. The removal efficiency of H_2S in different residence time

As can be seen from Figure 5, in the inlet H_2S concentration of not more than 100 mg/m^3, the residence time from the 32s ~ 192s, the system processing efficiency is 90% or more, especially when the residence time in the 84s, the system's processing efficiency is basically 100%. When the residence time in the 84s and above H_2S inlet concentration of 1000 mg/m^3, the system can achieve 90% removal efficiency. Residence time of for the 192s, the system is barely 90% of removal efficiency.

In addition, residence time from 32s to 84s, the system of H_2S removal efficiency increase quite significant, wait until after the 84s, with the residence time increases, the rate of removal efficiency change is not obvious. The reasons are likely to be on this system, the residence time in the 84s when enough time to H_2S degradation by bacteria. Even increasing the residence time, removal efficiency of the system is not much improvement. Therefore, this system is scheduled to stay in the 84s the economy and reasonable. In order to maintain more than 90% removal efficiency, H_2S inlet concentration of the maximum can reach 1200 mg/m^3. The device is suitable for handling the low concentrations of H_2S gas.

2) The Length of the Fill and the Impact of the Removal Efficiency

In the production practice, the gas concentration of pollutants along the length of the fill and the distribution of bio-trickling reactor design are important parameters, which directly determines the length of the filter packing layers. Residence time is 84s, select 5 different concentrations of H_2S to study the effects of the length of filling, the experimental results shown in Figure 6.

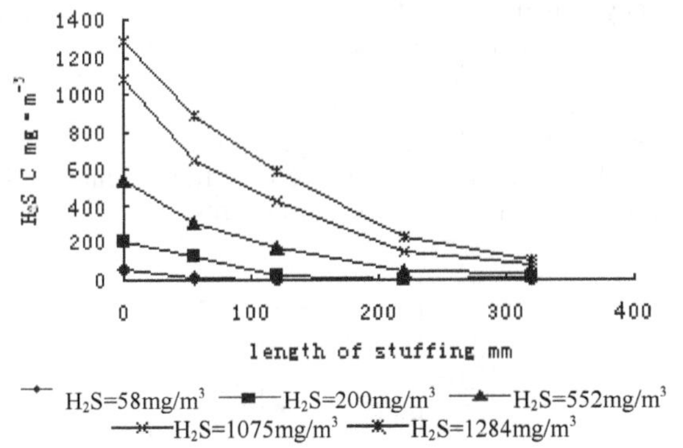

Figure 6. The relationship between the extent of stuffing and the concentration of H_2S

Figure 6 reflects the needs of all the removal of H_2S by the length of the fill. When the H_2S inlet concentration of 58mg/m^3, the length of the fill and requires only a 60mm; When the H_2S inlet concentration of 200 mg/m^3, the length of the fill and the need 120mm; when the inlet H_2S concentration of 552 mg / m^3, the length of the fill and the need 220mm; when the H_2S inlet concentration of 1284 mg/m^3, the length of the fill is 320mm at the time, H_2S removal efficiency of 92.2%, to achieve a higher removal rate need to increase the packing length. Despite the increase in the length of the fill and purification efficiency can be increased, but the actual operation should be based on the quality of inlet gas concentration and purification required to design the length of the fill, otherwise improve the purification efficiency is not simply increasing the tower's economy.

3) Vertical Adverse Flowing Biotrickling Filter

The vertical adverse flowing plexiglass biotrickling filter was developed for biodegrade H_2S as Figure 2. The effective filling height 300mm can be filling packing. Experiments use gas-liquid countercurrent operation.

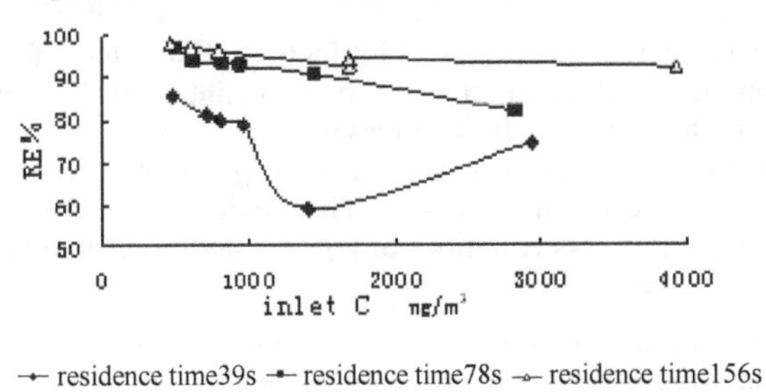

Figure 7. The removal rate of H_2S impact of different inlet concentration

H_2S removal efficiency decreases with the inlet concentration increases, the reducing trend especially of larger gas flowing. Inlet concentration of 478mg/m^3, the gas flowing is 2L/min (residence time of 39s), the removal rate was 85.8%; Import levels increased to 800mg/m^3, H_2S removal rate was 80%, continue to increase the concentration to 1403 mg/m^3, removal rate was only 59%.Vertical biotrickling filter is more suitable for high concentration and the large flowing with higher cost than cross-flowing type Biotrickling Filter.

4) *Washing of the Biofilm*

As the growing of biofilm limit the removal of H_2S are actually, uncontrolled thicker than the thin biofilm is not greater desulphurization capability. Regulation of liquid flow 10L/h flushing packed tower 10min. it is found a large number of perish bacteria out, this are because some bacteria can not get enough to fill the internal layer of oxygen in the anaerobic phase. After washing 1day, we can see from Figure 8, H_2S gas removal efficiency increased substantially.

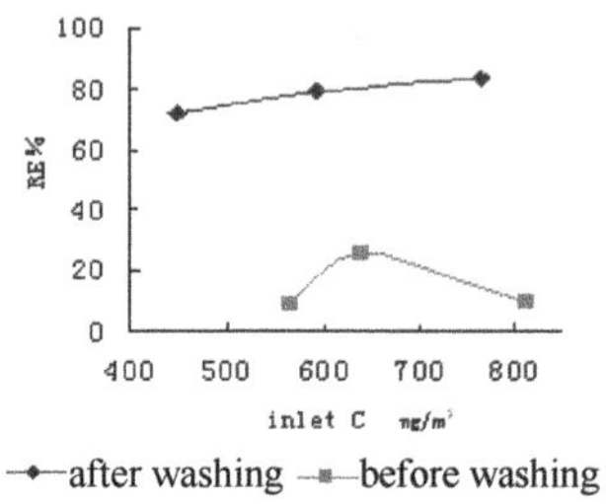

Figure 8. Impact of backwash on the removal rate of H_2S

Biofilm thickness is usually ranging between 0.07 ~ 4mm. When the film thickness exceeds this range, it is reduce the oxygen transfer rate and grow slowed down. In this case, the filter should immediately stop running and washing.

IV. CONCLUSION

Artificial selection of the relating bacteria before biofilm has advantage of high activated can be used directly and it does not require a long period of domestication. In order to achieve stable industrial technique, microorganisms isolate first is a better way.

Compared with other ceramic, modified ceramic by 1.5mol/L $FeCl_3$ as a modifying agent, keep calcination temperature 400℃, calcination time of 3 hours has better removal capacity of H_2S with good microbial attachment and growth.

Cross-flowing horizontal biotrickling filter has faster removal of pollutants in the reactor to maintain the smooth flow of the reactor. It is pressure drop smaller, power consumption is small. The horizontal height of the trickling filter low, ease of maintenance. The device is suitable for handling the low concentrations of H_2S gas and in no area limit place with suitable length.

Vertical adverse biotrickling filter is more suitable for high concentration and the large flowing with higher cost than cross flowing type Biotrickling Filter in large flowing and area limited place.

When the biofilm thickness exceeds this range, the filter should immediately stop running and washing.

REFERENCES

[1] National Environmental Protection Agency. The People's Republic of China National Standard GB14554-93. Odor pollutant emission standards. Beijing: China Standard Press, 1993.

[2] Ramírez, M., et al. Removal of hydrogen sulfide by immobilized Thiobacillus thioparus in a biotrickling filter packed with polyurethane foam. Bioresource Technology, 100(21), 2009,11: pp 4989-4995

[3] Kimberley Tang,et al. Bacteria of the sulphur cycle: An overview of microbiology, biokinetics and their role in petroleum and mining industries.Biochemical Engineering, 2009, 44(1): pp 73-9.

[4] Georg A. Parks. The Isoeleetrie Points of Solid Oxides, Solid Hydroxides and Aqueous Hydroxo Complex Systems. Chem. Rev, 1995,65: pp 177 ~ 198.

[5] HU Fang, WEI Zaishan YE Weijun. Study on purification of the odor by using biological packed column[J]. China Water and Wastewater, 2006, 22(13): pp77-81.

[6] HUANG Bing, LI Xiaomei, SUN Peishi. An approach to purification of foul-smell gas containing H2S in low concentration using a biofilm packing column[J]. Environmental Sciences and Technology, 1999, 87(4): pp17-21.

[7] SUN Peishi, HUANG Bing, HUANG Ruohua,et al. Special bacteria and their function for biopurifying waste gas[J]. China Environmental Science, 2002, 22(1): pp28-31.

[8] CHO K-S, RYU H W, LEE N Y. Biological deodorization of hydrogen sulfide using porous lava as a carrier of Thiobacillus thiooxidans[J]. Journal of Bioscience and Bioengineering, 2000, 90(1): pp25–31.

[9] XU Xiaojun,GONG Lei,YANG Hong . Biological purification theory and technology of odor gas [M]. Beijing: Chemical Industry Press, 2005: pp63-65.

[10] PARK D-H, CHA J-M, RYU H-W, et al. Hydrogen sulfide removal utilizing immobilized Thiobacillus sp. IW with Ca-alginate bead[J]. Biochemical Engineering Journal, 2002, 11(2/3): pp167–173.

[11] XU Jixian, LI Suyan LI Simin. Analysis on the generation rate of sulfur in biological desulfurization process[J]. China Water and Wastewater, 2003, 19(13): pp96-97.

[12] HO K-L, CHUNG Y-C, LIN Y-H, et al. Microbial populations analysis and field application of biofilter for the removal of volatile-sulfur compounds from swine wastewater treatment system[J]. Journal of Hazardous Materials, 2008, 152(2): pp580–588.

[13] CHO K-S, HIRAI M, SHODA M. Enhanced removability of odorous sulfur-containing gases by mixed cultures of purified bacteria from peat biofilters[J]. Journal of Fermentation and Bioengineering, 1992, 73(3): pp219–224.

[14] LIU Yanfang, XIA Yu, YU Ying. LIN Bo. Development of Water TreatmentNano – modified Ceramsites Filler With High Specific Area. Jiangxi Chemical Industry, 2004(6) : pp:91-96

[15] HAI Jing, HUANG Shangdong, ZHANG Fan, CHENG Jiang, LUO Xiaohong, DAI Songlin. Preparation ofModif ied Polypropylene Biof ilm Carr ier and Its Application. Environmental protection of chemical industry.2006,26 (4):pp333-336

[16] Foldesova, M., Tomkova, V., Lukac, P., Dillinger, P..Study of physical and thermochemical properties of modified zeolites.Journal of Thermal Analysis,1996,46(2):pp565-571

Brief Analysis on Featured Truism Innovation of Mongol-Yuan Culture

Li Jiao
College of Huaxia
Wuhan University of Technology
Wuhan, China
Email: 13600496@qq.com

Abstract—**Mongol-Yuan culture is an important component of the prosperous tradition culture of the Great China, as it has absorbed the essence of Northern and Southern China's cultures, but as well as accepted the essential parts from both the East and the West, thus forms its featured cultural appearances. So this essay is to dig out the true and pure soul of Mongol-Yuan culture, to utilize its potential cultural value as the source for landscape design, to make the past serve the present, to make the innovative designs, thus to contribute some experiences for nowadays tourism landscape featured innovations.**

Keywords—Mongol-Yuan Culture ; Tourism Landscape ; Featured Innovation; History and Culture

As the ongoing integration of world economy, the Tourism has stepped into the top of global industry, as it is recognized as the attractive "sun-rising industry", the most promising and vibrant in the tertiary industry. At present, the international tourism has a tendency of preference in cultural tourism, and thus the folk custom tourism that is to carry forward national culture and to reveal the local customs is increasingly drawing attentions.

I. Mongol-Yuan Culture

Mongol-Yuan culture is an integration of grassland culture, the Central Plains culture, and the exchange culture between China and foreign countries, it also generated from farming culture, nomadic culture and frontier fortress culture through processes of aggregation, integration, transmission and accumulation. Thus these cultures then grant it with apparent local characteristics and folk features, forming a system of inclusive local culture. Among the Mongol-Yuan culture, Mongolian imperial culture, court culture, festival culture, folk culture are the core elements.

A. Fundamental Spirit of Mongol-Yuan Culture

Generally speaking, the fundamental spirit of Mongol-Yuan culture can be showed in the following four forms: First is the heroic spirit, which forms an important part of Mongolian traditional culture, which is sublimed from the hero worship, because era demands hero, and Genghis Khan modeled his Heroic Era. Second is the spirit of freedom, which is an extraordinary feature of the nomadic people in aspects of society, economy and culture life, which mainly is reflected in nomadic lifestyle and everyday life of literacy and arts. Third is the pragmatic spirit, which for the Mongols, is to stand in realities and adjust themselves to fit in the nature, then to choose the most appropriate way of surviving. Fourth is the pioneering spirit, which involves two aspects, one is the land exploitation in daily nomadic life for surviving, and the other is the open-up of wars for political and military purposes [1].

B. Diversity of Mongol-Yuan Culture

The feature of culture diversities determines the open mind of the local people, who show out the tolerance to foreign cultures, and actively learn from and absorb the outstanding cultures of other nations. Mongolian Yuan Dynasty in ancient China history is the dynasty of least taboos on mental culture, compared to Ming and Qing Dynasties, which have dozens of times taboos. And Yuan Dynasty is the only one that enhanced freedom of religious believes ever in history. At that time, the Mongolians not only believed in Shamanism, but also accepted Buddhism, Islam and Christianity, and then formed a comprehensive religious system with its own regional and national characteristics.

Mongolians were good at learning from other peoples. Genghis Khan created the Uighur Mongolian text by learning from the Uigur people, Kublai Khan created the Basiba Mongolian text by learning from Tibetans, and Erdene created the Manchu language text from Uighur Mongolian text [2]. Mongolian used to be the first international language, and back to thirteenth century, the world-wide famous work "The Secret History of the Yuan" written in Uighur Mongolian has been well kept till now, which is a unique study of ancient nomadic civilization as the precious historical material. Since books are the carriers of culture, in the Mongolian Yuan Dynasty, the Central Plains Culture has been widely spread in the ethnic areas via books, such as the Confucian classics that have been translated into Mongolian language publications; a variety works of Buddhist Tripitaka and ethnic languages books are also published in succession. The flourishing of publish industry also reflects the prosperity of the Mongol culture from other aspects. Moreover, the Yuan drama (Verse and Drama) is also formed within this comprehensive cultural environment. The birth and prosperity of Yuan Drama is mainly due to the moral influence of the northern ethnic minorities and liberal cultural policies of the Yuan Dynasty.

The design of architecture has accepted the Han-style palace, pavilions and grassland accounts carpet style in both, fully reflecting the characteristics of the two kinds of architectural cultures. The building form and its basic structures of Mongolian palace have completely absorbed the Han-style traditions, but also have new creations in building technology, further structure, materials and architectural ornaments [3]. After the Ming Dynasty, the architectural art was influenced by the Han and Tibetan styles, as many of the new Buddhist monasteries and temples also absorbed the essence of Tibetan architecture, which formed an organic integration of the Buddhist statues, murals, Thangka decorations with the Mongolian, Han and Tibetan sculptures and paintings.

II. Tourism Landscape

In the 70 and 80 years of 20th century, "Landscape Theory" was introduced into tourism science, and thus the concept of "Tourism Landscape" appeared in the rising of Tourism Geography. "Tourism Landscape" concept can be interpreted as: In the regional areas, there form the attractive objects to tourists through the interaction between nature and human beings, and the positive phenomenon of cultural spirit which can make the tourists feel about [4].

In general, the tourism landscape, according to their attributes, can be divided into two categories of the Natural Landscape and the Cultural Landscape as shown in Figure 1. And the Natural Landscape mainly refers to the naturally formed attractions, such as the polar, mountain, climate, birds, animals, deserts, swamps, tropical rain forests, some other nature reserves and etc; while the Cultural Landscape, in comparing with the Natural Landscape, refers to the tangible or intangible attractions in the long river of human history through the continuous deposition of the landscape , such as architectural landscape, Garden landscape, heritage monuments, and culture & arts, etc; and in real tourism activities, the tourism landscape is often a complex of the natural landscape and cultural landscape in out-come.

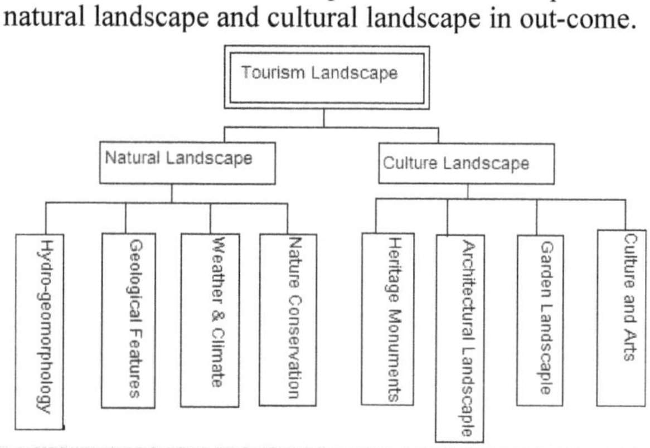

Figure 1 – Categories of Tourism Landscape

A. Features of Tourism Landscape

Participation

The participatory feature of tourism landscape concludes two aspects: One is the participation of tourists in tourism sports or projects; two is the participation of services from local residents and administrative involvement from local government.

Diversity

Different types of tourists expect more various planning types of tourism landscape for more spiritual enjoyment. Moreover, the existence of diversity plays a rather important role to ensure the stability of landscape ecosystems, and to give a buffer to the environment from interference of tourism activities.

Regions

Regional cooperation is the inevitable trend of tourism development around the world. That is to establish barrier-free tourist regions, and to enhance close cooperation among the different tourist regions, so that to create a favorable environment for the tourists as so called "I think, therefore I travel", which makes the travel routes flowing and clear.

III. Featured Innovation

This essay, in the process of analysis on the characteristics of the Mongol-Yuan culture, has also gotten to understand in some extent of its inner connections in geographic spaces– Geography Vein, of its historical traces in evolution - History Vein, and of its cultural root causes of the formation – Culture Vein [5], which is to say, the protection and development of the three veins resource can be reciprocally combined with the needs of the modern tourism landscape, and let the past serve the present, which leads to the possibilities of innovative designs, thus to accumulate experiences of featured innovations of today's tourism landscape. The principles of featured innovative design are shown in Figure 2.

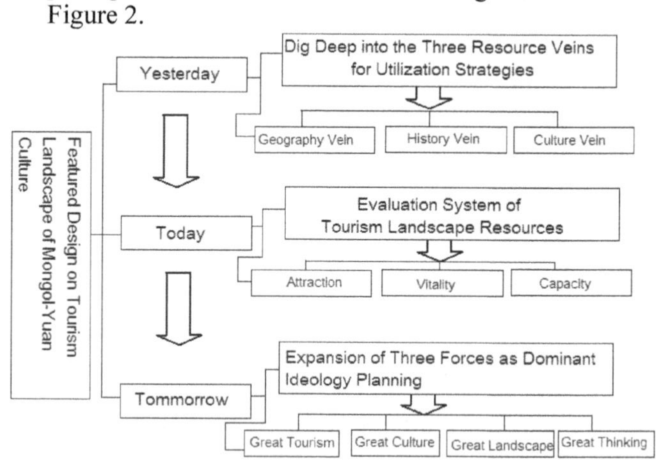

Figure 2 - Principles of Featured Innovative Design on Tourism Landscape of Mongol-Yuan Culture

A. Yesterday - Dig Deep into the Three Resource Veins for Utilization Strategies

Tourism landscape design will become "a river without water" or "a tree without root" if we put aside of its traditional culture background where it generates and develops, which is to show that culture is the foundation of tourism design and the soil of designing resources, so to dig historical information of local tourism regions and refining geographical history and culture should be the basic work for landscape design.

In process of the landscape design, via the study of the local traditional culture, historical evolutions, historical allusions, folklore, historical sites and etc, to exploit its cultural attributes and make them into the landscape design as integration, so as to combine the overall framework of landscape design, design principles with the geographical history, maintaining regional historical continuity, so that the landscape can have its unique vitality. The key elements of the featured innovations are shown in Figure 3.

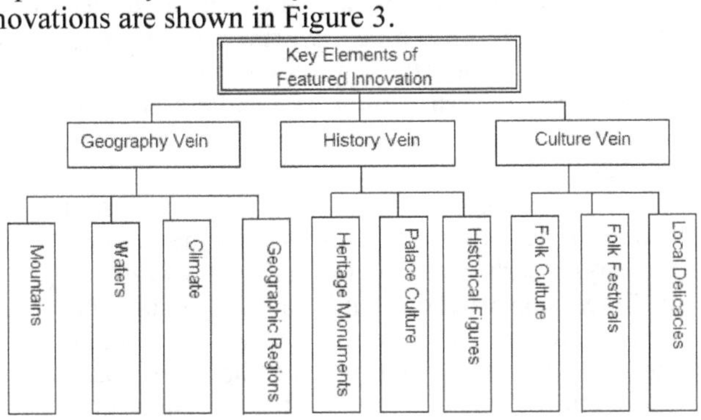

Figure 3 - Key elements of featured innovative design on tourism landscape of Mongol-Yuan Culture

Geography Vein

The geography vein is a complex combination of topography, climate, waters, soil and biological and other natural complex combination of factors is the formation of geographical landscapes of the substrate and background. The topography conditions have a decisive influence on the macroscopic appearance of the landscape, such as the delicate southern cities and the decorous northern cities.

History Vein & Culture Vein

The history vein is the historical connections of regional development that has a unique value, which is the basic foundation to study the regional characteristics of change and development and to carry out the featured innovations. Culture vein can be narrowly interpreted as "cultural connections". So to combine the history vein together with culture vein as the same value can grant every city with its own history & culture content, which can be reflected in the intangible patterns via language, ideas, and tangible patterns via buildings, streets, sculptures and other elements of the physical landscape. Such as cultural relics, precious old trees, historic sites, former residence of celebrities, urban patterns, historic streets, city squares, green system, figures, costumes and other physical substances.

B. Today -Evaluation System of Tourism Landscape Resources

Attraction

From the perspective of geography vein, the Inner Mongolian grassland, distributed on the Inner Mongolian plateau, has about 86.67 million hectares of natural grassland, where waves of undulating plateau, fresh air, blue sky, white clouds are all natural beauties to attract tourists. The unique attractions of prairie landscape and desert landscape from east to west have unparalleled appealing in the national system of natural tourism resources.

From the perspective of the history vein, the legend of Genghis Khan who led the Mongol horsemen galloping about Eurasia, has granted the mystery of miracles here, and attracts the curiosity of tourists throughout the world.

From the perspective of culture vein, here have lived the Huns, the DongHu people, the Xianbei people, the Khitan people, the Mongolians and many other northern nomads, which deposit a long-established nomadic culture, forming colorful folk customs. Mongolians are good at singing and dancing, and also have a rich variety of competitive sports, entertaining activities. The clothing and food there are unique as well, and all these features have been a major appealing to the tourists in the cultural circle of the Central Plains farming.

Vitality

The vitality notonly refers to the developing and expanding capabilities of scenic and tourism resources, but also refers to the surviving abilities of the scenic places in aspects of environment, society, economy and so on. Inner Mongolia is rich in tourism resources for its featured folk culture, and it is ready in economic, technical, policy and management aspects to develop the folk culture tourism, while the Inner Mongolian tourism's economic structure and regional market capacity also need to utilize folklore resources for a better balance, so as to evolve into a tourism project of strong attraction, thus promoting the regional tourism economy of Inner Mongolia.

In the development and utilization of grassland tourism resources, it should continue to make the tourism activities in harmony with grassland ecological environment from the point of view that has been long formed in the nomadic culture of nature and man in one. The grassland tourism activities, including eating, housing, traveling, viewing, shopping, entertaining and etc can be organically integrated with the nomadic habits of local residents, which can not only improve the service quality, satisfy the curiosity of tourists on nomadic culture in a greater extent, but also can protect original pastoral lifestyle of the local people.

Capacity

The capacity of the ecological environment in tourism regions refers to the maximum extent that the natural environment can afford for tourism development, and a well-maintained natural environment is the necessary basis for development planning of scenic spots.

The Inner Mongolian grassland has a relatively simple ecological structure and is sensitive to changes in the outside world, or to say it could be quite weak that it is very vulnerable to external intrusions. Since the global environmental changes and unscientific agricultural activities, the grassland has been experiencing the degradation. So if the tourism activities do not follow the natural laws of environmental development, it will bring greater damage to the grassland ecology and grassland landscape, which will finally reduce the appealing of grassland tourism resources. In other words, tourism development must be based on the capacities of the ecological environment in line with the local economic development and social ethics.

Therefore, in addition to enhance the ecological awareness of tourists, the planners should give careful and serious considerations when making design planning, such as road planning, infrastructure planning and site selection, so as to minimize the destruction to natural environment, and also should consider the might-be side effects of completed tourism constructions, so as to give early consideration for a reasonable solution.

C. Tomorrow - Expansion of Three Forces as Dominant Ideology Planning

Great Tourism

First, to establish the ideological concept of large regions, which means to enhance regional cooperation oriented by resource features and market demand. It not only refers to integration of tourism and its transportation, communications, commerce and other planned services, but more emphasizes on improving the correlation between tourism and local primary and secondary industries in tradition.

Second, to establish an ideological concept of great resources, which highlights the features of regional tourism, bringing its content into vaster areas, actively exploring various source markets, expanding the scope of tourism products to attract the market, so that to make a more diversified trend of tourist attraction.

Third, to establish an ideological concept of great lead, which can not only meet the demands of tourists, but also can promote the increase of local people's income level while improving their living conditions in the process of tourism development. In other words, it is to make an infinite circle with multi-driven benefits from tourists, suppliers, local residents and other parties.

Great Space

The establishment of great space tourism landscape has to integrate the commons and features of the regional spaces. On one hand, there is something in common among different tourism regions or resources, which is the basic foundation for regional integration; on the other hand, there is also something distinct among in various regional landscapes, which could possibly stimulate the tourists with motivation to travel around, so as to achieve a balance of advantages compiled from various regions, the sharing of tourists, and concerted efforts of the complementary relationships, thereby constructing the great integrated tourism space.

Accordingly, the products that could be developed from grassland eco-tourism include: grassland natural scenery sightseeing, grassland national customs experiences, grassland sunbathing, picnicking camping, grass skiing, horseback riding in long-distance. Also, the desert as an eco-tourism resource is irreplaceable, which can be carried out in following activities: the desert adventure, expedition, tour travel, tourism viewing, sand-boarding and so on.

Great Culture

From a global perspective of tourism developing trends, the competitions between tourism destinations, among tourism enterprises, have been rising as the competitions of culture. The one who can react on demand from the modern tourists, make full use of local cultural features, utilize modern means of performance, so as to develop the tourism products which allow them to feel worth talking, worth listening, worth sight-seeing, worth having fun, and thus provide them with unique experiences, refresh feeling, and endless aftertaste of the tourism services, then the one will be able to win the market.

Thus, it is very essential for a tourism destination to determine its culture-oriented development. In particular, the larger parts can be the tourism spots development, facilities construction, and smaller parts can be the service quality and clothing of service personnel, and all these need to fully embody the theme of local culture and features, which is to make the tourism with a unique soul, a main line and a good taste.

Great Thinking

In the beginning of China's tourism development, the competition for tourism mainly focuses on tourism resources. The fame or name recognition or the monopoly of tourism resources in a region or a city constitutes the competition core of tourism development. With the deepening development of China's tourism industry, till nowadays, the regional competition has shifted from the industrial competition to destination competition, which refers to the integrated competition among cities or regions, where includes types of tourist attractions, service facilities and reception system , but also include geographical locations, economic conditions, the overall environment and so on.

So, for those regions with rich tourism resources, who want to make these resources into tourism products, they need to carry out a process of digging, planning, developing, presenting and packaging, which is also an integration process of thinking. At this stage, the main concern of regional tourism development is to construct a perfect supply system. That is the "Six Elements" of tourism: food, accommodation, transportation, travel, shopping and entertainment.

This essay has taken the Mongol-Yuan culture, tourism and landscape as the foundation of featured innovations, and it shows the relationships among the three objects not only from the macro view or micro view, but also from the external planning and design, or from the internal factor analysis. Therefore, only a correct interpretation of the relationships among them can grant a good starting for a qualitative and well-oriented the design, so to make the featured innovative research on demand of the Time and People.

REFERENCE

[1] WU Yunbatu, GE Gengaowa. "On the Basic Spirit of the Mongolian Traditional Culture [J]." Social Sciences of Inner Mongolia (Chinese language version). Jun.1997:70.

[2] JIN Feng. "Nomadic Civilization Being Precious Treasure in Chinese Culture." Culture Website of Yuan Shang Du. 26th.Dec.2006.

[3] OU Jun. "On the Diversity of Mongolian Traditional Culture [J]." Social Sciences of Inner Mongolia. Marc.1998: 56.

[4] ZHOU Xiangpin. "Globalization and the Expansion of Landscape Planning and Design [J]." Urban Planning Forum 2nd Ed. 2001.

[5] LIU Binyi. The Historic Cultural Landscape & Planning and Design of Tourism. China Architecture & Building Press. Apri.2003.

[6] LIU Binyi. The Commemorative Landscape & Planning and Design of Tourism. Southeast University Press. Jan.2005.

[7] Committee of Professional Teaching Steering. Development and Innovation of Landscape Education. China Architecture & Building Press. Sep.2006.

[8] LU Xiaoqin. "Research on Tourism Development of Famous Historic Cultural Cities." Master Theses of Central South Forestry University [D]. Jun.2005:1

[9] QI Qige. "Research on Development of Inner Mongolian Original Eco-tourism System." Master's Theses [D] .May.2005.

[10] LI Mengling, JIA Yinzhuo. Art & Design of Landscape. Huazhong University of Science and Technology Press. Sep.2005.

[11] TAN Ying. The One out of Three in Harmonious Circle – Innovative Research on Composition of The Landscape. China Architecture Press. Sep.2006.

[12] DING Feng. Aphasic Culture – The Challenge for Chinese Landscape Planning and Design. China Architecture & Building Press. Sep.2006.

[13] HOU Xin. Theory of Urban Space Based on Cultural Ecology. Southeast University Press. Oct.2006.

[14] HUANG Yinong. "Exploration on Aesthetic Principles of Tourism Attractions Image-Design." Journal of Social Science of Hunan Normal University. May.2003.

[15] FU Zhiyi. "To Create the City Landscape with Local Features."Journal of Nanjing Arts Institute - Fine Arts and Design. Apri.2006.

[16] CHEN Bateer. "On the Formation, Changes & Features of the Mongolian Traditional Culture." Journal of Inner Mongolia University.No.36.Vol.3.May.2004.

Characteristics Design with Organic Link on the Environment Landscape of Campus

Yin Jing
Shandong Jiaotong University
Jinan, China
Email: yjing78175@163.com

***Abstract*—In this article some ideas and ways of the characteristic formation of the environment landscape in campus with the local conditions based on the environment landscape of the Eleventh National Games Boxing Venue built at the campus of shandong Jiaotong Unversity are discussed and then summed up, and give some references to the behavios pursuiting the characteristics blindly in the current construction of the environment landscape, and promot the sustainable eco-construction.**

Keywords-Campus;Sports Venues;Environmental Landscape; Organic; Features

I. INTRODUCTION

The National Games Venues built in the university campus helps to integrate the resources of education and sports. It also helps to operate after the match and reduce the financial burden. The Eleventh National Games Boxing Venue is built based on pre-built the Sports Venus in Shandong Jiaotong University in Changqing campus, and some strategies about it are determined:

In the view of sustainable development, compensate the lacks of the sports venues with the campus environment, meet the transportation plan during the tournament and after the game, integrate the role of of leisure, sporting events from a wider scope, enhanced the quality of campus environment.

Excavate and strengthen the landscape characteristics elements, enhance and Integrate the regional environment identifiability elements to form a complete and organic landscape system.

Protect the ecological environment of the Venues, follow the 3R circular economy principles, minimize the needs of material and energy resources, protect the earth's mineral resources in a sustainable way.

II. THE FEATURE FORM WITH PRGANIC LINKS

The environment characteristics can not be far-fetched, they are formed by discovering the local characteristics ements, and the feeling of reality and perception. First, the use and experience of the site are determine, followed the right form and quality are designed consciously. They are the best service to express its function and its purpose, and seeking planning experience. In this project the modern way of life, natural and cultural heritage are combined perfectly with protecting the natural environment and pursuit the ultimate demand that the human advocates nature and return nature.

A. The Ecological Landscape Characteristics with Protecting and Using Existing Local Elements Adequately

A continuous green background of natural ecological environment is formed with the natural landscape elements rationally, and echoing with the existing topography, planning and integrating landscape organically.

The Sports Venus is located in the northeast loop of the campus, there are two trenches for city flood discharge passing through southern and northeast. Excavating highly the elements feature, using the natural terrain and green environment landscape fully and innovatively, an organic ecological wetland landscape is formed.

Fig. 1. Preliminary Design Style

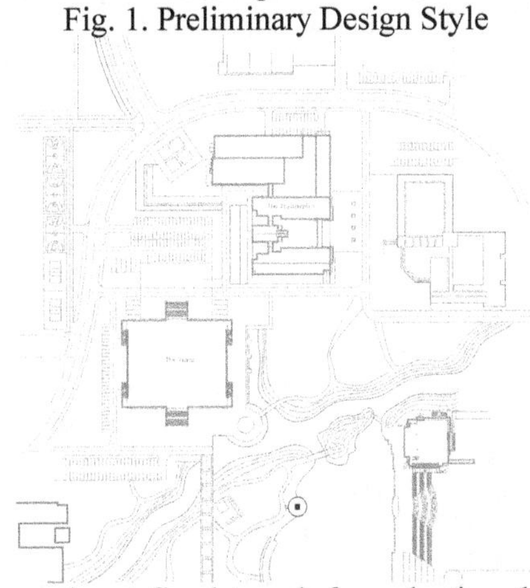

The northeast flood trench for colecting the flood from the northern mountains is planned as the natural landscape with a lot of trees and the terrain and a lot of stones, And the Center Square is formed with the natural green barriers and the Sports Venues and the Civil Engineering Building. In this region , the rainy season is short, the dry season is long, the patten of mountains and hills are formed by the shaped terrain which integrated with the distant rolling hills. During the rainy season, the low-lying topography is for the flood drainage passing through and into the landscape lake. During the dry season, it is green landscape with the natural topography and green plants, and huge stones.

The channel in the south of the Venues is planned as the river landscape across the campus. The river in the southeast of the venues is extended and become the campus landscape lake which integrates the Library Building in the southeast. The Sports Venues and the Librariy Building are surrounded by the wide lake and they are organizied as an whole landscap(Fig.1).

Make full use of the natural landscape and consumpt the lowest resources, finance the favorable personnel training, and build a hormous evoroment with the architecture. River, slopes, green land, gardens and cultural landscape as a whole, a hormous environmen are formed between human and natural.

B. The Characteristics Landscape with Specific Functionality

A complex feature formed should base on the local conditions realisticly and meet the functional requirements. The design of parking lot site of the Sport Venues conects the leisure, fitness, green square capabilities and forms an unique green parking landscape. Not only it meets the requirements for the traffic flow parking functional during the National Games event, but also meets the needs of life experience for the students after the game.

The Central Square adopts the grid of 6.0 m* 6.0 m. At the every grid node trees are planted. It forms 'forest Square' with trees-covered square. The way both meets the environment requirements for the large-scale physical construction of wide, open and warm stylel and it supplies the demand for the campus of leisure and fitness function with friendly and lively environment space .

Fig. 2. Adjusted Layout Style

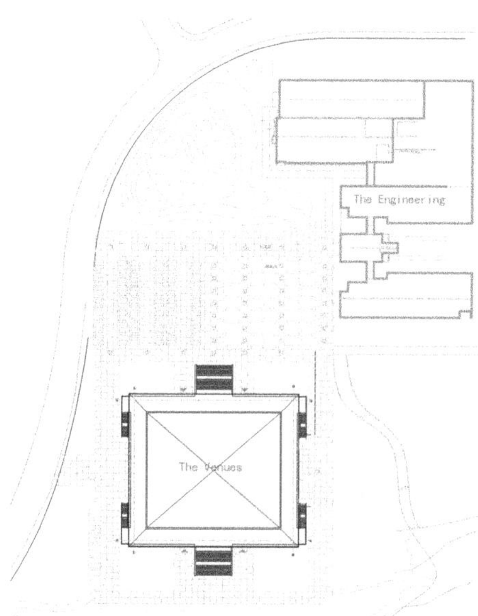

The roads in the Central Square are designed according to the lines and the numbers of traffic flow. The main roads are east-west which leading to the Sports Venues and the Civil Engineering Building. The former is for passengers 9 meters wide which is before the entrance of the Sports Venues, The later is for entrances of parking cars 6 meters wide which leads to the Civil Engineering Building . Between the two main roads there is a 9 meters wide and north-south trunk road in the middle of the Sports Venues. And there are eight 3.5-wide roads that is the north-south are arranged between the two east-west main roads. And the roads of 'forest plaza'extend to the stadium around and form a ring roads linking sub-channels to facilitate to car and fire-fence (Fig.2).

The parking unit of 4.8m* 4.8 m with green inlay grass brick are beside the secondary road. Each parking unit can accommodate two cars, which there are some blocks for car wheels are set in the edge of the parking unit, and some green belts are retained between the unit and the unit. A lot of same scale facilities are arranged to study and be relax for the students in the Parking unit, students may read in the morning, enjoy the cool air in the afternoon, or study in the evening.

The parking area in the 'forest parking lot' is primarily served for the athletes and officials during the competition. The temporary parking cars for the society's temporary vehicles are in the north-south road which is 18m wide ring of the road and the training ground nearby . The layout improves the space of the campus utilization.

A rolling development ways are adopted in the Forest Square which lowers down the one-time capital investment, and improves the utilization of sports facilities. It also leaves the free space for designers to creat next. So it achieves the benefits optimization in economic, social and environmental.

C. The Characteristics Landscape with Simple and Reasonable Green Allocation

The inertia thinking and behior of purchasing the plesure and old trees are breakthrough in planting configuration of the Venue landscape, We combine the technology, the arts and function perfectly. The planted with low-cost of the younger maple trees are the protagonist of the The Forest Parking space. These younger trees are from the 'green bank' seeded in the campus. Although they have slow the effects first, it helps long-term landscap development. A lot of yonger trees are planted in the free space of the campus in the early stage. It is a trees-bank of the campus with a green planting garden and it is an oxygen bar too. It soon becomes a part of the campus environment. When the trees are needed, a approach of 'transplant seedings interval' is adopted. This way can provide more growing space for left trees. The ecological approach is worth advocating. The yonger trees as a living organisms are constantly changing with the seasons and the growth process, the plesure trees or the older trees can not match with the yonger trees of a dynamic growing process. The unique curve, and the diverse form, the moist texture, the rich colors and the vivid light and changing shadow of the yonger trees are not shaped by any one sculptor either. The yonger trees is very ornamental value. Trees are Verdant in spring, full-bodied in summer, rich in autumn, in winter even if the leaves are wither, their grace graceful branches strech out horizontal or slanting ramp, seem to wait for the spring. The person's emotional experience also are accompaning by the growth of yonger trees with it(fig.3).

Fig. 3. The parking plaza landscape

III. SAVING AND USING RESOURCES COMPREHENSIVELY

When determining the width and materials of the pavement, we adopt the local materials and the low-cost, operational techniques to meet the asks of the functions use innovately abandoning some high-cost and harmful techniques. We avoided the formalization and absolute tendency in the current road construction. It is the idea of saving and eco-campus campus and it is a book the environment is in harmony with people.

A. Reduction Harden Pavement and Share of the Land

With the low cost, simple construction, and durability use, the concrete use is widely. But its drawback is that it is hard to move once laying. So in the roads and squares pavement, we adopt the blocks of granite with wide gaps paved the ground. The maintenance is facilitates and the material is easy to replace, the material service life is extended, the materials recycling is facilitate and the pruduced waste is less.

Create ecological landscapes by increasing soft pavement and changing the materials from hardened pavement into permeable pavement. The embedded straw bricks are paved in the parking area, the open areas as well as the edge of the road which both meet the infiltration of rainwater enhancing environmental benefits. Ecological beauty with the more weeds in embedded straw bale is showed. From the perspective of nature and sustainable, focusing on local, rainwater harvesting and conservation of groundwater and other characteristics, the pavement forms of the parking units with the landscape of large area of grass create a great beauty, giving person more refreshing feel. It helps to shape the students a broad-minded, tolerant character of physical and mental feelings in a subtle role.

The open space is reserved imposiblely outside the paved road and parking space to increase the green space and create a better environment. We adopt the slope-shaped green planting to transition the elevation difference between the car parks and the loop which forms a good square effect. And the rainwater absorpts through the ground surface.

B. Refining Pavement Design

The full utilization of energy resources is also reflected in regard to geographical specificity and timing of the pavement characteristics and significance. The width, linear as well as the patterns and colors of pavement design should be taken full account of.

Pavements covers a large area, and they create a great impact to the whole atmosphere of the venue. A general way is emphasis too much on road planning and simple traffic organization features. Not only we consider the long-term large-scale pavement which is the keynote with the grass embedded in brick and white granites, but also consider small-scale density and delicate texture of the pavement with the formation of granite of a wide seam block and brick inlay grass texture . The road layout with through-type, grid-like layout is easy to contact the building entrances and it is convenient to traffic and unobstructed. The secondary roads with 4.2 m width, on both sides of which there are brick inlay trim grass lay , it reduce plaque area. It is both easy to use for the two cars meeting, improving transport efficiency, and reducting the invalid path. It is also benefit to ecological environmental protection,

The pavement surface uses the patterns of three elements point-line-plane: the tone of the Venue site is green with green grass brick inlay and green open spaces. The linear decoration form that the granite paved the road gives people the sense of stability. At the same time, it has a strong directional role in guiding the traffic flow and pedestrian flow. The secondary roads which is perpendicular with the trunks and between the trunks appeared intervally as some straight lines which depicts a lively rhythm sense. The primary and secondary roads are both paved with the material of granite of 600mm* 600mm*60mm, giving the feeling of quiet and orderly, The pavement design is neat and uniform integrating the environment. The paving patten and mosaic with the Shidao red lies in the cross of the road and the entrances of the building with the role of tips and guidings. Its texture with wide joints form of massive granite is more clear and delicate, while saving 3% of the granite material. The wide seam form not only may keep rainwater in existence, but also facilitate the material maintenance, repairing and recycling.

The pavement colors adopt white zhenzhuhua granite, green inlay grass brick and red Shidao granite which matches with the construction. The roads are paved with white zhenzhuhua granite, the parking units are embedded with the green grass bricks, and the mosaics are dotted with red Shidao granites. The colors match between the well-coordinated to create a very attractive space. The effect s of the pavement color become more pronounced ,when the green grass brick inlay are paved beside the road of the white zhenzhuhua granites.

C. The Site Feature Integrated Pavement and Drainage

At present, the rainwater pipe is mainly pre-buried to organize rain . Altough the vertical design in this way is relatively simple, but the costs of pipes and the earthwork volume of the dark rows are high, and it is easy to plug, and it is unconvenient maintain in the future. Using the road to resolve the site rain drainage, the vertical design is complexity which primarily needs to solve the design of vertical elevation of the site, the slope organizations of the rain flowing, the elevetion relationship between the site, the building and road around, but it is simple and easy to maintain. In Jinan the rainy season is shorter, and there is rare rain and less rainfall frequency, but the rainfall is often large, so it is economical and suitable with the road drainage as the drainage channel.

The elevation of the Sports Venus site is located below the elevation of the loop around, and there are flood discharge trenches in east and southeast, so we take out drainage saving the cost of earthwork and pipeline. The rainwater in the site is collected by the ground elevation design into the south-eastern rivers. During the rainy season, the rainwater is stored in the penasitble ground and the excess rainwater along the surface of the roads is organized into the landscape river, and at the site of the inlet location which the rainwater flows into the river, the viewing platform is paved with the local freestone which extends into the river, both as a viewing platform, and as the channel of rainwater collected (Fig.4).

Fig. 4. The landscape flat as rainwater into the river

IV. SUMMARY

YouThe land area of the Venue environment is about 1.5 hectares, the afforestation rate is about 70% afforestation rate, the harden area is 5,000m2, and the permeable pavement area is about 4000m2. The total cost of the environment is about 150,0000 yuan, that is 100 yuan / m2 or so. It has saved money about 40% compared with the others. It achieves both good social and economic benefits.

The theme of the environment construction of the Sports Venues is 'natural wisdom' . It shows highlighting the natural charisma, broad, fragile and precious. And it also emphasises on caring for human beings emotions and a recycling society. Environmental problems become increasingly severe today, the project reinforces the theme for us: how to return to nature and in harmony with nature.

REFERENCES

[1] Arata Isozaki, "About my Method,", unpublished.

[2] James Wines, Green architecture, Koln; New York: Taschen,c2000.

[3] Sutton I. Western Architecture. London: Ttrasnes and Hudson, 1999.

[4] Roger Johnso, The green city, Macmillian Co. ofAustralia, 1979.

Comparison and Analysis of Urban Central Districts' Land Use Mode and Transport System

Guo Liang
School of Architecture and Urban Planning,
Huazhong University of Science and Technology
Wuhan, China
Email: paoren111@163.com

***Abstract*—In view of the conflict between current land development, use of urban centers and transport facilities construction and operation, this paper selected a few representatives of China's urban centers, compared the different characteristics of their land development, transportation facilities, and their operation ways and so on, so as to obtain experiences of urban central district's land use mode and transport system development from this comparison.**

Keywords- Urban Central Districts; Land Use; Transport System; Comparison

Urban center districts are the most prominent locations with the characteristics of complicated land use and inconsistent transportation, their healthy development are inseparable from the support of around transportation facilities. Strengthening the coordinated development of transport system in central districts plays an important role in improving central districts' accessibility and promoting the fairness and efficiency of transportation.

I. Comparison on Transportation and Land Use between Several Urban Central Districts

From the perspective of the correlation between urban central districts' land use and transport facility system, we can analyze what impacts their respective characteristics of each city have on their other facilities. Generally speaking, the intensity degree of development and mixed-use of urban central districts have impacts on traffic efficiency; while the layout and operation of transport facility system(including urban road network, cycling and public transport systems, parking systems, etc.) can cause impacts on their own operational efficiency and fairness, and the impact on the urban central districts' accessibility as well.

Public transport-oriented development of urban land use provided the experience for China's urban land use. Because of the complexity and diversity between each city's urban development realities, we should adopt different measures according to different type of cities in the process of promoting public transport-oriented development. For comparison purpose, follows selection of several China's cities to proceed the comparison between their urban central districts' land use development and transport facilities, operation status, in order to obtain experience of urban central district's land use mode and transport system development from this comparison.

Main comparisons are characterized by the land use features (including land use structure, land development intensity, mixed-use land degree), public transport features (including bus lines, bus stations, public transport operators), road transport and infrastructure (including road structure, road running, packing facilities) and so on. Among them, the survey of land development degree can be analyzed from Google-earth images, the main criteria for judging the land development degree can be well reflected from whether the high-rise buildings are "clustering". If yes, so the land development degree is "High"; if high-rise building are comparative less, but multi-layer and low-rise buildings are comparative more, it can be assorted in "Average"; if the high-rise building are numerous but dispersed, this situation can be assorted in " Above-average"; As to the analysis to mixed-use land degree, it can be measured through "Land-Use Fragmentation Index"(The amount of spatial layout index reflects disturbance degree of human activities when in land use, generally, with the development of city, and the increase of mixed use degree, the land use's diversity, evenness, and fragmentation increase accordingly.), land-use fragmentation index (C) usually is used to measure the land fragmentation degree, namely, the number of patches of different types of land, and its expression is as following:

$$C = \sum n_i / F. \qquad (1)$$

(F stands for the total land area.)

II. Several Relevant Characteristics of Urban Central Districts

This paper selected Daliang central district of Shunde, Dingzikou of Zunyi City, old urban district of Meishan City, Three Gorges Square in Shapingba of Chongqing City these four cities' central districts as the research objects. It set the land use situation as the research base, combined with surrounding transport facilities layout, and preceded a comparative analysis of current traffic conditions, in order to understand the relevant characteristics of coordinated development between urban central districts' land use development and surrounding transport facilities.

A. Daliang District of Shunde

The scope of land use in Daliang district of Shunde is relatively small. Although the district has been separated by the main road to west and east parts, but in fact the major commercial activities are concentrated in the west, it is probably relevant with its good spatial layout of its own "Non-Accumulation" feature. In the status that with little cross-road traffic in the central area, the surrounding road traffic still showed good order. Of cause, the number of bus lines cross the central area accounts 11, which provided the guarantee for people's easy gathering and distribution. Meanwhile, because the number of motorcycle, car ownership is higher, the layout of the public parking lots played a certain role in mitigating the status. (See Figure 1)

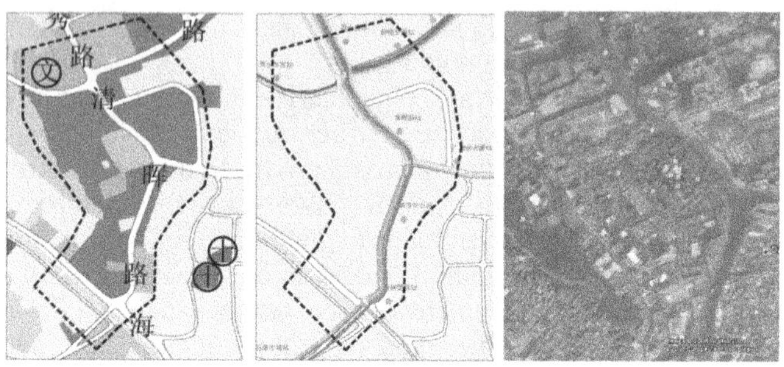

Figure 1. General Land Use, Public Transportation and Development Status of Daliang, Shunde

B. Dingzikou district in Zunyi City

The region of Dingzikou District was developed from the original urban central area of Zunyi City, so the surrounding commercial, office, residential area were mixed with each other. Because the roads mostly come off from hilltops, the development intensity degree is relative higher on both two sides of the roads, with narrow space between the two sides, but population density is quite high (about 2-3 million / km); Due to the "Fragmentation Index" is higher, therefore, the mixing degree of their functions is higher accordingly.

Dingzikou District is the center of the whole Zunyi City. Due to the scarcity of roads, almost a large number of public buildings are constructed along the main roads. Therefore, the main roads often bear the transportation, commerce and other comprehensive functions, what's worse, with a large quantity of building development on both sides of the roads; it increased a lot of pressure to the traffic in the roads. In this case, the development of public transportation has become an inevitable choice. However, due to the limitation of land use conditions, buses always stop at the road. When bus traffic is heavy, it will cause great disturbance to road status. At the same time, the commercial buildings located in the two sides of roads provide the reason for people crossing the roads. However, the solution of walking across streets further increased the traffic jams. (See Figure 2)

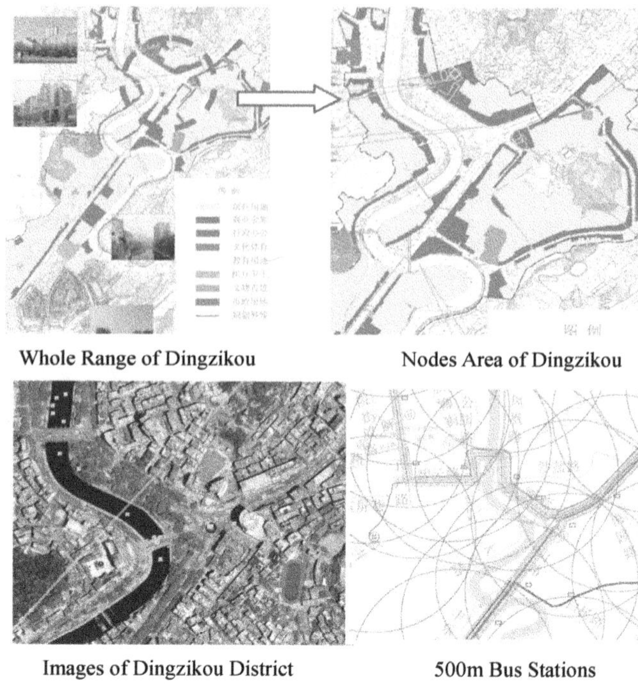

Figure 2. Land Use in Nodes Area and Public Transport Status

C. Original urban area of Meishan City

The features of Meishan central district are different from the form mentioned two cities. Meishan central district was developed based on the basis of the original urban area, and this feature is common seen in other small cities. The roads in original urban area are intensive, but the roads' width is narrow, which provides the reason for reducing bus entry. When the scope is not too large, relying on the bus lines and stations (300-meter radius to cover the entire central area) outside the original urban area can solve the travel in this region. From the perspective also shows that this area has good accessibility. However, when the private car is concerned, there still are a lot of cars access to the narrow roads, which arises the challenges to the internal traffic conditions and parking methods. (See Figure 3)

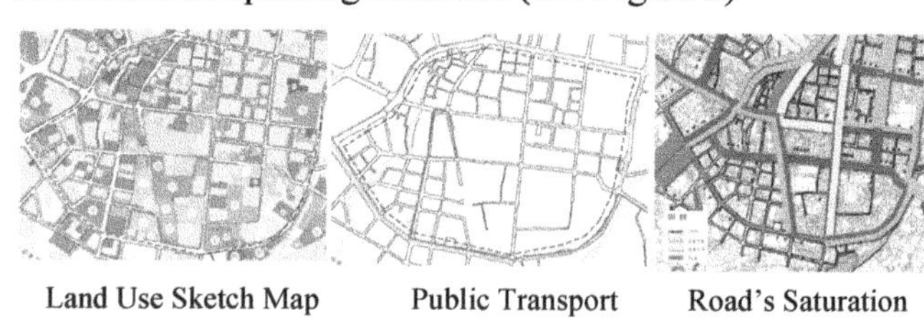

Figure 3. Land use and public transportation status of the original urban area in Meishan

D. Three Gorges Square in Shapingba of Chongqing City

The central district of Shapingba, Chongqing is located at the present Three Gorges Square. See Figure 4, Figure 5.

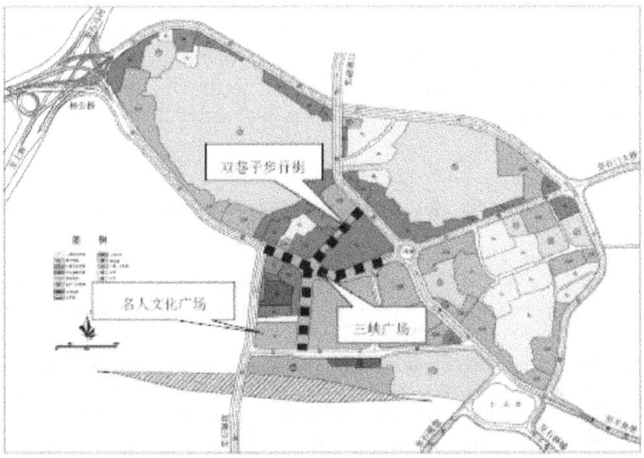

Figure 4. Walking street in Three Gorges Square

Figure 5. Saturation of Shapingba Central District

The average saturation of Shapingba central district's major road sections is 0.62, which is in C grade level. Of which, Hanyu Road(Shimen Bridge- Longquan Road), Shixiao Road, Yubei Road(Tianli Road- Yanggongqiao Overpass), Jialing Road, Tianli Road have greater loads, which is above 0.8, and is close to saturation when in peak periods.

The construction of Three Gorges Square attracts a lot of popularity and promotes the commercial prosperity, but at the same time, the traffic in Shapingba's central district become more and more congested, the high-intensity development inside the district is the direct cause of the large traffic volume, especially the major trunk roads have been cut off due to lower density of road network, and the resulting proliferation of traffic detour is the "culprit".

III. Summary to the Land Use Layout of Urban Central Districts

Through the analysis of the status quo to the above several urban central districts, the current urban central districts' development can be classified, shown in Figure 6.

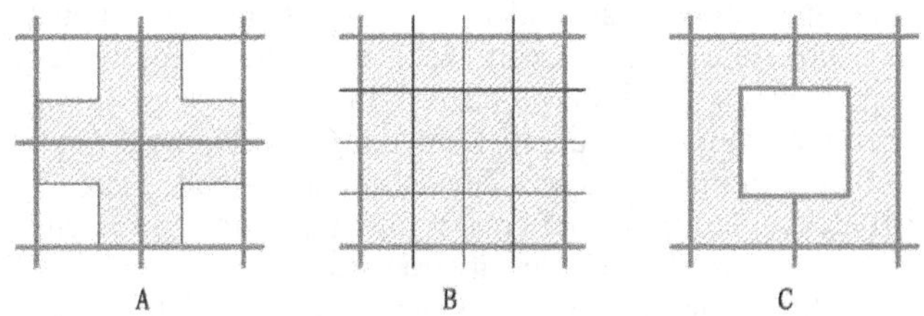

Figure 6. Several Functional Neighborhood's Development Mode

In which, Shunde and Zunyi are mainly based on A mode (actually Handan is the same), while Meishan is B mode-based, and Shapingba is in C mode range. The main features of each mode are as following:

(1) A mode is currently more common urban land development mode, but the negative effects are quite great, although the public transport lay along the main road would be more developed, and will also help address a significant portion of travel. When the buses are too many, the lack of Harbor Point Road will lead to excessive throw-style parking, coupled with the large number of passages across the street, all these can be the negative impact to normal roads running order. Therefore, if not compelling, such layout approach of urban central districts should be "abandoned" as soon as possible. However, considering many urban central districts' development in China is based on this mode at the beginning, it becomes very significant to explore the improvement approaches to this layout mode.

(2) B mode can be often seen in old regions of small cities. Although the neighborhood has a good scale and density of road network, in today's rapid development of motorization, it has also had to endure an increasingly intense shocks. Especially for street itself, to increase its traffic accessibility is very important.

(3) C mode usually comes from the building of "Walking Street" under the influence of the wave shape in the late 1990s. Admittedly, the benefits that this approach brought are great, but the blind construction in spite of construction conditions brought many negative effects that can not be overlooked.

IV. Conclusion

Combination of various modes of transport strategy and the corresponding effect, we can come to conclusions as following:

(1) Taking Meishan and Shunde, Chongqing for comparison, Zunyi and Shunde, Chongqing for comparison, we can draw the basic trend for the mixed degree of land use: That is, cities with large scope, their functional zoning feature is even more evident, the smaller the cities are, the higher the mixed degree is. Due to the higher mixed degree

has played an important role in reducing long-distance travels, there still exists much space for promotion of land use. Of course, this may be related to the too larger scope of overall land use in urban land's development process.

(2) From the view of development intensity, higher intensity of development will bring a lot of people, traffic, and create strong pressure to surrounding traffic. Shunde's experience shows that, under certain development intensity, more bus lines facility can meet the travel requirements when in high level of motorized time. But, Chongqing's experience shows that the high intensity development relying solely on public transport development is inadequate, but high density of road network is an important approach to meet the needs of high intensity transportation.

(3) Meishan's narrow road network in the old town is difficult to withstand the impact of motorization, but if the old town has a larger scale, accordingly the internal transport accessibility of old town will become worse, which requires the appropriate introduction of public transportation, or even adopt a one-way traffic method to resolve the issue of motor vehicles' running.

(4) Irrespective of which land development modes, public transportation developments are essential, but too many public transport sites in the central district and the surrounding road disorderly seriously affect the layout of the traffic operation, nor is not conducive to people's transfer. Therefore, building transportation hub in large-scale cities is very urgent.

(5)The configuration and layout of parking lots is very important to urban central districts' traffic orderly operation. However, to the current existing areas, which have formed a basic structure, a more workable approach should be looked for with the consideration of difficulty in parking lots' construction.

REFERENCE

[1] Guo Liang, He Hui. Countermeasures and Research of Behavior-based Demand for Urban Transport Planning Objective[J]. Urban Planning. 2009 (1) : 67~73

[2] Guo Liang, He Hui. Comparison of Association between Urban Transport Structure Optimization and Land Use Mode [J]. Urban Planning Forum, 2009 (5): 77 ~ 82

[3] Zhao Yanfeng. Evaluation of Spatial Pattern in urban land use and impact to urban residents[C]. China's Urban Transportation Planning 2006 Annual Meeting And The 22nd Symposium, 2006 : 617-621

[4] Shanghai Tongji Urban Planning and Design Institute. Comprehensive Transportation Planning in Urban Central District of Zunyi City[R]. 2006

[5] Shanghai Tongji Urban Planning and Design Institute. Special Planning to Road Traffic in Shunde District, Foshan City.

[6] Shanghai Tongji Urban Planning and Design Institute. Comprehensive Transportation Planning of Meishan City[R]. 2005

[7] Shanghai Xu Xunchu Urban Planning and Design Consulting Co., Ltd... Transportation planning of the core area of Shapingba, Chongqing City[R]. 2004

Computer Aided Optimal Design of the Pictography of Marriage Custom of Naxi Nationality

Zhang Chao & Zhu Xiaojun
College of Arts
Guizhou University
Guiyang, China
Email: zhangchao1232780@sina.com

Wu Xingkuan
School of Art & Design
Wuhan University of Technology
Wuhan, China
Email: wxk1952@126.com

***Abstract*—This paper intends to use the Adobe Illustrator, modern high-tech computer vector graphic processing software, for artistic processing, treatment and optimal design on the several excerpted Dongba pictography with regards to the marriage customs of Naxi nationality, and mainly aims to discuss how to reflect the "marriage custom" etiquette of Naxi people through the computer optimized Dongba pictography. Firstly, it chooses such typical Dongba characters as "wedding ceremony, blessing, kowtow, and deep love and so on" from the pictography of Naxi nationality, then employs the formal beauty principle of artistic and font design, and modern computer aided optimal design to express the happiness, excitement and deep love of Naxi people by the pictography. It tries to better apply this law of optimal design to the artistic design or applied design of Dongba pictography and creates more and better artistic works.**

Keywords- Nationality; Marriage Custom; Character Graph; Optimal Design

I. Introduction

The computer aided optimal design of graphic characters refers to the creation and design process combined with the computer drawing software and artistic design law, which uses the computer as the platform. This paper uses the vector drawing software-Adobe Illustrator as the main fabrication platform, summarizes the ideal of graphic design and creation, formal beauty principle and character design law from the perspective of computerized graphic artistic optimal design, and then applies the software to carry out the graphic artistic creation on Dongba pictography, and subtly blends the culture loaded in Dongba characters into the optimally designed characters through the specific design intention and fabrication process to make the graphic characters after the optimal design vivid and full of interest.

A good character design can be judged from different aspects, for example, the coincidence of character itself with its meaning; easy for speaking and distinguishing; good aesthetic demand to the audience, and therefore, to do a good graphic character design, many factors should be taken into consideration. This paper considers not only the advantage of the computer aided optimal design, but also the features of the character itself, culture background, meaning and so on. On the basis of observing the principle of character design, the authors of this paper try to make an innovative computer optimal design on the visualized characters.

II. Pursuit and Courtship

Figure 1 is excerpted from Dongba pictography of Naxi nationality, which means: "pursuit and courtship". As the marriage is a big event in life, in ancient times, commonly the man would pursue the woman he loved initially and it is the same for Naxi nationality.

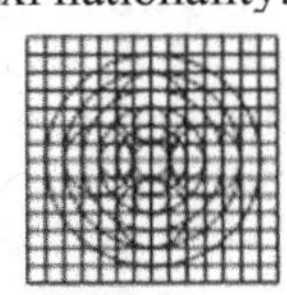
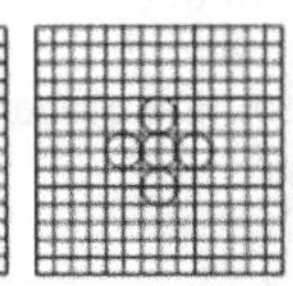
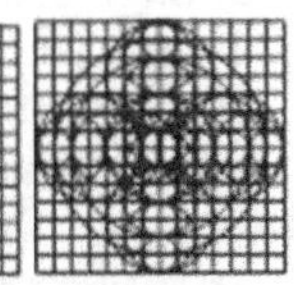

(Figure 1) (Figure 2) (Figure 3) (Figure4)

- Draw a 14×14 square on the print page of the drawing window of Adobe Illustration, and take the side length of the square as the radium to draw a circle, and then draw one circle that tangent to it at its up, down, left and right position respectively (see Figure 2)
- Draw several concentric circles centering on the center of the central circle in the square, and then take the diameter of one circle as the radium to draw four equal circles, see Figure 3, and draw circles on other directions in the same manner, see Figure 4.

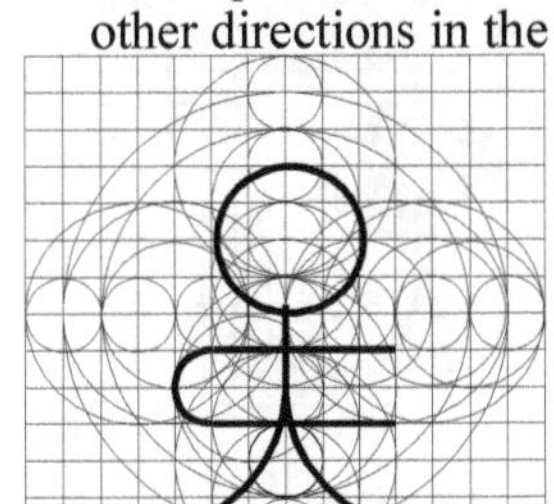
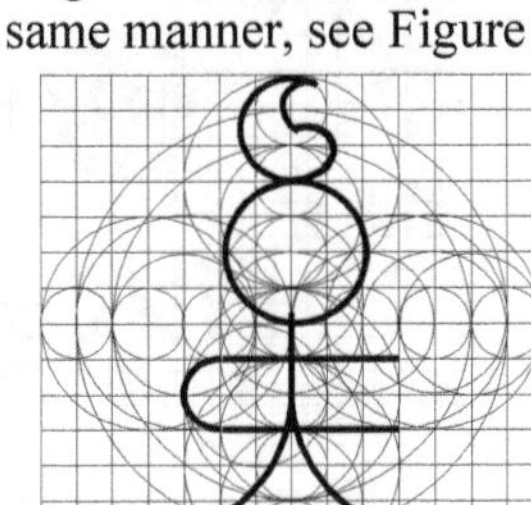

Figure5 Figure 6

- From the morphological feature of the word "pursuit and courtship" of Naxi pictography, we can see that the word "pursuit and courtship" is consisted of a man and a woman of Naxi, and the woman is ahead of the man. Use the circles in the square as the basic components and cut out the needed arcs from the circles, and then combine and assemble them orderly. In addition, the meaning of the word "pursuit and courtship" shall be considered, thus the early form of the computer aided design can be seen already, see Figure 5 and 6.

- Thickening the contour of the graphic character drew to make its form salient, and we get the Figures below (see Figure 7)

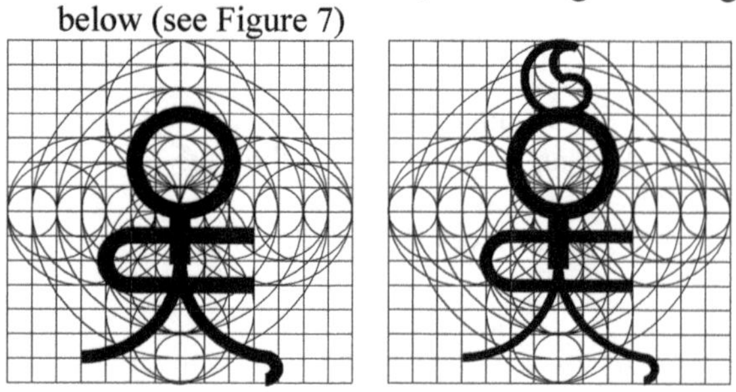

Figure 7

- After changed the fineness of the humanlike character and adjusted several times, we determine the form in Figure 8, which seems flexible and nature and adds the funny and the visual aesthetics to the graphic character. After that, add color to the graph based on the colors loved by Naxi people, through continuous practice, we finally decide to use the red and the blue as the main tone of the graphic character to make the whole graphic has strong visual impact.

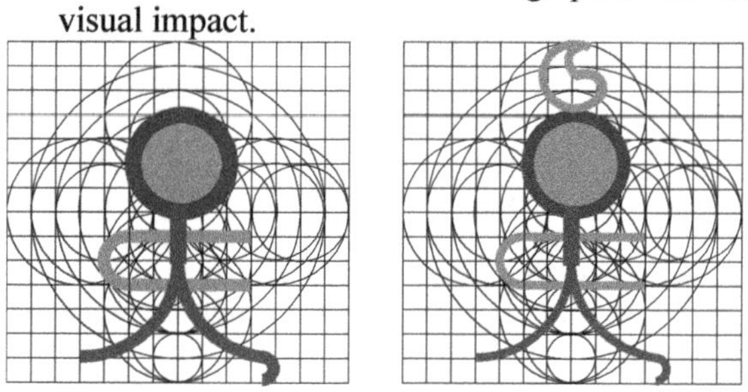

Figure 8

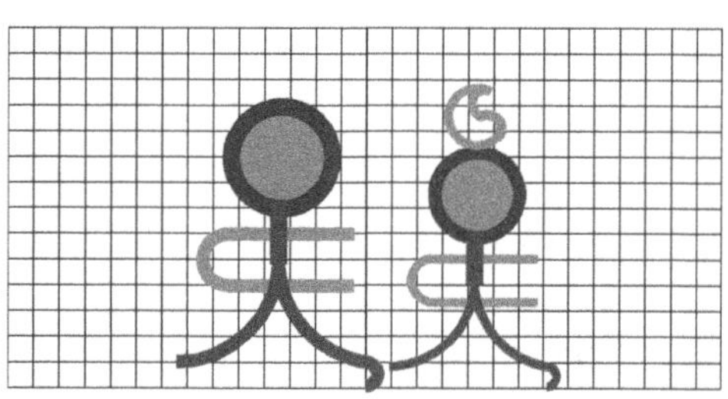

Figure 9 Pursuing and Courtship

- Figure 9 is the result got after we assemble and adjust the two exquisitely designed characters in the grid. The optimal design of the graphic character shows the "hardship" and happiness of the courtship process of Naxi people visually and interestingly. The whole optimal process assembles the geometry lines orderly just like a girl runs ahead, and a boy runs after her. The "pursuing and courtship" story of two lovers of Naxi seems to be embodied by one graphic character.

III. DEEP LOVE

The character in Figure 10 expresses the deep love between the man and the woman of Naxi. First analyze the structural feature of the original pictography when apply the computer software, and then consider the uniform and coordination of the whole design style, and besides, the congenial and happy feelings between the two lovers contained in Dongba pictography should be considered also. Depending on the accuracy of the software design, we purposely changed the freely written pictography into a new graphic sign full of modern feature, which is a completely new creation, and it aims to emphasize the meaning of the character itself and highlight the "deep love", which is the holy connotation of love.

Figure 10

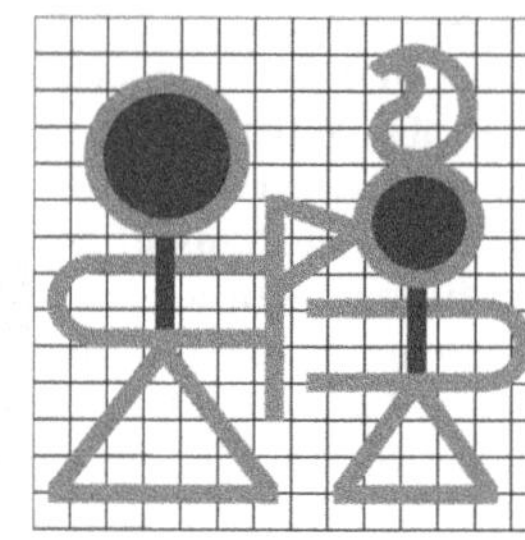

Figure 11

(Figure 11) and Figures below are the computer aided optimal design of the Dongba pictography by using Adobe Illustrator, which is the final sketch obtained following the steps from Figure 2 to 9 according to the design method of the pictography in "Figure 1 pursuing and courtship". The design of this graphic character bases on the feature of the original Dongba pictography in (Figure 10) and considers the feature of this drawing tool so as to combine the geometric curves with the oblique lines through the artistic method, which closely links the man and the woman together and embodies the heart love of the happy marriage. The symmetrical form embodies the equality between man and woman, the visual balance embodies a kind of peace, and the tangency and intersection implies the intimate and tacit relation between the couple, and the entire graphic character gives you a sense of concise and fluent, simple and nature.

IV. WEDDING CEREMONY

Figure 4 represents the process of "conclusion of marriage" in Naxi people's marriage custom, and the image of  in this figure represents the Dongba wizard of Naxi nationality, 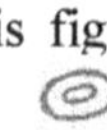 represents the ghee, besides the two pictography, the two images on the right represent the kowtow postures of the newly married couple. The whole font style reflects the ceremony of Naxi people, that is to say, they will spread the ghee when they get married. Use the computer vector software for Adobe Illustrator and follow the symmetric and balance principles of beauty, the pictography fully shows the jubilant scenes in the wedding ceremony of Naxi nationality, and this process implies the wish for good luck and happiness, prosperous family and perpetual happiness to the new couple.

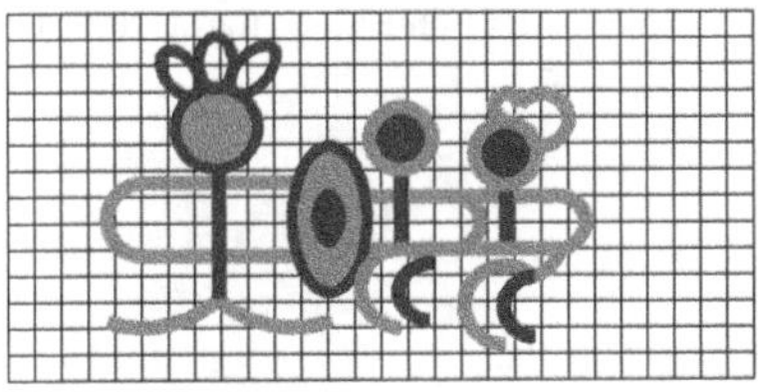

Figure 12 Figure 13

Naxi people value the marriage very much, and Figure 5 is the design based on Figure 4, the pictography of "conclusion of marriage", namely, "the wedding ceremony" of Naxi, and the design firmly grasps the form feature of the original Dongba pictography and the wedding ceremony of Naxi people. For the design of the crucial step when Dongba wizard applying the ghee onto the foreheads of the couple we use the arc, which has a fluent and natural visual effect and foils the boisterous scene of the wedding ceremony through the festive color.

V. KOWTOW:

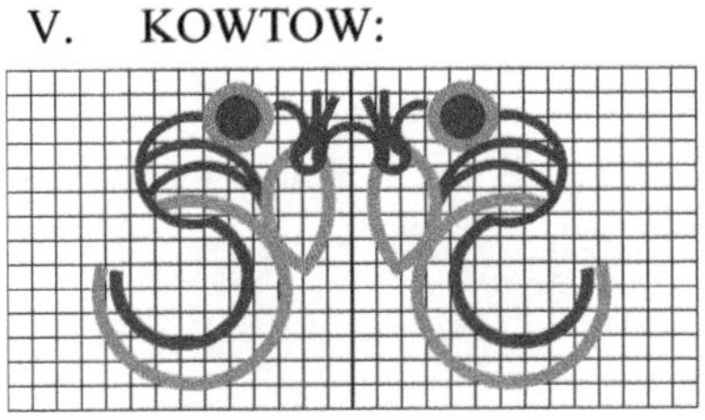

Figure 14 Figure 15

Figure 14 means "kneeing for worship, kowtow and kneeing with incense". Because Naxi nationality believes in divinities, when they held the wedding ceremony, Dongba people will blow the trumpet before the wedding ceremony officially starts, chant Dongba Classic, *Welcoming the God*, written in Dongba characters and let the newly married couple kowtow before the altar, and then pass the sacred arrow and tower in the bamboo basket to the bridegroom, and then pass the wool yarn that ties the sacred tower and god of the residential area to the bride, who will use the yarn to firmly tie the sacred tower to the arrow in the hands of the bridegroom, and then put them back into the basket. Through this process, the bride's life and soul has been closely linked with the bridegroom, and since then, their marriage is acknowledged by the society. What's more, at the end of the wedding ceremony, the bride and the bridegroom will serve disserts to each other, and use the bamboo chopsticks to serve the glutinous cake and preserved fruits to the mouth of each other, and all people around will bless and acclaim for the new couple. Hence, the kowtow ceremony is regarded as an important step in the marriage custom of Naxi people. Blending these details into a graphic character that seems to be mechanical symmetry and based on the computer aided optimal design better highlights the affection of mutual respect, equality and love between the new couple. The bright and contrast colors add the festival atmosphere and foil the theme of the pictography.

Figure 15 is the design of Dongba pictography meaning "kneeing for worship, kowtow and kneeing with incense" in Figure 6. The entire design still uses the arc line, which makes the graph more full, brief and full of rhythm. The whole symmetric form shows the husband and wife's bow to each other, and the entire design embodies a kind of aesthetic effect. In choosing the color, we choose the bright and festive color because it represents the wedding ceremony. The shape of the arc is very exaggerative, which shows the piousness of the new couple. The solemn and sacred process is expressed vividly by this brief graphic character design.

VI. PRAYING AND BLESSING

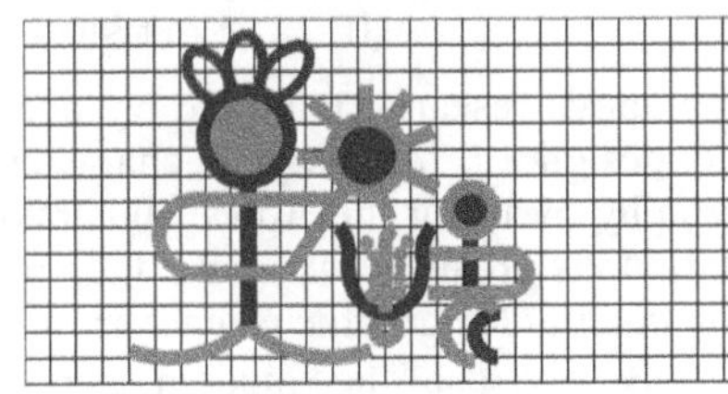

Figure 16 Figure 17

Figure 16 is Dongba pictography means "bless the prayer, and the Dongba wizard bestows corn, rice, wine, meat and so on to show the rich blessings." Just like other nationalities, Naxi people value the marriage event very much, and it is very jubilant in the wedding ceremony. They will invite the master of Dongba to preside the traditional ceremony-"God Si Worship". "Si" is the god worshiped by Naxi people, who can guard the house and bring luck to the family, and so it is named as "family god", and also called "life god". Holding the ceremony of god "Si" is to let the bride become a member of the family formally and pray for blessing from the god, and hope god "Si" to bless the prosperity of the family. This praying and blessing is vital for both the new couple and the whole family, and therefore this pictography is one of the necessarily required characters to try on the computer aided optimal design in this paper.

Blessing is a kind of traditional and admired religious behavior of Chinese people, and it is also an important ceremony in the wedding custom of Naxi people. (Figure 17) is the result of computer aided artistic processing and optimal design based on the original meaning of the pictography (Figure 16) of Naxi and the common principle and method of font style creation. Use the arc as the basic design element for the style of font, which is good for identification and blends certain connotation and artistry in it, so the whole character graph is complex but not in disorder, brief and compact. In addition, they will hold the seeing off ceremony for the god of marriage.

Through the above research and trial of the optimal design on the recreation of Dongba pictography with the computer software, we get some knowledge of the marriage customs of Naxi minority. Meanwhile, we apply the computer aided optimal design to make a creative design on Dongba pictography, which is very crucial to understand the marriage customs of Naxi people and adds its interestingness.

We can further review this precious heritage handed down from the ancestors of Naxi from the perspective of arts and optimal design. The graphic character after computer aided optimal design will present the audience a totally new appearance of the graphic character, integrate the thoughts and artistry of the original pictography, and make the conveying of the visual information in Dongba pictography more brief. Therefore, as the artistic designer, we can utilize the modern media and tools courageously to rethink our country's old traditional culture and make innovations so that a better inheritance and application of the excellent national culture on the basis of the good protection of it can be realized.

VII. CONCLUSION

The above research is a bold try and exploration on the artistic design of the graphic character of Naxi Dongba pictography by using the Adobe Illustrator, and on the basis of continuous creation, we make the following analysis on the computer aided graphic character:

- Boldly use the modern new computer drawing software, Adobe Illustrator to make a try on computer aided optimal artistic design and make the structure of the graphic character more precise, make it more visually logic, more psychologically in order and more balanced in proportion.
- From the design plan, we can see the character variation method and rule of the computer aided optimal design. The combination of the computer aided tool and the transformation method of the character graph, together with the tangent and intersecting rule of the geometric circles and the integration of straight lines and curves make the graph full of variation.
- Combining the traditional "marriage custom" of Naxi people reflected by Naxi pictography with the character graph and making the content and form uniform not only endow rich cultural connotation to the graphic character, but also facilitate the creation of more excellent works of graphic character, thus further extend the space and field of the design thoughts and greatly improve the thoughts, artistry, flexibility and diversity of the character graph. By this way, the feature and function of the graphic character can truly play their role in the development of the national culture and feasible method for the inheritance and further development of national culture are provided.

REFERENCES:

[1] Li Xiaobin. Artistic Design of Computer Two-dimension Animation, Beijing: Qinghua University Press, 2006.01

[2] Fang Guoyu. Naxi Pictography Genealogy, Kunming: Yunnan People's Press, 2005, 12.

[3] Zhang Ruilin. Black and White Decorative Pictures of Ancient Chinese Character. Anhui. Anhui Arts Publishing House, 1989.5

[4] Xi. Museum Essays on Lijiang Dongba Cultural Festival, Yunnan People's Press, 2003. 3, 1st edition

[5] Zhao Jingxiu. Notes on the Practical Words of Naxi Pictography, Yunnan People's Press, 2002, 9

[6] Chen Nan. Graphic Design and Innovative Thoughts, Nanchang: Jiangxi Arts Publishing House, 2005.1

Crystal Orientation Optimization of Nickel-based Single Crystal Turbine Blade

Qingwu Wang
Department of Power
North China Electric Power University
Baoding, China
Email: wangqingwu2004@163.com

Mao Pang
School of Mechanical & Automotive Engineering
Zhejiang University of Science and Technology
Zhejiang, China
Email: palmy@126.com

Hongyu Qi
School of Energy and Power Engineering
Beijing University of Aeronautics and Astronautics
Beijing, China
Email: qhy@buaa.edu.cn

Abstract*—This paper presents a method to optimize crystal orientation of Nickel-based single crystal turbine blade. Total strain range, crystal orientation and temperature range have effect on the low cyclic fatigue life of Nickel-based single crystal superalloy, on other invariable thermo-mechanical cycle conditions. The experiment data of many concerned literatures were analysed and how crystal orientation effect single crystal blade damage and fatigue life was derived. The evaluating method for fatigue life of single crystal superalloy was proposed. The 3 dimensional Nickel-based single crystal turbine blade was modeled in MARC program and this finite element model was capable of accounting for primary and secondary crystal orientation variation. Effects of variation in crystal orientation on blade fatigue life were studied based on finite element model with 68 material coordinate systems, by constitutive equation based on crystal slip systems. Result presented demonstrates that control of primary and secondary crystal orientation has the potential to significantly increase turbine blade resistance to fatigue without adding additional weight or cost.

Keywords -optimization; fatigue damage; single crystal superalloy,turbine blade; crystal orientation

I. INTRODUCTION

Under the same other thermo-mechanical cycle conditions , total strain range, crystal orientation and temperature range have important effect on low cyclic fatigue(LCF) life of Nickel-based single crystal superalloy. Under the same other conditions, if crystal orientation is different , the stress in single crystal superalloy turbine blade may vary, the strain distribution may vary, damage distribution may vary, the location where failure occur may vary and the turbine blade life may vary. Turbine blade geometry was determined by aerodynamics under certain working environment corresponding engine performance, so it has no range to change. Longitudinal orientation, known as blade profile laminated orientation, was controlled in the casting process of turbine blade. The other two landscape orientations were not controlled and stochastic. The two landscape orientations affect the natural frequency and fatigue strength of turbine blade. Although one crystal orientation was controlled in the casting process, the longitudinal orientation can not identify with this crystal orientation for technologic deviation. The included angle between the longitudinal orientation and crystal orientation was controlled within a certain range. Research on the crystal orientation effect on fatigue strength of turbine blade, optimizing crystal orientation, improving mechanical characteristics and prolonging turbine blade life by controlling longitudinal orientation and two landscape orientations in the manufacturing and designing process is importantly technically meaningful.

II. DIRECTIONAL PROPERTY OF SINGLE CRYSTAL SUPERALLOY

Single crystal superalloy has evident directional property on thermo-mechanical fatigue damage and life, under the same thermo-mechanical cycle condition.

Ding et al. carried out LCF life experiments of DD3 alloy. Experiments included multiple strain amplitude and load ratio under 680 ℃. The relationship of young's modulus between an arbitrary [hkl] direction and [001] orientation was studied. Power function of total strain range was used for fitting of fatigue life curve of DD3 alloy and modificatory crystal orientation had litter effect on fatigue life. The related coefficient between fitting curve and experimental curve, attained 0.8769, far above one between stress range fitting curve and experimental curve. [1]

Yue Zhufeng, et al. carried out LCF life experiments of DD3 alloy on [001], [012], [112], [011] and [114] 5 crystal directions. They concluded that LCF life of DD3 alloy was power function of young's modulus, as shown in Fig. 1. [2]

LI Ying; et al. carried out LCF life experiments of DD6 alloy and investigated the relationship of $\Delta\varepsilon_T E_{[hkl]}$ to experimental fatigue life. Results showed that fatigue life of DD6 alloy was simple function of total strain range corrected by young's modulus. Results under 760℃ were shown in Fig. 1 and the conclusion was alike under 980℃. [3]

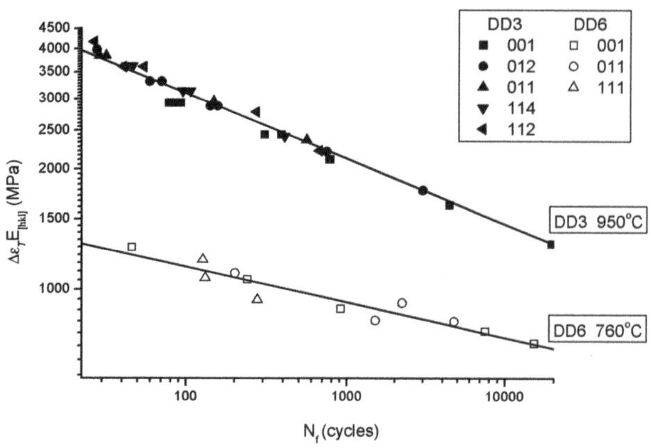

Figure 1. relationship between strain range and LCF life of DD3 and DD6

A great of experiments and research showed that directional property of single crystal superalloy reflected in thermal mechanical fatigue life may be expressed by young's modulus in the direction of load.

Thermo-mechanical fatigue life was related to temperature. D. H. Anton researched LCF characteristics of single crystal superalloy affected by temperature and he concluded cycle life was the power function of temperature. [4]

All the relevant experimental data served as the basis for model of thermal mechanical fatigue damage.

III. ANALYSIS METHOD FOR CYCLE DAMAGE OF SINGLE CRYSTAL SUPERALLOY

Fatigue damage model of single crystal superalloy was derived by those of directionally solidified superalloy or equi-axed grain superalloy. [5][6][7][8][9][10]

It was found that relationship between crack growth rate and strain range in single crystal superalloy was applied to power equation. Crack growth rate temperature depended satisfies with following equations:

$$\frac{da}{dN} = A_1 \exp(A_2 T)(\lambda \cdot \Delta K_\varepsilon)^{A_3} \tag{1}$$

where A_1, A_2, A_3 are empirical constants, T is the max cycle temperature, ΔK_ε is shape factor. And

$$\lambda = E_{[lmn];\alpha,\beta} / E_{[001]} \tag{2}$$

Where $E_{[001]}$ and $E_{[lmn];\alpha,\beta}$ are young's modulus related to [001] and [lmn] crystal direction. l,m,n express an arbitrary direction and [001] is reference direction. The value of α,β depends on direction.

The expression of strain intensity factor was substituted in (1) and after integral the following could be got:

$$N = \frac{\int da/F(g)}{A_1 \exp(A_2 T)(\lambda \cdot \Delta\varepsilon)^{A_3}} \tag{3}$$

Uniform geometrical shape is assumed, then

$$\int da/F(g) = C \tag{4}$$

where C is constant and

$$N = C/A_1 \exp(-A_2 T)(\lambda \cdot \Delta\varepsilon)^{-A_3} \tag{5}$$

This expression includes temperature and is coincident well with experimental results. Definition

$$\phi = A_1 \exp(A_2 T)(\lambda \cdot \Delta\varepsilon)^{A_3} \tag{6}$$

Get

$$N = C/\phi \tag{7}$$

ϕ may be defined as local anisotropic damage factor(LADF), a parameter related to local temperature, strain range and crystal direction. Its value relates to location in structure. ϕ is a relative value, dimensionless and serves as the basis to quantify local damage amount. From the equations it is clear that there is direct reciprocal relation between ϕ and life, so ϕ may be used to analyze the life situation.

IV. CRYSTAL ORIENTATION OF TURBINE BLADE DAMAGE

Single crystal superalloy has anisotropic property in [001], [011] and [111] 3 directions. [001] direction is very fit for Z axis so as to turbine blade has optimal thermal mechanical fatigue strength on Z axis. [001] direction, called No.1 crystal orientation, is not exactly as it is wanted, varying in a 15° range, as in Fig.2. Turbine blade has anisotropic property in the X-Y plane such as young's modulus.

If crystal orientation is inadequate in single crystal superalloy, greater strain tends to arise. Studies had shown that LCF crack related to crystal orientation. Crystal orientation affects location where crack arises and rate at which crack grows. In order to increase blade life, crack should be avoided. Researchers introduced method of crystal orientation control to obtain the best crystal orientation.

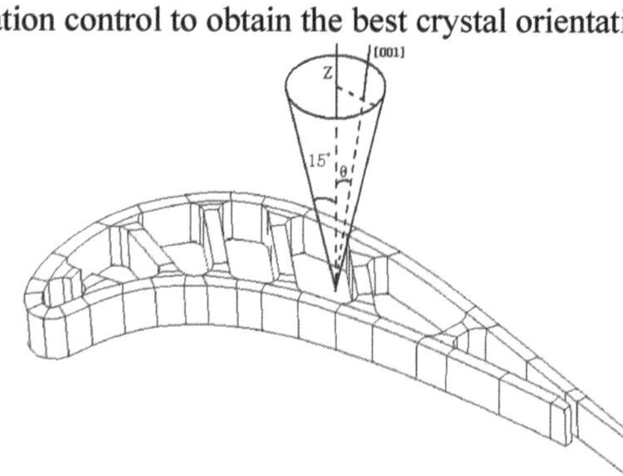

Figure 2. [001] diredtion is oriented with respect to Z axis

V. FINITE ELEMENT CALCULATION OF SINGLE CRYSTAL TURBINE BLADE

Z axis of blade casting coordinate system is in turbine disc radial direction, Y axis in axis direction and X axis in tangential direction.

This article calculates damage and life with multiple crystal orientations, [001] and [010] directions according to certain rule. Angle between [001] direction and Z axis is classified as 0°, 7.5° and 15°. [001] direction could revolve about Z axis remaining angle between them unchangeable, classified as 17 kinds in terms of spatial form, shown in Fig. 3. Vertically view, [001] direction and Z axis coincide where is labeled No.0. Where angle between [001] direction and Z axis is 7.5°, No.1 to No.8 are labeled, and angle15°, No.9 to No.16.

After [001] direction is decided, 0°, 22.5°, 45° and 67.5° (angle η) are selected to decide [010] direction. [001] direction and [010] direction together determine two axes of 68 material coordinate systems and the third axis is determined according to right-hand rule. By damage analysis in all material coordinate systems, the optimal crystal orientation was obtained.

Finite element model of turbine blade was build, which considers displacement restraints, centrifugal force and temperature load. Cyclic plastic constitutive model developed by Qingwu Wang was used based on crystal slips[11].

Based on the results by finite element model, fatigue damage analysis was carried out on Nickel-based single crystal blade.

VI. DAMAGE ANALYSIS OF SINGLE CRYSTAL TURBINE BLADE

The nodes on section outline are numbered by S and P, as in Fig. 4. The distance from a point to point O along outline is express by arc length. Section temperature profile along outline was shown in Fig. 5. Take No. 0 material coordinate system, η=0°, η=22.5°, η=45° and η=−22.5° for example, fatigue strength were analysed. Strain range profile on section along Z axis is shown in Fig. 6 and damage value in Fig. 7.

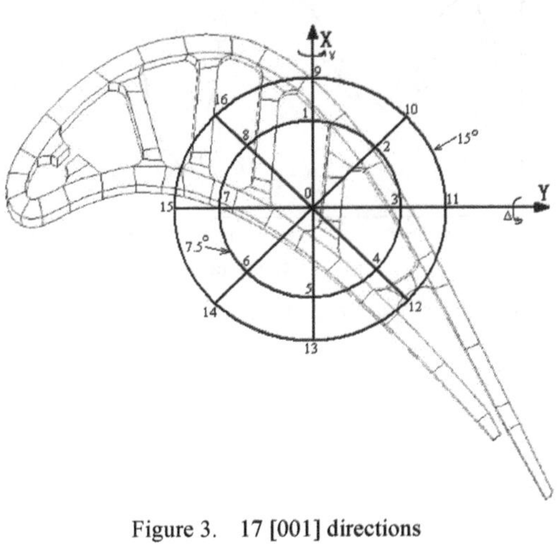

Figure 3. 17 [001] directions

Strain range profile on section along tangential direction is shown in Fig. 8. Damage value on section perpendicular to Z axis is shown in Fig. 9. The damage results in X-Y plane is related with [010] and [100] direction. Keep [010] direction constant, for example η=0°, to receive how blade damage are affected by [100] direction varying in all 17 cases.

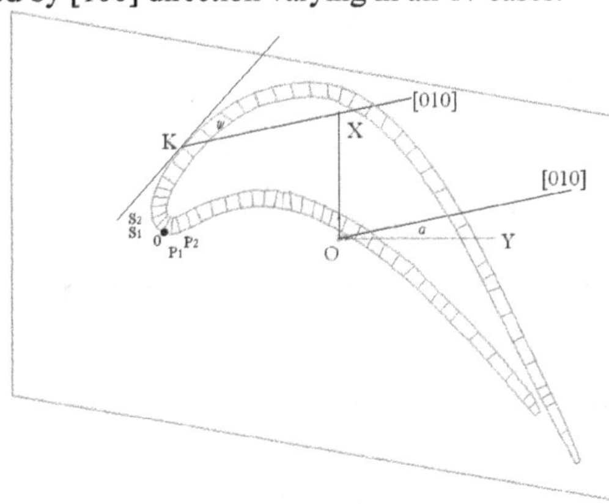

Figure 4. the definition of tangential angl

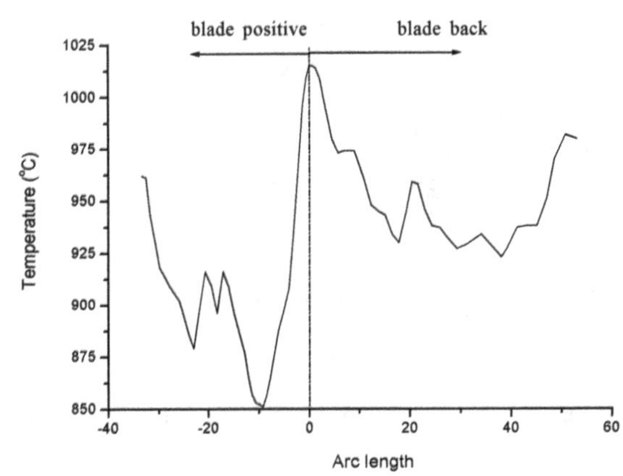

Figure 5. temperature of all points on section outline

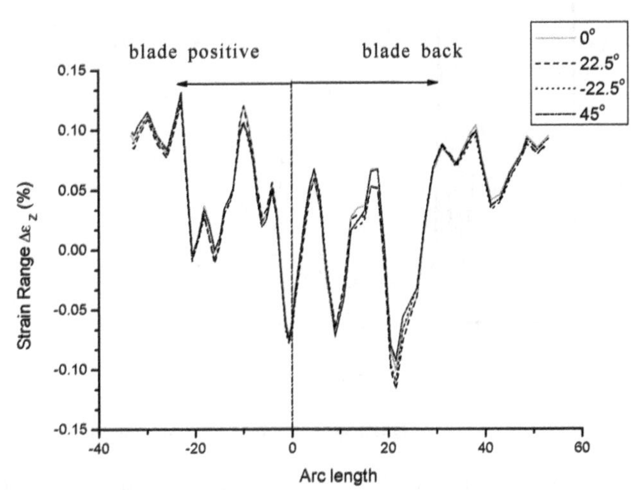

Figure 6. Z axial strain range of all points on section outline

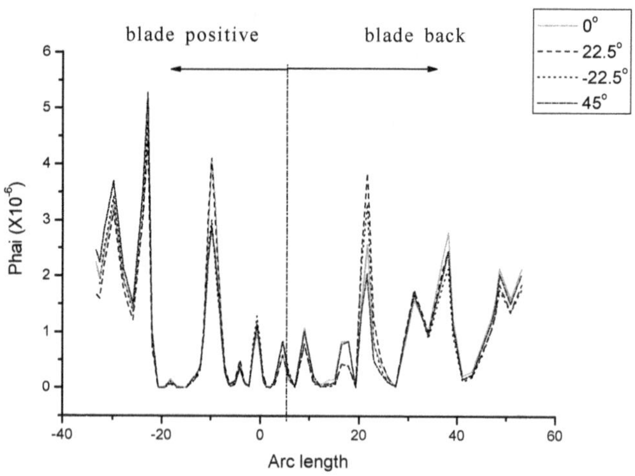

Figure 7. value of LADF φ at all points on section outline

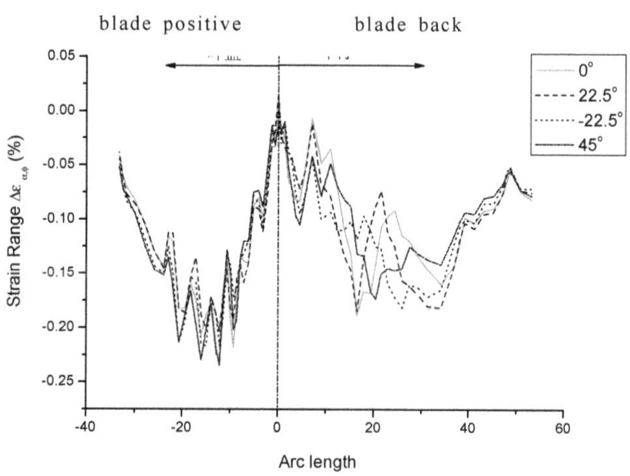

Figure 8. tangent strain range of all points on section outline

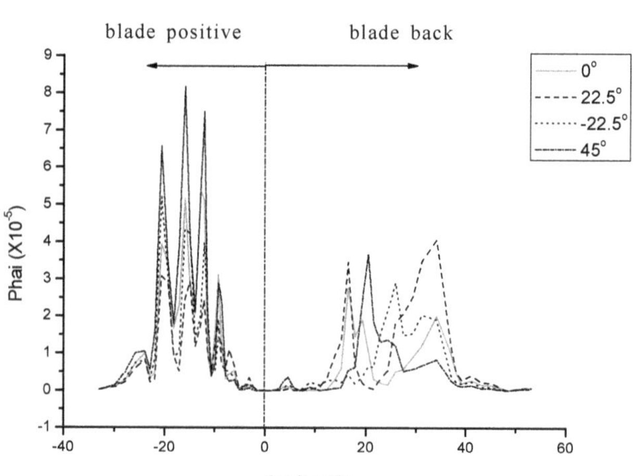

Figure 9. value of LADF φ at all points on section outline

As shown in figure, there are 4 damage peak points along all arc length, labeled P_{25}, P_{12}, 0 and S_{20} respectively on blade section. Damage values at the 4 points were further analyzed to find how damage are affected by [100] direction varying, as shown in Fig.10, Fig.11, Fig.12 and Fig. 13. It was observed that damage in Z axis direction possesses very strong regularity while [100] direction varying, to its maximum at No.14 and minimum at No.2.

Take No.14 material coordinate system, η=0°, η=22.5°, η=45° and η= − 22.5° for example, to receive fatigue strength, as shown in Fig. 14.

Damage profile was calculated where [001] direction is labeled No.2 and No.4, as shown in Fig. 15.

How blade damage perpendicular to Z axis direction are affected by [001] direction varying in all 17 cases could be received. There are 3 peak damage areas in each case where 4 maximum damage points were chosen, each 2 points in blade back and positive. How damage values at the 4 points are affected by [001] direction varying is shown in Fig.16, Fig.17, Fig.18 and Fig. 19. It is observed that as in Z axis direction, damage perpendicular to Z axis direction possesses very strong regularity while [001] direction varying.

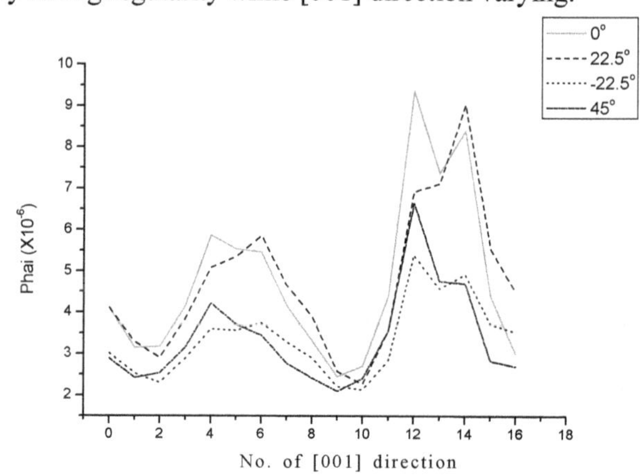

Figure 10. damage value at point P_{25} varying with [001] direction

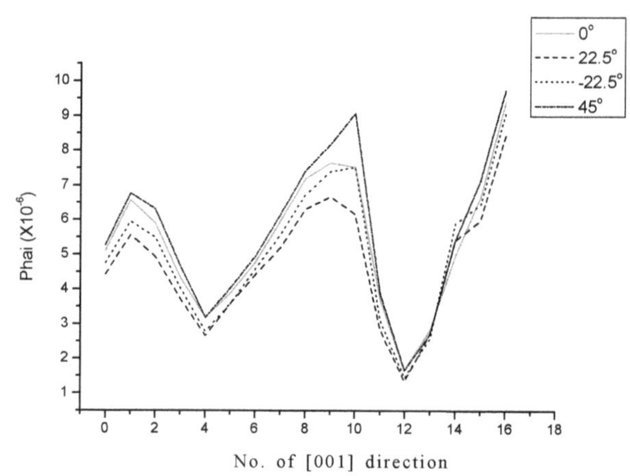

Figure 11. damage value at point P_{12} varying with [001] direction

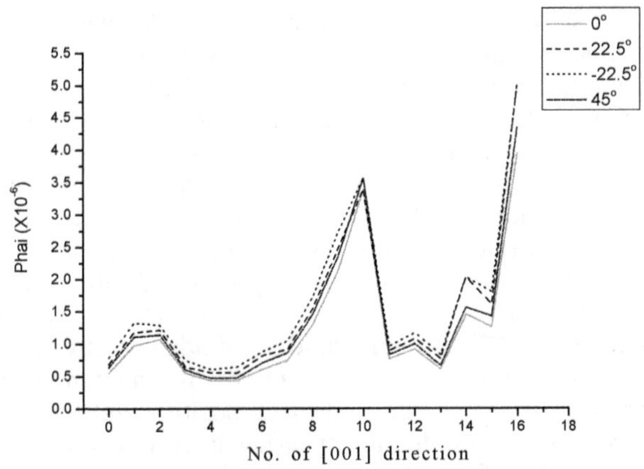

Figure 12. damage value at point O varying with [001] direction

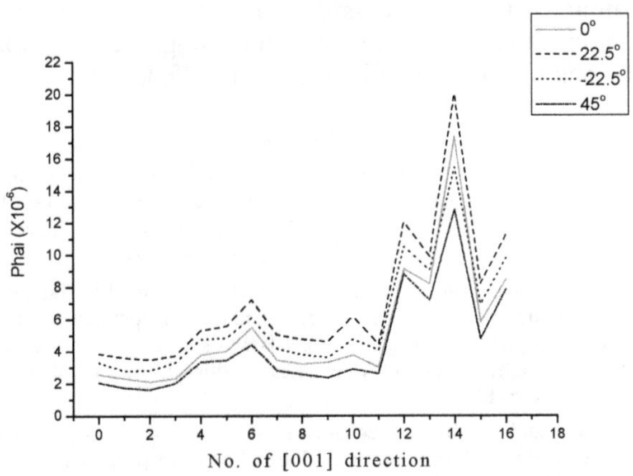

Figure 13. damage value at point P_{20} varying with [001] direction

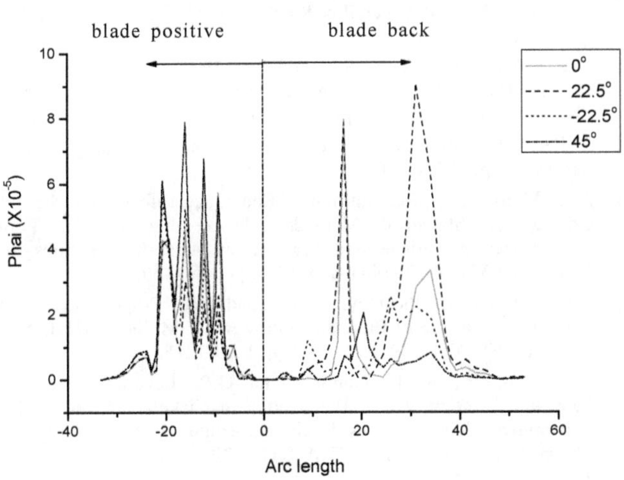

Figure 14. distribution of LADF φ value at all points on section outline

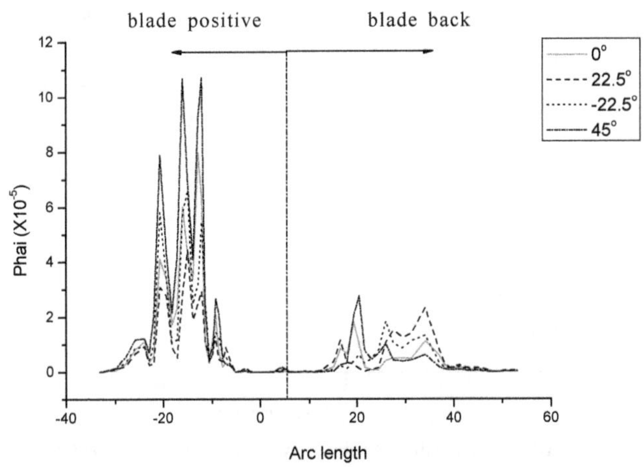

Figure 15. distribution of LADF φ value at all points on section outline

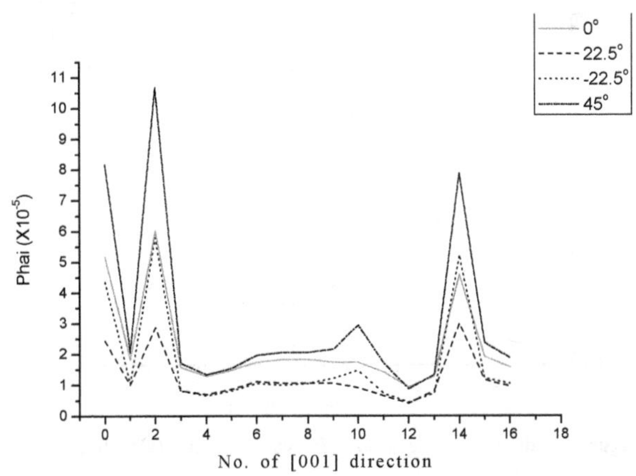

Figure 16. damage value at point P_{19} varying with [001] direction

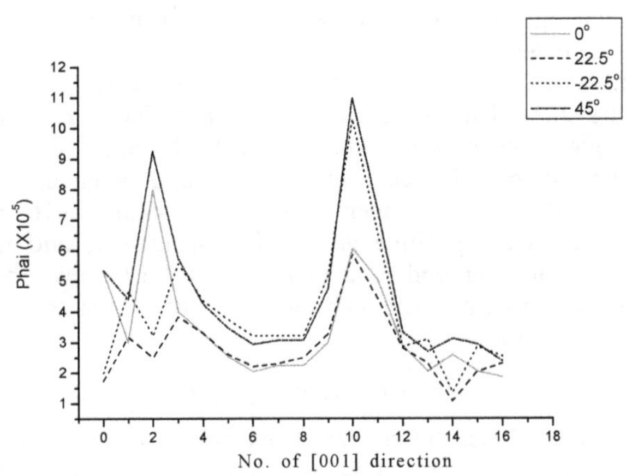

Figure 17. damage value at point P_{16} varying with [001] direction

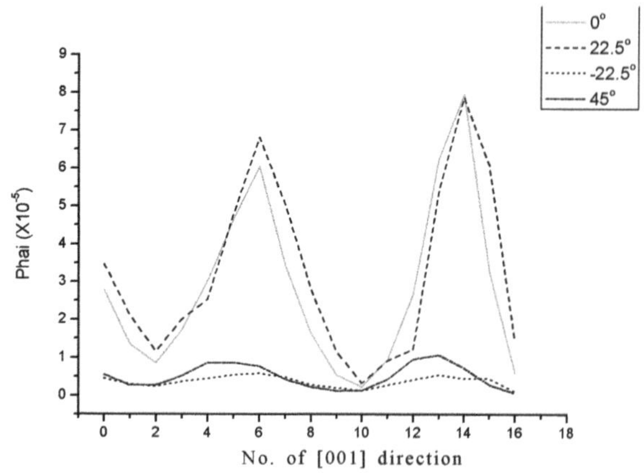

Figure 18. damage value at point P_{25} varying with [001] direction

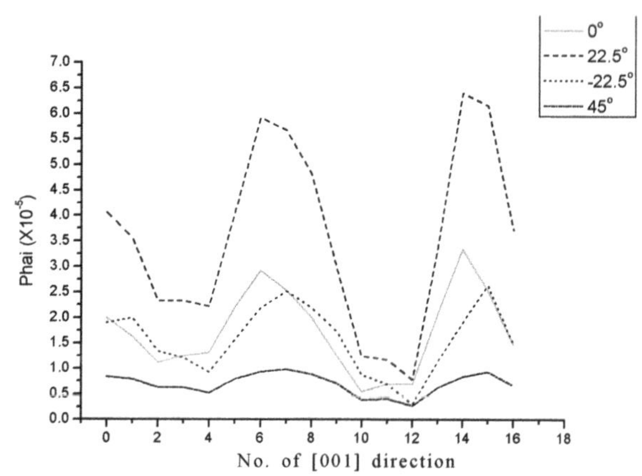

Figure 19. damage value at point P_{14} varying with [001] direction

Tangential damage changes with [010] direction at cross section are the same as vertical damage changes. There are 3 peak damage areas along the whole arc. Results show that Z axial damage value is the lowest under coordinate system No.13 and No.2, in blade positive and back respectively. Results show that damage value perpendicular to Z axis is the lowest under coordinate system No.13 and No.12, in blade positive and back respectively, that Z axial damage value is the highest under coordinate system No.16 and No.14, in blade positive and back respectively, and that damage value perpendicular to Z axis is the highest under coordinate system No.10 and No.14, in blade positive and back respectively.

RESULT AND DISCUSSION

On the whole, the change in material orientation has a great influence on damage value of single crystal blade. [001] direction has a far greater effect than [010] direction on damage value. [001] direction is dominant and [010] direction subordinative. The research reflected a trend: while [001] direction rotating on Z axis, Z axial damage show wave ridge-trough-ridge periodic change similar to a sine or cosine curve. Damage perpendicular to Z axis has the same regularity. Z axial damage range increases with raise of angle between [001] direction and Z axis, which shows Z axial damage becomes larger meanwhile. The results conformed to earlier studies. On the other hand, Z axial damage at many points fluctuate in the very close cycle, which indicates that a angle exists between [001] direction and Z axis, maximizing or minimizing Z axial damage. The change law of damage perpendicular to Z axis is much similar. Results shown that the area where the Z axial damage maximum and minimum appear keep unchanged with [010] direction varying, which has a little influence on Z axial damage distribution, as subordinative direction only affecting amplitude not location. [010] direction varying has some influence on damage perpendicular to Z axial but could not fundamentally change damage distribution, which only effects the maximum and minimum position in detail on blade positive and back, not the maximum and minimum appearing areas as a whole.

Cycle life of single crystal blade relates to crystal orientation. While the other condition keep same, it could opt for optimal material crystal orientation to obtain the maximum fatigue life. In designing single crystal blade, casting angle should be effectively controlled.

REFERENCES

[1] DING Zhi-ping; CHEN Ji-ping; YIN Ze-yong; and YANG Zhi-guo, "Low-Cycle Fatigue Life Prediction of Single Crystal Nickel-Based Superalloys under Asymmetrical Cyclic Loading," Journal of Aeronautical Materials, vol. 26, Aug. 2006, pp. 6-10,

[2] Yue Zhufeng; Tao Xiande; Ying Zeyong; and Li Haiyan, "A Crystallographic Model for the Orientation Dependence of Low Cyclic Fatigue Property of a Nickel-Based single crystal superalloy," Applied Mathematics and Mechanics (English Edition), vol. 21, Apr. 2000, pp.415-424

[3] LI Ying and SU Bin, "Mechanisms of Low Cyclic Fatigue of DD6 Alloy at Elevated Temperature," Journal of Aerospace Power,vol.118, Dec.2003, pp.732-736.

[4] D. L. Anton, "Low Cycle Fatigue Characteristics of <001> and Randomly Aligned Superalloy Single Crystals," Acta Met., vol.32, Oct.1984, pp.1669-1679.

[5] A.V. Kumar and C.Yang, "Study of Work Harding Models for Single Crystals Using Three Dimensional Finite Element Analysis". Int. J. Plasticity, vol.15, Jul.1999, pp. 737-754.

[6] S. Pearson, "Fatigue crack propagation in metals," Nature, vol.211 Sep.1966, pp. 1077-1078.

[7] A. J. McEvily, "A Redefinition of Endurance Life Design Strength Criteria by Statistical Methods". Proceedings of Air Force Conference on Fatigue and Fracture of Aircraft Structures and Materials, (AFFDL 70-144), Sep. 1970, pp. 451-459.

[8] B.H. Kear and B.J. Piearcey, "Tensile and Creep Properties of Single Crystals of the Nickel Base Superalloy MAR M-200," AIME MET SOC TRANS, Vol. 239, Aug.1967, pp.1209-1215.

[9] C.A. Rau, Jr., A. E. Gemma, and G.R. Leverant, "Thermal-Mechanical Fatigue Crack Propagation in Nickel- and Cobalt-Base Superalloys Under Various Strain-temperature Cycles," Fatigue at Elevated Temperatures, ASTM STP 520, American Society for Testing and Materials, Sep.1973, pp. 166-178.

[10] A.E. Gemma, B.S. Langer, and G.R. Leverant, "Thermomechanical Fatigue Crack Propagation in an Anisotropic (Directionally Solidified) Nickel-Base Superalloy," Thermal Fatigue of Materials and Components, ASTM STP 612, American Society for Testing and Materials, Aug.1976, pp.199-213.

[11] WANG Qing-wu, YANG Xiao-guang, QI Hong-yu, and SHI Duo-qi, "Crystallographic Constitutive Models for Single Crystal Nickel Base Superalloys," Failire Analysis and Prevention, vol.3, Feb.2008, pp. 67-73.

Cultural Landscape of Catholic Churches of Northern Shaanxi in the Pieriod of Late-Qing Dynasty and the Early Republic of China

Wang Li
School of Architecture
Xi'an University of Architecture and Technology
Xi'an, China
Email: superwli@126.com

Li Xiaoming
School of MEI
Xidian University
Xi'an, China
Email: superwlim@gmail.com

***Abstract*—This paper inquires into the development and the present feature of catholic churches in north Shaanxi, which were built during the period of late-Qing dynasty to the republic era, based on the field investigation data, analyzes Influence of Chinese and western culture and buildings values, to provide basis for cultural landscape protection and usage.**

Key words: late-Qing dynasty and the early republic of China ; northern Shaanxi ; cultural landscape ; church buildings

I. INTRODUCTION

Shanbei(northern Shaanxi) is located in the northern part of Shaanxi Province, to the south is the Guanzhong in Shaanxi, southern Shanxi and Henan Northwest, to the east is Jinzhong, to the north is the Ordos grassland and Mu Us Desert, and so, the unique topography, constitute a relatively closed and independent geography and environment, Tongzhi eleven years (1872), the Catholic Church became as a large-scale spreading to northern Shaanxi province, the Christian culture had an impact on many aspects of northern Shaanxi, in 1875, first Catholic church in northern Shaanxi was built, which is the Jingbian Xiaoqianpan church buildings. Since then a large number of church buildings built in the northern Shaanxi region, forming a fairly complete cultural landscape of northern Shaanxi. In this paper, the data and basis for conservation, heritage and sustainable development of Catholic cultural landscape is provided, which is acquired by field research of the remains of church buildings in the late Qing and Republican period.

II. THE DISTRIBUTION OF LATE QING AND REPUBLICAN CHURCH BUILDINGS IN NORTHERN SHAANXI

In 1872, the Belgian CICM missionaries Yemaozhi came from the outside edge of the Suiyuan (now Inner Mongolia) to Ning Liang town of Jingbian County in northern Shaanxi and carried out missionary activities, and built Xiaoqianpan church buildings in 1875, at the same time he also purchased land near the farm, subleased to local people looking for opportunity to spread his holy church[1]. From then to the the past several decades of Republic period, the spread of Christianity in northern Shaanxi region showed a trend of a considerable scale of development, almost affecting all regions across northern Shaanxi.

According to Local Records and Statistics of northern Shaanxi region, in late Qing and Republican period, in northern Shaanxi, two government two states (Yan'an Prefecture, Yulin Prefecture, Suide state and Lu State) built 36 Catholic churches and 25 Protestantism churches in 14 counties. So that Roman Catholic mainly spread over a wide range in the northern Shaanxi region, and followed by Protestantism, and the spread of the Orthodox almost had no documented record, in which the number of churches in different government counties is as follows: 18 Churches located in 5 counties for Yulin prefecture,6 Churches located in 3 counties for Lin prefecture, 8 Churches located in 3 counties for Suide prefecture, 54 Churches located in 9counties for Yanan prefecture, and there are a total of 86 churches.

Due to social unrest in late Republic, the spread of Christianity was severely affected, in northern Shaanxi church buildings are also subject to serious damage. According to the data archives of Shaanxi Province, the government reported questionnaire data shows that only eight Catholic Church and seven Protestant churches remain in northern Shaanxi region till 1936.

Although there may be some errors in these survey data [2], however, those data can reflect the drastic reduction of the churches' number in northern Shaanxi. The author have entered northern Shaanxi several times in 2005 to carry out field research on the existing Catholic Churches built in late Qing and Republican period, and classified as following categories.

III. EXAMPLES OF THE EXISTING CHURCH BUILDINGS IN NORTHERN SHAANXI

A. Romanesque style church

Yan'an Qiaoergou is the Roman Catholic Church architecture style. It was built by Spanish missionaries Yi-Xinghua in 1931, and completed in 1934. And now the churches become indoor training venues of Yan'an Lu Xun's art school.

The church is of the Basilica-style rectangular plane, north to south is 36.2m, east to west is 15.8m, the entrance locates at Southern front (Figure 1), Altar locates at the northern end, the height of the central hall is 11m, height of aisle is 6.3m, the roof is flat style. Clock-tower is located on both sides of the southern entrance, 21.9m high, Church facade decoration and detailed decoration adopt Roman-style arch, showing typical Roman style (Figure 2). Combining Chinese and Western Interior, the central hall is the arch structure, aisle gallery and the central hall are separated by arch, Altar space is unique form, the top covered with a circular dome, the altar stone decoration adopt the traditional plant style: plum, orchid, chrysanthemum, Lotus and Lung, The lower part of building adopt the materials of brick.

Figure 1. Entrance of Catholic Church Qiaoergou in Yan'an

Figure 2. The side elevation of Catholic Church Qiaoergou in Yan'an

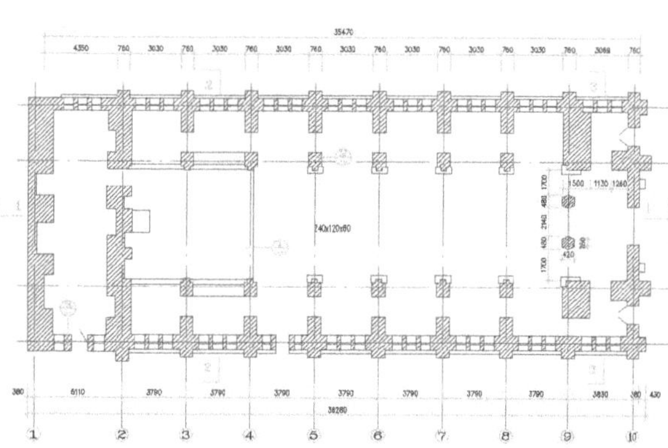

Figure 3. plane of Catholic Church qiao'ergou in Yan'an

B. Gothic style church

Yan'an Ganguyi Catholic Church is of the Gothic architectural style. In 1931, the Spanish missionaries came to Yan'an Ganguyi, employed 300 people of local labor to build the Church, and three years later, it was completed [3]. The internal facilities of the church were destroyed in the "Cultural Revolution", and now is Ganguyi granaries of Yan'an city, and now all the windows of the church is pluged for stored grain (Figure 4).

Figure 4. The side elevation of Catholic Church ganguyi in Yan'an

Figure 5. Entrance of Catholic Church ganguyi in Yan'an

The church is bacilica plane form, Length is 30.9m, and width is 12.7m. The arc-shaped altar locates at northern end of the church, the central clock tower is located at the southern entrance, which is 14.6m high, consists of the inner stairs. With a tall bell tower as the center, Façade is in well balance, the detailed decorations of stigma, roof, windows and doors are all of pointed arch, showing the tall features of Gothic church (Figure 5). The church aisle and central hall are separated by continuous Arcade. The northern wall of altar, there are six high windows with pointed arch and bright light, and there are brick decorations of local characteristics on the pointed arch window.

C. Integrated Chinese and Western architectural style Church

The Catholic Church in Jia County Tanjiaping is the one integrated Chinese and Western architectural style. Belgian missionary Nie finished the construction in 1909. layout of the church is rectangular, length is 15.8m, width is 15.2m, with two Pitched Roof, the flat roof on both sides is priest residence, two-floor-building in the east, is slightly lower than the church roof in the middle, and one-floor-building is in the west side, which was not completed due to priest return(Figure 6).

Figure 6. Roof of Catholic Church tanjiaping in jiaxian

Figure 7. Entrance of Catholic Church tanjiaping in jiaxian

Figure 8. The stairs of Catholic Church tanjiaping in jiaxian

The structure of the church adopt cave dwellings in northern Shaanxi, and it was built entirely using the traditional method of hoop kiln to form an independent brick cave churches, so that the church highlights its local characteristics. The main entrance is located in north facade with Gothic-style pointed tickets windows and doors, both sides of the brick pillar symmetrically set four small bell tower of different Height, from 10.1m to 14.2m, and there is not any stair in the clock tower (Figure 7). .A 1.5m outer gallery and L-shaped outdoor stairs is in the west of the two-story church (Figure 8), detailed decoration using the traditional plaque, peace doves and plant styles.

IV. CULTURAL LANDSCAPES VALUE OF THE CHURCH CONSTRUCTION IN NORTHERN SHAANXI

A. Historicall value

Catholic Church buildings in northern Shaanxi is not only spread religious culture, but also to promote a Western technology and culture in the spread of northern Shaanxi. Through the construction of churches, schools, clinics and other social institutions, Catholic groups have changed modern social life and ideas in northern Shaanxi to a certain degree. According to the Statistics of local chronicles in northern Shaanxi: in the period of late Qing and Republican, the church organizations had built seven hospitals and clinics, church schools up to seven, kindergartens up to four, and one nursing home for the aged, etc., and now those buildings are so far very few left, however, these incidents are recorded, which is Important witness of modern history and culture of the northern Shaanxi.

B. Technology Value

In northern Shaanxi, the western Catholic Churches are outstanding representatives of the more advanced materials, corresponding to their structure and construction methods. First of all, the church buildings are in early use of cement, steel, glass and other materials of construction in northern Shaanxi region. Secondly, the structure used in the church in northern Shaanxi adopts masonry arch structure, which is different from the traditional structures of cave building in northern Shaanxi, not only rugged durability, wind-fire, but also light, and the ventilation problems are solved. In addition, in northern Shaanxi, the church carried out the earlier use of outdoor drainage pipes, and achieved an organized drainage.

C. Landscape value

First appeared western-style buildings in northern Shaanxi, is the church building, which shows a strong religious overtones, and western culture, forming exotic landscape. The architectural style of Northern Shaanxi churches covers Roman and Gothic, and other classical architectural vocabulary, including classical column, pointed arch and a large number of Western-style decorations. These tactics have been absorbed into decorated houses, shops in the northern Shaanxi, such as Fung shopfront and Girls School in Mizhi County of northern Shaanxi, as well as other buildings, have some Western-style decorated traces, reflects the Easten and Westen cultural collision and blending in the process of exploration

V. CONCLUSION

Cultural Landscape is an important landscape resource. In the visual, the shape of church building is high and large. The art style and facade design is different from Shanbei cities and towns, which condense different features of social culture of Shanbei's cities and towns, representing those special stages of time and space. Those process of civilization formed in early 20th century has a typical material significance, and also show abundant historical culture accumulations and certain profound connotation, cultural landscape of church buildings become historical memory in shanbei. On the other hand, because that Shanbei is effected under traditional Chinese culture for thousands of years, those church buildings have a high historical and cultural value, and so, from the point of "cultural landscape continuity", the research of church buildings, will play a positive role for Shanbei in promoting the historical status and cultural value.

REFERENCES

[1] The committee compilation of Jingbian local chronicles. Jingbian County annals .xi' an : Shaanxi People's Press , pp. 429–435, 1993

[2] Archives of Shaanxi Province. questionnaire of Cathedral church and Mosque in shanbei, xi' an, Archives of Shaanxi Province Fond , No.9 Dir.4 Vol.1038.

[3] The committee compilation of yanchang local chronicles. yanchang County annals. xi' an : Shaanxi People's Press , pp.574–583, 1991.

Design for Natural and Intuitive Interaction

Touch-Based Human-Computer Interaction

Chong Zhang
College of Art
Zhejiang University of Technology
Hangzhou, China
Email: zoe0127@gmail.com

Yonghai Yu
Interaction Design Institute
Zhejiang University of Technology
Hangzhou, China
Email: yuyh@zjut.edu.cn

Bing Xu
Interaction Design Institute
Zhejiang University of Technology
Hangzhou, China
Email: xubing@zjut.edu.cn

Qiong Liu
Mobile Business Unit
Arcsoft Information Technology Co., Ltd.
Hangzhou, China
Email: qliu@arcsoft.com.cn

***Abstract*—The touch technology provides possibilities to realize more natural human-computer interaction which results from the elimination of the extra medium [1] of communication to a large extent. Meanwhile, the requirement for an intuitive operation that much meets expectation of human-beings appears as well.**

This paper will mainly discuss the "intuitive interaction" in two aspects: the requirement on exhaustive feedbacks from machine, and the discussion about current misunderstanding on necessary of reality simulation. The first aspect will aim at the comparison between the traditional human-computer interaction and the touch-based one. The second one will demonstrate the reality simulation from the data visualization which is the theoretical foundation of it.

Finally, a new concept of "sense on earth" will be proposed. Rather than reality simulation, it interprets the implementation of "intuitive interaction" more precisely.

Keywords-touch; intuitive interaction; feedback; reality simulation; sense on earth

I. INTRODUCTION

A. Background and Problems

The comprehensive employment of the touch technology is an inevitable future for the development of electronic interface. Obviously, this technical innovation offers new soil for more possibilities of human-computer interaction design. However, it also demands more solutions toward the interaction mode. It is substantiated by the obsolescent but still prevalent interaction model--the mouse-and-keyboard-based interactive mode. Rather than merely incompatible, the traditional interaction approach is obstructing the new technology put to use its inborn advantages indeed. It is the advantage that without the extra medium like mouse or keyboard, the distance between human and computer is shortened, and the communication relationship is also crystallized as well.

However, while many unique interaction approaches supported by touch pop out, there can be seen a chaos of numerous inappropriate application of the technology. Some cases appear as much too emphasizing on the visual effects, and that would probably cause user confusion in practical operation. But most cases attribute to the overwhelming successors' wrongly understanding toward some existent interaction models. These followers' immature alteration to the superior design finally results in the technology being inappropriately applied.

Therefore, the implementation of "intuitive interaction", if not been detailed, though making sense in general, is insufficient in lending guidance to design practice. This paper would discuss the change of interaction way which is directly brought by the touch technology, then would correct the overstating trend of reality simulation.

B. Terminology Definition

The concept definition is the logical premise of any research work. Thus, in order to make clear the "intuitive interaction", and to avoid the unnecessary dispute which is only generated for different jargon using, it is meaningful to describe this terminology by some analogous and historically used ones at the beginning.

As the "daily task" [1] called by Donald A. Norman, and "natural user interface"[2] named by Bill Gates, and the idea of "the inevitable and un-designed design" [3], which is raised by Johnason Ive[2], the similarity among these thoughts can be drew. It is the requirement toward user interface that is characterized by "less software sense but more human ways of communication". This includes demand for reducing the massive buttons and sophisticated menus

1. It is represented by the combination of mouse and keyboard versus monitor.

2. Jonathan Paul Ive, CBE (born February, 1967) is an English designer and the Senior Vice President of Industrial Design at Apple Inc.

crowded in interface. The concept of intuitive interaction embraces analogous terms as natural user interface and instinctive interaction, inclusive but not exclusive.

Additionally, it would be helpful to mention the artificial intelligence. In my point of view, this concept is more likely a fundamental design rule in the context of user experience, rather than one technology merely. That is to say, the machines that people face everyday, though not embedded advanced intelligence, they need be devised according to user's habitual thoughts. In this way, the interaction process flow can be bettered, and can be better presented through the user interface as well. Just as a responsible service person in the real world, the machine reacts to any behavior from the user through "language response", "appropriate eye looks", or even suitable "body languages". Thus, the feedback clarification would be discussed at the first place.

II. The Clarification of Feedback From Machines

A. To Build Correct Mental Model

First of all, the definite feedback from machines is the determinant of fluent human computer interactive experience.

Feedback received by users during operation, if exhaustive, would not only help user establish a correct mental model, but also help generate a sense of trust between human and computer. With reference to the unavoidable disadvantage of machine's insufficient means to react, the user interface should more emphasize the 100% intact feedback. Even when facing some merely minor misoperation which may not necessarily cause reversal in result, it still should be highlighted. Because it is the unambiguous perception of "what stage I am in" or "what process I have got through" that gets users be informed of one's own operation behavior without perplexity, not to mention astonishment. To witness typical case in action, one need look no further than the software shortcut key operation in Microsoft Windows. Once the shortcut key combination for copy (ctrl+c) is pressed, the machine would not give any response until the final result is displaying on the screen. The interrupted feedback directly contributes to the suspending of communication flow. This unpredictable mode finally turns the originally definite operation result to a fortune-based one.

B. The Input and Output Incorporated Interface

This demand of exhaustive feedback is especially crucial in a full touch circumstance. The reason is directly related to the distinctness of the new conditions which the touch technology brings.

Comparing to traditional operation mode --one that represented by mouse and keyboard versus monitor--, the employment of full touch inevitably leads to an integrated interface by merging the input devices with output devices into one operation platform. In other words, "how to maintain at least the original usability in the amalgamated interface" would be a far more key point in the transition period from the current conditions to the Full Touch era. In addition, the operation confusion is not well solved yet, even under the current circumstance in which neither the input nor the output is exclusive, and separated as well.

With respect to the current solution, the pop-up soft keyboard is rightly the visual signal of state switch from output to input. This virtually existent output device facilitates users operation, because it helps distinguish the two states clearly. In other words, user interfaces of most current touch products are still traditional indeed, as far as the input way is concerned.

However, when comes the extensive employment of handwriting technology in prospect, the state switching confusion would be particularly apparent. And that would probably obstruct to fulfill a fluent human-computer interaction experience.

The transformation from the input device as one-key-one-function to the one that is composed by varied gestures with multiple functions complicates the interaction process. And this would have a negative impact on the operation accuracy and the efficiency as well.

Furthermore, the learnability of the gesture learning process would be in doubt as well, since individuals have to remember more. The sophisticated operation attributes to the "less keys (only ten fingers) but more tasks". Even though a positive circumstance can be seen toward the touch product available in the market—users do not complain them more if compare to the traditional products--, we still need keep an eye on the future.

This is almost an inevitable process which could refer to the "development of new technologies, U shape curve theory - the complex to simple, and then again to complex" [4]. And that is the exact reason explaining why the current touch gestures are not as intricate as the aforesaid. Because yet the touch technology locates in the intermediate development status now, in which they are still not completely developed, and are aided by many user traditional methods.

To sum up, the exhaustive feedback is crucial in any form of interaction. Considering the distinctive disadvantages that the new technology embraces, the intact and exhaustive feedback would hardly be overemphasized in order to meet a more natural and fluent communication between human and computer. In the circumstances that artificial intelligence technology is not popularly employed, designers should already keep a mind of designing the interactive flow in a predictable way, even if the machine cannot "think" yet.

III. Data Visualization and Reality Simulation

The upgrading graphical interface is not analogous to reality simulation, and the reality simulation it does not mean an intuitive interaction. It is the intuitive interaction that is the ultimate goal, while the data visualization and reality simulation is only intermediate task to achieve it.

A. Data Visualization

The purpose of data visualization is to use perception for amplifying cognition [6], of which the essence is to ease the perception way of information reception. It does not mean that data visualization needs to look boring to be functional or extremely sophisticated to look beautiful. To convey ideas effectively, both aesthetic form and functionality need to go hand in hand, providing insights into a rather sparse and complex data set by communicating its key-aspects in a more intuitive way. [5] Yet designers often tend to discard the balance between design and function, creating gorgeous data visualizations which fail to serve its main purpose — communicate information. It is because they have forgotten the original intention – for information to be more easily perceived. A great deal of ambiguous or even idea wrongly conveyed icons can be seen on cell phones, websites, and computer applications as well. And it has to be admitted that there are some occasions that it is the words or text that can better express the meaning in a comparably accurate way, while the visual materials, whether photos or intentionally designed graphics, are insufficient to present. In these cases, the visual work is invalid, or even counterproductive, so that the obstacle is set within the interaction process. This "original intention" thought would hardly be overemphasized, since misunderstanding of visualization is widely existed. There is no rule for cell phones' tictactoemidlet interface design, or for criticizing text button as perfunctory design. It is things which are designed for merely decoration that are not in line with the fundamental starting point for data visualization.

B. Reality Simulation

Contrary to prevalent views, the visualization reality has been overemphasized and there is a trend of excessively exaggerating the theory, as I see it. But not fault for its own sake, rather the inappropriate interpretation leads to this contemporary result. In the context of touch technology, information ought to communicate directly since the extra medium has been eliminated. However, with reference to the already highly developing three dimensional display technology with the mutual support from hardware storage and software compatibility, more opportunities and probabilities that have been offered. These supports ignite designers' eager to build fascinating interface with great visual impact that is eyeball attracting.

However, is it truly necessary, the lifelike environment, or the reality simulation as realistic as possible? If the exception does exist, where do we draw the line?

In the book *Emotional Design*, Donald A. Norman rebutted his previous idea given in the Design of Everyday Thing--usability is overriding. He thinks that things with emotional impact, perhaps not easily using, could win products which are well functioned to some extent. So, does that mean the visual impact overrides the usability? I don't think so. And from my perspective, this pivotal line lies on the boundary of products' classification. If, for example, the product is mostly used for tackling with problems which asks for effective functions, the usability would be considered in the first place; on the contrary, while the product is not so essential to serious assignments but rather focus on individuals' entertainment needs, then the emotional impact as well as the visual effect would score.

Figure 1. The interface of Bob.

A remarkable case in point is the Microsoft Bob(see *Fig. 1*). Bob was a stark departure from the menu-based, text-intensive interface of Windows. In a precursor to today's customized wallpaper and screensaver programs, Bob is portrayed as the inside of a house, with different rooms to correspond to common real-world room styles such as kitchen and family room. Click on the various items in the room --Rolodex, calendar, checkbook-- and Bob would start up the appropriate program.

Bob was intended to provide a new, non-technical interface to desktop computing operations.[7] However, while the basic goals of Bob were sound, it didn't meet its goals all that well, and it actually made the world more complicated to navigate. In other words, since the program Bob is a function-based electronic tool rather than an application for entertainment, the user experience design which is not helpful but annoying directly led to the failure.

Thus, when it comes to the interface or user experience design, the ultimate goal need be clarified at the very beginning. The challenge for programmers and designers is to learn from Bob's successes as well as its failures. The initial purpose of Bob is valuable, and it has well presented the future user interface. And the computer industry should be careful it's not "throwing the Bob out with the bath water."

C. "Sense on Earth"

Compare to visualizing the objects from the real world, simulating the physically feeling that human perceive on earth is much useful while with respect to intuitive interaction. Here, "sense on earth" is created to describe the condition.

The computer language or the unique computer icons maybe seem alien one decade ago are mundane stuff to current users. That is, the knowledge towards these former aliens has been widely established already. Even if some unique terms that do originate from professional computer languages in the past, the current situation is that they are things with which people are quite familiar. Users knows "sandglass" as they knows the word "wait". In this sense, the necessity of extra visualization should be reconsidered, because it is not a final goal but rather a design means only.

However, the effect of adding the "sense on earth" into design is different. Attaching importance to this issue could greatly improve the interactive experience, for it is the dominant difference that people perceive in the real world that is different from the virtual computer one. The real world offers numerous kind of physical force, such as gravity and resistance, while the virtual world does not. Therefore, if we can simulate in the virtual world the real-world physics, which comes from human experience, and utilize them in appropriate ways, users' learning process toward machine operation would be shortened and simplified greatly. One convincing example is the iPhone. The orientation sensor built-in acceleration due to gravity sensor automatically detects the handset vertical and horizontal direction, to achieve image browsing or video player under the screen automatically rotate. Besides, the press and drag action within the unlock operation is also expressive. A simulated resistance can be felt when the external force to unlock the phone is weak. Though iPhone's GUI is well designed, the factor that works here is the virtual experience of resilience but not the visual appearance. Because even if the unlock valve is composed by two simple rectangles, as long as the sense of resistance unchanged, the expressive learnability would probably sustain the same.

This interactive mode which emphasizes the real world physical sense is an effective design in line with the demand of intuitive-or natural- user experience. People are quite familiar with the operational principle—the gravity, for example--, thus once users have tried to operate, a correct mental model could be established. These detail designs, such as aforesaid self-adapting of screen orientation and resilience simulation does bring users joy and familiarity due to the "sense on earth".

IV. Conclusion

In the final analysis, with respect to the input and output amalgamated interface due to touch-panel, the exhaustive feedback is rigidly demanded. Otherwise, confusion in the state switching between input and output would cause the using barriers to occur. Meanwhile, the reality simulation is mere one intermediate means to fulfill intuitive interaction, but not the final goal. Thus, it is necessary to reconsider the fundamental purpose of the design task. Last, simulating "sense on earth" is more effective in enhancing usability as well as user experience, since it is the essential difference between the real world and the virtual one.

References

[1] Donald A. Norman, "Emotional Design: Why We Love (or Hate) Everyday Things", Basic Books, 1 edition, May 2005

[2] Bill Gates: 'Vision and Touch and Speech and Ink, All of those I Call Natural User Interface' http://news.softpedia.com/news/Bill-Gates-Vision-and-Touch-and-Speech-and-Ink-All-of-those-I-Call-Natural-User-Interface-77123.shtml

[3] http://www.edibleapple.com/jonathan-ive-speaks-on-the-brilliance-of-apple-design-video/

[4] (US) Donald A. Norman, the Psychology of Everyday Things, translated by (P.R. China)Qiong Mei, Citic, 2003, pp.31

[5] Cameron Chapman, "Data Visualization and Infographics", Graphics, Monday Inspiration, January 14th, 2008.

[6] Stuart K. Card, Jock D. Mackinlay, and Ben Shneiderman, "Readings in information visualization: using vision to think", Morgan Kaufmann, 1st edition, February 1999

[7] Cowley, and Stacy, "Ballmer Analyzes Microsoft's 'One Big' Vista Mistake". CRN.com. July 31st 2006 http://www.crn.com/sections/breakingnews/dailyarchives.jhtml;jsessionid=ZEQ0I0LMYJC1MQSNDLPCKHSCJUNN2JVN?articleId=191600739.

[8] Pak Chung Wong, and J. Thomas, "Visual Analytics", IEEE Computer Graphics and Applications, Volume 24, Issue 5, Sept.-Oct. 2004, pp.20-21.

[9] Alan Cooper, and Robert Reimann, About Face 2.0--The Essentials of Interaction Design, Wiley; 2nd edition, March 17, 2003

[10] Alan Cooper, Why High-tech Products Drive Us Crazy and How to Restore the Sanity, Sams, 1st edition, March 23, 1999

[11] Pacific Northwest National Laboratory (PNNL), "Cognitive Informatics research and development in human information interaction", July 1st, 2008.

[12] Michael Newman, "Bob is dead; long live Bob". Pittsburgh Post-Gazette. May 23rd, 1999.

[13] Frits H. Post, and Gregory M. Nielson, and Georges-Pierre Bonneaum, "Data Visualization: The State of the Art", Springer, 1st edition, December 2002, 4.0 4.1 4.2 4.3 4.

Digital Art

A New Form of Artistic Expression

Gong Huilan
Art & Design Department
Zhongshan Polytechnic
Zhongshan, China
Email:ghl2000@tom.com

Abstract—**By discussing the integration of art and digital technology, the author elaborates a new art form--- "Digital Art" which produced by the integration; In addition, the author also discusses the status of digital art in art and design and explores this new art form from a dialectical perspective.**

Key words-Digital; Digital technology; Integrative; Digital Art

I. INTRODUCTION

The concept of "Digital" is developed with the emergence of computer technology, because the working principle of a computer is digital coding, encoding and decoding a variety of information. Nowadays, "Digital Technology" has been applied to our lives in every aspect. With the development of digital technology, the art form is moving towards the direction of digitalization, and thus a new form of artistic expression--- "digital art" emerged.

There is no need to argue if "Digital Art" is a form of art or not because art form is diversified, and it is enriched gradually with the changing times. If there is no film and no television, film art or television art would not be developed; likewise, it is reasonable that digital art is developed with digital technology, so discussing its "birth" would be meaningless. Whereas it is meaningful to discuss the integration of digital technology and Digital Art, and its status in art and design. Discussing these two issues will enable us to correctly understand the development trend of Digital Art and to guide the diversified development of artistic creation. The following discussion is dedicated to explore the integration of digital art and digital technology, as well as its status in art and design.

II. THE INTEGRATION OF ART AND DIGITAL TECHNOLOGY

In terms of human senses, art can be divided into: visual arts (painting, sculpture and architecture), auditory arts(music), audio-visual arts(dance, drama, opera, film and television)and ideological art(poetry and literature),etc. While what choices are these art forms facing in an era of entering into digital technology? This is a question that every artists needs to think about.

What arts and digital technology will turn into? Bearing this question in mind, we have to analyze the confusions of art. One of the confusions is the relationship between art and science. Is art and science opposite, or interdependent or complementary to each other? Are they belonging to a same culture, or two different cultures? Shelton Richard, an American artist, made a systematic exposition on this issue in his book On Art in Scientific Culture. He pointed out: In Renaissance, art and science are co-existed as two equal friends, since the rise of modern science, they began to separate. Since then, people treated science and art as two separate, unrelated fields, but is it true? In fact, science and art are two interdependent and interrelated systems. Art provides fuel for science and science ensures art to realize. This point of view has been properly applied to modern design, for example, product design is an organic integration of science, technology and art. After making clear the relationship between art and science, the relationship between art and digital technology become clearer too. Digital technology is the product of science and technology development, so art and digital technology will not repel each other like science and art, instead, they can integrate and should integrate. The only issue is that in which way and what extent they can be integrated? This is worthy of exploring.

Take visual arts for example, paintings originally express artists' emotional imagery through canvas and pigments, and now, through digital mapping software, painters can express their emotional imagery as well, such as drawing with (painter) mapping software to create an image effect of watercolor, gouache or oil painting. The only difference is that the image shows a virtual image, which can be copied over and over again. Of course, resorting to virtual imaging technology can turn images into real images and fulfill the same function as paintings, which finally achieve the conversion from "non-material image" to "materialistic image" conversion. However, are these digital paintings the same as traditional paintings? The answer is no, why? The reason ascribes to the different expression material and imaging methods that visual images relied on. To produce an image, digital mapping dependents on the imaging principle of digital technology, while traditional painting is the performance of painters' direct experience and skills, which needs not any other help. However, apart from artists' direct experience and skills, digital paintings have to rely on digital equipment to realize visual images, so people have to understand that digital technology is no substitute for manual techniques, which also make people clearly understand that

Digital Art in nature will not replace any arts, but emerges as a new art form - Digital Art. This is a bit like that camera technology does not replace painting arts but developed into photographic art; film technology does not replace theatrical art but evolved into film arts. Understanding that, one will neither totally repudiate the integration of digital technology and art, nor one-sided advocate the fully integration of digital technology and art, which respond to the above question of in which way and how much art and digital technology can be integrated.

The manner and extent of the integration of art and digital technology can be investigated in two ways, first, to find a way of integration with digital technology according to art form, say the art of painting: digital painting software enables to present various manifestations effect. In this case, its integration with digital technology is much comprehensive. As to the ways of integration, digital technology can be used to process the material to achieve the desired results as a painting reference, or act as a drawing tool to reveal painting effect, or re-process and re-create the traditional paintings. Therefore, the ways of integration are various. If it is an architectural art, the degree of integration with digital technology is relatively weaker compared with painting art, but it does not mean that they can not be integrated. In fact, using digital technology to display three-dimensional architectural design and construction space has become a means of artistic expression of modern architecture, but this means can only be reflected in a certain stage of the architectural art (architectural design stage), which is why the form of art determines the extent and manner of integration. Next, deciding the degree and means of integration should refer to the quality and requirements of the art, such as modern commercial art and design. Due to commercial purposes, the requirements to commercial design is not as high as pure art, producing a commercial poster needs no more, as before, artists' or designers' hand-drawn. At this time, using digital technology to solve the problem would be very efficiently, which at the same time determines the degree of the integration of this design and digital technology. Because of that, to produce a commercial poster, you can rely on digital technology to achieve this purpose. In other words, totally integration is possible. However, if using designated material to create a poster, for instance, by piecing or copper engraving, the outcome would be quite different. Digital technology, in this case, can only produce the effect of posters. Posters, after all, have to rely on traditional means, or on a combination of digital technology (CNC engraving technology), etc, which change the extent and means of integration fundamentally.

III. The Status of Digital Art in Art and Design

The integration of digital technology and art creates a new art form - Digital Art. Digital Art is a comprehensive art which essentially fuses various art forms, especially in modern art design, the role played by digital arts is becoming more and more important, from the advertising graphic design, architecture interior design, fashion design to animation film design, digital art and design is everywhere. People began to rely on this new art form, and controversy caused by it rose too. However, it is an indisputable fact that digital art enjoys popular support. And digital art makes art no longer the artist's patents, for it allows ordinary people to enjoy artists' joy of creation, which perhaps is the reason why digital art can be widely accepted and used. As a form of art, what role digital art performs in modern art and design? Questions like that are often answered in an extreme way, sometimes its function is exaggerated and sometimes belittled, why is it so? This is because people don't look at its nature. In fact, art develops itself all the time, especially its' artistic form of expression and techniques which are changing and enriching with the progress of science. When a new art form emerges, exaggerating or belittle its role is very common and inevitable. From the current application in art and design, Digital Art does hold a curial role, or we may say, play an indispensable role. But from the perspective of the nature of art forms - "a meaningful form performance with good taste" to regard Digital Art's function in art and design, we can thoroughly comprehend the role that digital art plays in contemporary artistic creation process, as well as digital art itself, which is a new art form and a new means of artistic expression. Art, fundamentally, lies in the emotions and ideas of artistic expression, lies in the artist's ideology; therefore, the significance of artistic thought comes first, it always has effect on forms regardless of art forms.

By exploring the integration of art and digital technology, we can perceive that digital art is a new form of artistic expression and a combination of science and art. By discussing the status of digital art in art and design, we can adopt a dialectical view in inspecting the role of today's Digital Art in art and design, which enables us to understand how various art forms serve arts and design activities and not to be fooled by the exterior form.

References

[1] Langer, S., K. Problem of Art. Nanjing, Nanjing Publishing House. March 2006,pp.54-57.

[2] Teng Shourao.The Sociology of Art. Nanjing, Nanjing Publishing House. May 2006,pp.125-134.

[3] Teng Shourao.Aesthetics and Design in the Age of Knowledge Economy. Nanjing, Nanjing Publishing House.Aug 2006,pp.312-334.

Discussion on Display and Exhibition Design of Modern Museums

Zhang Bo
School of Finearts
Nanjing Normal University
Nanjing, China
Email: zb238088@yahoo.com.cn

***Abstract*—this paper attempts to begin with the exhibition design, focuses on the discussion of modern museums and puts forward the necessity of the design and scientific structured approaches.**

Keywords- Modern museums; Display; Exhibition; Computer Items

I. Introduction

The space of every museum is the totally new untried experience for designers. They not only have to act as the director, they also should have rigorous and rigid perspectives as well as bold and unruly ideology. They should decorate the space with novelty and stunt, think over the details for many times, and express their life experience via the art literacy constructed by morality and cultural belief. In this way, it gives people visual pleasure as well as life inspiration. Consequently, those who have experienced it would sense space artworks of modern museums characteristic of humanity, artistic quality and socialization. (See picture 1)

II. Basic Rationale and Objectives

A. Basic Rationale

According to the conception of Mass Media, the design should focus on audiences and cover the brand new decoration means with the emphasis on the mainstream thoughts of subjectivation and individuation. The design should also fully present us the core information of museum exhibition features by means of multi-segment and manifold. In this way, the exhibition hall will be filled with the experience of scientific scene where stories are told slowly and emotions are bound to be affected by the situation. Besides, the exhibition hall will become a vivid exhibition context and communication platform where people would be eager to have a try. The exhibition design of museums deepens the rationale to achieve the target effect ultimately.

B. Design Objectives

- The exhibition design aims to perfect the combination of content and form and integrate the ideological, scientific, artistic and intellectual elements. In this way, the exhibition design would become the window of exhibiting, promoting and studying museum collections and a base for science education in order to meet the multi-directional, multi-level needs of audiences.
- The exhibition design attempts to grasp and express the theme, represent the resource advantages of museums precisely, and manifest the urban transformation and advantageous antiquity resources fully in the river of the long history. It employs various pattern of manifestation and makes use of new technology and new means to compound the effect. In doing so, it aims to achieve the multi-sensory design on the interactive level and exemplify it in order to inform and inspire the future generation.
- The exhibition design aims to coordinate the overall form and partial details; it aims to integrate the historical significance and brilliance and construct a dignified moment by means of fine-tuning; it aims to make eternal pictures where it tells of the formation and the making of excellent antiquity as well as implied stories, by means of our thorough understanding. It would move above or underneath and it would be smooth or rough with proper selection and representation so that it make us set foot on the holy palace filled with the sense of history, technology, environmental protection and artistic elements.
- The decoration aims to coordinate the decoration signs and the overall layout, and the decoration vocabulary and functions. In this way, when people enter the museum, they would sense the visual elements of strong local culture and the thick breath filled with civilization sediment.
- The exhibition design of items such as scenes and models aims to coordinate the overall decoration form and style. It also aims to optimize individual item and equipment adoption, material and craft, investment and cost as well as artistic effect.

III. Design principles and Bright Spot

The exhibition means of modern art museum are put forward on the basis of the construction tenets of museums and they cover several aspects as follows.

Reflect human-centric principle precisely with priority put in all visitors and staff. The selection of decoration materials conforms to requirements such as functionalized orientation, green design, energy saving, environmental protection. Lighting and illumination should be accompanied with intelligent control and air regulation. The construction technology of decoration furnishing should provide the descendants with the convenience of maintenance and renovation and the possibility of recycling equipment and materials.

Promote the technological content. Modernize the manifestation and technology of exhibition art. Exhibit all artworks of the exhibition hall in forms of high technology, such as special photo-effect, fiber optics, computer programming, and multi-media technology illumination adoption. Make use of computer intelligent identification system, response system, explanation system and management system of diverting tourists.

Make use of new system design theory and pursue the unity and wholeness of artistic features. In order to achieve the coordination of indoor environment and every level of indoor artistic styles, adopt the system control method of motif language and the material texture and eliminate all unharmonious factors and obtain variety from the simple scene.

Combine the eternity or timeliness of the craft aspect. In term of things that will be used for a long time, perfect the optimization of craft materials when taking into account the success at one time. Renewable considerations, perfect and forward-looking design attribute to the plausibility and reasonability of rebuilding and repairing.

Conception of "Highlights" in the display

a) Dig from culture ----- macro- and microelements. Reflect the broad culture in space building and scene designing. For example, scenes of civic cultures can be extracted from life. They are lively and kind. The audiences feel like going through time and space and are deeply moved.

b) Combine art with technology perfectly-----Scene representation, multimedia panorama sandbox, combination of life scenes and multimedia projection reflect the shocking effects of technologies and art. The graph and writing list and relic protection layout, etc can be made into audio, optical and electronic image versions. The visitors can touch the flashing image region with hand and background commentaries will appear.

c) Enhance the display effects with the latest display technologies-----Utilize the latest display technologies like HOLO screen, virtual imaging, phantom imaging, multimedia and virtual effects, etc, which are totally different from the mono static display form.

d) Enhance audience participation, interaction and interest ----- such as the interaction of dialects. The audience says some specific words. The computer captures voice messages and turns out the corresponding dialect. Through the voice contrast of the left and right channels, the audience can directly feel the characteristics of dialects. Also there are interactive games, interactive multimedia query systems, etc.

e) Exquisite craftsmanship ----- Differ from ordinary exhibition halls by details. The model recovery should reflect the beauty of technology and craftsmanship.

IV. THREE THE EXPRESSION AND CREATION OF THE CONTENT AND FORM

Design of venues is the soul for the design of the museum display. Every museum should have its own individualized style for displaying as well as a specific orientation of art. The design lays the emphasis on that the museum which is the base of cultural education should be different with other surrounding areas, and it should have an initialized characteristic of the region. The display techniques should breakthrough previous singular and static means or the conventional means for assisting the display in other museums. Each shining chapter for the stories in the museum should be fully revealed with a multi-media mean which is tridimensional, interactional and individualized.

Therefore, the main purpose of the display content should be understood in the design of the display. The spiritual connotation of the display should always run through the design of the display space. Through modeling different space nodes to make the spectators immerse in the experience, and then they will step on the experience tour with urgent emotion.

A. Space planning and design

The space is the basic source of the design. The original idea should be simple while facing an unprocessed space in the museum. The space should be regarded as a kind of place where people could express their behaviors and thoughts, and study from the simple framework that inside and outside of the room. The pencil should be used for expressing a jumping concept and thread that based on pursuing the aura of the space. The scheme is not certain at this time. Any petty or fragmentary spark is enough to lead the overall scheme. Pay attention to create the artistic conception at here. Different spaces could arouse different emotions, and different emotions will lead the users. Thereafter the basic pictures should be divided through the dynamic and the vocabulary of constitution mode that could enrich the space with a direction of function. It should be mentioned that the functions of independent spaces with different individualities should be linked together at this stage. The designer could imagine that he/she is just in the space and wandering along the thread. He/She could connect different components for all of the spaces with his/her own moving line. At last, a harmonious layout could be achieved. The next work that should be carried on is to make the direction of the style clear, reflect the individuality, and perfect the refining and the integration of the scheme. All of the decorations of the surface, the water-air-electric power, the psychology and etc. should be seriously considered. The stage for adjusting the scheme is the last stage. Further verification could be achieved through exchanging the views and ideas with the first party and fellowtraders. Of course, if there is a more mature idea that obtained at this time, it should be blended into the scheme bravely.

B. Design of sequence hall

Sequence Hall is not only a place for the gathering and separation of visitors, but also the identity of museum, which needs us to express the history in it through professional vocabulary and practices.

Sequence Hall is an important transport hub, which should take simplicity, generosity and clarifying the aim and theme as the main principle. The open and square layout is beneficial for the smoothness of audience sight line. In order to allow the audience has a profound overall impression on the museum at the latest time from the aspect of time and space, the design of fixture should be enough to take the space of sequence hall with representativeness, focus on the expression of shocking, and combine with the characters of sequence hall to elaborate the theme. The overall of ground decoration should make use of stone materials, which has mellow and profound temperament, and is in line with the atmosphere demand of cultural venues.

C. Design of exhibition hall

a) Follow and embody the requirements of "four modernizations".

- Firstly, art design can be seen everywhere, uphold the principle of combining realism and romanticism together, use a variety of artistic means, add aesthetic sentiment in the subject matter, and guide the audience in the loved situation step by step.
- Secondly, it's the user-friendly display design. The design of exhibition lines should fully take into account the psychological and physiological responses of audience, and intend to create circuitous twists of direction; the design of exhibition project should consider about the density and urgency of rhythm as well as the appropriation and change of information amount; increase the audience's "initiative" and avoid their visual forcing and aversion of information congestion; especially design the recreation space aimed at the much longer exhibition line, where audience can perform together with the guide (Actor) for physical and mental regulation; the necessary recreational facilities and afforestation are manifestation of humanity, and will also become an integral part of the landscape display.
- Thirdly, in terms of personalization, the design should make full use of the specific information symbols of museum, and highlight the regional characteristics through various forms of artistic expression.
- Finally, design turns to high-tech. The use of high technology should not go too far. The means of high-tech are primarily used in the aspects of materials, technology, lighting, sound and multimedia, so we should take the appropriation, and distinct the important one and unimportant one. For example, we can demonstrate the process of earthquakes in the Museum of Geological Environment through the technologies like sound, light and electricity, etc., explain to audience the changing reason of geologic hazards like land subsidence, active faults, ground collapse, etc., and deepen the audience's impression on the scenes of geological disasters that are hard to see through personal experienced landscape simulation in order to strengthen the audience's awareness of environmental protection.

b) Change visit to tour:

The entire display design should break the traditional model of missions and instilling, and focus on creating pretty space and the interactive exhibition project. Audience can visit everywhere in the exhibition project, or even interact with the person and display in the project; it contains both display projects and reality show performances, which is hard to tell the true or false and has great fun.

For example, when designing a streetscape rehabilitation, in order to show the marriage customs, festivals, temple scenes, and other unique local characteristics, we need to exhibit the details of life in front of visitors, in which part of the exhibition project involve the audience, and combine with many interactive dialects and other intangible cultural heritage, so that the visitors can feel more three-dimensional and more intimate about this city. Virtual scene should be combined with the multi-media scene, and make full use of sound, lighting, electronic technology to create a real-life atmosphere, and enhance the intellectual, cultural and interesting features of display. Through the display of arch, Folk Street, etc., we can provide a deduction platform for audience to understand the content of folklore, bring audience the immersive experience, increase the telepresence of exhibition and meet people's tracing back of history.

c) Avoid the average use of force, focus on importance, promote the entire area by a spot, set appropriate visual center in each unit and make them into landscape and attractions enhance the audience's excitement and deeper impression.

d) Take the right design style, express the historical and geographical content, shape, material, texture, color, layout, production processes, etc. with modern techniques, which are all rich in flavor of times.

Generally maintain the series and uniformity of style in each unit to (for example, the small sequence hall in each unit can all use the common abstract symbol, meanwhile, highlight the unit's personality with the thematic abstract innovative image on the contents.

D. Computer-assisted design

At an early stage of the design, the work of dividing different areas of the space in the museum could be done with the help of computer. Analyze the architectural drawings with AUTO CAD software. The analysis includes the analysis of the space function, analysis of lighting, analysis of items, analysis of water-electric-heating-ventilation and etc.

The SketchUp software could be used to draw the initial sketch. The means of body interspersion are mainly used to divide the space to obtain a basic understanding of the space. Tape Measure could be used for measuring the data of the target to obtain a precise numerical value and make sure that a reasonable guarantee for the division of the space could be obtained on the basis of an optimized design. A simple picture could also be used for previewing the final result.

The 3D MAX software is very important to design the concept effect for the space in the museum. There are mainly three steps:

- Model building—nowadays all kinds of required effects could be basically revealed by the software. Wall faces, the bump change of wall, include the complicated exhibition furniture, exhibits, plants, and even the person in the space, all of above could use the model to finish the initial building. The POLY model building is mainly used for some complicated models.
- Cover material—the setting of the material could be edited with Material Editor.
- Embellishment of the lights—Vray is a kind of extension software. It could be used for embellishing the picture through setting the lights and regulating the parameters.
-

V. UTILIZATION OF TECHNICAL DISPLAY ITEMS

Acoustics, optics and electronics, computer multimedia and devices have become the essential techniques of displays in museums. But it needs to have focuses, meanings and should be the right choice. All hi-tech techniques should serve the display theme and should never be distracting. Here some representative display items are picked for analysis:

A. Introduction of the e-book display item:

Authentic exhibits can only be displayed in a glass case with one page open. The approach breaks this restriction and the audiences are closer to contents of the exhibits. They are projected virtual books and audiences just need to wave their hands on the projected pages to turn the pages. The whole system is mainly controlled by the high-precision projection system. The programmable control system manages the power supply, inductive source and projector of the entire display system. (See Figure 2)

B. Introduction about the display item of roaming control with simulated controls:

The interactive project will match the images of the projection screen and produce a certain degree of shaking according to the scenes designed by the computer. Through the simulated flights (control the handle), the audiences can adjust gravity and get the feeling of roaming in the city with the combination of winds and sounds. (See Figure 3)

C. Introduction about the display item of fog screen imaging technologies:

The fog screen imaging method is similar to projecting but no screens are needed. "Fog screen" mainly applies the particles that exist in the air and let the static or dynamic images show on a very thin air layer. It brings unprecedented visual experiences to audiences and can substitute traditional writings and images. Plus, it can play image materials like slides and videos. Fog-screen gas is non-toxic and odorless. It has no pollution and won't cause any harm to the human bodies. The fog won't stay in the space and will immediately disappear when the power is off. (See Figure 4)

D. Introduction about the display item of holographic projection and e-guestbook for audiences:

Above the central showcase is the slowly moving coil ball suspending in the air produced by three-dimensional digital projection technologies. The holographic and transparent projection system above the showcase has the function of real space imaging with vivid color, high contrast and clarity and strong feeling space and transparence. Entities can be combined in and the combination of image and entity can be realized.

The display item has a touch screen computer and a camera. The audience may not only enter his opinions about the item but also offer comments for the museum and display. This is totally different from the traditional paper guestbook. Exhibit structure: This display item is composed by the holographic system above the showcase, computer, touch screen, camera and microphone. (See Figure 5)

E. Introduction about the display item of e-charts:

This chart applies luminescence film technologies and microcomputer control programs. Right now, luminescence film technologies are seldom used in domestic museums. Since luminescence can be made into different colors, we set different colors of luminescence on the same luminescence film and control the lighting of the different colors of luminescence by the controller. Therefore, the changes of each relic in different eras can be displayed on the same background image. (See Figure 6)

F. Introduction about the display item of the control system of universe creation:

This display item is carried out with virtual demonstration techniques.

Prelude 1, universe in chaos

Prelude 2, the creation of universe: The laser simulates the flying particles. The optical and electronic balls simulate the shocking scenes of exploding particles and simulate the shocking sound effects of exploding particles.

Prelude 3, the generation of universe: The laser simulates particle collisions. The optical and electronic balls simulate the early nebulas from particle aggregation. The projector shows the generation and expansion of nebulas on the upper and lower screen. The particle sound effects are simulated.

Prelude 4, the present universe: The laser simulates all kinds of cosmic particles. The projector shows constellations and galaxies on the upper and lower screen. The video broadcasts the science knowledge of the Big Bang and optical fiber shows the common constellation composition. (See Figure 7 and 8)

G. *Exhibition project Description of Holo Screen Technology:*

Features focused on a layer of advanced holographic technology film, which can display the images from the 35 ° projection, and all other light is ignored. The result is that even in the bright environment, the image is especially bright and clear. Semi-transparent display allows viewers to watch the image on the screen, and can also see through the screen, which is very useful for the exhibition, for the screen has become a part of the museum exhibition items.

Advantages and applications:

Semi-transparent holographic screen; can effectively resist the surrounding lights; screen sizes reach up to 60 "; display video, DVD, TV, computers; all is compatible with in 35 ° projection angle. (See Figure 9)

The display design of museums is a process of continual deepening, continual grinding, continual picking and continual upgrading. It must combine the different elements like history, culture, science, materials and the construction technology level. And it offers the audiences with displays that are worth experiencing with high qualification, level and information amount.

References

[1] Cai Qin, "A New Field of Museology", Zhejiang People's Publishing House, August, 2003.

[2] Yan Jianqiang, "theory and practice of the museum", Zhejiang Education Publishing House, September, 1998.

[3] Chen Shuonan, Shen Yonghua, "Basic Knowledge of Modern Museology Q & A",Museum of Natural Science Society of China, 2000.

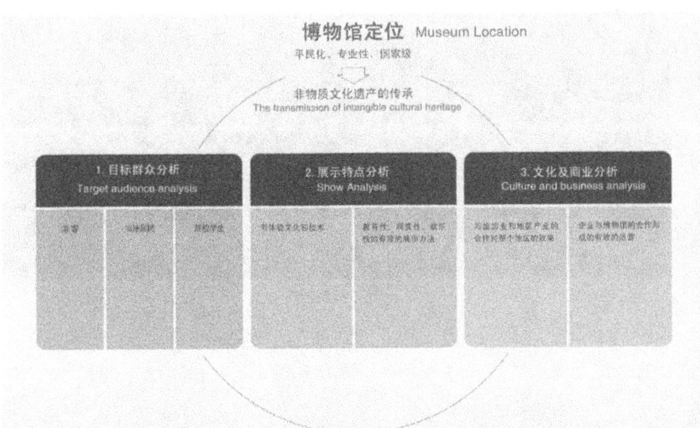

Figure 1

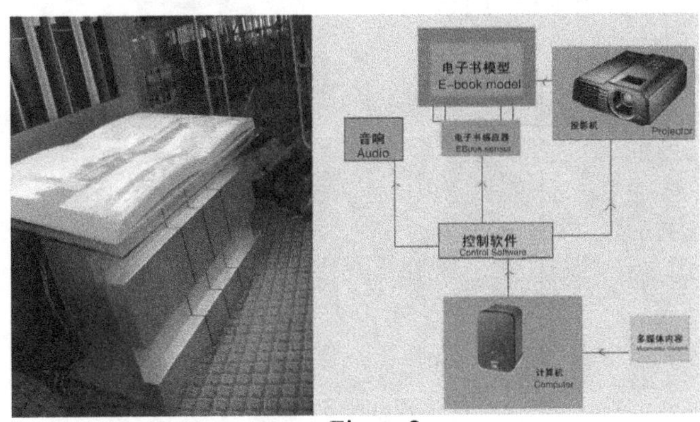

Figure 2

Figure3

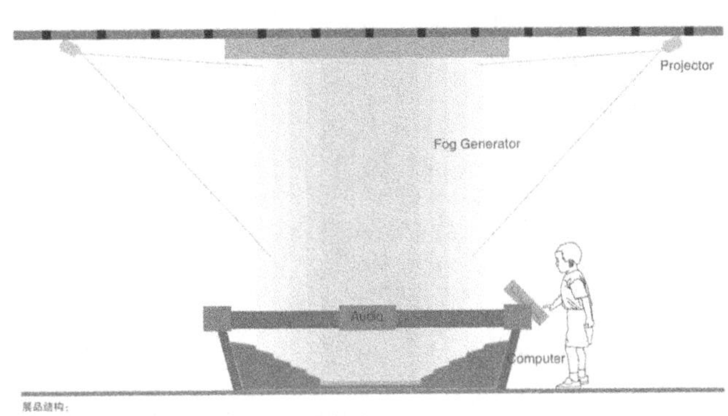

Figure 4

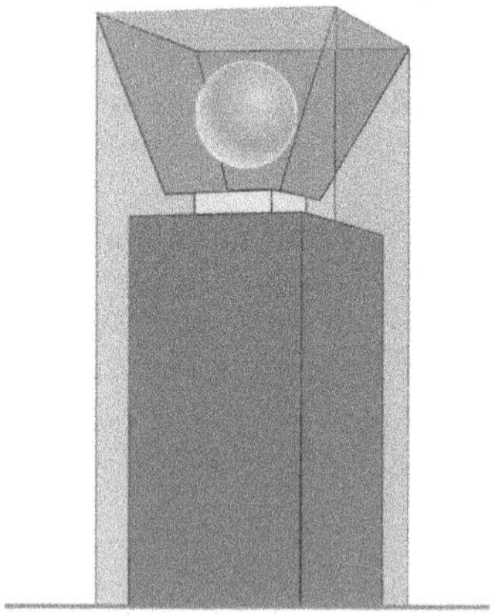

Figure5

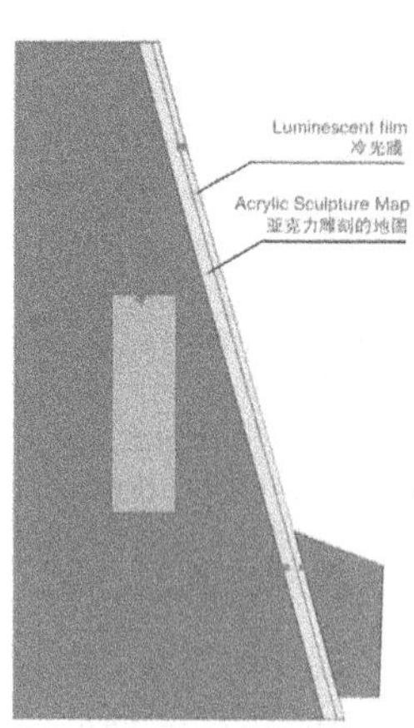

Figure6

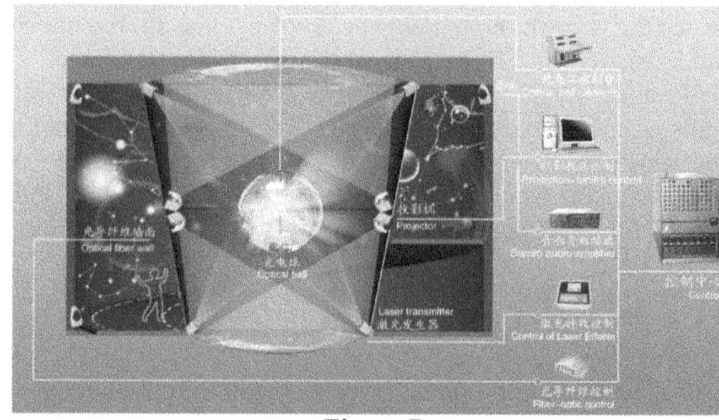

Figure 7

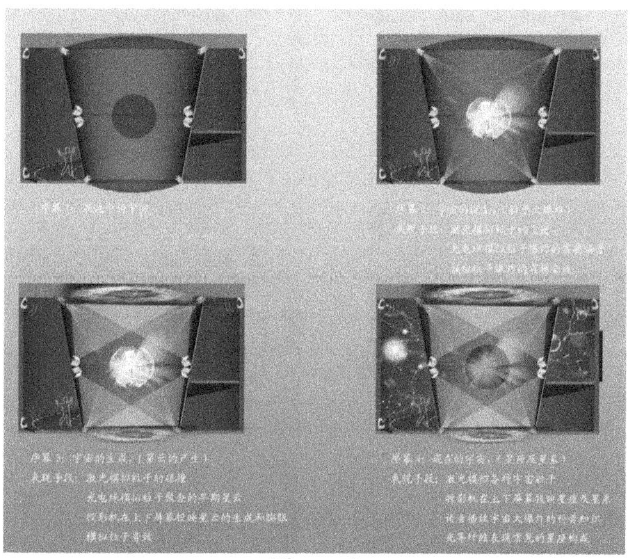

Figure 8

Figure 9

Ecological Landscaping Optimization Design

Zhou Xiao
Modern Art College Environmental Design Office
JiangHan University
Wuhan, China
Email: 455204044@qq.com

Zhou Yuanyuan
Modern Art College Environmental Design Office
JiangHan University
Wuhan, China
Email: 3217066@qq.com

***Abstract*—how to optimize the design of ecological landscape is a prominent question at the moment. The ecological landscape does not service for a single background, but the complex multiple ecological environments in nature. So the non-uniqueness design of landscape ecology, requests us to optimize the landscape ecology and to be systematic. The obtaining of ecological environment is not simply planting some more trees.Landscape planning and design essentially should be ecological planning and design on land and outdoor space.**

Keywords-analysis;ecologica;,landscape;optimization design

I. INTRODUCTION

When we are proud of the 2008 Olympic Games, we are amazed of the wonders of Olympic constructions, we dispute about the new materials of Water Cube, while our vision is not limited to the surroundings. The most distinguished feature of Britain's "the Garden of Eden" project is nothing better than the two cellular structured round roof which looks like a cut half grapefruit covering on the ground. There are all kinds of ecological community, huge greenhouses in which growing many tropical rainforest plants, even the Mediterranean farmland. The Water Cube uses the same film material ETFE as the Garden of Eden does. The design of "the Garden of Eden" apparently has an extremely futuristic and organic style, and its connotation is environmental protection. "Although it is a cliché, one of our goals is to change the world anyway." The founder Schmidt said "We are here to help people realize that if they can do a little bit small, multiplied by 60 billion, the result is great achievement." Urban development damages the ecological environment which affects our daily life to a certain extent, so how to optimize the design of ecological landscape is a prominent question at the moment.

The ecological landscape does not service for a single background, but the complex multiple ecological environments in nature. So the non-uniqueness design of landscape ecology, requests us to optimize the landscape ecology and to be systematic. At present, the requirements of ecological landscape optimization develops with the process of all kinds of ecological architecture practice all over the world, in order to make a harmonious development of the ecological city construction.

II. T HE ANALYSIS OF ECOLOGICAL ARCHITECTURE

In architecture, professor Burkhart from Kassel University of Technology in Germany criticizes the actions which interpret ecological thinking in formal and just come to the surface. He points out the concept first came into being in the 70s of last century, and gradually turned into a business trap. To build such an ecological apartment, one should be rich and self-contained; he should live alone in the remote desolate countryside but at the same time owns all the advanced technical equipments. He should be a most radical contemporary scientific explorer, but also has sufficient artistic accomplishment, in order to make his apartment provide reporters of the architecture magazine an ecological form which they can tell at a glance. When this idea spread to Germany, it satisfied the need of those farmers who were worrying about the selling of the traditional countryside apartment. So those with "ecological" brains began to build showy small commodity houses, such as with a greening restroom in it and so on, to make business profit. In addition, there were some so-called luxurious "ecological architectures". In order to build them, people use many times of money than a common same functioning house, consume more energy and resources, pay some more for the maintenance fee. This kind of architectures focuses too much attention on appearance, so professor Burkhart call them "post-modern ecological architecture". When British architect N. Frost talks about the development of construction which based on renewable solar energy, he sharply points out "it is definitely not for fashion, but for living." All these show ecological architecture is not a simple concept of space forms. In fact, like the process of operating ecological city, one of the important features of ecological architecture is its internality. That is, the key to ecological architecture lies in the internal operation system has changed.

From the existing building system transit to the ecological architecture, there are two paralleling and compatible thinking. The first one is gradual development thinking. According to this, to save energy and resources and reuse of ecological ideology of modern architecture, based in the functional established standards, norms and relevant index to rearrange and correction or update, and based on this new architecture function and principle, application of technology with economy, can achieve the goal of ecological architecture creation. In addition, there is another radical or advanced thinking. It pays more attention to the future pointing of the technology, materials and forms. Take the entirely depended on solar energy "self-sufficient type" architecture for example, the day lighting, ventilated, heating, garbage and sewage processing and daily use of energy consumption all come from the self-equipped solar switching system.

Either the gradual or the radical thinking, the produced ecological environment constructions have their apparent internality. In this respect, they are in the same line with the modern architecture, and the successor of the latter in new century. Therefore, we need not deliberately create a new set of "ecological recognition system" to distinguish the ecological architectures and non-ecological traditional modern ones. If the ecological architecture has some special form, it is caused by its internality. That is the feature the modern architecture owns from the moment it exists. But rising from the traditional modern architecture to today's ecological architecture is a natural and continuous concept circulation. Be similar with modern architecture, ecological architecture also emphasizes the space efficiency. But it is not trying to establish a set of index system of independent space, but more closely with the environment and people's life itself. If we say modern architecture focuses on people's life from the angle of space, then ecological architecture from a more macroscopic environment and resources angle. Take the sunshine for example; modern architecture considers it mainly from the hygiene respect, the reasonable sunshine space problem, but on the foundation of the formal one, ecological architecture considers the problem of obtaining and using the resource and energy. To achieve this, the materials and technology which are suitable to certain social economic level and can help to accomplish the goal should be first selection. Therefore, ecological architecture not only focuses on the efficiency of space, but focuses more on the efficiency of environment and resources.

III. THE ANALYSIS OF ECOLOGICAL CITY

The obtaining of ecological environment is not simply planting some more trees. Such as according to city or area resources reasonably determine the energy consumption of the standard, the water supply, treatment and recycling of waste produce standard, and the recovery and processing standards and methods of land use and transportation mode, standard etc, they are very crucial questions. Some cities can solve these problems within its scope, while others need to do it in a wider range. This is because in the sense of city, the characteristics and environment bearing capacity is one of the key factors which decide the reasonable size of a city, industrial structure, social space and economic space. Therefore, to promote the ecological city construction, the first thing should be renewing the concept and optimizing environment design.

Here the so-called resources is a generalized understanding, which includes three major parts, they are social resource, material resource and human resource. Social resources include social structure, organization, culture, history, tradition, the economic basis, knowledge, technology, system, etc; material resources include natural resources and environment, city, construction, etc; but human resources include such as population, labor, creativity, etc. The three kinds of resources overlapping in a certain extent, the sustainable development of any kind of resources depends on human's behavior and affect the living condition of the other resources. The scale of an urban ecosystem unit is decided by formation of human resources in a certain level. It can be a city, a community, and also can be an urban agglomeration. Anyway, it should have the following characteristics: 1. Local environmental creation, namely, it should have certain dependable local economic resources, such as enterprise, mining, commercial outlets etc. which can solve the employment problem. This resource forms a series of attractive culture, industrial center, also constitutes an important basis of future development.2. In the process from exploring to using, the urban resources implement the maximum saving principle, reuse principle and recycle principle, and the degree and strength of the explored and used material resources should be adapt to the recycle ability of the used resources, to make the recycling of the waste gas, water and garbage solved on spot. Therefore, the scale of urban ecological unit is proportional to the resources it used and its ability to protect resources. Its rationality of economic scale is decided by making maximum use of the available limited human resources, and making the wasted resources and environment pollution and destruction to the minimum. But it is the rationality of the society in scale to achieve these goals can provide the optimal system framework, knowledge framework and industrial framework.3. Establish a management system which adapt to the operating system of an ecological city, so that from now on we can lead every aspect of the development process to the sustainable development track.

IV. THE ANALYSIS OF ECOLOGICAL LANDSCAPE

Before If we interpret landscape planning as an implementing process which consists of analyzing anything concerning with outdoor space and land problem, putting forward methods to solve the problems and supervising it, then the duty of landscape planning designers is to help human to make people, construction, community, city and human's life living harmony with the earth. Landscape planning and design essentially should be ecological planning and design on land and outdoor space. Ecological principle is the core of Landscape Architecture. From a deeper sense, landscape planning and design is a human ecology system one. It is a minimum design by utmost using the natural power, also a regenerative design based on the renewal self-ability of the natural system, which to change the input and emission mode of the existing linear and energy logistics, build a circulation among the source, consumption center and flow. The creation of the landscape is a kind of sustainable landscape.

The general principles which urban landscape planning and design should follow:

A. Local principle.

That means design should have its root in the where it was. As to any design problem, what the designer first to think about should be where they are, what the nature permit them to do and what the nature can help them to do. We are always amazed at the layout of far from the madding crowd Chinese villages and the lovely folk houses. In fact, most of them were not created by designers, but the locals who live in harmony with nature in the long experience with the local places and based on their deep understanding of nature, design these reactionary works. Following the principles mainly showed as respecting traditional culture and local knowledge, be adapt to the natural process, as well as places of local materials, plant and materials

B. Protection and conservation of natural resources.

The earth's natural resources can be divided into renewable resources such as water, forest, animals, etc. and non-renewable resources such as oil, coal, etc. To achieve the sustainable development of human survival environment, it is necessary to protect and reserve the non-renewable resources. Even the renewable resources have limited regenerative capacity, so we should adopt a reserving and profit-making but not a killing the goose lays the golden eggs way to use them.

C. To respect nature and show nature.

The modern urban residents is becoming farther and farther from the nature. Nature elements and natural process is gradually invisible. The skyline of the long-distanced mountains, the horizon at the foot can be almost classified into abstract nouns. As the natural process disappeared from the public eyes in the traditional design, the supporting system of city life is also often hidden. Sewage treatment works, refuse landfill, power station and transformer substation are all consciously hidden as ugly objects. With the disappearing of the natural landscape and the process of it, the structure and process of the urban life supporting system, people show no cares of the present condition and future, not to say the cares of ecological environment and control the daily activities. Therefore, to make everyone involved in design, care environment, the natural process must be recovered. Let the townsman feel the freshet after the rain, the surface runoff stream into ponds; let them feel the natural wind through wavering leaves; let them notice the change of season by seeing flowers blooming and fading; let them see the process of nature decaying and degrading from the withering and growing of the leaves. In Thayer's word, landscape is a kind of language showing ecology.

In addition to the above principles, landscape ecological planning design also stresses that everyone is a stylist; everyone should be involved in design process. Landscape ecological planning is not only cooperation between human and nature, but also a process of cooperation among the humans. Traditional design emphasizes the individual creation of designers and regarded design as a pure and elegant art process; but modern landscape ecological planning stresses that everyone is a designer, because everyone is making decisions to his life and future, all these will directly influence the common future of him and others. From the daily means of transportation to choosing furniture, decorate materials, water using, selecting food, processing of garbage even the use of packaging, they are all ecological problems, because they have a profound influence on the whole community and environmental health. Everyone's decision should be the content of the landscape ecological planning design, so it is involved in all daily activities of the individuals. To the professional designers, this means their work must head for the public and society, and contains public knowledge in design. At the same time, the designers should optimize their ecological design ideas and purposes, to be accepted by the public, thus to become everyone's design and everyone's behavior.

V. THE INTEGRATION OF ECOLOGICAL ARCHITECTURE CITY AND LANDSCAPE

After People who have radical ideas describe the concept of ecological city as "eco-topian" compared with "Utopian" thought, to express the pioneering and originality of this concept. They think there are great discrepancies between the ecological city and the existing one and it's difficult to switch from the latter condition to the formal one, so it is necessary to build a totally new city on the foundation of ecological thinking. Although from reality point of view to achieve "eco-topian" ideal is still very far away, the existing technical development directs a more practical way for us. That is to build ecological city by building ecological community. At this level, "the green movement" proposes some principle can be more inspiring and practical.

A. Community self-supporting.

This principle is influenced by Schumacher's "intermediate technology" and "appropriate technology" theory. At present, it is more economical to achieve certain degrees ecological balance in community rather than achieve this goal in a whole city. For example, to set up the wastewater treatment facilities in the community can have advantages such as recycling use the water resources and comprehensive utilization, investment saving and easy to generalize. The community also has the scale of energy saving measures in the same benefits. Furthermore, to find a balance between living and employment not only can promote the fully economic development of the community and social, form a stable social structure, but also save a lot of investment which using for urban transport construction. Community self-supporting may prompt city to ecological step by step.

B. Be harmony with nature.

As "phototropism" to insects, human also have the nature-approaching instinct, and it should be satisfied during the creation and decoration of the living environment.

C. Decent realization and diversity.

The huge developing city and centralization will inevitably lead space environment to be compressive, the social efficiency, space efficiency oriented policy made this, making a result that all the big cities showed the same mechanized character. The dispersed living condition helps to achieve the goal of diversified environment.

VI. CONCLUSION

Anyhow, it is essential to reform the existing way of resources using and the management system in order to build an ecological city and optimize ecological landscape design. There are various ways to achieve this. Every city should select the appropriate constructing way based on its own resources and environment construction. Further, in the community level the city should form an ecological development strategy which in accordance with its characteristic, and select the appropriate ecological technique and method, to carry out the ecological principles absorbed into every single part of city and landscape design.

ACKNOWLEDGMENT

The completion of the article depends on Optimization Design of International Conference on (ICOD) of the upcoming. Also thanked Jianghan University Institute of Modern Art, Environmental Art Design my colleagues for their kind assistance, as well as Digital Technology Co., Ltd. Shanghai Heng Man Virtual Reality laboratory construction program.

REFERENCES

[1] Yang Donghui, China needs to Landscape Design - - the practice of landscape architecture from the United States look at our landscape garden, Chinese garden, 2000.

[2] Liu Bin-yi, the modern landscape planning and design [M] (first edition, second edition), Southeast University Press, 1999,2005.

[3] Wu Fei. Garden undergraduate professional training model Analysis [J]. Anhui Agricultural Science, 2007,35.

[4] Wu Liangyong. Habitat Environmental Science and Landscape Architecture Education [J]. Chinese garden, 2004, (1), 2-5.

[5] [Landscape Gardner. Chinese Construction Industry Press, 2007].

[6] [Wang Yujing. Plants Scenery of City Park. Liaoning Science and Technology Press. 2003.].

[7] [Simons.J.O. Landscape Design. China Construction Industry Press,2000.].

Ecological Water Landscaps Practice on Campus Landscape

Yin Jing
Shandong Jiaotong University
Jinan, China
Email: yjing78175@163.com

***Abstract*—In this article, the exploration on water landscape has been carried out with ecological collecting and utilizing water landscape pattern in the condition of the natural landform. The pattern has resolved the problems that the rainwater existing, harbor, water quality keeping, and comprehensive utilization. It also provides a further designing platform that the water can be used in interior and outside landscape, in temperature cooling and so on. During this process, I hope the ecological design thoughts may be put into the whole progress of campus landscape of designing, building, running, maintain and managing. I hope it may hoist the quality of campus landscape. I also hope it may provide some worthy utilizition experiences of landscape water through my ecological water landscape prictice.**

Keywords-water landscape; ecology; collect; purify; Utilization

I. Introduction

The scarce of water resource already becomes the worldwide problem. When the quantity of water source has already no way to be increased with tradition developing ways, collecting and using rainwater for landscape becomes one kind of ways both economy and pragmatic. The water for landscape has the multiple functions that relieving water resource crisis, improving an ecological environment by groundwater penetration, cutting down and slowing down rainwater, bringing down the rain flood disaster; providing water for inner landscape of building etc. During the collecting and fully utilizing water for landscape of Shandong Jiaotong University, some beneficial trials have been carried out and some fundamental ecological principles are sufficiently showed.

II. Collecting and Purring Water

A. 1.1Collecting and Harboring

Geology of Shandong Jiaotong University is for gabbro. By the long efflorescence the surface configuration appears stone around grit. And then it becomes a pit by mining grit. The rubbish has been dumped in it. Here the environment is very bad during initial stage. After the rubbish is cleared up, the physical feature appears high in south , west and east, low in north. It becomes the concave type physical feature naturally. The ground of No.1 Teaching Building is the highest place, the southern pond which is a big pit formed by digging the grit is minimum place.

The rainwater often discharges to lay the pipeline under the ground. In this project, it is organized through the shallow ridges with vegetation on the surface which is the passage road the flood and water away combining with the natural caves. And it is organized to the southern low-lying pond in the south. We reserve its low-lying physical feature of the place, with local swamp and expand appropriately the pond.

Collecting and storing the rainwater with natural condition that brings about a another problem that the used water yield is far less than collecting and storing yield. Therefore, the water function is sufficiently developed. The shallow ridges with vegetation being the rainwater discharges are also the stream landscape and the water irrigating channel. The collecting water is for interior and outside landscape, It also has the functions such as defencing fire, cooling tempreture ets. By the technology flowing by itself with transporting pressure, the using water spot and storing water spot are being connected mutually, they form comparatively stable ecological chains.

B. Water Purifying

Water landscape with large area confronts with the problem is how to keep water quality. It is important to adopt proper purification technology. At present, the rainwater treatment adopts much handicraft with the film. Its cost is higher. Some ecological purification ways has done in this project: The technology is simpler; and the purification results in well reduced prices; it is suitable in run, defends, management. Project construction cost and operation cost are low. Water can well loads the external changes. First the water is collected through the earth's surface, and it is organized to draining to the pond, then the water is bumped from the pond with a water pump to the water box constructing in the high platform of the No.1 Teaching Building. The water with high water level in the water box falls again along the steam with the shallow vegetation ridges into the pond. There are some water flavors when the water spraying, drops in the steam increase the chance getting in touch with water and atmosphere which the contents of dissolving oxygen in the water are increases. Water cycles between pond and wet land unceasingly eliminate the pollution effectively and keep the water quality. Water is purified moving back and forth. It becomes flowing water. The campus landscape becomes especially intelligent and elegant.

Rainwater first runoff after the processing such as precipitayion, fliteration, penetration, sorption and living things degradation, it The low greenbelt shallow ridges are lower 20-40cm than the road surface(Fig.1). It can reduce obviously rainwater runoff, dislodging rainwater pollution in initial stage.

Embankment is adopt various the slope filters belt. According to landform, road , hydrophyte, purifying function and landscape, we choose the styles of Embankment such as the sward entering water, ladder shape growing hydrophytes and so on.Through the side slope belt, pollution in rainwater gets degradated and hold-up , thereby the rainwater is purified before it enters the pond.

The plant island is set up in the pond may improve the ability of the pond self-purity. The aquatic vegetation has very good physics, reduces suspension, absorbs large amount of the nutrition salt and reach purifying water quality. Plant floating island provides a necessary space to the various hydrophyte and birds. And it can improve the lake landscape. The wet land is the transition belt between wet and dry land. It is one kind of ecological systems, and has unique ecological structures and functions. It can protect biodiversity and improve the natural environment.

Without concrete structure, some permeable method are adopt, such as laying the stones in the base and in the bank of the pond, planting the trees around the shallow ridges and the pond. With the opening circulation river system, it appears a living water landscape. In low water time, it shows the stone and vegetation landscape. This wet land can realized the water treatment and keep water quality with low cost of handled, run , and defended.

III. Water Landscape of Comprehensive Utilization

A. *Water for Landscape Outside*

The water in the pond is pumped into the box beried in high level of the outside platform through a pump and some steel tubes. It becomes source of a brook. The water in the box is leading to the plant shallow ridges with the function of collecting and purifying. It is planed according to the person route and the brook form and natural campus feature low in north ,height in south. The brook copies natural forms and likes a partner along the student advancing route, hiding or appearing in people's sight.The step stone in the brook across the traffic road. The brook wall is constructed using the distribute stone and the riprap rock of location .The stone are also create the landscape; adjust water speed and streams direction; form minor drop and brook noise; make water produce diffluence etc.The plants of brook is cultivated combining with wild, local plant and the aquatic or wet land .This cuts down the artificially scenery trail. It also acts as organizational line connecting the other campus space. The natural brook has been extracted and reproduced in campus. It increases campus scenery.

With rainwater uneven changing, the pond water yield during four seasons of the year is changeable. In this project the way keeping the water level of the pond does not adopt pouring water into it. The water level is kept according to natural water which creates the natural landscape changing with time and season. The process of water going up and down can make students build up the perceptual knowledge of lake change. There are two pieces of stone islands in the pond centre usualy which major part is located under the water level. So put a passage to connect the stone and the bank. In dry season, people can go to the stone island to sit or think of. In water season, people feel cheerful and contented with the duck and goose live on it. The stone island echoes with the northwest giant stone over a distance. We use valuable stone of reservation to record local natural culture, form effective identify. We also lay the road, the ladder shape close the pond on the basis of reserving original stone bank to keep persons closer to water. The arbor, bush and grass are planted suitable to local pond. With breeding some animals such as duck, goose, fish, natural pond landscape are formed(Fig. 2).

B. *Water Landscape for Interior*

The interior waterscape is attracting scenery within limited space. In general, the interior water landscape often uses running water or middle water. In this project , the pond water is adopt as the interior water landscape. It consumes low energy to keep the water quality, it has saved water source, the maintained cost and there are economic effect. First , the pond water is purified and lead to the pool for fire in the basement of the Library. And then by the water pump and buried the pipeline under the wall, it is lead into the pool of the Cafe in the secend floor, when the pool is full, the water flood along the east wall forming waterfall landscape to the pool in the first ground. Then the water floods to the brook under the glass ground , and along the brook the water floods to the west pool where there are one meter high tube. As the assembling water is super to the tube altitude , the water returns to the fire pool under the basement. So the water goes and returns in a circle. Circulating water adjusts air dampness and humidity interior. In summe it is hot and dry in Jinan. It makes person feeling cool. The sound underwater has decreased the noise interior, moving water has adds the contents of oxigy of air. The waterfall on the east wall, the stream covered glass plate,the plant around the waterscape strengthen the ecological environment effect in the Cafe(Fig. 3).

C. *Water for Adjusting the Interior Temperature*

The pond water can adjust the temperature interior by cooling down the structure of building outside. It is beneficial to promot natural ventilation. When the water evapors, it lowdown the temperature. The south wall of the Cafe in the Library is an entire glass curtain. In summer, the sunlight brings about large amount of radiation to interior. The purified water coming from the pond enters the pool in the basement of the Libray. It is also pumped to the top of glass curtain through the level pipeline buryed in the wall. And then the water pours down along the glass curtain outside and it decreases the radiation of the sun. This procedure by pond water cooling glass curtain surface outside is very effective. The surface temperature of glass structure outside in summer has been lowered to enough to ensure that interior is comfortable, at the same time the water becomes the element of the building landscape.

The water in the pond as a medium is also used to exchange heat. Under the water temperature about 2m deep is not much changible. The air-condition cooling system usually adopts a cooling tower as the cold source of air-condition. In this project the steel-made set tubes are as the heat exchanger which buried in the bottom of the pond pool. The water in the steel-made set tubes carriy out heat exchange with the pond water outside. The pond water provides the low-grade cooling source for the air-condition system which decrease the cost of air-condition equipment, construction and running. It lifts the quality of landscape and the beneficial environmental protection.

D. Underground Air-vent Windway Combining with a Pond

It is one of the important tactics that transport natural wind to building interior by the windway underground. However its detail design of air-vent is often short of thorough thinking. In fact, that arranging a reasonable place for air-vent of underground windway can promotes natural ventilating, gains fresh and comfort and natural wind making use of landform and configuration of location and the local climate. At the same time it can create the comfortble space. The ventilating duct is arranged to close the pond water of the Library. The water ability of heat accumulation is strong. In summer, the pond is a colder source than the land, the different temperature between water and land promotes the air flow. Through the water surface and the vicinity afforest moist, the air which is rich in negative oxygenion ensures the air humiture and cleanness. It provides the air quality to interior and brings health and delighted. Thingking of the current road situation, the selection of air-vent faces the pond and lies lower than the road. The valley feature can also promot the natural draft ability. The lieu with silence, nature, comfort is formed with the sinking ground, flight of steps, the newspape wall on the ventilating duct, the road piling up with locality stone, being on verge of the pond and vicinity afforestation (Fig. 4).

IV. SUMMARY

In this article, the beneficial exploration has been carried out, not only in water comprehensive utilization, but also in low project cost of running, defending, administration. In the project rainwater discharging is effective decreased. Circulating water system of campus and the landscape exploration benefits improve the tiny climate of area, reduce the source of water, embodys the usage function and culture functional coordination, creat the ecosystem campus. Ecological water landscape also becomes one kind of teaching. It stretches teaching in class to the environment education harmoniously. It plays important role in developing the ecologyical technology and improving people water-saving consciousness .

Purifying and keeping the water quality are treated by biological technology. the physical chemistry gives place to a self-purification of water in order to prevention and cure eutrophication. The consumption is low, the energy conservation is very effective.

REFERENCES

[1] Treatises on histrory of architecture,VOL.17.

[2] Scott R. Physical Geography. St. Paul: West Publishing Company, 1992

[3] Doshi. New York: Whitney Library of Design, 1998

Figure 1. The shallow ridges by the low waysided with vegetation have several function: it is a landscape brook, a pipe for plant irrigrate, for drainage, and also for collect rains

Figure 2. Natural water landscape

Figure 3. Water landscape in the Cafe

Figure 4. Air-vent Underground not only the place being ventilated, but also a silence, nature, comfort lieu

Enery-Saving Measures in HVAC System

Hongying Zhang
Department of Architectural Environment & Equipment Engineering
China University of Mining and Technology
Xuzhou, China
Email: zhykd@126.com

Rui Xiong
Department of Architectural Environment & Equipment Engineering
China University of Mining and Technology
Xuzhou, China
Email: 583702738@qq.com

***Abstract*—The energy consumption in China's building is increasing year by year, while the HVAC systems accounts for a large proportion in energy consumption of building. The article expounds energy-saving measures mainly from the selection of cold and heat source, the design and operation of the HVAC system.**

***Keywords*- HVAC; energy-saving; ecology**

I. INTRODUCTION

At present, in order to meet the indoor temperature and humidity requirements, the energy consumption of HVAC system probably account for 50% of the annual energy consumption in buildings. So optimizing the environmental performance of buildings, minimizing building energy consumption and using energy resources reasonability and effectively is an arduous and urgent task to us who is engaged in HVAC.

II. ENERGY-SAVING ON COLD AND HEAT SOURCE IN HVAC

The selection and configuration of cold and heat source of HVAC system plays an important role in building energy-saving.

The use of green energy, the use of energy from step to step and the use of renewable sources is the direction air-conditioning industry develop to.

At present, the selection of cold and heat source of air-conditioning system are tending to be diversified. It mainly focus on solar energy, geothermal energy, natural gas, nuclear energy, hydropower and so on. Natural gas and nuclear energy are both of green and efficient energy source, and have bright prospects, but they are non-renewable energy and reserves are limited. On the contrary, solar air-conditioning systems, cool storage air-conditioning systems, ground source heat pump air-conditioning systems are effective means of energy-saving and environmental protection.

A. *Solar Air Conditioning*

Solar energy is the fastest-growing of utilization in China and the world at the moment and it has been used in air-conditioning system. The energy required for solar air-conditioning system in part or entirely from solar energy. Solar air-conditioning system is on the base of solar radiation energy to refrigerate. Through the use of solar energy air-conditioning, it not only can make up for power supply, but also can contribute to create the Environmental Protection Model City. Promoting the use of solar energy air-conditioning neither can bring about urban heat island caused by using electric air conditioning nor pollute the atmospheric environment. Therefore, solar air-conditioning is truly a green and energy-saving air-conditioning. At present the main type of solar air-conditioning are: solar electric refrigeration air-conditioning, solar heating and cooling air-conditioning, solar heat pumps, solar liquid desiccant air conditioning. Solar Refrigeration and Air Conditioning include: solar vapor compression refrigeration and air conditioning, solar absorption refrigeration and air conditioning, solar ejector refrigeration and air-conditioning.

China's first solar-powered air-conditioning was successfully made by Stead in Zhejiang Haining in September 2004. The solar air-conditioning is based on solar's infinite light and thermal, using chemical energy to strengthen multi-level receiver to collect and store large quantities of energy, then through solar adsorber, heat and cold equipment to meet requirement in all kinds of weather. It achieved full optimization and efficient operation of the day and night, and is characterized by high heat value, heating and cooling quickly, no open flame, no smoke, no pollution. Oxygen which is produced efficiently by light energy is used to cooling and heating, and it can save more energy than the traditional air-conditioning system around 60%. In addition, in case of continuous cloudy and rainy season, the equipment can use the photoelectric conversion system automatically, in order to achieve continuous operation capabilities.

B. Ice-storage Air-conditioning

Ice storage system is mainly based on the phase change of water release to reserve cold. Since cold is reserved by phase change of water is much larger than the cold is reserved by temperature drop, so the storage capacity of per unit volume of ice is much more than water storage and more economical. As regards to alleviate the contradiction of supply and demanding power, the ice-storage air-conditioning is effective measures to balance urban power supply. At present, the ice-storage air-conditioning application is still in the initial stage in China. So far there are only more than 200 ice-storage air-conditioning projects, the transfer of the total electrical load of 1.0×105 kW, equivalent to the United States' 1/50 or Japan's 1/30. Ice-storage air-conditioning cooling stations of regional ice-storage system development become a trend at the moment. At present, the principle of inside ice storage equipment to ice and melting ice is still hot to study, and storage performance of equipment is still much more to be improved, and evaluation indicators and evaluation methods of the storage device (technology) are not perfect, does not truly reflect their economy, energy and rationality.

C. Ground-source Heat Pump Air-conditioning

Ground-source heat pump air-conditioning system is a using renewable energy, energy-efficient, environmentally air conditioning system. In the winter, it can absorb energy from soil, groundwater, surface water and other natural resources to heat buildings; in summer, it can release energy to natural resources to cooling the buildings. According to the form of natural resources Ground-source heat pump system can be divided into ground-source heat pump system, groundwater heat pump system and surface water heat pump system.

- The ground-coupled geothermal heat exchangers is the core of ground source heat pump system. Ground source heat pump is based on the earth as a heat source to air-conditioning buildings. It can improve the lower energy on the ground through the heat pump to heating buildings and storage cold to for summer use at the same time. In summer it can transfer energy from buildings to ground through heat pump to cooling buildings while reserving energy. Ground Source Heat of the original system is very popular in Europe, the United States, Canada and other countries because of its' energy-saving and water-saving performance. China has not application examples about it at the moment, but there are already some researchers note the system. In the south of Yangtze River and the southeast coast of China, people are hoped to be supplied heat in winter. Therefore, it has great development prospects and higher economic efficiency to research ground source heat pump technology in China.
- Groundwater heat pump system is divided into open and closed forms. The open system directly supply groundwater to the heat pump, and then recharge well water to the ground; the closed system transported groundwater to the plate heat exchanger, and need secondary heat transfer.
- Surface water heat pump systems are similar to ground source heat pump system ,both of them use associated plastic pipe in the groundwater to replace the soil groundwater heat exchanger heat exchanger. In Hunan, Lingtian science and technology use water source heat pump air conditioning systems and ground-source heat pump air conditioning, both of them use water (or other liquid media) as a medium .Water source heat pump is used widely at present, and ground-source heat pump is still researched at the moment .Both of its cooling or heating coefficient can reach 3.5 ~ 4.4 more than air-source heat pump (air - air or air - water heat pump) 40%.

D. Urban Sewage Heat

Sewage source heat pump air-conditioning system with urban sewage as a heat source, it can extract energy from sewage through the consumption of a small amount of electrical energy in the winter to heat building; while in summer, it can extract energy from inside building and then release it to sewage to reduce the room temperature to achieve cooling effect.

The primary energy utilization rate of urban sewage heat is greater than 1, with a very high energy efficiency. In addition, the urban is much rich in sewage resources. Urban sewage formed by the industrial wastewater and domestic sewage is the heat-based renewable clean energy sources, therefore, with urban sewage as HVAC energy has a unique advantage.

First, the urban sewage water occupied urban water supply more than 85%.In 2006 China's annual sewage discharge reached 53.68 billion tons, if 50% of urban sewage energy can be used, it can meet newly built a whole year energy use of heating and air conditioning and has a great development potential. Second, the urban sewage water temperature is high and stable, with warm winter and cool summer characteristics, the temperature is in the 10~25 ℃throughout the year, and is suitable for heating buildings in winter and cooling buildings in summer. Furthermore, compared to air-source and ground-source , urban sewage heat source has high heat transfer efficiency and high operation efficiency, and the heating factor COP can reach up to 5.20 and cooling factor can reach up to 4.25.

The use of urban sewage as a renewable energy, not only saves high-grade energy like coal and power, but also reduces the urban waste heat and CO_2, NO_x, SO_x, dust and other pollutants. With urban sewage as cool source in summer, the heat of cooled substance is not directly emitted into the air and can reduce the urban heat island effect. Compared to solar energy and terrestrial heat, the urban sewage heat made the greatest contribution to reducing the greenhouse effect.

III. Air-Conditioning System Design for Energy-Saving Measures

A. *Water System*

According to air-conditioning load change to control water system play an important role in energy-saving:

- With frequency control loop to replace control damper in the cooling water, and reduce the cooling water flow when in part of the load ,in addition ,with on-off control or variable speed drive of cooling tower fan will significantly reduce the energy consumption of the cooling water system;
- According to the actual operating characteristics, different working methods cooling chiller unit use different control strategies, combined with the microcomputer automatic control to control the cooling water pumps to achieve the best energy-saving operating conditions;
- With variable flow control to chilled water can effectively reduce operating costs of providing unit of cooling capacity , so the two pump variable flow system is widely used in project at present , but also use a pump system directly change chilled water flow in the evaporator .

B. *The Use of Frequency Control Technology*

In HVAC energy consumption, 15% to 25% of the heating energy consumption comes from the pump, 35% to 45% of air conditioning energy consumption come from the fan and pump .In this part of the fan and pump energy consumption, more than 40% was consumed by various of adjustable dampers. When in the partial load, the requirement of water or air flow will reduced through regulating the valve opening, the regulating role of the regulating valve is based on consumption of fan or pump energy consumption. Therefore, the energy consumption and investment costs will be reduced through regulating pump fan speed to control water or air flow .At present, the price of variable-speed fans, pumps are similar to electric air valve, water valve with same flow, so the initial investment will not be too high. Therefore, frequency control is a better choice, it can control speed with changing rotating magnetic and can realize stepless speed regulation. The use of frequency control to regulate flow not only can save energy, but also reduce noise.

C. *The Use of Large Difference in Temperature to Rationally Regulate the Proportion of Fresh Air*

There is a area made by temperature and humidity called comfort area in which the human body fell did not change significantly, but the system's energy consumption has a more substantial change, properly increase difference in temperature between air supply and air return can reduce the capacity of water or air road to decrease water pump and fan power.

Increasing difference in temperature between air supply and air return(water supply and water return) and using cold air are the two means to increase difference in air supply, cold air supply can be combined with ice-storage system. Both methods require air-conditioning terminal equipment to make corresponding changes to adapt new heat transfer conditions and cold air supply should solve problem of condensation at the outlet.

D. *The Selection of Fresh Air Volume*

If air-conditioning unit handle too much fresh air, it will increase its load and increase power consumption while dealing with too little fresh air will affect the air quality of the environment; so we should regulate difference in air supply temperature and fresh air ration in different situations. Such as shopping malls, stations , the hospital lobby where there are a lot of people we should properly increase fresh air ration while the indoor temperature does not necessarily adjusted too low. As for the hotel rooms where indoor temperature is explicitly required while having not many people, we can properly control the fresh air rate.

E. *The Use of Heat Recovery and Heat Exchange Device*

Since the introduction of fresh air, air-conditioned environment must row part of old air out .There are some difference in temperature between out air and indoor air, with cooling time as an example ,if the indoor temperature is 28 ℃, outdoor temperature is 35 ℃,when the indoor air was discharged into the atmosphere will bring the energy loss, using heat recovery exchanger to precooling fresh air , its temperature will decreased so that the load of the fresh air handling units will be reduced, this device is much more suitable for situation where can concentrate to row old air while needing large amount of fresh air.

F. *The Economic and Social Benefits of HVAC Operation*

To increase building economy, there is a great energy-saving potential from the management to formulate indoor temperature. For example, Japan is now promoting indoor temperature at 28 ℃ in summer, no higher than 20 ℃in winter, and there is a great saving in hot and cold load.

The use of hollow concrete slab which can form air channels in the evening to cool or heat room and make regenerative materials reserve energy at the same time, in the day time ,you can control air supply parameters and save operating costs by 15%. At present, with the electronics, telecommunications and automation technology use in building.

"Green ecology" and "Intelligent" will be combined, so we can depend on the weather, temperature, humidity and air conditions to automatically adjust the opening and closing and air-conditioning situation of "smart shell" which also largely reflects the economy in building.

IV. CONCLUSION

"Green" and "ecological" have become the trend of modern architecture. On the basis of environment-friendly the modern architecture is aimed at "function" and "comfort" in which the development of HVAC technology becomes an unavoidable factor. In this context, how to successfully apply modern HVAC technology to buildings according to evaluation criteria of green ecological architecture is undoubtedly a new challenge. We believe that HVAC technology must achieve greater leap forward based on the professionals' efforts in the near future and make us closer to natural.

REFERENCES

[1] Jiang Yi. China's building energy situation and the effective energy-saving means [J]. HVAC, 2005,35 (5) :30-40.

[2] Bu Zengwen, Liu Junyue. The technical measures of achieving green building in HVAC design [J]. Refrigeration and air conditioning, 04,4 (l): 7-9.

Environmental Conservation of Historical Architectures and the Application of Digital Technology in Universities

Dai Fei
Art College of Hubei Normal University
Huangshi, China
Email: dodo1997511@163.com

Abstract—**With the rapid development of science and technology and the prosperity of digital techonology in China, we have been studying the origin of the digitization and the environmental culture in universities. This article mainly talks about the application of digital technology in the conservation and utilization of the historical architectures and enrivonmental resourses.**

Keywords-**Conservation, aplication, digital technology, 3DMAX**

I. INTRODUCTION

Due to the increasing enrollment in higher education in China, every year thousands of students enter university, which meets the needs of cultiviation of most high school students for higher learning thus providing enough human resources on one hand, and also putting the pressure on the universities for sharing the equipements on the other hand. In this case, the institutes are undergoing a bran-new integration---pulling down and rearchitecture houses, establishing new ones in other places altogether with the relocation and alteration of people. This gives an impact to the historical sites in these institutes and their phisycal forms are facing the danger of despair.

II. REASONS OF THE DEFICIENCY OF THE HISTORICAL ARCHITECTURES AND ENVIRONMENTAL CONSERVATION IN UNIVERSITIES IN CHINA

Historical architectures and environment of the institute witness the prosperity of its development, embody its characteristics and the countinous culture of the institute, which should be protected and inherited and their chacterteristics should be developed and utilized. The tangible assets on campus consist of historical architectures which relate to the development of the cultural history of the institute representing certain historical chacteristics; campus environment that is naturally formed and mainly peculiar to that institute also including the manmade lanscape gardens with historical meaning, which are the important resources for teaching and research, and the hemerophyte of the institute culitivated in certain ecological environment. The historical architectures are the real picture of the development of the institute. Once ignored or impropriately dealt with, they will depair afterwards.

Besides there are other reasons. First, as the development of higher education, almost all institutes are expending their enrollment that makes the existing education resources be far from meet such needs. Second, the college cities are located far away from the downtown for the sake of the constuction of the city and province. The campus is relocated to the college city, while the original ones are bidded for real-estate and public places instead. In this way, the historical sites are destroyed. Thirdly, another reason owes to institues themselves, those who are lack of sound management system and ignore the particular conservation.

III. THE SIGNIFICANCES OF THE CONSERVATION AND UTILIZATION OF HISTORICAL ARCHITECTURES IN UNIVERSITIES

A) In the continuation of the culture and history of the university

To pretect the historical architectures and the environment of the institute is not only to preserve the exsiting campus, but to keep the culture continuous. They play important roles in teaching and scientific research, with irreplaceable value in forming the campus culture. The ancient classroom buildings represent a school sprit.

B) In the integration of universitiesg and the socioeconomic performance

Cultural heritage, a developing "assets", \will promote the economic development of the city, for people show much affection to the profound history, campus culture and unique human landscape. The distinctive architectures and the nutural environment can be utilized for tourism resources with a unique reputation for itself. Wuhan University integrates the ancient architectures on the hillside with the landscape, which attracts tourists with cherry blossom in full bloom.

C) In the integration of universities and the environmental performance

Now we need to reconsider the utilization of architectures and the environment. It will bring much negative effect to be eager for quick success and instant benefit by pulling down and rebuilding houses. The limited resources on campus can be fully used if protected properly. The conservation of historical architectures and the environment will reduce the negative effects caused by newly-built campus to the ecological balance. It also provides better social value in saving resources and reducing pollution.

IV. THE NECCESITIES OF THE CONSERVATION OF THE HISTORICAL ARCHITECTURES AND ENVIRONMENT IN UNIVERSITIES AND THE APPLICATION OF DIGITAL TECHNOLOGY

At present, the digital technology of art design in teaching architecture and environment has gradually come to maturity. It is a trend for modern institute to apply such digital technology to protect and utilize the historical architectures, which transforms altered virtual space into reality and brings new challenge and change for the traditional architectures. Constructs in different times could be integrated with the environment well in technological forms combining the old and the newly-built campus to make up for a deficiency of the original function and the exterior and interior space of the architecture itself. In this way, the elements and ways of rebuilding can be under better control at the first stage and the opportunity of public participation is increased. Teachers and students can vote for the plan. The application of digital technology consists of three phases as follows.

A) Proposal phase.

It includes the analysis of foundations and the transformation of drafts. As to the former, it involves the information research for the construction sites, analysis of the figures in GIS, Photoshop and systemic assessment of the current environment of the architecture. Ttransformation of drafts refers to transform the drafts into digital models and combinate them with construction and installation. Take Paul Andreu as an example. He designed the China National centre for the performing arts. He also transformed the his drafts with smooth lines and ever-changing space into digital documents.

B) Design intensification phase

It consists of making models as a visual reference, imitation of environment and the structure analysis. Digital techonology mainly contains CAD, 3DMAX, CAM and Rapidsketching etc. Environmental imitation supports the inspection of the physical environment, such as energy, light, air and sound insulation. And the structure analysis mainly concentrates on some special requirements for design, especially those public places with large span and high-rise buildings.

C) Submit of image phase.

This phase refers to report, communicate and to convey the meaning of design. The former talks about the communication of the entrusting party and the designer on the showing of visual expression and digit. Besides the mock-up, there are othre ways to convey the meaning of the design in different ways, including 3D and 2D cartography, explanatory note and even virtual techonology, to completely express and imitate the characteristics and the real effects of the program in CAD and rendering software.

V. THE CONSERVATION OF THE HISTORICAL ARCHITECTURES AND ENVIRONMENT IN UNIVERSITIES AND THE APPLICATION OF DIGITAL TECHNOLOGY--- 3DMAX SOFTWARE

3DMAX provides indispensable technology in the submitting phase and creates favorable conditions in selecting and optimizing the proposal, assessing campus landscape and coordinating the relationship between the engineering and the environment It can build virtual model in variety levels and with plentiful details and is also an exact reproduction of the project being established. It can imitate man's visual entering into the interior part of the building to examine the unifunctional pipeline, the rationality of the arrangement and the macth of colour, etc. we can better learn the entire effects of this locality, consider the reqirement for the sun light, direction and the property line and also the organization of the interior part with the exterior roads. It is better to research the reasonablity for the public feasilities and the grassland system. 3DMAX makes it possible to optimizing the campus architectures with the accessible softwares. 3DMAX in the alteration of Art Building in Hubei Normal University is a classical example to apply animation to the mentality of designing with 3D still life and animation. The specific steps are as follows.

Step	Content
Data Gathering	Investigation of the base, height, stories, location, color of the art building and picture taken
	Draft of the geometric plane surface, elevational drawing, section drawing and scenogram; data processing with GIS about the surface features; three-dimensional model as the actual scale in AotoCAD
Model of the Building	Input the data into the computer with AotoCAD、3DMAX
	Data correction ： an overall plan with exact datas
Map	Direct & indirect processing of digital film
	Partial tmap according to the materials
Lighting Design	Scene lighting in different areas as central and secondary activity area, background area, ambient or visible light
	Type of luminaire: pointolite, surface light source, ambient light, colour intensity
Render Output	Render with simple, 2-Sided, transparent materials one by one
	Use ordinnary shadow image with suitable resolution and light
Animation manufacture	Make a virtual object and portray its path
	Link and reorganization the path, key frame，design the path for the total images
Overall modification and experts assessment	Design the greening and environment, adjust the constract ratio and brilliance, and modification with the experts's experince

The historical architectures and environment are paid much attention in 3DMAX. So the materials must be the real materials of the architecture. Those used in the animation are taken by camera by restoring, cutting afterwards due to the weather and the environment and even the limitation of shooting angle, passers-by, parking cars, ect.

A) Direct processing of digital films

a) Deal with the superfluous parts. It is hard to avoid passers-by and vehicles in the picture actually which need processing with Photoshop. Picture 1 （a） is the original one of the art architecture in our college. We can copy a replicated pictures to the target area in order to wipe off the bush in front of the building to get picture 1(b).

1(a)

1(b)

b) Color revision. Every picture may not be of the same color due to time, angle, light, ect, which need modifying in constract ratio, brilliance and color balance with Photoshop. Make a contrast with the pictures before and after the revision.

2

c) Stretching and transfiguration.

The architectures and tall trees nearby will bring trouble for taking picture of high architectures in a long distance. Such kinds of photos need transforming in orthophoric angle with order edit- transfiguration-Free Transform. The next picture is the right side of the art architecture.

3(a)

3(b)

B） Indirect processing of the digital films

a）Transparent Map. As to making models of flowering shrubs, tree and sculpture, it takes a lot of time to imitate these landscape with 3DMAX directly. While tranparent map could look much real. Picture 4(b) show the application of the sculpture in picture 4(a).

4(a)

4(b)

b) Map. We can get the most elements from digital camera, but not all. We need to use Filter→Distort→Ripple in Photoshop to modify the color and the water wave.

c) In order to deincrease the number of the pictures repeatly, we usually adopt sole parterns and map them in the form of array like 5(b) not a whole one as 5 （a） .

5(a)

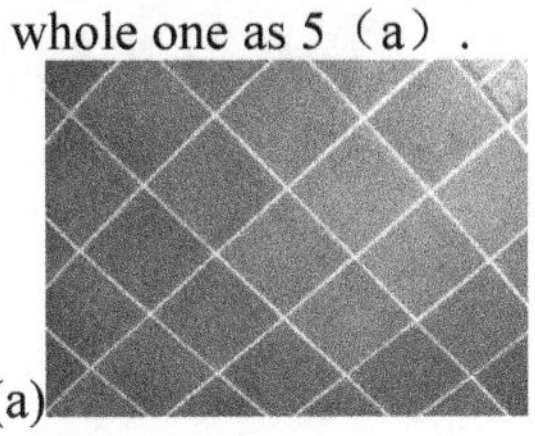
5(b)

CONCLUSION

Nowadays we have witnessed the emergance of so many modern architectures, digital libraries and labs. As the view of sustain development is emphazed, the conservation and utilization of historical resources on campus is a way for the harmonious development of the whole campus environment. This is a formidable task in the long run. The application of digital technology makes us pay much attention to the transformation of plane and static state into mobilism and integrating. The coexitance of modern architectures and the virtual environment is to inherit the campus civilation with digital technology, which is of particular characteristics of times. It aslo presents students virtual experience in learning with meticulous design in the functions of interior and exterior space. The modern technology and our concerted efforts will put into full play in the development of universities.

REFERENCES

[1] Song Zefang and Zhou Yihu, Campus Programmes and Architectural Design, Beijing: China Construction Industry Press, Sept. 2006

[2] Zhang Peng, The Environmental Design on Campus Visual Culture, Guangzhou: Lingnan Art Publishinghouse, Nov. 2005

[3] Zhang Jian and Lu Chao, Conservation and Value of Outstanding Historical Architectures in Shanghai Universities, Huazhong Architecture, vol. 2, 2007

[4] Ni Hongrao, Disappearance in university historical heritage and its conservation, Journal of Technology College Education, vol. 12, 2006

[5] Wu Xiao and Yu Gang, Rethink on the Application of Digital Technology to Contemporary Architectural Design: Thoughts on the Contemporary Architecture and the Digital Design Process. Architecture Journal, 2006, pp.95-98

[6] Guo Baofeng, Exploring 3D Architectural Animation, Journal of Huaqiao University (Natural Science), vol 7, 2002

[7] Chen Jinchan and Zhan Weijie，Research on Surface Mapping of 3D-Sences in Virtual Campus, Journal of Donghua University (Natural Science Edition), Vol 4, 2005

Exploration on a Model of Sticky Knowledge in the Process of Organizational Knowledge Transformation

Gang Kong & Liwen Tan
School of Economics and Management
Wuhan University
Wuhan, China
Email: konggang10285@126.com

Shengtai Zhang & Qian Li
School of Management
Shanxi Normal University
Linfen, China
Email: liqian1983625@163.com

Abstract—**The paper firstly summarized the existed relative researches as well as defined the concept of the sticky knowledge. Besides, it made the systematic analysis on the mechanism of the formation of sticky information and put forward a model of sticky knowledge in the process of organizational knowledge transformation. At last, the paper advanced the strategies to manage sticky knowledge according to the model.**

Keywords-knowledge management; sticky knowledge; knowledge transformation; model

I. INTRODUCTION

In the age of the knowledge economics, knowledge is the source of gaining sustain competitive advantage and to be regarded as the most resource of strategic value (Nonaka, 1994; Grant,1996; Simonin,1999). The capability of enterprise to explore, maintain and cultivate competitive advantage depends greatly on the ability to create, transit and apply knowledge, which called knowledge management capability (Drucker, 1999; Hoopes and Postrel, 1999). Therein, one of the most important linkages is successful transferring knowledge. However, in the transferring process, knowledge cannot successfully transfer from sender to receiver because of kinds of obstacles (Yi Wang, 2005), therefore, stick on the topical circumstances (Anderson,1999. The academe defined it as sticky knowledge.

The existence of the sticky knowledge inevitably calls forth transferring cost in the organization. Accordingly, study on the mechanism of the sticky knowledge's emergence and how to cut down it is extraordinary valuable for reducing the transferring cost, accelerating knowledge diversion and thereby acquire and sustain core competitive advantage. Especially for the enterprise shortage of knowledge resource, how to realize successful transformation among the inside and outside of the organization as well as create value has become the key to reply the dynamic competition under the increasingly uncertain circumstance.

II. REVIEW ON THE EXISTENT DOMESTIC AND FOREIGN REASEARCH RELATIVE

The researches which were about sticky knowledge were mostly originated from Eric von Hipple (1994) in his study process of technology innovation. He definitely advanced the concept of 'sticky knowledge' and 'information sticky'.

According to Prof. Yi Wang, since the 1990s, the research about sticky knowledge mainly arose from three aspects as follows, the process of technology innovation, the internal transfer of the organization and the external. Correspondingly, there were three basic different schools.

A. School of technology innovation process

Representation as Eric von Hipple, they main focused on the influence of sticky knowledge to technology innovation process and explore how to deal with sticky knowledge so as to achieve efficient innovation as well as prompted some innovation management tools.

B. Rchool of Knowledge Transformation Inside the Organization

Representation as Cabnei Szulanski, they concentrated in the formation reason and impact factor of internal sticky knowledge, the key blocking factors on knowledge transfer inside the organization and the measure to improve it.

C. School of Knowledge Transformation among the Organizations

School of knowledge transformation among the organizations representation as Bernard L.Simonin made their attention on the formation reason, impact factor and expansionary action of sticky knowledge existing in the process of transferring among the organizations. Simonin(1991) advanced the impact factor leading to sticky knowledge as follows, elusiveness, Asset Specificity, previous experiences, complexity, the tend to self-protecting, the distance between the organization and cultures.

In addition, some scholar put forward the measures to over come the sticky knowledge such as, Szulanski et al verified that the template was very effective to reduce the sticky knowledge by the marketing practice in the European department of Xerox. Sidney 和 Szulanski brought forward deal with sticky knowledge through replication.

Furthermore, the domestic researches on sticky knowledge started comparatively later. Existing and fewer study contain, Yi Wang, Guisheng Wu (2001) defined the concept of sticky knowledge and put forward a model about knowledge transfer in the process of combine of producing, learning and researching. Lingling Zhang, Hongming Luo (2006) explored the sticky knowledge among the knowledge-based enterprise. Qin Chen (2008) discussed the sticky knowledge within industry cluster.

To conclude, the studies existing discussed the sticky knowledge within the internal enterprise, industry cluster, the allied enterprises and the cooperative firms of producing, learning and researching. However, the studies ignored the logic relation between knowledge transformation and competitive advantage. Moreover, the researches mostly focused on how to reduce the sticky knowledge as well as ignored the positive impact of sticky knowledge. In fact, reasonable sticky knowledge is extremely valuable to protect core competence especially in the increasingly dynamic and uncertain circumstances. The article will make a great attempt to make up for it.

III. THE IMPLICATION OF THE STICKY KNOWLEDGE

A. To Define the Sticky Knoeledge

Stickiness originally as a concept in the Physics Science refers to the resistance to obstruct relative movement among the different molecules flowing levels when the liquid moves, which we called sticky resistance. None other than the sticky resistance influences the normal movement. In fact, any liquid would bring stickiness simply the different extents.

Eric von Hipple (1994) defined the sticky information as extra cost when some information in some way transferred to special distance. There existing a positive correlation between extra cost and the extent of stickiness. Afterward, some scholars else described the sticky knowledge as internal stickiness or knowledge ambiguity.

The article described the sticky knowledge as the knowledge which is hard to transfer or would bring additional cost. The stickiness of knowledge is mean to blockage or slowness appeared in the whole progress of absorbing, transferring, sharing and new knowledge emergence because of the preservation of sender, the weak ability of receiver and the ambiguity of the knowledge and information.

B. To Weigh and Measure the Sticky Knowledge

According to the above expound, the stickiness which is to measure the difficulty level of knowledge transfer depends on the cost arisen in the transfer process. As a matter of fact, the sticky knowledge is easy to weigh but hard to measure clearly. The current and excellent research on knowledge measure should attribute to Karl E. Sveiby and his Konrad Group(Swedish) in his classic book <knowledge tango-the art to management and measure knowledge>(2007). However, even in the excellent research, knowledge cannot be clearly measured for its inherent nature. The article will explore it in the future study.

C. The Characteristics of Sticky Knowledge

As a matter of fact, sticky knowledge is relatively hard to transfer other than by no means unable transferred. That is to say, it has characteristics of relativity and dynamics. The relativity is described according to specific receiver, if the receiver changes, the stickiness would change. The dynamics is mean that the stickiness probability and range of convergence may generate dynamic development with the changing of the characteristics of sender and receiver, the relations between them and the medium of knowledge transfer and so on.

IV. ANALYSIS ON FORMULATION REASON OF STICKY KNOWLEDGE

In the internal organization or among the organizations, there are different reasons. Some scholars analyze the sticking factor from the internal organization such as: Szulanski（1996）put forward four kind of factor contains traits of knowledge, resource of knowledge, knowledge receiver and the transferring circumstance; Reagans 和 McEvily（2003）discovered that the intensity of cohesiveness inside the organization and the extent of relational network can affect the sticky knowledge.

Other scholars considered the reason as strategic intend (Hamel,1991) 、 organizational capability(Lyles and Salk,1996 ; Pucik,1988) 、 partner choice(Makino and Delios,1996; Tiemessen et al.,1997) and trust(Aulakh, Kotabe and Sahay,1996; Dodgson,1996; Inkpen,1997)。

Obviously, the researches existing mostly analyzed the formulation reason of sticky knowledge from the resistant factors in the process of knowledge transformation inside and among the organizations. However, the article consider that it necessary to add the view at actively enhancing the stickiness. That is to say, strengthening stickiness can help maintaining core competence. Consequently, the article tentatively put forward the model(as in the pic.1) about formulation mechanism of sticky knowledge based on prof. Yi Wang.

A. Cognitive Factors for Transfer

Cognitive factors for knowledge transferring include the characteristics of knowledge, sender's ability of coding and transferring and the ability to decode and absorb of receiver.

According to the informational special model of Boisot（2004）, it is an important factor that the knowledge can whether be decoded. Generally speaking, the more implicit, complex and fuzzy, the knowledge will become harder to code and successfully transferred, this stickiness of knowledge will become larger.

Simultaneously, the sender should make the knowledge easier to understand and accept through flexible manner and methods according to the diverse features and requests of receiver other than only sending the knowledge.

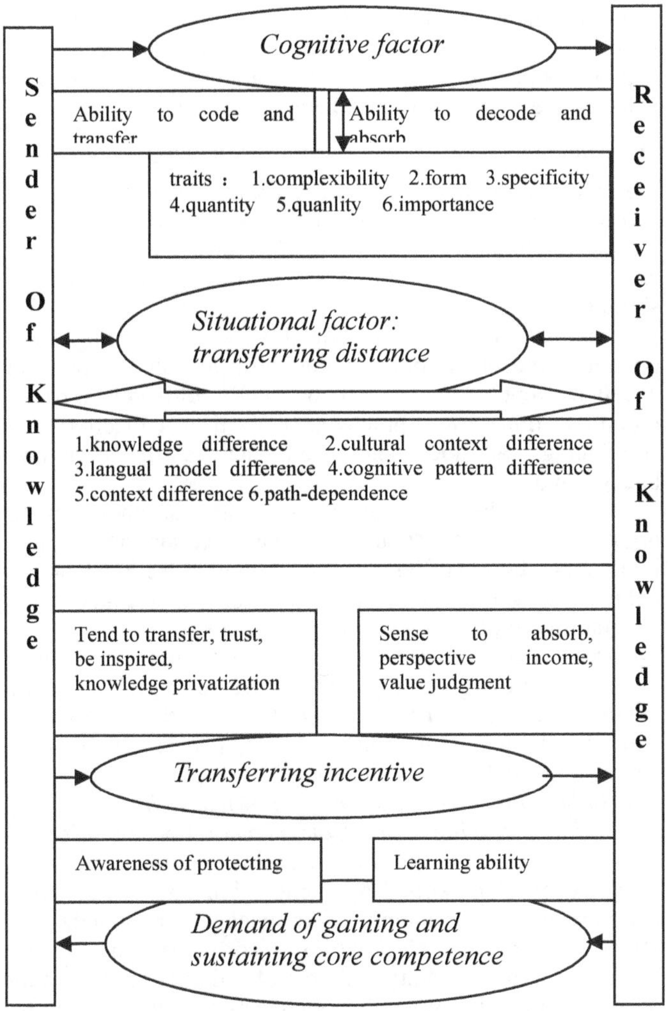

Pic.1 the formulation mechanism of sticky knowledge

Otherwise, the weaker the ability to absorb is, the longer the transferring time would be, the quantity of knowledge transferred would be smaller at the same time unit. That is why the capability of organizational learning has unprecedentedly become so important and emphasized unprecedentedly.

In a word, only the knowledge transferred be conserved, can the transformation be effect and valuable.

B. The Situational Factors—Transferring Distance

The differences of knowledge background, cultural backdrop, lingual model, cognitive pattern and path dependence made up of the situational elements of knowledge transferring.

For the most part, the larger the scale of organization is and the more complex the organizational structure is, the higher the degree of sticky knowledge would be. Besides, the organizational culture in more favor of transferring knowledge can diminish sticky knowledge degree.

When comes to the transformation among the organizations, the important context factors contain as follows, cognitive or geographical approachability, communicating channels, relations with each other, strategic similarity, knowledge setting, lingual style, cognitive pattern and so on. In general, the transformation between two similar organizations can deduce the extent of stickiness.

Apart from this, the path dependence of knowledge transformation can block the transferring of other skills and knowledge for the self-reinforcement.

C. Transferring Incentives

Essentially speaking, the source of the stickiness can be attributed to the incentives of transferring which is divided into the motivation of sending and the aspiration of receiving. The transferring intention, trust degree in each other, extent of being inspired, tends to private knowledge of senders and the Sense to absorb, perspective income, value judgment of receivers consist of transferring incentives.

The researches existed have pointed out that there existing conflict between private rationality and collective rationality so that the owner of knowledge might have not the intentions to share it.

In addition, the receiver will actively filter and judge the transferred knowledge though analysis on perspective profit and cost.

D. Demand on Attaining and Preserving Core Competence

The concept of core competence first put forward by C.K.Prahalad; G.Hamel (1990). They consider the essence of the concept as knowledge in the organization with the features of value, scarcity, hardness to imitate and replace. Based on it, the article thought that it is key and imperative to intentionally enhance the stickiness of knowledge to conserve the core competence.

To owner of core competence registering as knowledge, the sender will beyond no doubt improve the conserving awareness to prevent leaked on their best especially in the strategic alliance. That is to say, the stronger the conserving awareness and ability, the level of the sticky knowledge standing for core competence would be larger.

The receiver commonly as the competitor of the organization will attain and learn the core competence of the owner at any cost, which we called organization learning, a hot study field for academe and practice. Therefore, the capability to learn of the receiver influences the sticky knowledge.

Moreover, the dimensional embedment of knowledge formulates a special element for the emergence of stickiness. Under the wave of globalization and economic liberalization, the source of organizational competitive advantages depends on the knowledge dimensional embedment. The emergence of dimensional embedment depends on the tacit knowledge and organization internal social capital. It means to a cost in the flowing process of knowledge within the space. From this, to preserve core competence, the emergence of sticky knowledge has become inevitable.

In a word, a creative point of the article is having advanced a distinct view that it is necessary even inevitable to actively enhance the stickiness to preserve the core competence.

V. To Management the Sticky Knowledge

It is especially critical to realize successful and effective transformation of knowledge for the increasingly dynamic competence under uncertain circumstance. The inherent traits including relativity and dynamicity make it possible to achieve effect management to sticky knowledge through controlling the influential factors. Consequently, based on the above analysis on formulation mechanism of sticky knowledge, the article advanced some relational countermeasures.

A. Build Perfect Platform to Communicate and Share the Knowledge

Perfect platform can promote the normalization and standardization of knowledge and increase the capability of coding and decoding to reduce the stickiness brought out by cognitive factor. The modern information and internet technology can help conceiving an effective system and mechanism to transfer knowledge then realize the valid exploitation for existing knowledge and explore for the new ones.

B. Construct the Organizational Culture and Values Propitious to Transfer Knowledge in and Among the Organizations

Based on the context factors, the excellent culture and values can help diminish the difference of culture, knowledge, lingual model, cognitive pattern, context, path-dependence and so on. The cultivated culture should emphasis on individual self-innovation, development and collective, inspire both competition and cooperate, actively both explore, absorb and apply internal and external knowledge, attach importance to private and collective profit.

C. Manage the Sticky Knowledge in the Organizational Internal

For the cognitive factors, the private person and organization both should eliminate path-dependence and laziness to learning and enhance the awareness of sharing knowledge by effective and diverse communication as well as improve the comprehensive diathesis and capability of learning, absorbing, coding and decoding.

1)Emphasis on the Human Resource Management: Organization should provide effective and pointed training plan to perfect the knowledge structure of staff and heighten the relative capability. Simultaneously, to inspire the positivity and motivation of transferring, there strongly demand an effective mechanism of knowledge-based reward, performance management pointing to knowledge and so on.

2) Advance the Ability of Organizational Learning: As has been said before, organizational learning increasingly become a hot topic both in the academe and practice field. Only sustaining study can deepen the capability and of understanding, shoring and applying then successfully acquiring the core competence of competitor.

To heighten the capability of organizational learning, we should allow the existence of diversity, encourage try and explore, build the mechanism of tolerating risky as well as enhance the team-learning and emphasize life long learning.

D. Manage Sticky Knowledge Among the Organizations

1) Actively cultivate the trust mechanism and effectively shorten transferring distance among organizations: According to the view of New Classical Economics, the arising of transactional cost and expense can be ascribed to shortage of trust.

Trust is a mutually beneficial which is coming from opening atmosphere, built on long-term repeated games and excellent reputation. It can be well cultivated through repetitive transaction practice, perform a collective and convincing attitude and construct similar cultural background.

To shorten the time and dimensional distance among organizations, we should increase collaboration and intercourse, build sufficient trusting relation, seriously study culture and values of other organizations and enhance the adaptability to other organizational culture.

Another channel to reduce stickiness is to form an alliance or net with the organizations who processes high knowledge potential energy. Though effective profit distribution mechanism and valid mutual development mechanism, it will accelerate the transformation of knowledge.

2) Enhancing Organizational Knowledge Innovation to Maintain Sustaining Core Competence: For the inherent value increment and infinity extensity of knowledge, only actualize constant knowledge innovation and advance the complexity and dynamicity of knowledge can organization effectively prevent leakage of core competence by heightening the difficulty and cost of imitation and learning of competitor. Thereby, the organization can successfully achieve enhance sticky core competence and valid maintain and preserve their core competence.

On the contrast, for the receiver private and organization of knowledge, only enhance the capability of learning and rapidly response can they successfully attain the core knowledge of competitor then apply it into strengthening the core competence themselves. This can be found in the enterprise of Japanese economic miracle. None other than it can they formulate their core competence fleetly in a short time and overtake the traditional strengthening corporate at last.

VI. Conclusion

Sticky knowledge result from knowledge transformation may reduce the positivity of individual or organizational communication and makes them shortage of the capability of adapt and reconstruct organization internal and external resource so as to match the dynamic competition under the increasingly uncertain circumstance. However, on contrast, the existence of sticky knowledge can effectively prevent unexpected leakage of valuable and core knowledge.

Consequently, study sticky knowledge can help understand and explain organizational flexibility, potential value of M&A, successful opportunity of strategic alliance, profit from technology assignment and how to win sustaining and dynamic competitive advantage by managing organizational knowledge validly.

Review from the relative research existed, it is still an emerging academic field. For the matter of fact, there are various stages and level of the transferring process of sticky knowledge and the formulating mechanism is also diverse in every stage. Nevertheless, it is a pity that the article has not explored on this aspect. Undoubtedly, it deserve more exploring and analyzing as well as the issue of sticky knowledge in the process of knowledge innovation, which we will delved in the future research.

ACKNOWLEDGMENT

The paper can be accomplished successfully first because of the strong support by the two programs of the National Natural Science Fund of China, which are named 'research on the model of tacit knowledge transformation'(Confirmation Number:70871072) and 'research on the strategy of upgrading and appraising international competence of made-in-china based on the global value chain'(Confirmation Number:70772046). Also we have got the energetic support of the program of the National Soft Science of China, which is named 'research on the human capital management of the resource-based enterprises' (Confirmation Number: 2008GXQ6D149). Furthermore, we should be deeply grateful for the school of economics and management at Wuhan University as well as the management school of Shanxi Normal University.

REFERENCES

[1] I.Nonaka, "A dynamic theory of organizational knowledge creation," Organization Science, vol.5, May, 1994,pp.14-37.

[2] R.M.Grant, "Toward a knowledge-based theory of the firm, " Strategic Management Journal, vol.17,Nov.1996,pp.109-122.

[3] B.L.Simonin, "Ambiguity and the process of knowledge transfer in strategic alliances," Strategic Management Journal, vol.20, Aug.1999, pp.36-52.

[4] P.F.Drucker, Management Challenges for the 21st Century, New York: Harper Business,1999, pp.156-162.

[5] D.G.Hoopes, S.Postrel, "Shared knowledge glitches and product development performance," Strategic Management Journal, vol.20, Jan. 1999, pp.837- 865.

[6]Y.Wang, "Review on sticky knowledge transformation research," Science and Research Management, vol.3, Mar.2005, pp.34-46.

[7]Anderson, H.Poul, " Organizing international technological collaboration in subcontractor relationship: an investigation of the knowledge-stickiness problem," Research Policy, vol.28, Jul.1999, pp.625 - 642.

[8] E.v.Hipple, "Sticky Information and the Locus of problem solving: implication for innovation, " Management Science, vol.40, Aug.1994,pp.429-439.

[9] G.Szulanski, "Exploring internal stickiness: impediments to the transfer of best practice within the firm," Strategic Management Journal, vol.17.Winter Special Issue,Dec.1996, pp.27-43.

[10] W.Sidney, G.Szulanski, "Replication as strategy," Organization Science, vol.6, Mar.2001, pp.730-743.

[11]Y.Wang,G.S.Wu, "Research on formulation mechanism of sticky knowledge in the process of cooperation of production, learning and study," Science and Research Management, vol.16, Aug.2001, pp.56-72.

[12]L.L.Zhang, H.M.Luo, "A model of sticky knowledge transferring in the knowledge intensive corporate," economics management, vol.11, Dem.2006, pp.35-52.

[13] R.Reagans, M.Bill, "Network structure and knowledge transfer: The effect of cohesion and range," Administrative Science Quarterly, vol.48, Aug.2003, pp.127-142.

[14] G.Hamel, "Competition for competence and inter-partner Learning within international strategic alliances," Strategic Management Journal, Summer Special Issue, vol.12, May.1991, ppy.65-91.

[15] M.Boisot, A.Canals, "Data, information and knowledge: have we got it right?," Working Paper, vol.12, Aug.2004, pp.46-78.

[16] C.K.Prahalad, G.Hamel, "The Core Competency of the Corporation" Harvard Business Review, vol.3, Mar.1990, pp.79-91.

Exploration on Concept Planning of the Historical and Cultural Town under the Background of Tourism Development

Set the Concept Planning of Sanhe Ancient Town for Example

Gu Dazhi
College of Architecture & Art
Hefei University of Technology
Hefei, China
Email: xzandgdz@126.com

Xu Zhen
College of Architecture & Art
Hefei University of Technology
Shanghai, China
Email: gdzandxz@126.com

***Abstract*— Under the background of tourism development, the sustainable development of the historical and cultural town is a worthy of research. Based on the concept planning of historical and cultural town of entrance area of Sanhe Town in Hefei City, the author introduces the main contents of the planning, and summarizes the main aspect of concept planning, pointing out that on the basis of overall layout, the management of planning, content planning, and land development about the development in the future, is the important content of concept planning.**

Key words: historical and cultural town; concept planning; sustainable urban planning; management of planning

I. Chinese Historical and Cultural Town under the Background of Tourism Development

A. *The new trend of tourism*

Today, tourism has become one of the basic ways of Chinese people's life; the tourism industry has become one of the pillar industries in China. Meanwhile, "leisure life" becomes one of important parts of urban social evolution trend. To dwellers living in a city, weekend, the short holiday, has a high proportion of consumption in daily life.

Therefore, it is important to strengthen the tourism products and implement the tourism strategy. In the city and surrounding areas, planning some function of leisure culture, holiday, business conference and so on, can adapt to the promotion of tourism development of leisure life, and become one of the important functions of urban development.

B. *The new development of historical and cultural town*

At present, there are 143 national historical and cultural towns in China. As an important carrier of the history and culture, the town has a special mission to inherit the historical heritage.

But these historical cultural towns also meet a common problem that is how to solve the problem between the protection of traditional culture and the economic development. It is the contradiction between protection and development. Combination with the smokeless industry-tourism economy is important means of sustainable development in almost all the historical and cultural towns.

Therefore, **in this context**, the concept planning of historical and cultural town should embody the different characteristics and guidance.

II. Concept Planning of Entrance Area of Sanhe Town

A. *Project profile*

Sanhe Town lies in Hefei City. It's a national historical and cultural town. Its tourism develops gradually just based on the background tourism development in China.

1) The orientation of tourism development of Sanhe Town—"a bit, a ring, a circle" .

a) "A bit"— to make Sanhe Town become a very important tourist town on the tourism line from Jiuhua Mountain to Huangshan Mountain in Hefei.

b) "A ring" — to turn Sanhe Town into an important part of the Chaohu tourism annulus.

c) "A round"—to make Sanhe Town become a tourism leisure culture park based on the characteristics of "thousand-year-old town, ecological water town"(Fig. 1)

2) Location analysis.

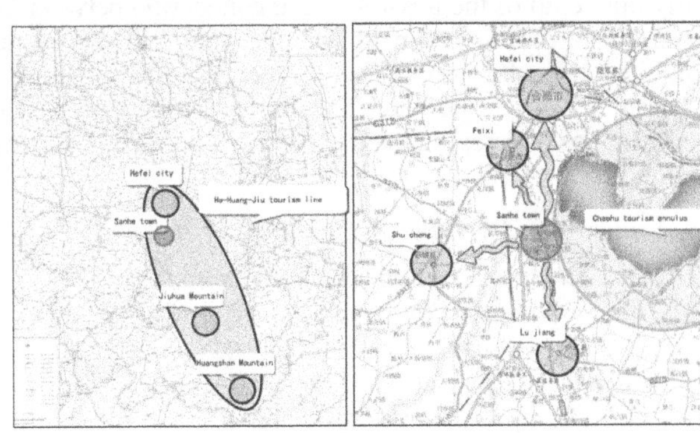

Figure 1. Orientation of tourism development of Sanhe Town

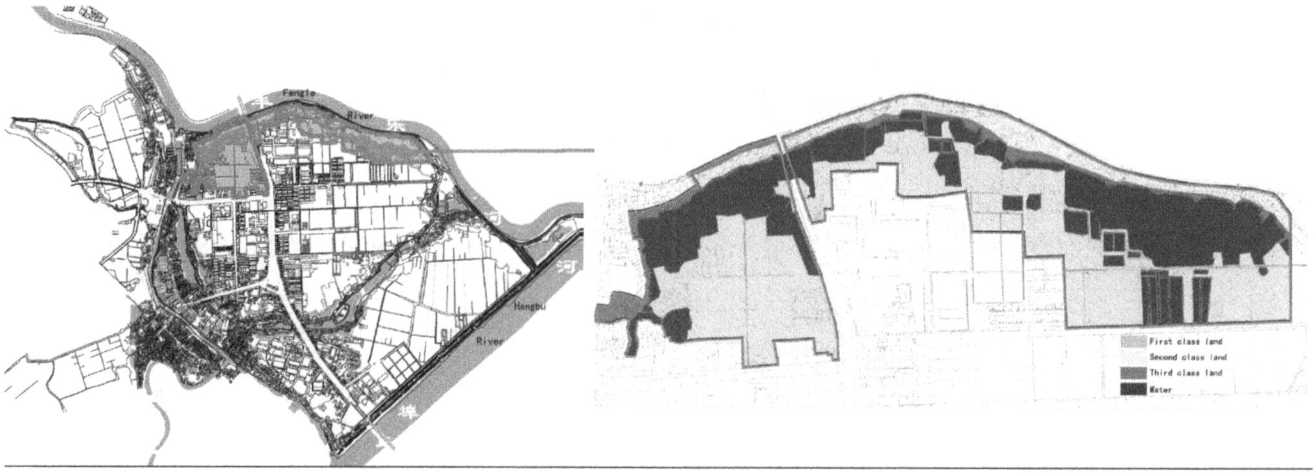

Figure 2. Entrance area base situation of Sanhe town

Sanhe Town is located to the south of Feixi County in Hefei, the capital of Anhui Province. It's 40 kilometers north away from Hefei, 7 kilometers waterway away from Chaohu Lake which flows into Yangtze River, Fengle River, Xiaonan River and Hangbu River around the town, that's the reason why it was called "Sanhe Town". The area of town is 4.71 square kilometers. As a national historical and cultural town, Sanhe Town has a long history, an outstanding folk tourism and rich waterfront cultural tourism. Since the leisure products of village and town become more popular, it has brought the new opportunities of development to Sanhe Town.

3) Base situation.

Sanhe Town belongs to the hilly region. Sanhe Town is from water, and the water becomes the soul of Sanhe. All the rivers in town are navigable. Bridges, rivers and houses make it into a typical "beautiful scenery in the south of China" in Anhui Province.

The base is located at the entrance of the Fengle River and the provincial highway which goes through the town. The base covers an area of about 0.41 square kilometers, and nearly one tenth of the town. It is the connection between the western area of protection of ancient style and the eastern ecological agricultural park. So, the geographical position of the base is very important. The base crosses the Yuceng River and has an irregular shape that two ends are wider than the middle part. (Fig. 2)

B. Analysis of development condition of the project

1) The advantages of project development.

a) The advantage of location resources: Sanhe Town has a unique geographical position. It is a satellite town to Hefei and one of the important gateways in the southwest area in Hefei. Its geographical advantages are superior.

b) The advantage of traffic resource: Taking the road traffic into account, it has Shanghai-Chengdu expressway, Hefei-Jiujiang railway and Hefei-Tongling highway.Taking the waterway traffic into account, Fengle River and Hangbu River flow into Chaohu Lake which is connected with Yangtze River. It is only 40 kilometers away from Hefei. So, Sanhe Town has an outstanding advantage of traffic resource.

c) The advantage of humanistic resources: Sanhe Town has a history more than 2500 years and a rich historical and cultural heritage. During the long history, it has formed its unique folk and numerous cultural relics. Humanistic resources become an important advantage to promote the tourism development of Sanhe Town.

d) The advantage of natural resources: Sanhe Town has typical "beautiful scenery in the south of China", and the ecological balance of the town is very well.

e) The advantage of architecture resources: The basic pattern of Sanhe Town is based on fractal road system, and it has traditional local-style dwelling houses, skeleton street, traditional space structure. Spatial relations between streets and houses are relatively complete. "Small bridges, water flowing and beautiful houses" is so artistic.

Therefore, the planning and construction of this area of town will further promote the integration of those advantages above.

2) The adverse conditions of project development.

a) Tourism system of Sanhe needs to be integrated.

Mainly lies in the following aspects:

First, the content of tourism is scattered and simplex. Sanhe Town of tourism has existed within a certain scale, but the type is simplex, it is mainly developing on the protection of the ancient town. Although the ecological agricultural park has been developed in the east town, the contact between the east and the west isn't close enough. So,the town hasn't formed a complete tourism system.

Second, the landscape system is incomplete, and lacks facilities. At present, the tourism area of Sanhe Town is dispersive, and has not formed a complete system of tourism landscape, wasting a lot of precious natural tourism resources. At the same time, the facilities and environmental quality need to be improved.

Third, the road system is not perfect. The existing road system does not fit the standard of tourism development both in class and quantity in Sanhe Town. Now, the road system of town has been reforming and consummating.

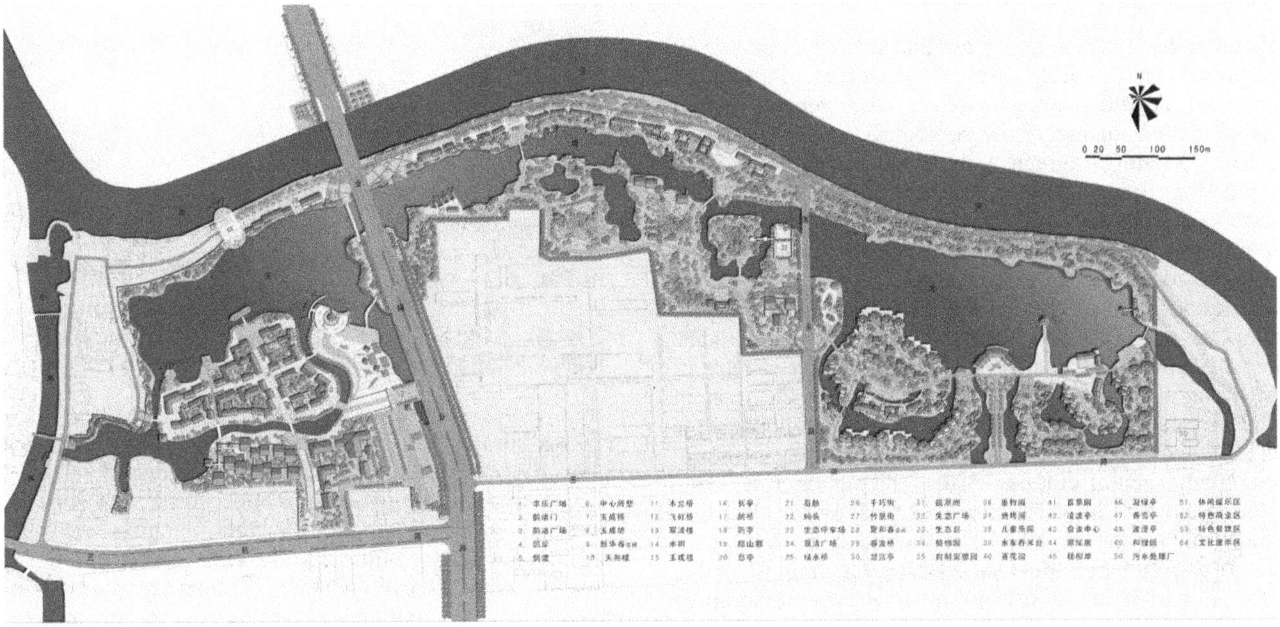

Figure 3. Overall plan

b) External tourism system of Sanhe Town remains to be strengthened.

Though tourism of Sanhe Town has been development to a certain scale, its contact with external town is limited. Although it wants to be an important tourist town on the tourism line from Jiuhua Mountain to Huangshan Mountain in Hefei and also a part of the tourist area around the Chaohu Lake, its traffic links and tour routes are not clear enough and cannot attract many tourists from different places. Therefore, it is important to form a complete tourist system by making the tourist routes on the tourism line of "Hefei-Jiuhua -Huangshan" and the Chaohu Lake.

So, the planning and construction will combinate the western pretection area of ancient style and the eastern ecological agricultural park, and the project will play a positive role in promoting the road system, tourist routes and facilities of Sanhe Town.

C. Project positioning and ideas

1) Project positioning.

The project positioning is to create tourism leisure culture industrial park, and turn the base into a tourism service zone involving tourism, entertainment, leisure, holiday, reception, business function and etc, and become an important part of the tourism leisure culture industrial park.

As a portal position, the planning should reflect the "thousand-year-old town, ecological water town" cultural characteristics. At the same time, the planning should improve the tourism system, combining the western pretection area of ancient style with the eastern ecological agricultural park, and shuld fully exert the existing resources advantage, further excavation and integration of the advantage of resources, culture, insisting on the fusion of tourism , economic and ecological aspects.

2) Planning ideas.

a) Combining with Sanhe Town culture: Based on the unique cultural environment, the town will integrate the ancient architecture structure and modern style, and also create the artistic character and ancient cultural landscape.

b) Combining with tourism industry: The planning area should be constructed into the demonstration area of Sanhe Town and the auxiliary services area of ecological agricultural tourism zone, realizing the connection with the western protection area of ancient style and the eastern ecological agricultural park.

c) Combining with ecological regulation: The natural landscape with "water" is prominent. Based on the favorable natural ecological environment, the planning should provide traditional folk leisure projects that are close to nature, recreation and return to the field.

d) Combining with the design theme of "Feng • Ya • Song": New interpretation of the town, the planning would have a new look to interpret the design theme, namely the ancient amorous feelings — Tourism new atmosphere, Elegant town, Beautiful environment, Praised passion and the famed ancient town.

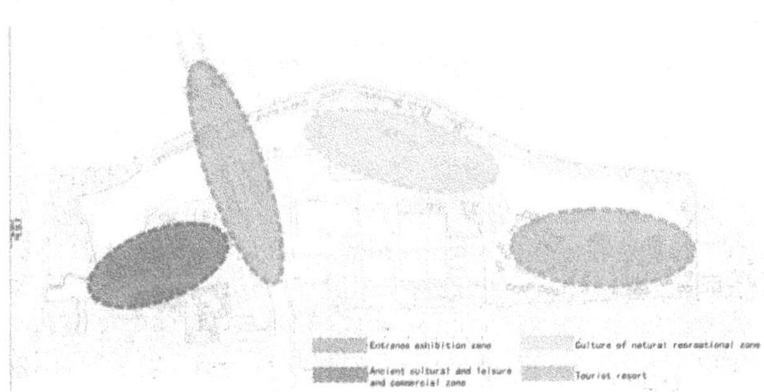

Figure 4. Planning structure

D. Overall plan

Through the analysis of the base and surrounding areas of geographical environment, natural resources, humanistic feature and development conditions, the general layout format four areas, namely "the ecological corridor", "slight praise of Quezhu", "green wild and gentle wind" and "suitable for living beside beautiful water". From the function, corresponding to: entrance exhibition zone, ancient cultural,leisure and commercial zone, culture of natural recreational zone, tourism resorts. (Fig. 3)

Four zones contact one another closely. "The ecological corridor" gives tourist first impression through the elements display of "culture of ancient town" and "water culture of town". "Slight praise of Quezhu" connects with the western protection area of ancient style closely, and publicizes "the culture of ancient town". Based on the natural ecological environment, ancient cultural,leisure and commercial zone use the elements of cyan stone, small bridges, cruise, pavilions and so on, to show the unique characteristic of Sanhe Town, and also develop various entertainment that visitors can participate in. It is not only for the entertainment project planning, but also provides a natural ecological green leisure park for the people of the town. "Suitable for living beside beautiful water" can attract many family weekend leisure activities, short-term vacation activities of various groups and related business conferences and activities, etc.

1) Overall planning structure(Fig. 4).

The "Hefei-Tongling" highway goes right through the north-south direction in the base, in which Yuceng River flows from west to east. So the planning has naturally formed the general layout morphology of "two axis and four areas".

Two axis are "Hefei-Tongling" highway and Yucheng River. It is not only the main traffic carrier but also an important landscape corridor.

Entrance exhibition zone is located in the way that goes into Sanhe Town and has a strong image display function. Ancient cultural, leisure and commercial zone is located in the west to "Hefei-Tongling" highway and in the south to Yucheng River, and connected with the protection of ancient town by the Xiaonan River. The culture of natural recreational zone is located in the east to "Hefei-Tongling" highway and in the south to Yucheng River, and connected with the ancient cultural, leisure and commercial zone under the Fengle Bridge. Tourism resort is located in south to Yucheng River and in the east to the culture of natural recreational zone, with the connection between Yucheng River and ecological agricultural park, tourism resorts and ecological agricultural park will contact one another closer and closer.

2) Traffic planning.

Except the trunk road and the sub-road which have been determined in the overall planning of Sanhe Town, roads in planning area mainly are divided into three levels: driveway, walkway and waterfront leisure walkway. Roads in each scenic spot are connected with the urban road. Waterfront leisure walkways connect various scenic spots through the walkway in the form of ringing the Yucheng River, and the connectivity of distinct traffic system has formed easily.

Figure 5. Entrance exhibition zone

The ancient cultural, leisure and commercial zone is the maximum flow distribution in the planning area, green parking are set for about 110 vehicles parked there. In the traveling peak, it can increase temporarily more than 30 parking spaces on entrance plaza. Besides, at the entrance of the culture of natural recreational zone and tourism resort, many parking spaces are set in order to avoid the influence of large public transportation.

3) Layout analysis.

a) Entrance exhibition zone(Fig. 5).

It consists of Sanhe Gates, Fengle River Bridge and Hetong Landscape Belt.

Sanhe Gate refers to the square group on both sides of the Fengle River. By transforming the green modeling, it builds up the spatial form of adduction and divergence of visual corridors. There are abstract sculptures with image of ancient archway on North Engle Square. Two carved pictures, one is ancient town picture and the other is water town picture, are symmetrically arranged on both sides of the highways.

Fengle River Bridge was reconstructed, using adornment plant to cover bridge column and part of the bridge, emphasizing the highlight of the night light effect, which could increase good landscape effect.

Hetong Landscape Belt refers to the road landscape from the Sanhe Gate to Sanhe Gate. On both sides of the road, thick shades of green trees give people a deep first impression. In the west, under the background of rich

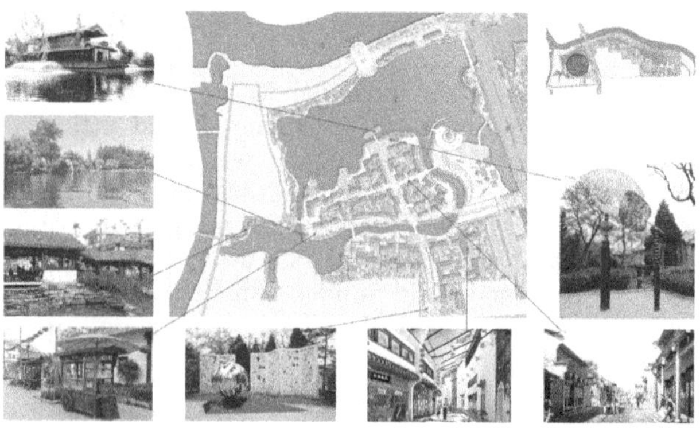

Figure 6. Ancient cultural and leisure and commercial zone

architectural features, great activity square extends gradually into the water. In the east, the miles of winding river shore lush and pavilions were seen from the shadow of the trees or boats on river. It is quiet beautiful pleasant scenery.

b) Ancient cultural, leisure and commercial zone(Fig. 6)

Go down from the Hetong Landscape Belt, and across the pontoon bridge above river, you can arrive at the gate of scenic spot — Sanhe Gate. There is a winding river with commercial shops on both sides. It is the commercial center of the ancient cultural, leisure and commercial zone. At the end of the street along the river, Magpie Pavilion and Magpie Bridge are opposite to the landscape. Go across the Magpie Bridge, go through the mountain along the river corridor, and walk along the Xiaonan River, you can reach the protection area of ancient style of Sanhe Town.

Yucheng Fang, north to the street along the river, is a group of buildings which combines the ancient elements with modern architectural elements, which has traditional forms and rich block space changes. Tea center is around the stone boat along the river. There is a shore of Magpie in the east to the stone boat, like water curving it naturally. It becomes more and more friendly and pleasant with music fountain. At the same time, it is a good wharf as well as a good place in which people people gathered.

There is a group of buildings, which is a culture center, having exhibition hall, museum, archives, etc., in the south to the street along the river. Shuangcui building lies in the southeast in this zone, which has a series of recreation composite function of living, food and bathing, etc.

c) The culture of natural recreational zone(Fig. 7).

The culture of natural recreational zone mainly consists of two parts: water town custom zone located in the north to Yucheng River and natural recreational zone located in the south to Yucheng River.

Water town custom zone begins with the tabor ship **wharves** on Fengle River, until the green water bridge, it is called "Thousand-bridge Street", reflecting the manual characteristics of Sanhe Town such as craft workshop, feather fan workshop, etc. The street space of "Thousand-bridge Street" is set along the Yucheng River, and changes according to the changes in the water. If you walk eastward, and go through "Thousand-bridge Street", "Zhuli" street is in front of you, which has a hydrophilic form and link with Yucheng River terrain closely.

Natural recreational zone mainly consists of natural green landscape. After Spring Wave Bridge, you would reach the rippling pavilion, named after "bubbling streams", which is located in the botanical garden. It is beautiful scenery. If you walk westward through the botanical garden, you will find the fishing garden, buffet barbecue park, children's park and yacht dock in turn, at last, you will arrive at Hundred- turn Bridge, which connects with Magpie Bank of ancient cultural, leisure and commercial zone. If you walk eastward through the botanical garden, there are ecological islands, homemade clay sculpture park, Lancui Pavilion and Nanjing Road, which is connected with tourism resort.

d) Tourism resort(Fig. 8)

Tourism resort area also can be divided into two parts: the recreation green belt in the north to Yucheng River and the tourism resort in the south to Yucheng River.

The recreation green belt continues the landscape of water town custom zone and is connected with a series of symbolic landscape such as Wave Overflowing Pavilion, Snow Incense Pavilion, Green Together Pavilion and the willow bank.

The tourism resort in the south to Yucheng River is also divided into two parts: Garden with Hundreds of Flowers and Garden with Hundreds of Herbs. You can enter the tourism resort along the original Wangci Road. There are waterfront passageways on both side of the road. Business center is along the Yucheng River, and combine with the water closely since the middle part of business center is hollow. Conference center is the core space of Garden with Flowers, concentrated green space and villas are arranged with concentric rings form. Entertainment food center is the core space of Garden with Herbs, in which cottages are built near the water with high appreciation.

III. OVERALL PLANNING MANAGEMENT OF CONCEPT PLANNING OF SANHE TOWN PORTAL AREA

For the concept planning of the historical and cultural town, it is not enough to focus on the function layout only, at this time, it is particularly important to take the overall planning management into account. The overall planning management is an important guiding for the next level

Figure 7. Culture of natural recreational zone

Figure 8. Tourist resort

planning and construction as well as an important content for ensure spirit program of town continued. The following aspects are mainly considered in the planning.

A. artistic scene

1) Amorous feelings of ancient town.

Sanhe Town is a thousand-year-old town, and human resources are rich. So reflecting the amorous feelings of the feeling is a key in the planning. Streets pattern, architectural style, function arrangement and so on are the main aspects that should be considered carefully, and some modern elements integrate some modern elements appropriately. The purpose is to reflect the ancient artistic style scene.

2) style of water town.

Sanhe Town has abundant water resources. So reflecting the style of water town is another key in planning. The main consideration lies in the regulation of environment, the integrating of the bank shape, construction of the street along the river, landscape of water and the construction of garden, etc in order to reflect the unique ecological environment and cultural features of the style of Sanhe Town.

B. Adjusting measures to local conditions

1) Function arrangement.

According to the function of the surrounding area and traffic flow, reasonable functional partitions are set up. The ancient cultural, leisure and commercial zone lies in the west of the base, which has a waterfront square. The culture of natural recreational zone lies in the middle part of base with many kinds of plants and complicated bank line. Tourism resort is located in the east of the base, because it is near the natural ecological leisure zone and the ecological agricultural park. A complete tourism inner circle is formed with the main tourism area in Sanhe Town.

2) Roads setting.

Roads are divided into driveway, walkway and landscape walkway. The driveways are set along the boundaries of each zone, and traffic transfer spots are set in order to reduce the influence of motor traffic. The walkways are set according to the ancient road structure. The landscape walkways are set along the river, and the landscape effect is noticeable.

3) Architecture Setting.

Single building form is mainly from representative symbol of Sanhe Town, and is integrated with the function, structure and materials today, realizing the architectural aesthetics fusion. At the same time, the form of staggered floor and overhead construction are used for the building near the Yucheng River, and scenic focal point is paid more attention to.

4) Greening configuration.

First, selection of trees species must comply with the requirements of the function and the ecological water landscape. Second, trees, shrubs and grass are used comprehensively together to form the multi-layer greening system. Third, the comprehensive consideration of seasonal changes of landscape appears alternately .

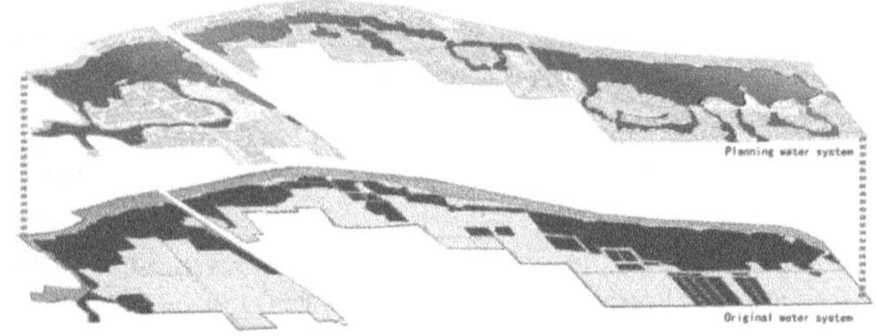

Figure 9. analysis of water system

C. in accordance with water(Fig. 9)

1) The rectification.

We make the water way of Yucheng River clear and complete, and connect the water system of the whole Sanhe Town. Those make the Yucheng River become an important carrier between each functional partition.

2) Water utilization.

In some areas, we bring water into the function of ecosystem, and layout half illicit activities and illicit activities around it. It is not only the success of different types of activities isolated, but also produces different space sequence of landscape, progressive and fascinating.

3) Building with water.

Based on the relationship between buildings and water, different waterfront processing modes are formed. In different functional partition, different activities are divided by water, people's public activities around the large water surface and open space as well as people's private activities and half illicit activities around small water surface.

D. Space positioning

There are five kinds of space in the planning.

The first kind is traditional architectural space form, mainly lies in traditional cultural, leisure and commercial zone. There is traditional architectural style with some modern spaces such as large plaza and platform on the river, etc.

The second kind is the natural ecological space form, mainly lies in the culture of natural recreational zone. Considering the ecology, natural green space is given priority to the space of the buildings.

The third one is the natural villages space form, mainly lies in the tourism resort. On the basis of natural green space and traditional village settlement, people in the city have the opportunity to make the most intimate contact with nature.

The fourth kind is water space form, mainly lies in the Yucheng River, which links the above-mentioned three space forms into a whole space through the interpenetration of the space.

The last is green space, it is the modification of water space, and it is also the important content of contacting with the above-mentioned space.

E. Open and closed

There are four zones in the planning, and each functional partition in shape is different.

The entrance exhibition zone has an emphasis on the effect of space exhibition and forms a series of rhythmic landscape elements, which meet the people's demand of landscape under different conditions such as in the garage and walking on road.

The ancient cultural, leisure and commercial zone has an emphasis on the open space. Dynamic commercial space is given priority to the static display space — the most representative and intense scenes of Sanhe Town culture appearance again in order to make visitors have a real and intuitive understanding of Sanhe Town culture in a short visit.

The culture of natural recreational zone has an emphasis on the green open space and the construction of the building blocks. Participatory experience chance is provided to visitors. Some small half illicit close space and illicit close space are set up and try to embody the characteristic of the Chinese traditional garden.

The tourism resort provides the function of conference, residential, business and leisure, etc but focuses on the closed space with a certain amount of half open space and open space. The closure of inner resort and the open of landscape along the Yucheng River are integrated perfectly.

Though four zones have different functional orientations, they are still united in the whole image planning of portal area of Sanhe Town and embody the culture and characteristics of Sanhe Town.

F. Analysis of development

1) In space, Four zones planning and development are in parallel, but the emphases are different.

The entrance exhibition zone: Provides the image display function of the entrance to satisfy people's different requirements of landscape in a car or walking on road. The main content is small square and sculpture, etc.

The ancient cultural, leisure and commercial zone: The zone is closely related to traditional culture, it mainly provides activity space that contains entertainment, leisure, traditional catering, festival party, etc. It creates modern buildings with traditional appearance and features, extracting the form from street pattern and architectural form of ancient town.

The culture of natural recreational zone: The zone is closely related with water culture of Sanhe Town, it mainly provides the beautiful natural scenery and arranges various manual workshops which visitors can participate in. Large natural scenery is important to enhance the quality of the entrance area and to promote the following development of tourism resort. It creates various characteristic environment of landscape through the different water treatment such as water affinity, paddling in the water, listening to water and so on.

Tourism resort: It is developed on peculiar natural ecological environment and human environment of Sanhe Town. It contains leisure, reception, meetings, and communication functions and so on. It can be used as family weekend vacation land as well as a business meeting land. The illicit close space, half illicit close space, the public space connect and transit one another naturally.

2) In time, one zone developing first and scrolling development of others.

During the course of development, we arrange the leisure commercial zone to develop first, because it links with the protection of style area closely. It can bring economic benefit and the brand effect and drive the development of culture of natural recreational zone. So we can create a good environment and improve the brand effect, forming the far-reaching influence.

3) In model , Independent development and mutual reliance.

The entrance exhibition zone using image planning mode, a mount of display elements become the beginning of the whole image shows of Sanhe Town. The ancient cultural, leisure and commercial uses tourism business model under the commercial mode development strategy. It builds business environment and inherit culture of Sanhe Town. It is the leading development of entrance area. The culture of natural recreational zone using natural ecological park pattern, it emphasizes on both the development and protection and focuses on formation of water culture atmosphere to improve the quality of entrance area of Sanhe Town. Tourism resort uses estate development mode. It combines the traditional estate business development and tourism economy as well as drives the development of real estate market, creating a new appreciation potential.

G. Tourism routes, self-sustaining, into the syetem of town.

In the concept planning, planning of tourism routes is combined with tour routes system of Sanhe Town. Ancient cultural tourism lines, natural ecological tourism routes and water tourism routes are three major tourism routes in planning, which can occupy one day or two days. So the internal tourism line is complete and is integrated with the tour routes system of the town, which improves the internal tourism system of Sanhe Town greatly.

IV. EPILOGUE

As a concept planning, it has a great significance on the following planning and construction. Therefore, planning orientation, planning management and content planning should be interpreted more clearly beyond the overall layout.

The sustainable construction planning of the historical and cultural town is a huge project. This article discusses the related content of the concept planning of the historical and cultural town under the background of tourism development. It is still plain. The author really hopes that the article can attract more attention and research on this historical and cultural town.

REFERENCES:

[1] GuChaoLin, "Concept planning: theory, method and examples," (second edition), China architecture &building press, 2005

[2] Wang Jinghui,RuanYisan, "protection theory and planning of historical and cultural city," China architecture &building press, 1999

[3] Zhao Yong, "Protection theory and method of Chinese cultural and historical town village," China architecture &building press, 2008

From Tujia Minority People's Residential Buildings to the Application of City Design

Jiang Meng
School of Art and Design
Wuhan University of Technology
Wuhan, China
Email: jiangm911@163.com

***Abstract*—Tujia architecture, a kind of unique civilized aesthetic culture with strong local characteristics owning to its special aesthetic judgment as well as its unusual structure, influences other regions' architecture designing or even a city's "blue-map-making" in China, which embodies Chinese traditional religious, philosophic ideology: the combination of nature and human, integration of universe and human, respect to God and ancestors; while its molding patterns as well as its decoration expresses Chinese traditional cultural essence in the aspect of art: place feelings on objects. Based on the study of the existed architectural sceneries and the city developing zones, this thesis takes the architectures of Enshi city as an example to conduct a study of the designing application in a city's construction.**

Keywords-Tujia Minority People; architecture; Landscape System

I. Introduction

Qingjiang River crosses Enshi City, and the city's landscape system named "Qinshui Corridor" is constructed along this river. "Qinshui Corridor" can be divided into four sections: the second section is Fengyu Bridge, which embodies Tujia People's customs and practices; the third section shall be the key competence of this city is from Selenium Square to Plaza of Nations, which inherits and enhances Tujia spirit; the last section is the ancient city wall and the Connected Corridor of Tujia Diaojiao Building, which covers a small part of the city while its architecture intension concludes strong Tujia cultural characteristic while discloses the essence of local people's spirit, and this is the end of this landscape system. Over-viewing the whole landscape system, the beginning section contains few elements of local characteristics; the construction of this part in form of art shall combine the specific characteristics of Tujia People and the economy consideration. So the needed work is to decorate the existed buildings while avoiding building big sculptures or other architectures and the best choice can be the cable stayed bridge (on building). So the author proposes some suggestions: place concrete sculptures along the bridge; apply embossments in relation to Tujia People or introduce the historic evolution of Enshi city on the bridge tower to display amorous feelings, culture, worship, natural scene, life convention and architectural style; so the bridge tower can be designed as Tujia People's totem post or monumental works of the whole city that embodies newly understanding of the local manners and feelings in the most charming, vivid , primitive forms.

II. Exploration of Traditional Tujia Architecture Landscape

Wherever Times is specified, Times Roman or Times New Roman may be used. If neither is available on your word processor, please use the font closest in appearance to Times. Avoid using bit-mapped fonts if possible. True-Type 1 or Open Type fonts are preferred. Please embed symbol fonts, as well, for math, etc.

A. Landscape Point in the Second Section: Qinhe Palace

Qinhe Palace, located in the second part of "Qinhe Corridor" landscape system, and its main architecture is Fengyu Bridge. This bridge combines Tujia customs and practices: concrete pier, steel approach, wood floor, and the whole bridge is covered by wood roof with rows of tiles. The tile roof contains upper and lower eaves with four claws, five tiers strew, which embodies Tujia thoughts "Five Sacred Mountains Pointing to Heaven". The bridgehead follows the width of Qingjiang River; both sides are balustrades; parallel with the balustrade are long-benches. This building's design is mainly inspired by primitive Fengyu Bridge's structure and by the historical architecture of Tusi Chieftain City.

1) Fengyu Bridge

The embryonic Fengyu (wind and rain) Bridge depends on it's functionality; it appeared firstly in mountain creeks or over rivers long time ago, which was used to be shelters in busy seasons for peasants or hunters on their way working or home, so people named it "Relaxing Pavilion Bridge". While with the maturity of bridge-making technology as well as the progress of people's aesthetic consciousness, this special bridge's construction evolutes, the structure is more complicated and exquisite, and its decoration becomes rich and delicate, the model is more unique, magnificent and conspicuous with local characteristics. The bridge body usually is divided into two parts: bridge body and bridge pier as a common bridge, while it's characteristics is the dyadic lifting beam, and the beam is connected by Mortises and Tenons, which looks abnormal elegant and pretty grand; sculptures and colored drawings are made on the beam column and eaves mouths, which is Tujia inherited architecture art form without exception. The pier is piled with stone bricks, and it takes forceful momentum in appearance; combined with the bridge body, the whole bridge forms a contrarily image of lightness and heaviness, falsehood and reality, strength and beauty.

2) Tusi Chieftain City

Tusi Chieftain City locates in En Shi suburb. There are undulating hills, watermarks on green hill, green pines besides tall bamboo, which takes on conspicuous majestic momentum, manifests vigorous strength of Tujia People. The main building "Jiujin hall" (Hall of Nine Terrace) is the politic and cultural centre of chieftain in the old time, which complies with glen of strategically location and difficulty of access, follows the physical feature of a place to rise one by one step, and the building has obvious functional division that manifests sense of wholeness and functionality. Combined with the forceful city wall surrounding along the mountains' peaks and valleys, the whole city can be a miniature society for Tujia People to enjoy life while stands a strategically located place. Although the Tusi Chieftain's time had gone forever, this special architecture as well as its spatial distribution and the building form has been conserved; even in some of modern architectures, Tujia People also takes its form. The typical application is in Hubei University for Nationalities. This university is built along a hill: it's constructing process, whole distribution, outward appearance, space feeling, and all expresses directly or indirectly this special architecture concept which embodies distinctive culture connotations.

B. Landscape Point in the Third Section: Plaza of Nations

Plaza of Nations locates in the third section of "Qinshui Corridor" landscape system, is a successful example of Tujia culture's application in modern city design. How to apply the traditional elements to reverse history essence as a city's memory to call up visitors' resonance while posses innovation to obey the new times' functionality as well as the aesthetic values in a city's construction? The key lies in the way of making use of and the reform of the environment. China is a country of rich culture, and a city's construction can not discard tradition, contrarily, a city's buildings or landscape systems should borrow scenery from the city's history to be a city's enrichment, complement and enhancement. It is only considering the specific customs, practices and the nature for the establishment of a city's buildings or scenery, the man-made works can be melt into the wholeness of a landscape system.

1) Ancestral Temple of Family Lee

Ancestral Temple of Family Lee locates in the suburb of Li Chuan City, En Shi, which covers more than 4,000 square meters, including 24 patios, 174 rooms, and the basic frame is wood-making. The front courtyard is the combination of Western and Tujia architecture patterns, while the entrance is a traditional Chinese opening door; the decoration of the columns as well as the semicircle-shaped eaves mouth of the main storied building takes western style; while the rooms on both sides is traditional "Diaojiao building" pattern. This family temple combines traditional and modern methods in its construction form that fresh people's eyesight. This peculiar architecture product's formation based on the combination of traditional Chinese culture and western culture in late Qing Dynasty: the host of this house was rich, and his eldest son was lucky to study architecture abroad; after World War II, the eldest son back home but the house ruined, so he decided to reconstruct it. The front courtyard is paved with flat bluestone; the front corridor column is thick and straight with gorgeous relief; and the diaojiao building's beam is interwoven with colorful paintings.

This family temple takes characteristic architecture concepts: "Zouma Corner House", "one post connected six beams", and "one post connected nine beams", which was highly praised and widely copied. And the application of western architecture patterns added new elements to Tujia architecture culture that attracts and stimulates designers' inspiration in the development of national architecture.

C. Landscape Point in the Fourth Section: Ancient City Wall, Connected Corridor of Tujia Diaojiao Building

The ancient city wall of Enshi City and the connected corridor of diaojiao building, locates in the fourth section of the landscape system, which is a typical harmonious unification of Tujia buildings in this modern city. The windows and doors sculpture and decoration of the corridor is vivid and changeful with various subjects, which mutually effects the aesthetic feeling with blue tiles and with wood wall that produces elegant and luminous beauty. The window and door sculptures also manifests Tujia carpenter's delicate wood carving techniques, which combines functional and auspicious pursuits together that displays Tujia People purity and passion of life.

1) Diaojiao Building

Diaojiao Buildings, representative Stilt Style Caroline Moreno architectures, are tradition residential houses of Tujia People, which are usually scattered around a mountain foot and are well-arranged in groups that creates grand effects in aesthetic aspects as a whole community. Two turrets stand parallel on both sides that creates symmetry

effects while the natural employment of apace arrangement embodies the harmonious cultivation of the special mountainous environment for required lebensraum. This special architecture form, actually, is the local people's best choice under such precipitous mountain areas. As for external form, most Diaojiao buildings are built with nine posts falling to the ground and with beams crossed excessively, which can be classified to dyadic buildings: the building shall be divided into two floors; and the corridor rounds the turrets with its post suspending in the midair like a man standing with one foot--such dyadic wooden building always is in the structure of five posts with six or seven hanging pillars that can be constructed in any dangerous steeps. The tight structure of this kind of building is very unified and solid without one nail or rivet and the advantages of this architectural merit are: clear functional distinction between force bearing structure and sheltering walls, that means the whole building's wooden framework burdens the force bearing while the walls just separate the building into several small rooms; secondly, easy adaptation to various climate condition, height of the building, thickness of the wall, and the size as well as the face of the windows and doors are arranged according to specific geographic conditions; thirdly, the special characteristic of wood and the building's joints can minimize the damage of earthquake that even the walls fall apart, the building would stands safely; fourthly, the fabrication can be conducted on the local resources in no need of brick-making or material-transportation.

Diaojiao building attaches importance to dragon vein in respect of traditional Fengshui theory that embodies rich cultural connotation. Besides its random complying with terrain conditions as well as its co-existence of people with Gods ideology, it also emphasizes the notion of architecture space to be universalized. The constructers of Diaojiao buildings usually make terraces along flights, foster post upon holes, so the building can be set in hills, or over waters, or on plains that fully employ the limited land while the buildings can be merit into the surrounding environment, which combines complement of human and nature, metaphysical sense of heaven and earth for the Cottage. So Dijiao building is ecological building of China's ethnic minorities' people with distinctive local characteristics.

Diaojiao building groups are nation's arts fruits of architecture, painting and sculpture, and also are outstanding representative in Tujia architecture history. The consideration of arts decoration and sculpture concerns the ridge of a roof, the pillars, beams, corridor balustrades, posts-bottom, eave boards, doors and windows that condenses wisdom of Tujia People.

III. THE CONSTRUCTION PROJECTS FOR THE CITY'S ENTRANCE

A. *The Necessity to be a City's Landscape*

In this part, the main project is a cable stayed bridge. This cable stayed bridge holds a peculiar important position in En Shi city for its projects crossing the Qingjiang River: firstly, its location is eminent; the first scenery that comes into people's eyesight is this bridge when visitors enter into the urban even from every road. Secondly, this bridge is abnormally huge that can make extremely profound impression to its visitors. Thirdly, the landscape establishments for this bridge is available because it is on build that the construction of its function and landscape facilities can be conducted simultaneously. The cross over the river of this bridge unifies the landscape from both sides to be an entirety and the graceful structure of the bridge tower can be one of the city's signal landscapes. From this aspect, people's feeling of its existence proposes essential significance without appreciating it via visiting a museum or gallery.

B. *The Inspired Roots of Architecture Arts*

In feudal society, Totem post is a kind of commend building erected to eulogize outstandingly meritorious deed , section mansion, benevolent rule and loyalty, filial piety, chastity and righteousness and also to offer sacrifices to ancestors, heaven and Gods while such activities were held, Tujia People would participate in hand-waving dancing to celebrate them. If the bridge tower is designed and constructed as a Totem post, enormous effects will be given to the citizens of this city. Tujia People in their tradition worships large-scale stele: Panlong Mountain, an overhanging cliff, located in the middle reaches of Qingjiang River, on the steep peak of which a stele temple stands tall and upright. At the foot of this mountain, there is a big cave elongates to the mountainside while leading a 238-grades-stone-stairs rounding up the way along the slope of the mountain to the peak. At the peak, among the dark green, old pines stands vigorously and forcefully while the wild flowers blossom. Yumu Stockade, a lonely stockade citadel made of stones and wood, sitting at the Peak of Panlong Mountain, in which numerous stone tablets were caste in rows. Yumu Stockade experienced tens hundreds of years, is a perfect combination of wood and stone in architecture form with flying-shape eaves that is covered among dark green. The Yumu villagers lead their life complying with the dangerous location of the mountain for cottage that embodies that man is an integral part of nature. As a representative of Tujia stockade culture, this group of architecture footmarks the development of Tujia architecture art that fully condenses the local people's wisdom as well as their creating abilities.

C. *Fengshui Consideration*

From the aspect of Fengshui Geomancy theory (that emphasizes "cause and effect"), the bridge tower establishment of Enshi Cable Stayed Bridge as a signal Totem post conforms to the theory of Geomantic "vitality" (or cause): the bridge happens to be in a position surrounded by mountains and girdled by a river, so this embossed building is not only able to exorcise evil spirits, but also to gain effects of "Feng Hiding and Qi Gathering". The bridge tower locates at the entrance mouth of the city's dragon pulse while the bridge floor faces to the upper reaches of Qingjiang River, so the emboss design can be proposed according to the "Qi leading" theory that comforts to "the Qi comes while the river leads" and so the whole city would be timely wind and rain.

D. *Permanent Landscape*

Talking about the permanency of a piece of art works, its long-lasting existence of physical form is not enough, the un-evaluated art values are more important. So the birth of such an embossed architecture in this small city may be long-lasted while the appreciation of the local people to it even may be forced. The face-to-face art communication of this art product with local residents exerts subtle influences to develop the citizens' moral sentiments as well as to advance their cultural quality.

The successful construction of the bridge tower in form of art as a node of this landscape system symbolizes the city's culture quality that takes irreplaceable place in the inheritance of Tujia culture, which stands upon Qingjiang River from dawn to night emitting art brilliance forever. It is a commended cultural symbol of a city, and even of a nation that promotes conveniences to the contemporary people while leaves perpetual footmarks to the future.

[1] John O. Simens, "Study of Landscape Designing," Beijing: Press of Chinese Architecture Industry, 2000.

[2] Manford L., "History of City Development: Origin, Evolution, and Future," Beijing: Press of Chinese Architecture Industry, 1989.

[3] Tang Qiong, "Architecture Arts of Tujia Minority People," in Decoration, Vol 7, pp 24-26, July 2006.

[4] Zhang Liangquan, "Tujia Diaojiao Building," Nanjing: Jiangsu Fine Arts Press, 1994.

[5] Man Yide, Xie Yaping, "Exploration of the Inhabitants architecture Arts of West Hubei Province: Take Three Villages of Enshi Autonomousdistrict as Examples," in Architecture, Vol 11, pp67-72, November 2007.

[6] Science History Study Institute of Chinese Acadamy: "The History of Chinese Ancient Architecture Techniques," 1984.

[7] Shen Fuxi, "The Cultural History of Chinese Ancient Architecture," Shanghai: Shanghai Guji Press, 2001.

[8] Li Renkui, "Explanation to Fengshui Concepts," Beijing: Quanyuan Press.

[9] Li Kaiqu, "Suggestions to The Construction of Downtown Sculptures in China's Main Cities," Beijing: Accociation of Chinese Artist, 1982.

Function, Scale and Level

Investigation of Multicentre Developmental Patterns of Megacities

Yongying Tian
School of Architecture
Tianjin University
Tianjin, China
Email: tyy_tj@163.com

Yukun Zhang
School of Architecture
Tianjin University
Tianjin, China
Email: zyk.tj@163.com

***Abstract*—At present, China is undergoing a rapid development of urbanization. The amount and scale of megacities are increasing at an amazing velocity, resulting that the single centre developmental pattern cannot meet the requirement of megacity. The present paper, in this context, elaborates the concept and characteristic of multicentre developmental pattern of megacity employing the fundamental principles of biology. It further investigates the issue, how to achieve the multicentre developmental pattern of megacity, so as to optimize the layout design of Chinese megacities.**

Key words-multicentre; function; scale; level

I. INTRODUCTION

China, as well-known, is undergoing a rapid development of urbanization. The ratio of urbanization in nationwide reached to 45.68% by the end of 2008. Especially the increasing velocity of the amount and scale of megacities appears to be amazing. It is predicted that the amount of the megacities with million populations will exceed to 150. A lot of challenging problems, however, become more and more acute with the enlargement of megacity scale. China faces a high tension on the urbanization due to a big population in countryside and the lack land resource of construction. The multicentre pattern, therefore, has the promising advantage in the optimization of urban structure. At present, a couple of cities such as Beijing, Shanghai, and Guangzhou are planning to form multicentre megacities with new structural districts for residence community and industrial community. It is noted that the premise of application is the clarification and comprehension of the concept and basic properties of the multicentre developmental pattern. Having this knowledge in mind, this paper is devoted to investigate the fundamental concept and properties of the multicentre developmental pattern, and to propose some practical suggestions on how to achieve the multicentre developmental pattern of megacities. It is a tentative step to optimize the layout design of Chinese megacities.

II. FUNDAMENTAL CONCEPT

In the process of urbanization, the developmental patterns of megacities are exclusively the sharp alteration of the inner system and the rapid extension of the outer profile. They both result in the multicentre pattern of cities. For example, Tokyo in Japan composed by three sub-capitals, i.e. Shinjuku, Shibuya, and Ikebukuro; London in UK, having 8 new districts around. Although the pattern seems to be the simple aggregation of multiple cities, it is, however, essentially different from the original pattern. They are not analog to the smaller elephant and the bigger elephant, but to the single-cell creature and the multi-cell creature. Obviously, the difference from the extended pattern and the original pattern is not resulted from the quantitative variation, but from the qualitative variation. Hence, the multicentre pattern of cities in this paper refers to the typical districts in the urbanization area, which are scattered in space, but are connected in logic. Each district in this area, moreover, undertakes a certain or some specific functions, and satisfies the basic requirement of daily life of urban inhabitants.

III. URBAN LAYOUT BASED ON BIOLOBY

As the father of ecology and humanism, Park, has noted that the urban spatial order is finally the production of ecology order, and human society is organized in two aspects, namely biology and humanism, resulting in that human being, similar to the creature, evolves in a way of competition. Therefore, it is practical to investigate urban layout using the fundamental principle of biology. The key point is to construct suitable perspective of investigation. It should obey the fundamental principle of biology, and reflect the basic properties of city as well. According to this principle, we investigate the multicentre developmental pattern of megacities based on three key words of biology, which are capacity, scale and level, respectively.

A. Urban function

The function of cells in biology is that the noble cells include the all genetic information, exhibiting the capacity of developing into new individual under favorable conditions. Moreover, the function of each cell does not only govern the health level of new individuals, also affect the evolution result. Similarly, in the process of urban development, the improvement of function underlies the health development of cities, and the structural evolution essence of urban space is that the structural space intelligently varies so as to meet the urban function. The development of whole city or urban area, therefore, has similar behavior to the organic evolution if each subcentre of city is viewed as a cell, which can grow into a new individual with the urban construction, and the urban function is viewed as the genetic information. A multifarious and dynamic city then can be formed by efficient combination of urban function. In fact, the difference in developmental stage and strategic orientation will lead to different levels of function of urban subcentre. The levels are divided into totipotent centre, multifunctional centre, and professional centre in principle. As was mentioned, the problems resulted from the development of city can be viewed as the disease parasitizing in the living creature. For example, some cities set out large-scale residence area in the suburb of cities in order to relieve the housing tension of townsfolk. Some new problems, however, arise such as the traffic jam though the residence requirement of the subcentre cored as the residence area can be satisfied, due to un-neutral attention paid on residence function, resulting in the lack of other facilities and conditions, e.g. education, commerce, and jobs. This layout actually does not settle down the housing problem; instead consume massive labor power and financial resources. The built subcentre is just like the new tissues generating from abnormal growth of cancer cell in organism. It occupies the space of health tissues need and consumes the resources of organism since its abnormal performance.

B. *Urban Scale*

Scale usually refers to dimension units or time units in describing certain phenomenon or investigate certain matter. It is also able to assess the dimensional range or frequency inherent in the phenomenon or matter. We explore the urban layout employing the scale concept in biology, as follows. First, the generation and growth of melanin will make the skin of nearly all mammals form homogeneous spots and figures. It is the result of creature evolution. We take the snake as an example. Its figures obviously distribute along the length of body due to its waistline being significantly smaller than its body length. Actually, the quantity and distribution of urban population and commerce aggregation have significant effect on the development of cities. If the population and commerce area are viewed as a kind of figures, the city will acts as a special creature having this kind of figures. We thus can figure out the integration properties of urban development, as shown in Fig. 1. It is seen that Shanghai has relatively concentrative figures since that Shanghai belongs to the spatial structure with high density and single centre. Other cities, however, have relatively scattered figures. As statistics, the spatial configuration of 31 megacities in China appears to be compact from 1990 to 2000. Second, the size and ratio of body length, weight and skin area do not only affect the appearance of animals, also affect the metabolism and thermal losses. They are further related to the organ function of animals. For example, digested food and respiratory oxygen both get into body-internal-circulation from the outer; the strength of shank relies on its cross-section. As usual, function is in direct proportion to the dimension of body. It will be, however, limited when the dimension reaches to an extent. The only approach, in this context, to prompt creature evolution is optimizing the inner organs. This principle is also suitable to the plant, and is, of course, suitable to the building and the city. When construction area is limited, optimizing urban spatial structure, not simple extension on urban scale, is the valid method to realize the sustainable development of city.

C. *Urban level*

Homogeneity or order is one of the common characteristics of living entity. Whether cells or their organic composition, such as tissues, organs, system, individual and even biotic community, all forms compose of highly homogeneous structure as certain levels. This kind of multi-layer, network and high-dimensional structures provides a smooth channel for delivering parent information, greatly improving the ability of self-control and response to outer environment. Therefore, reasonable level assessment is very important to the regular service of vital movement. A city, especially when it develops into a stage, will undergo a series of processes, e.g. the generation and distribution of new tissues, the generation of new cores and reconstruction of multiple cores. These processes exhibit multiple layers and structural levels, as shown in Fig. 2. The city or urban area as the mode of multicentre development is thus an integration of multiple subcentres with different function, different level, and different scale. These subcentres are closely related and interdependent, and can be divided into urban center, district center, suburban center and neighborhood center.

IV. SUGGESTION TO MULTICENTRE DEVELOPMENTAL PATTERN

It is seen that the multicentre developmental pattern exhibits the properties of getting from junior to senior, getting from simplicity to complexity. This evolution process does not stress the quantity and velocity of development, more importantly; stress the quality of development, especially the correlation between various centers. For an illustrative purpose, the totipotent of development is investigated. We thus propose a series of suggestions on how to carry out the core scale, and how to increase the core level so as to achieve the sustainable development of city.

A. *Corporation and Complementation, developing pole nucleus of regional function*

Nowadays, the urban function approaches to be complex. The individual centre of the multicentre developmental pattern should pay attention to the efficient combination between organic decentralization and function diversity so that the principle of corporation and complementation can be achieved, developing the powerful core of district function, and having the merit of unity better than summation of segments. The remark does make sense to many cities in China since their existing strong cohesion to the central area due to original layout principle of the single centre. In addition, the parasitic city should be prevented in the process of urban construction, which is often resulted from the lack of subcentre function, and the level discrepancy between main centre and subcentre.

B. Breaking up whole into parts, communication promoting regional development

There has the reasonable scale in the urban multicentre development. The scale relationship between various centers always changes and regulates in the different districts and different developing stage. In fact, many abroad cities use the railway to form the traffic network between centers so as to utilize the resources and serve the city. However, the city could become more crowded if we ignore the developing stage of centers and the distance between them, and only rely on the rapid traffic facilities, such as the countryside residence community mentioned before. The current situation in China is the unbalanced development of centers. We thus should use various traffic facilities with different levels according to the function of centers and their distance. Besides, the urban function with different scales should include the living schedule of residence folks, and make the distribution of urban space more reasonable.

C. Multipole development, increasing levels of centers

With the extension of urban scale, the level of urban centers should be adjusted and improved correspondingly. Especially the center with significant economy-increasing pole, e.g. the new district under construction located at countryside. Various centers at different developing stages with different levels will appear to decentralization characteristics, including massive infrastructure, location, politics and resources. These characteristics rely on the distance between subcentre and main center. It should be noted that the new city cannot play the expected role in some cities with multicentre due to the big discrepancy of scale between new city and central city. Therefore, the level of city development with multicentre must be controlled in an extent, which can be achieved by reasonably designing function of various centers, or by reasonably increasing centers with different function and different level.

V. Conclusion

John Baldwin has noted that each system has its own optimal scale, and within this scale, the bigger has better properties; beyond this scale, otherwise, the bigger has worse properties. Although it reveals that it is necessary to substitute the multicentre pattern for the single centre pattern, the single centre pattern is yet accepted at some situations. The single centre pattern, in fact, still plays an important role having significant effectiveness of function integration when it is used in a reasonable way. Besides, the geographical position, climatic condition, natural resource, population and developing level of economic society of various cities are obviously different, therefore the decision of choosing developing pattern of urban space needs comprehensive consideration and oriented survey. There are a lot of questions that need further investigations on how to set down a reasonable direction of urbanization.

References

[1] J. X. Zhang, Outline History of Western urban planning thought, Nanjing: Southeast University Press, 2005, pp. 180.

[2] S. L. Jiang and J. F. Cui, "Research into the evolution of urban spatial structure in information age," Planners, vol. 7, pp. 94–96, July 2006.

[3] X. S. Wang, J. Y. Liu, D. F. Zhuang, and L. M. Wang, "The spatial characteristics of the morphological change in mega-cities," Geographical Journal, vol. 3, pp. 392–400, March 2005.

[4] J. M. Thompson, Urban layout and traffic planning, W. X. Tao, W. Y. Ni trans. Beijing: China Architecute & Building Industry Press, 1982, pp. 245–248.

[5] H. L. Leung, "Planning strategies inspired from ecology," City Planning Review, vol. 5, pp. 51–54, October 2000.

[6] J. Duan, Urban spatial development theory, Jiangsu: Phoenix Science Press, 2006, pp. 149.

[7] B. Alain, The Spatial Organization of Cities: Deliberate Outcome or Unforeseen Consequence? Institute of Urban and Regional Development University of California at Berkeley, 2004.

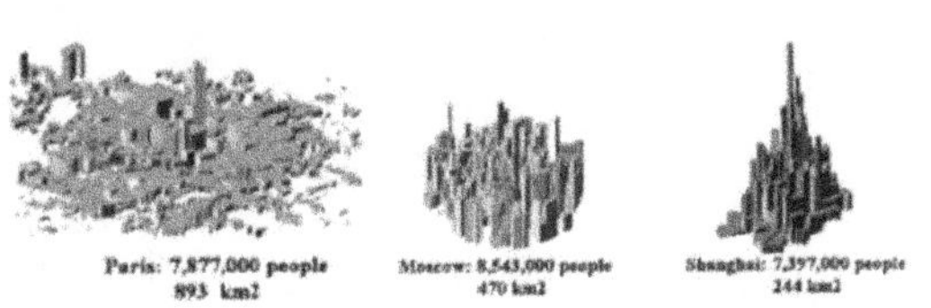

Figure 1. Graphs on population density of three-dimensional in Paris, Moscow, Shanghai.

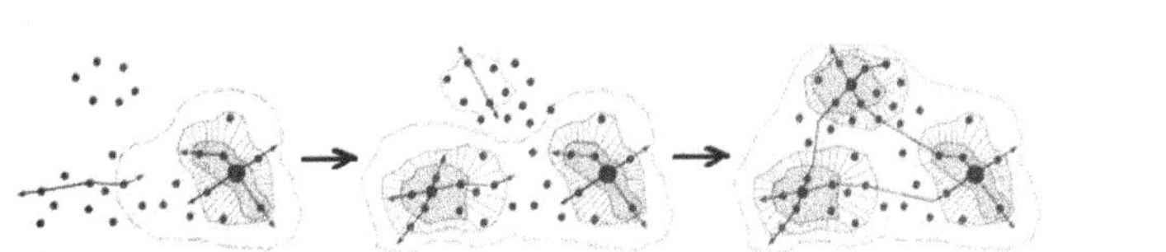

Figure 2. Schematic of growth model of urban space.

Further Study on Optimization Design of Multi-Dimensional Form of Books

Cao gang
Art & Design Institution
Wuhan University of Science & Engineering
Wuhan, China
Email: cgang@vip.qq.com

Li haiyan
Teaching and researching of art
Wuhan Experimental Foreign Languages School
Wuhan, China
Email: 277148900@qq.com

Abstract—**In order to meet the requirements of modern information dissemination, the design of modern books starts to seek multi-disciplinary integration of multiple technologies based on the design of traditional books. The characteristic of modern books is no longer the planar structure to transmit knowledge unilaterally, but a series of horizontal, vertical, multi-directional, and reflective new researches, which are implicative of each other, comprehend by analogy, and contain the traditional and modern research, the inner and outer part of book layout composition, the material composition and color communication, multi-dimensional structure, etc. Since modern books have changed from the inherent three-dimensional structure to multi-dimensional form, designers should strengthen even more on the optimal design of elements of books like structure, layout, colors, materials, etc., meet various reading needs of readers on the books, enhance visual, readable, convenient, and pleasant nature of books, extend the optimal design of multi-dimensional form for modern book, and develop the cultural values.**

Keywords: ***books, structure, layout, color, material, optimization***

I. INTRODUCTION

All The multimedia communication we faced in today's information age have had great impact on the patterns of books, and the laws used in the design of book design in the past are gradually broken. The forms of book design are gradually extended to the three-dimensional or multi-dimensional level, and the performance measures such as interactive technology, dynamic images, sounds and even smells and temperature are increasingly applying in the creation of book design field. Morden book design starts a series of new research on multi-dimensional structure, the inner and outer part of book layout composition, color communication, and the material composition.

II. OPTIMAL DESIGN OF BOOK STRUCTURE UNDER THE MULTI-DIMENSIONAL FORM

Book is the container to hold information. Generally speaking, people are very interested in space. Through multi-sensory experience, they can change the small physical space of books into multi-dimensional spiritual space.

The multi-dimensional forms of books in this digital era can be divided into two types: the first is the multi-dimensional form of books formed by the application of two-dimensional plane in three-dimensional space or even four-dimensional space; the second is the multi-dimensional form of books formed by the integration of various media advantage and the expansion of various sensory channels. The former is achieved by time and space factors, while the latter is the integrated use of various techniques. The expansion of multi-dimensional forms of books is not only confined to the narrow sense of "three dimension", but also contains the "interactive book" and "movable book" that interact with readers. Book itself is a time pattern contains the gradual process of reading. The famous Japanese book designer Kohei Sugiura described the overall form of books by body structure. The senses of levels, dynamics and four-dimensional space on books concerned by the design of modern books enlighten us to re-think and re-understand books from the points of multi-dimensional space and three-dimensional structure.

Modern book design has divided the printing space from the space structure of book ontology, and strengthened the spatial extension as well as the three-dimension, space, sound, figure or image extension on the aspect of book structure to increase the reader's understanding and to stimulate and expand the imagination. Book design is no longer limited to the single material or a simple structural design, but also contains the optimal design model of multi-dimensional structure, and is optimized to the changing structural designs that are interactive and operational, and can be turned, pulled, pushed, rotated, expanded, closed, extended, and spread, etc. It's aimed at the multiple interactive experiences, and stimulating the limitless imagination and creativity of readers. Structural Optimization of books enables them to have both functions of reading and gaming, and particular emphasizes on the using of hands and the participation of readers, which enables the books to have more possible forms, and accepts the information dissemination in the inspiration process of imagination and creativity. Fro example, the three-dimensional book "The Wizard of Oz" has combined with the application of paper materials and various producing printing technology, whose complex and fine structural optimizationdesign is spectacular. A large number of three-dimensional graphics in books described the main line of story. When opening the cover, a big tree will be erected on the book page, which highlights the central figure. In addition, there are a number of small devices to supplement the story, such as: grooves and pulleys, and the three-dimensional part can slide in the grooves to interact with readers and result in resonance. It is another performance of spatial thinking.

III. OPTIMAL DESIGN OF BOOK LAYOUT UNDER THE MULTI-DIMENSIONAL FORM

The purpose of layout design is to present clear structure relationship through clean-up of text context, and try to bring out the readability, convenience, and comfort of reading, which can not only convey information, but also generate sensory aesthetics. A good book layout design can clearly show the nature, style, structure and hierarchy of the original text, and be perfect harmony with the Folio, binding forms and the cover illustration style. With the space extension and strengthening of books outside plane graphics, the multi-dimensional form of books should be optimized to increase the readers' understanding as well as stimulate and enlarge imagination. The optimized design of book layout can determine primary and secondary relationships and changes in rhythm through different form, size, distance, direction, and color elements according to creative needs, express the forms in different sense of power, and adopt simple and clear fonts, graphics and symbols in design to allow the mutual harmony, mutual exchanges and mutual promotions of information, and jointly build a pattern and new concept of book layout design.

The design of modern multi-dimensional form books makes extensive use of digital media, which not only enables the reform of contact means between people, but also directly participates in the standard of out real life framework, and creates the novel and various layout optimization of design space. With the various kinds of dealing methods of image synthesis, thoroughly stack, direction rotation, image filters, etc. expressed by computer, it can form an optimized layout with multi-dimensional space. This optimized constitution method makes layout no longer a simple and single constitution relationship, but the three dimension with multi-view and conflicted space levels, thus to stimulate the audience and result in an unprecedented art form. This is also the optimal design of book layout under the multi-dimension.

The optimal design of book layout needs virtual magical use, white magical use, in-depth and detailed appreciation and creation, as well as unlimited exploration in the nuances. We should apply the four-dimensional (time, space) concept to the layout design, use the "black" and "white" to form the point, line and surface rich in rhythm, and lay them throughout every side of books' multi-dimension and every page inside the passage. So the whole book becomes a changing room for imagination and a wonderful black and white world.

Details not only have a role of finishing touch on the optimal design of book layout, but can also generate the fascinating charm through careful characterization. The row spacing is set for promoting the circulation of passage, while the spacing black and line prints can set off with each other; the determination of row spacing depends on the fluency of passage and the clear lines. Meanwhile, the treatment of relationship between white border and center layout should be adjusted according to different types of books; as for the organic composition of rows, columns, comments, top margin, and page number, etc., we should start from the contents and structure, and emphasize on the optimal scientificity within various factors as well as the relationship between them and the sights.

As for the optimal design of book layout, we often combine the chaotic graphics together according to certain sense of form, and organize the various form factors within unified structure. We can also array clutter part into certain canonical form, and change the irregular shape with lots of detailed changes into geometric basic forms, so that changes in various sizes, color, right and wrong can be arranged in gradual sequence, which are adaptation to the special visual psychological tendency.

Of course, during the optimal design of layout, we should also pay attention to the cultural background and consumer objects, and avoid cognitive impairment. If we mix the symbolic language with different patterns together, people will lose the basis for identification. In addition, when using new forms of symbol manifestations, we should make them easy to learn, easy to remember, and can never forget with one sight. Combine different presentation styles and smartness together to make them in line with the current artistic trends. Good works prove that only those layout which use spatial thinking to optimize the design of books contents are beautiful.

IV. OPTIMAL DESIGN OF BOOK COLORS UNDER THE MULTI-DIMENSIONAL FORM

Color is a polygon prism to express the feelings and transmit the feelings, and is an important visual element in books. Color is the extension, progression, complement and deepening of the artistic language in book design. Through rational allocation within colors, as well as the arrangements of interrelation within colors, we can achieve harmony of colors. As for the book design under modern multi-dimensional form, the important purposes of using colors are not only to bestow forms with the visual aesthetics, but to rationally choose and match colors through the understanding of color emotional characteristics, and create the works with profound moods and rich bookish mood. After the interaction between books and readers under the modern multi-dimensional forms, the use of colors should be based on the theme of books, and keep the affective personality of color in line with the theme contents, keep a appropriate degree of using colors, fully optimize the colors, and use color associations to achieve the best visual effects.

Optimal Design of color should also consider about its position on the layout, the impact on the layout when mixed-using as well as the orderly arrangement. For example, the complementary color is the most antagonistic within the mutual antagonism, while once it gains the harmony and forms the unity of oppositeness, it can presents a stirring sense of harmony with strong intensity. Color is instability, while the various elements of colors are interdependence. In the optimal design of color, and in the process of flexibly using various forms of law, we must grasp the overall effect, pay attention to the combination of theory and practice in order to optimally design a variety of color aesthetic forms.

Books have a wide variety and varied contents, and face all levels of readers. The symbolic of color and the emotional associations caused by it are often preferred by readers with different ages and different cultural backgrounds. The features such as the contents of books, the age of readers, cultural levels determine the use of color optimization, and different books have different focuses. For example, children is easy to be attracted by the crisp and bright colors, thus the cover of children's books often use orange, yellow, green, etc. colors that symbolizes youth and vitality as the main colors, which are suitable for the innocence and liveliness of children. While the colors such ad brown and gray can bring people the feelings of coolness, calmness and elegance, so they are fond by the middle-aged readers, especially scholars. Theory and academic works, etc. often use plain colors, namely, rational colors, which have stable visual sense, good potential, and are easy to be popularly supported; old books can use the colors with low purity, calmness, and harmony, so as to reflect the ideological connotation and artistic conception of the books, which are consistent with the character traits of our readers; bright and vibrant colors are often used for children's books, which have strong first feeling, and are easy to meet the visual pleasure; according to the contents, the colors of young people's readings have more widely choices. Optimized color language has its own characteristics and functions in the book design — simplicity, decoration, symbolic, psychological role, lyrical role, associative role and the media role.

The color optimization in the modern book design is not equivalent to the color optimization of other commercial products, for it is deeper, and has more cultural quality. The color optimization of books should be very careful: the colors must be penetrative, and can move minds, while the message they conveyed should be heavy and rich. A clear and simple optimization of color has strong symbolic, which can effectively unify the general features and highlight the overall appearance; gentle and subtle colors are implicit and inward, which enable people to taste carefully, and have a sense of reverberation proudly; pure colors make books have clear personality and strong visual impact; rich colors are symphony, which can significantly meet the needs of visual aesthetics.

V. OPTIMAL DESIGN OF BOOK MATERIALS UNDER THE MULTI-DIMENSIONAL FORM

The creation and breakthrough of book design concepts in multi-dimensional forms has orientation meaning and space expansion on the recognition of book design undoubtedly. The materials of modern books are not limited to papers, and any things can be the materials. Only if they meet the theme, they are our best choice. So during the process of optimizing book design these aspects come first: accurately grasp visual and touching feelings of materials; grasp the personality of materials, and fully express the contents of materials; diversified portfolio of materials, fully express the potential power of words through multiple kinds of materials.

The book design under multi-dimensional forms requires book designers to stand on the high place of system theory, and cut into the creation of book forms so as to focus on the inner spirit and external form of books, macroscopy and microscopy, literature and images, design and process, etc. such a series of issues. Since readers' appreciation on various parts of books can not be completed in a moment, but to dynamically form the knowledge of overall artistic effects in the process of reading and in the extension of times. In this reading process, the most important is the book materials. Therefore, the design of books should not only deepen into every space of the books, such as the cover, lining papers, flap, title, text, illustrations, etc. but also needs to consider about the different features of book materials such as softness and hardness, and fully play the handiness, harmony and correlation between graphic main bode and the materials like paper, leather, wood, etc. and among colors, as well as the rhythm beauty of extension of space forms in the dynamic books and the flavor formed with the transference of attention.

Under the promise of meeting book design theme, we need to seek for more colorful external morphology expression that is nearer to the contents, as well as various materials that have different real feelings. In addition to strengthening the use of materials and fittings besides ordinary papers, the use of different materials such as the stones, plastic, wood of the combination of traditional paper and various materials can produce a variety of visual effects. They can be attached and embedded into the page or hidden under the cover or other different materials in order to increase the readability and interests of reading. For example, at the same time of seeing the external shape of fox, people can feel the smooth texture of fox fur on the tail, thus have deeper understanding to identify the animals. In the books of introducing fruits, we can use the corresponding flavored materials to books to increase the understanding of information besides the sight, and reach the diverse transmission of information. The diverse choice of these materials makes the multi-dimensional shape of book design more diverse.

The beauty of books is determined by the beauty of materials to some extent. The optimal design of materials is also a creation of theme on the space shape. Such as Mr. Lv Jingren's "Thousand Character Classic list book of Zhu Xi", when creating the form of book, he thought that the characters of Zhu Xi are free and easy, so we have to not only keep the original flavor, but create a refreshing forms. Finally, carve the thousand characters on the Paulownia board, and imitate the printed wood edition of Song Dynasty. The whole passage is linked by the leather belts, locked by the Ruyi wood, and constitutes the unique pattern of books. Through the multi-dimensional design with mixed materials, we can bring books the direction and position, convey the antique cultural flavor, and expand the space cultural values of books.

In the virtual world of digital media, everything is composed of 0 and 1. The feelings of digital media audience on digital media can only be limited to two aspects: visual and auditory. The traditional books allow readers to feel a book from visual, hearing, smelling and touching aspects. The book designers should correctly look at the different features and benefits of different media, for the ultimate goal of optimal design is to convey information more effectively, and the optimal design and trade-offs of all the book materials should be focused on such a theme.

VI. CONCLUSION

The essence of book design is to take the initiative to design and convey knowledge and information, and the designers of modern books should not be satisfied with the only conveying means of letter symbol. According to the book information, modern book design has its own understanding and interpretation, and has developed from the unilateral act of writing to the multi-dimensional shape with multiple mode of transmission. No matter the paper-based books, e-books, pop-up books, concept books, or the more diverse books combined with the new media, and however the shape changes, the attribute of books as the material carrier for people to express information, disseminate knowledge, and accumulate cultures has never changed. The optimal design of modern books in multi-dimensional shape displays this information to readers with certain form of attractive and acceptable.

REFERENCES

[1] Lv jingren "Theory on Book Arts" [M] Beijing. China Youth Press, 2006,pp.36-70

[2] Li dingfang, "the Significance of material language in the bookdesign" [D], Tianjin. Tianjin Polytechnic University, 2007,pp.6-10

[3] Qiu yuan, "the "even" and "Uneven" in graphic design" [D] Beijing. China Art Research Institute, 2008,pp.7-12

[4] (days) Kenya Hara, "the design of the design" [M] . Zhustartled translation. Jinan. Shandong People's Publishing House, 2006,pp.20-40

Graphic Designs Based on the Modern Cognitive Theory

Peng Tian
Zhengzhou Institute of Aeronautical Industry Management
Zhengzhou, China
Email:tianpeng791128@163.com

***Abstract*—Graphic designs are the co-results of thinking and expression. During the whole course of graphic designs, innovation is the most significant footstone. This paper, which is based on the cognitive theory and makes visual perception as a basic element, applies the inductive method to put forward several elements causing and influencing visual perceptions from the point of information communication and acceptance and accordingly makes analysis and demonstration about the optimization and innovation of graphic designs. Through the researches, this paper attempts to construct an interpretable graphic design theory suitable for analysis for reference.**

Keywords- graphic designs;vision perception;optimization; innovation

I. AWARENESS OF THIS ISSUE AND IMPORTANCE OF THE STUDY

Graphic designs are a process of pure thoughts in which innovation is the most significant cornerstone. Innovation can break the old inertial thinking, help you to surpass yourself and get rid of formats and thus brings other people a band-new experience.

Since the significance of the study is obvious, how to find an effective way to achieve the optimization and innovation of graphic designs has become an important task in front of the designers. In this paper, I will elaborate on the understanding and awareness of this issue starting from the factors affecting visual perceptions.

II. THE UNDERSTANDINGS OF THE CONCEPT OF VISUAL PERCEPTION

Visual perception is a process which starts when the eyeballs receive visual stimulus and then the stimulus is transmitted all the way to the brain for acceptance and identification. It includes visual stimulus retrieval, visual information organization and appropriate reactions. Therefore, visual perception contains the basic elements of visual reception and two main parts of visual cognition. In plain terms, seeing and perceiving the existence of light and objects are related with visual reception. While to understand what the things we've seen are, whether it is meaningful and how the brain makes explanations belong to the part of higher visual cognition. It is important to add visual perception concept and to clarify that, in terms of graphic designs, visual perception is a kind of interaction between designers and observers rather than the unilateral performance in the traditional sense.

III. VISUAL PERCEPTION ELEMENTS IN MOTIVATING GRAPHIC DESIGNS

A. Contradictory space in the practical application of graphic language

Just like the contemporary art, graphic designs have already entered into a multi-dimensional creation space. Graphic design artists are no longer satisfied with expressing their ideas and feelings just in a two-dimensional plane of space. Instead, they extend their wisdom and originality to a three-dimensional and even ultra-dimensional spatial extent. In comparison, the graphic language in the contradictory space has become a form of high taste among the designs because of its specificity in structure kernel. It is not only with unique ingenuity, liveliness and interest but also full of wisdom and philosophical implication.

Then how to reflect the concept of "contradictory space" in graphic designs? The core is the application of "visual illusion" principle. Visual illusion is the illusion with the experience of the people's judgments based on knowledge and the observed morphological characteristics in the middle of real contradictions. The special beauty generated by the visual illusion phenomena has added a wonderful new area to people's aesthetic activity. The designers studying visual forms are making their utmost efforts to create outstanding works that can guide visual effects.

Several forms of contradictory space:

1) Closure form of expressions

Closure in this case refers to a cognition inclination which complements the gap or space present in the incomplete vision picture. The picture, at first glance, seems to be a fragmented collection of various unrelated forms. However, in a closure condition, it may become a kind of graph that can be identified and recognized and thus becomes a meaningful visual form.

2) The reversal form of expressions

Some geometric figures can present a three-dimensional sense and visual image with visual magic effects in a two-dimensional plane space. This performance allows space reversal and transformation of the effect of a graphical form. From the extraordinarily simple lines to complicated faces,, as long as there is a particular and appropriate inspiration, there may appear an unexpected graphic effect in terms of space arrangement relations.

3) Overlapping form of expression

Many works seem to have a reasonable form at first glance, but after a closer look, we find them contain a key point of new graphic performances in dealing with the spaces with two contradictions. Under normal circumstances, a screen can be changed or stressed when the size, shape and location of a basic unit have been transformed. When people are staring at a specific part of the screen, the isolated part will present a temporary sense of stability and order. However, a long residence time in a particular part will easily cause the eyes fatigue. At this time, one form of expression often overlaps with others, making the original visual image complicated, and then a new image nature and a second image characteristic may occur.

4) The progressive form of expressions

The performances of unlimited spatial visualization will often depend on the sequence patterns, which helps people effectively perceive the special visual illusion effects produced by the depth of three dimensions.

B. The relationship between graphs and backgrounds

Referring to graphs, we must talk about backgrounds because they are closely related and any graphs we have seen are graphs under a certain background. The relationship between graphs and backgrounds can sometimes be interchangeable. People always select certain parts with graphical characteristics within their visual field as a percept while consider the other part as a background.

The relationship between graphs and backgrounds is put forward by a school called "gestalt" in psychology. This school also believes that the correlations among various parts of the graphs are the most important attributes in graphic perception rather than the graph's location and size. They develop a point that that people will organically link every part of the graph according to the following principles of organization.

Proximity principle - people tend to regard proximal components in a graph as a group when looking at a graph.

Similarity principle-people tend to regard the similar components in a graph as a group.

Closure principle-people tend to recognize closed or complete graphs.

Good principal of continuity -People tend to recognize regular and continuous graphs.

Symmetry principle-people tend to recognize natural, balanced and symmetric graphs.

C. The impact of the observtion and graph direction on perception

As to some familiar forms in our daily life, even if we change the direction of our observation, the outcome will be the same. But if we are not familiar with a graph, our recognition will be greatly affected by the changes of the direction of observation and graphs. Some people have done such an experiment, which firstly let the subject sit upright looking at a graph in front of him. Then the direction of the graph or inclination angle of the subject's body is changed, and then the participant is required to identify the earlier graph from several similar ones. The result shows that as long as the direction of a graph is consistent with the direction of the retina of the subjects, their recognition reaction time is short and the errors are fewer. But when the direction of graphics and the retina are not consistent, the picture recognition time is longer and the number of errors also increases.

In addition, even if the participants are familiar with a complex graph, the change of directions of the graph will also affect our recognition because we have been accustomed to observe them from a particular direction. For example, the inverted Chinese characters can not easily be identified and so is the inverted face.

D. Illusions

Sometimes people will also be "fooled" and make mistakes on the graphic perception. Illusion refers to perception errors that are determined under objective conditions and are difficult to overcome. Illusions are often difficult to be perceived by people. In other words, people often consider the illusions as true perceptions.

Visual illusions are generated mainly in the cross and contrast of lines. People's appreciations and judgments of beauty can not be satisfied with stopping at the dim and fuzzy level. There are desires and attempts to quantify beauty. This process reflects people's efforts in making the fuzzy and random phenomenon accurate, quantitative, as well as the attempts to establish some correspondence between them. In fact, the relationship of fuzzy and precise, random and order is not diametrically opposed. In many research areas of the modern natural sciences, scientists are also trying to prompt quantitative rules hidden behind a large number of disorder and chaos phenomena in nature.

E. Percolation theory and visual balance

The concept of percolation is comparatively more subtle, so let us draw a simple analogy to illustrate the image; please envisage a square sand table covered with uniform dry sand and a person is spraying water droplets to any parts of the sand table at random. As this process continues, wet sand area in the sand table will gradually expand and connect with each other. After repeated trials, the scientists find that when the wet sand area increases to 59.3% of the total area, the left and right side (or upper and lower side) of the sand table can be fully connected by the wet part. Only at the instant moment when the "wet area" is connected, people get the best visual sense of balance. Originally, it has been incredible that such an irregular and random phenomenon as percolation is mastered by a precise figure. What is more unbelievable is that the figure actually goes with the wonderful golden ratio and has some internal relations with the human visual psychological balance, which can not but arouse a sense of uncanny workmanship.

F. Revelation from board games

Scientists introduces the concept of entropy when trying to make quantitative descriptions of the order degree (or perplexity order) of a system. Entropy is related to the state number owned by system. If the system has smaller state number, the entropy is smaller and then the order degree of the system will be high; in other words, the system will very neat. If this system has a large state number, then the entropy is larger and the system has a high degree of disorder, or we can say that the system is confusing.

When a neat composition transits to a messy one, people's mental state undergoes a spontaneous process from low entropy to high entropy state, and, therefore, people take a normal psychology to accept it; On the contrary, when the composition transits from the mess into a neat one, there is a non-spontaneous process from high entropy state to a low entropy state. People usually present unusual rejects. People's vision experiences are a process of tension - relaxation - re-strained.

G. *Fractal structure and plane composition*

In nature, there are a lot of phenomena with "self-similar features".

This fractal structure and "similar figures split" in plane composition can be said to be different in approach but equally satisfactory in results, which at least illustrates two problems: First, fuzziness and mess can be described quantitatively, even though this description is still very "elementary", but people are making efforts to make it more accurate and more comprehensive. The second is that there are many places in exploration areas of science and art where they can learn from each other. This is because both science and art are exploring the mysteries of mankind and nature. Modern science pays more attentions to the chaos, instability and breaking. The essence of chaos is neat; the order can be generated from the disorder and a balance can be achieved through the unstable. All above are consistent and interlinked with modernist art in the pursuit of harmony in disharmony, non-rhythm in rhythm and the sense of wholesome in the clutter.

IV. CONCLUSION

In conclusion, the paper attempts to deliver the following views:

- The generation and formation of graphs are the reflections of not only the human beings' cognitions but also their rational thinking.
- A better graph is an effective combination of meaning, enjoyment, formations and expression method in order to enable the graphic language with clear superficies and profound meanings rather than a simple form and expression method.
- There are many elements influencing human beings' visual perceptions. Only through blending various elements reasonably into graphic designs can we get rid of formats and give other people a brand-new experience.

REFERENCES

[1] R.Arnheim, "Art and Visual Perception (new) (Fine Arts Education Translation Series)," Trans, Meng Peixin, Hunan Fine Arts Publishing House, August 2008.

[2] Zhang Liguo, "Modern Design Aesthetics,"Henan the Fine Arts Publishing House, July 2007.

[3] M.W.Aisenke,M.T.Kean, "Cognition psychology," Trans, Gao Dingguo, East China Normal University Publishing House, February 2004.

[4] Gu Daqing, "Design and Visual Perception," China Building Industry Publishing House, January 2000.

Green Architectural Technology Optimization Design

Longgang High School's Canteen Design as an Example

Peng Yue
College of Architecture
Xi'an University of Architecture and Technology
Xi'an, China
Email:Wingyue@163.com

Jiaping Liu
College of Architecture
Xi'an University of Architecture and Technology
Xi'an, China
Email: liujiaping@xauat.edu.cn

Bin Zhou
Dept. of Foreign Studies
Xi'an University of Architecture and Technology
Xi'an, China
Email:zhoubin8282@yahoo.com

Yixuan Lu
College of Architecture
Xi'an University of Architecture and Technology
Xi'an, China
Email: luise18@126.com

Abstract—**Problems such as poor indoor environment, pollution to the surrounding environment and waste of water and power universally exist in the school's canteens. In view of this, in the design of Hanzhong Longgang High School's canteen, the pre-research was firstly made to discover and analyze the existing environmental problems and their causes; then the design goals were brought forward to lower the environmental pollution, improve indoor environmental quality, and reduce energy consumption. After that, green technological measures were selected and optimized to be applied to the architectural space design. Lastly, architectural scheme was created. In the architectural design, the green technology has been applied to achieve the proposed goals in order to create a healthy, comfortable dining environment for the students, and at the same time accumulate the experience of optimized design on green buildings.**

Keywords-green technology; canteen; optimal design

I. INTRODUCTION

In recent years, the enrollments of various types of schools have been continually growing; as a result of this, there are many projects either for setting up new campuses or for expanding old campuses. The canteen buildings in the school have such properties as the large capital investment and a large number of people using them. What's more they have a high impact on the campus environmental quality; for example, the soot emitted from the canteen pollutes the environment, the noise of exhaust fans in the kitchen disrupts the surroundings, the dining hall lacks ventilation and lighting and the indoor thermal environment of kitchen is bad. Because these problems commonly exist in the traditional campus canteens, it is urgently needed to improve the situation through the application of green technology. In addition, as a large-scale public building, the canteen's space form, function zoning and moving line organizations also deserve in-depth study and exploration in order to meet the requirements for the development of multifarious functions.

II. DESIGN REQUIREMENT OF THE LONGGANG HIGH SCHOOL'S CANTEEN

Longgang High School is located in the city of Hanzhong of Shaanxi Province. It is expected that there will be 4000 students. The school's goal is to set up an excellent model school in the province, so it has a higher requirement for its planning and architectural design. This paper chiefly describes the design process of green technology optimization in designing the Longgang High School's canteen.

The functional requirements of the campus canteen are as follows:

- The dining hall should accommodate 4,000 students at the same time, and each floor can be independently operated.
- In the architectural design, green technologies should be adopted to reduce environmental pollution and reduce energy consumption so as to create a healthy and comfortable dining environment for students.
- Every functional room should be arranged according to the cooking operation process to make moving line short and convenient.
- The canteen should contain an indoor playing field basketball court、 a table-tennis playground、 a fitting room and a sports equipment warehouse. The playing field could also be used as a cultural performance spot, so the stage and some related rooms should be set. In other words, this space is a multi-function hall.
- The relative position of three main functions of boiler room、 dinning hall and the multi-function hall should be reasonably determined. Some outdoor space should be set apart as the Vehicle Park or temporary depot in the entrance of the boiler room and the kitchen.

III. INVESTIGATION AND SURVEY BEFORE DESIGN

Before starting our design, we investigated many types of students' canteens of the same scale such as Nanjing Jinling High School、Hangzhou Yuhua High School、Hangzhou Qingyuan School 、 Shanghai Yuying Experimental School、Shanghai Qingpu High School、Suzhou Lanying High School、Wuxi Liyuan High School、Yancheng High School、Taizhou High School and Dafeng High School. Taking into account the similar features and size of High school and college students' requirements in the use of canteen, in order to obtain adequate reference, we also visited six college students' canteens in Xi'an, namely, Xi'an University of Architecture and Technology、Xi'an Jiaotong University 、 Shaanxi Normal University 、 Northwest University of Politics and Xi'an Petroleum University.

A. The Investigating Contents

In the survey, many items were investigated, such as main functional zoning 、 the main moving lines 、 the kitchen functional organization、the space forms of dinning hall、diet sale way、dining hall and kitchen area ratio、indoor air quality 、 lighting conditions and ventilation way 、 water and electricity using situation 、 the type of cooking energy use、wastewater discharge ways and the impact on the surrounding buildings, etc.

B. The Main Problems Found in the Survey

Through the survey, we have found a severe problem, that is, the school's canteen building design and its use have been broken off. Recently great changes have taken place in the canteen's equipment and processing methods, but the current design codes and design specifications have seriously lagged behind the requirements for use, resulting in a number of canteen projects with unreasonable design 、 large investment and resources' waste and serious pollution. There are a number of specific problems, which are as follows:

- According to the current *Code for Design of Dietetic Buildings*, the ratio of areas between the dining hall and the kitchen is 1 to 1 [1], which is completely inconsistent with the actual functional requirements. Consequently, there is no scientific basis to follow when designers determine the proportion of areas between the dining hall and the kitchen, which further turns out that the area of the kitchen is often so small, making the kitchen crowded、disorderly and of poor environmental quality.
- It is quite common that the space form in the dinning hall is very monotonous.
- Since problems such as insufficient lighting and poor ventilation commonly exist, the air quality of the dining hall is poor.
- The indoor environment in the operating rooms of the kitchen is very bad and the thermal environment is very poor, especially in the summer. For example, the indoor temperature and humidity are all very high.
- As the kitchen lacks effective measures of disposing of oil stains, therefore the canteens have seriously polluted the school's environment.
- In the operating area of the kitchen, there are only rooms in one side which have natural lighting, and those in the other side next to the dinning hall are dark rooms.
- The operation of the kitchen's ventilation equipment is less efficient and resources such as water and power are seriously wasted.
- The functional partition in the kitchen's operation area is not professional; as a result, the moving line doesn't go swimmingly.

Through the survey, we have found that the problems mentioned above universally exist in the school's canteens, and meanwhile we have a deeper understanding of the urgency of the application of green technology in the design.

IV. THE MAJOR PROBLEMS TO BE SOLVED IN THE DESIGN

According to the previous survey, we have put forward the key problems to be solved in the design:

- Determine the reasonable "area ratio between the dining hall and the kitchen" to suit the modern cooking pattern、kitchen equipment and students' dining way. Based on the analysis and study, we put forward a feasible area ratio between the dining hall and the kitchen that is 1.5 to1.
- Adopt the effective oily fume disposing technology to reduce environmental pollution and inefficient、ineffective energy consumption.
- Improve the kitchen and dinning hall's indoor air quality through organizing the effective natural ventilation and the necessary mechanical ventilation.
- Make a full use of natural light and improve the indoor luminous environment through the design of architectural space and form.
- Change the single form and enrich the indoor and outdoor space of dining halls.
- Change the kitchen's traditional extensive water-use patterns and reduce water wastage.

V. THE ADOPTED MEASURES OF GREEN TECHNOLOGY OPTIMIZATION DESIGN

The project is a practical opportunity for applying green technology to architectural design. The technical measures of green architectural design include natural ventilation、ultra-low power consumption envelopes、natural lighting、heat pump air-conditioning、use of solar energy、green building materials 、 intelligent control 、 wastewater treatment and recycling and roof planting, etc. In terms of various problems existing, we are planning to adopt the following technical measures in the design:

- Follow the green-building design principles "healthy, energy-saving and environmental protection in building's full life cycle" [2] in the site planning of

Longgang High School、building space design and construction design of the school's canteen.

- Adopt indoor and rooftop greening technology to improve indoor and outdoor air quality and thermal environmental conditions [3].
- Apply wind pressure and hot pressure ventilation principles and technology, optimize the vents design and effectively organize natural ventilation to improve air quality, create a comfortable environment and reduce air-conditioning energy consumption [4].
- Apply the thermal insulation technology of walls、windows、roofs and other building envelopes to create energy-efficient and comfortable indoor thermal environment [3].
- Employ natural lighting techniques to make full use of natural light in large space [3], such as the dining hall and the multi-function hall.
- Use such green building materials as 3R as much as possible i.e. reduce raw material (reduce) 、re-use (reuse) and recycle (recycle) items, and non-toxic materials [5].
- Adopt recycling techniques of disposing rainwater and wastewater to save water resources [3].
- Properly use intelligent control technologies, timely automatically adjust ventilation 、 lighting 、 shading、 cooling and other technical measures according to the changes of natural weather conditions [2].
- Make full use of solar energy to reduce non-renewable energy use [2].

VI. THE GREEN ARCHITECTURAL DESIGN OF SCHOOL'S CANTEEN

After the economic benefits and environment benefits of the above-mentioned technical measures were estimated according to the specific functional requirements for the buildings, the green technology was optimized to ensure its efficient and economical effect. The following technical measures were adopted to solve those problems identified:

- The kitchen's device noise is fully considered from the building's layout to avoid disturbing the nearby buildings, as is shown in Figure 1 and Figure 4.
- The general plane relationship between the dinning hall and the kitchen is that the dinning hall occupies a better position facing the east and south and the operating room faces the north and west. This is good for the dining hall to get adequate lighting and the kitchen placed in the north is suitable for food storage, as is shown in Figure 1、Figure 2 and Figure 4.
- The vertical lighting shaft is arranged in the kitchen to raise the level of natural lighting inside the kitchen, as is shown in Figure 2 and Figure 4.
- Set a number of ventilation openings in the dinning hall, and the vertical shaft to use the chimney effect to enhance the dinning hall's ventilation thus improving indoor air quality; at the same time, these shafts enhance the natural lighting, as is shown in Figure 2 and Figure 3.

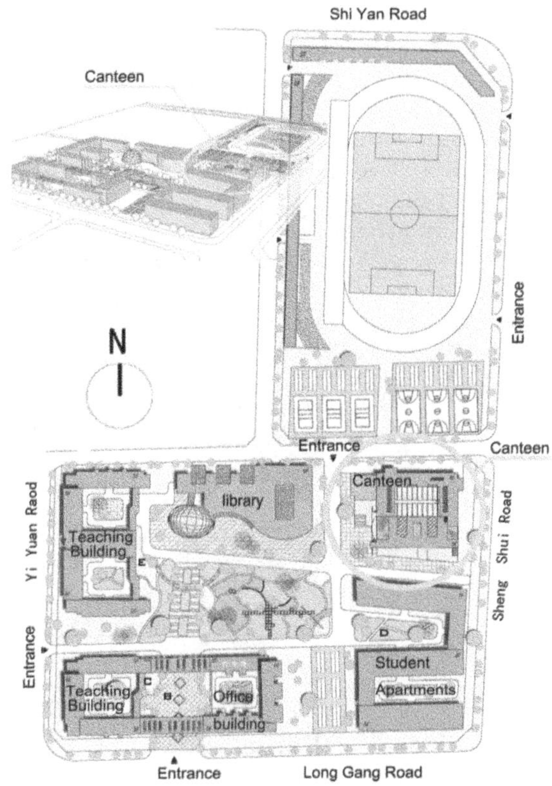

Figure 1. Schematic diagram of the overall plane.

- The sunshade louvers are used in the windows and glass curtain wall and water spray is used in the light top of the dining hall for cooling in summer.
- Set up water pool at the entrance to lower the indoor and outdoor air temperature in summer and improve the thermal comfort, shown in Figure 2、Figure 3 and Figure 4.
- The building greening is designed adequately, for example, using the areas of apron、roof、veranda and platform to increase greening covering so as to improve the visual quality and air quality of the environment and reduce the surrounding temperature in summer, shown in Figure 3、Figure 4 and Figure 5.
- Kitchen's cooking fumes' treatment is the key to solving the pollution problem. As to the cooking equipment, an effective absorbing and exhaust cooking oil device should be installed. Its working principle is: at first, the water spray absorbs the soot, and then water and oil are separated in the collection box, water can be used circularly.
- Use rainwater collecting and water recycling and treatment systems for greening、roof water spray cooling thus saving water resources.

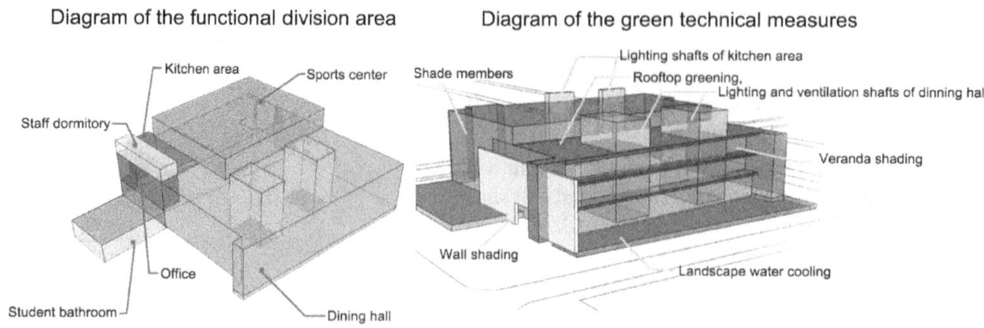

Figure 2. Schematic diagram of functional division area and integrated measures of green technique

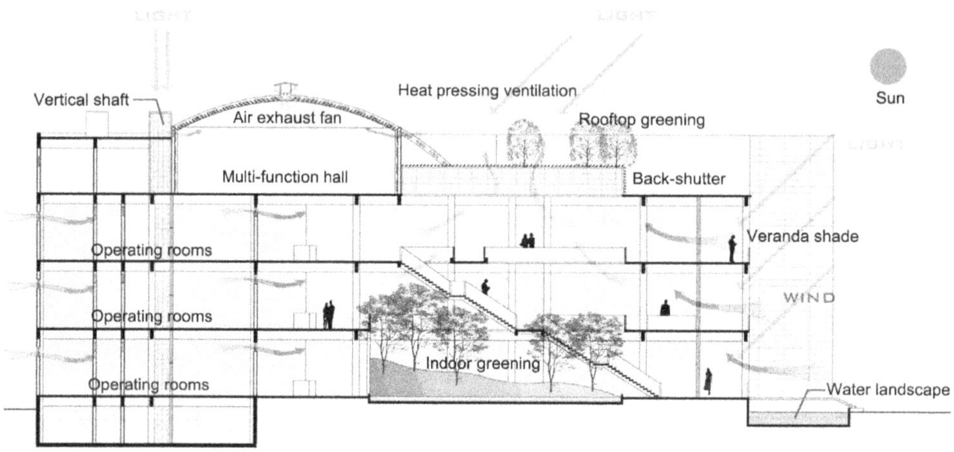

Figure 3. Schematic diagram of natural lighting、ventilation and sunshade.

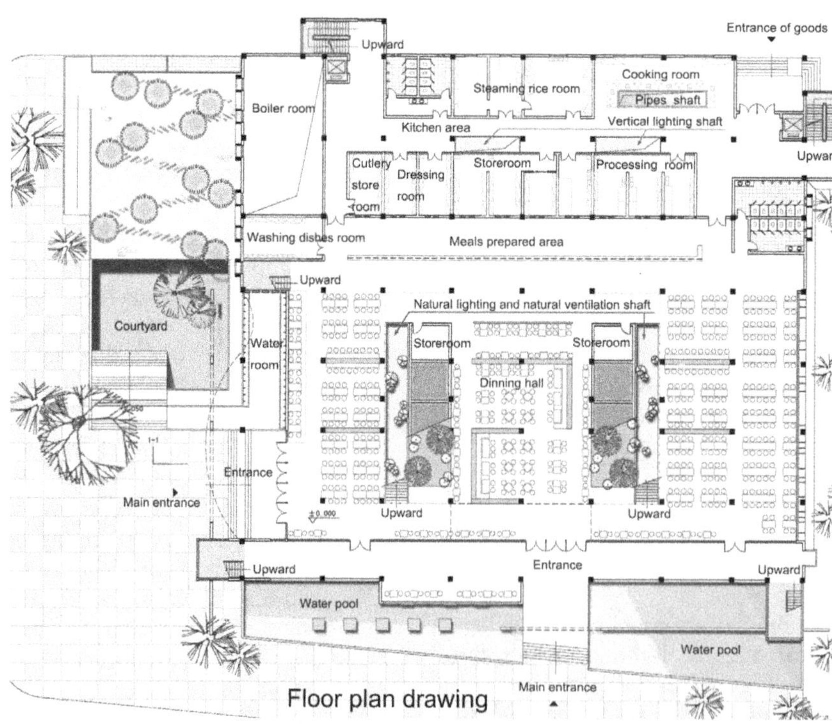

Figure 4. The position relation between the dining hall and the operation rooms of the kitchen.

- The building's elevation and shade structures are designed to meet the shading requirements for different faces, as shown in Figure 6.

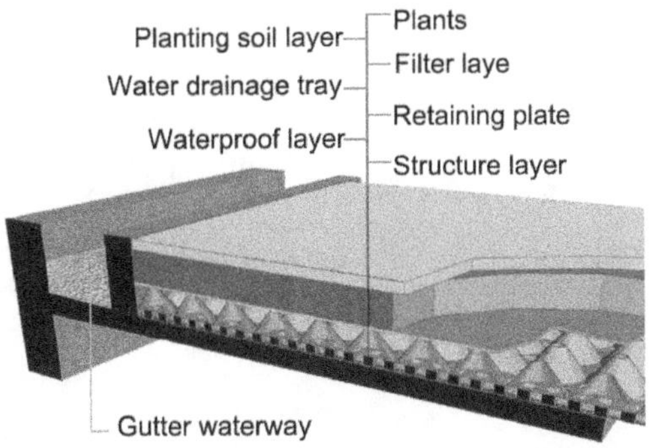

Figure 5. Schematic diagram of the rooftop planting.

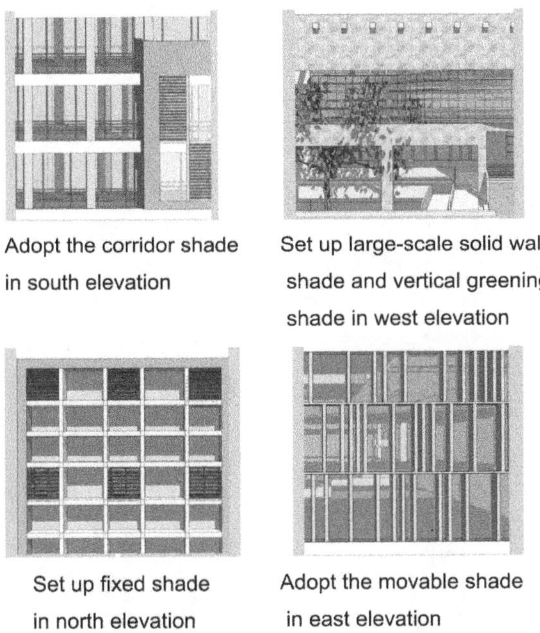

Figure 6. The elevation design to meet the shading requirements for different faces.

VII. OTHER PROBLEMS IN THE PLAN DESIGN

- The students' dining time is relatively concentrating with a large number of people going and coming, so sufficient buffer area should be reserved near the entrance and in front of the table of food sale to prevent congestion.
- The mechanical ventilation is used in the kitchen and the natural ventilation is used in the dinning hall. The ventilation of kitchen and dining hall to be organized into different parts respectively should be taken into account. At first the kitchen's ventilation installations draw out the fume and steam, which are then treated and purified, and in the end discharged from the roof. The kitchen should have unobstructed air openings. Compared with the dinning hall, the air in the kitchen should be formed into negative-pressure to prevent the kitchen area's air from going into the dinning hall.

VIII. CONCLUSION

"The campus canteen green architectural design" aims at employing the green technology in the architectural design to solve the problems universally existing in the campus canteen such as dysfunction 、 environmental pollution and waste resources, etc. First of all, surveys are conducted to get to know the existing problems and their causes, then analysis and study are made to define the main problems in the architectural design, and then the green technological measures to be fitted with building's operation functions and economic costs are put forward; finally these technical measures are carried out in the design of architectural space and structure.

REFERENCES

[1] JGJ 64-1989, Code for Design of Dietetic Buildings [S].Beijing: China Architecture and Building Press, 1987.

[2] Xi'an University of Architecture and Technology Research Center, Green Building [M]. Beijing: China Planning Press, 1999

[3] Jihong Han,Yan Jiang, Eco-Building Demonstration Project of Shanghai [M]. Beijing: China Architecture and Building Press, 2005.

[4] Runming Yao, Koen Steemers, Baizhan Li, Sustainable Urban and Architectural design [M]. Beijing: China Architecture and Building Press, 2006.

[5] Yun Xia, Ecology and Sustainable Development of Building [J]. Architectural Journal, 2002, (7) :8-10.

Identity of Space and Information Experience in Display Design

Hu Yiping
School of Art and Design
Wuhan University of Technology
Wuhan, China
Email:hyp333@163.com

Abstract—**The display design is a subject with a rich content, and a wide range of areas involved with the development of the times. It also constantly enriches the content of the subject. The design of the display space is at the soul status, and it is the key to correctly handle the relationship between space and human. Based on the display space and the effect of emotion and experience of information, the paper thoroughly analyze the application of identity of space factor in the display art. The paper also point out that making rational use of "space" language is necessary to show the real significance and vitality.**

Keywords-display art; space design; information eperience; information-oriented system

I. INTRODUCTION

Design Art has been always associated with the development of social and economic .As the integrated application of a variety of design techniques, display design is also in full swing along with this trend. The early display design is only limited to the goods on the store shelves to be placed and simple Layout display, intended to cause customer's attention and induced to buy. With the development of economic and needs of foreign exchange, display design has evolved into a comprehensive spatial visual communication design. In fact, the display design integrated product design, environmental design and graphic design and many other factors. It is based on space design form for the purpose of the information communicated in a specific time and space. Display design fully mobilize the audience's sight, hearing, touch and even smell and taste and all perception by the means of the props, colors, materials, lighting, audio, text, illustrations, images, and exhibits. It also makes full use of new media, new technologies and new results in order to develop the interaction between "human" and "objects".

II. SPACE COMPOSITION AND EMOTIONAL EXPRESSION IN DISPLAY DESIGN

A. Display Space from the "No" to "Yes"

From a sense to see, space design is "something out of nothing". Fully aware of the characteristics of exhibition space is great significance to the success of exhibition space design. The main features of the exhibition space are in the following areas: Four-dimensional nature of space-time, combination, mobility and opening of Space. It is difficult for anybody to competent exhibition space design who is short of an accurate grasp of the exhibition space or imagination. A famous British fashion design once commented on those fashion "artists" who lack the concept of human body: clothing design is not a painting but a sculpture and designer must see the body inside the human. By the same token, the exhibition designer's mission is not to be a painter, but to be a designer who is creative and predictable and has a good idea of exhibition space.

Exhibition space is a "virtual form" and the "field" in people's vision formed under the definition of visual display entity. If visual display entity can be consider as a real physical space, the "virtual form" can be regarded as virtual space. Thus the final realization of purpose and function of display design is based on certain occupied space as a prerequisite. It effectively conveyed information by the object display, layout, lighting, props, audio, color and other multi-media. Therefore, the design of display space is activities created by man-made environment; the exhibition space planning is leading actor and core elements of the display design. On the one hand, exhibits and a variety of display props must occupy a certain degree of physical space; on the other hand, the interaction between the space and audience psychology and space effect formed by interpersonal interaction come into being the potential psychological space. So the continuity and approach of the two kinds of spatial relations is the key to display space design.

B. The Principle of Display Space Constitution

The principle of display space constitution can be summarized as the following: to provide the best display location (function), efficient use of space, and the optimum visual perception (aesthetic), the economy and efficiency principles (money, goods, time), and a unique style. Whether it is successful or not can be measured with this. Display composition design, that is o create a prefect visual and psychological environment through the organic mix of exhibits, display props, display space and the color in order to achieve the established exhibition intention. Display constitute must meet the functional requirements, the intention of expression and aesthetic requirements; pay attention to space variety, the division of the regional and functional implementation. Therefore, it requires the designer to be rich in imagination and rigorous scientific attitude and economic minds.

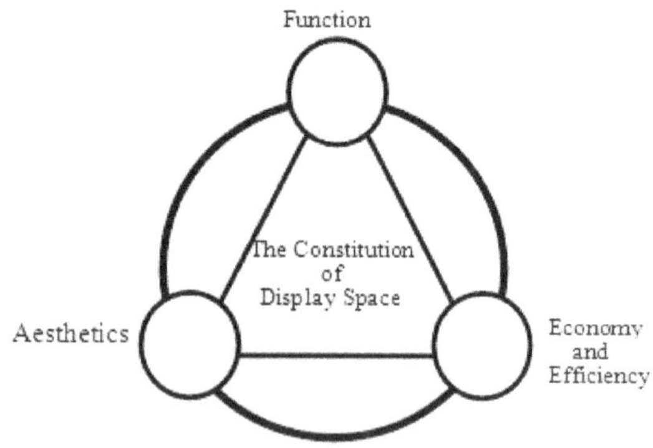

Figure 1. Factors of Display Space Constitution

Place structure, path structure and the field structure consist of the basic structure of exhibition space. Place structure is the basic property of space because the place reflects the fundamental relationship between people and space. It is human give the property of fourth-dimension to the display space, which embody the human subject, form the "force" and "field" effect by relationship of the center, the direction and the regional synergies so that it can be from the imaginary state to showing reveals reality through human action in the display environment. At the same time, human accesses to all of the psychological feelings on the course of this experiences about the space. Therefore, the human is the ultimate object served by exhibition space. Human factors must be considered in the display space and display space should provide better services for human.

Figure 2. Shopwindow display: in a limited space, the performance of a rich composition form through many different permutations and combinations

While the display methods are diversified, comprehensive, but any exhibition must follow certain principles and has some common application method. The flexible use of constitution methods will effectively maximize the utilization of display space and quickly and accurately deliver information to the audience. For example, it can express different emotions and visual effects by using the concentration of point, extension of line, wholeness of surface and the weight of block. So the combination of point, line, surface, block and various geometric shapes can plays a subtle role on the applications and information presentation in exhibition space. It also can establish different design styles according as different display contents, such as the channel type, courtyard, garden-style, center type, pitch type and so on. In addition to the text font style, the forms of display constitution are various including echelon, tower, sphere, cube, cylinder, and the changes in a variety of combinations, all of these can maximize space utilization and increase information in the limited space. Exhibition space constitution involving horizontal and vertical (upper, middle and lower) design and open nature of the space including three major aspects such as closed, semi-closed and open. In the specific design, it must take into account overall these three major factors, survey the program in multi-angle. Only in this way, is the design reasonable and perfect.

C. *Display Space Design and Emotional Expression*

In addition to human engineering, the elements of display design should furthest satisfy human's psychological and emotional needs which is the basic basis of display space analysis. Comfortable and harmonious display environment, prefect effect of sound and color, information-rich display contents, safe and convenient space planning and thoughtful services and facilities, etc., all of these are spiritual requests for human in the display design.

This requires designer to carefully analyze the behavior of visitors and give full attention to the ergonomics with a scientific attitude. The shape, size of display space must be compatible with the size of human; the proportion of the various parts of space, must assort with human's perception and action, which is the most basic space requirement. At the same time, human should be in a comfortable environment for the activities. If it can not create a kind and warm psychological space, it is only an indifferent machine shell, even using the most advanced display means. So only a humanistic exhibition space can be called a "sensible" and "reasonable" design.

During the display design, although the prominent exhibits are the ultimate goal of the entire design, but the visitor's experience and feelings to the exhibits and the overall display are the key to the success. Promoting the communication between human and object with a variety of media, adjusting human psychology by using a variety of spatial combination, all of these design skills are in the view of person's emotional needs and psychological feelings in mind. While in the specific design, it must resort to some of the media in order to establish a bridge between visible objects (substance) and cognitive subject (human).The use of implication, contrast, metaphor and other ways can play intensification, dilution, dispersion, illusiveness and other psychological reactions. Human's space perception includes the shape, size, distance, and direction perception. The receivable information of physical organs and the physiological mechanism of the brain obtain qualitative and quantitative understanding relying on the several reference of environment. Characteristic of space such as open, closed, dynamic, static sense and rugged, elegant, quiet and so on, which can infect and stimulate the human. For example, increasing the gradation of the distance using a variety of methods of contrast, mirror reflection, mutual penetration between internal and external, fuzzy boundaries, changing perspective relationship, extension of an element, virtual-based visibility, forward or backward sense of color, tortuous deep and visual illusion caused by using intermittent or turning lines which obtain a sense of artistic conception, all of these skills can bring psychological and emotional associations to human.

III. The Use of Visual and Perceptual Psychology to Design the Visit Route

A. *The Law of Human Vision and Perception*

The display formation of exhibits constitutes a variety of visual elements, which affect people's vision in varying degrees. Observation method of the human has certain commonness, researching these common characteristics is a basic starting point of studying vision law.

Design is associated closely with the human vision and perception. Consciousness includes both the human brain's direct action and the sense organ's overall reporting to objective things. The consciousness object is composed of many parts and every part has different characteristic, but people always feel a unified object. Because of gravity, person always watch from top to down and from left to right in flat space, which is the same direction of arranged text. So, in the limited flat surface, viewers focus on the left and the top at first. In other words, this different area in the same flat surface has the different attraction to viewer correspondingly. According to attractive degree, the optimum region is left top part and upper middle part and then right top part, left bottom, right bottom. The optimum region has great value in graphic design which is applied in layout design, advertising design, bill design, packing design and exhibition design.

B. *Human's Behavior and the Perceptual Experience*

In modern times, the vision psychology and art psychology created a lot of research results which were used widely in the modern exhibition design. Early in the 19th century, Gestalt psychologist Rudolf Von Labon Aenhaim (Germany) proposed "mentality field" and "behavior field". Field was the entirety of interdependent extant fact and individual psychological life space was his experience contents within a specific time, including his psychological environment and perceived environment. People's behavior was a function of his mental space forever. Meanwhile, the exhibition space is the product of interaction between people and the environment. Gestalt psychology pointed out that perception is greater than the sum of feeling. People don't regard an object as the separated parts but the integrated organism. Human's experience is a whole mentality about the exhibition action and human's participation is his variety of behavior, his direct or indirect relationship with the object. During the exhibition design, audience should be guided to the role participation and action participation. So, there is a direct communication between subject and object.

C. *The Design Rules of Visit Route*

When watching the objects or the exhibits, people who stand immobility prefer to move step and change visual angle in order to obtain multi-angle, multi-faceted perception. Time is a dynamic annotation way. People in showing space, facing the time efflux and the space change, obtain the entire experience. The space temporality is an objective factor in showing design, and this "fourth dimension" wielding sufficiently is foundation of creating dynamic space form. It is also an only way to create a "floating beauty".

The maximal characteristic of showing space is to have very strong flow ability, so showing space design should combined the subject, content, rhythm, primary and secondary with space division, signs and introductions to plan a visit route . At the same time, we must also consider the people's visual and perceptual psychology.

The first basic rule of space design is that showing form should be dynamic, serialization, rhythmic. This is decided by the character of showing space and human factors. People in showing space always move to visit, then gain experiences in motion and ultimate space feeling at last. So, according to this foundation, we should design the most rational visit route in showing space. People moving on this route can participate in the all action economically, little or as much as possible without taking the repeat route. The whole space design should be logical, like smooth, cadence, clearly delivered music melody. In addition to functionality, people feel the charm of spatial variation and the infinite fun of flexible design. With the changing spatial directions, the form of spatial display is more smooth, rhythm and people may enjoy a full range of space in constantly changing visual composition.

In this regard, Chinese garden art is different in approach but equally satisfactory in result with show art, from which we can get some inspiration. For example, Chinese gardens emphasize turn and transfer, brightness and shade in space array, and constant scene changes and emotional changes in visit route, which are valuable for the form of display space.

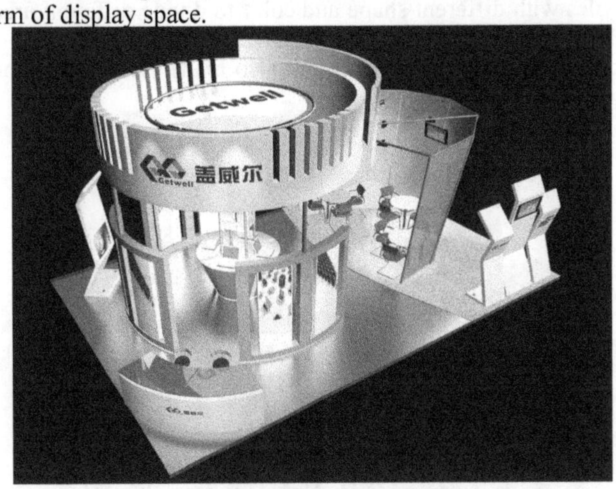

Figure 3. To guide tour streamline and enrich the space through the semi-enclosed and transparent approach

IV. The Conveyed Information of Oriented System in the Exhibition Space

In general, the design concept of oriented system is divided into two parts: one part including graphic symbols is used to indicate the direction and the region; another part is the performance of symbols in the environment space. The starting point of former study is visual communication, whose concern is how to use simple graphical symbols to express the precise meaning and be instantly distinguished by everyone without language. The latter depends on the environmental design, focusing on the material, appearance, location, artistic expression and other factors, whose concern is how to use the graphic symbols embedded in the whole atmosphere of the environment. Therefore, the oriented system design is related to the visual communication design and environmental design. Two fields are not only cross but also independent each other. The oriented system design of exhibition space start from spatial organization and the audience perspective, and spatial orientation has a series of personalized form for the audience. Used serialized-oriented form, the scattered exhibition space is all connected and becomes a comprehensive and unified exhibition field.

A. The Oriented Identity in the Exhibition Space

The legibility and orientation of the environment space are two sides of an issue. Well-oriented design of space makes people easy to identify. The successful application of oriented system can promote friendly interaction between the audience flow and exhibition space. The audiences walk smoothly and visit harmoniously in the full and uncrowded space. Space resources are integrated and space efficient is promoted, which is a humanistic consideration for the exhibitors and audiences. The display form of the oriented identity is that the same identity element is repeatedly applied in space, and the audience in a different space can consciously visit with the same identity element which includes the same layout, materials, installation forms, etc. For example, in 2005 Aichi World Expo, there were sign poles with different shape and color to divide different areas. They became the special landscape in parks, which not only beautified the environment, but also made visitors find their own pavilion fatly.

Figure 4. Aichi World Expo Park: using sign poles with different shape and color to divide different areas, such as: Oceania, South Asia, Africa, America, etc

In the exhibition hall of CHIC2005, the duplicate flags marked with hall number were hung above the passage, and guided the audience along the path to visit. Through the same logo or the regular number, the different spaces organize an organic sequence entirety. It is very easy for the audience to visit from the previous space to the next one. The unified CHIC2005 LOGO was distributed in every corner of the exhibition space. Visual text symbols appeared repeatedly in each space, or complicated conversion place, and they displayed clearly every space identity, no matter where the audience was. In addition, the monomer space identity is a point of the entire exhibition space. Through the oriented system s or instructions, many different points link a visit route which guides and implies the audience imperceptibly.

B. The Oriented Form of Color Management

Color achieves space management by color communication and color distinction. The same colors appear repeatedly in different space, the same property of the space will be identified. On the contrary, difference space can be distinguished by different colors. Color management is a good helper to define space, such as different regions distinguished by different colors series. Red is the standard color of CHIC2005, its special form of color management is that the trunk was wrapped by a red cloth. It gave the fresh impression that the trunk is also one part of the space logo or display props. It rendered consumedly the overall atmosphere of the space. The production of oriented identity system used a lot of inkjet canvas and the blankets, being economic and practical. With using red, white and black, the combination had a great tension, a strong visual impact and a good spread of brand image.

V. CONCLUSION

From a macro view, the display design can be called the art of spatial mobility. But, its performance in space is different from the general living and entertainment environment. The uniqueness of the exhibition space design lies in not only meeting the person's physical and mental needs, but also promoting and provoking a sense of people's participation and communication. Therefore, the display space is a four-dimensional blend including three-dimensional space and time. It is an art form which can be accessed, heard, touched, and observed from a different angle. The fundamental purpose of display design is the efficient transmission and exchange of information. In this process, the design and planning of exhibition space, and audience's feeling for information experience is essential.

REFERENCES

[1] Rudolf.Arnheim, "Art and Visual Perception," Sichuan People's Publishing House,1998.

[2] Lin Fuhou, Ma Weixing, "The Display Art Design," Beijing University of Technology Press, 2006.

[3] Guo Hao, "Exhibition Design and Practice," Shanghai Jiaotong University Press, 2005.

[4] Donald A.Norman, "Design Psychology," CITIC Publishing House ,2003.

Improving University Administration Under the Perspective of Tacit Knowledge Management

Ruihong Zhang
School of Art and Design
Wuhan University of Technology
Wuhan, China
Email: jiujiu_zj@126.com

***Abstract*-Administrative management is a key component of universities' daily operation. It is the foundation of ensuring the successful operation of research and education. With the advent of knowledge based economy, universities are increasingly faced with challenges, especially in elevating the overall quality of education and building a harmonic campus environment. This research explores the deficiencies in university administration from the perspective of tacit knowledge management, and presents some suggestions to improve the efficiency of the administration in universities.**

Keywords- university; administration; tacit knowledge

1. DEFINITION OF TACIT KNOWLEDGE

Knowledge can be classified into dominant and tacit knowledge. Dominant knowledge can be expressed and spread via symbols. It can be heard or read and comprehended, such as mathematics. On the other hand, tacit knowledge exists only in minds, hence hard to express via symbols. Even if they are expressed in symbols, audience can not fully comprehend the full meaning of the knowledge, which was developed via experience and has not been defined explicitly via logical tools.

Polanyi (born in 1958) was the first to propose the concept of tacit knowledge. He believed such tacit knowledge was silent, could only be perceived rather than expressed, coming from individual's perception and understanding of the external world. Such knowledge can't be separated from the perceiving individuals. Management guru Peter Drucker believed that compared with dominant knowledge, such tacit knowledge is in essence a skill, of which the existence can only be demonstrated, as an experience and perception process. Tacit knowledge originates from experience, similar to specific skills and mindset, with roots in an individual's action and mindset, also in his/her life goal and value system. As a result, such knowledge is highly personalized, hard to expressed and propagated via symbols, shared with others, such as experience, subjective opinion, emotion and culture [1]. Tacit knowledge can be obtained only via personal experience, constituting a large portion of human being's knowledge.

2. IMPORTANCE OF TACIT KNOWLEDGE MANAGEMENT IN UNIVERSITY ADMINISTRATION

2.1 Definition of tacit Knowledge Management

Knowledge management regards knowledge as a basic resource in organizations, a key object of organizational management, and hence integrates it into all aspects of organizational management. In organizational activities, the origination and consolidation of knowledge occur throughout the whole organizational activities. Knowledge management aims to facilitate the creation, propagation and application of knowledge, via effective consolidation means. From this perspective, knowledge management is the guiding line of core activities, becoming the core management strategy for long-term organizational growth and competency [2].

The subject of tacit knowledge management is tacit knowledge: via extracting, organizing, sharing, transferring, converting, applying, creating and evaluating organizational tacit knowledge, ensuring the fully exploitation of the value from such organizational tacit knowledge. In summary, organizational tacit knowledge management is the effective management of organizational tacit knowledge.

2.2 Main Issues Faced in University Administration

1) Lagging concept and system

Compared with the western researchers, researchers in China are lagging in the research of scientific administration. As a result, our university administration often lacks the guidance of scientific and systematic approach, leading to university administration falling behind of society development. At the same time, the regulation is not fully defined even though it is the symbol of modernization for university administration. Consequently, universities in China often focus on research and education, de-emphasizing on administration [3]. The manifestations are: the regulations can't meet the requirements of administration; lack of monitoring in daily administration; excessive administrative meetings; difficulty in coordinating departments, all leading to loss of valuable university resources, impacting the growth of universities.

2) Inflation of organizations, non-optimized organizational structure and low efficiency

University organizational structure was developed under the planned economy era, hence leading to highly centralized model, and as a result, with similar problems faced by government. The typical structure is with the party committee and president at the top, there is usually the hierarchy of university, college and department [4]. There are designated interfaces at each layer, which presumably lead to high efficiency. However, the drawback is the overabundance of layers, organizations and personnel [5].

At the same time, the quality of administrative personnel is not up to the standard, lacking professional qualification. Many of them do not possess the necessary knowledge, leading to random and subjective behavior. In addition, the age distribution is out of proportion, lacking a strong core middle-aged management layer, but with outnumbering junior and senior personnel [6].

3) Low utilization of information technology (IT)

In recent years, the advent of IT leads to significant advance in university administration, because IT is the basic requirement for expedited and highly effective administration. However, for many universities, the lack of IT knowledge and skill prevent them from fully deploying IT. For example, the computers are network remain lowly utilized in many places. Many tasks are still processed in the traditional way.

In summary, there are obviously many areas for improvement in university administration, which deters the full development of universities. At the same time, the most critical factor impacting an individual's efficiency is his/her tacit knowledge, leading to the phenomenon that once a person leaves the position, it takes a long time to train the replacement person, which wastes precious time and resources [7]. Therefore, the quality and efficiency of administrative departments are adversely affected.

3. Adjustments for University Administration under the Perspective of Tacit Knowledge Management

University administration needs to break through the limit of obsolete concepts and systems, with many areas to be improved. However, with the time it takes to change the overall system, it is important for now to focus on improve the conversion of tacit knowledge to dominant knowledge, as an intermediate way to improve overall administrative efficiency.

3.1 Transform Concept and Establish the System for Tacit Knowledge Management

Based on the specific nature of the current university administration, it is important to focus on the conversion from tacit knowledge to dominant knowledge, and improving the overall knowledge management by steps. As a result, the knowledge based will be open and transparent, leading to the efficient propagation of knowledge. Universities can also build its own knowledge and expert base, as a store for tacit knowledge, via converting the experience and results into symbols such as text, sound and visual information properly categorized and managed. In the process of building the data base, emphasis should also be placed on facilitating the sharing and flowing of such information.

A. Flatten the Organizational Structure and Improve Inter-person Communication

Flat, open and flexible structure facilitates the sharing of information. At the same time, such structure expedites the spread of information, rotation of personnel, leading to the fast propagation of tacit knowledge. For example, if a person in function A is transferred to function B, he/she will naturally spread the knowledge in the new department while learning the new knowledge there. In addition, the transfer forces the functional departments to define the roles and responsibilities of each personnel, as well as working process, in order to make the transfer viable. As a result, such a transfer eventually led to the full spreading of tacit knowledge within the whole organization. Even if one person leaves, the overall efficiency will not be affected.

3.2 Build a Full Knowledge Management System and Cculture via Modern IT

Culture is often defined as the belief, value, regulation and behavior specific to an organization, i.e., unwritten rules. A culture based on shared knowledge is built on such belief, attitude and habits. To develop a sharing based culture depends on leadership and continuous sharing and positive feedback/acknowledgement [8]. University administration should create an environment for learning and sharing, thus facilitating the flow of tacit knowledge. All sorts of means should be employed to enhance the management of tacit knowledge, such as via Internet, neural network, to categorize, update and display information, especially the most valuable ones. Personnel should be able to freely discuss the optimal solution or work process for an administrative issue, and reach consensus via such interactions, thus leading to effective sharing. Such learning, sharing and propagation should become an integrated part of university administration. The tacit knowledge management system should be constantly tuned and improved, in order to gradually build a culture matching the unique nature of university administration.

4. Conclusion

Compared with dominant knowledge, tacit knowledge is a key factor impacting the efficiency of administrative personnel. Because of the lack of emphasis on knowledge management, the research on tacit knowledge discovery, storage and propagation has been rare. Since it will take time to improve the overall university administration system, it becomes effective and hence important to emphasize on tacit knowledge management in order to increase the overall efficiency of administration. University administration should build a tacit knowledge database, and utilize modern IT to manage and maintain the database, so that the tacit knowledge management becomes an integral part of university administration culture. In such way, the blueprint of harmonized, fast and stable growth of universities can be achieved.

References

[1] Y. X. Zhang, "The mechanism of tacit to dominant knowledge conversion," Journal of Liaocheng University, 2002, 4, pp.124-126.(In Chinese)

[2] J.J. Chen, "Research on enterprise knowledge management system architecture," Science & Technology Progress and Policy, 2009, vol. 26, pp. 128-131. (In Chinese)

[3] N.Guan, "Issues and solutions in university administration," Theory Horizon, 2008, 3, pp. 248-249. (In Chinese)

[4] C.Z. Zhang, "On the Present Situation and Reform of Administration in Universities and Colleges," Journal of Fuqing Branch of Fujian Normal University, 2008, 1, pp.84-87. (In Chinese)

[5] M.H. Liu, "Discussions on university administration," Science & Technology for Development, 2008, 4, pp.31-32. (In Chinese)

[6] B.L and A.F. Zhang, "Issues and ways of resolutions in university administration," Journal of Hunan Financial and Economic College, 2008, vol.24, pp.141-143. (In Chinese)

[7] X.D.Teng and J. Kong, "Applications of Knowledge Management in University Administration," Journal of Beijing Union University (Humanities and Social Sciences), 2006, vol.4, pp.89-92. (In Chinese)

[8] Y.J.Guo, "Sharing of tacit knowledge in organizations," Information Studies: Theory & Application 2004, vol.27, pp.130-133. (In Chinese)

Interior Optimization Design

Analysis on Sustainable Development in Interior Design

Ying Hu
Wuhan University of Technology
Jiangxi University of Finance and Economic
Nanchang, China
Email: huhuhugogo@hotmail.com

Abstract—**The sustainability development in interior design is a theory concern about specific interior space environment and characteristic of usage by using technology and artistic features to create a safe, clean, comfortable, scenic living condition in order to satisfy the physical and mental needs of modern people. It is an artistic design by using interior to provide the function of substantial life and spiritual life. The design theory is derived from the main purpose of protect environment, cherishing life and focus on spiritual needs as a green action. It is an artistic reflection on the thinking of the relationship between environmental, social and individual. Sustainability development is a new trend developed in interior design aspect which can be applied to creating a green interior space environment.**

Keywords-Interiordesign;Environment;Sustainable; Development

I. Introduction

In recent years, rapid industrial development brings prosperous economy and society development, but also leads to problems like damage on world natural environment, the ecological unbalance, growing population, global warming, ozone depletion, decrease of tropical rain forest, environment pollution, species extinction, species extinction, and the conflicts between human and environment becomes acute too. People are paying more and more attention on global environment problems. Material consumed in fields of architecture design, construction and application occupies nearly a quarter of total human resources consumption. As people's awareness on environment deepens, they gradually realize that the importance of natural sight in environment. Beautiful scenery and fresh air may improve work efficiency and also perfect men's spiritual status. No matter greening and greening space are in or out of buildings, no matter they are in private residence or public environment, a grace and changeable natural sight may influence people in the long run. At present, many people are sparing no effort to bring natural plants, water and stones into environment design. They add reinvention on natural view in people's living space. How to use limited resources, reduce influence on environment and save resources, energies, land, water and material effectively during healthy, comfortable and safe residence is a problem concerned by the architects worldwide and the whole world. Hence, architectures and interior designers have to consider bringing human activities into ecological system and reconsider the relationship among man, architecture and environment. Ecological construction emerges as time required. At the same time, the ecological interior design, as the significant composition of ecological construction, emerges too. And it becomes an important part for human's sustainable development. Since people hope to improve living environment through design and then put forward the strategy of sustainable development. Interior design is a bridge connecting spiritual civilization and material civilization. When reflected in interior design activities, it is called "ecological design" or "green design". Like the urban planning and architecture design, it mainly aims at harmonious existence of nature and people, and the sustainable development of human being and the earth. They make efforts to design interior environment with high science and technology content as well as humanity affections.

Design with sustainable development concept is a new trend. It undertakes environment protection as a duty, so the environment and ecology concept shall be embodied throughout the while design process. Compared with traditional design, the interior environment with sustainable development concept has not only universal interior characteristics, but also many unique characteristics. In general, there are systemic integrity, ecological organism, scientific of system regulation as well as the proximity of natural environment and design openness. The sustainable development interior environment requires reducing consumption of material and energy, eliminating discharge of harmful substance, and requiring the related products and material shall be able to classified and recycled or reused.

II. THE INTERIOR DESIGN WITH SUSTAINABLE DEVELOPMENTCEPT

A. *Integral environmental view*

The integral environment proximity focuses on the balance relationship between human being and nature. So the environment benefits shall be considered thoroughly in the design. It is required that to create interior environment more scientifically and rationally in stead of traditional superficial creation. In the context of guaranteeing comfortable functions and grace appearance, the life cycle shall be prolonged as long as possible and try to harmonize the relationship with nature. Design intention, conceiving, style, and atmosphere creation shall be understood by integral concept and interpreted by systemic method. Make both the "interior" and "exterior" be interactive with buildings, parks, and cities. A successful interior design shall be done after thoughtful consideration of ambient landform, landscape, and buildings. In that way, the interior environment may combine with ambient environment organically. The environment proximity may be furthered and human live harmoniously with nature. The integral environment concept is now represented.

B. Excessive design

Under consideration of functions and performance of interior environment, try to complete by the most suitable design language and approach. Deviate from design theme and randomly pile up designs may cause waste due to excessive design, and this shall be prohibited. Along with the science development, people's aesthetic standard changes from complex style to simple style. Designer shall master the change of people's aesthetic trend in different time since it is not only a problem of aesthetics but also morality. Because to much design will use plenty of construction material, and this may lead people to an opposite side, i.e. advocate waste. Designer shall undertake the leading role to instruct an appropriate value view, promoting environment-friendly standard and lead the society to accept correct aesthetic concept.

C. Green moderate consumption

Moderate consumption is advocated in commodity economy. Interior decoration is not only creating artificial environment, but also a kind of consumption, and it is very important to residence consumption. Although ecological interior design aims to "create comfortable and graceful residence environment", different from before, the ecological interior design advocates moderate consumption and economical living way, disapproving luxury style. It tries to maintain the production and consumption within the acceptable scope of resource and environment, and ensure a sustainable development. This presents a totally new ecological cultural and value perspective.

III. SPECIFIC EXPRESSION OF GREEN AESTHETICS

Compared with traditional aesthetics concept, in practice of instructing interior design the new aesthetics conquers many limitations and negative factors. It is embodied with more integrity and constructiveness. The design value of sustainable development is mainly represented on harmony between nature and cultural environment. It advocates social reliability and realism, opposes environment pollution or distortion of nature. It aims to express "ecological" beauty. In addition, it establishes new thoughts and approaches of green scientific and technical invention based on more extensive ecological network as well as promote Chinese traditional ecological aesthetics.

IV. SPECIFIC EXPRESSION

A. Energy-saving and high efficiency aesthetics

Save regular energy is a significant direction for interior ecological design. The residence interior design, as the section consuming most resources, it is possible to reduce resources waste and pollution through reasonable and effective design. Meanwhile, the progress of modern science and technology provides more possibility for energy-saving design. For example: the circulation of water resource, application of illumination devices with high energy effect, and clean energy as solar, etc. Taking the energy-saving and high efficient use as part of green aesthetics is helpful for people to improve people's recognition so as to apply energy-saving and high efficiency in specific design.

B. Setup the aesthetics with green cultural concept

Actually, the ecological crisis confronted with is cultural crisis and concept crisis. It is aroused due to unreasonable value, behavior and social and economic factors. "Human centrism" is the root of crisis, thus establishing harmonious green ecological culture between human and nature is the key to conquer green crisis. The aesthetics founded only on the base of green ecological culture may infiltrate into people's deep souls. The primary problem solved by green ecological culture is the green ecological c consciousness. The green ecological practice has close relationship with environment characteristics of various fields. Green aesthetics depends on regional environment and humanity thought and relies on high technology. To be exactly, it relies on high technology of environment protection. Therefore, green aesthetics is a culture mode with high level and high-tech cultural information. During the whole process of interior design, the green culture understanding, application and promotion of designers may embody their social responsibilities of being a member of the society.

C. Aesthetics of taking territory as friend

Interior space belongs to architectural space. As a part of micro-aspect of architecture environment, it has close relationship with the architecture itself, ambient environment and humanistic environment. Firstly, it shall be the integration with architecture environment, trying to extend the architecture on functions and form, avoiding excessive structural modification and function replacement. Secondly, it is the coordination with ambient environment. Apply ambient natural environment, regional climate to bring more benefits to interior environment. Thirdly, it is to extend regional culture and realize the sustainable development of humanistic environment. Interior environment is a natural grower of specific environment. Local material and technology represent the indigenous culture and the cultural characteristics of traditional residence environment. It has green aesthetic value to protect, improve and develop local characteristics.

D. Integral harmonious aesthetics

From interior design to construction and to application, the course is an open dynamic integral system with input and output of energy and material. In addition, it has life cycle. During its whole cycle, it is characterized as high efficiency, low energy consumption, free from pollution or small pollution.

V. GEEN DESIGN

Green science and technology means the application of greening of science and technology, i.e. the applied technology has distinct meaning of environment protection and satisfies the demand of ecological protection. For example: the application of intelligent science and technology and humanity science and technology. The aesthetics with green science and technology embodies the thought of humanism and benefits the development of human being. The green science and technology aesthetics in residential interior design may better represents the aesthetics essence of scientism as well as the science view and value of ecological spirit.

A. Elements for interior design of sustainable development

- During interior design, climate, environment, and energy shall be included in consideration at anytime and anywhere. Sustainable utility of natural resources and energies shall be disposed properly, including sunlight, ventilation, water, timber, and various natural resources. As for space structure, decoration shall adopt nonpolluting and natural material as much as possible and create a living and office environment by sustainable spirit.
- According to ecological requirements, material selection for interior design shall pay attention on non-toxic, fire-proof, dust-proof, mothproof and antipollution factors, including applying the degradable building material.
- Interior decoration shall minimize the consumption of energy and resources. It is necessary to consider the regeneration and reuse of them.
- Interior designer shall engage in ecology design and research, increasing the nature proportion in design, so that the interior decorate may further closes to nature.
- Chinese interior designer shall breakthrough traditional "old form", abandoning excessive decoration, not purchasing superficial luxury or so called "fashion". Thus, visual pollution, and waste of labor force or material may be reduced.

B. Interior decoration with outside style

People like to close to nature and designer will decorate the interior as outside through introducing nature. This design method is called interior decoration with outside style. By the means of construction design or reconstruction, penetrate the indoor and outdoor; or open partial walls to integrate indoor and outdoor environment. An open flowing space is created, and residents may enjoy more sunshine, fresh air and landscape. Due to the technology progress of building structure and material, the movable construction with huge roof is available now. And the interior and exterior may be connected as a whole.

C. Introduction of green plants and sight

Live creature may be applied in interior design, such as potted plant, bonsai, waterscape and ikebana to obtain natural feeling. At this aspect, there are abundant wealth may be use from Chinese traditional bonsai arts. The method of drawing landscape is inspirited by hominid petroglyphs. The landscape mural, plants and flowers, clouds and water add natural effect and art atmosphere. Through making use of interior garden creation, such as Four Seasons Hall, build large natural sight in public space. These round-the-clock ecological gardens are very popular.

D. Nature simulation

Emphasize the application of natural texture in interior design, so that users may feel the natural material and have a feeling of coming back to natural. Designers think much of surface material and disposal. They try to break through frame through texture emphasis and function hints. The cement surface, wood and metal material will be exposed intentionally to present their original face. Some material are rough, some have been polished to be smooth. This is called "Seeing color from hearing, and hearing voice from sight" artistic expression. In this way, people may find their underlying affection of back to nature. Sound and smell simulation in interior environment, such as birds twitter, flower flavor, wind or wave sound, may bring people nature-like scent and hearing. Advanced science and technology make such creation become possible.

VI. APPLICATION OF NEW TECHNOLOGY IN SUSTAINABLE INTERIOR DESIGN

A. Environment-friendly decoration material

Ecological environment-friendly material is in the process of clean production and product ecologization. They won't harm human body or ambient environment during production and application. The used material replaced from interior are easy to be degraded and transformed and may be reused to produce new product as renewable resources. This is a developing trend for all building material. At present, the produced non-toxic paint and renewable wallpaper embody the realization of such target. Since most products couldn't reach such requirement, so the first choice for decoration material will be non-toxic, nonirritating, non-radioactive, and low CO_2 exhaust material.

B. Save regular energy is a significant direction for interior ecological design.

Anti solar glass, privacy glass, and thermal insulation wall have many superior performances. If they could be combined with interior design, then both thermal insulation and lighting may be achieved and energy will be saved greatly. Set the floor panel heating as an example: the principle is to heat floor to 18-23℃ by the heating piping-PAP or contact tube, which is buried under floor, and radiate heat evenly. Advantages of floor panel heating are: 1 Floor temperature is even, and the heat will descend from bottom to top gradually, so it is very comfortable; 2 Cross-ventilation is weakened: reduce dust; 3 Save energy 10%-20% when compared with other heating methods, and it may save 2%-3% indoor usable floor. 4 Sound insulation effect is good. In addition, application of energy-saving lamps and water-saving parts may reach the effect of saving regular energy. Saving of lighting device power is an important aspect too. There are mainly two sides: one is use lighting device with high energy benefit and the other one is to reduce the use time of lighting device. Besides that, automatic system may be applied to save energy too. For example, infrared or ultrasonic sensor may be installed to detect human movement in office, passage or stairs and start lights automatically when necessary to save power.

C. Clean resource

Solar energy and wind energy are inexhaustible energies, and they are renewable and non-polluting. Thus, the solar and wind energies will influence interior design inevitably, especially for the appearance and ventilation system of building. Then people have new understanding of external surface and natural ventilation: introducing sunlight, natural ventilation, thermal insulation, shading, and anti-glare, appropriate application of solar and wind energies.

D. Modern ecological technology

It adopts all-around modern ecological technology. Since plants may absorb CO2, formaldehyde, benzene and bacteria in the air, so a healthy interior environment may be built up. It acts as ecological aesthetics function. Therefore, it is a key item to introduce interior greening and garden for interior ecological design. Now, the technologies of humus generation, waterproof treatment, and soilless culture may provide technical support for interior greening. Interior greening is multi-layer, It could be at any floor, or placed at balcony or roof. The multi-layer greening has compensated the deficient ground greening, and connects ventilation with outdoor atmosphere. Natural lighting disposal becomes a significant link of interior design and it may improve the isolation status of interior and nature.

CONCLUSION

The sustainable development concept in interior design is an internal space design arts by using scientific and artistic means to create safe, health, comfortable, grace internal space as well as satisfying human physiological and psychological demands and the physical and spiritual requirements according to different environment and use purpose. The design idea and approach aim at environment protection, life concern, and spiritual demand. It presents the interior design consideration on relationship among nature, society and human being.

It's a fashion to create green and sustainable interior environment interior designer, it is very necessary to implement the sustainable thought and persist in green design. Depending on all successful human achievements, try to contribute by self exquisite design.

REFERENCES

[1] LI MING ZHANG, ENGLISH In Architecture and Construction ,Chinese Architecture dusty press,2002

[2] Matthew Line, Book of Design A Complete resource of interior style Weidenfeld &Nicolosn Wellington House 125 strand London WC2R 0BB.

[3] Ambasz, E.(ed) The New Domestic landscape New York Museum of Modern Art.

[4] Glenn Porter, Raymond Loewy-Never Leave well Enough Alone, The Johns Hopkins University Press, London 2002.

Investigation of Design Optimization on Inner Roof Structure of Nansha Stadium for 2010 Asian Games

Ran Zhao & Demin Wei
Department of Civil Engineering,
South China University of Technology
Guangzhou, China.
Email: zhaoran1129@yahoo.com.cn

Wenbo Sun & Haoliang Zhu
Architectural Design Research Institute,
South China University of Technology
Guangzhou, China.
Email: swb_scut@yeah.net

***Abstract*—Inner roof of Nansha Stadium for 2010 Asian Games is investigated in this paper, a comparison on the optimization of structure design scheme is performed. Double-layer reticulated shell and radial beam string are chosen as two acceptable schemes for the inner roof, through the comparison on the aspects of structural mechanical characteristics, steel consumption and visual effect. The results show that radial beam string scheme has better mechanical characteristics, more beautifulness and less steel consumption**

Keywords- Asian Games Stadium, reticulated shell, radial beam string, optimization design

I. INTRODUCTION

Nansha stadium is dance and martial arts stadium for 2010 Asian Games. Total GFA is 41200m^2, shape of the stadium like a conch, plane size of that is about 270m×115m. Stadium roof is comprised by outer part and inner part, as is shown in Fig.1. Outer part of the roof is one-layer surface shell structure; inner part is a circle with a diameter of 98m. There are two choices for the inner part of the roof could be accepted: double-layer of reticulated shell system and radial beam string structure system. Based on the two choices, detailed analyses are performed for optimization design.

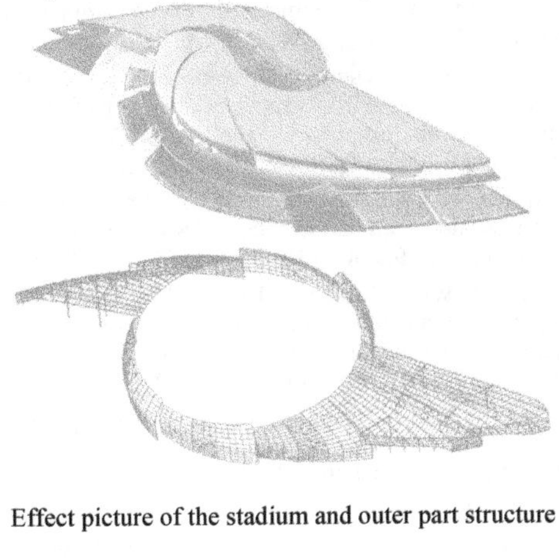

Figure 1. Effect picture of the stadium and outer part structure of the roof

II. CHOICES OF THE INNER ROOF STRUCTURE

Two design schemes are shown in Fig.2. and Fig. 3. The first structural scheme is a double-layer reticulated shell system. Thickness and the height-span ratio of the shell are 2.6m and 1/13 respectively. All the elements are steel pipes, the joints are welded balls. The structure below the inner roof is concrete beams supported by 36 columns. To release the boundary stress, some small plate elastic rubbers are adopted in roof bearings.

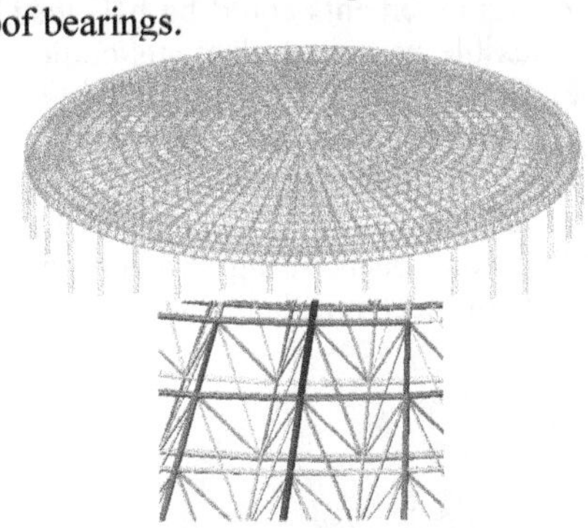

Figure 2. Integral and local structure of the first structural scheme

The second scheme is a radial beam string structure system, the upper elements are welded steel tubes and cross-braced pipes, middle elements are steel pipes, lower elements are tensioned cables. The structure below the inner roof is the same as the first scheme. Bearings of the inner roof are sliding in the process of construction and hinged after lower cables are all tensioned.

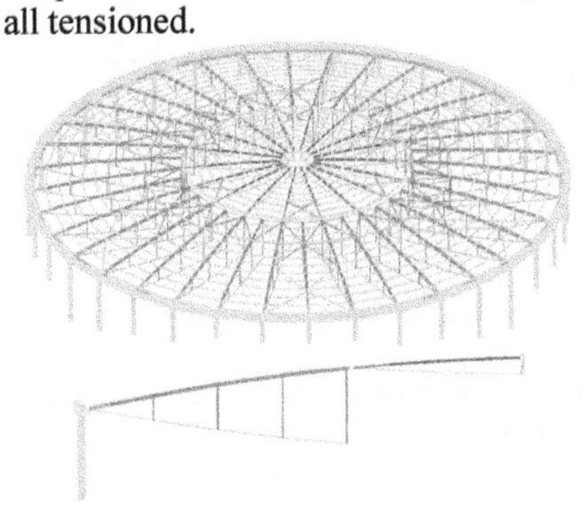

Figure 3. Integral and a radial element of the second structural scheme

III. COMPARISONS FOR OPTIMIZATION DESIGNS

In this chapter, mechanical characteristics, visual effect and steel consumption of the two schemes are compared for optimization design.

A. Comparison of mechanical characteristics

Mechanical characteristics are the first considerations in choosing structural scheme. Double-layer reticulated shell system is an excellent type of spatial structure [1], however, because of its special shape, the height-span ratio of the structure is only 1/13, mechanical characteristics of the shell are discount a lot [2]. Furthermore, displacement of the structure is very large because of the sliding bearings [3]. If rigid bearings are adopted, the large thrust of the shell edge cannot be balanced by itself. It will result in large stresses of the edge steel elements and the concrete substructure. It is difficult to resolve the contradiction.

Radial beam string structure is hybrid structure consists of flexible cables and rigid elements [4]. The enough tensile strength of cables could be utilized, and rigid elements could ensure stability of the structure. Moreover, the thrust caused by the upper rigid elements could be balanced by tensioned cables, and flexible and rigid elements could work together perfectly. FEM software named strand7 is used to analyze the two schemes, geometric nonlinearity need to be considered in the second scheme [5]. Table.1 is the deformations of the two schemes. From the results, it is found that the radial beam string structure is better than the former in terms of mechanical characteristics.

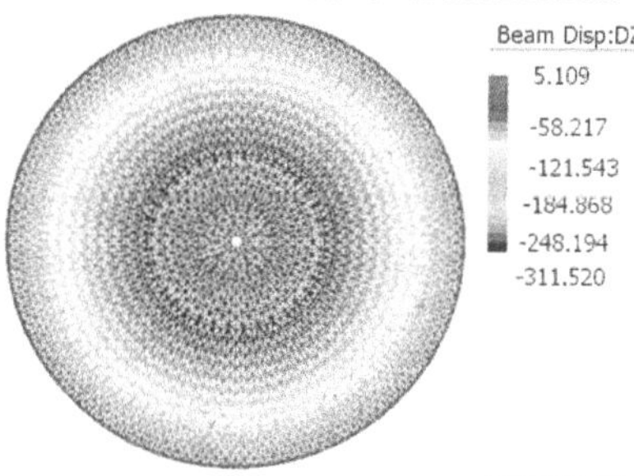

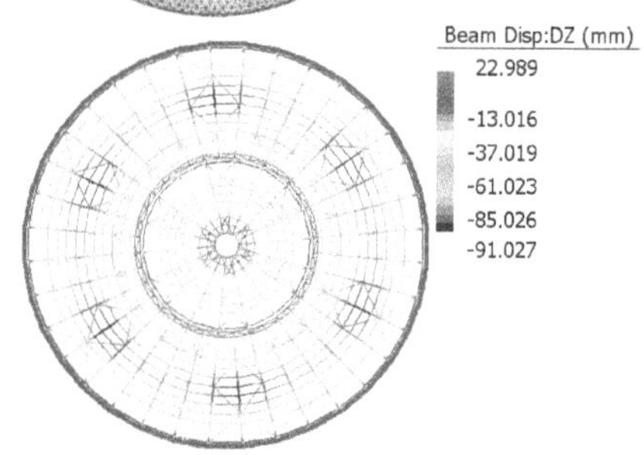

Figure 4. Displacement contours of the two structural schemes

Table.1. Maximum dispalcements of the two design schemes for the inner roof structure

	Dead load and full-span live load	Dead load and half-span live load
Scheme.1	312mm	273mm
Scheme.2	91mm	115mm

B. Comparison of structural visual effect

The visual effect is often ignored in engineering practices [6], however, if the visible elements of steel structures are too many and unsightly, structural engineers have to change the structural scheme. The advantage of long-span steel structures is simple and light, however, because of the length of appropriate grids of reticulated shell is about 3m, these elements of the first scheme are excessive visually. In contrast, as shown in Fig.4, these elements of the second scheme are considerable less than the former. The flexible cables and radial rigid elements make the structure light and concise, the structural effect is better obviously.

C. Comparison of steel consumption

Under safety conditions, the steel consumption is controlled by the design of steel structures. Table.1 gives the steel consumption of the two schemes. General, the steel consumption of the long span steel structure is about 70kg/m^2~90kg/m^2. From table.2, it is found that consumption of the first scheme is 70kg/m^2, and that of the second scheme is only 58kg/m^2. It is evident that the second scheme is more economical

Table.2. Steel consumption of the two design schemes for the inner roof structure.

	Elements (kg)	Joints (kg)	Total weights (kg)
Scheme.1	437000	86000	523000
Scheme.2	408000	27000	435000

IV. CONCLUSIONS

Inner roof of Nansha stadium is a long-span steel structure, because of its importance and complexity, careful design is the necessary. In order to obtain an optimized structural scheme, comparative analyses of two acceptable structural systems are performed in the paper. The results show that the radial beam string structure system is better than the double-layer reticulated shell system in many aspects, including appearance, mechanical characteristics and construction cost.

REFERENCES

[1] S. L. Dong, Luo, R Z, and Y. Zhao, "Analysis Design and Construction of New Space Structures". China Communications Press, Beijing , pp. 447-482, May 2006. *(references)*

[2] P.K. Umesha, M.T. Venuraju, D. Hartmann, and K. R. Leimbach, "Optimal design of truss structures using parallel computing". Structural and Multidisciplinary Optimization, Vol. 29, Feb. 2004, pp. 285-297, dio: 10.1007/s00158-004-0420-1.

[3] Y. Zhong, "Displacement Control of Prestressed Structures". Computer Methods Applied in Mechanics and Engineering, Vol.144, May. 1997, pp. 51-59, doi:10.1016/S0045-7825(96)01164-4

[4] Y. Hangai, M. Wu, "Analytical Method of Structural Behavior of A Hybrid Structure Consisting of Cables and Rigid Structures " .Engineering Structures, Vol. 21, Aug. 1999, pp. 726~736 , doi:10.1016/S0141-0296(98)00027-3.

[5] X. F. Yuan, S. L. Dong, "Nonlinear Analysis and Optimum Design of Cable Dome". Engineering Structures, Vol. 24, May. 2001, pp. 965-977, doi:10.1016/S0141-0296(02)00017-2.

[6] Y.Q. Wang , L. Huang, "The structure technology and visual art of architecture". Journal of Nanchang Institute of Aeronautical Technology, Vol. 6, Oct. 2004, pp. 87-91.

Investigation on the Concept "Design"

Liu Hainian
Wuhan University of Technology
Wuhan, China
Email: dhlhn@163.com

Xu Yonggang
College of Law & Business of Hubei University of Economics
Wuhan, China
Email: xuyonggang1@yahoo.com.cn

***Abstract*—The concept "Design" is ever-changing and ever-developing. Putting forward the concept "design" is aimed to clarify and construct the corresponding design activities. The earliest concept in China, developing in the process of manual production, is quiet different from its counterpart in Europe, which is contained in the context of art. Chronologically, the author in this paper will deliberate the design concepts and their differences between in China and Europe.**

Keywords-design, sustanbility, time-featuredness

I. INTRODUCTION

The word "design" is often used in our daily lives, such as "clothing design", "packaging design", "landscape design", "store design", "environmental art design", "Urban architectural design" and so on. In order to re-clarify and re-construct the concept "Design" from professional perspective, the paper integrates definitions and explanations of "Design" from many distinguished works on art design, which will help to understand "Design" more deeply.

II. THE ORIGIN OF "DESIGN"

A. The Origin of "Design" in China

Kaogongji, China's first compilation of handicraft technologies, is a classic literature about ancient science, technology, design and production. Although it doesn't put forward the term "design" explicitly, but there are words "zaowu" and "chuangwu" [1], which are the earliest explanations for "Design". It is said in *Kaogongji*, "Wise men make creation and skillful craftsmen inherit the technology. The soul should be passed to the later generations forever."(Xi 2008, 3) *Encyclopedia of Cihai,* published in 1936, a well-known Chinese lexicon and character dictionary, gave such a definition to "Design", also named "Pattern", "before the creation of art crafts and architectures, we must consider their shapes, structures, colors, decorations and how these elements can be put together. The pictures and drawings that reflect the consideration are called patterns, including architectural patterns, dyeing patterns, ceramic patterns, metalwork patterns, decorative patterns, advertising patterns and so on." The second edition of *Modern Chinese Dictionary*, published in 1983, explains "Design is methodologies and drawings developed to satisfy certain purposes and requirements prior to a project." "Design" can be spelled in Chinese words "设计". The character "设" contains following meanings: 1, set up; 2, plan; 3, suppose; 4, assume (Lu 1983, 1013) and "计": 1, calculate; 2, measuring instrument for time and temperature; 3, idea, method or plan; 4, intend or plan to do something; 5, a surname. (Lu 1983, 532) The Ministry of Education in 1998 incorporated the discipline of "Design" into Curricula of Regular College Course, named Arts Design, and Academic Degrees Committee of the State Council promulgated and named it "Design Arts" in Curricula for Post-Graduate Education. The two are the official names of "Design" in China.

B. The Origin of "Design" in Western Countries

"Design" originates from the Latin word "Desigara", which means proofs or media such as badges and tokens that distinguish things from the other things.

"Design", originating from the Renaissance, in English means "to reach a certain state, or form some sort of plan through action." The meaning is "the intention in artist's mind for creation". The word "Design" at that time still remained a title within the scope of art.

The interpretation of "Design" in the 1786's edition of *Encyclopedia Britannica* reads as: "harmony reached in artwork's line, shape, proportion, dynamic state and aesthetics". Later "Design" gradually widened its boundary and grew out from the context of art.

III. THE DEVELOPMENT OF "DESIGN"

"Design" shares different definitions between different cultures. The following parts will display the definitions both from China and western countries in order to investigate the origin of "Design".

A. The Development of "Design" in China

In Industrial Design Guidebook, (Zhang 1993) the editor Prof. Zhang Daoyi defined "Design" as: 1. The basic meaning of design is to achieve a particular purpose to envisage, plan or advance program. It is a dynamic process of thinking and creating, and its result is eventually delivered in some symbols, such as language signs, drawings, models and etc. 2. Design, a word with multi-meanings, shares

[1] zaowu, chuangwu: creation

extremely extensive references. When used, generally it should be added with proper modifiers to express the complete and accurate meanings, for example, clothing design, packaging design, environmental art design, and computer program design. Unless used in a specific context, the single word "Design" can be confusing and meaningless. The changing definitions of "Design" reflect social advancement and demands, and also its connotation changes in accordance with human's changing demands for design activities.

B. *The Development of "Design" in Western Countries*

In Western countries, the definitions of "Design" also vary with social development in various degrees. In the *Encyclopedia Britannica* over 200 years ago, "Design" was used to distinguish paintings from applied arts. In the 19th century, "Design" shared the same bracket with "Pattern", bringing on a strong sense of decoration.

After World War I, German's university Bauhaus applied "Design" to some courses, such as "Metal Design", "Furniture Design" and so on, which made "Design" rise from pure artistry and creative practice and attained a lot of rich contents. The extension of the meaning made related design activities more complicated.

At the early 20th century, the connotations of design appeared to be diversified.

1 "Design" is the simulation of something that we want to create before we are confident of the final result.

2 "Design" is a kind of creative activity - creation of unprecedented, innovative things.

3 "Design" is an activity that makes artificiality change.

4 "Design" is an imagination transition from existing fact to potential possibility.

Former Soviet Union Aestheticians proposed in *Concise Dictionary of Aesthetics*, "Design is a kind of creative activity (including its production) and its purpose aims to form and adjust object, space environment. Its function and aesthetics achieve unity in this process.

In *Concise Encyclopedia Britannica,* "Design" (fine arts) usually means the process of planning, and it can also refer to concrete plan kept in mind or made in draft. Product design first refers to the relationship between the components of an unfinished product, which is often subject to four factors, including material properties, material processing methods, the combining degree of individual components, and the influence that the overall product effects on the audience and users.

IV. TIME-FEATUREDNESS AND SUSTAINABILITY OF "DESIGN"

"Design" is a very complicated concept, which has very complex forms and room for further development. Therefore the understanding for the concept should be diversified and dynamic. Each era shares its unique development style, so is the concept "Design", which obviously brings on the characteristics of the time. Design activities change with the change of society and the concept "design" is its external manifestation and the static expression of the development process.

A. *Time-featuredness and Sustainability of "Design" in China*

The industry of art design is expanding very rapidly in China. The majority of universities have set up the major Design. According to up-to-date information, there are more than 700 formal educational institutions for design. The number of design schools, design majors, and people engaged in design activities are in the increase along with the expansion of design field. Modern time is a time which needs design, and design is becoming a more and more important creative activity. In the curricula of these design institutions, some focus on industrial design, some lay emphasis on urban architecture design, and some regard graphic design as critical. Therefore, the concept "Design" is a moving, diversified and time-featured concept. In China, there were argumentations for concepts such as "decoration arts", "crafts design", "utility design", etc. With the development of society and industry, people began to switch their importance from these concepts to the distinction between "art" and "design" and re-estimate the relationship between them, which is inheritance and development of European Renaissance's design within the field of art. Social development and industrial revolution are objective conditions decisive to the independence of "Design".

B. *Time-featuredness and Sustainability of "Design" in Western Countries*

Before European Industrial Revolution, the design was understood as production technologies and decoration techniques in the process of industrial planning. The low productivity at that time determined that craftsmen must depend on manual power to complete products. After the revolution, Design mainly pointed to industrial planning and its purpose was to satisfy people's practical demands. Drawing-oriented and mass-production-oriented designs became popular and were distinctive changes after Industrial Design Revolution.

ICSID, established in June, 1957, organized many experts to define the concept "Industrial Design" several times. In 1964's International Conference for Design the organization put "the definition of design" as the meeting's priority to deliberate. President Maldonado of Supreme Molding School of Ulm in Germany believed that design should not be regarded as an aspect of industrial production but an independent activity, which was published in *Design Education.* The definition of "Design" was concluded as follows, "design is a creative activity, whose purpose is to decide product's form properties, although form properties includes the outside features of a product, it lays importance to the inter-relationship between its structure and functions. The properties turn a product into the unity both from producer's and consumer's sides." Later, the scope of industrial design got further expanded. In the 11th annual conference of 1980, the organization re-adjusted "Industrial Design" to "'As far as mass-manufacture industrial product is concerned, it is called design to equip material, structure, color, surface processing and decoration with new qualities and qualifications by virtue of trained techniques, knowledge and experience.' According to the situation, industrial designer's work should concern some or even all of the aspects said above. Furthermore, when needed, industrial designers should exploit their techniques, experience and visual assessment ability to solve the problems in packaging advertisement, display and market development. Such activities also belong to the field of industrial design." (Li 1991, 73-74)

C. *Prediction of the future changes of Design*

With the advancement of digital and information technologies in modern society, the world is becoming smaller and smaller, and thus the connotation of "Design" attains new elements. In the future, design activities will represent features greatly different from those before. Information Age will revolutionize design activities as Industrial Revolution did hundreds of years ago, which is bound to make the activities deeper, more personified and even super-personified. Future design activities will pay more attention to human's abstract thinking abilities than to the concrete thinking, which will further extend the concept of "art design". Design will change physiological relationship between people as the proceeding of some design activities such as Japan's introduction of design investigation to study of sexual behavior on Internet. In the near future, design will change all kinds of human relationship such as individuals, family members, groups, societies, time, space, universe, vision, olfaction, gustation and perception. Design will no longer be a conceit of physical change for tools or materials, and it will change the conscious content of other fields.

V. CONCLUSION

In modern society, with the great advancement of productivity, digital and information technologies, "Design" obtained more vivid and up-to-date connotation. Design activities nowadays tend to pay more attention to ecological environment and sustainable development, for modern industry has excessively exploited natural resources. Design's connotation and extension is in such continuous motion and its definition is increasingly bringing on the color of humanity and science.

Design makes proper adjustments with the advancement of society, time, and productivity. The adjustment is the abstract manifestation of design activities. Its practical, appreciable, cultural, creative and scientific connotation is the mark of social advancement. With continuously higher ability to understand and change the world, design will have a more scientific definition in the future.

BIBLIOGRAPHY

[1] Cao Tianquan, *Introduction to Art Design.* Shanghai, People's Fine Arts Publishing House, 2005

[2] *Concise Encyclopedia Britannica*, Beijing: Encyclopedia of China Publishing House, 1986.

[3] Hu shou hai, *Introduction to Design.* Hefei: Hefei University of Technology Publishing House, 2006.

[4] Jin lei, *Introduction to Design.* Hebei: Arts Press, 1997.

[5] Li Yanzu, *Introduction of Industrial Arts.* Jilin: Fine Arts Press, 1991.

[6] Lu Shuxiang & Ding Shengshu, *Modern Chinese Dictionary.* Beijing: Commerical Press, 1983.

[7] Tian Zibin, *History of Chinese Industrial Arts.* Beijing: The East Press, 2001

[8] Xi Chuanji, *Classical Arts Literature.* Shanghai, People's Fine Arts Publishing House, 2008

[9] Xu Xiaogeng, *Modern Design Arts Wuhan*: Yangtze River Literary & Arts Publishing House, 2003.

[10] Yin Dingbang, *Introduction to the Science Design.* Hunan: Science & Technology Press, 2008

[11] Zhang Daoyi, *Industrial Design Guidebook.* Jiangsu: Science & Technology Press, 1993

[12] Zhu yu, *Introduction to Design Arts.* Hunan: Hunan University Press, 2006

Landscape for Educational Technology

The Importance of Virtual Reality Technology in Landscape Design Teaching Method

Liu Jing
Modern Art College Environmental Design Office
JiangHan University
Wuhan, China
Email:3938533@qq.com

Chen Li
Modern Art College Environmental Design Office
JiangHan University
Wuhan, China
Email:20018285@qq.com

***Abstract*—With the development of the society, landscape designers should be required to have long-term vision and comprehensive ability of balance, gradually finding the way to foster talents of landscape designer. To achieve that simulating the development and change of landscape design and various, changeable natural conditions with advanced technology in the way of teaching is one of the means to improve the accuracy of predicting the practicality of the designed works by landscape designers, compared to the effects of initial drawing by hands and the manufacture of 3D static effects, which can more likely realize error- free designs with better use of technology. At a macroscopic angle, the introduction of virtual reality technology is also the true effective measure to the save of natural resources and the sustainable development.**

Keywords-landscape design; landscape designing major; virtual reality

I. Introduction

Landscape designing dates back to 1858, the father of landscape design of America, Olmsted realized it and insisted on naming the occupation he undertook “landscape architecture” in stead of the temporal “landscape gardening”, thus explored a broad field for the development of the major and subject of landscape design, lasting for 100 years. As well as architecture, the occupation of landscape design formed before the landscape architecture. It is the landscape architecture that developed and improved the theory and method of landscape design on the base of large quantities of landscape designers’practice.

II. THE CURRENT DEVELOPMENT SITUATION OF LANDSCAPE DESIGN EDUCATION

A. The beginning of landscape designing major

In the 1980s, the appearance of the name “environmental art design” began a discussion of arts category in education industry. Introduced from Japan, the name rapidly replaced that of the famous department of interior design at Central Institute of Fine Arts Industry, becoming the first choice of numerous universities’ naming of the major. Actually the category of environmental art design is too extensive for the interior design to contain. Several years later, many experts and professors who had deeply analyzed and understood the name thought that environmental art design does not equal to interior design. The extension of the concept of environmental art cannot be covered by interior design. Therefore many universities began to review the category of environmental art design. Like industry division, as the developing degree of society grows with people’s feeling of beauty, industry division of detailed becomes a trend. Environmental art design, existing as a major in universities, also needs very detailed division to reflect the intention of the name. Yu Kongjian went back after graduation from Harvard in 1995. Landscape architecture Academy in Beijing University was founded in1997 with two masters programs, master of landscape architecture and landscape gardening. With these great events about the detailed division of environmental art design education industry occurring, landscape architecture came to China. From this point of view, the entering of landscape architecture to china began with college education, so landscape architecture education became one of the most important methods to lead the development of the occupation of landscape design. As landscape design becoming an occupation that approved by the design industry in China, the plans and methods of fostering the talents of landscape design and various teaching methods drew more attention in the industry, which directly affect the quality and level of occupation of landscape designer.

B. Current situation of landscape designing major

Throughout the arts colleges and arts major in universities, landscape architecture has become an indispensable training orientation. To strive for greater market of art students, most universities describe the perspective of the major gorgeously. Of course, it is undeniable that many professionals are optimistic about landscape architecture because of the glory of its environmental protection and sustainable development. This current situation also makes many universities offer core courses of landscape architecture like landscape design, garden botany, techniques of presentation, and computer aided design. But in the view of the professional characteristics, objectively landscape design is influenced by regionalism and climate change; subjectively not only are the students required having good scale relation and sense of style, but also they should have the thought of sustainable development. Most of the art colleges in China provide the major of landscape design, but many graduates cannot make best use of the knowledge obtained on campus on their positions and cannot bring their specialty into full play because their professional knowledge's range is not wide enough and the working position is unsteady. And most landscape designing major can meet the current problem that students on campus have a good ability of hand-painting that can paint the design paper carefully with unique style and arrange the plants species well and colorfully on paper. On the other hand, when the designs of the students are brought to market, the practicality of the gorgeously design paper is almost zero.

III. THE CURRENT TECHNOLOGY AND METHODS OF LANDSCAPE DESIGN EDUCATION

A. The current situation of landscape education technology

Whether in the arts colleges or arts major in universities, teachers often teach knowledge of landscape design by multimedia, while students work on large design plans by hand-painting. The other way of design is by computer aimed design, using software like Auto desk, CAD, or 3D MAX to accomplish floor plans, effect graphs, or dynamic cruises. As to technical level, it can realize the exchange of information in teaching and learning, but as to the teaching effect, students can only know the teachers' experience and relevant information that teachers get from teaching materials and reference media. There are two parts lacking of verifiability, factuality and making the students know the changes of landscape design with the changes of natural condition and man's factor. During the teaching process, students can only accomplish the landscape design with static and timing representations. The whole teaching process is made up with teachers' lecturing and students' exercises, unitary in method. As to effect, the majority of colleges focusing on landscape design courses can only make a judgment of the students' standard of obtaining the professional knowledge by the ability of producing drawings, rendering skills of drawing, and color matching.

IV. THE ANALYSIS OF SUPERIORITY AND PRACTICALITY ON INTRODUCING VIRTUAL REALITY TECHNOLOGY

A. The influence of virtual reality technology on teaching

Virtual reality can realize the reconstruction of all spatial relationships with 3D technology in teaching in colleges. It can increase the students' understanding of the standard of spatial scale, construction technique, seasonal variation of plants, and design of circulatory water, which assess the rationality and action ability of landscape design. Virtual reality needs to be equipped with software assistance and hardware environment. Comparatively speaking of teaching, virtual reality should be one part of laboratory for professional of design. For teachers, the introduction of this technology to teaching can accomplish the scene setting that landscape design needs with basic 3D manipulation software.

In addition, by placing accomplished models into virtual reality system and multi-channel 3D display, teachers can simulate possible changes about the description of scale relation, the explanation of plants selection and collocation, and the change of seasons after the landscape design, thus can lecture landscape design knowledge more convectively. The applications of virtual reality in landscape education not only have the advantage of reusable cases, but also help the students without good ability of spatial thinking during the process of teaching.

In the process of real-life operation, virtual reality can identify the general-purpose 3D software, Sketch up documents, and platform compatibility. It can also give the teachers numerous technological choices of cases in the process of teaching. In the view of real teaching situation, virtual reality has more reliability in the necessary software and hardware compared to many other teaching technologies. Meanwhile, it is unnecessary for the teachers to bring students to landscape job location, which may cause an accident. Teachers have collected some landscape materials; they are not able to be shown to students in school who cannot come to the site by themselves. Virtual reality makes it possible for the students to know the image formation of classic landscape.

B. The evaluation of students' information by virtual reality technology

The main features of virtual reality technology reflect on creation stage of landscape design are as follow. The space experience in design has more interactivity and flexibility.

The depth of field is expressed as much as possible and the authenticity increases greatly. It makes the landscape spatiality fully expressed in this stage, which provides visualized and vivid teaching demo for the teaching and learning of landscape designing major. In landscaping design, virtual reality technology combines the advantages of departed media, possessing unprecedented figurativeness, authenticity and vitality. The observers experience personally, observing the landscape in various methods, wandering indoors and outdoors, and feeling the material quality and airflow with the aim of virtual reality tools. And it can obtain relevant system information of the case with hyperlink. For example, the seamless connection with Google earth can blend the landscape in system into the surroundings, having better effect of virtual display and experiencing more reality of the design. At the same time, it can get virtual scenes of all kinds of architectures and gardens successfully, making the students have a deep understanding of classic architectures and garden landscapes all over the world. Students can learn in the situation that they can participate in and control by themselves. According to the plan of designing paper, the feasibility of all the designed elements in landscape design can be examined. The students can grasp the action ability of the case on campus.

V. THE VITAL FUNCTION OF FRIENDLY INTERACTION OF VIRTUAL REALITY HAS ON LANDSCAPE DESIGN EDUCATION

The characteristic of interaction is not only reflected in professionals in education to education objects. In the level of education technology, interaction is also a stage that design education technology is experiencing or will experience. The technology changes the teaching methods and the information that the students obtain as well. It also changes the approving degree of the students' designs when entering to landscape design industry. The application of virtual reality technology can improve the quality of teaching in many aspects.

VI. CONCLUSION

Landscape designing is developing in China, which plays an important role either in city construction or in residential property development. It calls for a good supply of designers. In the educational technology and teaching with the teaching conditions of development of experiments being to grow at the same time, the landscape of educational technology to improve the existing landscape design can make up the standard of teaching can not be achieved diversity, interactive reality. Landscape design-based platform for educational technology is even more necessary in the teaching process rational use of computer-aided design.

Virtual reality technology in teaching to better highlight the gap between landscape design space large, content-rich features of the scene can be a more intuitive visual effects, participating students to measure the real effect of raising awareness of the relationship, and learn.

The landscape of education determines success or failure of China's landscape development of the industry's prospects. Landscape design university training in the use of object-teachers make good use of technology, multi-faceted integration of state, will be better able to improve the practice designers design level. Therefore, as the medium of virtual reality technology, the more possible means should be given the full attention of various art institutions.

ACKNOWLEDGMENT

The completion of the article depends on Optimization Design of International Conference on (ICOD) of the upcoming. Also thanked Jianghan University Institute of Modern Art, Environmental Art Design my colleagues for their kind assistance, as well as Digital Technology Co., Ltd. Shanghai Heng Man Virtual Reality laboratory construction program.

REFERENCES

[1] Wang Chengwei, Gao Wen and Wang Xingren, "The Theory, Realization, and Application of Virtual Reality Technology,"Tsinghai University Press, Guangxi Technology and Science Press, 1996.

[2] JohnVince, "Virtual Reality Systems, " Addison-Wesley Pub. Co, 1995.

[3] Grigore Burdea, and Philippe Coiffet, "Virtual Reality Technology," A Wiley-Interscience Publication, John Wiley & Sons, Inc,2005.

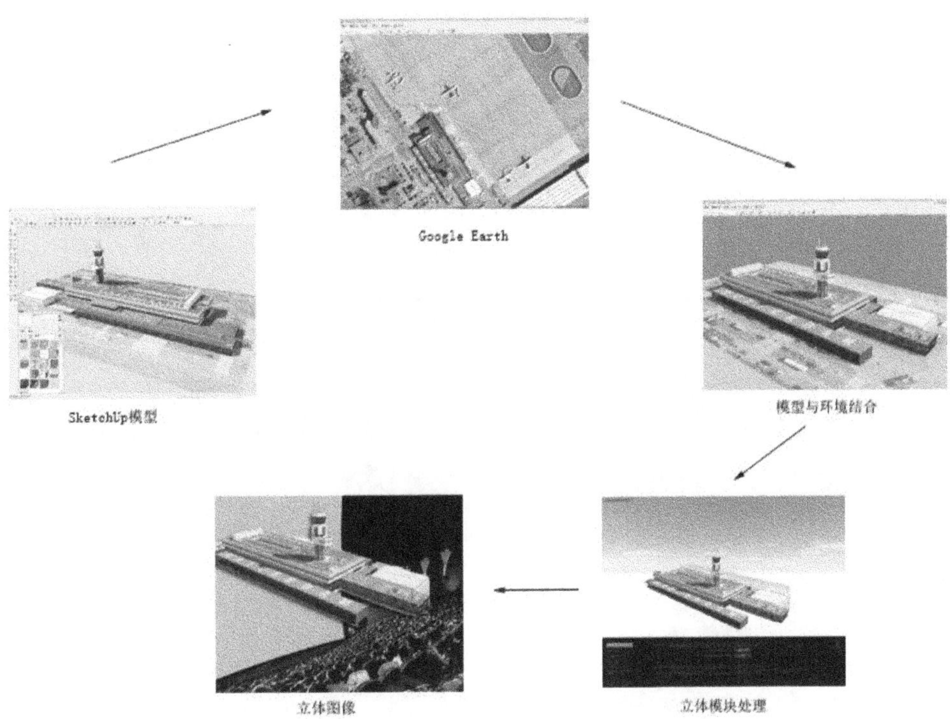

Figure 1. Sketch Up Model → Google Earth→Combining Model with Environment→Three-dimensional Module Process →Stereo image.

Landscape For Educational Technology

Landscape Design with Practice Teaching Thinking

Zhou Yuanyuan
Modern Art College Environmental Design Office
JiangHan University
Wuhan, China
Email: 3217066@qq.com

Zhou Xiao
Modern Art College Environmental Design Office
JiangHan University
Wuhan, China
Email: 455204044@qq.com

***Abstract*—With the development of information technology, teaching content, teaching methods, teaching patterns are changing rapidly, Teaching and Training Program on the ever-changing. Teaching involves a lot of professional design issues, such as square landscape design is not simply copying the style of several ground-mosaic called square, and the need to investigate the project's geographic, cultural backgrounds, such factors as the use of the object. Links Landscape educational practice is to study the use of the profession to be the focus, we have to realize in practice, in practice, the use of innovation in practice.**

Keyword-educational technology; practice teaching ; practical significance.

I. INTRODUCTION

Landscape Art and Design started in the eighties of the last century in China, continue to develop.The current landscape of education undergraduate courses, including theoretical teaching and practical teaching. There are general education courses, professional basic courses and specialized courses in theoretical teaching. Practice Teaching Practice has focused on innovation and practice of extra-curricular links and links. But the practice of teaching can not be completely separated the division, in addition to purely practical classes, there are theory and practice of theoretical courses and practical courses, which illustrates the infiltration of practice teaching in general, environmental art and practice of curriculum design needs to improve the design depth and contact practical.

Symonds has, in its classic passage lay bare the landscape architect's work and the corresponding definition of discipline, "We can say that the landscape architect's life-long goal and job is to help human beings, people, buildings, communities, cities and their lives in harmony with the life of the Earth. " This requires the teaching of the course is not a mode of teaching, require integration of life, do not cross-industry combination, should be out of the classroom to the practice exercise books.

II. THE BACKGROUND OF THE PRACTICE CAPO

The formation of the landscape more a reflection of our people, man and society, people and culture, and even human beings in the past, present and future thinking. Teaching involves a lot of professional design issues, such as square landscape design is not simply copying the style of several ground-mosaic called square, and the need to investigate the project's geographic, cultural backgrounds, such factors as the use of the object. As well as complex commercial landscape design, industrial landscape design and residential landscape design, content will need to investigate more.

There are many practical courses, clearly the purpose of practice while also exploring practical approach. For example, the current curriculum to go out collecting folk songs reflecting the lack of this class, former students curriculum requires students to paint landscape, architecture etc. This is definitely able to exercise the students drawing skills. However, the design for the students also requires experience life, experience the community and experience the culture of different regions to understand the different lifestyles, habits and values. Insights to be good at human spirit, find out the local cultural background, history, customs, symbols vector, so that design will meet the local people living habits, and aesthetic taste of the public space. This design is part of the local people belong to local people's life, belonging to the local culture and history. Thus human landscape design should express and embody the essence of property is a kind of human spirit, which is in the design process should always be the focus of attention is the design form of the inner pillars. Such as footprints and roads, gates and bridges, walls and fences, sculptures and squares, architecture and urban, are described with regard to human beings, man and society, the relationship between people and local culture. I put the concept of"Nature as Teacher, experience culture," into landscape educational practice, called for students to start before the landscape design of natural phenomena, natural beauty, history and culture, customs and other aspects of the manpower to find design ideas, determine the design focus of attention After the design concept and performance coupled with human point of view. In this way, so that is no longer an empty classroom to teach the skills, but more to the students thoughts back to earth, back to history, back to our culture and life, so that they learn and design during the process to appreciate more creative fun, make up for their lack of experience and understanding of natural defects.

III. PRACTISE THE PRINCIPLES OF CURRICULUM DESIGN

The last century, 80 years of reform and opening up of China's art and design has entered a new period of development, with the country's economic construction and development, people's living standards improve, the community has an increasing demand art and design, art and design products in developed countries and cultural trends are also with the Chinese and foreign cultural and art exchanges and profound impact on the development of Chinese contemporary design. With the quality of education in full swing, teaching goals are also killed by instilling knowledge, students turned to the comprehensive capabilities. Facing the new situation, an important part of teaching - layout of major changes have taken place in schoolwork. Previously, a large number of written assignments arranged mainly practice-based, but now has turned to academic arranged practical application of knowledge. Practice-based homework how to design, that should follow the following principles:

- The targeted system. Teachers in the design of tasks before, we must thoroughly understand the outline of this class is fully aware of the specific objectives and requirements of students and teaching students to understand the specific circumstances for these requests, the situation is to design this class of schoolwork.
- The guiding principle. Against the design premise of schoolwork, but also to enable academic instructive. In the academic ago, it can on the academic requirements, and methods to make a brief description, if necessary, to give practical demonstrations.
- The interest in principle. Interest is to do a power thing, so when in the design of tasks and students must bring their knowledge and interest in the use of combined effect until the completion of such tasks better.
- The principle of moderation. From a psychological point of view, the attention of students always remain at a certain period of time, so the design should be suitable for students at academic age, how much homework should be moderate, degree of difficulty should be moderate, neither too easy so that students do not pay attention also should not be too difficult to provide students with fear or frustration, not to make too much homework a burden of students.
- The operational principle. Practice-based tasks aimed at training students in practical ability and application of knowledge to solve practical problems.
- The principle of diversity. Type of practice is designed to academic forms. There is no single form of selective, not necessarily in line with the characteristics of all students, and thus can not give full play to the creativity of students.
- The principle of objective evaluation. Schoolwork is completed, teachers should be taken seriously, especially in practice-based tasks, I believe that can not simply be evaluated with right or wrong. To give an objective conclusion.

IV. PRACTICAL SIGNIFICANCE OF CURRICULUM RESEARCH

Attempt to analyze China's current higher education landscape art and design professional development programs put fear into the arrangement with the adjustment of clear practical courses, courses in practice, a clear link, as well as graduate design of the introduction of real issues to achieve standards. Teaching the structure of interlocking done without duplication, the students skills development and the market can be closely integrated.

From the three aspects of the Practice Course Category description: grade classification (degree students), different grades from low to high practical course arrangements implemented progressively. Curricula, design courses in the two thematic convergence is not consistent, students accept the need for a transitional link. Course themes and designs of the market based on real practice issues, including the design and analysis, terrain studies, terrain reconstruction planning, space to create and layout of the road system, environmental assessment, the ideal design of assessments, a reasonable choice of materials, garden ornaments in the rational use of and so on.

A. *Theoretical significance*

1. the students in the learning process can clearly course goals, methods, and applications.

2.teacher capability requirements, to avoid the scripted, the overall quality and scientific research requirements.

B. *Practical significance*

Students learn in school is able to in-depth, rather than a form of design patterns. In terms of employment can quickly get started, and after the creative has a vast space for development.

Theory to practice; practice to theory; practice sublimation theory; theory to guide practice again, this is a progressive cycle.

C. *National teaching and research of landscape design*

1) the advanced nature of modern American landscape education

The United States is the birthplace of modern landscape architecture. In 1900, a small Foley was Fisk • LO • om (FLOlmsted, Jr.), And Shurcliff (AASharcliff) opened a landscape design at Harvard courses, formally established its status in modern subjects. One hundred years, the American Society of Landscape gradually improved education and development to a scientific discipline system, and an open attitude, to meet the changes and challenges of the times, forming a multi-level education system, both to provide a good vocational education but also on the basis of research work. Landscape design from the gardens extended to the vast space systems, such as urban public space, residential areas, waterfront areas, parks and green channel, scenic road system, even including national parks and scenic tourist areas. Landscape planning theory not only to comply with the adaptability of the land, but also attach importance to the needs of human life. colleges and universities set up the Department of Landscape Architecture, American Society of Landscape Designers Association has 48 chapters, nearly 13,500 professionals.

a) American Society of Landscape Design characteristics of curriculum

- to design courses as the core, to develop students creativity

Landscape architecture is a comprehensive cutting-edge cross-disciplinary, it uses the principles of aesthetics and science of space in the layout of the venue, students need to know a lot of ecological, social and artistic knowledge, and coordination with land planning and design capabilities. Landscape design course is the core of education, curricula were organized on different scales of landscape design, aims to nurture students to use a variety of knowledge and creative problem-solving abilities. Design course to take given the typical assumptions, or the combination of the actual site, students observe, imagine, thinking and problem-solving ability.

- focus on a comprehensive interdisciplinary education and establish a wide range of values

Some modern American landscape architecture in the establishment of the College of Agriculture, while others are set up in the Institute of Art and Architecture. In the design disciplines, architecture, landscape design and urban planning professionals all three side by side, thereby strengthening the link between the three professions. Harvard University, the University of California at Berkeley, and so to adopt such subjects setting, the Department of Landscape Architecture and Architecture students often, urban planning department of the exchange of teachers and students, which has trained them to think in terms of habits. Students completing the compulsory core of the theory and core design courses, you can elective own areas of interest.

- teaching practice, improve their practical ability

As human society is moving away from the natural environment, answer Landscape Design no longer be an enlarged version of the construction yard, but a practical topics covering almost every aspect of the relationship between man and the natural environment, the practice of discipline, its theory and social responsibility also will be expanded and changed. Teaching with practice, you can quickly enhance students practical abilities. Design lessons are often topics of practical subjects, students have the opportunity to exchange the local population, so designed as to solve practical problems, an effective way to improve the situation.

2) studying the situation in Japan

In Japan, the garden Education and the Institute of organizational development review process and the status quo on the basis of the analysis, first of all general education in the Japanese garden, "six-oriented" and "nature": "6" of the garden of education, including international, socialization, character, the name of diversification of the garden, the garden's scientific research in the field of quantitative as well as the breakdown of the garden; "nature", including the garden of education systemic (system-based), developmental, educational links of practicality.

a) mature landscape of the education system in Japan: The Japanese Landscape educational "6" of

Garden of education internationalization. Garden education, socialization. Historical heritage nature of landscape evolution. Specialization. Garden name of diversity. Quantitative research garden the garden in the Japanese emphasis on scientific research and quantitative research. LA garden area of education and research segments of the Japanese Institute of Landscape With the development of interdisciplinary area of research carried out in accordance with the following papers.

b) Japanese garden educational "nature"

The development of education of the garden. Garden systematic education. Garden education, manifested in the systematic research (teaching and research) to set the systematic, faculty and curriculum with systemic several key aspects of the systemic, while the most important thing is research (teaching and research) to set the systematic .

c) in the practical aspects of teaching.

Landscape Architecture for practical subjects, in the garden of teaching, particularly in undergraduate teaching should attach importance to practical education.

V. COUNTRIES OF OUTSTANDING LANDSCAPE DESIGN EDUCATION INSPIRATION

A. American Society of Landscape education landscape to Education in China

The landscape of education in the United States on the subject on the basis of a scientific orientation, the development of the times to maintain a high degree of sensitivity, the teaching process to design lessons focus on, through the multi-level, multi-disciplinary elective courses to enrich student's knowledge structure, in order to cultivate a generation after generation of creative, and social responsibility of designers to promote the completion of the project the balance of the environment and natural systems. Ten years a tree, takes a hundred years become the human beings.

Thelandscape is put into the cultivation cycle of long, slow bear fruit only when we adopt a pragmatic approach, face up to the status of education of China'slandscape, learning the advanced Western countries, landscape planning theory and science of teaching system, proactive, committed to reform, in order to keep pace with the times, and take the major responsibility of higher education to the community.

B. the Japanese education system in the landscape of China's educational development landscape Inspiration

- for the community, appropriate adjustments to the direction of interdisciplinary education and research education and research should face the society and the development along with the development of society.

- new ideas, the development of modern gardening subjects according to their different characteristics at various schools, in education research should be in a different direction characteristics; various scholars, according to their different learning research background, a different perception on gardens, and engage in different aspects of the education and research.
- emphasis on the garden's scientific research and quantitative research to ensure the proper development of Landscape Construction and Japanese garden with a high level of quantitative scientific research, compared to our plants from the garden planning and design studies are somewhat insufficient, especially in the planning design far behind.
- a reasonable set up teaching and research gardens, improve the education and research system and schools should be based on their own characteristics, reasonable to set teaching and research, education, common garden is reached, based on the characteristics of the creation of school teaching.
- at different stages of the cultivation of different abilities should pay attention to undergraduate education should pay attention to basic skills training and practice teaching links, MA, PhD postgraduate training should strengthen scientific research capacity.
- garden plants and correctly handle the relationship between planning and design to make up for a real garden belonging to the content of education and research disciplines, the majority of our school garden plants, horticultural education and research to the tilt, while the planning and design aspects of the city planning or building tilt, garden plants and planning and design out of line, really belongs to discipline the use of garden plants to create a space and environment there is a vacancy in this area.
- garden plants, research should be to apply mainly to avoid excessive micro-oriented direction of garden plants should pay attention to the use of plants to create an environment and space research, urban greening techniques, garden green ecology research.
- in achieving "common" schools based on the pursuit of "characteristic" At present, Landscape Architecture can be divided into two categories: one category is agriculture and forestry colleges in the education system, focusing on garden plants and garden planning and design, emphasis on the natural the transformation of landscape and ecological environment;In addition, there are a number of art schools set up the professional landscape design focuses on environmental art, is not comprehensive, can not do co-ordination to grasp.

C. *the trend of educational development of China's landscape.*

Although China has the "mother of gardens" reputation, but in the development of the modern landscape but at the apparent backwardness. Modern Landscape education system has not been established. At present, there are many colleges and universities have been gradually began enrollment in professional landscape design, but in teaching theory and methods, there are still many problems. Such as in teaching theory, still the traditional Chinese garden design ideas for the dominated, the lack of modern landscape design theory, the introduction; in teaching methods, the core of the design lessons attach enough importance to the depth and breadth of curriculum is not enough, most of the Elective courses are not breaking the College of restrictions on cross-disciplinary elective courses less.

In recent years, as the art of doubling enrollment, in order to protect the quality of teaching and the overall level of new faculty and school match the overall level of existing teachers, art colleges and universities to recruit new teachers, except special expertise outside the standards are set by is a master's degree or above. Teachers no practical experience, in the grasp of the overall curriculum is completely self-taught, such a self-bound students with each session of the run-in the teaching process to complete, meaning that the term, two students who slowly find ways to .Some schools set up both practical courses, are also more technical courses, such as construction technology, professional bidding, budget and other purely practical course. A lot of misunderstanding of the real issues that will limit the design thinking play, it is not true, we should understand the entire process of real things to do for the program design services to provide more inspiration.

Although the landscape has a profound understanding of the many problems in education, but the proposed system is far from enough to guide teaching theory can only be explored in advance. For now, need to combine into the practice of teaching, how binding is also the need to continue in-depth study.

VI. CONCLUSION

Links Landscape educational practice is to study the use of the profession to be the focus, we have to realize in practice, in practice, the use of innovation in practice. Domestic undergraduate education, the second, third and fourth grade at different stages of learning into practice to the appropriate link. Arrangements can not simply some social programs euphemistically of the market, this will only make the students into stereotypes, but to integrate into the practice to design the curriculum links, so that designs are based on including history, environment, background, standardized and so on, is not a simple function of aesthetic research.

ACKNOWLEDGMENT

The completion of the article depends on Optimization Design of International Conference on (ICOD) of the upcoming. Also thanked Jianghan University Institute of Modern Art, Environmental Art Design my colleagues for their kind assistance, as well as Digital Technology Co., Ltd. Shanghai Heng Man Virtual Reality laboratory construction program.

REFERENCES

[1] Ding Shao-gang. China's agriculture and forestry colleges of higher professional status and Garden Education Reform [J]. Chinese garden, 2001, 17 (4): 15 - 17.

[2] Kongjian Yu, LI Di-hua, landscape design: professional, academic and education, China Building Industry Press, 2005, P8. P128.

[3] Liu Zhongmin. The United States the development of landscape architecture, architects, 35.

[4] Yang Donghui, China needs to Landscape Design - - the practice of landscape architecture from the United States look at our landscape garden, Chinese garden, 2000.

[5] Pat Symonds, JO, Landscape Architecture, China Building Industry Press, 2000.

[6] Wu Liangyong. Habitat Environmental Science and Landscape Architecture Education [J]. Chinese garden, 2004, (1), 2-5.

[7] Bin-Yi, landscape studies three major areas of discipline and direction of [A], First International Conference on Education Proceedings landscape, China • Shanghai 005.10.28-29 [C], China Building Industry Press, 2006, Beijing.

[8] Liu Bin-yi, the modern landscape planning and design [M] (first edition, second edition), Southeast University Press, 1999,2005.

[9] Ding Shao-gang. China's agriculture and forestry colleges of higher professional status and Garden Education Reform [J]. Chinese garden, 2001,04.

[10] Fan Yi-rong. Design class teaching reform of thinking and of [J]. Ningbo University (Polytechnic edition), 2004,06.

Low-Cost, Energy-Saving and Ecological Optimization Design for Rural Housing

Guohui Fan
Department of Architectural Engineering
Henan Polytechnic Institute
Nanyang, China
Email: fgh19801214@163.com

Jiding Zhu
Department of Architectural Engineering
Henan Polytechnic Institute
Nanyang, China
Email: zhujd777@sina.com

Abstract—**Based on the status of development and the features of local production environment in the less developed rural areas of Central China, an exploration of the energy-saving design for Rural Housing is made in this thesis. With the principle of utilizing the rural waste and the local resources , measures such as the new style wall, chimney effect, energy-saving curtain, ground insulation and so on are taken to realize the low-cost, eco-friendly residential energy-saving optimization design goal.**

Keywords-rural residential; energy-saving; optimization design; ecology

I. PREAMBLE

With the arrival of "low-carbon economy," residential energy conservation has become increasingly popular. It has long been widely promoted in urban construction. However, in rural areas, the development of residential energy conservation is still at an early stage or even a blank in some areas. In Central China's less developed rural areas, its promotion is even more delayed, which has a serious impact on the overall process of residential energy efficiency.

II. STATUS OF RESIDENTIAL ENERGY CONSERVATION IN THE LESS DEVELOPED RURAL AREAS ,CENTRAL CHINA

The less developed rural areas of Central China refer to Henan Province as its representative, including Hubei, Shanxi, Jiangxi and some other regions. It is hot in summer and cold in winter In these regions, where the average temperature is 0 ~ 10℃ in the coldest month, 25 ~ 30℃ in the hottest month. Residential energy-saving design must meet the need of stopping heat in summer and thermal insulation in winter.

In these areas, residential facades, roofing, windows, doors and other structural systems still adopt conventional practice: solid clay bricks are used for the most building envelopes(a part of them is made of soil); doors and windows are mostly made of wood and a small amount of them single-layer steel; the only function of roof is waterproof. Scattered coal stoves are used for heating in winter, and there is less refrigeration in summer and the ventilation is in bad condition. Housing envelope is in poor insulation. Heat transfer coefficient of exterior walls and roofs respectively are 50% more than the architectural design standards for energy efficiency, which results in energy waste.

Although lots of rural residential energy-saving researches are made in recent years, the promotion scope is very limited because the energy-saving measures divorce from farmers' basic situation. The reasons are as follows: the pre-input is high, so many people don't want to spend money; the construction technology of energy-saving measures is of complexity, due to being afraid of trouble, many people simplified it in practice; furthermore, there is no correct understanding of building energy-saving construction and building materials; many people are shortsighted who are not aware of energy-saving.

In addition, there are some drawbacks in the current practice, such as filling the center of the block (or hollow brick) with polystyrene board or polyurethane panels or some other chemical products to meet the building energy efficiency requirements. But as the raw materials of such products are crude oil, it will meet the requirement of building energy efficiency as well as increase the oil consumption and do not meet ecological requirements.

Therefore, it is pressing to explore a set of low-cost, eco-type, using local materials and simple energy-saving measures and promote it in the less developed rural areas in central China.

III. OPTIMIZATION DESIGN OF ENERGY-EFFICIENT HOUSING IN RURAL AREAS

A. *Optimization Design of Wall Structure*

Building envelope plays an important part in residential energy-saving of which the wall is laid extreme emphasis. According to the agricultural characteristics in the regions, a lot of crop straw can be effectively used to improve the wall structure. The new compound wall is constructed with steel as framework, plant fiber and cement composite board for the panel, (Using crop straw as the reinforcing material and cement as adhesive will not produce pollution.) and cob for the core layer. The wall makes full use of local crop waste and solid waste, which are local materials. The processing is convenient and more ecological.

The wall is of light weight, low thermal conductivity and outstanding energy-saving effect. Compared with 240mm clay brick walls, it can be about twice energy-saving. And it has the flame retardant characteristic, its life expectancy is 50 years or more, and after the expiration it can be recyclable to reduce construction waste. Under the same condition of the housing function, the cost of this kind of wall is very similar to that of the brick-concrete structure, so that it can effectively reduce residential energy costs [1].

In addition, some frameworks can also be set up on the wall to plant some climbing plants. This approach can achieve the three-dimensional greening effect, and using it as shade of the building can save energy [2].

B. Optimization Design of the Roof Structure

In winter, the roof contacts with hot rising air and heat is eliminated a lot, which results in poor thermal insulation; in summer, due to the strong solar radiation, radiant heat absorbed by the roof makes the roof overheating and leads to hotness indoor [3]. At present, sloping roof structure is commonly used in these regions, which creates an ideal thermal buffer layer for the top of the building, but the thin roof is a main reason for the energy loss. Making full use of crop straw-- binding and paving them into the inner roof and then paving plaster-- can be of good thermal insulation. And it is also an ecological approach.

An important measure of rural residential energy conservation is natural cooling in summer. It is a simple and effective method to transform the roof slope, set the exhaust chimney and use the “stack effect” to regulate the natural ventilation in the house. In order to avoid the "reverse chimney effect", wind-driven exhaust fan should be installed in the air outlet to speed up the indoor air emission, and meanwhile the air inlet should be set in the bottom part which is in the shade, ventilation to keep the supply of outdoor cool breeze. If "chimney effect" is smartly used, it will have a very effective role in regulating the indoor temperature and improving indoor air quality.

C. Optimization Design of The Door and Window

Windows and doors are the main thoroughfares for the cold air infiltrate to the heat source [4]. Generally speaking, for a residential building, the amount of heat loss through windows is more than one third of the total heat dissipation. However, it is difficult to have energy-saving windows. The common approach is to install steel or aluminum double-glazed windows and doors. But in rural housing construction process, restricted by the local construction techniques and product quality, energy-saving effect is not satisfactory yet. Besides, the investment is so high that it cannot realize the purposes of low-cost and easy promotion

It is an economical and practical method to paste laminated transparent film. The transparent polyester film should be pasted on the inside of the window frame, and generally there should be 15 ~ 20mm air compartment between the film and the window glass. In this way, 60% of the energy consumption during the heat conduct through windows can be saved, and in winter the temperature can be increased about 2℃, and can avoid condensation on the glass [5]. This method can be also used for the summer insulation. Their transmission performance is better than ordinary plate glass and the price is much cheaper than glass.

D. Optimization Design of the Ground Insulation

In the low-rise building, the ground heat consumption accounts for about 6% of the envelope heat consumption, so the ground heat consumption can not be ignored [6]. The ground insulation has a great influence on people’s feeling about temperature, so it should be strengthened to improve living comfort. The proposal is to have 200mm thick dry slag under the cushion of the ground and external surface of the outside wall within 0.5m ~ 1.0m. (Dry slag is by-product of heating for local residents and it is treated in general as garbage disposal.) And meanwhile, in order to achieve the moisture effect, 20mm thick layer of plastic film or a layer of plastic film can be put on rammed soil to raise the ground heat storage capacity. This method is economically feasible.

IV. Conclusion

For the rural energy-saving housing project, promotion is the key and efficiency is the guarantee. In this paper, low-cost and eco-friendly measures are put forward which is simple by taking the local environment and local materials into consideration .They can improve the quality of the farmers’ living conditions, promote eco-energy saving of buildings, reduce the environmental loads and play a useful role in accelerating the new rural development as well.

References

[1] Yun Wu and Yulong Wang, “Requirements of Wall Material and Detailed Construction of Steel Structure Energy-Saving House,” Architecture Technology. Vol.39, No.11, 2008, pp. 881-884.

[2] Peng Di, Haojun Wu and Baiyi Li, “Housing Energy Conservation Strategy in New Countryside,” Industrial Construction, Vol.38, No.11, 2008, pp. 118-120.

[3] Yan Liu, Jianxin Zhang and Liyang Fu, “Design Proposal of Countryside Housing's Energy Saving in Hot Summer & Cold Winter Areas,” Construction Conserves Energy, Vol.36, No.2, 2008, pp. 34-37.

[4] R. de Dear and Gail Schiller Brager, “The adaptive model of thermal comfort and energy conservation in the built environment,” Int J Biometeorol, Vol.45, No.2, 2001, pp. 100-108.

[5] Zijiang Yang, “Doors And Windows Energy-Conserving Measures Of Small Town Housings In Hot Summer And Cold Winter Area,” Industrial Construction, Vol.35, No.7, 2005, pp. 19-22.

[6] Zhi Qu and Yanzhuang Wang, “Discussion on Energy-Saving Residence in Southern Area of Henan Province,” Chongqing Architecture. No.3, 2009, pp. 5-7.

Ocean of Knowledge & Heights of Science

Architectural Design of Guangzhou No.2 Children Activity Center

Xiong Dizhu
CAUP of Guangdong University of Technology
Guangzhou, China
Email: 38458020@163.com

***Abstract*: -----To respond to the discovery of science and technology by children today, the three building blocks are presented as science, technology and knowledge based learning respectively, while the whole project is captured as a symbolic science mountain for future young generations of China to conquer.**

Keywords----- science, technology and knowledge

I. PROJECT PROFILE

To promote socialist cultural and ethical progress and offer more study and activity places for the children in Guangzhou, the municipal government of Guangzhou decided to build Guangzhou No.2 Children Activity Center in Zhujiang New Town, functioning as a large modern base for quality-oriented after-school education and fostering high-caliber talent reserve for the sustained development of Guangzhou. With a planned site area of 16,570 square meters, the Center is located on Plot J1-7, J Area, Zhujiang New Town (Fig. 01) and the southern part of Zhujiang New Town of the new city central axis, bordering on the Huajiu Lu on the south and Zhujiang Avenue on the east. As a key part of the public space development on the new city axis, the Project, together with other three large civil buildings i.e. Performance Center, Museum and Library constitute a quadrangle central public space, representing the newly appeared city center.

The municipal government of Guangzhou launched the International Architectural Design Competition for Guangzhou No.2 Children Activity Center in 2002, where SBA from USA won the competition. The Project was completed and put into use in 2005.

The building of the Project is designed into 7 aboveground floors (partially in 6 floors with one basement) with a total building height of 34.5m, hence is classified as Class 2 high-rise civil building. With a site area of 16,750 square meters and the capacity of accommodating 20,000 persons (including outdoor activity place), the project is planned with a building coverage of 6,540 square meters and a gross floor area of 45,887 square meters (including 9,938 square meters for basement and 33,234 square meters for aboveground structure except the open-up floors), and involves an estimated total investment of RMB294 million Yuan.

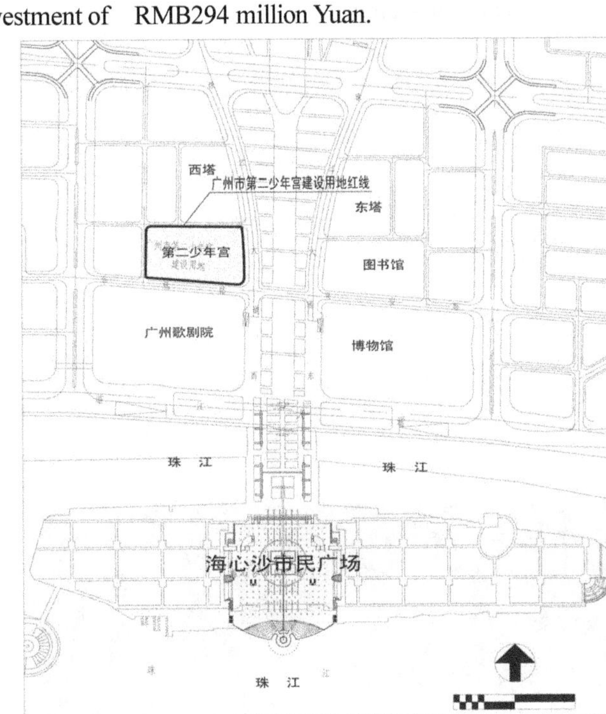

Fig. 01

II. SITE PLAN AND TRAFFIC DESIGN

A. Analysis of Building Layout

Based on the location on the central axis of Zhujiang New Town and particularity of constituting Guangzhou's Culture and Art Square with other three cultural buildings (Guangzhou Opera House, Guangzhou Library and Guangdong Museum), the concept stressed on creating a symbolic public space, which can organically integrate with the surrounding environment with unmistakable identity, distinct sign of the times, rich local characteristics and cultural connotations and strived for constructing a landmark building preferable for urban view and located in the base of urban visual center and on the city central axis. The main entrance was designed in the southeast of the Plot to constitute an enclosure with other three cultural buildings, as the integral part of Culture and Art Square (Fig. 02 and 03). The foursquare classrooms are provided in the north of the site, accommodating its function, whereas the curved Cultural and Performing Arts Center in streamline form stands on the south, forming the Red Scarf Square through enclosed arrangement of romantic curves; the outdoor activity place is provided in the area between them. This way, the overall functional layout is organically coordinated with the environment, while the functional zone inside the singular building is rationally arranged.

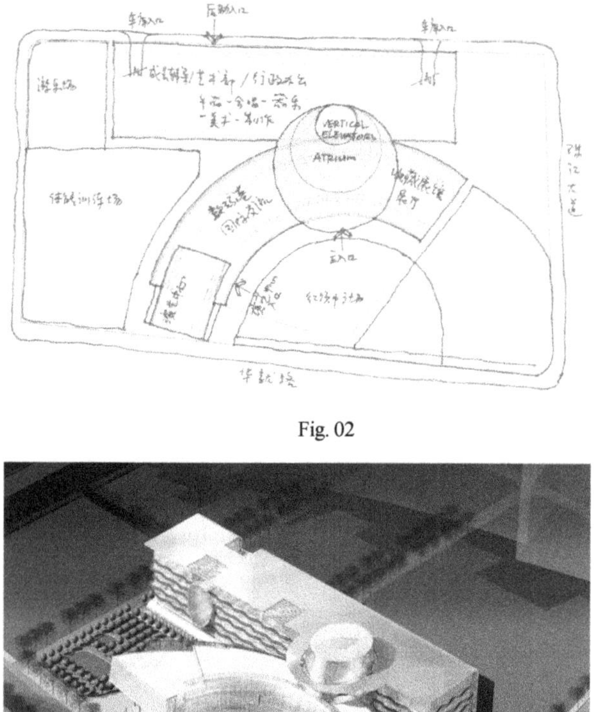

Fig. 02

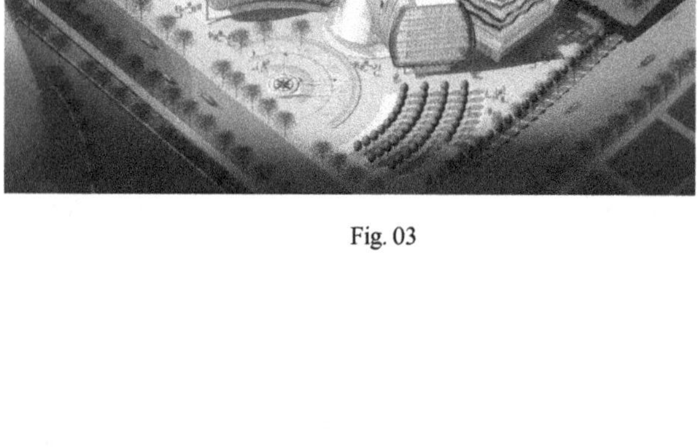

Fig. 03

B. Analysis of Traffic Circulation

Vehicles approach the building by the west and north roads, which are guided to the entrance of the north underground garage and semi-outdoor parking area, while bus circulation is arranged on Zhujiang Avenue (west) on the east and Huajiu Lu on the south. Thus, the vehicle flow is relatively concentrated without interference between the vehicle and bus.

Pedestrian enters into the site mainly through Huajiu Lu, the urban road and West Zhujiang Avenue, the central axial landscape road of Zhujiang New Town. The different functional zones of the building can be accessed through elevated space and various squares formed by building layout. The design has minimized the circulation cross, while ensuring the integrity of city central axis square.

C. Analysis of Underground Space

The underground space mainly serves as the underground garage and MEP rooms in combination with the civil air defense design. The entrances/exits of the underground garage are concentrated in the north of the site. The foursquare column grid for classrooms is used for underground garage, favorable for parking and vehicle flow direction. The civil defense shelter gives full consideration to the use for both peacetime and wartime and maximizes the efficiency of underground space.

D. Analysis of Landscaping Concept

1) Design for Public Square

In the landscape square design, the square is organized into a cultural and recreation space featuring rich connotations and harmonious entity with the building and surroundings through modem landscape design concept. The ground floor of the building is partially opened up through open, transparent and light architectural form, thus allows for the mutually penetrated indoor and outdoor spaces (Fig. 04). The parent waiting square takes after the architecture features and connotations of arcades in Guangzhou, serving as both the transition for building entrance and the connection between inside and outside squares (Fig. 05). The Red Scarf Square, the main entrance, culture and ecological square, parent waiting square and ecological square of outdoor activity base of the Center are all provided along the southeast of the site (Fig. 02). The ecological square of outdoor activity base in the west leads to the central elevated square, then reaches the Red Scarf Square, the climax of the whole square at the southeast end of the site. Connecting to the Urban Culture and Art Square via the eastern ecological square, the open space of nearly almost 10,000 square meters, represents the center of Children Activity Center, another highlight of square on the city's new central axis. Thus, the site square is organically integrated with the urban square. The openness of the singular building not only stands for the integration of interior and exterior spaces, but also a large window of the Center opening up to the urban space (Fig. 07).

Fig. 04

Fig. 05

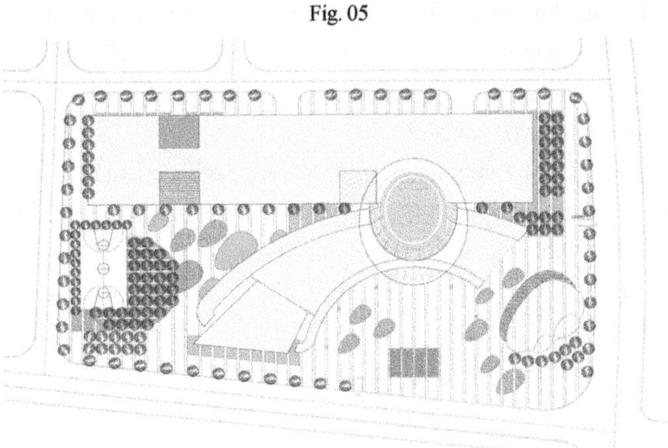

Fig. 06

Fig. 07

2) Landscaping and Environment Design

The landscape and environment design focuses on how to present Guangzhou's characteristics and cultural connotation of the Center. In fact, the comprehensive arrangement for the square spaces, the centralized green space and selection of tree species not only try to reflect the flavor of southern China and the charm of Guangzhou as a "Flower City", but also fully implement the ecological ideas. The charm of "Flower City" is highlighted in concise and succinct greening arrangement, while the environmental furnishing and leisure hard pavement are taken as the main environment design elements to add more interest to the city central axis in Zhujiang New Town.

3) Landscape Design for City Central Axis

The super high-rise on the north of the site, the Children's Activity Center and Guangzhou Opera House cascade in building heights toward the Pearl River, forming the musical notes leaping on the city central axis as well as the suddenly emerged skyline near the exit to the Pearl River.

III. SINGULAR BUILDING DESIGN

A. Planar Building Function Analysis

The main body of the building comprises of the unfolded curved public building (6 floors) in the south and regular quadrate building (7 floors) in the north (Fig. 08) with one basement.

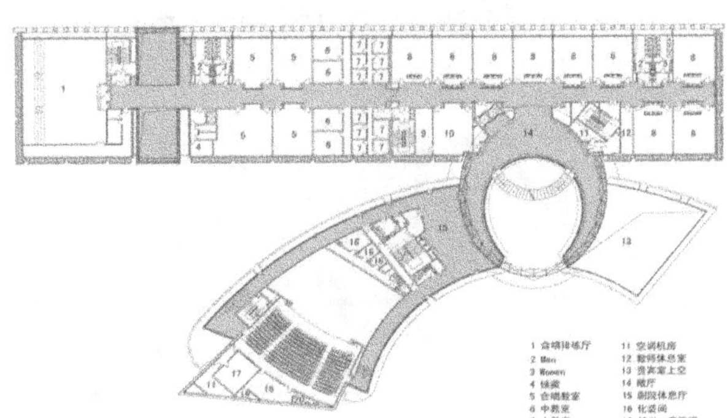

Fig. 08

The entrance lobby and theatre entrance lobby are intercepted into the elevated and open space on the south ground floor in a flowing form, while the northern main building is designed into Development Guidance Area in sequence.

The theatre and support functions are provided on F2-F6 in a centralized way in the southern curved building. The functional layout of F1-F7 in the north quadrate building is as follows:

F1: entrance lobby, exhibition hall

F2: classrooms for foreign language and computer training

F3: fine arts and calligraphy training center

F4: classroom for calligraphy training and chorus rehearsal room

F5: musical instrument training room

F6: dancing training center

F7: international exchange hall and musical instrument rehearsal hall

The basement serves as parking garage, MEP room and civil defense shelter (Class 2, Grade 6) combining uses in both for peacetime and wartime.

B. Analysis for Building Section and Elevation

The whole building is composed of curved 6 floors and quadrate 7 floors, where the middle oval glass penetrates through 7 floors and cuts into, connecting the south and north parts (Fig.03). The floor height is 5m for F1, 4m for F2-F5, 5m for F6 and 7m for F7. The south façade is clad with external metal mesh (Fig. 09 and Fig. 10), curved shape and some square and stripe openings, which not only function as breathable wall (insulation of noise and light pollution) but also create esthetical, novel and original design effect. The north facade uses lateral perforated metal plate as the ornament shelter and shade for the main AC machine units (Fig. 11).

Fig. 09

Fig. 10

Fig. 11

C. Traffic Organization

1) Planar Traffic Organization

The F1 entrance of the classroom area is arranged in the south of the building. There are 2 escalators inside the oval atrium leading to F2. The F2-F7 are designed with classrooms in similar plan layouts, i.e. organizing the planar traffic passage via the smoke preventing escape stairs and fire elevators at both ends and internal corridor.

The theatre entrance is arranged in south elevated F1. The theatre can be reached either via the stairs in the theatre entrance lobby to the theatre lounge, then through the outer corridors at both sides; or via another outdoor stairs on the open-up floor directly.

2) Vertical Traffic Organization

For the classroom area, there are 3 smoke preventing escape stairs and 3 fire elevators (those in the west are concurrently used as freight elevators), among which two escape stairs lead to the basement. 2 smoke preventing escape stairs are provided for the theatre with full consideration to the fire and traffic circulation requirements.

IV. ARCHITECTURAL DESIGN IDEAS

The architectural image reflects Guangzhou No.2 Children Activity Center's spirit, a palace for discovery. The facility has been envisioned as a learning device, which can be adjusted and reset for adoptions to future socio-economic changes and children's learning capacities. Through learning, children gradually discover the essence of this learning device and explore the fascination – the deep meanings of a place for discovery. The spatial dynamics and variety is inspired by a cutting-edge technological attitude. Light, both natural and artificial, is the essence of the design. Guangzhou No.2 Children Activity Center is luminous, not simply illuminated. Façades and roof act as a fabric, which moderates the natural and artificial light. They become a screen to which festival images are projected according to seasons or activity themes (Fig.12). With its characteristics of transparency, permeability to light, reflection and refraction, there is a constant change of images and effects during day and night, effecting not only the appearance but also maximizing the comfort and minimizing the energy consumption of the building.

Fig. 12

The Children Activity Center is captured in the fluid-like and continuous folding plane forms intercepted by a glass tower. The fluid-like form embodies flexible open public, social, performance, interaction and exchanging functions. The folding plane form embodies flexible organized learning and administration functions. The glass tower embodies flexible display and main circulation functions. This unique tripartite composition is envisioned for a cutting edge 21th century children activity place. The three building blocks are presented as science, technology and knowledge based learning [1] (Fig. 03).

Guangzhou No.2 Children Activity Center enters the competition among high rises in Zhujiang new town center for the best alternative with the sky. We design the tilted glass tower with a new strategy of visual dissolving top of building into the sky by and extending the glass skin beyond the top structure of the building (Fig 03 and 13). The tower is treated as a freestanding object settled onto the Honglingjin plaza, facing south while perfectly align with the center point of the plaza. In a unique way, it is this location and angle that gives the identity to this central element – a symbolic science mountain for future young generations of China to conquer. [2]

Fig. 13

The central tower form is the core body of public open and display spaces, and main circulation, which is made of steel structure and cladded with point-supported curtain wall system. The front curved fluid element curves free of the building's frame, symbolizes the ocean of knowledge and references the water of Pearl River rubbing against a passing ship, tying the building to its context. This unique element is cladded with woven metal mesh with changing perforation to conceal outdoor hung HVAC condensing units and cut down the energy consumption. At the elevated building base, simple circle and oval forms cladded in glass and arranged in grouping with stone walls and metal columns give the building a uplift energy that extends upwards to float the gently curved building element.

The folding plane concept for the Children Activity Center

transforms a porous building morphology via a series of programmatic and ecological functions. The intercepted overall building mass, which reference computer memory blocks, brings a dynamic quality, helps soften the building mass, and accentuates the main portal as it embraces the tripartite composition. These horizontal banding elements are perfectly correspond to all open out door corridors, as sun shading system and mechanical screens for HVAC units (Fig. 14). Two large scaled opening are designed to create vertical porosity in the block with a ruled surface system freely connected to control prints, plan to section (Fig 03 and 15). These large, dynamic openings corresponding to the internal public spaces are the ecological lungs of the building bringing natural light down and moving air up through the section (Fig. 16).

Fig. 14

Fig. 15

Fig. 16

V. CONCLUSION

Guangzhou No.2 Children Activity Center, from its international competition to operation, has created another children's activity place, a new urban public space and a up-to-date knowledge communication facility in Guangzhou.

ACKNOWLEDGMENT

I would like to express my gratitude to Mr. Don Deng for his generous help during the writing of this thesis.

REFERENCES

[1] Don Deng, "Design Brochure of Guangzhou No. 2 Children Activity Center", unpublished

[2] Don Deng, "Design Brochure of Guangzhou No. 2 Children Activity Center", unpublished

On Chinese Traditional Architectural Art and Eco-culture

Lu Xiaobiao
College of Light-textile Engineering and Art
Anhui Agricultural University
Hefei, China
Email: xiaobiao511@163.com

***Abstract*—In this paper, the eco-culture of Chinese traditional architectural art is analyzed from the angle of the relationship between architecture and eco-culture, examples are given and analysis is made on the typical form of architecture and the arrangement form of courtyard house in the Xi'an Banpo and the Yangshao culture site and the middle and lower reaches of the Yellow River in north China as well as the forms of folk houses in the northern and western border areas, and the enlightenment of eco-culture of traditional architectural art to modern urban ecological construction is proposed, that is, modern urban ecological construction should absorb such four major ecological viewpoints as systemization, naturalization, economization and humanization of traditional architectural art.**

Keywords-Traditional houses;architectural art; eco-culture; ecological construction

"Eco-culture is the material embodiment of material civilization and spiritual civilization in the ecological relationship between the nature and society. It is the culture for harmonious coexistence between human beings and environment, continued existence and stable development."[1] Whereas architecture is just one of the major carriers of human eco-culture, the change in human habitat conditions reflects the change in the relationship between human beings and the nature, and the style of architecture will naturally influenced by the factors of natural environment and natural resources and psychological factors of human beings in relation with the environment. Therefore, as the intermediary between human beings and the nature, architecture spans over all levels of eco-cultural system, no excellent architecture example neglects any item of them in the ancient and today at home and abroad. The styles of traditional architecture in different regions all represent the inseparable relationship between architectural art and eco-cultural system. Different styles of architecture are closely related with the natural environment and natural resources.

I. CHINESE TRADITIONAL ARCHITECTURE AND TRADITIONAL CULTURE

The characteristic of the Chinese traditional big-roof buildings has close relationship with the natural environment in China. The ancient China originates from the Yellow River basin, here the winter is very cold and the summer is very hot. The rainy season is concentrated and there are mostly storms, this results in higher requirements of Chinese traditional architecture for cold-proof, heat stroke preventing and rain-proof functions, therefore, the characteristic of big-roof construction with high and thick walls is adopted and attention is given to north-south orientation of the houses as well. There are dense forests in the Yellow River basin, plus the Loess Plateau abounds in loess, therefore, the earth-and-wood structure of Chinese traditional architecture is also determined by the environment. While the dignity of big-roof architecture and bucket arch as well as various decorations of flying dragons and leaping tigers are also related to the orthodox concept of Chinese traditional culture. They represent the traditional world outlook of Chinese people, which advocates the thought of harmonious development of people and the nature and the so-called "favorable climatic, geographical and human conditions".

The sociality of human beings determines the intrinsic character of architectural art, states under the background of the same culture may adopt different forms of architectures under different natural environments, but the there are many common points in the connotation of the architecture. For example, the courtyard folk houses in Beijing, pile supported bamboo building of the Dai nationality in Yunnan Province, Ke Jia clay building in West Fujian Province, vernacular dwellings in vicinity of rivers in water villages south of the lower reaches of the Yangtze River etc. all represent the architectural arts of Chinese folk houses with the ancient ecological culture of "the unity of heaven and man" as the kernel concept. Let's take the residential architectures of various nationalities in different regions with local characters as examples to analyze the relationship between architectural arts and eco-culture.

II. CHARACTERISTICS OF TRADITIONAL ARCHITECTURE IN THE YELLOW RIVER BASIN

With the Yellow River basin as the center in the north of our country, especially in the middle and lower reaches of the Yellow River, there are formed relatively typical forms of architectures therein. Ever since a long time ago here not only the styles such as cave dwellings evolved from the ancient troglodytism, stone masonry dwellings with timber frame created according to local conditions and flat-roof dwelling etc. have been remained, but also various kinds of manor residences with the courtyard dwelling as the basic construction have been built, and become the most basic form of the buildings in the north of our country.

The forms of temples, ancestral halls and assembly halls correspond with the courtyard dwellings, except that their layout is more magnificent, their structures are richer in variation and decorations are more luxuriant. Such buildings are the works of higher standard among folk buildings, they are the important places for folk activities such as folk belief, people-to-people intercourses and cultural dissemination etc. Another type of residence in northern region is the cave dwellings, they are mostly distributed in the loess regions such as Henan, Shaanxi, Shanxi, Gansu etc., more highlighted is the house and courtyard outside the caves, which still keeps the layout of the courtyard dwelling. The common folk houses in countryside in northern regions are mostly made of local materials, the walls are laid with bricks, stones and sun-dried mud bricks, the roofs are made into flat or slope ones with loess and line as the surface plastered with gray lime or covered with tiles or grass, the interior of the house is warm in winter and cool in summer, and some houses utilize heated brick beds to keep warm in winter. The folk houses in northeast regions still maintain the most basic style of northern folk residences, i.e. three-room arrangement is the basic form. Generally the house site is wide and the courtyard is relatively large for convenience in receiving sunlight, meanwhile the walls are thick and relatively enclosed due to heating requirements.

The forms of folk houses in northern and western border areas are relatively varied. In part of the minority areas in Inner Mongolia, Gansu, Qinghai and Xinjiang, the tents adapting to the nomadic life style are the common type of residence, among them the most dominant are ayils, people of Mongolian, Kazakh, Kirgiz, Tajik nationalities etc. mostly live in ayils; the ayil is also called Mongolian yurt, which is also a style of building adaptable to the mostly windy and sandy environment with big temperature difference between day and night in the northwest grassland, especially adaptable to nomadic life, such Mongolian yurt made of multilayer felt can be formed with partitions whose thickness can be changed with the change of air temperature in the four seasons, moreover, windows can be opened for temperature adjustment. Shepherds of Tibetan and Yi nationalities in the southwest regions also live in different forms of tent. The tent construction is light and easy to move.

III. Characteristics of Traditional Architecture in Yangtze River in China

The architectures in the middle and lower reaches of the Yangtze River have no longer copied the original pile supported characteristic since Han and Wei Dynasties, and apparently received the influence of northern architecture. "Since the Ming and Qing Dynasties, the regions such as Jiangsu, Zhejiang, Anhui, Jiangxi Provinces etc. which have developed culture and economy and dense population, have developed the folk houses characteristic of water villages in the areas south of the Yangtze River."[2]. The typical folk architecture built in vicinity of rivers in water villages in the areas south of the Yangtze River is representative of the folk houses in over 150 small and large towns spread over more than 900 square kilometers outside Suzhou City. These ancient towns are often built alongside rivers with small bridges and flowing water, the streets and alleys are deep and quiet, three sides surrounded by water and one side facing the harbors, river channels crisscross in the towns, hemmed in lake lights and water colors, the streets run alongside the rivers, the main streets are back to rivers and secondary streets face the rivers. Streets, alleys, stores, stored houses, pavilions, residences and water ports, water steps, jetties etc. are mostly arranged beside or by water, and spread along with the water system spaces, the combination of water with roads, bridges and buildings is very appropriate, starting from functional and artistic need, there are arranged small bridges, brooklets, small streets, small alleys and small buildings, thus constituting the artistic style of small and exquisite layout. In addition, the dwellings in the areas south of the Yangtze River generally take enclosed courtyard as the unit, which is called "Jin"; houses with several "Jin" are arranged along several axial lines, which are generally by south or southeast direction. Large residences are composed of various rooms to form a group of courtyards, and rich families have gardens, thus forming a clear and beautiful style typical of house gardens. In the areas in the lower reaches of the Yangtze River, more typical house garden architecture in Suzhou and Hangzhou and the architectures in the Huizhou region in Anhui and Jiangxi Province are representative of local style and character in these areas. If it is said the artistic beauty of the courtyard architectures is presented through central axial lines and symmetric balance, then the artistic beauty of "riverside dwelling" architectures is exhibited through ecological harmony between architectures and between the architectures and the surrounding water.

The Jianghan Basin and Dongtinghu Basin in the middle reaches of the Yangtze River, including part of areas in Hubei, Hunan and Sichuan Provinces and small part of areas in the south of Henan and Shaanxi Province are different from the regional culture of the Central Plains. The construction mode contains elements of northern architectures and are influenced by the architectures of the lower reaches of the Yangtze River, their styles are relatively plain and free, generally folk houses are of symmetric or rectangular layout, while urban dwellings are mostly narrow deep courtyards arranged by "Jin". The relatively characteristic residences are typical of Sichuan

Basin and southeast mountain areas where the residences are laid out according to terrains parallel to contour lines. In this way different styles of slope houses with different depths are formed, and the construction materials are selected locally and are put to good use.

IV. Traditional Architecture of Different Nationalities with Local Characteristics

Most areas of Yunnan and Guizhou Provinces, northwest areas of Guangxi Province and southwest areas of Sichuan Province are one of the major colonies of national minorities, their houses have distinct regional features and national features. In such areas more houses of pile supported construction with elevated infrastructure are remained, their planes are mostly of transverse long layout, they are composed of wood frames in upper and lower layers for different purposes. For example, the "pile-supported" bamboo building of Dai nationality in Yunnan Province make full use of the local bamboo resources. Bamboo houses are well ventilated, cool and damp-proof. Pile-supported folk buildings have a long history in our country, and are found on the prehistorical site excavated in Haimen, Jianchuan County, Yunnan Province. "Jiu Tangshu-Nanman Zhuan (the Old Book of Tang-Nanman History)" recorded that "in the mountains there are toxic grass and louses and pallas pit vipers, people live in buildings, they enter the rooms by means of a ladder, such buildings are called 'pile-supported house'." The features of the pile-supported houses mentioned here are: "the dwelling layer is a house raised by support posts on the ground with ladder access to the rooms, it not only adapts to hot and rainy climate and is favorable to waterproofing and moisture-proofing, but also can prevent attacks from wild animals and toxic worms."[3]. People of Tujia nationality in Wuling Mountain areas can only live in selected lands beside mountains and water, because there are rare flat dams. Here is the subtropical mountain area, so the climate there is changing and the rainfall is abundant. The suspended building of wood frames is not only accessible flexibly and adaptable to the rugged geographical environment, but also suitable for the air temperature and climate conditions here with the concurrent damp-proofing function, and greatly saves building materials, makes full use of local resources, forming an architectural form which is level in the air though the ground is uneven and occupies the air but does not occupy ground. Some of stilted houses of Guangxi Zhuang nationality and Dong nationally are as high as three stories, the roof is made into an attic with ingenious construction. Various temples, stockaded villages, balconies and bridges have different characteristics, among them the drum towers of Dong, Yao and Zhuang nationalities and Dong villages in Hunan and Guizhou Provinces etc. and the theater stages and folk temples of Zhuang and Dong nationalities as well as watch towers and village gates of Miao nationality etc. are representative of them. In the areas in the plain or hilly areas where people of several nationalities such as Han, Miao, Buyi, Yi etc. live, three-bay courtyard houses with wood frames are dominant, but building materials have their respective characteristics. All these buildings effectively merge the buildings with surrounding natural environment in a natural and harmonious way in the geological environment with good ecology and aesthetics and reflect the philosophy of "harmony of heaven and man" and "telepathy between human beings and nature".

The regions in Guangdong, Guangxi, Fujian, Hainan and Taiwan are anciently called Lingnan regions, except for the border and the minority areas, most of these regions are more greatly influenced by the Central Plains and have merged the styles of the areas south of the Yangtze River, thus formed the construction system with regional features. The most of grand courtyards are arranged symmetrically to the central axis, and the structures are mostly of stone masonry and wood frames, similar to those in the areas in the lower reaches of the Yangtze River. Common folk residential courtyards are relatively free, mostly particular are the Kejia residences in the mountain areas in Fujian and Guangdong Provinces, the clan life style formed large residential communities. Their layout is similar to the ancient folk buildings in the Yellow River basin. Influenced by the natural environment and climate, the building materials are taken locally, for the building siting, sunny and wind-proof places facing water and close to roads are chosen as the building sites in order to facilitate production and life. The compacted soil is more than one meter thick, thus can prevent the intense heat from attacking in hot summer and can shield from cold wind attack in cold days, thus a microclimate is formed inside the building which is cool in summer and warm in winter. The thick earth wall has adjusting function which other wall bodies cannot compare with, under the climate conditions with too big change in outdoor dryness and humidity, the thick soil layer can keep the humidity suitable for human body. These ideas all have relationship with modern sciences such as geography and geology, architecture, ecology, aesthetics etc., particularly they advocate harmonious coexistence between human beings and nature from ecological angle.

V. Conclusion

The transition of traditional form of architectural art from enclosed type to semi-enclosed type and to open type with time reflects the gradually changing relation between human beings and nature and society, and also exemplifies the contribution and effect of changes in eco-culture on architectural culture.

The experience of ecological destruction and the ruthless retaliation of the nature make human beings recognize that "man can conquer nature" is impractical.

Only by learning coexistence with nature, following the will of Heaven and seeking harmony of heaven and human beings, can human beings maximally recognize nature, utilize nature and remodeling nature. To improve utilization ratio of resources through competition, to promote fairness of ecological relationships through coexistence and enhance sustainable ability of system by itself are the three major means for achieving harmonious coexistence of man and nature and sustainable development.

Therefore, modern urban ecological construction should absorb the eco-culture of traditional architectural arts and should have such four major ecological views as "systemization, naturalization, economization and humanization."[4].

Systemization: introduce the principles of ecology and compound ecological system in Chinese traditional culture according to the trend of reductionism in urban construction, and pay attention to the features such as landscape integration, metabolic cycle, sensitive feedback, technical crossing, system integration and continuity of time and space etc.

Vitalization: create a comfortable settlement environment, have stronger survival and development mechanism; lay emphasis on the features such as mobility, smoothness, vitality and adaption of human beings to nature etc. as well as lower value of risk.

Economization: accommodate more ecological functions in smaller space, exchange smaller ecological cost for higher economic benefit and exchange smaller traffic volume for higher traffic volume in order to improve utilization efficiency of resources as far as practically possible.

Humanization: better satisfy various social needs of residents for health of both body and mind, and induce and inspire various merging and sublimation of people.

Urban ecological construction does not mean returning to primitive ecology, but actively developing ecology. It should enhance people's life quality, including the contents such as dwelling conditions, community environment, infrastructures, service facilities etc. It should mainly reflect reasonable utilization of land resources, safe utilization of water resources, reasonable planning and layout of towns and communities as well as harmoniously and friendly managing infrastructures and service facilities for ecological environment with a view to providing residents with convenient, healthy, safe, rich and comfortable life and working environment

REFERENCES

[1] Wang Rusong and Zhou Hong, "People and Ecology", [M] Yunnan People's Publishing House, 2004: 225

[2] Qin Geng, "folklore in literature", [M] Anhui Literature and Art Publishing House, 1993:225

[3] Tang Jialu and Pan Lusheng, "Science of Chinese folk arts", [M] Heilongjiang Fine Arts Publishing House, 2000: 84

[4] Ma Shijun and Wang Rusong, Compound Ecological System of Society, Economy and Nature. [J] Acta Ecologia Sinica. 1984:(1)1-9

On the Interior Environment Optimization Design in Digitization Age

Deping Ge
Art Department of Zhaoqing Science & Technology Polytechnic
Zhaoqing, China
Email: gedeping1980@163.com

Abstract—**In 21 century, as the application and popularization of digital technology as computer and internet, the society has gradually entered a digital time of diverse information, which is totally different from the past industry times. Interior design has been transformed for the sole physical practice into the optimization of various construction element to create an update interior environment that meets the needs and development of human.**

*Keywords-**interior environment; construction element; optimization; combination***

I. INTRODUCTION

As the application and popularization of digital technology like computer and internet in 21st century, the society has gradually entered a digitization age of diverse information, which is totally different from that in the past industry times. This will thus change people's life style and behavior and effect the traditional interior space and environment. Compared with the simple production of object, interior design has been much more creative. It is an optimization of various information and a systemic, scientific and coherent design of the relevant constructive elements for creating suitable interior space in line with the progressing times and the need and development of humanbeings.

II. THE CURRENT SITUATION OF INTERIOR DESIGN

According to a report in the First National Seminar on the Qulity of Room Air and Health on April, 2003, every year more than 111 thousand people die of interior air pollution and 4.3 million are in emergance case while 220 thousand call for outpatient service.

Statistics shows that the energy consumption of construction materials takes up half of the total number, most of which are nonrenewable. "A large propotion of the environmental pollution in construction is caused by the production, construction and reconstruction of the interior decoration materials. [1]"

The living environment mainly consists of the man-made environment with architectures. People spend more than 90 percent of their lifetime living, working, studying in interior space. Thus the quality of the room air is so important in living a healthy life. Especially in the 1990s, as the development of the economy and the awareness of environment conservation, the interior design has become a top task to improve the interior environment. At the same time social and environmental problems do exist in ecology, humane, time and benefit, which will effect the life quality and destory the harmony of ecological and social environment, and even the sustainable development of the society.

Another fact is that nowadays designs with Chinese and western charateristics begin to crop up across China, like *Aegean Sea, Eurasian Flavor, Spain Coast,* which are just of simple imitation and mechanical application.

III. CONCEPT OF OPTIMIZATION

In mathematics and computer science, optimization, or mathematical programming, refers to choosing the best element from some set of available alternatives. In the simplest case, this means solving problems in which one seeks to minimize or maximize a real function by systematically choosing the values of real or integer variables from within an allowed set. More generally, it means finding "best available" values of some objective function given a defined domain, including a variety of different types of objective functions and different types of domains (Wikipedia).

Interior optimization design is to satisfy people's comprehesive need for interior environment with modern technology information on base of mathematics optimization therories, which imitate the interior elements for a best available design to some purpose under certain conditions. It will be much people-oriented to reduce the waste and pollution of resource and keep the balance of ecology and sustainable development as well as provide comfortable environment for living and studying.

IV. THE ELEMENTS OF OPTIMIZATION

A. The elements of interior environment

"Environment" has different meanings in different circumstance. In modern digital time, due to the economic development and the high energy consumption, the concept of environment refers to nature, society, economy and culture and the like in general not only to space or room. The traditional concept on the specific and material environment has been become much more comprehensive in macroscopical and systemic view. [2] The interior elements mainly contains the hard environment like space, furniture, docration and display and soft environment like lighting, heating and ventilation, sound, ecology, etc., which are enbodied as shape, color, nature and culture in design involving different groups, the proprietor, designer, enterprises, government, ect. Their relationships are as follows (See Picture 1).

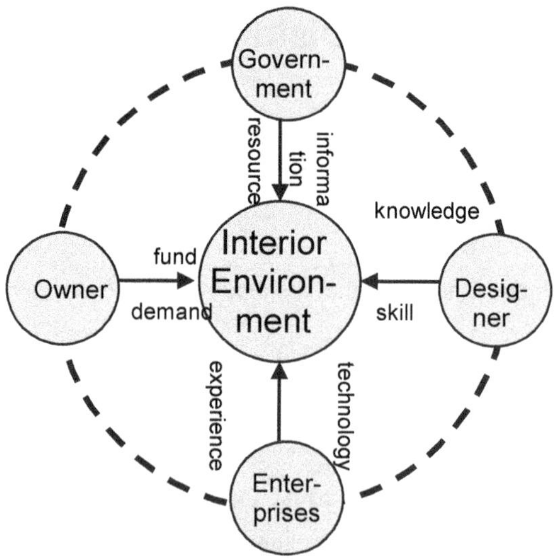

Picture 1 The realationships between the resource groups

B. Design Process analysis

- *Design objective*

The process is mainly subject to the target and design objective to optimize various resources in certain information technology with full consideration on the relevant participants and elements, thus creating comfortable environment.

- *Ways of design*

According to the law of systematology, optimization is a systemic and organic composition of all elements. Nowadays, any interior design program involeves interdiscipline and all works of lifelike architecture, materials, furniture, lighting, construction and management. The interdiscipline such as sociology, aesthetics, design phychology, human engineering, economics, ecology usually effects the results of the design. So the designer should first embark on the whole ideally and optimize the overall situation systemically of all groups and elements.

- *Principle of design*

Because the elemens of the interior environment is multivariated and prodess of design is long and complex, we need to follow the principles.

① principle of sustainable development

② principle of harmony

③ principle of energy saving

④ principel of humanistic care

- *The process of design*

a) Information gathering.

It is to gather information about the programme sites and the surroundings by means of IT, including terrain, traffic, vegetation, infrastructure and layout and sort out the original data and pictures with details. Tools such as network or digital camera are adopted to investigate and consult the relevant information for actually understanding. To meet the need of the users. Information also should contain people's needs, tastes, hobbies and the cost. Various construction materials like stones, lumber and emulsion varnish and lighting are the main factors consisting and effecting interior design. So we need to collect information released from the government and enterprises about the construction material, techonology and new energy through the internet and media to avoid the waste and pollution.

b) Analysis and imtation

First, consider people's need for the space and analyze the shape, size and color of the the space and furniture with the gathered information. The interaction of people and the material environment could convey the emotion and idea for the space through these factors like color, shape and sound which also effect and determine people's behavior and awareness. Whereas we can find out the best size and shape with the analysis.

Secondly, analyze the room light, thermal environment and interior air and their mental and physical effects. Nowadays, with the highly energy-consuming and inundant pollution, the natural factors like sunlight, wind, water, plants and the fresh air play active roles in interior soft environment, energy saving and pollution abatement. Patrick Geddes (1854～1932) proposed in his book *Cities in Evolution* that we not only need "coal and tap water" but also "sunlight and air". [3]

Thirdly, architecture, a carrier of culture, can be much further optimized by developing its humanistic elements. As American designer Poulos said people always considered there were just three aspects of design, namely, aesthetics, technology and economy, but the most important one should be human nature. This refers to the humanistic care. The design in the 5th Garden of Vanke Group in Shenzhen adopts the primary hue of white and black in traditional Hui style, for example the white walls and black tiles and horse head shaped walls, which represent profound culture and rich architectural deposits. (See picture 2)

Picture 2. A photograph of the 5th Garden

Based on these analysis and information, a preliminary imitation or demostration can be conducted about the program by means of computer. According to the functional need for space, we can apply Auto CAD（Auto Computer Aided Design）in 2D modling to divide the interior area in the form of plan and elevational drawing. By this way, the ratio of space and the furniture, greening, display and the arrangement can been easily seen. Second, we fully use the Auto CAD to design technical drawings including hydrographic chart, circuit diagram, caliduct erection drawing, profile pattern, ect. by orders like map, edit and dimedit. Better understanding the equipments and arrangement in advance can make the resources optimized at the most in virtual forms.

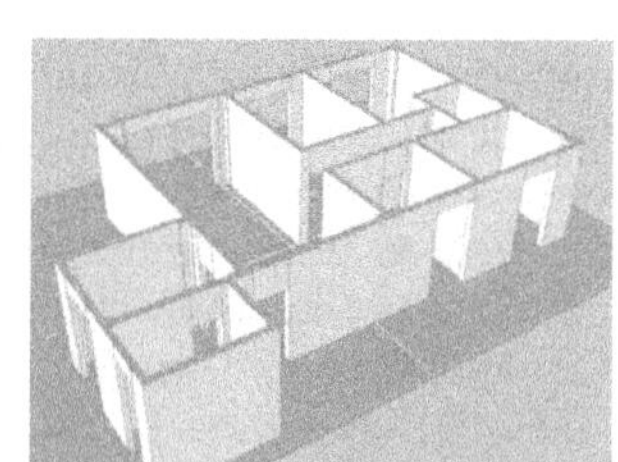

And then we can utilize 3DMAX to intimate the modling with startling reality by 3D modling, material edit and render. This provides a intuitionistic understanding of the results before decoration.[4](See picture 3 above)

For the development of Lightscape and Vray, the visual pictures of the structure, texture and color of the space, furniture, display, greening become much real. We can set up the color, illumination intensity and color temperature accurately while rendering and try different ways for admitting light and light source match. In this way we can do all count on the energy comsuming.

We can also use architecture animation to better understanding the combination results of design elments in the whole, and even experience with computer science in 4D space. At last, as to the construcion cost and running cost of the program, we can calculate the the way, frequency and maintenance of various kinds of materials furniture and energy with computer aided software like Excel in order to reduce the economic,social and ecological cost as well as satisfy people's need and tastes.

V. THE OPTIMIZATION OF THE CONSTRUCTIVE ELEMENTS

The construction of the interior environment is a systemic and complex engineering thatinvolves more than one field and different subjects. The draft design, is an imitation and analysis of relevant elements and resources. While the whole program refers to the implement of the plan, the use and maintanence of the space. It involves all kinds of groups as talked before. With the interaction amd support of tthese groups, the constructive elements can fully bring into role and optimize the resources and information at the most.

Living Tomorrow in Amsterdam is a classic exmaple of interior environment design in digitization age. It applies the lastes technology optimizing the future life and the working environment and effectively uses solar energy and cross-ventilation to control the heating exchange. The materials used can be recycled for environmental protection. [5]

CONCLUSION

Interior environment design is a giant systemic and technological engineering. As the rapid development of the times and the progress of science and technology, it requires that our designers should better understand the most important and relevant elements with close relations and cooperate with technicists from other works of life. This is to make full use of those elements to effectively promote the quality of interior environment under the condition that the design satisfys the needs of function and taste.

REFERENCES

[1] Hu Junhong, Lead Ecology into Interior Design[M]. Art and Design, Vol.2, 2002

[2] Wang Haisong and Shi Lili, From "Environment System" to "comprehensive Design"- The Trends of Contemporary Environment System[C], Analects of China Environment Art Design. Nov. 2007

[3] Bao Shixing, Research on Urban Landscape –A Study of Chinese Landscape Complex[C], Analects of China Environment Art Design. Nov. 2007

[4] Zheng Qingrong and Yang Wei, 3ds max/VRay[M]. Sciene Press, Aug. 2008

[5] Qiu Jianwei, Study of Residential Patterm om the Digitization Age[M], Art and Design,July 2006

Optimized Color in Web Page Plane Design

Yu Lan
College of Art and Design
Wuhan University of Science and Engineering
Wuhan, China
Email: 89219032@qq.com

***Abstract*—Compared with plain, colorful pages are more attractive in web design, because web design, to some extent, is a kind of art design, which not only requires exquisite and fluent patterns, but also innovative and proper colors. Conveying message through colors is an effective way to reinforce its understandability and identification. Language with theme color promotes the conversation with viewers, and extends the viewing time. In color optimization design, according to the principles of harmonious, balanced and focused, we create fascinating pages by mixing and assorting different colors. Color, is the first and most lasting factor to impress viewers with the image of a website. With good framework and page arrangement, as well as the optimized match of colors, a successful design will come out.**

Keywords- web page; color; optimized design

I. INTRODUCTION

A piece of outstanding graphic design work demands not only excellent patterns, but also excellent colors. Similar to post advertisements and television images, layout design in web pages should also be done with eye-catching and appropriate colors, in order to focus on the theme and serve viewers with the appreciation of beauty. Color optimization plays an important role in web design, and it's a significant way to adjust viewers' visual psychology and to draw viewers' attention. Web design is completed by computers. The color system in a computer provides red, blue and green as the Three Primary Colors. The creation of varied and visual excitement stimulating colors by mixing the three primary colors can achieve the goals of drawing people's attention, conveying information and winning users. As the old saying goes, "colors impress in distance while patterns in close look", so color has great impact on people's feelings about surrounding. The emotion features color shown by should be identical with the characteristics of pattern, thus making the best of colors in web design, so that pattern and color can resonate with each other under a same theme.

II. THE RELATIONSHIP BETWEEN WEB DESIGN AND COLOR

In the web design, colorful pages always outweigh uncolored ones, and are more attractive than them, because web design is a kind of artistic creation to some extent, so it not only requests graceful and smooth pattern design, but also novel and proper color. Meanwhile, the utilization of color can produce a strong visual effect so that it makes the pages more dynamic. Moreover, the first impression is given by the color design of the pages when explore them. Undoubtedly the color is a direct impact on the audience's interest in browsing.

The color visual language in website design has the specificities of idealism, symbolism and compatibility. The idealism of color visual language is that the designer can use the color subjectively, and change the color of natural scenery to generalize them. For instance, according to the demand of pages, the designer can color the sun into blue or green to meet the effect of depth and calm ness in design. As long as it is the requests of pages, it is the ideal design of designers. The symbolic of color design is based on the character of its own color. When using the color, designers should fully demonstrate the symbolic and allegorical nature of color and make it similar to idealism, in order to create effects that beyond its own image, such as the black and red in Beijing Opera, the colorific boldly applied in modernistic design and painting, and so on. It is also true for the symbolic in literature, the symbolic in color could play an extremely effectual role in the design. The harmony of color produces peculiarity and form a particular way in using color through a coherent and coordinated use of color. According to different design contents, the performance of such color harmony determines the amount of levels and hue in order to stress the key point and present the theme.

Every color has its "unique" expression, but also has varied meanings. For example, the blue and green in website design can relieve people's visual stress, making people feel calm, natural. healthy, peaceful and relaxing, and also give people a feeling of solemnness and regularity; yellow and red represent the expression of warmth and ardency, which are the memory or knowledge closely related to the character of website design and the acquisition through psychological association, and also are the color of caution that have a function of warning and suggesting. So the final impression on people vision and conscious given by color is a consequence achieved in virtue of people's experience, memory and knowledge, as well as psychological association.

III. THE ROLE COLOR PLAYS IN THE WEB PAGE DESIGN

In the web page design, the prime colors of hue have an attractive power to catch to convey emotions. And arose people's psychological resonance. In the color optimation design, different colors are assembled to form a splendid page layout according to the principle of harmony, balance and emphasis. Color design, which is the important element in the webpage-design is not only the language of wed design, but also the method to convey visual information. Followings are the main roles that the color plays in the web design:

A. the division of visual region

The most important function of web page is conveying information. Color is the important element to create ordered flow of visual information. Designers classify the different kinds of information and differentiate the visual region of web page by using color distribution. By using different color distribution, it brings readers different psychological effects so that primary and secondary orders can be differentiated well, and web page is easy to read and convenient to guide. Therefore, ordered visual flow can be formed.

B. the emphasis of main theme

Network is the sea of information in which readers seek for information required. The content and transmitting way of web page are uniform, which is one of the requirements makes the designing works a success. The expressive force, emotional effects, and the aesthetic psychological sense of the different colors are used in the web page, which combine its content and form organically. By using intrinsic power of the color to beautify the web page, the web page can be made graceful, dynamic, fashion, and striking. Besides, foiling and giving prominence to the theme makes it possible to communicate with readers.

C. the attraction of view line

There are hundreds and thousands of websites on the Internet, so it is not easy to make our own site outstanding. In design, a variety of beautiful web pages can be made to meet the readers' visual demands by using the function of color. In the web page design, no matter what kind of optimization form are used, the main purpose is to beautify the web pages, catch readers' attention and improve the attraction of the web page.

D. the enhance of artistic quality

Color is the way to convey visual information as well as the language of art design. Color design brings the works of web page vitality, and plays a decisive role in the art quality of the works, which is not only reflected on the sight, but also on the mental influence and symbolical significance. Good color design can strengthen the artistic quality of the works of web design.

IV. KEY POINTS OF COLOR OPTIMIZATION IN WEB PAGE DESIGN

Two main points of color optimization in web page design: overall style and color matching.

A. Defining the overall style of the website.

The style is abstract, which refers to the integrated impression among viewers provided by the overall image of a website. The "overall image" includes the CI (logo, color, font and slogan) design of a website, layout design, browsing method, interaction, character, mood, content value, existence meaning, website prestige, and so on. Websites of different styles give people different feelings. The three basic intentions of color are brightness, harmoniousness and plenty.

The beauty of brightness---- the color structure is mainly focused on bright colors can create the aesthetic feelings of lucidity, flamboyance, strength, light, irritation, excitement, diversity, glitter, markedness, freshness, and so on.

The beauty of harmonity ---- the color structure mainly focused on harmonic color can create deep, thick, soft, subtle, reserved, calm, beautiful, elegant, plain, refined, old, and other feelings of aesthetics.

The beauty of plenty---- the subtle changes and proper combination of all kinds of colors create aesthetic feelings of richness and luxury.

B. Optimization match in colors on web ages

No matter in graphic design or in web design, color is always the most important thing. When we are far from the screen, what we see is not the beautiful layout or pictures, but the colors of the page. Color, is the foremost and ever lasting factor to impress viewers with the image of a website. With nice framework and page arrangement, as well as the optimized match of colors, a successful design will come out. The methods of optimization are as follows:

1) Color distribution on web page

- The content of web page is the room for information, so it requires bright background color, dark characters and high contrast. Generally, it's black and white, and if the background is black, the characters can be gray or other light colors, light enough to be ignored.
- The heading of a web page is the place for its logo and other key information, and it should be dark. High contrast makes users to locate more conveniently in the website. Titles usually have different "appearances" from other parts on the page, and characters or colors exceptionally different from content can be used, or simply the invert colors of the content.
- The area of navigation menu should have dark background color, and then we can distinguish menu from content exactly by using patterns of high contrast or unique font.
- Sidebar, although not used in all the web pages, is still a useful way to show additional information. It should be divided from content clearly, at the same time to be convenient to read.
- Footer, as a light role on a web page, shouldn't be ever emphasized. The colors same as the sidebars' or other darker colors can be used in footer.
- Utilization of some bright colors to guide viewers.

2) Color balancing

It's a conservative principle of supplementary color matching. It's hard to handle if many colors coexist, so they should be unified. For example, we can use the color phantasm phenomenon, and the weight phantasm is practical in daily design, which means the lighter things look brighter while the heavier ones look darker.

The visual balance of colors includes the positions of colors, the proportion of each color, the area of each color, and so on. Bright colors should take smaller areas; otherwise people will feel uncomfortable and dazzling. That is a balanced color matching. We can't shed too much light on one color; otherwise the whole page will get imbalanced. For example, if the large central area of the page is one color, the surrounding area must be filled in some different small areas of colors; if the left area is bright, the right area can't be all gray or blank, which means there should be bright colors to resonate with the left. If too many bright colors are employed then a little black or some dark colors can be adopted to make the page heavier, and to achieve color balancing. In web design, balanced colors will meet most people's color psychology. The visual balance of colors is the last adjusting problem of identifying the main color of a website, so it should be identical with the prerequisite of presenting a completed visual impression.

3) Identification of the main color, by following the principles of primary and secondary

It's the most stable principle of processing colors. In fact, red, yellow, green, blue, purple, each of them represents a serious of specific colors, which is also known as hue, the appearance of color, and the name to distinguish different types of colors. On every page, apart from white as background color, the color slathered is the main color of the website. If too many colors are used on one page, if handled improperly, it'll be hard to tell which one is the main one, so that the whole page will seem dazzling and messy. Therefore, the area of auxiliary colors shouldn't be too big, and it should be limited in a certain area ratio, to follow the principles of primary and secondary.

Certainly, post-modernism designers strengthen individuality and symbolization, and often utilize unique color methods, combining history and reality with symbolic, humorous and allusive decoration methods, to create many wed sites with distinct characteristics. A outstanding designer should always follow the fashion trend, and sense the impression on the color emotion given by the aesthetic changes of colors incisively, as well as to identify the unique color style of web site by combining the use of colors and his/her own appreciation of the beauty.

V. SIMPLE DESIGN METHODS OF COLOR OPTIMIZATION IN WEB PAGE DESIGN

There are so many theories about color, and the designer is not designed to describe them precisely due to the limited time. During the process of matching colors on web page, if the designer is not very confident about the application of colors, then there're some simple skills that can be helpful.

- To use single one color. Here it means selecting one certain color at first, then by adjusting the transparence or saturation, the web page can then make the colors of the page seem united and rich in layers
- To use two colors. Referring to selecting one certain color, then choose its contract color. Pay attention to the area ratio and the priority.
- To use one color scheme. In brief, using one sense of color, such as light blue, light yellow and light green; Or dusty yellow, dusty grey and dusty blue.
- To use similar colors. Colors in the same hue with different purities and lightness are easy to match with each other, like using several kinds of green, and therefore, normally there won't be many problems on the page; to choose similar colors, congener colors, and adjacent colors in the hue or color cycle to match if possible.
- To use the color library in wed site design software. When we open the color box in Color Dialog, if the colors are grouped, and there are 6 colors in one group, and each group is a nice color scheme but a little bright, then we just need to exchange 2 of them with dark colors in the same hue, to avoid being too bright as a whole.
- To use fresh and bright colors, setting against white, and to process the details with gorgeous and distinct colored bands or blocks. This kind of match is young, fashionable and bright.
- To strengthen the visual effect by using contract colors with different purities, and to assuage vision by utilizing different kinds of grey with different lightness. To use more grey to coordinate in assistant colors, transition similar colors and adjacent colors; to use gradient color effects.

During the process of matching colors on web page, keep some misunderstandings in mind.

Don't try to use color as much as possible. Three to five types of colors will by available.

The color match of characters should be separated with the background. There should be an obvious color comparison between the setting and the content (Setting should not be too complex. Never use intertwined traceries as the setting, just keep it simple and easy), so that the main content could be highlighted.

VI. CONCLUSION

Color is a critical and charming visual language in web page design, with the help of which we can create a unique visual effect. Only by owning a rich knowledge of colors, fully using color optimization methods, and an understanding of state culture can create a good web page design work.

In conclusion, the key to web design is to transfer the information, thoughts and intention conveyed to some visual manner, so that the web page viewer could easily get the required information, and also the appreciation of beauty at the same time. In terms of web page design, the innovative process not only embodies the understanding of the visual language, but also means the upgrade of it. A successful web page designer should fully understand how to employ these elements to convey there design idea in a comprehensive way. A designer should make visual detailed adjustment to the original color match through using the methods and skills of color optimization on web page, so as to acquire a balanced and harmonious sense of beauty, and eventually by analyzing the psychological outcome of the viewer from the perspective color and emotion, to enhance the cultural taste of colors on web page.

REFERENCES

[1] Zhang Zhusheng , computer-aided color design (dissertation), Northwestern Polytechnical University, 2002

[2] Li Junfeng, the traditional colors of the new understanding of the Gansu Science Education 1997 (3)

[3] Gu Daqing design of visual perception-based interior design and decoration 2002 (9)

[4] Chen Liannian Chongqing, Southwest China Normal University Publishing House Color Design 2001

[5] Liu Ming,Cui Tianjian color constitute the performance of computer-aided techniques Posts & Telecom Press 2002

[6] Liu Hong, Resolution Graphic Design in the colorful language Decoration 2003 (4)

[7] Ye Bei page color art with the design of computer and network 2004 (10)

[8] Tian ShaoXu digital color Exploration of Shenzhen University, 2001 (3)

[9] Wang Yu, Sui Yong visual elements on the Web Journal of Beijing Institute of Graphic Design 2005 (12):

[10] Wang Qingbin, Zhang Fuchang modern web color design, Wuxi University of Light Industry 2001 (6)

[11] Yu Ji Zhen, Du Lijun on the development of digital art [J] Decoration 2006 (6).

Optimization Design of the "Axis" as the Characteristics of Urban Landscape

Liu Jin
Hubei University of ZhiXing College
Wuhan, China
Email: 171347627@qq.com

***Abstract*—Urban axis usually refers to a spacewhich connect the elements of linear space in the city. "Axis" as a character of the urban landscape link the urban environment, people always get the first impression from it. As a result, Axis planning is the kernel content of city's configuration. According to axial landscape as a starting point, this article take urban historical feature as a researching background. Discuss how to inherit and renew the urban axial in system. The content including: analyzing the creation and development of axis; investigate the influence of culture ; research the composing of urban axial landscape; though classical cases, in search of how to Optimization Design of the "axis" as the characteristics of urban landscape.**

Keywords- Urban axis, urban axial landscape, the composing of urban axial landscape design, optimization design

I. INTRODUCTION

City is the most wonderful, most advanced, most complex and most profound product which created by human being.

The development of civilization and city updates, our living space made more demanding. As a carrier of modern civilization, cities are rapidly changing adapt to the changing times.

For the urban environment, a harmonious and pleasant atmosphere of the landscape will further promote the development of the city. Internally it can increase people's sense of identity, pride, helping improve the quality of the people and strengthen urban cohesion; cities in dentification of foreign enhance, improve the cultural quality.

City axis as an organization of urban space, itis one of the ancient and effective method has been well received by city planners. In today's urban planning, focusing on city image, heritage and revitalize our urban landscape environment is a high-profile study. How effective and innovative on the basis of tradition would be the path of the future development of the urban landscape.

II. BACKGROUND

In Primitive stage of human culture, primitive people recognize the space and environment relys on sunrise, people describes the sense of space for the initial external impulse, widespread in all cultures of the sun worship which constitute the original understanding of the axis.

As an organization of urban space, cities axis is a design technique can not be replaced, as the completion of the space entity, it is also an important way worth urban planners to study.

In August 2008, Bayannao'er government invite Professor Wu Xingkuan Studio design Bayannao'er channel landscape planning. This design covers Ulat Street Center, the regional landscape and Yonggang, Yongji drainage along the drainage area.

The planning area is the important place of the city's political, economic, cultural, scientific, educational center, show the urban landscape , features and characteristics as the first window. From the planning we know it's a long narrow channel, the landscape topography, water consistent with a distinct axis of landscape features. I participated in the project master plan and thus take part in ICOD meeting, which is the source of this research.

III. URBAN LANDSCAPE EMBRYONIC AXIS

With painting, music, sculpture, and even different forms of architecture and art, urban space and the formation of the axis takes a huge human, material and financial resources, and often experienced a long period of development. A fairly long period of time within the generally small size of cities, urban residents and nature and more frequent contact. As the relatively low level of industrialization, urban eco-environmental quality can be better protected. Relying on the basis of the axis of the axis of the landscape more to help the natural environment, such as a green corridor along the river, dense riparian vegetation, etc., artificial traces performed at this stage is not very clear.

In addition, there is a class of factors shape the city axis, which is the "urban context", which is a long-form belonging to specific geographic areas and periods of collective unconscious, including the world outlook and values. Philosopher Michel Foucault (Michel Foucault) indicates that the space is the relationship between human society and the projection of power, to initiate "bio - power" to illustrate the space. His power here, covering the period of human ignorance, theocratic, feudal period, the throne, after the European Renaissance by advocating the human rights and so on.

Through investigation and study found that in the early stage of urban development, urban planning is probably inseparable from the axis of two ways:

A. Natural conditions

Natural conditions is a prerequisite for the formation of urban axis. It is the city's background and relying on, but also the natural basis for urban development. Any one of the urban environment through the building are made of the historical accumulation. Urban design, urban architecture and urban spatial pattern of harmony with nature is to maintain and promote the city's characteristic important content.

B. Power relationships

No matter what kind of government are faced with how to make their own power long-term effective rule of this issue. Therefore, it must be through a variety of media indoctrination and the dissemination of its ideology and values, thereby legitimizing their own, so that it generated a priori ruled that the trust doctrine. The magnificent city of the axis space as vital to the public and the media, a natural power, "deification" an effective tool to sign on the different regions and different nature of the regime in this way are valid. In the physical space and psychological space, the axis has a strong symbolism, representing national spirit. In order to "power" materialized into the urban space on the axis of the capital city of the West adopted a similar approach, which Beijing and Washington axis of comparison cities is particularly clear: to create a sense of geographic center, which is self-centered human primitive instincts and the concentrated expression of imperial power.

IV. LANDSCAPE FEATURES AXIS

Urban axis formation, regardless of the length of its development is a historical process of development.

The cultural level, the character of a city can be, and the spirit of an axis are closely linked, so that people have a concept of the dominant spirit.

On the physical structure, the urban landscape connectivity axis of urban green networks in the city's two-dimensional space, can be visually reflect the various types of green space between the functional and ecological processes on the contact situation.

Landscape planning is an axis of urban spatial texture, and the most intuitive reflection of its history and culture, urban development and urban landscapes of the same strain axis.

Throughout China's traditional architecture and cities, the presence of the axis is very clear, China's ancient capital in the north-south direction, multi-axis and the city exists mainly within the urban centers. This lives in the city center in the axis of the creation and construction is deliberately seeking, and it is closely related to the layout of the entire city. Ancient China put great emphasis on the ideological orientation and position of hierarchy with power and closely linked. "Lu Spring and Autumn" and this is also written: "The ancient king, and chose among the founding of the world, choose the country being written for the Palace, Palace choose among the Li Miao" 4 "and choose the" principle of urban planning in ancient China already in existence, from the current view of the city remains of the ancient Chinese city planning in the direction of a large number of north-south axis, with the "centered", "Home North" and "south side" closely related to ideology. Western civilization is a civilization in different places at different times of the continuous supply of the development forward, this growth has created its rich and varied background, democracy and tolerance characteristics. The pattern of Western cities is a concentrated expression of this civilization, rather than clan domination. One axis of development of the system, then more of a practical, spiritual and aesthetic pursuits. Throughout Western civilization, you can learn the West axis of the landscape with a few distinctive features:

- The overall layout relying on the road and was accompanied by large areas of greenery, water surface; appeared smooth, bright. Constitute the use of a wide straight road to link up the perspective scenery, more use of the state and thus complete the axis of turning, full of changes eclectic.
- Plant configuration emphasis on flower bed design, with floral patterns constitute a variety of flower beds, ground-rich content;
- Trees pruned into a geometric shape, or composition of the hedge, tree walls, could serve as a construction; the use of pools, canals ,fountains and other forms, waterside plants, architectural sculpture etc.; various elements were in strict accordance with the geometric symmetry are homogeneous graphic format layout.

V. URBAN HERITAGE AND UPDATING AXIS

As life needs care, the air needs cleaning, the machine in need of repair, any city is a living organism and always in a state of metabolism, so it is always in a state of change and development. Since the reform and opening up the past three decades, the acceleration of economic globalization, in a free market and called for humanistic care of today, cities are increasingly transformed into an economic entity and spiritual home.

Over time the composition of the city's roads and construction groups continue to aging and outdated, the city's structure, network and transport development is inevitable. How to protect updates and the protection of urban heritage is the outstanding issues.

Take Beijing for example, the city's axis with the development of the city landscape is also constantly updated. One of the biggest change is the urban nature of the changes. The establishment of New China, Beijing has a feudal dynasty ruled by the center to become the capital of the People's Republic has become the new China's political and cultural center. Since the liberation of Beijing, the biggest change in the axis than the original city center ------ Forbidden City, Tiananmen Square, the conversion and retreated with the subordinate position. So that the latter became the new city center. This is a requirement of social progress, and, from the spirit of the area changed fundamentally the meaning of Beijing's central axis.

With the expansion of the scope of Beijing urban planning, in order to growing urban transport and accommodation, sightseeing, tour, the needs of the city's 8km long axis from South Garden has been developed, from the south to the north, North Fifth Ring Road of about 30km range. In the context of this 30km long, arranged in line with the axis of the image of the landscape of urban public space is not a simple continuation of the axis, which require designers to use many of the paragraphs of the vision of sustainable development, planning and design. Axis function in the city is no longer positioned to reflect the monarchy services, but the addition of the content of popular participation, open the Forbidden City, Jingshan Park, alteration of the Tiananmen Square and so on.

VI. DESIGN CASE: INNER LANDSCAPE PLANNING BAYANNAO'ER CHANNELS

Bayannaoer Department of Mongolian, meaning "rich lake." It is located in northern China, western Inner Mongolia Autonomous Region, in north latitude 40 ° 13' --- 42 ° 28 ', longitude 105 ° 12' --- 109 ° 53' between. Design object:

1. Ulat Street Landscape Planning
2. Yonggang sewers, drainage landscape transformation Yongji

A. *Planning Analysis*

Both exports as a road and city construction and transformation project, the center area through lighting, landscaping, and building a harmonious and rich in national geographic characteristics of urban landscape environment. Yonggang sewers, Yongji drainage of water from the lake to the bridge Mato Yongji first check gate length of 7.2km. In accordance with the requirements reflect the farming culture, right from the opening line of channels, starting with landscape layout 100m wide, 30m wide landscape Avenue.

B. *Planning Objectives*

Based on the national culture of Bayannao'er regional characteristics for planning subject to grassland culture and the spirit of the Yellow River culture as a guide, based on local terrain conditions and according to the different characteristics of the channel water features landscape design, line string in the form of points, architectural art and culture, to be a lively show. Bayannur channels, landscape planning and distinctive features, advantages stand out:

- From a functional point of view, to maintain the integrity of the historic city, optimizing the city's functions to meet the city's growing material needs and spiritual needs;
- From the landscape point of view, the use of a continuous axis running through the region would be artificial axes with the combination of natural landscape to form a commercial development and urban nature of the organic integration of texture, harmony.
- From an economic point of view, through the planning and landscape remodeling, the beautiful environment and perfect facilities will attract more visitors.
- Therefore, the future development of urban axis reserved for the future development. Determine the axis and the nature of the land on both sides of the city in general on the functional orientation of China's urban axis construction and the most important issue facing the development.

VII. URBAN LANDSCAPE AXIS INHERITANCE AND UPDATE STRATEGY

Heritage, traditional culture is to overcome the negative factors and retention of positive factors, learn spirit of the times to meet the development of the times. In order to promote the development of new things. Establish a correct sense of awareness with the correct guidance of the practice, maintaining city image.

Update means change and change give a new life to the old cloth new. Urban axis updates are nature-based updating; the people-oriented update; to civilization-based updates. To create a favorable image of the urban landscape, bringing together the overwhelming majority of people. Therefore, the landscape in the city axis, we should note the following updates:

Urban Landscape axis updates to take into account historical and cultural protection, and coordinate the relationship between protection and development;

With full respect for the city's historical and cultural environment, protect historical and cultural context, for some cultural and historical neighborhoods and river landscapes rational development and utilization of protection, and restoration of the political landscape in the urban axis to "the entire old as before," "in order to keep its true" to "longevity" instead of "rejuvenating." Mr. Wu Liangyong in the "old Beijing Hutong children with Chrysanthemum," a book on the update made the following definition, that the update mainly includes three aspects:

First, the transformation, alteration or re-development (Redevelopment), means a more complete remove some aspects of the existing environment, the purpose is to explore space, adding new content to improve environmental quality,

Second, the remediation (Rehabilitation), refers to the existing environmental regulation of the use of reasonable, generally refers to a partial adjustment or minor changes;

Third, the protection (Conservation), is the protection of existing patterns and forms, and to maintain the general not allowed to make changes. Thus, the updated content is more abundant.

VIII. OPTIMIZATION URBAN AXIAL LANDSCAPE DESIGN

Urban Landscape Planning axis goal: to safeguard the ecological balance, improve environmental quality, beautify the urban landscape to facilitate the masses for the purpose of recreation, planting green plants conservation, management as a means of local conditions to ensure the continuity of the axis visual system, strengthen urban axis Landscape system in the overall urban planning in the position and role, set up scientific and complete urban green space system.

The "axis" as the characteristics of the landscape as a network connecting urban style is the most direct city image display window. For the residents living in the city, the city as material and spiritual home, and create beautiful city is the landscape design objectives, regional culture and history is a source of cities. Therefore, the axis of the transmission and update constitutes the core of urban spatial planning of the most content.

Axes of space as a collection of elements that it involved a wide range of positive aspects that we should be reinforced so that it can play an effective role in the planning were.

First, the axis of space control should be reflected in the limited guidance, rather than a mandatory order. In line with the axis of the function will only be accepted by urban residents and urban history.

Secondly, in the axis of landscape planning, emphasizing the public nature and public participation. Rather than the needs of landscape, ultra-scale axis we must be cautious to use, otherwise they will become a mere formality, as the axis to axis.

Third, the axis of the design can also import more methods, it should be more emphasis on its regional impact, not only in physical form, is more of the urban culture and urban charm highlight;

Fourth , the continuity ,history is a focus of our attention for the value axis of space, and we should actively use its planning into new areas.

Axis of the transmission of urban landscape and update on the urban economy, culture, environmental quality, urban aesthetics, and so plays an important role. It is more metropolitan landscape structure optimization provides a new way of thinking.

References

[1] G. John. Ormsby. Ximeng Zi (a) .21 century garden city----the creation of livable urban environment. Liaoning Science and Technology Press, 2005.5

[2] Meng Gang Li Lan Li Rui Dong Wei Shu compiled.Urban park design. Tongji University Press, 2005.9

[3] Wang Hong-cheng Lv Chen forward. Streetscape urban garden design. Tianjin University Press, 2003.1

[4] Dai Zhizhong Liu Jinchuan Lihong Lie with. Urban intermediary space. Southeast University Press, 2003.9

[5] Health and spring the car. Urban green space landscape structure analysis and ecological planning ---- A Case of Shanghai. Southeast University Press, 2003.3

[6] Xi Zhou Li Zhan, Dong compiled. ----- Eco-design of the new theory of reflection on the eco-design and re-recognition. Southeast University Press 2003.6

[7] Yoshinobu Ashihara. Street aesthetics The Aesthetic Town space. 2006.10

[8] Ideal space --- the ideal space for urban planning in the cultural strategy and actions. Tongji University Press, 2006.6

[9] Ruan Yisan, Wang Jinghui, Wang Lin eds. Historical and cultural relics protection theory and planning. Tongji University Publishing .1999.8

[10] Yu Kongjian LI Dihua. Urban landscape of the road ----- with mayors to exchange. .2003.1 China Building Industry Press

[11] Wan-yong book. Old harmony updates. .2006.9 China Building Industry Press

[12] Liu Binyi forward. Landscape Planning and Design of Urban Roads. Southeast University Press, 2002.3

[13] Dong Pingyang, city monsoon, Oriental Press, 1994

[14] Zhaoyong Ge, WANG Ya-nan, a hundred years the city's changes, China Economic Press, 2000

[15] XiongYue, the Shanghai History, Shanghai People's Publishing House, 1999 [18] Wu Liangyong, UN-Habitat Environmental Science, China Building Industry Press, 2001

[16] Zhang Hongyan, invasive and succession --- a new theory of urban social structure changes, Southeast University Press. 2000 .

Figure 1. Layout plan of Baroque Rome.

Figure 2.

Optimization Design on Positioning Model of Destination Image: A Case Study of Nanjing

Xiong Yuanbin
School of Economics and Management
Wuhan University,
Wuhan, China
Email: ybxiong@163.com

Chai Haiyan
School of Economics and Management
Wuhan University,
Wuhan, China
Email: chaihaiyan0391@163.com

***Abstract*—Destination image has been shown to be an important influence in the tourists' behavior. The traditional positioning approach of destination image has many shortcomings in tourism planning and marketing promotion, such as multi-level and variety, product-oriented, reduplication. The purpose of this paper is to provide an optimization model of destination image positioning, ascribed it as "four context theory", including geographical context, cultural context, marketing context and stakeholders' context. Nanjing is used as a study case. The paper points out that Nanjing's destination image should be positioned as "ancient capital of mountain and water, modern life".**

Keywords-destination image; optimization design; Nanjing

I. INTRODUCTION

In an increasingly competitive tourism industry, one of the most significant marketing challenges for destination marketers is the need for an effective destination positioning strategy. A key component of this positioning process is the creation and management a unique destination image to differentiate themselves from competitors. The existing literatures suggest that destination image plays an important role in influencing tourists' behavior as well as their decision to visit particular destination [1]. So image is the key construct in destination positioning. A major objective of any destination positioning strategy will be to reinforce positive image already held by the target audience, correct negative images, or create a new image. But most of the previous studies have followed the traditional approach to positioning that is based on image creation using a number of attributes that reflect the destinations' most attractive products. Traditional approach is supply-oriented or attribute-oriented and not account sufficiently for the consumers' demand and stakeholders' opinion, which lead to destination image homogenization and non-particularity.

The purpose of the article is extend traditional positioning theory of destination image by analyzing destination image formation and influencing factors, proposes the optimization model of image positioning. The model creates the positioning analysis framework of destination image, based on "four context" including geographical context, cultural context, marketing context and stakeholder context. The article selects Nanjing, Chinese famous tourism city as a case, discusses marketing application of "four contexts" theory.

II. THEORETICAL BACKGROUND

A. Destination image

Destination image is the sum of beliefs, ideas, and impressions individuals have of attributes and/or activities available at a destination and is the formation of overall mental pictures (imagery) of that destination [2]. Image formation is defined as a construction of a mental representation of a destination on the basis of information cues delivered by the image formation agents and selected by a person [3]. Gunn thought that destination image formation could be separated into organic (i.e., newspapers, magazines, the opinions of friends) and induced (i.e., advertisements, commercial promotions).Fakeye and Crompton believed that actual visitation created a image more realistic than existing prior to visitation, separated this type of image from Gunn's organic image and ascribed it as the third level, "complex" image [4]. According Pike and Ryan, destination image includes cognitive and affective image components. The cognitive image component consists of beliefs and knowledge about a destination, primarily focusing on tangible physical attributes. The affective image component represents feeling about a destination. Cognitive image is an antecedent of affective image [5]. The existing literatures suggest that unique destination image is formed depending on tourists' perceptions of the strength of cognitive and affective attributes. Induced and organic information sources have significant influence on cognitive image, tourists' motivations influence affective image.

B. Traditional position model of destination image and its shortcoming

Previous studies have concentrated on destination attributes that are seen to attract visitors, such as climate, scenery, and accommodation. Crompton suggested that the strong attributes that are perceived as important by visitors should be first identified for effective positioning for a destination [4]. Botha offered a critique of the traditional approach to positioning, based on image creation using the attributes of the destination's attractive products; preferring one based on measurement of what already exists in the tourists' minds, which could be positively reinforced. The traditional positioning approach based on attributes was introduced into tourism planning by Chen Chuangkang in china, ascribed it as "context", including natural geographical environment, historical and cultural tradition, social psychology and economic development level. Fan divided it into two factors, geographical context and cultural context. Geographical context includes climate, geological features, scenery and transport. Cultural context represents social and cultural factors, including heritages, social custom, political and economic factors. This theory is applied widely to tourism planning and destination positioning analysis.

With the increased number of destinations available and their competition for a limited number of visitors, effective destination positioning strategy becomes a key challenge for destination marketers. But traditional two-context positioning approach has a few shortcomings in increasing destination competition and tourism image marketing, as following:

First, destination's geographical and cultural contexts are multi-level and variety, which lead to vacillate in destination image positioning. For example, Hubei province lies in the middle of china, the culture of Chu, Three Kingdoms and the Revolution of 1911 form its cultural context. It is difficulty to select for Hubei's destination marketing organization, which lead to change continually on Hubei's tourism image, weaken Hubei tourism competition.

Second, traditional approach is product-oriented, resource-oriented or supply-oriented. Because cultural context of destination is high correlation, especially within one country or same culture region. Geographical context also exist resemblance. Additionally, marketers copy blindly famous destination positioning strategy. Thus many duplicate destination images are formed, based on traditional positioning approach. For example, there are a few "River City" and "Little Guilin" in china.

Third, the superposition of cultural context disturbs tourists' cognitive. There is a long history in china; the superposition of cultural context is common phenomenon. The situation that destinations contest cultural attribute is emerged in endlessly. There are a few battle for certain culture hometown, such as Ancient LongZhong (Zhu geliao's home) between Xiangfan in Hubei province and Nanyang in Henan province, Mulan hometown between Huangpi in Hubei province and Yuchen in Henan province, Chibi (old battlefield of Three Kingdoms) between Puqi and Huanggang. Because historical event may have relation to a few regions, these contests are almost inevitable. But tourists may appear cognitive maladjustment owing to lacking historical knowledge.

III. OPTIMIZATION DESIGN MODEL OF DESTINATION IMAGE POSITIONING

The literature reviewed pertaining to destination images and the shortcoming analysis of traditional positioning model enable one to create an optimization design model of destination image positioning as delineated in Figure 1. The model, which includes four factors, such as geographical context, cultural context, marketing context and stakeholders' context, reflects more comprehensive positioning approach. Marketing context is a core of other factors and also the most important and essential element of image positioning model.

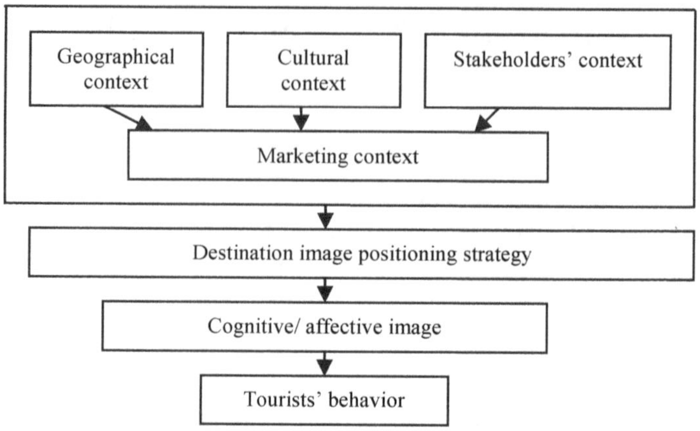

Figure 1. Optimization design model of destination image positioning

As can be seen from Figure 1, the positioning strategy of destination image should consider destination's supply-side and demand-side variable, especially marketing context analysis. Positioning analysis requires more than an understanding of a product's image in the mind of the consumer. What is also required is a frame of reference with the competition since a position is a product's perceived performance, relative to competitors, on specific attributes. Destination positioning strategy influences tourists' cognitive image by destination-oriented marketing promotion (i.e., advertising and guidebook) and affective image by affective quality of environment. Both cognitive and affective evaluations form the overall image of a destination. Destination image, both in terms of an overall evaluation and within its different components, has been postulated to influence various consumer behavior concerning not only before, but also during and after visiting a destination. Thus, as many researchers suggest, generating a positive image among potential visitors may be a very important process through which destination managers can generate demand.

In positioning model of destination image, geographical context includes climate, scenery, geological features, and transport. Climate, scenery and geological features are foundation of tourism industry and the primary elements of destination appeal. Transport is the life line of tourism industry, which influences destinations' accessibility and destination competition. Cultural context includes history, social custom, and heritage, political and economic environment. Reisinger and Turner argue that cultural differences are very useful constructs for international tourism promotion, and they can provide very accurate criteria for targeting and positioning [3]. As a result, destination marketers should take into account the cultural backgrounds of tourists when developing positioning strategies. Stakeholders' context includes host communities' residents and other stakeholders, such as news media, other government departments and special interest groups and so on. Resident attitudes are crucial in creating hospitable and appealing environment for tourists. Moreover, resident attitudes can directly impact tourist enjoyment. Tourism development that is subordinate to local character and identity as well as to local needs, wants, and priorities is the best possible guarantee against tourism saturation. Destination marketing organization also recognizes other stakeholders as being important, because they supply or facilitate funding, provide the tourism superstructure and product, participate in or generally support their programs, or influence governance [6]. Thus the opinion of residents and other stakeholders are important element in destination positioning. Destination image positioning and landscape planning must embody residents' cognitive and gain their identification. Marketing context includes tourists' cognitive/ affective image on special destination and competitors' positioning analysis. According positioning theory, the positioning of destination is the process of establishing a distinctive place of that destination in the minds of potential visitors. Thus, examining image that already exists in the tourists' mind can provide guidance and/or inform the decisions of the destination's positioning strategy. Most of the previous studies have examined destination image as a dependent variable, suggested that factors such as familiarity, tourists' socio-demographic character, could influence potential tourists' preference of destination image [7]. Therefore, the measurement of potential tourists' cognitive/ affective image and the satisfaction of the destination attributes, combined with the identification of the tourist needs and desires, can be seen as the basis on destination positioning strategy. Competitors' positioning strategy is a referenced frame of destination positioning; distinctive image can attract potential tourists' attention and increase competition.

IV. THE CASE STUDY OF NANJING CITY

In this paper, Nanjing, a Chinese tourist city is chosen as our case destination. Nanjing is the capital of Jiangsu, located at eastern a part of China, about 3 hours drive to the Shanghai. It has a long history and one of the eight great ancient capitals in China. In addition, Nanjing, with a favorable geographical location and picturesque landscape, is also a famous tourist city. As an ancient capital city, green city, cultural city, riverside city and city of Fraternity, Nanjing has emerged as one of China's most attractive destinations for investment and living. There is little research on Nanjing image as destination. However, several studies have made significant contributions to analyze Nanjing image positioning strategy. Xiu, for example, compared the expectation and confirmation of tourists; image about Nanjing through both structural and non-structural methods. The empirical study shows Nanjing is lacking of an integrative image, tourism image of Nanjing mainly focuses on "ancient", "green" and "rich in cultural heritage". The author suggested an image of "capital of six dynasties, cultural Nanjing" should be set up [8]. Other researchers suggested other positioning strategy, such as "capital of ten dynasties, cultural Nanjing". "A city of universal love" is Nanjing slogan and government positioning on Nanjing. However, these researches or suggests were supply-oriented or attributed-oriented, which did not account for the complexity of the positioning strategy, such as stakeholders' factor and competitors factor and so on. Furthermore, previous positioning is static and single culture, such as six dynasties' culture or the culture of the Republic of China, which lead to the battle between differential cultural contexts. Nanjing is a modem city brimming with dynamism and confidence and also the most important commercial city in the Yangtze River delta region after Shanghai. These factors reflect the current and future of Nanjing, which must be included in the positioning strategy.

The article applies the optimization design model to analyze Nanjing positioning contexts and reposition it. According table 1, mountain and water should be one of the cores of Nanjing image, which be identified with marketing and residents. In addition, ancient capital culture and the culture of the republic of china are the most prominent cultural context. Modern city landscape is the third factor, also more attractive attribute in future. Based on the above analysis, the article suggests that Nanjing image positioning should center on mountain and water, ancient capital, modern city and the culture of the republic of china, the slogan is "ancient capital of mountain and water, modern life".

TABLE I. FOUR POSITIONING CONTEXT OF NANJING

Geographical context	Cultural context	Marketing context	Stakeholders' context
Mountain and water(Yangtze River, Qinhuai River, Xuanwu Lake, Niushou Mountain, Qixia Mountain)	the culture of Six Dynasties, Ming Dynasty, the Republic of China	ancient green clean	Thriving and prosperous Ancient capital Mountain and water city

REFERENCES

[1] Woodside and Lysonski, "A General Model of Traveler Destination Choice", Journal of Travel Research, vol. 27, May. 1989, pp. 8-14

[2] Echtner and Ritchie, "The Meaning and Measurement of Destination image", The Journal of Tourism Studies, vol. 14, May. 2003, pp. 37-48

[3] Asli D.A., Tasci, and William Gartner, "Destination Image and Its Functional Relationships", Journal of Travel Research, vol. 45(4), May. 2007, pp.413-425, doi: 10.1177/0047287507299569

[4] Fakeye and Crompton, "Image Differences Between Prospective, First-time, and Repeat Visitors to the Lower Rio Grande Valley", Journal of Travel Research, vol. 30,Aug.1991, pp.10-16.

[5] Steven Pike and Chris Ryan, "Destination Positioning Analysis through a Comparison of Cognitive, Affective, and Conative Perceptions", Journal of Travel Research, vol. 42, May. 2004, pp.333-342, doi: 10.1177/0047287504263029

[6] Sheehan and BrentRitchie, "Destination Stakeholder Exploring Identity and Salience", Annals of Tourism Research, vol. 32, 2005, pp.711-734, doi:10.1016/j.annals.2004.10.013

[7] S. Baloglu and K. Mccleary, "A Model of Destination Image Formation", Annals of Tourism Research, vol. 26, Aug. 1999, pp. 868-892

[8] Xu Feifei, GuKai, Wan Xucai, LiuPei, "Study on Nanjing Urban Tourism Image", Geography and Geo-Information Science, vol.3, May.2005, pp.93-96

Optimization on Information System of Highway Natural Zoning in China

Yuxin Niu
Highway College
Chang'an University
Xi'an, China
Email: hicky@foxmail.com

Jinliang Xu
Highway College
Chang'an University
Xi'an, China
Email: gl15@chd.edu.cn

***Abstract*—Information system of highway natural zoning is a platform for data storage, data calculations and results inquiry. Optimization methods on information system of highway natural zoning in China are enumerated in the paper. On the basis of the analysis on the structure of information system for highway natural zoning, the optimizations of the system data and system performance are analyzed. By choosing the recovery model of "Simple" and "bulk-logged", using the API function, the speed of data loading and bulk data copying is significantly improved. By the introduction of DECODE function, SQL statement is optimized. Then, the temporary table is established to optimize the query cache. The contribution of this study is to reduce the system response time and improve the system accuracy, in particular to provide a systematic, integrated spatial decision support platform to raise the automation and quantification of highway natural zoning in China.**

Keywords-highway natural zoning; information system; database; optimization

I. INTRODUCTION

Highway natural zoning in China is the process of division of certain geographical regions based on the regional climate, hydrology, geology, topography and other conditions affected on the highway construction. The information system of highway natural zoning, abbreviated as HZIS, involve physical geography elements of topography, climate, vegetation, hydrology, hydro-geological and so on. With the conversion, calculation and renovation, the amount of data is more large and complex. As cross-impact of the zoning indexes, the phenomena of data mess, redundancy and damage in the information system, lowered the scalability and maintenance of the system. The article analyzed the structure of HZIS firstly. And then, on the basis of the construction analysis, the optimization is researched from two sides, the system data and the system performance.

II. STRUCTURE ANALYSIS OF HZIS IN CHINA

The main principle of HZIS is that by importing the initial data, such as DEM, meteorology and so on, the system calculated the data by spatial analysis model. The initial data and the results were stored in the database in order to export the results, provide indicators, detailed results to users.

According to analysis the GIS structure model and the characteristics of highway natural zoning, the mixed-mode network of GIS, C/S combined with B/S, is selected. Based on Oracle database management system and the SQL search engine, HZIS used Geodatabase as the spatial data analysis model. The structure of the system was shown in Figure1. [1]

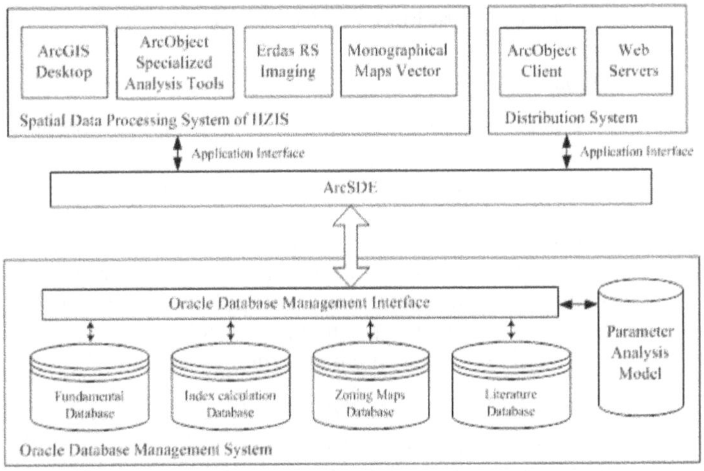

Figure 1. Structure of HZIS in China

III. DATA OPTIMIZATION OF HZIS IN CHINA

Data analyzing required for a long time and large storage space because of amounts of data and variety of data types. So, optimization on HZIS is needed to reduce data storage space, shorten processing time and the real-time query execution time. [2]Data optimization is discussed in the article through data importing, data processing and data storage. The purpose of optimization is to increase the running speed of HZIS by the data response time decreasing.

A. Data Importing optimization

There are many ways to accelerate data importing, such as selecting appropriate database recovery model and appropriate bulk importing methods, setting up clustered index. [3]When importing the initial data of the system, "simple" recovery model is chosen. And when importing incremental data, "bulk-logged" recovery model is picked as shown in figure2. The latter one can reduce the space of log caused by operation of large-scale replication.

The data involved in the process of highway natural zoning is huge. Take meteorological data for example, it is the data related to highway natural zoning in the past thirty years. So choosing a appropriate bulk importing method directly affects the speed of data entry. API function is a bulk copy function which can be used to achieve bulk copies of the system data. Considering the needs of data-read-back when data importing failed, the line of the data is one thousand. [4]

In order to solve the hybridity caused by data importing, clustered index was created in the system tables. Imported data are auto-arranged in the order of key values when data importing. And the order is recorded. The construction of clustered index could strengthen the accuracy of HZIS.

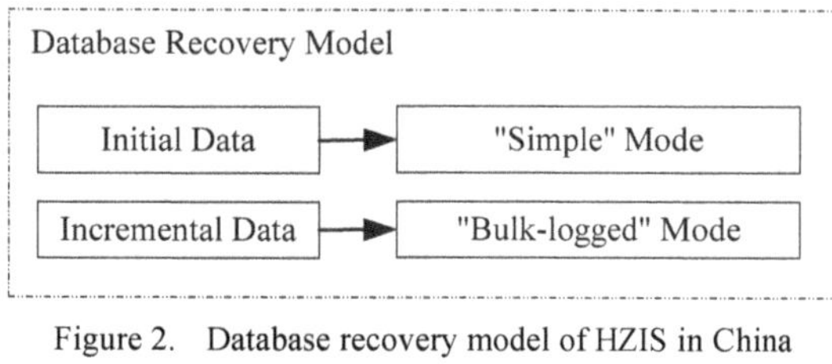

Figure 2. Database recovery model of HZIS in China

B. Independence of Parameter optimization

The data about natural zoning for highway are complex. Cross-phenomena were produced in the spatial analysis procedure. For example, climate zoning, hydrological zoning and geological hazards zoning are all related to rainfall of the geographical regions. If the algorithms of cross-data are stored in the database rather than the application, the independence of the parameters was enhanced. Though the administrator password, the algorithms and the parameters could not only be replaced but also modified.

C. Database tablespace optimization

The size of hard disk space occupied by HZIS and the speed of query were directly affected by datasheet.[5]The tablespace of HZIS is consisted by the Oracle system tablespace[6], rollback tablespace, temporary tablespace, ArcSDE system tablespace, index tablespace, basic data tablespace, indicator tablespace, mapping tablespace and analysis model tablespace. As Oracle system tablespace, rollback tablespace and index tablespace were active, these should be stored separately on different hard disk to reduce I/O conflicts between the tablespace. [1]

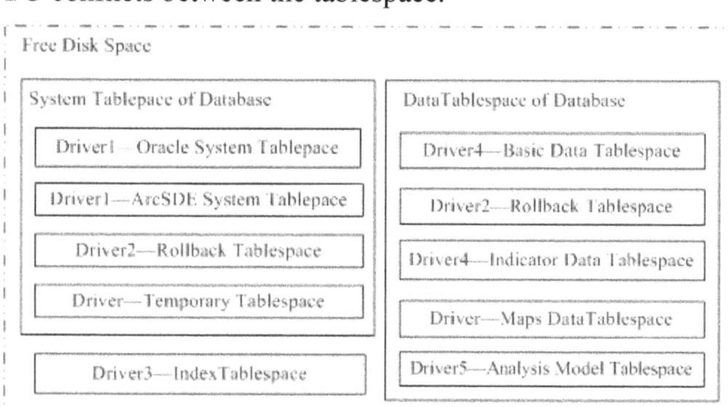

Figure 3. Storage location of Database Tablespace

IV. PERFORMANCE OPTIMIZATION OF HZIS IN CHINA

A. SQL Optimization

In the process of data consideration and inquiry, most of the application operations are achieved by parsing SQL statements. Therefore, optimization of SQL statement is critical to system optimization. DECODE function of Oracle is the function that can convert scattered data into practical table form. [7] For the convenience of inquiry or printing of original data as in figure 4, DECODE function is used to optimize the SQL procedures:

CREATE VIEW v_xsmx HS

SELECT Parameter, SUM(DECODE (TO_CHAR (Data, 'MM'), '01',Value,0))Value1,

SUM(DECODE (TO_CHAR (Data, 'MM'), '02',Value,0))Value2,

……

SUM(DECODE(TO_CHAR (Data, 'MM'), '12',Value,0))Value12,

FROM t_xsmx GROUP TM Parameter

Finally, the view of v_xsmx is selected for data inquiry or printing.

Parameter	Data	Value
Solar Radiation	198109	a
Accumulated Temperature	198507	b
...	...	...
Frozen Index	200205	n

Parameter	Jan.	Feb.	. . .	Dec.
Solar Radiation	a			
Accumulated Temperature		b		
. . .				
Frozen Index				n

Figure 4. Conversion of data between storage and inquiry

B. Cache Optimization

The establishment of the cache could improve system efficiency significantly and reduce the frequency of access to the database. [8-11] LOG_BUFFER is used to specify the size of redo log buffer in Oracle databases. The size of the data buffer is defined by DB_CACHE_SIZE and shared pool size is set through SHARED_POOL_SIZE. [12][13]

(a) When client sends a request to read the data in the database, application server check whether the requested data exist in the cache firstly. If there are, the data are directly sent to the client. If not, the application server will send the request to the database server. Then the requested data would be stored in the cache as shown in Figure5 (a). When there is the same requisition next time, the application server will return the data to the client directly from the cache, without visiting the database server again.

(b) When application server receives the client's request to update the data, it put the request into cache instead of sending the request to the database server immediately. All of the requisitions are submitted to the database server one-time by the application server until the user clicks "Finish" after entering all data requests. At this time, the data are renewed by the database server as shown in Figure5 (b).

(c) When the read-write commands are sent by the application server, the database server does not implement the operation immediately but collecting these commands together. Whether implement the operation or not depends on time-period of the system. The process is shown in Figure5 (c).

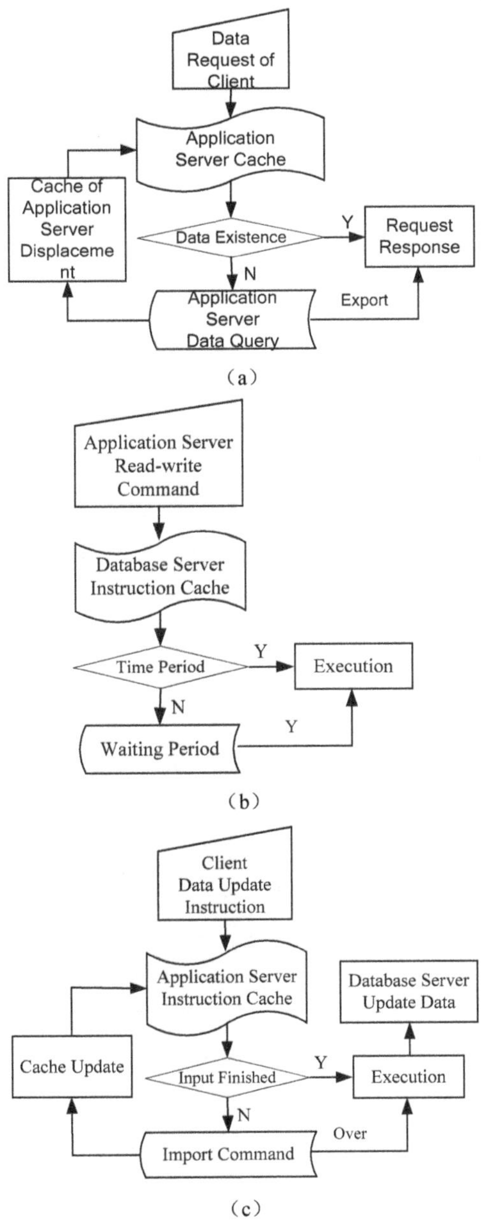

Figure 5. Optimized design on cache of HZIS in China

If the buffer is occupied by data fully and there is a new user-request, the data in the cache will be replaced based on FIFO principles.

Setting up cache among the client, application server and database server, could save the time of page transferring and reduce the number of read-write command. The efficiency of the system is holistically improved.

V. CONCLUSION

By studying the independence and loading speed of the data, the accuracy of the data calculation is improved. Putting the tablespace into different disk could reduce the data conflicts. Introduction of the DECODE function may optimize the SQL statement. And the access speed of HZIS is speeded by setting up cache between the client and the server. The optimization on information system of natural zoning for highway could achieve scientific and modularity in data collecting, data analysis and the results sharing.

REFERENCES

[1] Xu Jin-liang, Li Chao and Yang Hong-zhi, “Design of spatial database on natural zoning for highway in China”, Journal of Traffic and Transportation Engineering, vol.8, Dec.2008, pp.47-53,in Chinese.

[2] Ye De-qian,Lu Bao-chun and Li Guo-yi, “Research on Improving Synthetical Performance of OLAP System”, Computer Engineering and Applications, vol.26, Aug.2003, pp.198-199,in Chinese.

[3] Microsoft Corporation, “Microsoft SQL Server 2000 Resource Kit”, Unshu studio Trans. Beijing:China Machine Press, 2002,in Chinese.

[4] Huang Xue-ju, “Building of Multidimensional Conceptual Models and Designing of OLAP of Scientific Management Information”, Journal of Huzhou Vocational and Technological College,vol.4, Mar.2005, pp. 80-83,in Chinese.

[5] Harinarayan V, Rajaraman A and Ullman JD, “Implementing Data Cubes Efficiently”, Proc ACM SIGMOD’96,1996.

[6] Devillers R,Bedard Y and Jeansoulin R, “Multidimensional management of geospatial data quality information for its dynamic use within GIS”,Photogrammetric Engineering and Remote Sensing, vol.71, Feb.2005, pp. 205-215.

[7] HE Ze-heng, LU Jian-bo and LI Jian-jun, “Optimized design of information system based on ORACLE database”, Journal of Harbin University of Commerce Natural Sciences Edition, vol.18, Feb.2002, pp. 156-159,in Chinese.

[8] Wu Ji-dong, “The Working Mechanism of the Computer Cache”, Science and Technology Information, vol.33, Nov.2007, pp. 411, in Chinese.

[9] FENG Xiang,Liu Zhi-man and SHUAI Dian-xun, “A Novel Game Particle-Field Approach to Parallel Cache Resource Allocation of CDN”,Chinese Journal of Computer, vol.30, Mar.2007, pp.368-378,in Chinese.

[10] Le De-guang,Guo Dong-hui and Wu Bo-xi “Caching Technology in Network and Its Application”, Applied Technique, vol.30, May.2003, pp.33-35,in Chinese.

[11] Che Xiu-mei,Zhou Yuan-cheng, “Research and Discuss of Web Caching Protocol”, Journal of Shanxi Agricultural University, vol.26, Apr.2006, pp. 408-410,in Chinese.

[12] Teng Yong-chang, “Oracle9i Database Administrator Guide”, Beijing: Tsinghua University Press,2004,in Chinese.

[13] LoneyK. and KochG., “Oracle9i : The Complete Reference”, Beijing: China Machine Press,2003,in Chinese.

Optimizing Design of Landscape Symbol and Image-Building of City Culture

Wu Biao
Chongqing University of Arts and Sciences
Chongqing, China
Email:wubiao510@163.com

Xingkuan Wu
Wuhan University of Technology
Wuhan, China
Email: wxk1952@163.com

***Abstract*—The landscape symbol of a city is the visual language of the city's cultural right. It communicates with viewers the information of the city and displays the social status, cultural significance and time spirit of the city. Thus the process of designing landscape symbol is actually the optimizing designing process of image symbol which most represents the city culture. This thesis analyzes the relationship between a city's landscape symbol and its cultural image; it elaborates the importance of reinforcing the cultural rights of voice of a city to the establishing of a city's cultural image, from the perspectives of the form, the composition and the culture of landscape symbol.**

Keywords-landscape symbol; optimizing design; cultural right; city image

I. INTRODUCTION

Landscape is an important composing element of the culture of a city and its symbol designing represents directly the city's cultural soft power. A city's landscape is also a name card which communicates with the outside world its temperament and cultural property. It influences imperceptibly the spiritual needs of the public of the city and at the same time, the public are more and more willing to receive the mental enjoyment of it when the physical needs of them are fulfilled. "The core mission of modern designing is to construct symbols."[1] However, phenomena like blindly borrowing patterns and misusing language of designing symbols in the designing of it are common thus it frequently destroys a city's traditional landscape culture and weakens the city's recognizability, which causes the city to lose its speciality and hinders it from its sustainable development. Under these circumstances, it is necessary to re-examine and re-grasp the relationship between landscape symbol and cultural image of a city, and to optimize further symbolic designing of city landscape so as to strengthen a city's cultural right of voice and to establish its distinguishing cultural image.

II. CULTURAL IMAGE OF A CITY AND LANDSCAPE SYMBOL

Cultural image of a city refers to the specific perception, general opinion and integrate judgment of the public to the inner cultural strength and outer cultural vitality displayed by a regional city. It is an objective existence that can be felt constantly by the public and concentratedly shows itself in its cultural essence, nature, function, constructing level and cultural atmosphere. It is also an organic unity of impression deepest in the heart left to people which includes visual image that is physical and appreciable image that can be perceived in mind. No matter visual or appreciable, cultural image is a kind of psychological effect so it needs certain medium to stimulate people's mental response either directly or indirectly.

The landscape of a city is both an important composing element of its cultural image and an essential medium which makes it possible to show. It makes it possible for people to watch, experience, perceive and associate a city's developing history, cultural path and value so as to form a conscious judgment of some kind of stable cognition to that city. Therefore, landscape should naturally carry the information like developing history, cultural path and value, regional feature and so on of a particular regional city; besides, it should be the visual image of constructing a city's cultural information—landscape form.

The constructing of landscape form is a kind of symbolization based on fusing, refining, generalizing and then abstracting the complex multi-information of a city's objective reality. The result of this process is called symbolic form of landscape, which includes form, meaning, color and some other attribute essence of symbols. This symbolic process is defined as optimizing and designing. Thus, the optimizing and design of landscape symbol is exactly the process and result of generalizing and abstracting the complicated multi-information of a city. It displays the culture of a city in a most direct way and reaches the goal of convey the city's cultural image rapidly and accurately. According to the theory of semiotics by Saussure, "linguistic symbol is summarized as the signifier and the signified; the signifier is the external form of linguistic symbol, while the signified, the meaning of it; certain agreement exists between the two." [2] This view further proves that landscape symbol has both the attribute of meaning symbol and form symbol in it. Just as linguistic symbol, landscape symbol not only appears as the physical image of a city's culture, but also implies and influences the city's internal spirit and meaning.

Therefore, the symbol of landscape meaning makes effort to reflect a city's historical path and cultural connotation, while the symbol of landscape form and color displays the city's cultural atmosphere and appearance as well as its spirit and philosophy. They both help to construct the cultural image of a city.

III. THE CONNOTATION OF CONSTRUCTING CITY CULTURE BY LANDSCAPE SYMBOL DESIGNING

The designing of landscape symbol is a visual design of artistic form that borrows concise while profound symbolic language so as to reveal the regional culture feature and unique aesthetics of a city. It is a practicing process of people to acknowledge experience and continue the group spirit of a regional city, and constructing its city image. The constructing of a city's cultural image by landscape is completed by many symbols or symbol combinations which contain a lot of information of the city. But not all of the information can accurately and aesthetically express the city's cultural image, so it is needed to be optimized and be transformed into physical appearance. Thus, the designing of landscape symbol based on constructing city culture must adopt some optimized methods and approaches, which consist of optimizing of symbolic form, structure and cultural quality.

"Pierce's theory of semiotics divides symbols into three types, image symbol, indicatory symbol and symbolic symbol." [3] Image symbol represents an object by a similar picture which contains resembling content and form of it. Besides, image symbol can deliver directly the information of the object so it is easy to be understood straightly by the public. This is substantially the optimizing design of landscape image through which makes it possible to express characterizing, instructing and symbolic meaning. For example, Xiangzhou of The Humble Administrator's Garden, a classical Chinese garden, is designed according to Huafang of southern area of Yangzi River in China, both of which share some similarities. Vase gate of the garden is made with the inspiration of ancient vase which symbolizes safety. Round Moon-gate takes full moon as model that stands for satisfying life. The flower window with a shape of paper fan is made according to paper fan which in Chinese has the same pronunciation as "kindness." Another flower window with a shape of two ancient coins with a hole in each symbolizes wealth.

Instructive symbol is a kind of sign that has some direct and innate relationships with the object which enables people to associate correspondingly. As in the United States, Morase, the designer of US-Japan Historical Square, which is located in the riverside park of Portland, put a regular stone in which the name of the square is carved, at the entrance of the square. Thus this symbol is an instructive symbol.

Symbolic symbol refers to one that is based on factors such as cultural values and aesthetics of observers, which makes it possible for people to associate it with a group of symbols on the basis of fully understanding of the characteristics of the object. For instance, the color red is considered the representative of China; the traditional architecture caisson is conventionally seen as a symbolic sign of dignity. On the other hand, symbolic sign is given different meanings due to various subjective factors of observers.

Therefore, through optimizing selecting of these three types of symbols above and appropriate using of the characteristic, metaphorical and symbolic form of symbols, the design of landscape symbol can extract, generalize, reform and optimizing the natural and humane landscape of regional city.

The designing of city landscape symbol researches the structural form with which the symbolic meaning expresses correctly the regional feature and city spirit. Everything in the world has in itself some certain organizing structure that produces individual differences and similarities from one another landscape. The wide application of semiotic theory in landscape designing has provided theoretical support for further study of the formal elements and organizing rules in its study. Thus it is of great importance to examine the symbolic rule in landscape designing and how to express exactly its symbolic meaning.

"Symbolism of designing is the most primary characteristic of contemporary design because it serves not only the function of common symbol but also the rights even violence of it,"[4] which characterized the cultural epoch feature in a symbolic society. Symbol in modern design has developed into a form of right, mostly cultural right, which brings about social stratification. Whether a city has the cultural right decides if it is able to develop sustainably. Therefore, as the local culture is being crashed by the international culture nowadays in cultural internationalization, landscape designers should learn to balance the cultural factors of these two different values. In order to gain and strengthen the cultural right of a city, the displaying form of city's landscape culture should be optimized and innovated.

A substance in relation with a certain nation is most lively and influential in the process of spreading culture, and is most welcomed. That is to say the designing language with the characteristic of local culture can actually increase the cultural speech right of city landscape. The reason why Chinese brush drawing and Spanish bullfights are welcomed is that they are strongly characterized by their national features. I.M Pei, when designing the new museum of Suzhou, integrated perfectly the traditional and modern culture, which set a great example for the innovative development of Chinese designing culture. The new museum has adopted the color tones of pink and black for wall and tile, which includes both modern expressing style and traditional south China symbol. The garden landscape organically combined modern style and traditional style thus made it both with traditional architectural feature of Suzhou, and with modern feature of geometric model and deliberate geography. Therefore, the designing of landscape symbol must be based on the living style and aesthetic convention of certain group of people so as to build a unique structure of deep culture and life accumulation that potentially exists. To react to the crash of modern multi-culture by using the innovation of landscape symbol is in fact the optimizing designing of symbolic culture of the landscape.

Landscape is like a man, while symbol the language spoken by the man. In order to express a topic accurately, only some favorable words are not enough. Besides, some reasonable sentence structures which have cultural meaning are needed. Thus, if a landscape is wished to convey some information about the city to the public, it is necessary to correctly transmit the cultural meaning of it through optimizing the form and structure of symbolic language, and if possible, by using metaphor, symbole and some other rhetoric devices.

IV. Conclusion

To sum up, the designing of city landscape symbol should pay attention to the optimizing design of the type, structure and culture so that the natural environment and regional culture is able to be seen from it. Besides, the historical path, humane spirit and regional feature should all be included in expressing the significance of the city from the landscape symbol. At the same time, when encountering foreign culture, the designing should accept some rational elements to a certain degree with an open mind on the basis of inheriting the marrow of local culture so as to continue and develop the time spirit of local landscape symbol. In addition, the dignity of city culture should be displayed through it. These are all special emphasises on the designing and researching of landscape symbol. Designing culture is multi-developing nowadays. In order to assume the responsibility of raising the cultural taste of a city and building up a city image, it is necessary for the designer to call for the respect for and the protection of the sustainability of nature and regional city's cultural feature. Furthermore, rationally using of different cultural symbols for landscape designing practice is indispensable.

Acknowledgment

My deepest gratitude goes first and foremost to Professor Wu Xingkuan, my supervisor, for his illuminating and careful guidance. Second, I would like to express my heartfelt gratitude to Professor Chen Hanqing and Professor Li Yanzu for their literature point of view, from which I have benefited a lot and gained a good deal of inspiration. I am also greatly indebted to my beloved family for their self-sacrificing care in my study and life, which has strengthened my courage and confidence in overcoming difficulties during the course of the thesis. Last my thanks would go to all the specialists, teachers and my fellow classmates who take their attention on my study.

[1] LiYanzu, "Expansion of consumption, sociology of symbols and design," Journal of Nanjing Arts Institute (Art and Design Edition) ,April 2007, pp. 10.

[2] Saussure,Course in General Linguistics,Commercial Press.Beijing, 1980, pp. 4-32.

[3] WangMingyu, Linguistic Semiotics,Higher Education Press.Beijing, pp. 73-74, 2004.

[4] LiYanzu, "Expansion of consumption, ,sociology of symbols and design," Journal of Nanjing Arts Institute (Art and Design Edition) , April 2007, pp. 10.

[5] JiangHui, "City spirit and shaping the image of a modern," Business Times, October 2007.

[6] ChenHanqing, CaoYali, "Sign language interpretation of the design," Decorative, December 2005.

[7] XuXiaojing, ZhengQianli, "From the Kung Fu Panda to see genetically modified culture in the animated film role," Decorative, August2008.

Outdoor Space Optimization for Senior Citizens Based on Interactive Activities in Old Communities in Wuhan / China

Peng Xia
School of Architecture and Urban Planning
Huazhong University of Science and Technology
Wuhan, China
Email: supersunny7906@126.com

***Abstract*—With the great improvement of health services for senior citizens and rapid urbanization in China, more and more aged people living in old communities need to deal with the problem of disappearance of communicational space and facilities by the urban renewal movement and real estate development which greatly prevent elders getting outside. It is essential to get to know what interactive activities and the relationship between outdoor space and activities of the kind before we can find solutions. In this article, we take the case of one historic community Baoqing as an example to see how to improve outside activities for aged people.**

Keywords- aging society, outdoor space, interactive activities

I. INTRODUCTION

Generally speaking, urban planning and architecture design in China have showed little consideration for aging citizens who live in old communities because of the rapid urbanization movements, and the research foundations we referred to have been out of time already, so we will have to conduct further research on the daily activities of senior citizens including senilities and their spaces.

So we have explored senior citizens' real lives and observed their active space, listened to the vigorous stories and understood what kind of space and public facilities are needed for senior citizens living in old communities.

II. CASE STUDY ON WAN'AN COMMUNITY OF HANZHENG STREET, HANKOU COUNTY

Hankou County was initially formed in the 16th century and was named after its location at the junction of Han River and Yangzi River. During the Emperor Chongzhen period (1628～1644), Hankou was developed along the river with relatively stable markets. Since then, Hanzheng Street has become the main market street of Hankou County.[1] (Fig. 1)

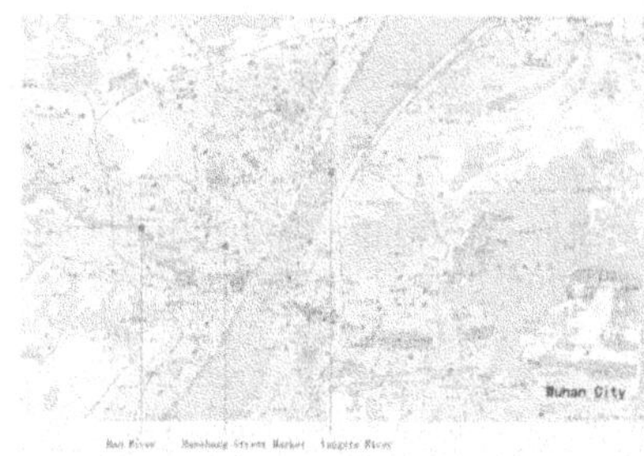

Figure 1. Location of Hanzheng Street district

Wan'an community is located at the south-west part of Hanzheng Street district and east to Wan'an lane. The whole area covers about 0.027 km^2, with a population of 2,902, 480 of which, a proportion of 16.54% are senior citizens. (Fig. 2)

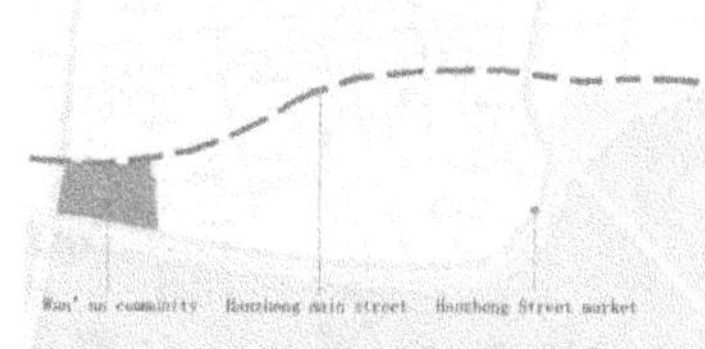

Figure 2. Location of Wan' an community

We choose the community because the living space and historic buildings are well-preserved, and life character could be quite representative of the whole area for the following reasons:

- The vertical lane grid patterns express the texture of Hanzheng Street district, constituted by three major mixed housing types: low storied row dwellings, multilayer compensated dwellings and high-rise buildings for both commerce and habitation.
- The proportion of aged people is 16.54%, higher than the average level of the whole city, which is 16.18%.
- The intercourse conditions are better than the other communities of the area as we know. So the types of outdoor space are diversified. And the daily activities we observed are persuasive enough. (Fig. 3)

[1] Rowe.William T (1984), Hankou: conflict and community in a Chinese City, 1796-1895, Stanford University Press

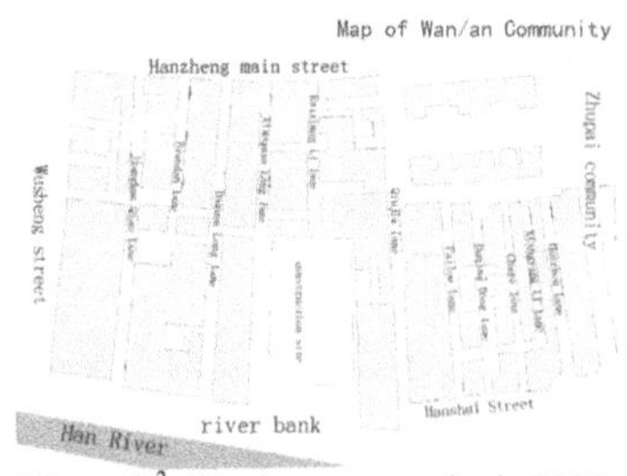

Figure 3.[2] Wan'an community in 2000

Main lanes above: Jianghan Qiao lane(L1), D Biandan lane(L2) , Daquan Long lane(L3), Xiaoquan Long lane(L4), Ruixiang Li lane(L5),Qiujia lane (L6), Wan'an lane(L7)

Sub-lanes: Pailou lane, Duqing Dongli lane, Chaye lane, Huizhou lane

III. TYPE OF LIVING SPACE

A. Low Storied Row Dwelling

Most two-storied buildings were built with a sloped roof at the end of Qing Dynasty, but some have been rebuilt or transformed years ago. The structures are generally brick-timber works with two patios located on the east and west. A few multiplayer buildings(<4F) are apartments constructed by owners' institutions or companies after liberation. Outdoor space of these two types is lanes and remained space.

Figure 4. Birds eye view of Wan'an community

B. Multilayer Compensated Dwelling

This kind of buildings was built for people whose houses were removed on the spot because of the commercial development and financial investment from the end of the 1980s to the middle of the 1990s. The buildings are 8-9 stories without elevators like the edifice Wan' an Tower C, and the ground floor of which is now used for commercial use. The second is used for storage and the third for traffic platforms or playing platforms for habitants.

Figure 5. Perspective to Wan'an Tower C

C. High-rise Building for Both Commerce and Habitation

This kind of building was mainly for real estate development during the late 20th century and early 21st century. Examples of this kind of building are Fortune Times (29 stories) and Band Garden (19 stories). The affiliated facilities for out space activities are well utilized and even a public security system has been installed and managed by professional companies .But bounding walls made the elders isolated from other districts and most people inside the towers conduct their interactive activities on the roof platforms above the annex .

IV. RESEARCH ON INTERACTIVE ACTIVITIES OF SENIOR CITIZENS

We make results by tables and figures below.

A. Step 1

TABLE.1 Relationship between daily activities and in/out space for senior citizens

Daily activities	Indoor	Outdoor Activities
physical movement	Having dinner/sleeping	
housework	cooking/cleaning/washing clothes	cooking (4) [3]/buying food (6)
Interactive activities	chatting (6) /listening to dramas (5)	looking around(2)/chatting (12) /playing cards (3) /exercising (11)
other activities	reading newspapers (5) /watching TV (10)	floriculture (2) /Sitting along (9) /taking a walk (13)

[2] The map of Fig.3 comes from Wan' an Community Administration Committee

[3] The numbers in brackets are numbers of interviewees living at the same place during the time of investigation

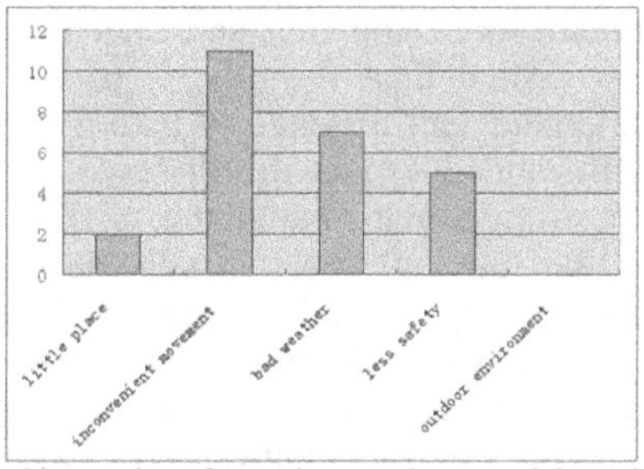

Figure 6. Obstructions for senior people to participate in outdoor activities

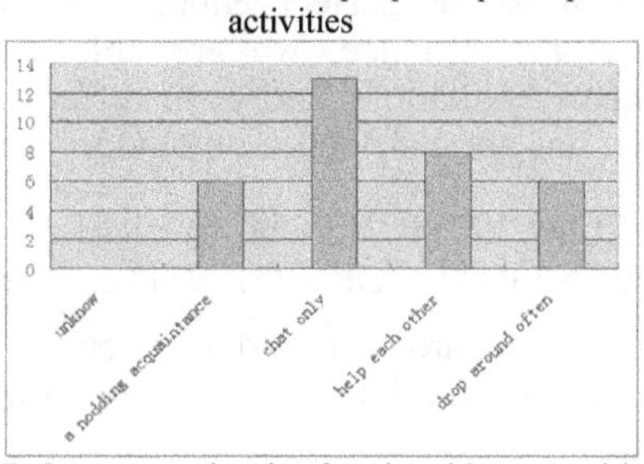

Figure 7. Intercourse levels of senior citizens (multi-choice)

Most elders have the desire to participate in outdoor activities, like to chat with neighbours and to relax themselves, but some of them still want to stay alone and would like to read newspapers and to enjoy the fresh air. Most of them attribute their lack of outdoor activities to their unsatisfied health condition, the inconvenience of movements or bad weather. However, we still have to place emphasis on activities such as chatting or dropping around. (Fig. 7)

From the statistics, we could see that the majority (68.42%) like to chat with neighbours, near half (42.11%) would offer assistance to each other like buying commodities for the disabled, helping them take more activities outside or wash large clothing, etc. What's more, about 31.58% elders dropping around show deep interactive actions, which could make harmonious atmosphere between neighbours.

We also see that most activities take place outside home (65.31%) where the types abundant as well (54.55%). Elders would rather communicate with people whose physical distance and social distance are almost the same. This kind of social network would be helpful for deeper interactions.

For all the activities we observed occurring at the remaining space, which include daily facility spots and crosses. And chatting with people and dropping around to see neighbours are two major actions that occur all day, and the remaining actions such as sitting along and looking around randomly were still frequently happening during the daytime.

B. *Step 2*

We can divide 18 people into three groups, each group live in one of the three types of buildings evenly. And we see the relationships between interactive activities and frequency /time/distance, etc.

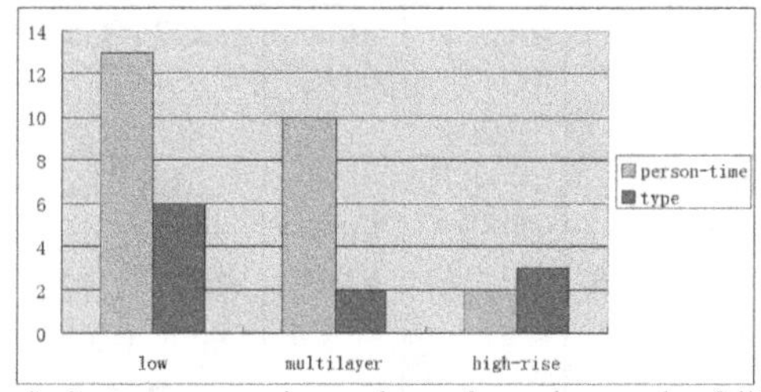

Figure 8. Stat. on person-time and type for senior people of different living space

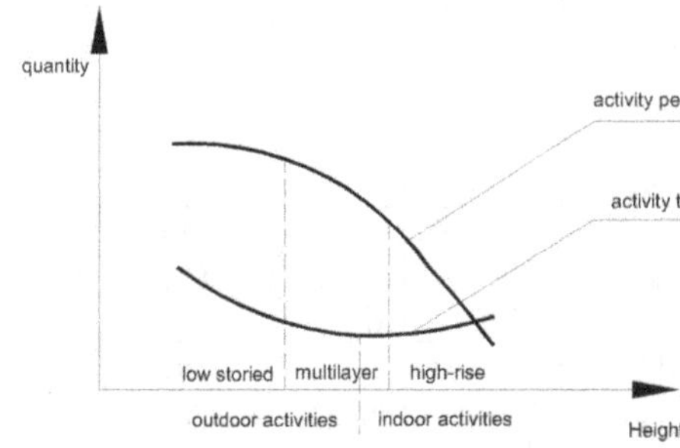

Figure 9. Stat. on person-time and type for senior people of different living space

TABLE.2 The relationships between area, time and frequency of outdoor interactive activities of interviewees

Daily activities / Area	Time	Frequency	Content
around residence (less than 100m) i.e. lanes or platforms in front of gates	07:00~10:00	1 time/day	have dinner (C4) /cook (C2、C4、C5、C9) /do exercise (C7、C10、C13) /floriculture (C3、C9)
	15:00~7:00	1 time/day	look around (C3) /play cards (C2)
	20:00~20:30	1 time/day	take a walk (C1)
	07:00~10:00 or 15:00 ~17:00	Several times /day	chat (C1-C12) /sit alone(C2、C3、C4、C5、C6)/take a walk (C7、C14) /pets (C9)
inside community (about 300m) i.e. lanes ,streets or riverdocks nearby	About 08:00 or 15:00~16:30	1 time/day	take a walk (C1、C2、C3、C4、C13)
	6	1	Do exercise

	: 00~7 : 00	time/ day	(C2、C5)
around community (about 500m) i.e. Changdi street, Guanyinge market, Carrefour supermarket	07: 00~9: 00	1 time/ day	buy food (C4、C6、C9、C10、C11、C12、C13) /take a walk (C1) /do exercise (C8)
Out of Hanzheng street community (more than 550m) i.e. Qingtai Park, Zhongshan Park(riding)	06: 00~7: 00	1 time/ day	Do exercise (C8、C9、C11)
	14: 00- 16:00	1 -2 times /wee k	study at universities especially for elders(C11、C12)

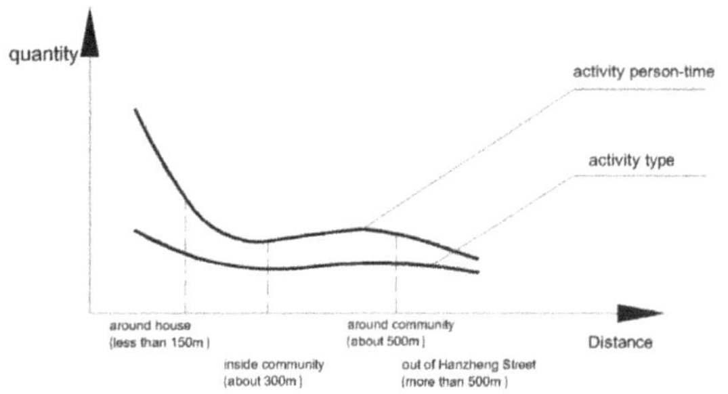

Figure 10. Relationship between distance and person-time(add up) for senior people

From tables and figures, we could see that the elders living at low storied row dwellings have most of their interactive activities outside the room (96.15%) and they have more types of activities than the other people who living in multi-storied and high-rise buildings (80%); the higher the building is, the less the interactive actions would occur and less helpful to the boost of relationship. On the other hand, the types of communicative activities that happen at middle-layered houses are the lowest among the three types of buildings comparatively.

The location for elders' activities living at low storied row dwellings are lanes, river banks and public spaces, which have more person-times (55.1%) and the space at multilayer compensated dwellings are platforms above annex with less person-times visiting (36.73%), even actions partly happening at home like sitting along, reading newspapers or watching TV; and that happens least at high-rise buildings because of the height above the ground.

Seen from larger area, outdoor interactive actions could be heavily influenced by the distance from home, elders who live within less than 100m from home take their outdoor activities much more frequently and their types of activities are more diversified than the others (63.27%,81.82%), so 100m is the convenient distance to take a walk by foot for most of them, and 300m is the distance to walk or drop around with stronger will, but 500m is the distance to do something like buying food, vegetables, the actions happening once a day, sometimes done by helpers or family members. Activities in a distance over 550m are commonly attached with specific intentions, such as doing exercise at parks or going to universities for study where the environment must be perfect enough to stay. However, regardless of the distance, we are able to see that the environment of habitations and even different weather and microclimate has great influence on outdoor activities.

V. CONCLUSIONS ON THE RELATIONSHIP BETWEEN OUTDOOR ACTIVITIES AND SPACE

Through the research on outdoor interactive activities of senior citizens, we could understand the optimized condition for them:

Firstly, good health is the foundation for all activities, and poor health condition is basically an obstacle for engaging in outdoor activities. Besides, we still see that there are other obstacles like bad weather and lack of facilities. And activities like playing cards, chatting with people or dropping around, which do not have specific need for facilities, should have serious consideration on the outdoor environment. And activities like doing exercise, sitting along and looking around should also be safe and enjoyable.

Secondly, the activities are communicational for most types, interactive actions between neighbourhoods could be strengthened step by step. And the more help elders get from other people, the more eagerness they would make. On the other hand, it also enhances the harmonious relationships among people living together.

Thirdly, no matter how far the distance to the ground floor, the higher the floor, the less outdoor and interactive actions occur for all three types of buildings. So platforms, roof gardens and open spaces above annexes or even higher are good places for communicating with each other. And for low-storied people, level distance by foot is still an important factor to be considered;

Fourthly, to satisfy the basic requirements by senior citizens, offering more non-obstacle establishments like ramps, handrails and chairs wherever they go and building more facilities or places such as medical treatment rooms will guarantee that they can go out without worry.

REFERENCES

[1] Rowe. William T, Hankou: conflict and community in a Chinese City, 1796-1895. Stanford University Press, 1984, pp.68–69.

[2] Leon A. Pastalan & Benyamin Schwarz, Housing choises and well-being of older adults, London: Haworth Press, 2000.

[3] Liu Ying, Research on the outdoor living space for the elderly, Wuhan: Huazhong University of Science and Technology, 2005.

[4] Sandra C. Hoglund, Design for Aging: Patterns of Use, Massachusetts Institute of Technology Press, 1980.

Planning of Tourist Resort Based on the Culture and Ecology Factors

Case Study of Jining North Lake Tourist Resort

Xiuqin Zhang
School of Architecture
Tianjin University
Tianjin, China
Email: suekin_tju@yahoo.com.cn

Paolo Vincenzo Genovese
School of Architecture
Tianjin University
Tianjin, China
Email: pavic@163.com

***Abstract*—Based on the case study of Jining North Lake Tourist Resort in terms of culture and ecology factors, this paper approaches the planning method of Jining North Lake Tourist Resort and raises some strategies for development. By the application of "Pan-Tourism City" idea, construction of characteristics road network, landscape ecological planning and spatial integration, the master planning of Jining North Lake Tourist Resort folds the exploitation of tourist resort, the integration of city function and the spatial structure as a whole to develop the culture and tourism of the city.**

Keywords-tourist resort; culture; ecology; spatial integration

I. INTRODUCTION

A. *Introdction of Planning Area*

There is a widespread Chinese poetry, "Jining is named another Suzhou, facing the mountain and holding the river. Numerous houses stand by the shore, countless boats float in the river", describing the Jining origin depending on the Beijing-Hangzhou Grand Canal. The case just locates the south of Jining, which is the only provincial tourist resort in the south of Shandong Province.

The site lies in the south of Weishan Lake, and in the north of the old district of Jining. It is surrounded by the Beijing-Hangzhou Grand Canal and the Guangfu River. While the Old Canal goes through the site. The total area of the waters is about 11 square kilometers. The North Lake, south in the site, is 3.3 kilometers wide from north to south and 3.1 km wide from east to west (Figure 1).

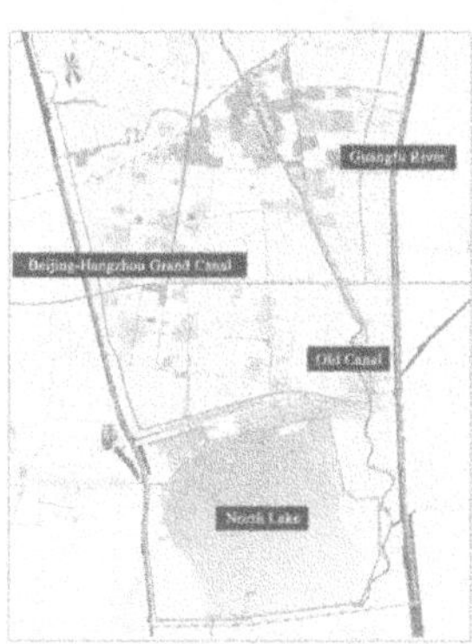

Figure 1. Actuality of water system.

B. *Analysis of Planning Element*

The key to the Jining North Lake Tourist Resort planning is analyzed from three aspects. Firstly, on economic aspect, only by finding the appropriate development pattern for both space and industry, and making tourism promoting urban development and city providing the guarantee for tourism, can achieve the goal of win-win. Secondly, on the cultural aspect, only by grasping the rich cultural connotation that Beijing-Hangzhou Grand Canal gives and determining the unique cultural orientation and achieving the interaction between culture and economy, can make it showing itself by surprise. Lastly and importantly, on environmental aspect, Jining North Lake Tourist Resort is rich in ecological resources. There are lake and river inside the site and around the shore not only planting large area lotus but also distributing broad wetland. Plentiful aquatic products teem in the lake such as fish, shrimp and crab. Ecological protection is premise and basis for the development. Only a solid foundation can help to achieve sustainable development.

So we can draw a conclusion that culture and ecology are the two major elements of this planning. Under the guidance of culture and ecology the planning feature that "ecology is primary, landscape is bone, culture is soul, the city and the park depend on each other" can be achieved. And the exploitation of tourist resort, the integration of city function and the spatial structure can be folded as a whole to develop the culture and tourism of the city.

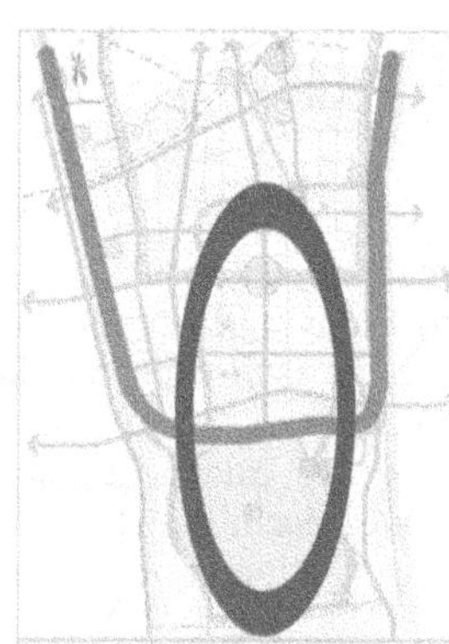

Figure 2. Structure of tour route.

II. TOURISM PLANNING

A. *"Pan-tourism city" Idea*

There are three rivers across the site and the North Lake hold extensive water surface .While the attractions are used to locating along the shore. So the scenic area is broad and the tour route is long. For this situation we put forward the "Pan-Tourism City" idea to construct the tourist city by city open space, special tour lines and themed scenic spots, which melts the tourism concept into the layout of everywhere and the life of townsfolk bit by bit.

B. *Organization of Tour Route*

Because of the broad scenic area, the organization of tour route is asked to abstain from turning back and single type, and to coordinate with the stage construction of scenic area.

Firstly, by analysis of actual road system and the road system in Jining Master Planning which is in revision, we mark off the road which can not be changed and which can be changed but keeping the direction as the basic of the new road system. Secondly, by excavating an artificial river between the Beijing-Hangzhou Grand Canal and the Old Canal, and by constructing roads and arranging attractions along the river to achieve the relative separation between tour route and urban transport line, and to further carry out the concept of "Pan-Tourism City" to obtain networked attractions and systemic tourist lines. Thirdly, except taking full advantage of land tour line, with the help of water tour line the tour route system constructs a ring structure ringed by "O-shaped" and "U-shaped" tour line (Figure 2). This structure not only combines with the city function but also join almost all attractions. The "O-shaped" tour line covers North Lake scenic spot, the Old Canal scenic spot and the center scenic spot of the tourist resort; and the "U-shaped" tour line covers the Beijing-Hangzhou Grand Canal scenic spot and the Guangfu River scenic spot. While through the "U-shaped" tour line can easily connect with the old distrcit of Jining. Lastly, for showing the cultural and ecological characteristic of North Lake Tourist Resort, the planning emphasizes the organization of city slow traffic system and tries to separate the slow traffic system from fast traffic system. Two methods help to achieve the slow traffic system. One is vehicle, including waterbus, solar bus, bicycle, and so on. Another is road section, by designing special path for tourist bus, bicycle and pedestrian, and increasing the quantity and width of the greenbelt, and minishing the width of fast traffic road in slow traffic system.

C. *Tourist Attraction*

Introduction of main attractions:

Tasting Vino and Enjoying Lake—combines the bank of the North Lake, takes the landscape of lake and country as backing, people can enjoy vino and chat. White Sand and Night Moon—lies in the middle of the Old Canal to construct the man-made beach. People can play in the water in summer, while enjoy the moon beside the shore in winter. Inn and Song—lies in the island of North Lake to plan the marina. People can rest in the inn, taste the aquatic product, and listen to the fishermen's song. Water Streets and Night Markets—combines the marina to plan the streets along the river, which be equipped with performing arts plaza, lantern shop, embroidery workshop and folk craft stores.

III. ECOLOGICAL PLANNING

A. *Ecological Environment Capacity Analysis*

The planning also does the ecological environment capacity analysis. The area, north to the North Lake, west to the Old Canal, south to the Rihe Railway, has high environmental carrying capacity. The relatively large-scale and high intensity urban development projects can be carried out in this area. While the low environmental carrying capacity area lies in the shore of the Beijing-Hangzhou Grand Canal, the Old Canal, the Guangfu River and North Lake. This area can be put the city function of ecological conservation, ecological recreation and eco-community in low-density (Figure 3).

B. *Application of Basic Spatial Principle of Landscape Ecology*

The best patch shape of ecological significance is the irregular one, which has a circular core, bending edge, and some finger-like salients. Its circular core is conducive to protect resources, the bending edge is helpful to the interaction between patch and matrix, and the finger-like salients do favor to the long-distance species dispersal. Based on the research of the relationship between patches and corridors of different properties in ecological network, the planning makes the natural and artificial ecological space interacting and integration. Considering the abundant water system and the region's cluster arrangement, the planning sets the natural landscape of North Lake as the core, takes the original Beijing-Hangzhou Grand canal, Old Canal and GuangFu River as the backing, makes the construction area and river corridor integration, and uses the "cell wall" pattern—parks and green space of all levels dispersed in the region, in order to construct the urban eco-system with distinct regional feature of "The lake as the heart, water as the pulse, ecological matrix, corridors and patches of multi-level combined".

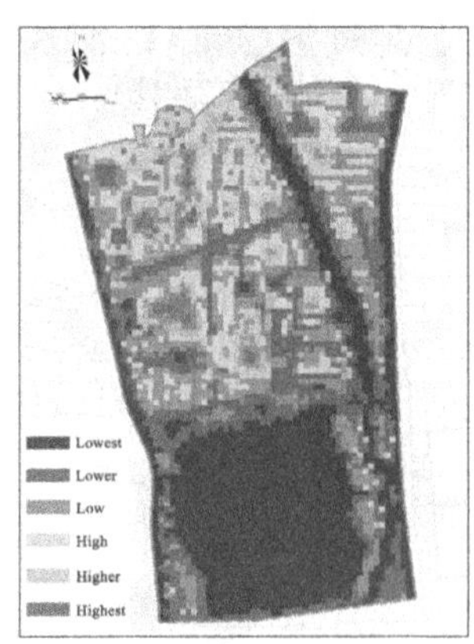

Figure 3. Ecological environment capacity anaysis

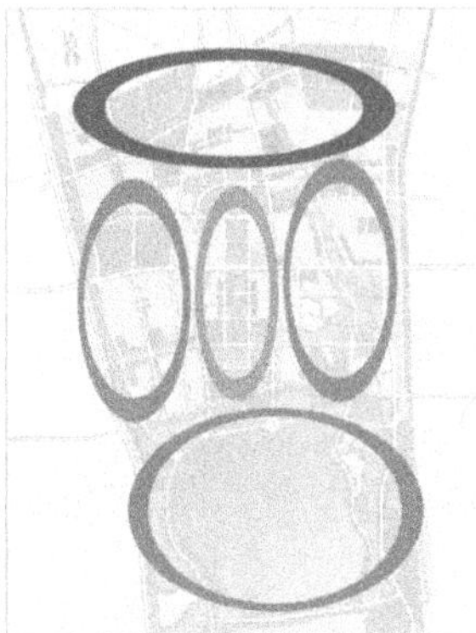

Figure 4. Layout structure

IV. SPATIAL INTEGRATION

According to the tour routes and the city main roads the layout is divided into five parts (Figure 4).

North Lake, including its surrounding areas, as a relatively complete scenic water features but without isolation, is connected with the Beijing-Hangzhou Grand Canal, the Guangfu River and the Old Canal by the rivers and the sight-seeing boats. The area between the "O-shaped" and "U-shaped" tour routes are the core of the region, carrying the functions such as travel services, administrative business, entertainment, cultural exhibition and residence. This area is the most important part with the vitality and development potential of the whole tourist resort. And we aim to create a regional tourism service center. Paralleling to the core area of east-west distribution lies two integrated tourism functional areas. The west one carries the functions such as the canal tourism, education and residence, and the east one serves as eco-scenic area based on the Old Canal and the Guangfu River.The northern part along the railway forms the urban industry functional areas on the basis of the existing enterprises through upgrading.

Each Area has its own planning of specific projects, which can drive the spatial development, according to the status quo, the functional requirements and distribution of the scenic areas (Figure 5).

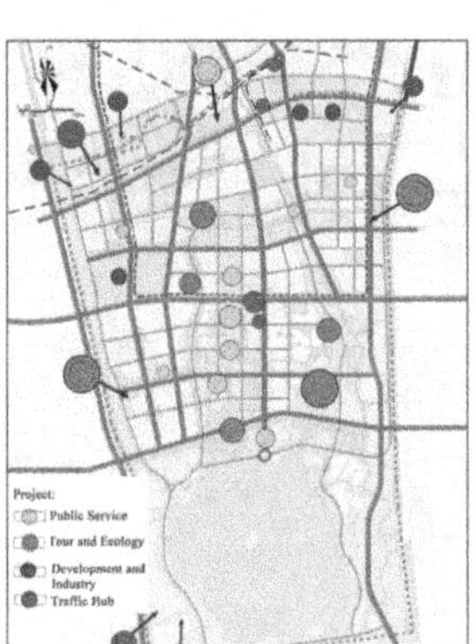

Figure 5. Projects that drive the spacial development.

V. DEVELOPMENT STRATEGY

A. *Star Project Initiation Strategy*

To construct landmarks and important tourist service facilities promotes the interaction between government and market. For example, Jining TV tower, located on the shore of North Lake, brought along the construction of starting area, coordinated with the core area, will become the focus of sightlines and important node of landscape. And administration building, as a model of development made by the government, simultaneously increases the value of the land and establishes the layout structure.

B. *Infrastructure-Oriented Strategy*

To take road and bridge construction as the dominant development pattern ensures the city started construction of high quality. For example, North Lake-middle road, which is a landscape avenue connecting the old district of Jining and the North Lake Tourist Resort, will greatly enhance the accessibility of the transportation and the aesthetic feeling of the landscape of the new tourist district. Chezhan-south road is an important road connecting the Jining Station, the new tourist district, the characteristic commercial street and the Old Canal landscape belt. And its construction will ensure the formation of Old Canal tourist axis.

C. *Market Demand-Oriented Strategy*

To invest in waterfront recreational facilities, marina, resort hotel, characteristic commerce and so on can induct the exploitation of the tourist resort, and become the "catalyst" of city development. For example, Lakeshore Hotel is not only a landmark of the region, but also a development flagpole of North Lake Tourist Resort. While providing high quality service, it perfects the logistical guarantee system. Waterfront Bar Street, develops characteristics tourism project, creates fashionable tourist services and enhances the grade of tourism services.

VI. CONCLUSION

In the past 20 years, the tourist resort construction developed rapidly in China, also leading to many problems, such as the protection of environment, the culture revival and the urban sustainable development. What should we do? Is the innovation of concept or the breakthrough of method? Is the excavation of connotation or the replacement of function? Or perhaps these are all perhaps. Actually, to master the idea of sustainable and to find the characteristic of the site are the right way.

REFERENCES

[1] Bojie Fu, Liding Chen, and Keming Ma, Landscape Ecology Principles and Application, Beijing China: Science Press, 2001.

[2] Jianguo Wu, Landscape Ecology: Pattern, Process, Scale and Hierarchy, Beijing China: Higher Education Press, 2007.

Preliminary Study of Energy Saving and Emission Reduction Strategy in Green Building

YuanYuhui
College of Art and Design
Wuhan University of Science and Engineering
Wuhan, China
Email: 105189368@qq.com

***Abstract*—Green building is an unavoidable tendency of the development of world architecture and the design of energy saving and emission reduction in green building plays an important role in realizing green building. This text sets about from the introduction of the concept and development of green building, and discusses the designs of building's energy saving and emission reduction, relatively generally focused on the way to realize the design of energy saving and emission reduction strategy in green building.**

Keywords-Green Building; Energy Saving; Emission Reduction; Renewable Energy Sources; Design

I. INTRODUCTION

Architecture is indispensable in human's life. Reviewing the history of human's architecture, from the initial shabby shelters to the architectural achievements today, a serious of severe troubles are in front of us, while we are enjoying the happiness and comfort brought by modern culture: a huge demand of architecture materials, the over exploitation of non-renewable resources, the air pollution caused by architecture activities, and so on, so the relationship between nature and human beings is in a precarious situation. Yet China is a country extremely lacking resources. Considering from energy supply and demand, ninety percent of the verified deposit of fossil energy resources in China is made up of coalmines, yet the per capita volume is only half of the world's average level, and the per capita deposit volumes of oil and natural gas in China are only 11 percent and 4.5 percent of the world's average level respectively. Considering the land and water resources, the per capita amount of arable land and water resource in China is only one-third and one-fourth of the world's average respectively. According to research, the energy consumption in architecture in China triples that in developed countries in Europe, but the whole urban ecological environment is much worse than that in European and American cities. The arduous fact has told us that the sustainable development and the development of green building are imperative for China.

II. THE RAISE OF THE CONCEPT OF GREEN BUILDING

Dating back to 1950s, there were foreign scholars raising the concept of human inhabitation architecture, combining the problem of human inhabitation, ecology and environmental sciences. In 1970s, with the emergence of worldwide resource crisis, the depletion of Earth's resource and the environment issue were becoming worse and worse day by day, so the green building related to energy saving started to draw attention, and there came a puff of "green" wind in architecture field. After 1980s, based on the fast development of the ecologically sustainable development theory, the concept of "green building" began to thrive in globe.

The concept of green building was firstly brought by American Italian architect Paul Soleri in 1960s, meaning the space providing people with healthy, comfortable, and safe living and working condition, at the same time realizing buildings with efficient use of resources and the smallest influence on the environment in the construction activities throughout the life cycle, and it's also called arcology. According to this, we can say that green building is a combination of architecture and environment, based on local natural ecological environment, using the basic principles of ecology and architecture, arranging appropriately and organizing the relationship between buildings and other factors. Fundamentally speaking, green design is not some kind of standard which can be defined simply, but a way of thinking about designing, and even a way of life. For architects and engineers today, green design means considering all the potential factors, like building, structure, high technology, material, and so on with brand-new concept, and studying new technology in a spirit of innovation, in order to make them harmonious with our environment. Now this concept has drawn a lot of attention, and has been put into practice in many countries, becoming a direction of world architecture.

III. THE ENERGY SAVING DESIGN OF GREEN BUILDING

Green building requests the efficient use of energy and the minimum effects on environment so that it can achieve the harmony of the three facors "human-building-nature" . Therefore, the energy saving design of building becomes the key of green building. Building energy saving means increasing energy utilization of the building, using the most limited resource and minimal energy consumption, and achieving the maximum economic and social effects. According to a rough estimate, 2/3~3/4 building energy consumption can be reduced by using thorough and efficient architectural technology. From the perspective of social and economic development, green building is the inexorable trend of the building development in the future, and the energy saving design of green building will also be highly valued.

A. *The site selection, planning and design of building*

The site selection, planning and design of building are the important parts of the architectural design as well as the key of realizing building energy saving. The improper site selection or unreasonable design will bring irredeemable losse in the running of the building in the later period. Therefor, the architect should start from the integrated design concept, and cooperate closely with the structural engineer, equipment engineer, energy analysts and environmental experts so that the site selection, planning and design can reach the demand of energy conservation and environmental protection.

First of all, the architect should take local climate, economy, culture, traffic, energy and any other factors into overall account and make a reasonable selection of the sites of building. Secondly, it is necessary for architects to adopt a reasonable design for the construction external environment, such as decoration of vegetation and water around the construction aiming at improving partial climate and avoiding heat island effect, in order to reduce the construction internal energy consumption. Finally, the architect should make reasonable construction shape design with the maximum use of natural light, natural ventilation, in order to reduce the architecture's dependence on equipment.

B. *Making full use of renewable energy*

The use of renewable energy plays a very important role in saving traditional energ in the energy saving design of green building. Nowadays, the energy crisis is paid widely attention to all around the world. The exploitation and utilization of unconventional, renewable and green energy such as solar energy, wind energy, geothermal energy, bioenergy become the development trend. Besides, it lives up to the concept of green architecture and the idea of sustainable development.

1) *Solar energy*

The sun is the most important resource provider in Mother Nature. Currently, the most widely used renewable green resource in green building is solar energy, and it's acknowledged by all to be the most potential renewable resource. In the design of energy saving in green residence building, we can make full use of the conveniences brought by the sun: for example, most of the lighting systems in green building can use solar energy, and the cooling system can be activated by the move of air, and the heating system can collect energy from human bodies and office equipments.

Solar energy is one of those early exploited energy saving technologies, and has been put into full application through the two technologies of active and passive forms. The passive sunshine greenhouse technology, which means that the sunlight filters in the glass warm house or the big glass window facing the south and one side of the transparent glass window absorbs the solar radiation, has been widely used. Currently, the most common active utilization of solar energy, like the solar water heater, solar thermal collector, and photoelectric cell, means using the electricity generated by solar panels. The solar thermal collector itself is waterproof and heatproof, similar to the function of roofs, so that we can use all or parts of the solar thermal collector to replace roofs, and thus cut cost. What's more, there is also solar absorption chiller, solar drying system, and so on.

2) *Wind energy*

Wind energy is one of the important renewable resources on the Earth. Wind energy is abundant, renewable, widely distributed, and non-polluting. Wind energy is, especially to coastal islands, very significant as a reliable way to solve the resource problem in producing and living. In the areas away from grids, the stand-alone power supply system becomes the most important power ressource for people there. Even in developed countries, wind energy, as a kind of efficient and clean energy, is receiving more and more attention.

Currently, wind energy is mainly used in the following field: wind power water-lifting, wind power generation, wind-aided sailing, and wind power heating, among which wind power is more and more becoming the main form of wind energy utilizations, draw great attention from all countries, and develops most quickly.

The main manifestation of green building using wind power generation is: in the ground floor, wind power generation units can be set in a relatively open place before construction starts, so that it won't undermine either the overall appearance or the use of wind energy, and can determine the capacity of the wind power generation units according to the number of users. Considering the fact that the wind energy resource increases as the height grows, it's feasible to install the wind power generators on the roofs. Among the buildings, many wind power generators can be set in the vertical direction in air channel, according to the distribution of wind field. That's why the use of wind energy in green building should be mixed with the idea of the new energy using technology at the beginning of designing, and be brought into the overall design of buildings.

Furthermore, the use of geothermal energy mainly includes: the high-heat geothermal used in power generation or directly in heating and hot water supply and the low-heat geothermal used by geothermal pump and the utilization of air through tunnel system. But in all, the scale of renewable resources is limited, its technology is not perfect, and the level of its commercialization is relatively low. It demands further study.

C. *To reduce building energy consumption*

As building's internal heat is circulated through the envelope the level of transferred heat is related to the surface area for heat transfer and sealing situation. In other same conditions, the building heating energy consumption increases with the volume coefficient proportionally. In order to save energy, we should control the shape factor reasonably. For the building envelope, we should improve the thermal insulation and sealing of the envelope. Overseas data indicates that improving the additional investment of building envelope thermal insulation properties can be compensated completely by the savings in energy costs. In regular conditions, the additional costs of improving the energy saving performance of building envelope accounted for only 3% to 8% of the total investment, with the energy saved up by 20% to 40%.

Wall is the main construction that protects the building. Single material for load-bearing or separation can't meet the high requirement of insulation. Therefore, composite walls have been developed rapidly. It not only solves the problem of interior and exterior thermal bridge, but also have the advantages of increasing indoor usable area and being convenient for interior redecoration. Now, the composite systems mainly used are molded or extruded polystyrene board, rock wool panel, glass cotton board, polyurethane foams and other heat insolating materials. Window's heat transfer ratio is much higner than exterior walls and protection structures. 35% of heat loss is due to the heat exchange between doors and windows and exterior heat. It's a great measure to reduce the heat consumption by using insulated glass, vacuum glass and low emission glass to replace single glass. Roof insulation technology mainly applies air-entraining concrete, and cement polystyrene board property, cement prelate, pumice sand or aerial roof insulation sometimes. The design of envelop enclosure is developing with the development of new materials. We can improve the building's function of saving energy by using thermal protection and insulation wall materials and comprehensive applied materials, efficient thermal radiation material, and new insulation and sealing materials.

IV. THE DESIGN OF GREEN ARCHITECTURE FOR POLLUTANTS EMISSION REDUCTION

The design of green architecture for pollutants emission reduction is mainly indicates that the buildings have minimum influence on environment, completely realize low pollution, low emission in their life period, and create healthy and comfortable living environment.

The rapid development of economy promotes the rapid development of urbanization and also brings sweeping changes for people's life. On one hand, the development of architecture technology and economy promote the comfort of people's life; on the other hand, they also have given rise to concern about people's health. With the rapid development of high-grade building and decoration and large-scale application of building decoration materials, the "invisible killer" has been quietly lurking around us. The sampling inspection of indoor air quality of office provided by experts indicates there are hazardous substances namely ammonia, formaldehyde, ozone etc. These substances can cause eyes, nose and throat dry, irritated and general weakness, fatigue and discomfort, nervous headache, memory loss and other symptoms, in common sense namely " volatile harmful substances allergy". More and more robust people fall ill in the office, and more and more odd symptoms appear in hospital.

Traditional building material industry is a major basic industry in national economy, which is the one of industries with the highest natural resources and energy consumption, the largest environmental damage on production and the most serious pollution of the atmosphere. And with the development of social economy, emands of people for materials are increasing as well as for its health and security. "Green Architecture Materials" was firstly raised in International Symposium on Material 1988, then numbers of countries have developed performance standards on "Green Architecture Materials". Therefore, people began to research "sustainable development materials" with high-performance and environmental protection, namely green materials.

The internationally recognized green materials have minimal impact on environment and are beneficial to people's subsistence and health in the processing of raw material selection, production, construction, use, waste disposal etc. Compared with traditional building materials, green building materials contained in green materials should have the following characteristics: save energy and resources, less use or no use natural resources, the large-scale use of industrial waste, no radiation and no pollutant. They completely embody their "health, environment protection and safety" in the using process.

Nowadays, people pay more attention to "Green, Environmental protection", and also attach great importance to the use of green materials, so that increasing green materials are applied to our architecture and interior design. Meanwhile, the encouragement of native materials and plants selection is indispensable in the using of green materials, because the management and maintenance costs of the local species is minimum and also regional, such as in the south areas with lush vegetation, we can widely use renewable resources like bamboo as raw material. Act according to circumstances, so that we can design regional architecture with local characteristics which carry the regional culture.

V. CONCLUSION

In contemporary society, green building is becoming more and more welcome among people, and the development of green building is the only way of world architecture development. In terms of energy, our country is big, but our architecture is an industry which is hign in energy consumption, material consumption and pollution. To develop green building in our country is the only way to realize a harmonious society, agreeing with the direction of the Scientific Outlook on Development guidance, having great prospect. Meanwhile, the energy saving and emission reduction of green building is the key to realizing green building, and it's a highly integrated subject combing many subjects crossed. Therefore, we should highlight the collaboration among different subjects and fields, design and build more green buildings, and make contribution to the sustainable development of society.

REFERENCES

[1] Collected Works on Green Buiding Technologies, China Academy of Building Research,China Architecture and Building Press 2005.

[2] Lin Haiyan, Building's Energy Saving in Sustainable Architecture Development, Architecture Science 2006（.9）

[3] Green Building.Research Centre, Xi'an University of Architecture and Technology,China Planning Press 1999

[4] Ecological strategy of Architecture Design, School of Architecture, Tsinghua University, China Planning Press 2001（9）

[5] Chen Zhipeng, Basic Understanding of Green Building, Architecture of Shanxi 2004

Product Innovation Based on Universal Design

Jie Zhu
Institute of Urban Design
Wuhan University
Wuhan ,China
Email: 8826707@qq.com

Qianghua Duan
Wuhan Office
Earthasia Design Group
Wuhan ,China
Email: danwood @163.com

Abstract—**With the continuous development of China's economy, improving of people's lives, Chinese society has entered the long-age and declining birth rate society. Economic prosperity let people's demand for product diversification. How do Chinese products to be recognized in the international market and to meet domestic demand in China's long-age society? In this paper, the concept of product innovation based on universal design, it broke the previous design patterns, proposed a new design concept, product is no longer just the pursuit of high-tech content and high economic returns but to improve the functionality of products, artistic and cultural, so that design becomes friendly, designed for consumer services.**

Keywords- product; innovation; universal design

I. INTRODUCTION

21st century is the pluralistic period on the design, it can be regarded as "the era of prosperity design", "humanization" design innovation among them becomes the light spot, and form a kind of irrevocable trend. "Universal Design'' is very good explanation of the "Humanization Design'' in the product.

Innovative products of traditional technology-based manufacturing model use the new scientific and technological achievements throughout the entire product manufacturing, they will transfer scientific results into technology manufacturing, science and technology is the winning weapon of product. However, these products often ignored the real needs of people. High-tech technology products is not necessarily a good product.

At this time people began to doubt whether the technology is real needed? Through numerous examples of the failure of high-tech products we can see that the answer is no, Let the Hi-tech products into people's real needs, enterprises must have design innovation. Now, with the product segmentation and market competition intensifies, traditional products of technology innovation turns into design Innovation, design plays an important role in entire cycle of the production.

II. THE TREND OF USER-FRIENDLY DESIGNING

With the development of more user-friendly design, "Humanization" design behaves as follows:

Firstly, pursue the interest and recreational of the products. The product has not only satisfied people's basic needs but also met modern's pursued easily, humorous, joyful psychological need. Some designer already put design feeler stretch to people in the abyss of the mind, through in metaphor and aesthetics design, let users appreciate thoroughly and feel more warmness.

Secondly, Pursue the model form suitable for the organization of human body. The designer still carries on research and discussion on how to make the products individualize and suitable for the human size ,even more, they hope the products can become a part of our body like clothes and jewel, let the design things to integrate in user.

Thirdly, The designer's humanization mind. The designer has also found a channel of communication with consumers, through listening to consumer's opinion, feedback consumer's information to reach coordination of design things and user, thus design the products making consumers more satisfied. Emphasize the communication, contacts and exchanges of both sides of designer and consumer, design the trend but not regarding designer and consumers as Opposite both ends. These have given the profound side of the humanized in fact. The designer's mind turns to the humanized level progressively [1].

Fourthly, Design brings enthusiasm into disabled person's articles. The designer concern with human nature has already been expanded to the field of the disabled person, become one side that the human industry design has humanitarian and human touch most.

Professor Guanzhong Liu said: Design is aimed to meet people's needs, and to make people's lives better [2]. Pay attention to humanized design is the noble ideals industrial design to pursue. Universal products, that is for human beings to create a more comfortable and better life and the environment. it's a new innovative ideas of contemporary made-in-China to the world.

III. THE BACKGROUND OF UNIVERSAL DESIGN

The rise of the concept of universal design originated from the 1950s, U.S. Pastor Martin Luther King(1929—1968) led the black civil rights movement. The United States has experienced two world wars, since the late 1940s, large-scale outbreak of poliomyelitis coupled with the subsequent Korean War, makes the obstacles faced by American society, veterans and persons with disabilities to return to social work, for the most people doing the design for these people reveal inconvenient.

Dr. Martin vocal advocate equality of all mankind, not only for blacks fight for the rights and dignity, but also encourage the society to address the issue of person with disabilities, greatly influenced the U.S. government after the formulation of social welfare law. These laws prohibit against discrimination persons with disabilities, and also to provide them with easy to use education, public housing, communications and transport facilities, to reduce obstacles in their living environment, Barrier-Free Design is also resulting.

In 1961 American Standard Association "Convenience of the disabled and the use of the building close to the standard" become the world's first "barrier-free standard". Barrier-free design at this point is only tend to focus on the body of people with severe disabilities, especially with regard to functional impairment of action of environmental design. Advances in medical technology and equipment make average human life expectancy increased, the advent of an aging society, and many other factors, making more and more emphasis on barrier-free design, The design process to become one of the factors that must be considered[3].

Initially, these designs tend to medicalization， that is mainly from the medical point of view to provide products for the disabled and the environment, the design is now more user-friendly. Through the creation of new products and new environment to make it more acceptable for the disabled, take them into the daily lives of ordinary people. To reduce the social stigma experienced by disabled. disabled symbol design started in the late 1960s international , and developed in the mid-80s, both services for health and the disabled persons in Universal Design.

Universal design rises in Japan, because Japan is one of the biggest long-age countries of the world. Every enterprise channel into the universal design concept, the market put out various types of products to high age. The intent of universal design is to simplify life for everyone by making products, communications, and the building environment more usable by as many people as possible at little or no extra cost. Universal design benefits people of all ages and abilities.

IV. UNIVERSAL DESIGN CONCEPT

Universal design concept one： find the point of maximum tolerance. The design is not to find average, but the point of maximum tolerance, the universal design minds design for all, pursuing the equal point of the society, enable everyone to use fairly. And permit to use by all kinds of methods, even the use by wrong way will not bring inconvenience either, and can lighten consumer's onus in use, that minds totally develop the products with consumer's demand and position.

Universal design concept tow：handy and esthetic. As society comes to high age, the level of universal design can be accepted raise constantly, but there have most different between barrier-free design for high ages and universal design. barrier-free design is for obstacle person or high age person, universal design is correspond to various user's standards, designer can consider different physiology and psychological need of user, let users have greater freedom, the good's handy that should pay more attention besides beautiful appearance. The universal design pursues both handy and esthetic.

Universal design concept three： derive new market. Universal design can create the good economic benefits also. The user’ s sound has not been noticed before, when the universal design concept incorporate into the products, consumers can realized potential demand, and through consumer's use, universal design is not only become a new prevalence but also achieve large economic benefit in potential market. For example, the famous design of ball-point pen "Hand Birdy" Fig. 1, Not only in 1999 was Japan's T-Mark Design Award, and is selling goods on the market until today[4].

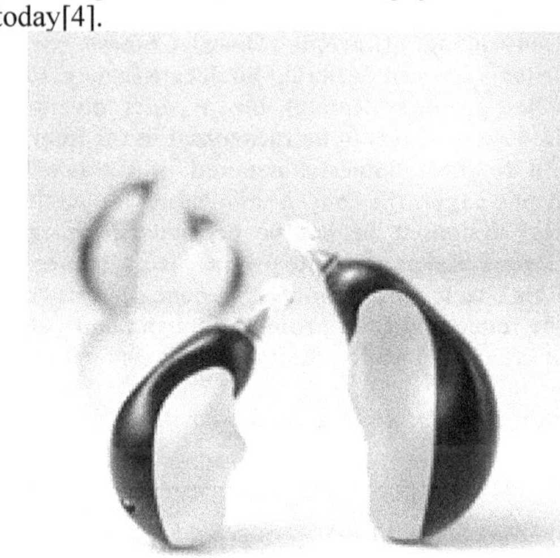

Figure 1. The famous design of ball-point pen "Hand Birdy"

V. THE DESIGN METHOD OF UNIVERSAL PRODUCTS

A. *Universal Design of product function*

First, universal design of product function must meet the man-machine engineering design method. As long as people use products, they should consider man-machine engineering, make the product more in line with people's movements and the using habits. In addition, under the premise of meeting the human body comfort, functions of the product design should also consider the use of different populations habits and special requirements, changes of using place and time, ect.

During the early stages of product design, research the product users is needed, at the same time, should conduct an investigation in special population's way of life to find the inconvenience caused by existing products and their special requirements . These requirements are not only physiological needs but also psychological needs. There are two kinds of simple design method of universal products through the investigation and analysis. One is the improvement of design based on the design originally.

For example, "spoon cutlery" design, it slightly changes the curvature of the traditional spoon handle, make the small spoon with a variety of grip ways. This change will meet different people's dining habits, including some hand disabilities or weaker hand the elderly, children, etc Fig. 2.

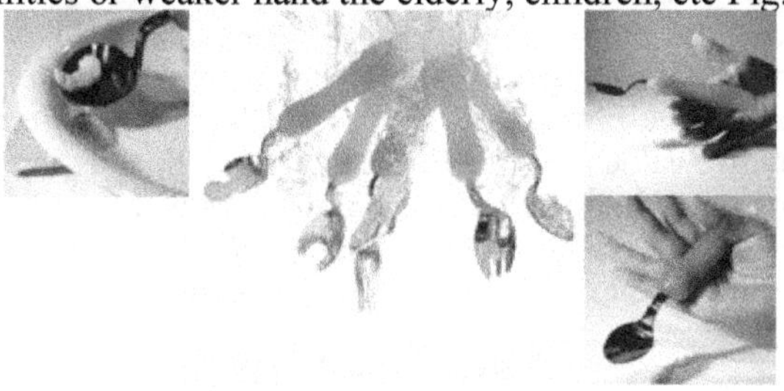

Figure 2. "Spoon Cutlery" design

Another is the invention of new function products for special groups or purpose, such as "handy wormy" design. People go to the supermarket is always with heavy shopping bags which are inconvenient, especially hands of persons with disabilities and elderly, designer designed a small handle to solve their troubles. This small handle used principles of mechanics to make shoppers save a lot of effort, and designed a number of hooks to lift up several shopping bags at one time Fig. 3.

The idea was inspired from the needs of the consumer, designers listened to their opinions and needs. "Problem solving" as the fundamental starting point to design products, products for the use of maximize crowd , ultimately consistent with the concept of universal design.

Figure 3. "Handy Wormy" design

B. *Universal Design of product Color*

Color is an important aspect of product design, color universal design is about the visual perception of color application. color universal design is taking into account the visually impaired, the needs of those suffering from color amblyopic or color blindness during the design process, improving the environment through product color design , allow varying degrees of color vision capable people easy identification and use product.

Today, about 7% to 10% of males and 1% of the female population are color-blind or color amblyopic. In addition, under certain environmental conditions, such as into the movie theater, at the fire scene, etc, even the normal visual users may also be temporary visual impairment, therefore, designers need to know how to avoid the use of ambiguous color as an important medium of communication consulting, inconvenience of the user or the worst result put the user into a high-risk situations. Designers have to be active in product design using knowledge of universal color, improving product design services for the majority of people.

World designer carry on different attempts to the concept of universal color in product design, Japan also has a color universal design organization, referred to as CUDO, specializes in the design of a universal color. June 2008 Olympus designed μ 1020 and μ 850SW Cameras and interface design of li-ion battery charger all took into account of the issue of universal color. Consumers are not sensitive to color (suffering from color amblyopic or colorblind) in the past, when using the user interface is not easy to see the pink fonts and images on the silver bottom. Therefore, the new camera interface, the icon will be speaking into a blue and dark red, brown black. Furthermore, the access indicator lights of memory card and chargers both with orange, to avoid the problem is that red-green blindness consumer can not distinguish use Fig. 4.

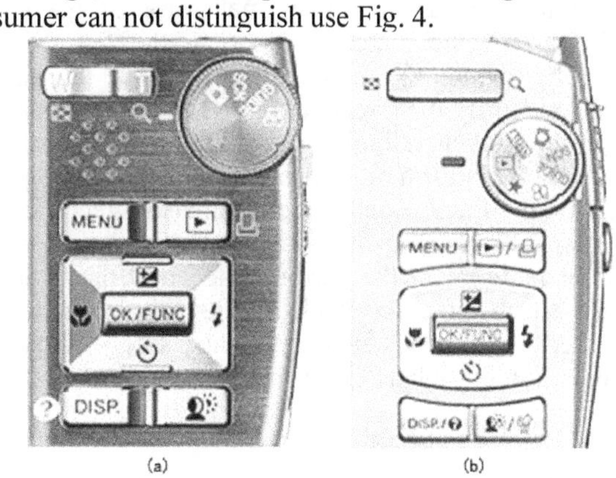

Figure 4. Olympus Cameras: (a) This is the old Camera Image. (b) This camera is to use the universal color design

C. *Universal Design of Public environmental product*

Universal design has been get people's attention in product design, it should be promoted more in public environmental product. Public environmental product is a public or street in the community, a variety of utility services system activities for people to certain conditions or quality assurance, and the corresponding identification system.

Public environmental product with multiple functions integrated service system. Members of vulnerable groups as part of the community, it should be like normal access to social participation and communication power, environmental product offered to the public for use, it is an integration of universal design in the design, that disadvantaged groups and ordinary people can participate in social activities, feelings of collective joy.

In the design process, first of all should fully investigate and understand the disabled, the elderly, women, children and other vulnerable groups in the public space in the behavior patterns, walking routes and special needs. Universal design should maintain the artificial environment of harmony with the natural environment, universal design should avoid or reduce the weakening of the natural character of open space, according to the size of open space, offers a variety of options as far as possible. Different accessibility of the environment should meet the needs of people with different abilities, emphasizing interoperability, to avoid "special place".

Environmental product of universal design should pay special attention to safety, because public space is a fusion of places, human, social, natural blending, for the vulnerable groups, there are many safety factors of instability.

For example, in some waterfront design, public product, we should pay special attention to their safety and security. Waterfront facilities prominent people and water in close contacts and exchanges. Contact with water for certain disabled or visually impaired people is unthinkable, but as long as do the safe handling of the details, can allows people with disabilities and ordinary people to enjoy water.

In the design of waterfront facilities, protective barrier often designed to seat or platform, where people can enjoy the water features. Design may make use of glass as a blocking part of the material, bump in the fence on the design of the texture. On the one hand facilitate the visually impaired to identify the direction, on the other hand also play a role in to prevent slipping. In addition they are often designed wood platform or wooden plank road, where people can walk on the Plank Road, people can also directly contact with water or play in the water. Design, the use of transparent material in the surrounding good protection, to install the touch direction of identifying the instructions on the armrest. So that everyone can safely play and communication in such an environment.

VI. Results and Discussion

Design has an important role in product, but it should be pointed out that the core values of technology issues, "Technology" is the core in design aesthetics, and it's the soul of the product. While we stress the importance of design, technology can't be ignored. Advanced technological achievements as an important means of design, it can plug the wings to the product. Successful product should be combination of perfect design and technology. How to combination the two things will be an important subject in contemporary product design. Technology is a rational, the design composition of the art is the sentimental reflection. Product has a rational core technology and psychological feeling of emotional appearance that closely related to people, and it's the comprehensive interdisciplinary, is the crystallization of the wisdom of all humanity.

Design concept was often with the designer's own intention or idea, in the relation between designer, products, user, and the user's demand is always neglected. "Universal design'' is a very good explanation of the "Humanization design'' in the product. It has broken the past design mode, can show value by "use", design should satisfied most persons need, but not fewer, should think over the "value of designed", but not consider by "sell well all over the world", the starting point of designing concept should be "handy".

VII. Conclusion

In the production of 21st century, design is not only to create new styles, but also to create wealth. Through the actual situation we have found that the design of the huge market space. At present, China strongly advocated "creative industries" as a representative, "creative" to create wealth and life-changing for a better world.

Universal design can be developed into a new business strategy for market competition, it is not only the design concept that concern for disadvantaged groups, but also could become a kind of marketing strategy for market expanding , time and intelligence resources and material costs saving, effectively enhance the competitiveness of products . Not only reflect the progress of a social and spiritual civilization, increasing corporate profits and enhance overall competitiveness, but also has considerable development potential. China is gradually moving into an long-age society and the globalization, we have to face the different peoples, cultures, customs and lifestyle of the ethnic group of users. Differences in characteristics of the user it's severe test that domestic enterprises have to face. Therefore, to vigorously promote product innovation based on universal design will become an important boost of global competitive edge of domestic enterprises.

Acknowledgment

The authors would like to thank colleagues of department of Urban Design, Wuhan University, each heated discussions gave me intellectual inspiration. Thank Wuhan University of Library and urban design the library to borrow books from the teacher for each patient to help, Professor Zheng Yang also kindly reviewed the best release and made many suggestions that improed the final version.

References

[1] Xu Wang, Emotional design in product, Art and Design, 2005, vol. 20, pp. 32-56.

[2] *Guanzhong Liu,Outline of science matters, CS:Central South University Press,2006, pp. 15.*

[3] Qun Huang, Universal design and Barrier-free design, BJ: mechanical industry press, 2000, pp. 6.

[4] Shengxi Fan, Behavioral and cognitive design/design of the humanized, beijing:china electric power press,2009.

Product Optimization Design Based On User Needs

Han Ji An
College of the Arts
Jiang Xi University Of Finance and Economic
Nan Chang, China
Email: hanja1021@126.com

***Abstract*—User needs is a source of product design activities, and meet user needs is the ultimate goal of product design. In order to design products that can meet user needs, there need to establish the design patterns which is product optimization design based on user needs. In the text, the product optimization design is elaborated, and it proposes to analyze the three elements of the design variables, objective function and design constraints of the model. Only by establishing a suitable objective function, selecting design variables and design constraints reasonably can control the complexity of the product optimization design at the appropriate level, and obtain the best design that meets the user needs.**

Keywords-user needs; optimization; design; product; model

I. Introduction

The ultimate goal of product design is satisfying people's needs, people's needs are different at different times. The early industrial revolution, the main criteria for user requirements is a good realization of the product function; With the continuous deepening of the volume of large-scale production, the products can only be used no longer meet the needs of social development, and the better use of products become the major consideration when users select products; With plenty of material products and the continuing rapid development of information technology, human consumption has entered an era of sensitivity, diverse needs is the main features of that stage. The diversity of needs leads to more factors of the product design should be considered, in order to achieve design optimization, we need to adopt the concept of optimal design to create the products which is the most suitable for user needs.

II. User Needs and Product Design

The ultimate goal of product design is satisfying people's needs, people's needs are different at different times. The early industrial revolution, the main criteria for user requirements is a good realization of the product function; With the continuous deepening of the volume of large-scale production, the products can only be used no longer meet the needs of social development, and the better use of products become the major consideration when users select products; With plenty of material products and the continuing rapid development of information technology, human consumption has entered an era of sensitivity, diverse needs is the main features of that stage. The diversity of needs leads to more factors of the product design should be considered, in order to achieve design optimization, we need to adopt the concept of optimal design to create the products which is the most suitable for user needs.

A. The internal and external factors affecting user needs

1) The external factors affecting user needs

Objective factors that affect the user needs are social, economic and technical factors, which does not change with a user's will. Social factors are the trend of social development which often result in the emergence of a new way of life, it makes existing products can't meet the needs of users so that need to develop new products; The economic factors are the purchasing power of the users, the products which originally elusive will enter into the consumers' life as the users' income changes, the popularity of computers is just a typical example; Technological factors are the main driving force of the products' upgrades and the new products' appearance, it will lead to the escalation of user needs.

2) The internal factors affecting user needs

The subjective factors that affect the user needs are the requirements which are led by the needs of user's own physical and psychological. These subjective factors include the following aspects: motivation, operation, situation, experience and expectation. Motivation is a inner agent to inspire and sustain people's actions, and it will make a target-oriented action in order to meet certain needs; When the user has motives, he would consider how to operate in order to achieve its objectives, which is the operational level of demand; Situation is the scene when the user uses the products, it requires careful analysis because the situation will affect the user's motive, action and expectation; User Experience, also known as "user habits", is usually very strong, there are some new products are fail just because that we don't do any sufficient research on user's experience and habits; he user expectation means the users' hope for the products. It will greatly increase the value of the products to meet the expectation of the users.

Under the action of internal and external factors which affect user, the user demand for the product is mainly reflected in the product's safety, comfort, aesthetics, accessibility, functionality, economy and product's shape, color, material, size, etc.

B. User needs and product design

Product design is a course which makes a certain purpose or need of human convert to a specific physical form or tool, to make the way of planning assumptions, solving problem expressed as a nice form by a specific carrier. It can be seen, the user needs is the beginning and the end of product design activities, that is to say user requirements is the root of leading product design, product design is the excavation and analysis of user needs, and eventually in the form of materialization to meet their needs. The analysis of user needs should be started with the internal and external factors that affect theirs' demand to get a clear demand and through organize to form product's elements (shapes, colors, materials, etc.) to provide products that meet customer needs.

III. PRODUCT OPTIMIZATION DESIGN

A. *Optimization design*

Optimum design is a new discipline emerging in the last century 60's. It is the most optimization techniques and computer technology applied in the design. The so-called Optimization design is the process of searching design optimal solution to achieve optimal design goals. Optimization design is to change the way which solely relied on human intuition, experience and constantly tests to achieve the best design goals in the past, because the method is often subject to the experience, time, environment, etc. So it is difficult to get the best results. Optimization design provides an important scientific design method for the modern product design and provides a basis for product design to find the best design plan in a variety of programs.

B. *Product Optimization design*

Product Optimization design is a method which combined optimization design concepts with design practice, to optimize the various factors which product design is involved in order to achieve the best product design goals. The best design programs of product design is the most satisfactory and most appropriate programs which is obtained by meeting a variety of design goals and some constraint conditions. The best design program of product design has relativity; its standards will change with the change of social development and design conditions. Product Optimization design is the process which play designer's initiative to obtain optimal design under certain constraints. Therefore, to achieve the goals of product optimization design must have two conditions: first, there is one goal of optimization design; second, there is multiple designs. In short, the product optimization design consists of two aspects, one is to build the product optimization design model; second is to solve the problem of product optimization design. To make the product design involved factors became the abstract optimization design problems and establish the design models that meet realistic requirements is the key. Only by establishing a reasonable model can we get a reasonable optimal program.

C. *Product Optimization design model*

Product optimization design model is taking a series of the optimization design of constraints condition which is the factors involved in product design as the design parameters, in order to provide the basis for the design specifications of optimal manner. Therefore, the product optimization design model consists of design variables, objective function and design constraints, that is the three elements of optimization product design. Design variable is the factors which will change in product design, these factors can be intrinsic factors (the user's expectations, motives, etc.), and external factors (social, economic and technical aspects), such as safety, comfort, aesthetics, accessibility, functionality, also the constituent elements (shapes, colors, materials, cost, etc.)of products ; The objective function is the goal of product design, it can be expressed as a function of design variables. The objective function is the criterion function to evaluate the performance of product optimization design; Design constraint is the choice of the best solution which is obtained under some conditions, these additional conditions are called constraints of product optimization design.

The optimizations of product design problems can be expressed as a set of priority selection parameters, making its design specifications which can meet a series of restrictions on the parameters achieve the optimal value. Product optimization design problem can be described as:

Demand variables (security x_1, comfort x_2, aesthetics x_3, accessibility x_4 ...);

To make the function optimization, $f(x_1, x_2, ... x_n)$, $X = [x_1, x_2, x_3]^T$ (The variables of product design involved are mutually independent);

Satisfy the constraint conditions:

$$g_u\ (x_1, x_2\ ...\ x_n) \leq 0 \qquad (1)$$

(u=1, 2...m) (Material properties, shape processing, color ...)

$$f_v\ (x_1, x_2\ ...\ x_n) = 0 \qquad (2)$$

(v=1, 2...p) (Production costs, technical capacity ...)

Therefore, product design issues should be described mathematically to adapt to using optimal design method for solving, this is the so-called mathematical model of product optimization design.

IV. PRODUCT OPTIMIZATION DESIGN BASED ON USER NEEDS

User needs is the basis of product design, no demand is no design, product design is the purpose of creative activity for meeting user needs, namely, product design is an important way to achieve user demand. It is the dialectical unity between the user requirements and product design, product design is meaningful which must be based on user needs, that is, Product optimization design based on User needs is one of the main ideas of modern design activities. The so-called Product optimization design based on User needs is that take the various factors which influence user need as variable (product safety, comfort, aesthetics, accessibility, functionality, economy and product shape, color, material, size, etc.) to get the best design. The specific design process as follows:

1) Analysis of design objects

Making a comprehensive and detailed analysis of the object of optimization design before product optimization design and explicit the requirements of optimization design to reasonably determine the objectives and scope of optimization design, in order to effectively enable the design problem can be resolved by optimization design. In the specific analysis, we must distinguish between primary and secondary issues so as to avoid affecting the purpose of optimization design because of overly complex models. Only according to the *orderliness* of optimization design and combination the features of product design, can design objects be adapted to use optimization design.

2) Determine the design variable

The factor which will change in the Product optimization design is the design variables, it will directly affect the design results and design specifications. As many factors will affect the design, in order to carry out optimization design effectively, when selecting the design variables , it is necessary to consider the following aspects: First, combination the nature of design objects to select the element which has a direct impact and can fully reflect the required of optimization problems; Second, choice the number of design variables reasonably, for any product design, the more variables , the difficulty of optimization will be increased, the fewer then the design freedom will be limited ; Third, the design variables should be independent each other, so that you can ensure that a design variable changed will not lead to all variables change in the process of optimization design.

3) To determine the design constraints

During doing product design, some constraints should be met, such as the size of the products can ensure that their chips can be installed and comply with the materials processing accuracy of choice, etc. These design guidelines is the basis of the product, in order to achieve the purpose of optimization, we should determine the constraints reasonably.

4) To establish objective function is the core of optimization product design

To create the objective function must select the target optimization first.In product design, common optimization index include that the safest, most comfortable, lightest, most compact, most accurate, the most affordable, most beautiful, most scientific, most sensual and so on. According to the most significant indicators that affect design requirements to establish the objective function.

5) Select the appropriate optimization method

After establishing the model of product optimization design, you need to select the appropriate optimization method, there are some common optimization methods, such as intuitive optimization method, experiments and explore optimization method, evolutionary optimization method and mathematical optimization method.

6) Optimization results analysis

After optimization, a comprehensive analysis of the results of solving is needed to confirm that whether it accord with the design requirements of originally envisaged or not, and select the satisfactory solution from reality. Sometimes, the results of optimization design may not be feasible, so we need to amend and adjust the optimization design variables and objective function until obtaining the satisfactory results.

V. CONCLUSION

Product design is the process of user needs materialization, so as to create products that meet user needs, that is product design derived from user needs, but also ends up with the user requirements. Product optimization design is a design methods that is fully consider the various factors which affect the design to obtain the optimal solution. The purpose of design optimization is to create the most satisfactory and suitable products for customers. Therefore, only product optimization design based on user needs can create end-products that meet the user needs.

REFERENCES

[1] Cagan, Vogel forward, "The creation of breakthrough products - from product strategy to the project finalized the innovation".Bei Jing: China Machine Press, January 2006.

[2] Alexandre Dolgui , Jerzy Soldek , Oleg Zaikin, Supply chain optimisation,"product/process design, facility location and flow control", Berlin: Springer,December 2004.

[3] Utah Rizzi, Eppinger book, De-Lin Yang translate and annotate, "Product Design and Development",4rd ed.,Ji Nan:Northeastern University Press Finance, November 2008.

[4] Bang-Chun Wen, Qing-Kai Han, Hong-Liang Yao, "The structural performance of products and dynamics of optimal design: the theory and methods of integrated product design based on engineering",Beijing: Mechanical Industry Press, October 2008.

[5] Donald A Norman, "The design of future products",Bei Jing:Electronic Industry Publishing House,June 2009.

[6] Kahn, K.B, "The PDMA Handbook of New Product Development", Second Edition,2rd ed., Bei Jing:Electronic Industry Publishing House,September 2007.

Renaissance and Inheritance of Regional Culture in Design on Style and Feature of Small Town

A Case Study of Bapu Yi Minority Town in Liangshan Region

Cheng Bin
Civil engineering and Architechture institute
Southwest University of Science and Technology
MianYang, China
Email: chengbin1971@126.com

Luo Neng
Civil engineering and Architechture institute
Southwest University of Science and Technology
MianYang, China
Email: superluoxiao@hotmail.com

***Abstract*—Design on style and feature of small town, is importaut approach to mold town image, embody town personality, and form town feature. Regional architectural culture is not only key element of design but also soul of design.The paper proposed that the cognition, refinement and expression are the important links to revitalize and inherit regional culture in design ,and thought that forming space, choosing scale and public participation is the important method to achieve this goal. With the Liangshan Bapu Yi minority town planning case,explained the specific application of this view.**

Keywords- Yi minority;Style and Feature of Town;Regional Culture;Cultural Renaissance and Inheritance;Bapu

I. INTRODUCTION AND REGIONAL CULTURE

Style and feature of small town is integrated with customs,culture,traditions, topography and architectural form,is reflected in the temperament of the performance relying on the material form of the town.The design on style and feature belongs to comprehensive urban design from its object,scope,and procedure,is an important approach to mold town image,embody town personality,and form town feature.The cognition, the refinement and the expression to regional culture is the most important link in the Process of design on style and feature,which refined cultural connotations.Regional culture is not only the important design element of the design style, but also the soul of the design style.The outstanding design always is the deduction,,renaissance and inheritance to regional culture,On the contrary,which lack of regional culture is pale and boring.The unique style and feature of small towns all have a very profound social and cultural connotations, it truly reflects the lifestyle of the residents,behavioral psychology and values.

The region culture refers to local material culture, system culture and energetic culture,Specifically,including the natural environment,climate,geology, topography,ethnic customs, history and culture,religious beliefs,as well as social needs and so on.To small towns in minority areas,the regional culture is not only reflected in the national ethnic nature of the era,but also reflects the level of development of productive forces at all times, life characteristics, form a unique architectural language symbols.The small towns,In the process of its formation,growth and development,experienced the baptism of history,the traditional symbols,such as Historical block,traditional festivals,architectural decoration and cultural heritage,become symbol of the spirit of regional culture,but also an important element for the designer to make use of.

II. THE WAY TO REVITALIZE AND INHERIT REGIONAL CULTURE

Although different cultural characteristics have different requirements in design on style and feature,the design methods and programs are the same.In general,there are two ways to revitalize and inherit regional culture,one is through designing the building and creating urban space to achieve,and second,through developping specific design guidance framework to achieve.The regional culture is not only abstract,is also the embodiment, manifests the integrity and the unity of regional culture, must closely hold the three links:cognition, refinement,and expression.At the same time,there must have the architects who understanded regional culture,combines the study of regional culture and architectural studies,and expressed in architectural language.

A. The Cognition of Regional Culture

Beyond the urban complex external performance,looking for the hidden order and the nature is the designer's an important work[1].We believe that the topography-related natural characteristics is external physical form of regional culture,while the residents way of life is the essence of regional culture.During the investigation stage,we can easily collect to the towns in all aspects of the physical elements,but the collection of cultural elements is not an easy task.To master the designs information,the first task is to understand surrounding geographic features,through thoroughly analyzes carried on the terrain landform features,definite physical character;Next is to come the comprehensive cognition regional culture by studing the regional life style, including architectural feature, people's life style,cultural pursue,cultural personal status, folk craft,traditional art and so on.

B. The Refinement of Regional Culture

By means of behavioral study, the geographical features and abstract way of life can be categorized, summarized and integration.They are both the protection and inheritance of regional culture, but also is the respect to geographical customary way of life and the guidance to modern new lifestyle; more is the concerns to social demand and crowd ecological environment.Through abstraction to the concept, the lifestyle can be summarized into the corresponding physical environment model of the space,and even regional culture can be humanized,urban morphology can be symbolized.This environment is a comprehensive environmental Characteristics,including all the elements of regional culture.

C. *The Expression of Regional Culture*

The expression of regional culture,including space, image,composition three aspects.The expression of space refers to how create a identity sense and belonging sense to the towns,by using the surrounding environment elements including pography, geology, vegetation,water and other materia elements,and the variety of urban spatial form.for example,many traditional buildings of Sichuan Province characterized by overlap long slope roof,with the moving out of place constitution way,combined with traditional practices such as Diaojiaolou,ChuanDou, EAVES to create a lively dynamic.The image expression is refers to the use of architectural linguistics methods, such as narrative,metaphor,metonymy,translation,semantic reset and so on,by means of behavioral,sociological approach,study the structural relationship under the times conditions,expresses the traditional pattern according to the new relations with the modern principle.The constitution expresses refer to one kind way using a detailed structural design to express regional culture. In this way,due to the use of the flexibility that can reproduce the subtle signs of regional culture.

III. The Method to Revitalize and Inherit RegionalCulture

A. *Scale Controlling*

Small towns have a macro-and micro-scale dual properties.At the macro scale,it focuses on the overall organization of space and function; on the micro scale,were concerned about a certain part of the public space,street landscaping and building forms,and even details such as curbs and street lighting.But regardless of from macroscopic and microscopic angle,when design,must analyzes from person's behavior custom and person's psychological feeling,these two kind of scale should be unified,and has manifested person's scale[2].This kind of scale relations are established between the person and the cities,the person and the building, the person and the spatial scale relations,as well as in small town between construction entity,spatial, entity and spatial one kind of harmonious criterion relations.

B. *Spatial shape building*

The spatial shape design is key point of the style design,as the expression of regional culture are mostly in the form of a space to express.Designer's main design measure includes that protecting the small cities space outline perfect,carryying on the commanding point the design, completeing the sight plane and the sight section plane control;based on this constructs the new landscape positively,provides the good landscape spot,the landscape line,the landscape surface. At the same time,must unify the landscape characteristic,the layout easy to distinguish and the sensation,the rich characteristic each kind of nature shape space,forms logical and the rich individuality overall spatial shape system.The spatial shape's creation needs to utilize each kind of landscape essential factor,forms variabilis and flavor spatial sequence,in satisfies the function request under the premise,created as many as varied activities for people to provide various content areas,created pleasant environment,including the square, green space,or walking the streets or interesting leisure space, take opens the space as the carrier to revitalize and inherit unique local conditions and social customs.

C. *Public Participation*

The design on style and feature will have one kind of social practice characteristic,the public participation will be an unevadable important question,which emerges from the 1960s in the Western country,and has manifested the multiplex value choice.The public participation taked the Community organizations as represents can maximum limit guarantee own benefit.At present,an increasing number of planning and designing public notice system was introduced,which to some extent, increase the public's participation.Of the many elements in design on style and feature,which can best reflect regional culture is the residents lifestyle,so residents participation is very important, and it is necessary.

IV. Case Study - Bapu Yi National Minority Town Planning

A. *Background and Objectives of the Planning*

Bapu town is located in the Meigu County of Liangshan State, is the largest inhabited by the Yi national minority,is also the birthplace of the Yi Bimo culture.For thousands of years,Bimo culture here,carried forward from generation to generation,well-known in the Sichuan, Yunnan,Guizhou,Guangxi and other Yi neighborhood,is the Holy Land of Bimo culture.The town is located northeast of Liangshan State,located in the heart of Liangshan mountain,171 kilometers away from the state capital,Within the terrain slope from northeast to southwest,with an average elevation of 2,000 meters above, is a typical Yi ghetto; is rich in regional culture.

Bapu Yi Zhai plans to build in the Meigu County nearby,its base,covers about 300 acres,the main planning objective is to revitalize and inherit Yi original cultural,historical neighborhoods,architectural patterns,in order to promote local cultural,economic and tourism development.

B. Physical shape design strategy

1) Street space

The street is important content in the Bapu Yi town design.Before the plan,we had once investigated and studied local some villages, has analyzed path's shape, the layout,in our planning,roads,is mainly in T-shape,the road shape conform to the mountain,in line with the traditional mountain road of the city's fabric.Consider the fire and the way of modern tourism,town roads are divided into 7m,5m,and 3m.7m for the trunk in the region,the formation of the loop,usually the vehicle no admittance, in the emergency case the fire engine may enter.

Ring road tandem with the landscape spindle,so that the layout is conducive to the commercial value of the street improvements,and easy to identify (Figure 1).The reason why the historical town gives people a warm feeling,mainly because of their buildings is a measure of the level of continuous and pleasant.Therefore, in our planning area,We designed a two-tier building,building height and street width of the basic ratio of about 1:1,so that can form the pleasant street space,planning construction area is larger than outer space coverage,which space has positive significance.

Figure 1. Example of Streetscape.

Figure 2. Designsketch of Tile Thatched Cottage.

2) Border

Yi residential buildings in history, there are three kinds of monomers prototype,soil palm room, stack Kibusa and tile thatched cottage (Figure 2).According to this three kinds of prototypes to be refined,we have designed a soil palm room area,stack Kibusa areas and tile thatched cottage areas,appropriate to do some relaxing stay in space.

3) Node

Node is the strategic focus, the node connection point is also a gathering point.Ten Calendar Square (Figure 3),toast House,the street of small squares constitute the primary landscape nodes in the region[3].Pay great attention to the design of space nodes connected on a line of sight;due to river longer,so in the local expansion of the formation of a small node to break the long river of space,eliminating the monotonous sense of space.In addition,a number of specially designed buildings also created a more interesting node.

Figure 3. Perspective of the Ten Calendar Square.

4) Area

If people can live in distinctive areas,it is always very happy thing.The planning area has land palm room area,stack Kibusa area,tile thatched cottage areas and toast the government district,the regional personality prominent, easy to identify.At the same time,pay attention to the entire region with sparse density and color. (Figure 4)

Figure 4. Perspective of Planning Area

5) Buildings

Architecture is the carrier of culture,we have used the traditional style in the architectural design,but carried on the massive changes to its plane design to meet the modern life need,caused the housing to be possible to form the three-dimensional commercial space with the downstairs. (Figure 5),and building also has its own unique style(Figure 6).

Figure 5. Spatial Relations Between Streets and Buildings

Figure 6. Perspective of The Single Residential.

6) Markers

In the town's planning and design, according to Yi traditional culture, we have extracted out the door (Figure 7)with national characteristics, and markers tower, also designed the nation's decorative patterns, which the town's residents and foreign tourists have a strong oriented.

C. Non-material Shape Design Strategy

1) Functional Classification of Small Towns

If all the land developed into a commercial,sales will face great pressure on the end we will build the public,commercial as a 1:2 ratio of settlements.Human spirit is the core of such a development to avoid flood of visitors during the day, an empty city at night.We carved out along the river,the landscape spindle, traffic trunk area as commercial facilities in the region,some relatively remote regions are designed as residential areas, so that can be achieved both tourists and residents to live together,but also to promote the traveling in-depth development.

Figure 7. Entrance Gate of Yi Zhai

2) Culture and Place

The culture is the accumulation and the inheritance of human life and the survival experience,the cultural inheritance and the development is the present era main melody,highlighting the special characteristic culture is inevitably the key to construction harmonious society.We have designed the front door,the tower as landmarks,according to the Yi national minority tradition's counts year way, has designed ten calendar squares (Figure 3),additionally builds Yi national minority's religious facility,these become the important component of culture.

3) Resident Participation and Business Planning

The small town is not only resident's life place,is also one kind of folk custom landscape,to promote the provincial economy,simultaneously attracts more tourists,we make the business planning to the small town street, like an antique street, a national accessories street and so on.

V. CONCLUSION

Under economic globalization's background,the cultural homoplasy is day by day serious,the region characteristic faces is flaw,the small cities style is easier to receive the external cultural impact,and is neglected in the urbanization tide. We have the responsibility also to have the duty to reappear and to to revitalize and inherit regional culture in design on style and feature. But the style and feature of small town is rooted in the broad regional culture,Only then pays great attention to preserve its simple architecture thought that the cultural connotation,can the style and feature enable to have the formidable vitality, realizes the towns sustainable development.

[1] XuXianfeng,ChenChangyong, "Cultural Renaissance-Based Buddhist Heritage Preservation Planning:A Case Study of Dananhua Regulatory Plan,Shaoguan City," Journal of Planers, vol. 25, June 200, pp.30–34

[2] BaiHui, "Return to the Tribe Luowu-An Attempt to the Protection of the Regional Style of the Old Townin Wuding ," NEW ARCHITECTURE,vol. A247, pp. 529–551, June 2006. pp.65–68..

[3] PengXiaolie,LiDaoYong, "Exploration on Landscape Planning of smal Towns – Take Laoguantuo of Liaozhong,Shenyang as an example," Journal of Shenyang Jianzhu University(Social Science), vol. 10, July 2008, pp.257–260.

Research of Jingzhou City Circumvallation Park Plan

Wang Yanan
School of Architecture
Xi'an University of Architeture and Technology
Xi'an, China
Email: kingwya@126.com

Li Zhiming
School of Architecture
Xi'an University of Architeture and Technology
Xi'an, China
Email: kingwya@126.com

An Yunhua
School of Horticulture and Landscape
Yangtze University
Jingzhou, China
Email: zjfdq@163.com

Zeng Junfeng
School of Horticulture and Landscape
Yangtze University
Jingzhou, China
Email: zjfdq@163.com

***Abstract*—Jingzhou city is also called Jiangling city, Jingzhou ancient circumvallation lasts the longest time, most dynasties and changes to be the only ancient walls developing from soil-made ones. It is the intact one among the few ancient circumvallation existing now in our country, and it is also the intact circumvallation in the middle reach of Yangtze River. It is famous both at home and abroad. On the base of protection, the relic protection and plan governor is confronted with the problems about how to update the conception of protection and usage, proceed to plan and design better, find more characteristics about its history and culture, play its comprehensive role fully, satisfy people's demand such as rest, travel, and culture education. This research makes an introduction about Jingzhou City Circumvallation Park and an analysis. Five communication such as protection and development, modern times and ancient times, inside and outside the city, upper areas and lower ones of the city, water and land are determined as plan principles.The circumvallation will be divided into core section, open section, buffering section and periphery. Aiming at different section the layout principle, layout target and specific measures will be researched differently. The vegetable slight is partitioned into two parts: the remained part and the new one. The new part is divided into Jingyuan chunhua, Fuhui xiayin, Churi qiushi, Shouze dongnuan, Sanyi garden, Tianwen lake. The detail design is made at last.**

***Keywords- Jingzhou;* City Circumvallation; Garden; Plan**

The Achievements of scientific research Project *(Q20091214)*, Education Department, Hu Bei Province

The Achievements of Human-social Research Project, *(2007ZS018/2008y135)*, Education Department, Hu Bei Province

I. INTRODUCTION

Jingzhou city is also called Jiangling city, Jingzhou ancient circumvallation lasts the longest time, most dynasties and changes to be the only ancient walls developing from soil-made ones. It is the intact one among the few ancient circumvallation existing now in our country, and it is also the intact circumvallation in the middle reach of Yangtze River. It is famous both at home and abroad. On the base of protection, the relic protection and plan governor is confronted with the problems about how to update the conception of protection and usage, proceed to plan and design better, find more characteristics about its history and culture, play its comprehensive role fully, satisfy people's demand such as rest, travel, and culture education[1-5].

II. THE PRESENT SITUATION AND ANALYSIS

Jingzhou is among the first famous cities of great history and culture in our country, and the ancient circumvallation is the most attracting historic remains. The existing circumvallation, having two cities such as eastern and western ones, dates from the soil-made walls built by Guan yu in the three kingdoms period. In Eastern Jin, Jingzhou governor, hengwen Shi, built soil-made city to connect the eastern and western cities, then Jingzhou circumvallation was preliminarily formed. The present brick-made city began to be built in Jingnan kingdom in five dynasties times, and was rebuilt many times. The existing circumvallation was built in Ming dynasty and rebuilt in early Qing dynasty.

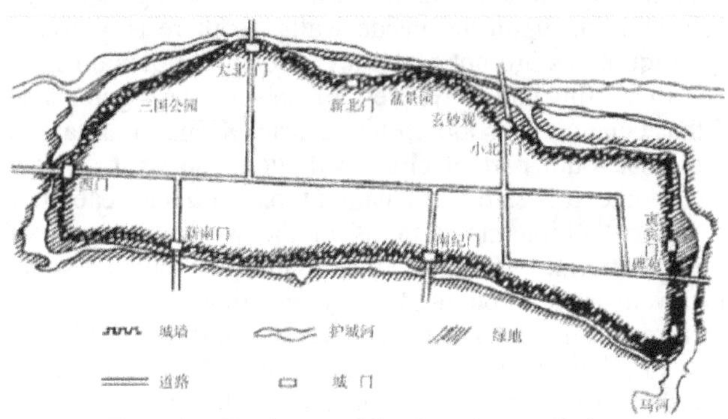

Figure 1. Sketch map of Jingzhou circumvallation area

The existing Jingzhou circumvallation and its moat are almost intact, and show irregular ellipse shape. The whole circumvallation is about 3.75km long in east west, 1.2km wide in south north and 11.3km in girth. The moat surrounds the circumvallation to form the blocked style with the border of circumvallation. The area inside the circumvallation is about 4.47 KM^2. The wall height is from 9 to 11m and its thickness is about 1 m. The soil-made wall is 10 meters thick with 6 paltforms,2 gate towers and 28 emplacements. To see form the sky, the ancient circumvallation stretches along the lake ,which is like a giant dragon laying on the land.(Figure 1,Figure2)

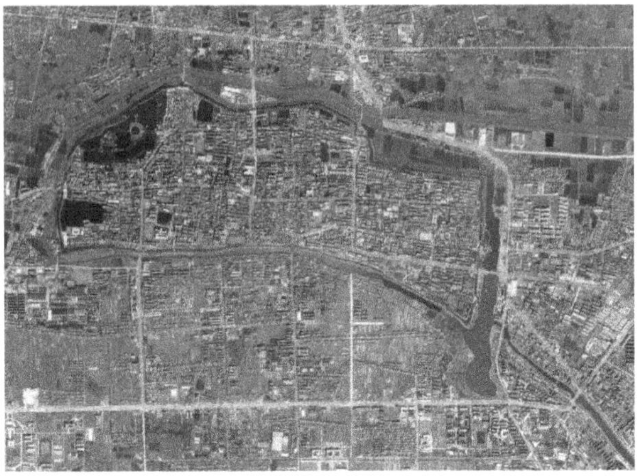

Figure 2. Airscape of Jingzhou circumvallation area

For better protection,developing its usage to play its role of relic exhibition and education meaning, and providing a comfortable visiting esplanade for tourists, the local government give money to build it. Now the open circumvallation park is built up with the ancient circumvallation as principal part, including soil-made city, brick-made city, water city and the area surrounding it.

In the macroscopical level, Jingzhou is the cradle of the Chu culture, and is also the important site of three kingdoms culture, many historic relics and architectures were remained. Jingzhou circumvallation area is the center of culture and history in Jingzhou city,the center of tours. The circumvallation park is the main carrier of this center. For instance,three mountains such as Songjia mountain, Xiejia Mountain, and Zhijia mountain are actually the big soil stage closely abutting upon the inside walls, so there is a saying "three mountains are not real ones". Songjia mountain is to the north of Jingzhou city, west of new western gate, it is said that Guanyu ever loosen his loricae. Xiejia mountain is in the west-south part of circumvallation, the west of new southern gate, it is said that Guanyu took off his loricae after one victory. Zhijia mountain is in the west-north part of circumvallation,it is told that Guanyu failed some battle ,went back through Mai city and finally were killed. Therefore ,the soldiers had to surrender and take off their loricae in Xiejia mountain. Besides this, "Jingzhou's three temples", "iron woman temple","Zhangfei Yidantu" are ranked as national key relic under protection.

At the same time, Jingzhou circumvallation park becomes esplanade for citizens and tourists because of its particular circumvallation sight and ample vegetable sight, playing the important role in city.

III. PLAN PRINCIPLES

A. *The communication of protection and development*

Circumvallation, ancient architectures need protection, moat and historic block need protection, the special circumstance of circumvallation need protection. At the same time, it is the key travel destination and history and culture traditional education base, many tourists and citizens being attracted to come here for a tour or rest. Therefore, on the base of protection, developing it properly and playing its comprehensive effectiveness is the key principle of layout.

B. *Communication of modern times and ancient times*

The existing architecture and relics should be protected actively, cultural relics should be kept as what they were, the temporary and illegal buildings should be removed, some ancient architectures should be properly rebuilt, and the regional features of Jingzhou circumvallation will be displayed adequately. At the same time, modern skills in displaying some architectures and relics will be used and for some architectures which are not suitable for exhibition or not rebuilt can be displayed in modern ways of dummy imaging, which is not only the intergrowth and communication of modern and ancient buildings ,but also the respect of ancient architectures and relics.

C. *The communication of the inside and outside*

Now there are many differences between inside and outside of the city. Outside shows the ring road ,moat and the area of grassland and wetland with the width of 20-300m, so it has vast eyeshot ,ample vegetables, birds and fishes, fresh air, the beautiful sight and resting place and so on ,which can give people a sense of delightful rurality. The inside of the city shows inside ring roads with the width of 10-20m, modern architectures. In some sections there are ancient buildings and blocks of history and culture ,where the dense of buildings is big ,the street is narrow, showing the life of the meeting of modern times and ancient times.

Therefore, the space of the inside and outside of the city should be handled differently according to the cultural route of space: In outside, the exhibition of the whole circumvallation, southern life and garden sight should be stressed. Using simple ways to make the circumvallation goes back to what it were in southern part in China is not only the inheritance of history and cultural, but also a creation of beautiful landscape for the whole city space; in inside, modifying the modern architectures and circumstance comprehensive managing in the area of Three Kingdoms Park, Sanyi street, Wen temple, Guandi temples, Gongan gate(peace gate), yanbin gate, "three mountains", "three temples", the poems and stories of famous people should representing the Jingchu culture ,three kingdoms culture and other ones should be fixed at the outside surface of buildings in some streets and blocks--which will create the circumvallation foot space which emphasizes the humanization and is equipped with good serving architectures---will not only make progress in rest, recreation, service ,but also conserve the main site of great culture and history for the modern citizens. Nine gates combine the inside the outside of the city organically.

D. The communication of the city's upper and lower areas

The vegetables in the inside slope of some parts of circumvallation should be made a sight, and the upper space of the circumvallation will be made a place used to show its history which will show its education, exhibition and recreation meaning. The lower part of the city will be made as the circumvallation park system with the greening as its main part, the service devices as the complementarities. It emphasizes the tour, rest, recreation, and is of connection in vision, organic union of two spaces through the wall ladder, soil slope way. The two spaces will form a fully unified and different space system.

E. The communication of water and land

The moat outside Jingzhou city is 20-200meters in width, and in some area of this moat its width almost reaches 600m. There are bridges ,islands and it is connected with Jingsha river and Taihu harbor. More than 10 lakes such as North lakes, dragon king pool lies inside the city. So some bridges ,islands, boats and the pavilion on the islands should be planned there to form the southern style of the union of land and water.

IV. THE PLANNING DIVISION

According to the demand of protection and development, combined with the facts, the circumvallation and necessary districts surrounding it will be divided into core section, open section, buffering section and periphery. Aiming at different section the layout principle, layout target and specific measures will be researched differently.

Circumvallation protection core section: this section should include the soil-made city are, brick-made city area, their nearby ancient architectures and relics are. In this section, the rigid protection plan will be performed, the object of which is to protect the ancient style in ancient circumvallation and nearby area. The specified measures are to constitute and perform the rigid protection plan, not allowed to build new architectures in principle. If it needs more building, then rigid examination and approving will be necessary.

The open common section: in the inside and outside ring ways of the upper and lower circumvallation, some parts should be separated as resting places for tourists.

The buffering section: this section has two parts. One is called inside buffering section which lies between the outside ring road and the outside of the moat, where some necessary common travel devices and garden sight devices will be planed to be built. The other one is called inside buffering section which lies in inside ring road, where the present architecture style should be planned to be kept.

The periphery section: the area stretches outwards from the buffering section and in this area the building should be controlled.

V. VEGETABLE SIGHT DIVISION

Jingzhou circumvallation park vegetable sight division is displayed as Figure 3.

Figure 3. vegetable sight division

The greening can be sued to make the whole sights with the sight motif 'two green banks, beautiful four seasons, Jingchu Three kingdoms, interacting to show its greatness'. In deploying, following the principles of making vegetable sights, paying attention to vegetables' color, shape, and season changes, and actively finding the vegetables' cultural meanings and Jingchu culture and Three kingdoms culture, are to form the vegetables sight section of its own characteristics [6-7]. The specific divisions are as follows:

The remained section the trees which grow well and is beautiful near the moat need conservation, for instance, the cedar, elm near the eastern gate; the big number of camphor trees in the area from new south gate to west gate, the metasequoia forest in the area from west gate to big north gate, and the willow in the area of little north gate need to be protected.

The new part: It is divided into Jingyuan chunhua, Fuhui xiayin, Churi qiushi, Shouze dongnuan, Sanyi garden, Tianwen Lake.

Jingyuan chunhua spot: In the spring, many tourists walk in the area of naji gate to west gate, or enjoy the sunlight and the sight of green growth and fresh flowers in Jingzhou. Therefore ,it is necessary to reinforce the spring sight and create and build Jing garden spring scenic spot .The sight vegetables which need to be recommended are prunus persica,prunu triloba which bloom in early spring, cherry flowers ,magnolia which bloom in middle spring, rhododendron,and rape pollen which bloom in late spring.

Fuhui xiayin spot: in summer, in the region of west gate to little north gate, through maple ,poplar with big shadow and vegetables such as pomegranate, hibiscus, lagerstroemia, albizzia, june snow which bloom in summer and the trees on the circumvallation, Fuhui xiayin spot can be created to make the tourists feel the cool sense of trees in summer, enjoy the sense the ancient circumvallation gives us.

Churi qiushi spot: in the region of little north gate to east gate, there are some everglade and in fields. Therefore, the sights such as great eye shot, many trees of different color, beautiful autumn, aromatic flowers, and ample fruits can be made here. The sight vegetables can be chosen as ginkgo biloba, sweet gum, golden-rain, four-seasons-cinnamon, tangerine and rice.

Shouze dongnuan spot:in the region of east gate to Nanji gate, in winter ,there are many tourists ,especially the old people enjoy the sunshine, walk here, so it is needed to create winter sight in southern district through cedar, mei flower, fealty bamboo.

Sanyi garden spot: on the north of west gate,the Sanyi garden can be created. There many peach trees are planted to form the peach garden, and statues of Liu,Guan, Zhang in peach garden should be made. The statue , the 'Zhangfei yi dan tu'and 'Guanyu three mountains' together can let the people feel the three kingdoms culture in Jingzhou.

Tianwen lake spot: The vast moat in the region of east gate can be used to build Tianwen lake. The scene spots contain Chu official Quyuan, dragon boat trace, Tianwen books, Jusong forests.

VI. DETAIL DESIGN

Jingzhou's ten scenes should be rebuilt ("Zhong xuan yu yi", "Hua shan yin yue", "Hu tian san bi", "Liu men chang ting", "Ju song ming yue", "Yu lie gao ge", "Jin feng yu ri", "Jiu long tu shui", " wang xi rang", "Gong an shui bo") .

VII. CONCLUSION

Through the careful designing of circumvallation, I hope it will give a chance of getting the citizens to know more about circumvallation history, fell the changes with the time going, and enjoy beautiful peach paradise.

Through analyzing the historical and cultural essences of circumvallation, the layout principle is made out, and the layout division, vegetable sight division and detailed design are done. A special esplanade for citiizens with historical culture and local traits is designed to renew the city, developing the function of rest, travel and cultural education,exploring the protection and usage methods of historical remains in new period.

REFERENCES

[1] the compilation committee of Jiangling county,"Jingling county annals",1990

[2] Ruan yi san,The theory and practice of history circumstance protection,Shanghai: Shanghai science and technology press, 2000

[3] Ning qian,Jingzhou ancient circumvallation defence devices' research and example analysis,Master thesis

[4] Shanghai Tongji Urban Plannging and Design Institute, Jingzhou Urban Plannging and Design Institute, Jingzhou historic city protection planning (2008-2020)

[5] Guide line of Chinese ancient circumvallation protection—Nanjing principle

[6] Xu Wenhua, An Yunhua, Qiu Cuiju, Jingzhou circumvallation park vegetables sight planning research,the Journal of Yangtze university 2008.6

[7] An Yunhua, Chen Youlin, Wan Xianfang,Jingzhou circumvallation planning research[J],Hubei agricultural college journal, 1999,19(2): pp.131-133.

Research on Climate Adaptability of Urban Environment Based on Simulation and Analysis

Hui Zhang
Huazhong University of Science & Technology
School of Architecture and Urban Planning
Wuhan, China
Email: zhhust@163.com

Hao Xie
Huazhong University of Science & Technology
School of Architecture and Urban Planning
Wuhan, China
Email: xiehao1978@hotmail.com

Jing Wu
Huazhong University of Science & Technology
School of Architecture and Urban Planning
Wuhan, China
Email: wuerna_777@hotmail.com

Abstract—**As an important part of urban environment, urban climate constitutes one of the key factors influencing urban spatial environment and urban planning. This paper takes Wuhan City as an example, and applies CFD simulation analysis to research urban environment. It takes Hanzheng Street of Wuhan City as its research objects, analyzes the environment of Hanzheng Street on the basis field study and CFD simulation, and presents effective measures to improve urban thermal environment, to create a comfortable human living environment and to promote urban energy saving.**

Keywords-urban environment; climate; CFD simulation; Hanzheng Street; energy saving

I. INTRODUCTION

Urbanization, as a necessary result of social and economic development, sharpens the contradictions between population and resources as well as environment. With the rapid expansion of urbanization in China in recent years, urban area and population increase gradually, and industrial and economic activities expand intensely. Accordingly, there are tightened urban land use, energy consumption and soaring anthropogenic heat discharge, leading to increasingly worsening urban thermal environment. Such a worsened thermal environment has a huge impact on urban microclimate and has aggravated urban heat island effect, deteriorating air quality and keeping accumulated pollutants, as well as posing a serious threat to the health of urban inhabitants. As the same time, increased urban environment temperature causes soaring citizen power consumption. According to research [1] findings from Hong Kong, citizen power consumption will grow by 10% and industrial and commercial power consumption will go up by around 3% when city temperature increases by 1 ℃.

In pursuit of sustainable development, it's essential for urban planning to feature balanced consideration and coordination of human living environment and urban overall layout. To be specific, it's important to discover how to create, from climate adaptability perspective, a good urban space and landscape design that helps improve urban climate because a favorable unban climate environment serves to enhance urban thermal environment and save energy in building and cities [2].

The rapid advancement of computer technology has provided powerful technological support for numerical simulation and analysis of atmosphere circulation. Computer simulation has become one of the key methods to research the relationship between urban microclimate and urban planning in a scientific, objective and systematic way, and to explore energy-saving design in urban planning. In this context, this paper uses Computational Fluid Dynamics (CFD) to explore the impact of urban microclimate on urban spatial layout by establishing urban digital models, conducting numerical experiments and computer simulation as well as analysis[3]. It carries a high value for research on urban thermal environment as well as urban thermal micro-environment, for exploration of climate adaptability of different urban conformations and street spatial layout, and for test of climate adaptability of different building conformation and urban spatial layout, as well as for energy saving.

This paper takes as an example the transformation of the old urban area of Hanzheng Street in Wuhan City to discuss climate adaptability in city transformation and its potential in urban energy saving. With the objectives of improving urban thermal micro-environment and urban thermal environment and of reducing urban heat island effect, it presents solid and reasonable urban spatial design approaches on the basis of quantitive research that takes the spatial transformation of the old urban area as its research object and features a series of simulations of the relationships between urban climate and urban space, interface as well as greening system.

II. RESEARCH THINKING AND METHOD

A. Research Thinking

Up to now, research on urban thermal environment has at least a history of 170 years. With the continuing and rapid advancement of scientific analysis method and computer technology, simulation experiment, computer simulation and mathematical model analysis have found their way into research of urban thermal environment and has produced remarkable results. J. E. Cermak, A. Soundborg and L. O. Myrup as well as K. Yoshio have made significant contributions in this field [4].

This paper introduces the CFD to research urban and building thermal environment by using various mathematical methods, conducting numerical experiments and computer simulation and analysis. Air current within a certain space can be simulated by establishing a mathematical and physical model. The temperature, velocity and concentration field generated by air flow within the space can be calculated and simulated according to provided boundary conditions and parameters. Based on the results, whether the results are feasible and reasonable can be analyzed and determined, therefore, the design can be continuingly optimized. At the same time, the laws and principles in urban planning and transformation can be explored and determined that can be applied to guide future design work [5]. Accordingly, it's vital to accurately simulate, from the perspective of building and urban planning, air flow and ventilation, air quality and heat exchange by air, along with human living environment comfort for the sake of research urban thermal environment and urban thermal micro-environment, of the exploration and test of climate adaptability of different urban conformation and street space and of energy saving.

B. Research Method

A city, as a complex material entity, comprises not only various buildings but also variable natural landforms. Urban spatial layout are influenced respectively by weather features, such as wind speed, temperature and atmospheric pressure along with solar radiation, urban characteristics, such as urban building density, population density and vegetation, and landform features. Therefore, a reasonable and accurate CFD urban digital model is the key to research the relationship between urban climate and buildings as well as cities by means of CFD simulation [5].

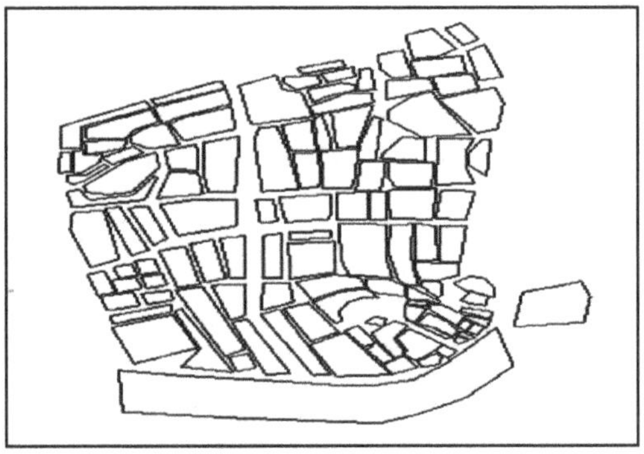

Figure 1. Hanzheng Street analysis digital model.

The research is conducted in a highly organized way. To start with, according to current urban layout, for the sake of establishing CFD urban digital model a city is divided into different area units by urban roads, mountains, rivers and lakes, etc.. Second, those units are graded according to the different property displayed in simulation and computation of CFD digital models containing building density, plot ratio, rough population density and greening. And CFD parameters are set differently according to the grade of CFD models. Third, define the most frequent wind direction as the initial condition on the basis of exhaustive urban climate analysis. Average wind speed and average temperature are adopted for calculation. Following this sequence, urban digital models are established and simulated to generate CFD numerical results for the sake of studying urban thermal environment [5].(the detailed application of CFD numerical simulation in urban digital models is discussed exhaustively in my related studies.)

C. Parameter Setting

According to climate information from the Observatory in Wuhan City and Green Building Research Center in Architecture and Urban Planning School of Huazhong University of Science and Technology, parameters are set as the following:

- Geological parameters: Wuhan City, north latitude 30.6°, east longitude 114.3°, altitude 23 meters, GMT+8 Time Zone, calculation month is August.
- Weather parameters: define southeaster as the dominant wind in summer, with a temperature of 32 ℃ and a wind speed of 15km/h.
- Material property setting: in the newed CFD models, regionalized modeling is adopted; the surface attributes of different blocks cannot be represented in models with relationship of quantity. Therefore, the comprehensive averages of different blocks are used in accordance with their respective actualities.

III. Analysis of Current Condition of Hanzheng Street

A. Current Condition Introduction

Wuhan, at the confluence of Yangtze River and Hanjiang River, lies in the east of Jianghan Plain that has a typical subtropical monsoon climate. The urban area of Wuhan City has a humid and hot summer and humid and cold winter, belonging to a typical hot-summer and cold-winter area. This climate characteristic is determined by special geological climate conditions, and the hot summer of Wuhan is intensified by urban heat island effect due to urban spatial conformation and city interface.

Hanzheng Street is located at the central area of Hankou District of Wuhan City, bordering Hanjiang River in the south and Yangtze River in the east (Fig.2). It's multi-fictional, offering not only business but also living places. It has a low building density but a high plot ratio. All these factors have a typical bearing on urban conformation and urban environment.

The studied area starts from Hankou Liji South Road in the west, to Youyi South Road in the east, and borders on Yanhe Road in the south, and Zhongshan Road in the north. (Hankou Liji South Road, Youyi South Road, Yanhe Road and Zhongshan Road are all names of streets). Research finds that the urban spatial conformation carries the following characteristics and contradictions:

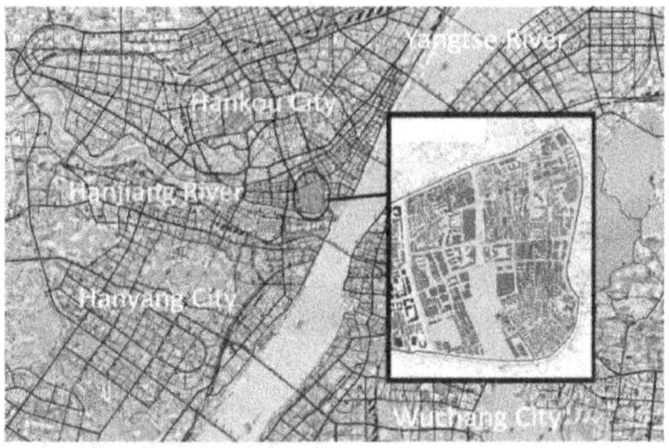

Figure 2. Location of Hanzheng Street.

- This area can be divided roughly into two parts by Duofu Road, namely the eastern part and the western part. The eastern part has a higher building density and a smaller average story number; business shops concentrates on main streets that is narrow, even only allowing one person to pass in some section of streets. The western part has more residences, fewer shops and businesses. Streets, some of which are obstructed, are broad there. Newly built high commercial buildings scatter at some turning points of main streets, which is changing the patterns of some roads in Hanzheng Street.
- Street space has becomes the uttermost pubic space because of Hanzheng Street characteristics due to historical formation, and because of the dramatic shrinkage of plaza space and courtyards due to continuing expansion of buildings. The streets play not only a transportational function but also various business functions, such as extension of shops, food markets, food and drinks street and cargo yard. Business activities have dominated these streets. Its impacts are multifold. To start with, it has caused human living environment to suffer. The transportation function of these streets is decreased dramatically and public activity space is shrinking. Second, it has deteriorated local environment due to tight spatial layout. Third, the urban thermal environment is worsen because heat produced by urban human activities fails to dispel itself and stays at these streets. This has a seriously negative impact on neighboring residential and commercial environment.

Field study on Hanzheng Street finds that summer daytime average street temperature stands around 40℃, and that air ventilation is poor with a long stagnation time. The rivers in the direction of dominant summer wind can serve to cool some buildings down the rivers. Therefore, the opening of Diyi Road will direct the summer wind into urban areas so that the temperature of main roads can be lower than that of other areas and that of some open areas.

B. Environment Examination and Analysis

Three different locations are examined and compared to analyze the environment of Wuhan City in the research. The locations are Hanzheng Street at Hankou City, Hong Kong Road Hankou City and Shenjiawan Street at Wuchang City. Temperature in continuous 48 hours in summer sunny days are recorded and compared as in the table1.

As it can be seen from the table, among the three temperature measurement locations in Wuhan City, the average day and night temperature of Hanzheng Street is higher than those of Hong Kong Road and Shenjiawan Street by 2℃to 4℃. And among the three places Hanzheng Street has the highest limit temperature that reaches 50.16℃ and that is higher than those of other places by 8℃ to 10℃. However, the low limit temperatures of the three locations are almost the same. The temperature at 12 at noon of Hanzheng Streets is higher than those of the other two locations by 3℃ to 5℃ while the three locations has little difference in terms of temperature at 0 at night. However, generally speaking, Hankou City has a higher temperature than other areas of Wuhan City, and Hanzheng Street is even higher, displaying a relatively poorer urban thermal environment. From the measurement results, Hanzheng Street of Hankou City has formed “urban heat island”.

C. Simulation and analysis of its current condition

From the above approach of modeling, establish the digital model of Hanzheng Street, set related parameters, and simulate and analyze its thermal environment. From the simulation results (Fig.3), it can be seen that Hanzheng Street at the interior of Hankou City has a high temperature and a radiation temperature reaching around 40℃ under the influence of southern dominant wind. From the wind speed simulation result, it can be found that ventilation within urban area is relatively poor, resulting in stagnant air and a less comfortable urban living environment. What’s more, the thermal environment condition has obviously formed an urban heat island effect. And Hanzheng Street is located the center of the heat island of central urban area, exerting an impact on the thermal environment of surrounding urban area. Compared these simulation results with those of field study, temperature, wind speed and environment comfort have demonstrated the same characteristics.

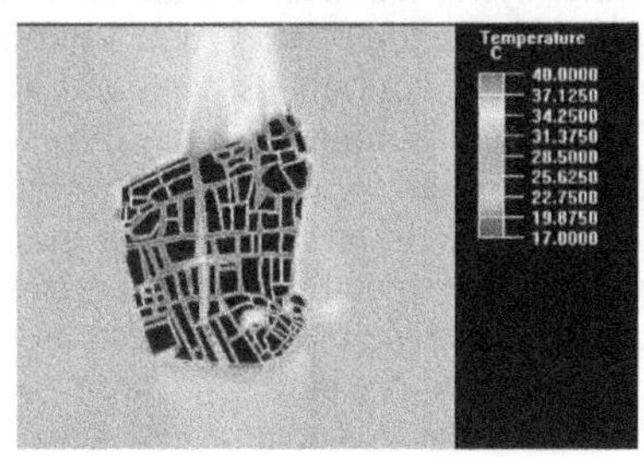

Figure 3. Thermal environment analysis of current Hanzheng Street.

IV. Optimized Design Analysis Based on Climate Adaptability

The improvement of urban ventilation constitutes an effective measure to lower urban temperature and to enhance urban thermal environment. There are two approaches available for urban ventilation: one is to direct wind while the other is to make wind. Directing wind is seen at some traditional cities where the urban patterns featuring the consistency of the direction of summer dominant wind and the direction of major streets sever to increase heat exchange between urban areas and their respective suburbs. The fundamental principle of making wind is to generate temperature difference between different areas that will cause a pressure difference. And a pressure difference makes air flow from one area to another. The most effective way to make low temperature area is to increase greening because the growth of plants themselves and the transpiration of their leaves can dissipate a great deal of heat from the environment, thus lowing local temperature.

Hanzheng Street is very suitable for directing wind because of its location. It's possible that relatively cooler summer dominant wind from rivers can be directed into the urban environment to accelerate air flow so as to improve urban environment and lower temperature. In the study, two areas are selected to conduct case study, optimized design and analysis (Fig.4).

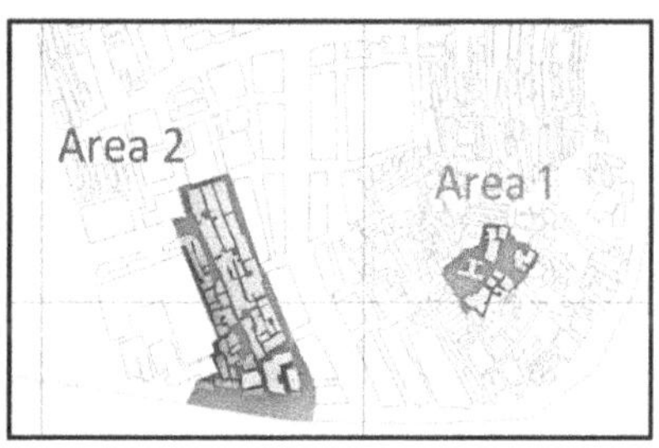

Figure 4. Transformation areas distribution.

A. *Transformation area one*

Research finds that upright road network is not suitable for directing wind while arborization road network is suitable. High buildings can change wind direction while the empty space on the sides of high buildings can increase the air volume of wind. According to this principle and in accordance of the layout features of transformation area one, it's recommended to broaden some streets within Hanzheng Street so as to form a continuous ventilation route in the direction of summer dominant wind. As a result of recommended transformation, wind from Yangtze River will function as a key factor to adjust the microclimate of Hanzheng Street. To be specific, according to existing road network, select and broaden a road from north to south, arrange direction of road within this area to direct wind deeply into the interior of this area; at the same time, reserve some plazas for improving microclimate, gathering and distributing goods, and human recreation.

B. *Transformation area two*

Transformation area two borders Yanjiang Road and is surrounded by high buildings. And some narrow streets cause serious transportation problems, and influence the business of shops in an indirect way. Therefore, it's recommended that three high commercial and residential buildings should be built to improve residential conditions of local inhabitants, and that businesses and shops should be centered to an effective wholesale marketplace with convenient transportation.

C. *Simulation analysis*

The model is modified on the basis of transformation approach from optimized design. From the results of simulated urban areas influenced by dominant wind, the temperatures and heat radiation temperature in the two areas have declined notably. What's more, in comparison to the actualities of the two areas, local temperature can be decreased by 3 ℃ to 5 ℃. At the same time, the wind condition in the two areas have enhanced, improving the thermal environment of the two areas and surrounding areas and the air quality within the two areas. In the transformation areas, air around buildings flows more smoothly, resulting in an increased wind speed within and around these areas, an improved living environment comfort and an relieved urban heat island effect, along with an augmented building and urban energy efficiency.

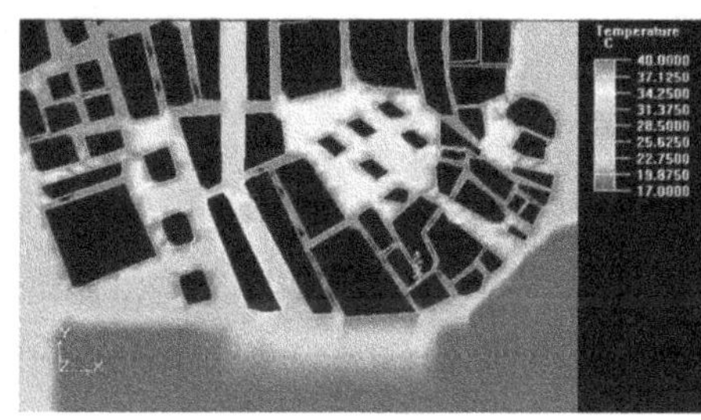

Figure 5. Analysis result of transformed urban area environment.

V. DISCUSSION

As a summary to the above analysis, this paper sheds light on the adjustment function of urban climate environment to urban ecological environment under the influence of climate environment. It's demonstrated that it's total feasible to apply CFD technology to urban planning to study urban thermal environment under climate influence [5]. CFD simulation should precede traditional planning methods in urban planning. Although it does not involve specific human living environment planning, it offers guiding directions and specification for following urban planning, drawing on advantages and avoiding the disadvantages and giving judicial guidance according to circumstances, and serves to create an ecological and climate-adapted human living environment using available ecological resources [6].

In this research, two upgrading methods are presented on the basis of Hanzheng Street current conditions analysis and of its thermal environment simulation and analysis. One is to adjust road network structure to form urban ventilation roads; the other is to adjust architectural form on the premise of keeping building area constant. The two ways both serve to improve business performance and regional micro-climate. The transformation of old urban areas involves many things, among which social, economic and political factors are dominant. With the deepening of urban ecology research, it's found that regional micro-climate is one of the keys to urban energy-saving. Therefore, application of simulation analysis and balanced consideration of all factors in pre-planning can help make a research scientific and accurate.

VI. CONCLUSION

To improve human living environment has become an internationally prevailing trend of thought, which calls for a high harmony of population, urban environment and natural environment. Urban environment, as a subsystem of the urban ecological system, involves human, climate system, and artificial environment as well as natural environment, all of which are part of the urban ecological system. The sustainable development of urban ecological system is vital for creating a desirable human living environment, enhancing living quality, and promoting urban and building energy-saving. Therefore, for the sake of improving urban ecological environment and creating a favorable urban human living environment, it's necessary to present solid principles and methods of corresponding planning and design on the basis of in-depth research on a specific zone or urban area by using computer simulation analysis technology in couple with related disciplines and theories.

ACKNOWLEDGMENT

Here, special thanks are dedicated to Prof. Zhuang Yu from the Architecture and Urban Planning School of Huazhong University of Science and Technology and to the National Natural Science Foundation in China (Number: 50578067).

REFERENCES

[1] C. Y. Li, C. H. Peng and Q. B. Zhao, Characteristic Analysis of the Abnormal High Temperature in 2003's Midsummer in Wuhan City. Journal of Central China Normal University (Natural Sciences), 2004, vol.9. pp. 379-382.

[2] G. T. Wang and X. Y. Wang, Numerical Simulations of Atmospheric Environments for Urban Sub-domain Planning, Journal of Tsinghua University (Science and Technology), 2006, vol.9. pp.1489-1494.

[3] G. T. Wang, Meteorology, Environment and Urban Planning. Beijing: Beijing Press, 2004, (in Chinese).

[4] T. Kazuya, and Y. Harunori, Measurement of Thermal Environment in Kyoto City and its Prediction by CFD Simulation. Energy and Buildings, 2004, vol.36. pp.771-779.

[5] Z. Yu and H. Zhang, Digital Model of CFD Simulation Design in Urban Planning, City Planning Review, 2007, vol.6. pp. 52-55.

[6] G. T. Wang, X. Y. Wang and S. G. Miao, A practice on the Modern Urban Planning Concept and Method: Modeling Analysis of the Atmospheric Environment Impacts in Foshan Town Planning. Urban Planning Forum, 2005, vol.6.pp.18-22.

TABLE I. TEMPERATURE MEASUREMENT AND COMPARISON

Temperature Measurement Locations	Average Day and Night Temperature	Limit Temperature(High)	Limit Temperature(Low)	Temperature at 12 at noon	Temperature at 0 at night
Hanzheng Street	37.5℃	50.2℃	30.4℃	42.3℃	32.0℃
Hong Kong Road	35.4℃	42.6℃	30.3℃	38.5℃	33.2℃
Shenjiawan Street	33.8℃	40.8℃	27.8℃	37.4℃	32.5℃

Research on the Elements for the Revolution of the Mobile E-commerce Driven by Cell Phones Terminal

Yonghai Yu
Interaction Design Institute
Zhejiang University of Technology
Hangzhou, China
Email: yuyh@zjut.edu.cn

Kaiyu Zhu
Interaction Design Institute
Zhejiang University of Technology
Hangzhou, China
Email: elongnet@yahoo.com.cn

Bing Xu
Interaction Design Institute
Zhejiang University of Technology
Hangzhou, China
Email: xubing@zjut.edu.cn

Jingsong Zhang
User Experience Department
Arcsoft Information Technology Co., Ltd.
Hangzhou, China
Email: jszhang@arcsoft.com.cn

***Abstract*—By fully integrating the business concepts in networking and communication industries, Apple's App Store formed a new e-commerce model and created a landmark scene in the history of the world's software industry. This paper makes a comparison between the traditional e-commerce and the new e-commerce landscape centered on App Store, analyzes the development of the several elements of the mobile phone platform and network platform, explains the nature of the new e-commerce landscape and its influence on the mobile e-commerce model and takes a glance into the future application of the function-enhanced terminals in the traditional e-commerce.**

Keywords- user experience; mobile phone terminal;mobile e-commerce;App Store

I. INTRODUCTION

In recent years, the market share of e-commerce has been growing rapidly, making a great impact to the traditional market model. Different types of e-commerce sites were transforming themselves horizontally or vertically to reach a more complete and professional level. This self-improvement seems to be the most effective means of attracting users except market promoting, until the newly-born App Store model successfully gives the possibility of another driving force. App Store, the software applications store for Apple's smart phone, iPhone, provides the third-party applications software sales service, a new e-commerce model Apple created by integrating network and mobile phones. With the development of new smart phones represented by iPhone and wide application of wireless networks, mobile phone played an important role in this e-commerce revolution in the software industry. Besides mobile phone's mobility and portability, iPhone's unique man-machine experience, platform scalability and Apple's brand radioactive makes it the driving force of the new e-commerce model. This paper describes the elements of the model revolution from different points of view and further elaborates on the combination of mobile-phone platform, the model elements and the traditional e-commerce.

II. THE MAIN VALUE OF APP STORE MODEL

A. *iPhone's product value*

The product value is the product's ability to meet the needs of the people. Potential human needs [1] can be divided into four basic levels: basic physical needs, needs of health and safety, needs of self-esteem and respect and needs of self-realization. iPhone stands unique in its design, and application, producing a powerful terminal driving force. The elements of its product value as the following:

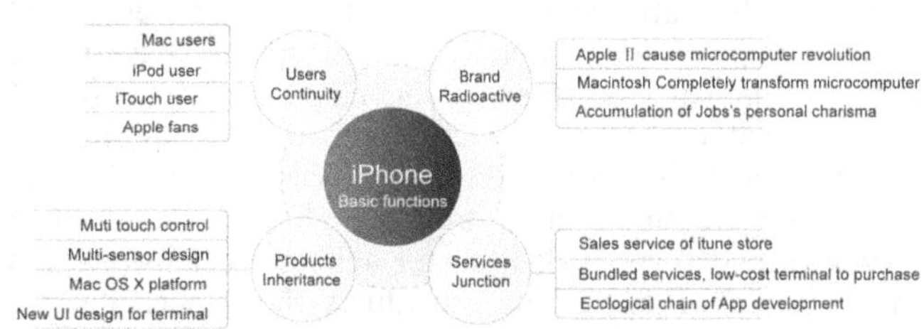

Figure 1. Main product value of iPhone

1) User continuity: The long roll of Apple's classic products, from Mac, iPod to iTouch, attracted many loyal customers through their product design concepts and precise product positioning, thus ensuring a broad user foundation for iPhone..

2) Brand radioactive: Through its creative hardware and software design, Internet technology and equipments, Apple brings some of the best experience to its users and constantly reflects the unique charm and personality in its products. All these contributed to the stand-out public image of iPhone.

3) Product inheritance: iPhone's operating system provides the basis for its function scalability and ensures its high performance. From the user-experience point of view, it endows the user with a most natural way of operation.

4) Bundled service: Apple successfully expands the traditional mobile phone sales service to a wider and broader landscape, and its approach of the application software revenue dividing enhances the ecological chain generated by this mode.

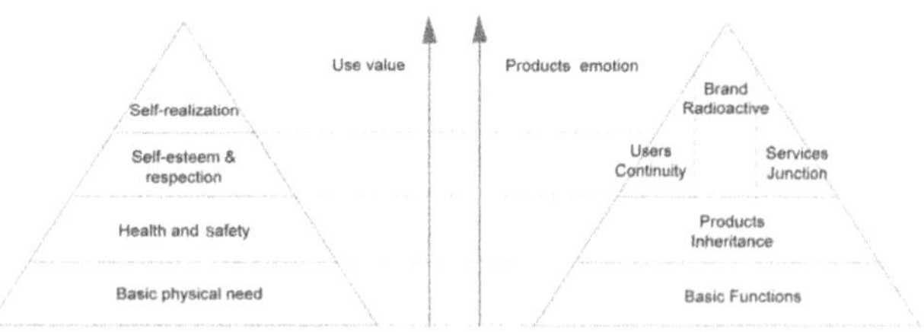

Figure 2. Potential human needs model with product use value

The model of user needs for the product can be expressed as a pyramid consisting of four levels, with each level the needs strengthen. The value of products is determined by how much it can satisfy the user's needs. Accordingly, the values of iPhone can be analyzed by which level of user needs it can satisfy. The basic function, phone calling and SMS that iPhone provides corresponds to the satisfying of the basic needs of a user from a mobile phone. With reliable basic function, the user needs reaches a higher level, which equals the needs of operation convenience, and even to the needs of pleasure acquired in using the product. These higher needs represent the more subtle usefulness of a product. Set aside some pure technical factors, this subtle usefulness of a product depends on transforming from simply solving technical performance to designing reasonable user experience, including the creation of convenient operating styles and allowing various forms of user-machine interface. Trying in this direction, iPhone's gravity sensing, multi-touch functions all enable another level of user experience and human-computer interaction [2], allowing the machine to adapt to human operation by developing an method of operation and control device compatible with human movement and also various display devices that ease the way human acquire information from it. The iPhone terminal system provides the scalability for software extensions and natural style of operation, constitutes the higher need a product can satisfy, corresponding equivalently to the human needs of health and safety. When the shallow basic needs are met its incentive function will be reduced, with deeper level of needs replacing them to become the main driving force.

Human needs of self-esteem and respect and needs of self-realization is the highest level of our needs. People who seek this level of satisfaction value friendship and family highly, tend to build a harmonious inter-person relationship with the surrounding people and are eager for the recognition by others and the society. So in order to satisfy the needs of this level, it's important for products to establish a distinctive product value and brand radioactive. The three product value elements that play a role in helping satisfying the highest level of human needs are user continuity, brand radioactive and service continuity. These three elements constitute the core of the brand building. The so-called brands are actually the product or service experience and some feelings arising from it, by consumers with similar consuming patterns on the needs of products and service. When consumers no longer consume brand products because of their product functions, they are actually consuming the added value of these brand products. People enjoy Pepsi because it represents youth and passion, people drive BMW because it represents success and wealth and people use Apple's products because they reflect some sort of taste and life attitude. In the process of self-realization, people enjoy a so-called peak experience emotionally [3].

As the value of a product satisfies the user's potential needs on a more complete scale and a deeper level, users also tend to develop an emotion with the product. iPhone's product design and brand management in meeting these needs made bold innovation and followed some long-term culture accumulation. As a mobile phone platform, it has become the driving force behind the mobile e-commerce revolution.

B. App network platform elements echoing with the mobile phone terminal

The most basic model of App Store is a platform for the distribution, downloading and review of application software based on a certain mobile phone operating system. In this prototype of App Store, besides the driving force of mobile phone platform, App network platform also has an important position.

App network platform is indeed a service delivering channel, a service provision joint for the more powerful functions of a mobile phone and the mobile phone industry. Based on Apple's SDK, developers create applications, enjoy the ease of avoiding complex business relationship and property rights disputes and could directly sell their products on the App network platform. Construction of this network platform manifested both the technological continuity and "radioactive" effect. From the "continuity" point of view, a single Web site can not directly imitate an App Store model and establish an ecological chain like App Store on the network platform, simply because a single site has neither a user-end terminal like iPhone, nor can it possess a position like carriers with a huge number of user groups and the ability to control mobile phone manufacturers. From the "radioactive" point of view, Apple's App network platform attracts a huge user base. And because of the network platform's cumulative effect inherited by the success of content selling of iPod and other brands, it's much easier for the App network platform to establish its reputation and popularity, as well as its acceptance among people.

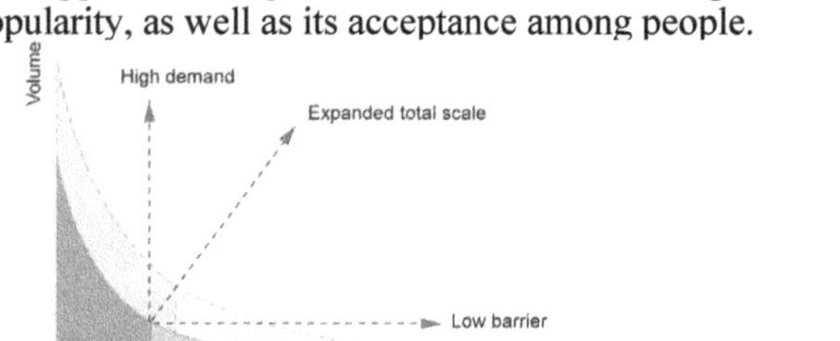

Figure 3. The long-tail effect to the scale of App retail

Despite these advantages, on the matter of the sales mode, App network platform still requires needs "low barrier" and "high viscosity" which represent a large user groups and a sufficient number of application developers. Because of a typical long-tail effect [4] appeared in the pattern of the sales of applications on the network platform, in order to continually expand the overall scale, an open platform is required, plus a low entry barrier for users and new developers. In the mean time, such a platform also needs to have sufficient "viscosity" which can be realized by using "revenue division" to enhance the stability of the number of different applications to achieve "high viscosity." At last, the combination of the platform and its brand radioactive could lead to the expansion of both the user base and the amount of applications, reflecting the channeling value of App network platform and also the platform's authority to guide and decide the content on it.

III. App Store Model and Traditional Mobile E-Commerce Model

A. *The ecological chain in App Store model*

In an ecological chain in which carriers takes the hold [5], the direct user interfaces only matters about mobile phone interface and telecommunication user interface. Among them, the mobile phone interface is a one-time interface for users in the sense that user who purchased the mobile phone wouldn't need the phone manufacturers to provide any further services or products. However, the carrier's user interface is a continuous interface in the sense that users who are using speech and data transforming services need to contact the carrier by using the carrier's user interface, which leads to a situation in which the carriers takes strong hold of the entire mobile applications ecological chain. In the traditional industry chain in mobile communication industry, the provision of services like CP and SP demands that users must use the carrier's user interface to receive the desired service, which means the carriers is at the core of the whole mobile added-value industry.

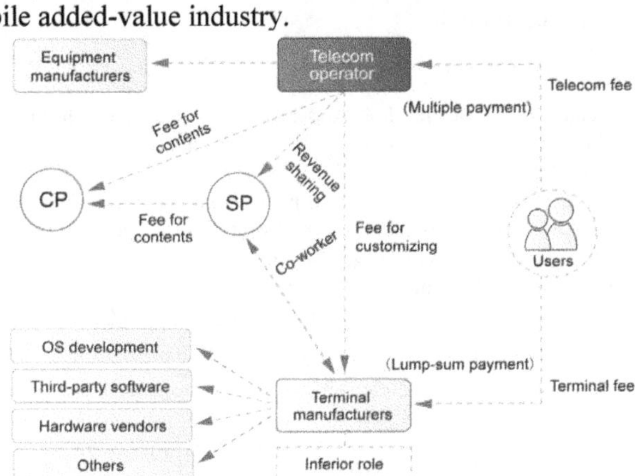

Figure 4. The ecological chain of traditional mobile e-commerce model

The App Store model presented is, however, built on the basis of a terminal-operator-centered ecological chain. iPhone's user continuity, brand radioactive, product inheritance and service junction constitutes its elements of product value and becomes the basic driving force of e-commerce in the software industry. Manufacturers themselves and mobile phones affect the users' choice of network acquisition and service acquiring. Services that CP and SP provides are sold in the form of applications on the platform, and due to the effect of relatively low entry barrier and high viscosity, both developers and users eagerly engage into this ecological chain, rendering the terminal interface a continuous one. Compared with the traditional e-commerce model, the core competitiveness of iPhone makes the ecological chain in the App Store model unique and make the App Store e-commerce model a revolution driven by mobile phones.

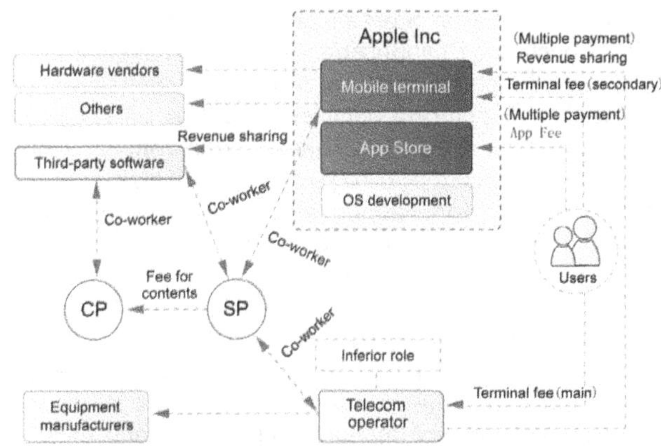

Figure 5. The ecological chain of App store model

B. *The developmental mode of mobile phone e-commerce*

1) *E-commerce based on mobile phones*

In the earlier phase, App Store is indeed the e-commerce in the software industry. The mobile phone market has been constantly exploring the best way to sell cell phone software [6]. Before the advent of the App Store model, the best way to make a mobile phone application popular is to persuade the carriers to pre-install this application. Obviously, this is a very difficult procedure. In the past, users rarely downloaded any application to their mobile phones, mainly due to two reasons: most mobile phones at that time are based on an enclosed platform and do not allow users to install other applications; another reason is that rare mobile phones are connected to the web, and those that are connected cost a high fee with a slow web connection. App Store model greatly eased users' access to mobile phone software; besides, App Store also provides those third-party application developers with a convenient and efficient software sales platform by dividing application sales revenue with these developers. In this way, as application developers actively engaged into App Store, users' needs for customized applications are met. This established a healthy mobile e-commerce ecological chain that could mean more than simply iPhone's application store in the future.

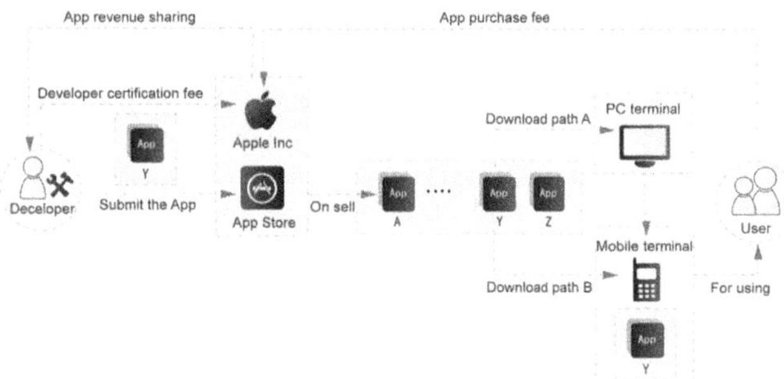

Figure 6. Mobile e-commerce chain based on mobile phones

2) Mobile e-commerce based on independent App

With the development of software store under the influence of the App Store model and users' transformation of the expectation of application software from entertainment function to functional service, those new application stores will insist their platform openness, meaning an even further expansion of the Wap concept to the Web concept, from the simple communication to the service of diversified network services. Some applications already jumped out of the previous model and appeared as the mobile e-commerce service entry providers on mobile phones. Business model like "In-App Purchases" such as depth-oriented allows users to use the application to make easy purchases with the aid of the application. As for the e-commerce that with the nature of goods freight, those third-party electronic payment [7] providers will balance mutual relationships and enrich the ecological chain.in the e-commerce in some kind of reasonable manner.Compared with the initial App Store, this mobile e-commerce has already passed the concept limitation in the software industry and developed in a service-oriented and entity-oriented trend.

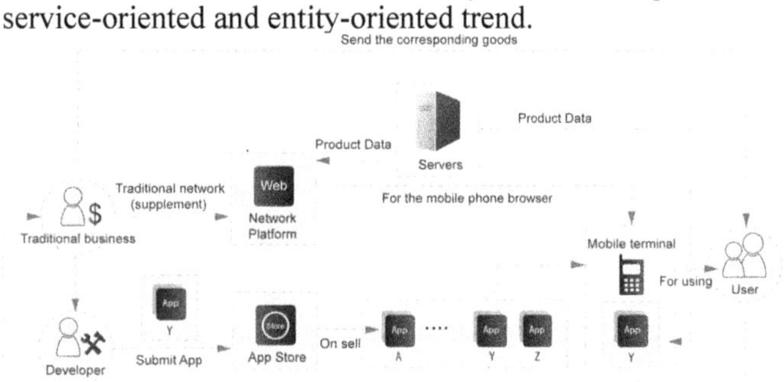

Figure 7. Mobile e-commerce chain based on independent App

3) Industrial Co-operation and in-depth customization on the mobile platform

Some traditional e-commerce platforms have obtained the qualifications to cooperate with terminal manufacturers to develop customized mobile equipments with apparent e-commerce characteristics. The combination of software and hardware technology along with the state-of-art embedded technology make the users approach the terminals in a more convenient way and expands the traditional Internet service to mobile Internet services. Therefore, those in-depth customized mobile phones should narrow the gap of functions and operations between them and the mature PC terminals. In addition, many of the traditional purchasing functions on e-commerce platforms will be transplanted to the in-depth customized mobile phones, like payment tools, communication tools, merchandise display systems and bar code scanning functions. Equipped with these functions, with a mobile phone users are allowed to use the video camera to identify the merchandise's bar codes and obtain its information like merchandise type and production time instantly on the mobile phone's display. In the mean time, an easy channel of conducting purchase is build for interested users to make the next move or perform real-time communication with the related sellers or manufacturers. This is the complete mobile phone for the mobile e-commerce platforms in the not so distant future.

IV. CONCLUSION

All this time, the ecological chain of the mobile phone industry has been taken hold by carriers, with mobile phone application design and configuration decided by mobile phone manufacturers. iPhone's success proves that, based on the convenience of users, the innovations in the operation and human-machine interaction can lead to further satisfy the user needs of a product,and eventually achieving the desired aim of improving human-machine usefulness, the ultimate feature that determines the product's market potential and even its potential to become a driving force for a new business model. The platforms like network can greatly increase developer and user interactivity, thus enriching the value of the whole industry ecological chain. The in-depth services aimed at application software and in-depth customization for the mobile phones would certainly improve and drive the new mobile e-commerce model's future development.

REFERENCES

[1] Ventegodt.,S.,Merrick.J. & Andersen, N. J.:Quality of life theory,Ⅲ. (Maslow revisited). The Scientific World Journal,3, 1050--1057. (2003)

[2] Preece, J. Rogers, Y. & Sharp, H. Interaction Design: Beyond Human-Computer Interaction. 2nd Edition. New York, (2007)

[3] Norman.D. A.. Emotion an design: Atrractive things work better. Interactions Magazine, ix (4), 36-42. (2002)

[4] Chris Anderson:The Long Tail.Wired Magazine Issue,10th (2004)

[5] Takeshi Natsuno:i-mode Strategy.Nikkei BP Planning Inc,(2002)

[6] Analysys International Inc: The report of Innovative business models and telecom industry chain, (2008)

[7] Zhang Kuanghai,Zhang Jing:Analysis of third-party payment.China Credit Cards,(2006)

Research on the Landscape Design of Urban Riverside Area from the Perspective of Historical Geography

The Case Study of the Conceptual Design of the Left Riverside Region in Hanzhong City

Feng Wu
Urban Planning
Xi'an University of Architecture and Technology
Xi'an, China
Email: 419572287@qq.com

Jingyi Fan
Urban Planning
Xi'an University of Architecture and Technology
Xi'an, China
Email: 297267949@qq.com

Fangxin Cheng
Urban Planning
Xi'an University of Architecture and Technology
Xi'an, China
Email: 326248154@qq.com

Jingtao Li
Urban Planning
Xi'an University of Architecture and Technology
Xi'an, China
Email: 71834664@qq.com

***Abstract*—"Integrate Culture Element into Design" is an important method in the space planning of urban riverside, which focuses on infusing humanistic factors into the design of natural landscape. In terms of some special riverside areas which have time-honored culture, the planner should discuss them systematically through the introduction of relevant concepts and methodologies in historical geography in order to emphasize its value. According to the result of analysis, landscape space is restructured after the screening and organization and under the guidance of core value. In the thesis, such method is tried and applied by taking the idea design of the left riverside region in Hanzhong City for a case study and it is expected that the method can be presented to give reference to the landscape design of urban waterside area with relevant cultural background.**

Keywords-historical geography; left riverside region; demonstration and screening; landscape reconstruction

I. INTRODUCTION OF HISTORICAL GEOGRAPHY PERSPECTIVE

The waterside of urban area is generally where a city originates and develops. It plays a significant role in shaping and carrying urban landscape and regional culture. Generally speaking, such area boasts superior ecological environment and rich cultural ambience. It is the focus and highlight for the creation of urban brands, manifestation of cultural connotation and improvement of life quality. Therefore, the area has emerged to be the focus in the building of waterside city for its unique and rich ecological, cultural and landscape values.

At present, with the gradual perfect of planning concept and accumulation of practical experience, the landscape planning of urban riverside area has presented an overall development towards being individual, valuable and daily. Apart from that, waterside leisure space is created by making use of natural resources, comprehensive functions such as entertainment, leisure, recreation and sightseeing are blended, and appropriate humanistic landscapes such as pavilions, towers, sculptures and bagatelles are shaped. The planning method not only creates rich landscape effect, but adds vigor and vitality to waterside area through the connotation exploration and facility construction.

However, in some areas, especially the riverside space which symbolizes the finishing point in the demonstration of urban landscape, such method of creating pleasant waterside space as stirring up feelings by sight is much weak in that it is lacking in macro and systematic thoughts and insufficient in depth and breadth exploration. Thus the core value of waterside space which is more important is ignored. To make up for the shortcoming of the planning method, we introduce the concept of "historical geography" and give explanation on the planning of left riverside region of Hanzhong City.

The historical geography research gives conclusion on the corresponding characteristics of regional culture from such three angles as historical natural geography, general historical social geography and special historical humanistic geography and through the observations of three time periods of past, present and future. The planning design researches and applies the historical geography from the following two aspects: the first is positioning study from "integrate culture element into design" to "content determines form", which takes the regional culture interpretation from the perspective of historical geography as the core; the second is the landscape reconstruction of riverside area from the perspective of historical geography, which takes the determination, acquisition and application of core value as the focus.

TABLE I. RESEARCH CONTENTS OF HISTORICAL GEOGRAPHY

Category	Subcategory	Branch	Contents
Historical Natural Geography		***Historical Geomorphology Geography***	Mountain Range, Desert, Seashore, Soil, Glacier, etc.
		Historical Hydrology Geography	River, Lake, Wellspring, etc.
		Historical Biology Geography	Vegetation Cover, Animal, Ecosystem, etc.
		Historical Climate Geography	Temperature, Humidity, Rainfall, Solar Terms, etc.
		Historical Disaster Geography	Earthquake, Flood, Drought, Landslide, Epidemic, etc.
Historical Humanistic Geography	**General Historical Social Geography**	***Historical Political Geography***	Settlement, Ethnic Group, Nationality, Integrated Planning, Political Area, Ancient Capital, Place Name, etc.
		Historical Economic Geography	Agriculture, Cities and Towns, Water Conservation, Traffic, Population, Handicraft Industry, etc.
		Historical Military Geography	Battlefield, Passway, City Defense, Beacon, Great Wall, etc.
	Special Historical Humanistic Geography	***Historical Cultural & Educational Geography***	Education, Talents, Style of Study, etc.
		Historical Religious Geography	Primitive Belief, Indigenous Religion, Foreign Religion, etc.
		Historical Custom Geography	Dialect, Folk Custom (Including Marriage, Residential Room, Diet, Clothing, and so on), Fold Way, etc.
Research on Historical Geography Theories and Literatures		***Research on Historical Geography Theories***	—
		Research on Historical Geography Literatures	Language Literature, Historical Map

Source: Drawn in line with relevant contents of Yong Lan's Chinese Historical Geography

II. POSITIONING STUDY OF SHIFTING "INTEGRATE CULTURE ELEMENT INTO DESIGN" TO "CONTENT DETERMINES FORM"

"Integrate Culture Element into Design" refers to the method of creating humanistic landscape by intervening human activities on the basis of natural environment, which is formed by overlapping cultural connotation and human activities based on natural landscape. The method proposes that, to form the cultural landscape of riverside area with characteristics, such landscape type which possesses both external and internal characteristics is needed. It lays emphasis on the infusion of intangible cultural connotation into tangible landscape space.

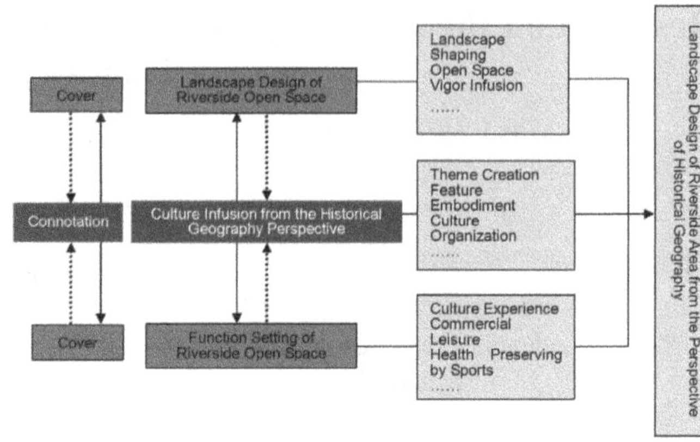

Fig.1 Study of Design Pattern

Generally speaking, the concept is of high guidance value in planning and design, just as we mentioned above. However, in specific lot, we deem that content is more important and the form is only a cover. Composition that various cultural characteristics integrate to one or complex of diversified impressions and perceptions of regional culture should be extracted systematically from the viewpoint of "historical geography", and the landscape will be further adjusted (Fig. 1).

A. Location Value Definition of Left Riverside Area of Hanzhong City

Hanzhong, situated at southwest of Shaanxi Province and the geometrical center of Chinese territory, is the capital of Emperor Gaozu of Han Dynasty Liu Bang when he was titled as Han Emperor. For a long time, Hanzhong remained the strategic place and region hub. With time-honored history and culture and profound regional culture, Hanzhong was the place where cultures from all directions intersects and blends, which can be described as history goes through Shaanxi and Sichuan and culture links all directions. Some people gave such statements: "just like Tang culture and Song culture are central to Xi'an and Hangzhou respectively, Han culture also plays an indispensable role in Hanzhong in terms of inheritance and profoundness of culture".

Han River, "source" of Hanzhong, is called Han Shui (literal meaning is water) and Mianshui River in ancient times. Including Yangtze River, Yellow River, Huai River, the four rivers are called "Jiang He Huai Han". 8 km of the river passes through the urban district of Hanzhong following a northwest-southeast direction. About 3.5 km in length, the left riverside area is located at the north bank of Han Shui and convex bank of river course, and thus the visual focus of the city. Compared with other section of Han River, this area boasts a nature-endowed advantageous terrain. Therefore, it is not just a place for leisure and relaxation in the city, but one that displays Hanzhong culture and city brands of Hanzhong. Based on the above orientation, we initiate our quest from the angle of "historical geography".

B. Interpretation of Hanzhong Regional Culture from the Perspective of Historical Geography

With the change of political, economical and geographical situations and region layout of city as well as the gradual convergence of modern culture, quite a number of cities' regional culture characteristics are weakening. For this reason, it is a necessity and a must to interpret regional culture from the viewpoint of historical geography to absorb the quintessence and discard the dross.

1) Interpretation of Core Conditions of Historical Natural Geography for the Evolvement and Development of Hanzhong

Hanzhong City lies to the southwest part of Shaanxi Province, the middle reach of Han River (the largest tributary of Yangtze River). It is backed against Qinling Mountains in north, shadowed by Bashan Mountain in south, and embraced by Hanzhong and Xixiang Basins in the middle. Due to the unique macro historical, natural and geographical conditions, Hanzhong has emerged to be the transition area of such three biggest historical, natural and geographical areas as Yellow River valley, upper reach of Yangtze River, and middle and lower reaches of Yangtze River (Fig. 2). It is the best place compared with others of the same latitude on the earth. Lofty Qinling Mountains and vast Bashan Mountain surround it and Han River, tributary of Yangtze River, winds through the area. All these factors contribute to the beautiful landscape of Hanzhong which is the same as that of regions south of the Yangtze River despite that the area is situated at the western region of China. In the area, mountain, water and city complement each other and are in perfect combination.

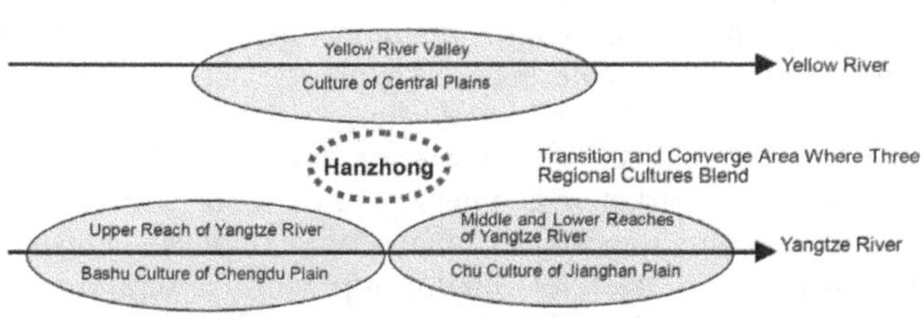

Fig. 2 Historical, Natural and Geographical Area of Hanzhong

2) Interpretation of Core Conditions of Historical Humanistic Geography for the Evolvement and Development of Hanzhong

a) General Historical Social Geography

Given the uniqueness of historical, natural and geographical conditions, Hanzhong used to be a place of strategic importance over the past dynasties, a place that bodes well in Han Dynasty, and the birthplace of Han culture. From the viewpoint of historical political geography, cultural elements such as "Han Ethnic Group", "Han Character" and "Han Culture" relevant to "Han Dynasty" all derived from the place, which can be concluded as "Culture from Han"; from the perspective of historical military geography, the historical fact that Liu Bang establishes the Han Dynasty relying on the geographical advantage of Hanzhong and a stream of policies and wars can be defined as "War Culture"; from the angle of historical economic geography, Hanzhong, as an important trade city in the middle reach of Han River in history, plays a role of trade hub, which can be defined as "Commercial Port Culture".

b) Special Historical Humanistic Geography

Second-to-none natural geography conditions make Hanzhong the transition and convergence area of the three largest culture regions like Culture of Central Plains at Yellow River Valley, Bashu Culture of Chengdu Plain at upper reach of Yangtze River, and Chu Culture of Jianghan Plain at middle and lower reaches of Yangtze River. The tolerant "Han Shui Culture" with unique feature is thus formed. In colliding with other neighboring culture, this culture, on one hand, demonstrates its innate charm in an unyielding manner; on the other hand, it absorbs the essence of cultures from other circle. It aligns with Bashu Culture, inherits each other with Jingchu Culture, communicates with Qin Culture of Shaanxi, and converges with central plain culture. In terms of ethos, it combines south and north; in language, it is a mixture of Qin and Shu so that the culture neither belong Qin/Chu, nor Ba/Shu. The culture displays vigorous life when developing in tolerance and absorption.

C. Determination of "Content Determines Form" Method on Left Riverside Area

From the above, we are able to conclude that the left riverside area of Hanzhong City is a core space to showcase the urban image of Hanzhong City (Fig. 3).

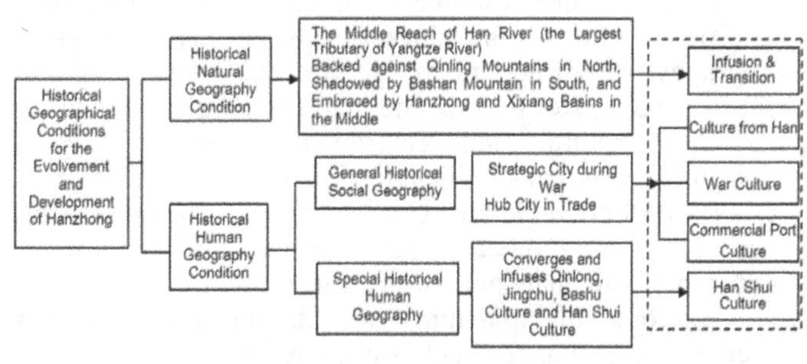

Fig. 3 Regional Cultural Characteristics of Hanzhong

Secondly, Hanzhong City is embodied with rich culture in itself with nature ecology being the base and regional culture being the trait. Consequently, importance should not only be attached to the ecology, landscape and life of the area, but also to the demonstration of regional culture value and the deduction & adjustment of landscape in content.

III. Landscape Reconstruction of Riverside Area from the Perspective of Historical Geography

Based on the above research and from a historical geography angle, we think that the landscape structure of a certain area in some lots can not only be a combination of natural landscape and human activities, but should be natural showcase of human activities orderly and being organized. In this planning with a historical geography viewpoint, it is expected that culture structure is established according to lot feature through gene acquisition of regional culture, green landscape features is blended and dynamic elements of lot are infused in light of the requirements of landscape design and function setting so that a new landscape structure can be set up (Fig. 4). In this way, regional culture is fully conveyed in content. Logic and emotion are translated from concept to actual landscape by the media so that the gradual and continuous integration process of culture landscape can be veritably manifested in time or space.

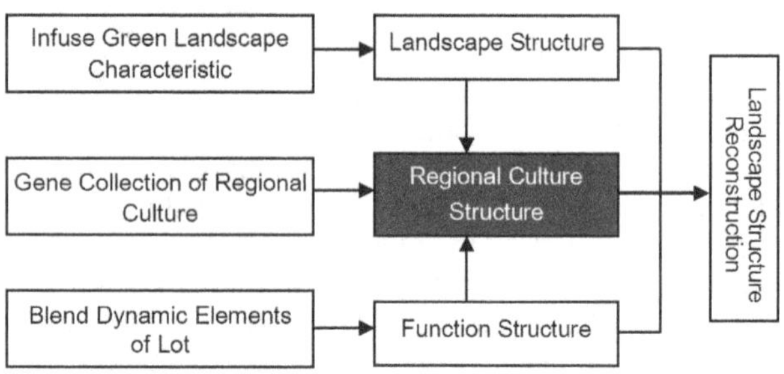

Fig. 4 Landscape Structure Reconstruction

A. *Gene Acquisition of Regional Culture*

——Take Advantage of Regional Culture in line with Local Conditions to Divide and Position Characteristic Culture

It is culture that distinguishes one area from another. During the scientific research and analysis of indigenous culture resource, it is suggested to fully explore and make use of the culture resource of a city in line with local conditions and under the concept of overall macro historical geography. According to the gene acquisition of regional culture, we make use of the current tangible resources (including natural and environmental resources) and intangible resources (including social and historical resources) of the region to screen systematically the area and give it an approximate culture orientation. The culture landscapes with characteristic are thus created.

Therefore, the planning has further complemented, screened and enhanced such aspects as "Culture from Han, Commercial Port Culture, and Han Shui Culture" based on the analysis from the angle of historical geography. The forming factors symbolizing Han culture and the reflected Han culture factors are finally determined. Through the sound organization of these factors, function division is launched in line with local conditions, which is used in the subsequent detailed design for reference.

TABLE II. GENE ACQUISITION OF HANZHONG CULTURE FACTORS

Forming Factors of Han Culture in Hanzhong	**Components of Han Culture Elegance and Simplicity of Han Architecture**	**Gene Acquisition of Han Culture**
Profound Historical Human Background	Feng Shui Perspective of Han Ethnic Group	The Source of Han
Superior Natural Geographic Environment	Tolerance of Han Culture and Collision with Surrounding Cultures	Nurture by Hanjiang River
People from Qin and Shu Mingled Together, Goods across the Country Converge Here	Trading Place, Sightseeing Place, Place to Remember	Trade through Hanjiang River
Abundant Han Culture Relics	Han Ethnic Group, Han Dynasty, the Hans, Han Clothing, Chinese	Comprehensively Mirror All Aspects of Han Culture

B. *Blend in Green Landscape Characteristic—Blend in Regional Culture from Multi-Angel; Improve Landscape Quality at Riverside*

The design of open space at riverside not only focuses on the control of the overall style, but also goes into the detail to expose the cultural image of the lot from all sides. Therefore, the choice of plant, the name of scenic spot, the sculpt, style and color of architecture and sculpture, land sculpture, street lamp and railing, etc. can all be blended into regional culture. If symbolic architectures can attract people in a short time, landscape articles of plant with regional characteristics and sculptures with cultural identities can render people appreciate the genuine culture connotation. Hence, after overall identifying the cultural characteristic orientation, we should infuse the culture content into details of landscape design from multi angles, reshape the style, and improve the landscape quality at riverside by integrating regional culture both wholly and partially.

C. *Infuse Dynamic Elements into Plot—Overall Showcase Regional culture, Form Driving Force of Regional Development*

Intangible culture needs to be carried and embodied by tangible behaviors and activities. On the basis of identifying cultural theme positioning, planning activities, such as performing arts, catering, shopping, accommodation, aesthetic enjoyment, experience, etc., is the major means to stimulate the vitality of open space at riverside. Besides the reference of market customers source and surrounding industries, more importantly, the setting of these dynamic elements should take its relationship with cultural image characteristic of the plot into account. For this reason, historical cultural element can be activated as a mean in modern times. Conflict between fashion and tradition forms a kind of harmony, which drives the development of surrounding areas, forming driving force of regional development.

IV. PRACTICAL RESEARCHES ON THE LEFT RIVERSIDE AREA FROM THE PERSPECTIVE OF HISTORICAL GEOGRAPHY

Based on the above exploration and analysis of Hanzhong culture from the angle of historical geography, in this landscape design of Hanzhong riverside area, three landscape areas are divided according to the plot characteristics and three cultural themes—Han Shui Culture, Han Trade Culture and Culture from Han have been identified by selecting relevant culture genes respectively. In accordance with the above discussed design mode and method, we finally get landscape structure of this design "cultural corridor, ecological band and axle of vitality" by establishing the relevant green landscape intention and project setting in light of different regional cultural climate(Fig. 5).

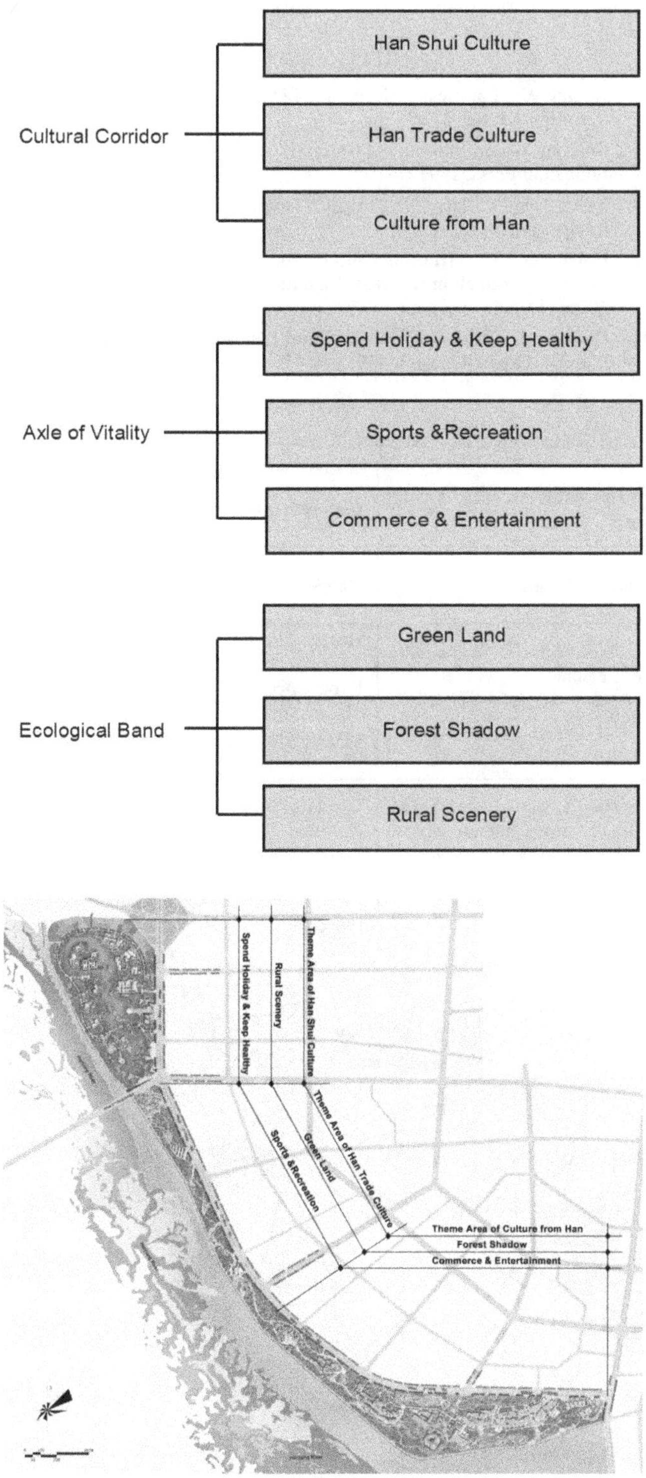

Fig. 5 Landscape Design Intension of Open Space at Riverside of Hanzhong

A. Blend Culture from Han into Design Theme of Core Area

"Culture from Han" is the monopolistic core competitiveness of Hanzhong City, The basic platform of its surrounding industry in theme area is the key point for developing the whole plot, which should be outstanding. Therefore, we divide the culture orderly into three subordinate landscape areas on the basis of element selection, namely "Source of Han", "Rhyme of Han", and "Elegance of Han", in which "Source of Han" is located at the axis, starting from "Milky Way" and with "Han Culture was the Fist Sign" as the transition, "Han Annexed Other Countries" as the extension, and "Han Was Located at Weiyang" as the climax. Artistic conception beginning with nothing and ends with nothing is thus created and the arduous process which the Han Dynasty rose from here and reached out to the world is manifested at the same time.

"Rhyme of Han" and "Elegance of Han" are the backdrop of "Source of Han". By using the technique of sculpture and clay sculpture, the former expects to create a folk custom garden showing the well-being and peace of Hanzhong people; while the latter prefers to make a cultural garden displaying culture achievements at that time by taking the advantage of "Han Ballad and Han Music". (Table III is put back at the end of the paper)

B. Han Trade Culture in Recreational Landscape

The culture theme area of Han Trade Culture is of long and narrow terrain with the narrowest part being only 85m, so focus should be attached to the creation of landscape effects complemented by corresponding service function. Not only does graphic design pay more attention to trees-lined shade, but more to echoing the planned theme. It emphasizes on the geometric plant of tresses and bushes. Repeatable line texture, like lines of sprays surging toward the shore, thus represents dynamic effect.

Two points are planned at north and south, which rely on the Baidu and Shangshuidu sites to establish two important ferry crossings activities and demonstration spaces. The same method is adopted between the two ferry crossings while several transit nodes are set. Characteristics of the nodes are demonstrated through changes of pavement. On the side of panel node, a certain amount of space for sport is reserved for the surrounding residents.

C. Inheriting Han Shui Culture in the Theme of Spending Holiday to Keep Healthy

Located in the Hanjiang Riverside, Han trade culture theme area is adjacent to the city center. People can overlook the mountains and enjoy the prosperity of the city from this area, which is regarded as "close to nature while still in the city". The area is well located and with advantageous landscape. So the plan takes advantage of such characteristic to build an international holiday area of health perservation and recreation with rural characteristic; and forge a Han Shui cultural exhibition area infusing different cultures by the water surface intensified by the quiet and peaceful effect.

In an effort to fit the functional image and Han cultural background, the plan emphasizes harmonious coexistence principle between human being and nature to forge a comfortable and pleasant landscape environment and create a landscape-creating conception with Chinese-style garden landscape. The environment will become the best place for recreation, spending holiday, holding commercial meetings and other activities. What's more, ground pavement and sculpture pieces embodying relevant culture are decorated to achieve the effect of interpreting Han Shui Culture from multi aspects.

V. CONCLUSION

City is the carrier of culture and culture is the spirit for the development of city. The landscape design of the urban riverside area has great significance for showcasing city characteristic and inheriting city culture. Therefore, the thesis takes Hanzhong as an example, from the perspective of historical geography, to demonstrate the mode and method of landscape design on the riverside which takes regional culture content as the core. It is expected to present reference to the landscape design on urban riverside area with relevant cultural background and guidance for relevant tourism landscape development through the application analysis on landscape reconstruction of Hanzhong riverside area.

REFERENCES

[1] Jia-hua Wu, Landscape Morphology [M], Beijing: China Architecture & Building Press, 1999.

[2] En-yong Wang, Introducing Cultural Geography [M], Beijing: Higher Education Press, 1991.

[3] Yong Lan, Chinese Historical Geography [M], Beijing: Higher Education Press, 2002.

[4] Hong-yan Li, Territorial Culture and Old City Renewal:The case Study on Protection of Xi'an Lianhu Historic Distric [J] , Planners, 2006. (12):34.

[5] Chang-wei Jin, Zhi-hao Liu, Cultural Landscape Design in Township Plan [J], Urban Research 1999.(4): 42–47.

TABLE III. ANALYSIS ON HISTORICAL CULTURE COMPONENTS IN THREE SECTIONS

Component \ Area		Cultural Area of Source of Han	Cultural Area of Han Trade Theme	Cultural Area of Han Shui Theme
Cultural Genes		Culture from Han	Han Trade Culture	Han Shui Culture
Activity Project		Culture Display, Commercial Recreation, Characteristic Catering	Folk Arts, Sports Recreation, Bar & Leisure	Health Preservation and Recreation, Accommodation, Folk Custom Experience
Space Environment	**Architecture Style**	Modern Elements in Traditional Style	Free Expression in Traditional Element	Mainly Showcase Traditional Style
	Landscape Element	Sculpture Landscape Sequence with Theme	Pavement Change Combined with Land-scape Characteristics	Display Sculpture Pieces of Relevant Cultures

Research on Wetland Landscaping Optimization Design in Wuhan

Qin Song
School of Arts & Design Wuhan University of Technology
Wuhan, China
Email: asongsongs@126.com

***Abstract*—Wetlands have significant function of "lung of a city", which are irreplaceable for metropolis eco-system maintenance, biological diversity, climate adjustment, pollution degradation, air cleaning and environmental improvement. It is vital to attempt wetland garden building in Wuhan. In this paper, a new optimized computerized analysis method combining the influencing factors "importance" with "direction" has been used to evaluate the effects of wetland garden landscapes in Wuhan. The results show that the actual effects of design could be reflected much more precisely.**

Keywords- Wetland;landscaping design;optimization;Wuhan

I. INTRODUCTION

It is generally believed that wetland is a kind of transition area between the land eco-system and aquatic eco-system [1]. The well-known definition of wetland is presented in "wetland Convention" as "For the purpose of this Convention wetlands are areas of marsh, fen, peat-land or water, whether natural or artificial, permanent or temporary, with water that is static or flowing, fresh, brackish or salt, including areas of marine water the depth of which at low tide does not exceed six meters". The wetland has significant function of "lung of a city", which is irreplaceable for metropolis eco-system maintenance, biological diversity, climate adjustment, pollution degradation, air cleaning and environmental improvement [2].

The wetland landscape is mainly classified as two kinds as natural wetland landscape including lakes, bottomland, rivers and artificial wetland landscape including paddy fields, ponds, artificial ditches, reservoir and pounds etc [3]. The lake wetland accounts for about 41% of the whole wetland area in Wuhan. Many lakes are faced with degeneration due to environmental pollution and excess exploration in recent years. In addition, rational utilization of these wetlands between aquatic and land systems has imposed an urgent task needed to be discussed.

There are two kinds of exiting wetlands in Wuhan. One refers to the river wetlands including the Yangzi River, Han River & their beaches, riverbeds, mudflats such as central bars. The other refers to the lake wetlands, which covers the majority of the whole wetland area including some lakes and lake mudflats within Wuhan city. There exits some success instances of wetland parks in US, EU and Korea nowadays. Sound effects have been acquired for trying to build wetland parks in various cities such as Chengdu, Nanjing in China.

More emphasis has been placed on the utilization of hills and lakes in the landscapes and gardens city plan of Wuhan. Meanwhile taking the distinguishing features of ample wetland resources into consideration, some wetland gardens have been developed. The exploitation of wetland gardens is favorable for protection of wetland, which can make full use of "lung of a city" and improve local environment, help relieve air pollution and noises [4] . At the same time, landscaping design of wetland can provide a unique light-spot of entertainment and attractive scene for tourism in Wuhan. Therefore, it is of great significance to attempt wetland garden building and it surely has bright prospective in Wuhan.

The principles of optimization of wetland landscaping design are shown as follows:

A. *Ecological principles*

The aquatic system, soil system and plant system are closely linked with environment in the city, which can not only exert the function of storage and infiltration of the wetlands, but also purify air, ease waste gases & noises at a large extent. To reserve primary plants to the utmost is a best way of dealing with the environmental problems in the cities in the aspects of both economic benefits and ecological requirements. On the one hand, it can be favorable to protect domestic species and prevent exotic biological invasion. On the other hand, it can create remarkable lakes feather and restore domestic plant system, which is in favor of restoration of native birds and fishes. To restore large-area plant system can provide many habitats and food for these animals and recover natural scenes to the maximum extent.

B. *Public principles*

The wetland scenes should meet with public requirement within the cities. For Wuchang as an example, there lie large-area resident districts near to the East lake, South lake and Sha lake. As a result, the wetland gardens should bear the features of regional Greenland for local residents besides for meeting the needs of "lung of a city". The wetland gardens should be designed to satisfy the material & spiritual demands coming from multi-aspects of public health, entertainment and aesthetics. The wetland gardens should provide a location of gathering, leisure and releasing for general public.

C. *Cultural principles*

The demands of culture should be satisfied for the wetland gardens. As a window of city, wetland gardens should have the function of cultural transmission. The wetland gardens should present for historical, ethical and cultural features of certain regions comprehensively as a carrier of regional culture while meeting public aesthetic requirement. The vivid characteristic of full of lakes in Hubei province is related to some graceful folklores, literature and arts. And residents of Hubei living in such an environment for generations have strong sentiment for the products from lakes and lake culture. Thus, the wetland gardens in Wuhan should give prominence to distinct styles and become a carrier for inheriting and spreading of Hubei culture..

II. STRUCTURAL ANALYSES OF WETLAND GARDENS LANDSCAPES

A. *Water landscapes*

The water landscapes are considered as long-distance landscapes in view of enjoyment. It is of great importance of the integrity of the landscapes to some large-area water. A mistakes usually made in the garden designs is to pursue the diversity of the plant and landscaping items and ignore the effects of landscapes as a whole. As a result, it seems disorderly. The astonishing overall landscapes can be designed through the feeling of distance stemming from waters landscapes whatever the specific processing methods are. In additional, the long-range harmonious view among the horizon line in the city, sky and birds should be taken into consideration. The design objectives need to be more pure due to the magnificent water landscapes and wild reference substance.

B. *Transition landscapes*

The transition landscapes belong to a kind of transition area between the water and land landscapes, which are composed of abounding factors. Therefore the mess effects could be more easily caused by ill-treated design. There are strip landscapes in the interim area between the water and land. The transition plants, which are near to the revetment or the plank roads, are usually regarded as short-range view. The integrated and partial effects should be considered comprehensively in the design. Moreover, crowd streaming line is an important factor in the garden landscapes. The garden view will surely be messy if the crowd is in a disorder state. The crowd streaming line is controllable in the design. The location of crowd completely depends on the road lines and the installation of the view-spot. The relatively rational method for dealing with the crowd is to utilize the embankment and the terrace near to shallow water area. Furthermore, some recreation such as fishing should be carried out by using the plank roads to improve the participatory in the garden.

C. *Shore-side landscapes*

The factors of shore-side landscapes are friendly and the details should be paid attention to. The species diversity should be emphasized in the plantation and the color, weather view and smelling are also important factors. It is of vital importance for the function of a variety of construction due to their direct touch with people. Other than the landscapes, the field of view is more important for the shore-side landscapes. The attempts of developing residential districts around the wetland gardens are worth promotion in Wuhan because the economic benefits of real estate increase while the crowd gathers in the garden. It is stressed that wind direction must be taken into consideration in the residential districts around the wetlands. Living at the upper hand of winds and waters is favorable unless the living standard is greatly influenced and the flows of water and air are retarded.

III. FUNCTIONAL OPTIMIZATION OF WETLAND ENVIRONMENT

A. *Optimization of the boundary*

The treatment of boundary of the wetland gardens is one of the perplexing problems. There are large-area residential districts near to every inside-city lake with rapid development of estates and the enlargement of residential areas in Wuhan. Although this specific way of gathering is favorable for maintaining the unique living culture in Hubei, the living by-products led by concentrated population could add heavy burdens to the environment of the inside-city lakes, which is one of the reasons for lake degradation and area decreases sharply. Some designs in the residential districts paid more attention on integrated effects in dealing with the shore lines by making the original wilding shore lines straight, filling in intrinsic channels or breaking up the small areas linked with large-area waters originally. The integrity and continuity of wetland environment are destroyed by those designs seemed to bring about preferable effects and the circulation of eco-system of wetlands is hindered, which would impose fatal influence on the biological species living in some shore-side areas and channels.

Therefore, it is critical to keep the original view of wetland shore lines furthest. Actually the linkage between the small water area and large-area waters can ensure good water quality in the residential districts and reduce the upkeep costs and protect eco-environment of wetlands. On the meantime, the liquidity of the shore lines, as well as the safety of the residential areas, should be considered.

B. *Optimization of plant system*

In addition, to utilize the environmental function has the characteristic of beatifying natural environment, which is unique for the other sewerage treatment. That is to say, the plantation of some ornamental water-born plant will make up unique attractive scenes in the treatment facilities and make these facilities harmonious with the landscapes in the gardens. The reaped plants can be used as fiber materials and feeds and can bring about comprehensive effects in the water purification projects.

In view of this point, it is not so difficult to choose those kinds of plant appreciate for the climate situation for there is much more options in Hubei. However, it is not pretty easy to present for regional features. It is worth to further research that ecological requirement should be met to suit for construction with distinguishing features. The scenes of wild ducks and lotus roots all round is typical description of "the land full of fish and rice" in Hubei, which is really attractive to the local residents. In recent years, there are many input species with frequent international exchanges and the species of plant used for wetlands are plentiful due to strong compatibility since Wuhan lies in subtropical areas. But some plants with strong regional features still need to be selected and become higher priority. The wetland plant is classified as three categories: water-born, transition and shore-side plants. Take water-born plants as examples, there are three categories: floating, emergent aquatic and submerged plants. The water-born plant system in Hubei is mainly composed of emergent aquatic plants. For instances, the lotus root is always the main emergent aquatic plant in Hubei. The reed commonly in Hebei, still popular in Hubei, is hard for large-area plantation and only can be used as transition installation because of poor economic values. The water caltrop and foxnut cover the majority of the floating plant in Hubei while the water shield can not grow there which is often seen in Jiangsu and Zhejiang provinces. So there is a limiting condition in plant diversity frequently mentioned.

C. *Optimized wetland landscaping design in Wuhan*

The Analytical Hierarchy Process (AHP) is a commonly used computerized method of systematic analysis and designing for wetland landscapes optimization. It was proposed firstly by professor T.L. Seaty from U.S in the 1970s. The computerized simulation optimization is to quantitatively describe the wetland based on AHP model according to the wetland systems of internal structure and external environment, using mathematical analysis method. Due to the optimization model of the precise solvability, flexible adaptability, quick processing and high efficiency, it plays a more important role the research on wetlands research. However, when it is used in comprehensive evaluation of the effects of wetland garden landscapes, the weight of actors originally assumed to be positive would cause deviation to the final results. In view of this drawback, a new optimized method is proposed associated with the characteristics of the wetland garden landscapes to reflect the importance and directions of the influence on the landscapes simultaneously. Specifically, two steps are adopted in building the judge matrix. The first one is to deal with the issue of "importance". For instances, the ecological and public principles are defined as exerting "very important" effects on the wetland garden landscapes and cultural principles is defined as "important". The second step is to take the "direction" (positive or negative of the effects) into consideration. The direction could be investigated by direct observation. For instances, the straight shore lines and the blockage of the waters in residential districts from large-area waters are surely regarded as negative factors in the landscapes design and building the plank roads near to lakes is of course a positive factor in the optimization design. By using this optimized analysis, more rational calculation could be carried out and the results have turned out that the actual effects of wetland gardens landscapes design could be reflected much more precisely.

IV. CONCLUSIONS

It is vital to attempt wetland garden building in Wuhan. Three principles of optimization of wetland landscaping design should be taken into consideration carefully. A renewed AHP method combining the influencing factors "importance" with "direction" has been applied to evaluate the effects of wetland gardens landscapes design in this paper. The results have turned out that the actual effects of design could be reflected much more precisely.

REFERENCES

[1] Xie Di, "Study of landscape design in artificial wetland in urbanized areas, " Master degree thesis in Southwest Jiaotong University, pp. 43–49, 2006

[2] Qiu hengjia, "Study on landscape pattern change and landscape ecology planning in the zone around Tailake, " Doctoral degree thesis in Nanjing Agricultural University, pp. 89–97, 2005

[3] Gong zhaoxian, Deng yi, "The spatial differentiation of urban landscape ecology and its optimization via architectural design, " Journal of Guangzhou University (Natural Science Edition), Vol 7, No 6, pp. 76-80 , Dec, 2008

[4] Wen Zi-juan, "Study of fuzzy evaluation for urban road traffic safety based on AHP method, " Journal of Min jiang University, Vol 30, NO 5, pp.106–109, Oct, 2009

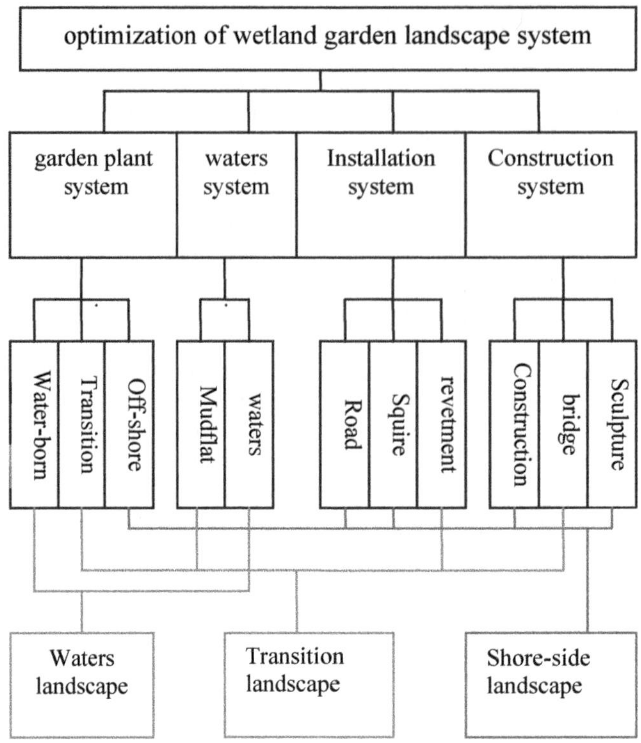

Figure 1. figure 1 Structure Diagram of wetland garden landscapes

A objective level B sub-objective level C realization method D optimization system

Researches on the Dynamic Changes of Wetland Ecologic Landscape Pattern Based on 3S Techniques

Shang Lei
School of Architecture and Urban Planning
Huazhong University of Science and Technology
Wuhan, China
Email: shangup7878@163.com

Xiong Huiying
School of Architecture and Urban Planning
Huazhong University of Science and Technology
Wuhan, China
Email:bearxhy@126.com

Abstract—**In this paper, taking Zhalong wetlands reserve, which is the largest habitat of red-crowned crane in China as the research object, based on 3S technologies (Global Positioning System (GPS)), Remote Sensing (RS), and Geographic Information System (GIS)), using the LandsatTM remote sensing images in 2005 and 2008, and a 1：50 000 topographic map as basic data resources, by applying theories and method on landscape ecology, it studies the dynamic changes of wetlands ecologic landscape pattern. Moreover, this paper conducted a quantitative and qualitative analysis on the utilization of land, the temporal and spatial evolvement of land coverage, which therefore provide basic data for wetlands protection and the researches on water fowls' inhabitation.**

Keywords- landscape pattern; dynamic analysis; land use

I. Introduction

A Geographic Information Systems (GIS), Remote Sensing (RS) and Global Positioning Systems (GPS), which are often abbreviated to Three-S technologies in China, play significant roles in many extensive integrated research related to space and time, and are valuable techniques and tools in obtaining, storing, managing, analyzing and visualizing ecological, water resource and socio-economic data for effective and efficient inventory and optimal policy and decision making. In China these advanced technologies started to gain attention soon after the open door policy in 1978, Three-S technologies often were applied in land use planning, land resource surveying, crop yield evaluating, disaster assessing, flood monitoring, fire preventing, and desertification controlling by interrelated research institutes, universities and government sectors. However, little research has been devoted to the use of Three-S technologies in the field of wetlands. The first part of this paper briefly describes the history of wetlands research, inventory and management in China; the second one describes the development and advancements of the use of the Three-S technologies in the field of wetlands; finally, In conclusion, limitations and needs for optimal use of the Three-S technologies for the sustainability of wetlands in China are highlighted.

Largest marsh area in China, have been converted to farmland (NWAP 2002).Of the 1，200 rivers and 50%of major cities have no drinking water that meets acceptable hygiene standards (NWAP 2002). In 1992 the Chinese government signed the Ramsar Convention and designated seven wetland sites as Wetlands of International Importance. The flood disaster in 1998 raised awareness of the government and public for the importance of wetlands conservation, and the government designated again fourteen sites as wetland conservation in 2002.The first wetlands resource investigation has recently been completed (1995-2001)and the results will be published shortly. The latest 2amount to 38.48 million hm², which include marsh wetlands of 13.7 million hm², coastal wetlands of 5.94million hm²,, river wetlands of 8.21 million hm², lake wetlands of8.35 million hm², and reservoirs and ponds of 2.28 million hm², Now the total of wetland with international importance amounts to 21, and according to the first wetlands investigation there are376 wetlands with national importance. The conservation and wise use of the wetlands of China have been listed as a priority area under China's Agenda 21.The strategy outlined under the National Wetlands Conservation Program(2002–2030)aims to establish 713 wetlands nature reserves, designate another 80Ramsar sites, effectively protect more than 90%natural wetlands and restore 1.4 million hm²of wetlands(NWCP 2002).The development and implementation of the National Wetlands Conservation Action Plan(2002-2010)and the 2002-2030 National Wetlands Conservation Program are preliminary outputs of the strategy. In China, wetlands research and inventory started in the early of the 1950s.The research topics mainly included inventory, institute in comprehensive limnology and watershed research. Geographical research is concentrated on the Yangtze River Basin, western region, and southeast coastal zone.

Land use is a direct reflection of human being's activity based on the nature. Land use is full of distinct spatial and temporal characteristics, [1] through remote sensing technology, periodic land-use data and powerful spatial analysis function of GIS can be quickly obtained, which will lay a firm foundation for the reproduction of land use spatial information. [2] Landscape spatial pattern and analysis of dynamic evolution are one of the core parts for landscape ecologic research. [3] The changes of landscape pattern are mainly represented by the changes of land use and land coverage, whereas the essence of the dynamic changes of land use pattern is a process of human being's constant adjustment of the land use in order to meet the demands of socio-economic development. It shows the development trend that human beings utilize the land for production and living activities. [4-8]Therefore, this paper takes Zhalong wetland area as the research object, being supported by 3S technology, it use landscape indices to conduct a quantitative analysis on the spatial and temporal evolution of landscape pattern in terms of land use. The evolving trend of landscape patter and its interior system are revealed. Furthermore, virtuous circulation and sustainable development of landscape ecologic system in this region will be realized. All these will provide theoretical guidance and technologic support to the project of wetland ecologic restoratio.

II. RESEARCH AREA AND RESEARCH METHOD

A. research area

Zhalong wetland is inland wetland mainly consisting of reed marshes. It is a nature reserve of water ecosystem. This region is located in the west of Heilongjiang Province, China and the downstream of Wuyu River on Songnen Plain. It is at the junction of southeast of Qiqihar City, Southwest of Lindian County and northwest of Durbote Mongolian Autonomous County. Its geographic coordinates is 46 ° 52 ~ '47 ° 32 'N, 123 ° 47 ~' 124 ° 37 'E. the climate there is temperate continental monsoon climate. This region is located in transition zone from sub-humid to semi-arid area, so its ecological environment is fragile. The average elevation in the region is 144.0 m and the annual average temperature is 3.9 ℃. Annual average precipitation there reaches 402.7mm. In addition to the dominant reed marshes, there are meadow grassland, farmland protection forest and lake habitat ecologic environments. The reed marshes occupy a larges area with a total coverage of 80% to 90%. Among the various birds, there are 35 genres which are state-level protective genres. Among the 15 kinds of cranes in the world, 6 types of cranes inhabit and breed there, in which the number of wild red-crowned crane occupies 1/6 of the total number in the world. There are other rare birds like Ciconiaciconia, Cygnus Cygnus, Platale leu-cordia, Egretta alba, as well as geese and gulls. It takes an important position for the undertaking of crane protection and wetland eco-system protection. Zhalong wetlands not only provides a habitat for cranes and other rare bird species, but also plays a significant role in regulating the dry weather in the western region of Heilongjiang Province, balancing rainfall, water storage, and flood diversion[9].

B. Data selection

This research based on the two Landsat TM pictures in 2005 and 2008, a standard 1：50 000 topographic map, a 1:100000 land use pattern map and a plants distribution map as well. It regards the GPS fieldwork research data as the basic data, and takes other word information and images as auxiliary data.

C. Classification of landscape

According to features of land use and coverage in the research area as well as the research demand, in combination with spectral signature of the satellite remote sensing image, and based on the land use types in the research area, its landscape is classified into farmland, woodland, grassland, residential and mining construction land, wetlands and water area etc.

D. Processing of remote sensing data

Supported by the processing software--ERDAS Image 9.1, the remote sensing images was undergoing the processing of geometric precision rectification, optimizing spectral region, radiation reinforcement and other processing, forming the basic interpretation of remote sensing data; at the same time, according to the established landscape classification systems and signs of remote sensing interpretation, by using AOI tool to select training templates. Then the possible matrix of the training template and the preciseness of classification template were assessed. Meanwhile, the remote sensing images were supervised and classified. In combination with the method of man-machine interactive screen visual interpretation, thematic maps of ecological landscape were drawn. Being supported by the software of geographic information system --ArcGIS 9.0, topology data processing, property assignment, and statistical computing was conducted targeting at the landscape types. Finally, the selected landscape indices were calculated and analyzed by using the analytical software of landscape..

E. . Selections of landscape pattern indices

In this paper, through selecting a set of reasonable index system, it is expected to reflect the individual features and overall features of the landscapes [10].Furthermore; it is also expected to reflect the spatial pattern and dynamic changes of landscapes.

- Percent of landscape (PLAND), which means the proportion of some landscape type to the total landscapes, and reflects the constitution of each landscape type in the research area.

$$PLAND_i = \frac{\sum_{j=1}^{n} a_{ij}}{A}$$

- Fractal Dimension (PAFRAC), which is the main application of dimensional Geometry in natural science. It is mainly used to detect the complexity of the patch shape in landscape ecology so as to quantitative describe the landscape patter. For the non-regular geometric shape of Africa and Europe, the relationship between its and perimeter (P) area (A) can be expressed in the following expression:

$$P^{\frac{1}{Df}} \propto A^{\frac{1}{2}}$$

In the expression, Df stands for the fractal dimensions of the irregular figure. Viewing from expression 2, the area and perimeter of the figure has such relationship with its fractal dimension:

$$\ln P = C + \frac{Df}{2} \ln A$$

$$Df = 2 \times \ln(P/4)/\ln(A)$$

In the expression, C is a constant, C=ln4

- Shannons' Diversity Index(SHDI), which is a reflection of richness and complexity of landscape. It describes the even degree of the proportion distribution of different landscape and the dominance of the main landscape element.

$$SHDI = -\sum_{i=1}^{m} p_i \times \log_2(p_i)$$

- Shannons' Evenness Index(SHEI)

$$SHEI = \frac{H}{H_{\max}}$$

$$H_{\max} = \log_2 m$$

- Dominance Index (DI), which is used to detect the dominant degree of one or more landscape element on the landscape. It is equal to the difference between the maximum diversity Index and its actual value. Its formula is:

$$D = H_{\max} + \sum_{i=1}^{m} P_i \times \log_2(P_i)$$

- Landscape Type Fragmentation Index (LTFI), which means the fragment degree of the landscape element, reflects the complexity of landscape spatial construction and the influential degree of human beings' activities on landscape structure.

$$LTFI = \frac{N_i}{A_i}(i = 1, \ldots, m)$$

- Landscape Shape Index (LSI), which is the weighted average of the ratio of actual shape perimeter to a round shape perimeter of the same area. The lager the value is, the more complex the shape will be. And the shape will be far from a round circle. The calculation formula is:

$$LSI = \frac{P}{2\sqrt{\pi A}}$$

III. Result and Analysis

A. *analysis of landscape element index*

- For the percentage of landscape area, area of the patches, numbers of such patches and character of landscape type patch, please refer to Table 1 and Table 2 . Distribution of landscape types is also shown in Fig 1 and Fig 2.

TABLE I.

Type	Area /km^2	Number of patch	Percentage	Fractal Dimension	Shape index	Fragmen-tation Index
Woodland (T1)	10. 77	56	0. 48	1. 286 4	11. 474 1	5.2003
Grass land(T2)	361. 06	121	16.05	1. 278 2	17. 501 6	0. 335 1
Farmland (T3)	254. 09	160	11. 29	1. 296 6	19. 903 0	0. 629 7
Saline and alkaline land (T4)	223. 46	129	9. 93	1. 311 3	22. 496 3	0. 577 3
Water land(T5)	107. 23	177	4. 77	1. 274 2	14. 239 9	1. 650 7
Marshes (T6) 1	1269. 16	51	56. 40	1. 234 8	13. 223 1	0. 040 2
Used land of residential and industrial area (T7)	24. 39	110	1. 08	1. 283 2	12. 550 5	4. 509 4

The character of landscape type patch in 2005

TABLE II.

Type	Area /km^2	Number of patch	Percentage	Fractal Dimension	Shape index	Fragmen-tation Index
Woodland (T1)	17.74	117	0.79	1.3410	15.5159	6.5947
Grass land(T2)	378.68	110	16.83	1.3064	23.2730	0.2905
Farmland (T3)	270.67	256	12.03	1.3158	24.2228	0.9458
Saline and alkaline land (T4)	198.92	166	8.84	1.3357	27.9019	0.8345
Water land(T5)	126.73	203	5.63	1.2829	15.7985	1.6019
Marshes (T6) 1	1232.46	57	54.77	1.2509	15.6035	0.0462
Used land of residential and industrial area (T7)	24. 95	124	1.11	1.2980	14.2828	49696

The character of landscape type patch in 2008

- From Table 1 and Table 2, the areas of various landscape types are in the following order: T6>T2>T3>T4>T5>T7>T1. Among them, Marshes occupied an advantaged proportion, with area percentages of more than 50%, which were respectively 55.34%, 56.40% and 54.77%. It was the basic landscape type in the research area, which controlled the circulation of energy and logistics in the whole area. The next was grass land. Woodland and Used land of residential and industrial area took up a relative small percentage. It showed that the natural landscape like marshes and grass land are in a dominant position, whereas the semi-natural landscape and artificial landscape (Woodland and Used land of residential and industrial area) only occupied up a small proportion, which was in a subordinate position. Thus, a landscape pattern dominated by marshes and supplemented by and enchased with grass land, farmland and other landscapes was formed. This was mainly related with the features of nature reserve, and local population and economic development degree etc.
- Fractal Dimension
  The fractal dimension can be applied in the process of identifying and maintaining the existing landscape patch shape. Meanwhile, it can also be utilized to identify that if the landscape patches are affected by natural regulation or human being's activities. [11] A larger the fractal dimension indicates that the landscape only suffered a few man-made interferences and the geometric shapes of the patches are more complicated, vice versa. Therefore, the value of by fractal dimension reflected the influence of human beings' activities on the landscape. The fractal dimensions in 2005 and 2008 showed that: the fractal dimensions of farmland and saline and alkaline land was larger, while that of marshes was the smallest. The large fractal dimensions of farmland and saline and alkaline land indicated that the patch shapes of these two landscape types was tending to be complex and irregular and they suffered large influences from human beings' activities, so that they were under the process of development. The small fractal dimension of marshes indicated that being influenced by human being's activities like exploitation of farmland and fishpond, their shapes were growing more and more regular, resulting in deterioration ecologic environment. The fractal dimension in 2008 grew larger than that in 2005, which showed that the complexity of land use was increasing during this period. The patch shapes also grew more and more complicated. But it should be noticed that the level of fractal dimension can not fully reflect the interference of human being's activities on the landscape, since there are other decisive influencing fractal dimension in addition to man-made factors, such as topography, geomorphology, and other natural conditions.
- Shape index
  In 2005 and 2008, the shape indices of each landscape types are arranged in the following order: T4>T3>T2>T5>T6>T7>T1. The orders of two different years are almost the same with similar changing trend and landscape proportion. Among them, the shape indices of farmland and saline and alkaline land were relative larger. The shape index of woodland in 2005 was the smallest. In 2008, the shape index of used land of residential and industrial area was the smallest. From the data of the two years, the shape indices of each landscape type was larger, appearing the same trend with the changes of fractal dimensions of each landscape type. This phenomenon showed that general boundary shape of the whole landscape grew more complex. Additionally, the larger shape indices of farmland and saline and alkaline land was mainly caused by human beings' activities. Suffering human's interference for a long term, the boundary became irregular and the patch shape was also growing irregular and complicated.
- Fragmentation Index
  Fragmentation Index refers to the fragmentation level of the division of landscapes, reflecting the complexity of the landscape spatial structure. To some extent, it also reflects the degree of human beings' interference with the landscape. The order of these fragmentation indices of different landscapes is: T6> T2> T4> T3> T5> T7> T1. The data in 2005 and 2008 almost appeared with the same changes. The fragmentation indices of woodland and used land of residential and industrial area were relative large, while that of the marshes was the smallest. Landscapes produced by human beings' activities like residential area and woodland also influenced greatly by human beings, resulting in sparse distribution, increased number of patches, and large degree of fragmentation. Distribution of marshes was tending to be gathered and exist in a large scale. They constituted the dominant ecologic landscape, thus the fragmentation index was relative small. Obviously, the research area was dominated by marshes and grass land, which were the basic landscape in the research area. Semi-natural landscapes and artificial landscapes (woodland, Used land of residential and industrial area) etc. were in subordinate position.

B. analysis of landscape heterogeneity index

Landscape diversity refers to the richness and complexity of landscape types, reflecting the proportion of a landscape type to the total area and all the landscapes. Landscape Evenness Index focuses on the unit evenness of each landscape type. [12] Dominance Index is utilized to detect the unit dominant degree of landscape structure by one or more landscape elements. As it can be seen from Table 3, the increase of Landscape diversity and Landscape evenness and the decrease of dominance index reflected the enlargement in the differences of each landscape proportion, and landscape heterogeneity was also significant. All these changes went against the stability of ecosystem in this area. This was mainly due to expansion of human beings' activities and the intensification of human's interference, which resulted in increase of landscapes like farmland, woodland, water land and saline-alkali land etc.. All these forced the lands use to be eave and diversified. Thus the integration of the region was damaged and the spatial pattern of the landscape was also changed. The information content and uncertainty of landscape cell were in a high level, which caused the landscape pattern to be diversified and complex.

TABLE III.

Year	Diversity Index	evenness Index	Dominance Index
2005	1. 8927	0. 674 2	0. 914 7
2008	1. 9461	0. 693 2	0. 861 2

The comparison of landscape heterogeneity

IV. CONCLUSION

From 2005 to 2008, in the research area, the number, area and position of each landscape patch underwent changes of different degrees. Generally, the area was in a landscape pattern that the marshes were the fundamental landscape, along with other main landscapes including grassland, farmland and saline-alkali land. The rest landscape types were patch-likely distributed in this region and their landscape pattern did change. Shape indices of each landscape type can be arranged in the following order T4> T3> T2> T5> T6> T7> T1. The shape indices of farmland and saline-alkali land were large. Fragmentation indices followed the order below: T6> T2> T4> T3> T5> T7> T1. The Fragmentation indices of woodland and used land of residential and industrial area were relative large and that of marches was the smallest. Based on the research result, it can be concluded that in the process of economic development, the land use should be planed based on some rules and principles in order to guarantee reasonable utilization and protection of land simultaneously, so that the sustainable use of resources and regional development will be realized. [13]

REFERENCES

[1] Cai Weimin, Tang Huajun, Lu Gang, et al. Application of Pattern Analysis and Transition Matrix in the Researches on Characteristics of Land Use [J]. China Land Science, 2006, 20 (1): 39-44.

[2] Liu Jiyuan. Remote Sensing Survey and Dynamic Researches on Resource Environment in China[M]. Beijing: China Science and Technology Press, 1996.

[3] Xiao Duning, Bu Rencang, Li Xiuzhen. Eco-space Theory and Landscape Heterogeneity [J]. Journal of Ecology, 1997, 17 (5): 453-461.

[4] Liu Huiming, Yin Aiguo, SU Zhiyao, et al. 3S Technology and Its Applications in Ecological Researches [J]. Ecological Science, 2002, 21 (1): 82-85.

[5] Liu Jiyuan, Buhe Aoseale. Researches on Temporal and Spatial Features of Land Use Changing Process --- Based on Satellite Remote Sensing Data [J]. Quaternary Research, 2005, 20 (3): 229-239.

[6] Liu Chunyuan. Remote Sensing Survey and Dynamic Researches on National Land Resource Environment in China [M]. Beijing: Science Press, 1997.

[7] Wang Zuocheng, Xue Lixia. GIS-based ITS System Structure [J]. Journal of Chongqing Institute of Posts and Telecommunications (Natural Science Edition), 2005, 17 (6): 772-775.

[8] Shi Peijun, Gong Peng, Li Xiaobing, et al. Methods and Practices for the Researches on Land use / Changes of Land Coverage [M]. Beijing: Science Press, 2005.

[9] Han Min, Cheng Lei, Liu Quan. Researches on the Extraction Model of Zhalong Wetland [J]. Remote Sensing of Land Resources, 2008, (1): 43-47.

[10] Xue Lixia, Wang Zuocheng, Li Yongshu. Researches on the Characteristics of Regional Landscape and Its Dynamic Changes Based on Remote Sensing --- A case Study of A new Area in North Chongqing [J]. Remote Sensing of Land Resources, 2005, (2): 64-68.

[11] Miller J.N., Brooks R. P. and Croonquist M. J. Effects of Landscape Patterns on Biotic Communities[J]. Landscape Eco. l, 1997, 12: 137-153.

[12] Ma Keming, Fu Bojie, Zhou Huafeng. Measurement of Landscape Diversity: Affinity Analysis of Pattern Diversity [J]. Ecology, 1998, 18 (1): 76-81.

[13] Gong Wenfeng, Fan Wenyi, Zhu Ning, et al. Landscape Pattern and Diversity of Outskirts of Harbin [J]. Journal of Northeast Forestry University, 2006, 34 (2): 50-53

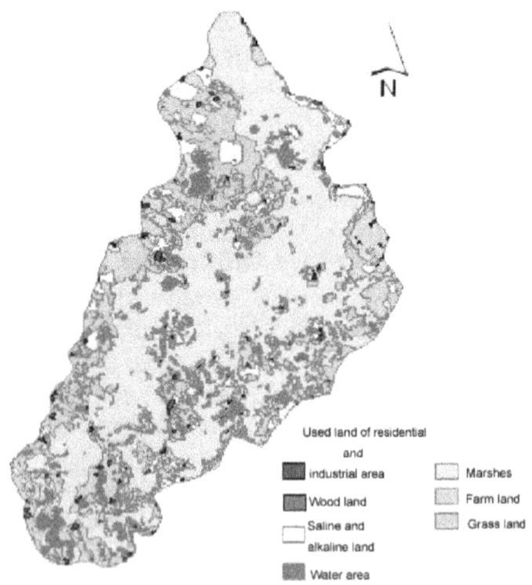

Figure 1. The landscape types distrribution map in 2005

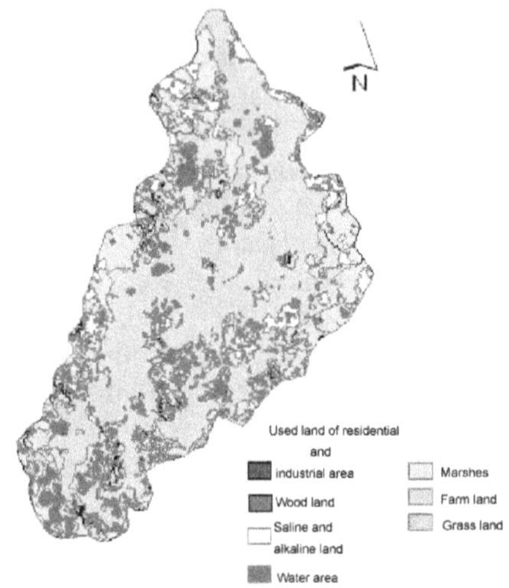

Figure 2. The landscape types distrribution map in 2008

Rural Landscape Compensative Design Based on the Overall Planning of Urban and Rural Development

Qin Zhang
School of Art & Design
Wuhan University of Technology
Wuhan, China
Email: superzhangq2004@yahoo.com.cn

Jing Wu
School of Architecture and Urban Planning
Huazhong University of Science and Technology
Wuhan, China
Email: wuerna_777@hotmail.com

***Abstract*—over the past 20 years, China has experienced rapid process of urbanization. With the growth of GDP and infrastructure construction in urban areas, lots of border areas of cities and rural areas in China will be transformed into cities, which bear the dual properties of urban and rural areas. How to create the connection of urban areas and rural areas and combine the high-quality of rural landscape and urban contemporary elements has become the key point of the urbanization. The paper approaches the high-quality of rural landscape and the disadvantages of the urbanization, discusses the meaning and feasibility of compensatory design of rural landscape and puts forward the ideal model of the coordinated development of urban and rural areas plan, including the green border, gray border, historical boundaries and interest boundary.**

Keywords-sustainability; compensative design; boundaries; overall planning

I. INTRODUCTION

Urban-rural coordinated reform areas such as Chongqing, Chengdu, Wuhan and Changsha are located in the middle west of China, having a varied topography with both flat lands and mountainous regions. There exist bounded edge districts between the core areas of cities and the typical rural complexes, benefiting from the mutual complement and combination of resources, while are probably harmed by the negative edge influence. So the structure and functions are indistinct and of permeability, having dual attributes of the countryside and the city.

The rural landscape, far less spoiled by industrialized manufacturing and “modernization”, has accumulated numerous amounts of histories, cultures and folk customs through thousands of years’ evolution. However, there are problems such as less efficient cycling of materials, energy, and the low speed of information transmission. The urban-rural coordinated reform is to set up more infrastructures and provide opportunities for the industrial structure of economy. At the same time the spread of urbanization inevitably leads to the damage of territorial ecology and the disappearance of folk cultural heritage and the drainage of social structure and belief system.

City industries, together with various demands and pollution, enter the countryside. The enhancement of rural economy is characterized by the building of broad roads, the destruction of farmlands for the construction of houses, and the filling up of damp earth for farming. Modern cities to be eliminated seize another breezing land in the countryside, breaking energy transformation and material cycle. The mingled contradictions of ecology, society and economy will become all the more complicated.

II. RURAL LANDSCAPE

Rural landscape is different from urban landscape and natural landscapes. Rural landscape is a kind of landscape with specific behavior, morphology, content and process. This settlement groups represent land use extensive characterized, lower population density and has obvious natural features of landscape garden area. According to the process of history development, the rural landscape record spatial distribution and time evolution with different cultures and natural records. On the one hand it reflects the current stage of human disturbance on the environment, on the other hand it preserve precious historical value of the human landscape heritage because of slowly development.

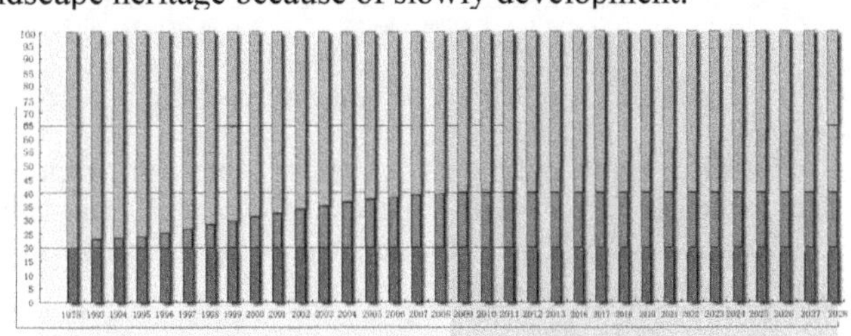

Figure 1. The Percentage of Urbanization

In terms of geographical scope, rural landscape refers to human settlements except urban landscape. In terms of composition point of view, the rural landscapes contain rural settlements, economic landscape, cultural landscapes and the natural environment. In terms of feature, rural landscape is complex landscape with cultural and natural characteristically.

“cultivate in sunrise, sleep in sunset, to relax between sky and earth area ..." reflects the leisurely and harmony life style in village. "The Death and Life of Great American

Cities," author Jane Jacobs called the building of the modern doctrine of human cities the building of the damage, disaster. Owing to economic confine factors, the rural landscape fortunately escaped this disaster and still maintained their freedom, fresh, simple atmosphere and integrity of the ecological context. The most basic elements such as farmland, farmhouse, fence, vegetable, fruit, mulberry, hemp, rice and wheat constitute pure, simple and natural garden style. Relatively, due to the emergence of air-conditioning in the city, it is no longer subject to weather constraints, due to the emergence of vehicles, it is no longer subject to the constraints of distance. While the unprecedented freedom bring city sprawl, elements of nature are pushed into a narrow space. However, the country is still affected by climate and distance, so people must live in the most energy-efficient, economic and harmonious way, forming a continuous vegetation cover and landscape-scale suitable for settlement. Exactly, these features adapt to ecology principles which people pursuit at present. Therefore, our research believes that the protection of these constraints in the development of the boundary settlement will be protecting the country's baseline and the essence of the natural ecology.

III. THE PROBLEM EMERGES

Since adopting urbanization and industrialization policy, china experience serious imbalance in development of urban and rural regional economy. This situation is constraining the overall economy improved and consumer demand expanded.

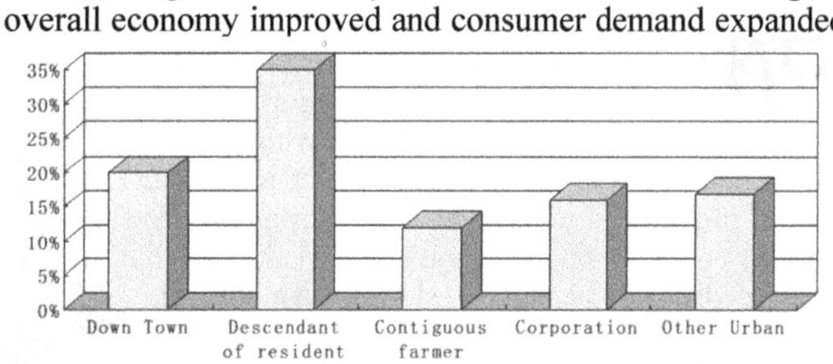

Figure 2. The emigration can be divided into 5 situations.

Finally, rural problems have become an important disadvantages factor for the success of China's urbanization. Due to lack of guidance, the process of urbanization in the landscape brings a series of problems.

A. The rural landscape does not have abundance aesthetic experience.

A large number of two or three layer "matches" rapidly emerge in the villages, which not only can't conform to requirement of sustainable landscape development, but also doesn't meet the production requirement of life, seriously affected villages visual image. This stereotyped phenomenon gave rise to the saying "sees the new houses but not the new feature".

B. Infrastructure can't satisfy the requirement of economic development.

Owing to lack of water supply, drainage, roads and other infrastructure facilities construction, rural areas will be confronted with seriously difficulties in transportation, sewage flow, and garbage collection, resulting in severe pollution and damage in water system, green vegetation and ecological resources.

C. Residential function

1) The scale of residential house is unreasonable and the pattern of residential function is imperfections. With rarely consideration the functional requirements and the architectural style according to village heritage, the newly residential housing design is monotonous, simply and cruder.

D. Style of construction

The style of construction in rural almost totally imitate the way in urban landscape. They build a large number of concrete roads through farmland. Natural River has been canals oriented and straight-oriented. As a result, the ecological communities during thousands years of history has been seriously destroyed. The traditional local characteristic of rural landscape is gradually disappearing.

E. Urbanized aesthetic psychology

People conceive urbanized aesthetic psychology. Rural residents would like prosperous and luxury life in the city such as high-rise buildings, private car. They hope transform there rural lifestyle into urban lifestyle through economic development. This psychological lack of understanding about the value of rural landscapes and treasure, would lead to further urban sprawl and huge destruction from modernism wave in major construction projects.

IV. TYPE STYLE AND FONTS

Set up digital model of stability assessment and ecological alarm system of rural landscape. The complementary design based on energy transformation and material cycle will form a powerful organic cycling system and be infiltrated into cities and countryside, and ultimately contributes to the sustainable development of the regions.

General Principle for rural landscape

- Provide the efficient street network for car, transit, bike and pedestrian movement.
- Provide open place as multiple uses.
- Restore and conserve the natural process and functions. Mitigate the impact of streets runoff on water quality and quantity to enhance local watershed health and treat storm water as a resource not a hazard.
- Preserve local open spaces and habitat.
- Create unique and attractive space to protect and enhance livability for commercial and residential.
- Interconnected landscape systems.

- Reduce traffic volume and speed on main driveway.
- Link green spaces with construction to create continuous, well connected green network.
- Easy to access green spaces.
- Encourage green infrastructure with street right of way and adjacent open space.

V. Four Ideal Boundary

A. *The green boundary*

Farmlands, pastures, woods, lakes, internal green spaces of the community and gardens form a continuous open space, a green boundary separating village communities and surrounding areas as well as a green belt connecting the communities.

B. *The gray boundary*

Establish rigid technological standards of public facilities and the long-term plans and laws and regulations. Taking advantage of the mountains and allowing nature to take its course, one road connects with an exterior one, joining all facilities including water supply systems and drainage systems, sewage treatment systems, garbage disposal facilities, power and energy facilities, communications and security facilities etc..

C. *The historical boundary*

To maintain primitive natural landforms with a history of thousands of years, that is helpful for preventing floods, land sliding and sustaining the growth of plantations and the underground water source reserve.

The restoration of architectural features and village humanities environment:

Architectural style in contour, scales, structure retention and multiplicity still remain as before. The construction takes advantage of the mountain, but not towering. The social atmosphere is still plain. The local character and style, Geomantic Omen Woods, Geomantic Omen Pond, boundary tablets, ancient roads, Dragon King Temple and the ancestral grave serve as the spiritual reposing carrier. The quadrangles with the crowing of cocks and the barking of dogs inside, farmlands, farmhouses, vegetables fences and fields, the fruit tree, mulberry, hemp and the rice paddy corn field all convey the retention of local character and style.

D. *The interest boundary*

The countryside is an interest whole, and the input and output of public consumption are to be increased, the renewal of technology, equipments and variety to be conducted, and the use of resources effective.

VI. Compensatory Landscape

Compensatory design differs from traditional design and ecological design. The former model hasn't account ecological and sustainable aspects. The later model emphasizes environmental protection, with the original intention to reduce the "negative interference". Compensatory design includes the two aspects, by the first part of this design represents the protection of ecological nature, and by the second part of this design represents redevelopment design. As discussed in the former segment, rural landscape conservation natural and history value, we must protect its precious properties. Also in order to satisfy overall planning of urban and rural development, the existing landscape pattern can't satisfy the increasing requirement of infrastructure. So it is necessary to expand and repair based on the existing construction. Other than development model in urban, compensatory design didn't need remove or protect the entire village and historical properties. It defines some boundary to protect and develop infrastructure and building using assemble and restore approach. Based on the principle of the boundary to retain its integrity structure of construction, while develop new landscape to satisfy increasing economic requirement. Finally, the life of the village has been maintained and extended. From the birth of a new countryside development, they have been given the spirit of the local, historical, and cultural.

Compensatory landscape must follow the principle of spatial layout

- Avoid blind expansion of construction and imitation phenomenon.
- Efficient construct artificial ecosystems, implementation land-intensive approach and protect farmland structure to create a pleasant and amenity rural Landscape.
- In addition, protect vegetation patches, increase the natural green corridors and scattered patches according to local conditions, compensate and rehabilitate landscape ecological functions.
- During the process of project implement, conserve land to rebuild a beautiful environment in harmony with natural systems.
- Respect for the geographical features, the original ecological pattern can only be adjusted not be changed.
- Respect for the geographical features, the original ecological pattern can only be adjusted not be changed. According to protect rural natural and history elements, look for Country Brand Features.

From the view of planning and design principles to create a living landscape atmosphere, remain more vegetation other than build more houses. Highlight the natural characteristics of rural landscape elements: Scarecrow as sculpture, open grass shed as "Pavilion", farmland as garden and so on. Rather than to mimic the style of artificial urban landscape structures such as: Pavilion, platform, floor and the porch. Design needs rustic characteristic to assort with nature. The rhyme of natural landscape often bears more attractive than carved geometric patterns. Emphasize local and national characteristics including building style, materials and species used in the garden to make sure different functional style of artificial structures in harmony with the surrounding environment. The style of architectural adapts China's geographical features. In order to highlight the unique local resources, Construction material is inclined to simple, elegant, true color. In terms of aesthetic content, our landscape design has ample culture and art characteristic nutrition from Chinese traditional history, formed its own distinct characteristics. The following principle is the essence of ecological aesthetics, and penetration of the Chinese culture, the aesthetic meaning and reflection.

Protecting local characteristics, respecting the rural functional orientation and maintaining the rural landscape pattern as well as the traditional culture environment are aimed at harmonizing the relationship between human and nature and fully embedding the features and characteristics of rural areas. Protecting the specific production processes, biological diversity, landscape and nature has a special value and protective function for the purpose of achieving integration between nature and Human and sustainable development.

VII. CONCLUSION

The protection and renewal of style are as follows: cultural, architectural and artistic environment in villages and towns should be paid attention to in the plan, and also culture should be integrated and reflected in construction. From the " angle of "special economic structure" and "ecological space" in Village, protecting and repairing special large-scale natural vegetation patches (Special green property), combining the general landscape principles of spatial planning, is based on special local climate and the special green and market-oriented products. Guiding and designing a natural corridor or a broken plaque with the Ministry of the form of decentralized control of human activities infiltrate the building lot or farm lot. Lay out a number of scattered residential premises according to the distance from the built-up areas and along the large scale vegetation area and architecture land. Some small patches of agriculture can be added in large patches of natural vegetation between the plaque and construction. Integrity of large-scale vegetation patches, as well as giving full play to its ecological functions, should be focused on the use of land for the purpose of ensuring the stability.

Finally with the establishment of rural landscape and ecosystem stability, evaluation of the number of early warning models, ecological, social, economic, material recycling and energy conversion compensatory designed to form an organic, has a strong vitality of the circulatory system, penetrate into urban and rural areas, eventually integrated into the sustainable development of fringe areas.

REFERENCES

[1] Appleyard D: Public Streets for Public Use – Foreword. Ed. Anne Vernez Moudon, ISBN: 0-442-26404-6, Van Nostrand Reinhold Company Inc., Printed in USA, 1987, p. 351.

[2] Kaplan A, Kucukerbas E, Ozkan B: Foca kentsel Yerlesiminde Gorsel Etki Degerlendirmesi Uzerine Arastirmalar. (In Turkish) (Researches on Visual Impact Assessment of Foca Coastal Town) Ege University Research Fund Project No:96-ZRF-009, Izmir, 1999;112:23–34

[3] Moughtin C. Urban Design: Street and Square. Architectural Press, 1992, ISBN: 0-7506-4274-2, 238:131..

[4] Blair GE: Visual Impact Assessment in Urban Environments (Chapter 13): in Smardon RC, Palmer JF, Felleman JP (eds): ISBN: 0471881848, John Wiley & Sons Inc Publisher, 1986;400:223–246

[5] Mccluskey J. Road Form and Townscape. Butterworth Architecture, Oxford, 1992

[6] Wright C. C. and Curtis B. Aesthetics and the urban road environment. Proceedings of the Universities Transport Study Group Annual Conference, Napier University, 3–5 January 2002 (unpublished)

Shallow Exploration on Modern Sports Leisure Park Landscape Design

On the Landscape Design of Honglian Lake Water Sports Training and Completion Base

Chen Li
Environmental Design Office
Modern Art College
Jianghan University,
Wuhan,China
Email: 20018285@qq.com

Liu Jing
Environmental Design Office
Modern Art College
Jianghan University,
Wuhan,China
Email: 3938533@qq.com

***Abstract*—The emergence of sports leisure parks is the inexorable trend of pursuing good quality of life. To satisfy the competition is one side of building a sports leisure park, what is more important is taking reuse into consideration, which attracts citizens do sports events in the park. Taking the landscape design of Red Lotus Lake water sports training base as an example, some features which need a full comprehension have been list as follow: participatory and cultural.**

Keywords-sports leisure park; participatory; reuse

I. THE INEVITABILITY OF BUILDING SPORTS LEISURE PARK

With the rapid development of economy and life-style, human is taking further away from the nature as a result of enlargement of metropolitan area and pollution problem. Almost 70% citizens stay in sub-health status proved by data. The problems come from working press and lacking exercise. Recently human pays more attention on health. The number of people taking part in all kinds of gym centre is increasing everyday.

A healthy life cannot leaves alone with pleasant ecological environment and active physical exercise. A good ecological environment makes people being carefree. Active physical exercise helps relieving the pressure. The emergence of sports Leisure Park satisfied the demand of material life and is being an inexorable trend of senior life.

Sports Leisure Park was post in 1990s, and has been admitted by many countries. The park of Grove in Sweden leaves a large free area for people to play around. With a swimming pool and series of water entertainments in the eastern part, it has a very high utilization rate in the summer. The western part provides winter sports. Different slope and the hillside topography in 30m hill make snowboarding convenient for teenagers.

The first sports park in Britain, Can piece - Hill Park, has swimming pool, Tennis field, boats, lakes, wharf and other facilities. In Britain, the lawn and laid groundcover area is mainly used for sports activity, which constitutes the special features of the park. These are the achievements of foreign sports Leisure Park.

With the Olympic Games held in China, China's cities also pay more attention to the construction of sports and Leisure Park in Beijing, Shenzhen, and Xian, also built Sports Park.

II. THE DIVERSITY OF SPORTS LEISURE PARK FUNCTION

The sports leisure park should meet the requirements in the sports events as well as function in post-match. Taking the self-running into consideration could not only satisfy the demand for pubic sports place of our citizen, but also avoid leaving it unused and reaching win-win. So during the designing of the sports leisure park, catering to the requirement competition and training should be taken into consideration. Besides, more attention should be paid to leisure and citizen's body-building entertainment. The function should be classified clearly.

As the water sports competition area of the Sixth National Urban Sports Meeting, Red Lotus Lake water sports training base aims at the international standard of water sports events and satisfies the technical requirements of the international competition. Meanwhile, it could be used as the state's Canoeing training center.

Considering it is one of the radiation areas and has marvelous view-sight for located on the bank of Red Lotus Lake, our government strives to build it into Water Park for the competition, training, tourism and leisure. It could be a good area for city residents to exercise and recreation in the free time.

III. THE TOPICALITY OF SPORTS LEISURE PARK

As for the landscape, it should reflect the parochial culture and the subjectivity of the park. The cultures can increase the taste of the park to attract more people. We can always find some scenic spots that have more fascination because of precipitate through history of long-term. The design of Sports Leisure Park can borrow these design methods to stress the topicality of the park. On the premise of according with the original master plan, the design should make good use of various landscape elements and organize the space to create elegant visual landscape, sports cape and culture landscape.

According to the analysis of planning and designing of Red Lotus Lake water sports training base, the whole base is made up with the competition part and the assisted part of leisure. The two parts, which will become the main frame of the landscape design, should be connected with a planning road. Our landscape design is spread out with "an axis and a chain". The axis is the landscape for competition watching, which can meet the function of competition and characterize the theme of water sports competition. The axis starts from the entrance, along the aqueduct for competition, penetrating through the whole base from the west to east.

The entrances in landscape planning of sporting places stress imposing manner and grandiosity. There are two purports of the entrance in design of water sports center. One is to manifest the feature of the sports center for water sports. The other is to improve the visual sequential space.

A. The gate of "Flying Rainbow over the River"

What comes into view firstly is the imposing gate, whose integral designed model looks like a flying rainbow across six pieces of red lotus petal. The red lotus reflects in the water. The waves comb; the rainbow flies across. It expresses the characteristics of the base as a water sports site. Meanwhile it symbolizes the hard training of the athletes and the aspiration that there must be a rainbow after the storm. It is how the name of the gate comes. Overlooked from the sky, the gate looks like a long shuttle. The banding openwork on the top of the gate makes it look light and handy and full of sense of modern. It also increases the changes of shadow under the gate. There is a bold idea of using materials and colors; the general color is warm yellow. The model of petals is beset with stripes of red materials. The gate also meets the requirement of fire-control and convenient transportation. The streamlined sides of the lodges located on both side of the gate make it seem open.

B. The Feature Wall of Canoeing

The scenicfocalpoint of the main entry door is a feature wall presenting the scene of canoeing with anaglyph. Since the space of the base is huge, in order to increase the visual hierarchy, a handling technique called "flange" in classical Chinese gardens is used, settling the feature wall here to form a good landscape visual hierarchy. At the same time, the pretty relief patterns actually increase people's feeling of beauty in the entrance area. In front of the feature wall stands canoe models sculpted by plants, which forms delightful contrast with the feature wall.

C. The Flying Olympic Rings

Walk on along the main road, the Olympic rings can be found made up with shrubbery mass of many colors. The postern round parterres are hung high by some color stripes as if they are golden medals in the competition.

D. The Wing-spreading Bridge

The bridge is named because it looks like a butterfly flapping the pretty wings. It symbolized that the athletes finally emerge from a chrysalis like a butterfly after hard training.

E. The Square for Rest in Athlete Apartment District

The square for rest in athlete apartment district is embosomed in trees. It combines grassland with paving floor. The whole square is a helical line spreading outside. The seats around provide a good place for people to have a walk and relax in the shadow.

F. The Natural Grass Slope for Watching Games

Taking advantage of the natural grass slope, the design combines the spectator seats with steps. Good grass seeds are planted in the lawn of the grass slope, which makes people can touch the grass directly. The slope result from the micro relief is like a carpet of flowers, which forms a good landscape. In such a way, the planting tree along the slope is the first stage in the landscape, the terrace is the second stage, and the grass slope is the third stage, thus the whole landscape is with rich hierarchies.

IV. THE PARTICIPATION AND JOVIALITY OF CREATING SPORTS AND LEISURE PARK

Modern Sports Park is not just to form urban landscape. More importantly it provides exercising and entertaining places for citizens. So it is essential to own the participation and joviality factors in it. The "first chain" of in the water exercise and competition base in Red Lotus Lake, as a landscape chain of natural landscape entertainment district, is just designed for attracting citizens to be there and do exercises after the match, to create a joyful and people-involved atmosphere.

Go straight by the left side of the entrance, along with the planning road, to the extending landscape avenue of the lake, you can see the three landscape regions connected into a circle in this landscape chain.

A. the open fire training region

It is a duty-bounded responsibility for a sports park to attract the public during the no-matching time and make them involved in the positive healthy sports activities. So how to attract the citizens during the no-matching time and self-sufficiently maintain the daily management is the main subject throughout our design. In recent years, open fire camping and expanding are becoming more and more popular among the public. Especially the youth and new rising enterprises would choose these activities as the best way to lead their weekends and for a company entertainment. Unfortunately, they have few choices, for Wuhan does not have many specialized bases for this kind of activities. According to the analysis of the base, the isolated region by the river which divided by the planning road, lying in the most northern part of the base is very suitable to design as a wild camping region. The landscape levels of the region can be more abundant by using terrain. The whole area consists of three major spaces, respectively, the bonfire square, campsites and expanding training camps. The bonfire square locates in the high point of the region. Spiral-lined road inlaid in the grass path extends to the tent camping and training area and connects with riverbank path, making people enjoy the wind caresses in face while doing sports in the nature.

Opposite to the wild camping region is the walking area which specially designed for the tranquility-loving people and the elderly. The forest-surrounding pathways are the best places for people to enjoy the fresh air.

B. Riptide cyclotron viewing region

1) central traffic island

As the planning road intersection, visitor flow rate is comparatively high, the traffic island is becoming an important landsacpe node. Firstly, the round-shaped traffic island sitting at the three-way intersectionm, its landscape will be visual focus. Here, we use a plant mode, such as sculpture landscape design elements, composition, asymmetrical manner to ensure that three directions have a better landscape visual field. The large sports theme sculpture "scale new heights forever" as a whole visual focus sitting on the ship-shaped framed flower bed, with a concise structure, was composed by the up-right cylinders. The arc on the top of cylinder symbolizes the water waves, aslo shows the characteristic of water training. The arcs of the two modelling groups stretch to two different directions, looking from a distance, like the eagle soaring in the sky, symbolize the positive athletes. The unfolding hands symbolize the friendship in contest. The growing trend of whole sculpture symblizes sports spirit of the constantly going beyond oneself, fighting tooth and nail, scaling new heights forever. The red and blue colors represent the color of national flag and water to echo with each other.

Under the ship waves mold lines and curves mold lines, symbolizing braving the athletes, forge ahead of the spirit of hard struggle, plus the mold lines around the spiral line was composed of sculpture – flower ship - mold lines of three levels of rich modeling.

2) Recreational game-watching square

Opposite the traffic island is the recreational game-watcing square which belongs to swift current circle round region. According to the current situation of the planning roads, the square of the flow line shows stretches out of square, with the main project riptide whirly curve photograph echoed. The whole square is divided into two regions in funtion and landscape level. The first level is horn square. The spiral curved horn looks like she is singing the triumph song for the athletes, also symbolizes the wispy ripple in the water. The pulling boat shaped picture on the road symbolizes the forge valiantly ahead spirit of athletes. In the north of horn square, there is a landscape wall which introduing the riptide activity, to show people the basic knowledge of the activity, popularize the general sports knowledge to the public. The second level is green square region. By combing the tree team, grass and pathway together, borrowing the method of mountain and water garden making. The region treats the greenland as water, main road as the wave, stone road as the boat, to create an over-crystal-blue-water landscape, also add certain interest to people's recreational tour.

C. Lakefront recreational region

The lakeline of the lakefront recreational region is designed in wave.along the lakeline there are natrual islands, yacht dock, water terrace and lotus pond.

Natural island loctes in the right hand of reception center, connecting with the land through the planning road, and have rich luxuriant plants on it.In the morning, the mist surrouded island on the water waves is just like Penglai reflecting into eyes, providing a best place to have a walk for the people who accommodate in reception center. What's more, the reserved wetland which correspoding to the natural island adds some wild factors here.

Following the riverbank path down to the south, then you will reach yacht dock and water terrace.Curve form echos with the waved lakeline, making the whole lakebank more consistent.the dense lotus crowds echoed with the natural island from afar, serves as the "period" of the riverline.

V. Plant Designs

According to the topography and the requirements of water training center, it makes every effort to maintain the pristine state and creates the visual value. Therefore, trees should be planted under the laws of nature to rebuilt phytocoenosis.

A. plant designs in competition terrain

Evergreen macrophanerophytes is planted as the background in the entrance to form well arranged cubical space and carry out a natural transition to the carpet flower bed. Macrophanerophytes and shrub are arranged as the background to the Sculpture Wall, which shapes the square space integration and displays heavy breathing.

The area behind the racing tract takes the advantages of the topographic relief and obedient to the natural distribution, constituting rich plant communities and clear transparent, shade relations. Standing on both sides of the border is the colorful bright flowers, which gives pleasant atmosphere. Evergreen trees form green barriers as a background. In order to guarantee the fairness and match with international level, the principle of ecology is applied to get the effect of strong ability of wind resistance, good hydrophobic through the composite structure of community.

B. Recreational Area Plant Design

Within the ecological island, the original wetland is completely retained. The new planting trees will increase with the original wetland as basic pattern. The slowing stream around the ecological island will be the main sedimentary area of pollutions. Therefore, the reeds and Typha orientalis Presl is immigrated to absorbing the pollution. Water resistant plants like ChiSha and metasequoia grow on the island.

Ornamental shrubs and feature wall in the sportsman apartment area construct a semicircle, bringing a closing feeling. Not only is the overall environment taken into consideration, but also the part decoration with colorful leaves. Thus, the whole give a few minutes lively.

In the river way, the plants prevent the slope being washed out by the rain and flood, preserving water and soil. As to the plants deploying, evergreen flowering suffruticosa plants or lianas whose root system is developed are chosen to form landscape features of lasting soil stabilization and perennial scenery. The tree species are periwinkle, common camellia, gardenia jasminoides cv.prostrata, buxus harlandii, sims azalea, Japan fatsia, China mahonia, loropetalum chinense （R.Br.） Oliver, procumbent juniper and jasminum mesnyi.

In the waterfront landscape area, the original waterfront environment and the species diversity should be taken into consideration. When design the planting, the original vegetational community should be combined with the artificial community to express the natural characteristics and to improve the ability of sustainable development on the lake bank. Planting weeping willow, Chinese tallow tree, cotton rose, metasequoia and cedar and water plants like cannas glauca, reed, fleur-de-lis, lotus, eichhornia crassipes, cattail and umbrella plant.

In the camping expanding area, maintain the original trees in the north part, design the plants in other areas according to the structure of plant community to preserve the original eubiosis. Along the garden, flowers that can grow in shade should be planted more to form natural habitat and colorful woodland landscape.

VI. Conclusion

Modern Sports Leisure Park is a multi-functioning modern one which giving priority to sports leisure, also focusing on urban landscape, popularization of science education and afforesting, etc. We should fully embody the theme and increase the cultural connotation in design, to meet the exercising and recreational needs of people. This design centers on the theme of water sports, integrate people's participation and sports culture into court, in order to construct a recreational sports park which you can enjoy nature and have an intimate conversation with nature.

References

[1] Li Weihong,"Research on Sports and Entertainment Space Landscape Design, "Nanjing Forestry University, 2006.

[2] Li Meijuan, and Deng Fengming, "New Concept on Modern Construction of 21st Century Park,"A cademic Paper of Hebei University of Construction and Engineering, 2006.24(1); 43-45.

[3] Ma Jun, and Meng Xiangbin, "Modern Views on Chinese Sports Park,"Chinese Landscape Garden,2005(4); 35-38.

[4] Meng Gang,and Li Lan, "Design of City Park, "Tongji University Press,2003.

[5] Wang Yujing, "Plants Scenery of City Park,"Liaoning Science and Technology Press,2003.

[6] Hu Changlong, "Landscape Garden Planning and design, "China Agriculture Press,2001.

[7] Zhen Xiaoxiang, "Connotation and Characteristics of Ecological Sports,"School Paper of Chendu Physical Institute, 2005.31(2); 43—46.

[8] Liu Binyi, "Three Major Fields and Direction of Landscape Subject,"China Construction Industry Press,2006.

Specific Geographical Waterfront Landscape Optimization Design

Jun Ren
Zhengzhou Institute of Aeronautical Industry Management
Zhengzhou,China
Email: luoyao810@163.com

***Abstract*—Owing to people's pursuit of spirit and natural love for the water, the waterfront scenic spot, as the unique environmental resource in the city, has been developing rapidly in the past few years. The Yellow River is the most valuable water resource in Zheng Zhou City, but the development of the tourism has brought about a lot of problems to the construction of the waterfront landscape in the Yellow River scenic spot. This article puts forward the ideas for the design of constructing the beautiful, pleasant and water loving landscape from the aspect of waterfront landscape, taking the tourists' aesthetic feature and behavior as the drawing dimension.**

Keywords-specific area; waterfront landscape;optimization design

I. INTRODUCTION

The water is closely related with life. However, once water is mentioned, Zhengzhou an inland city in the north of China would be embarrassed and heart aching. According the statistics, by 2010 Zhengzhou will be lack of water by 56,283million cubic meters. Production and living water situation is not optimistic, but it shall not quench people's passion for water culture and waterscape resource, and the water resources of the Yellow River scenic spots in the west city greatly satisfy the people's wish to be closed to the water and nature.

The Yellow River scenic spots are located 30 kilometers away from northwest of Zhengzhou. Since the 1980s, the construction of the scenic spots has never been stopped, and various entertainment and ecological projects have greatly met the needs of the urban residents. However, the development of tourism industry has also brought about a lot of problems for the construction and maintenance of the waterfront landscape in the scenic spots.

Firstly, the water ecology is destroyed. Overdeveloping the fishery industry and tourism are the main reason for the water pollution of the Yellow River resources. Owing to the reputation of the carps in the Yellow River, there are quite a lot catering vessels docking along the bank; because of the food processing all the waste water is poured into the Yellow River directly. In addition, some of the tourists' uncivilized behavior also causes water resource pollution to some degree.

Secondly, the construction of the waterfront landscape lacks of hydrophilcity. There are two scenic spots supporting by the Yellow River, Xinghai Lake and the Yellow River Bund. Xinghai Lake mainly conducts entertaining activities and part of the landscape facilities exist potential safety hazard. While the Yellow River Bund where people have to go does not have any construction. The bund has been severely air-slaked without any protective measures; speed boats under the bund are scattered; the roads on the bund have not harden already. The intimate contact between people and our mother river is blocked on the bund, even though some tourists go down to the shallow against risk, they still can not be satisfied. While as Zhengzhou is a city lacking of water resource people here are more eager wish to come close to the water than those from the seaside or region of rivers and lakes.

Thirdly, in terms of the mode of the landscape, the regional features and the audiences' aesthetic needs haven't been considered comprehensively, and some of the sceneries have reduced to formalized settings which can not arouse the audiences' interest. The customs and practices of the landscape are not profound enough. The regional feature of the folk-custom is very distinctive and can also have unforgettable impression on the new comers to the city. As to the present inheritance of the culture is only the restoration and indiscriminately copy without taking the new needs of the age into account.

A sound waterside environment can not only satisfy people's needs of tour, rest and entertainment, but also has good publicity for modeling natural and harmonious city image. The followings are the author's ideas for the optimization design of the waterfront environmental landscape in the scenic spot.

II. BUILD THE QUALITY ECOLOGICAL WATER ENVIRONMENT

First of all, ensure the continuity and completeness of the ecological pattern in the waterfront region including plan and use the land properly and protect the natural plants and the ecological safety jointing other plots. Secondly, maintain and improve the water quality and make sure the virtuous cycle of the ecological system. Thirdly, construct ecological greenbelt and enrich the planting landscape resource to form multi-layer plant community and organically combine the greening and the water. Areas near the water are suitable to plant damp tolerant plants such as weeping willow, dryland willow, Chinese ash, poplar tree, metasequoia and Chinaberry etc. partially with emerging, submerging and free floating plants to form the stereoscopic community structure with the water as interface.

III. THE CONSTRUCTION OF THE HYDROPHILIC LANDSCAPE

"The so-called hydrophilcity is that people can go the river side or a physical phenomenon that can be approached easily; in other words, it is a psychological phenomenon within the reach of the hands."[1]People's loves for the water because of their nature, and people of different ages have different feelings for the water. The exploration of the children is of curiosity and imagination and the adults are more implicit; they are likely to wander along the river banks chatting, taking group photos dynamically and statically. While the elderly prefer to stay alone fishing or sitting peacefully, which is quieter? Because of the diverse hydrophilic behaviors the waterfront region takes on the vitality and vigor.

A. Xinghai Lake is still water region with gentle slope needless of demanding flood prevention and control, and it can enable the tourists feel peaceful and quiet. Directed towards this features the followings shall be handled:

1) The bund

The bund is the construction nearest to the water, and it is interdependent with the water environment, especially within the tourists' vision range, therefore the construction of the affiliation and hydrophilcity of the bund should be taken into account. Sceneries can be set alongside or the shoreline can be built as tracing pattern to enable more points of sight of tourists to come close to the waters. Meanwhile, the curve both concave and extruding the lake water can bring in different psychological feelings for people. Moreover, the woods and stone materials can be added to the material for the bund construction. The natural color of the wood is plain of appetency, while the natural shapes of the cobbles and lakeside rocks and other stone materials are simple and unsophisticated and interesting .They can be combined with the water environment to form harmonious bund landscape. Meanwhile, the pattern of the water plants shall be reconditioned to increase the bund appreciation degree.

2) The bridge

The bridge is the expansion of the road. It is also a good way for the tourist to come close to the water. The location, the size and the style of the bridge are also important factors influencing the waterfront landscape. The surface of Xinghai Lake is not wide. The purpose of building a bridge over the water is for part of transportation as well as for the tourists to appreciate the sceneries and also for coming close to the water and splashing in the water. The style of the bridge can be the level bridge and the curved bridge etc. The major design of the level bridge should be brief and light and grateful the water or a bit above the water. While the body of the curved bridge is zigzag and changeable that can provide different angles for the tourists to appreciate, thus the bridge should be built in the scenery-rich environment.

3) Waterfront platform

The waterfront platform is a commonly-used means for the tourists to come close to the water. Standing on the platform, they can have a broad vision and can deeply feel the contact with the water. Build a waterfront platform of the length and depth about ten meters within step-reach to the water and of guardrails alongside set according to the safeguarding by the shallow waters along the bund, and on the platform facilities conformed to the environment such as the chairs and lamps etc.

4) The stepping stones on the water surface

The stepping stones are crossing water facility generally set in the shallow water for people to splash in the water and come close to the nature. They can be irregular natural stones set according to a certain footstep to form the natural lifesome water transportation. The tourists can paddle, stroll and rest on the stone.

5) flowing water landscape

Except sight and touch, the tourists' sense for the water there still exist hearing. The flowing water can also bring life to the environment. According to the resource condition water fall scenery of less fall can be constructed by piling up hillstones and priming water from the bottom and then flowing down into the lake from the top of the hills; or using the water wheel of the scenic spot to change the embarrassing situation as a decoration, and lifting the water by the rotation of the wheels and then pouring water into the lake from the water wheel.

6) Cultural landscape

The Yellow River is the origin of the civilization. The Yellow River culture developed from the Yellow River civilization is one of the most bonds of Chinese culture. The ancient people left their footprint of the development here, and also many historical legends and stories were circulated in this region. Xinghai Lake scenic spot is a newly developed one. It does not only enable tourists to come close the water but also provide a lot of entertaining facilities. Integrating its features when designing the cultural landscape infotainment and interestingness can be added .For example, when constructing the water fall scenery the pattern of the ancient pottery of the Yellow River can be combined---the water fall flows from the pottery into the lake, which will enable the tourists to experience the lingering charm of the Yellow River culture in the interesting pattern of the landscape.

B. Appreciate the Yellow River within a short distance on the bund of the Yellow River. The water of the Yellow River has become mild at this place. Standing on the bund the vision is broad and it can bring in a grand feeling for people. According to this feature, the followings can be processed:

1) the bund

For the mother river people are full of longing and enthusiasm。 "Ambition never dies until one reaches the Yellow River" also conveys the same eagerness. The bund of the Yellow River is at the state of aboriginality piled up with natural loess and the bund wall which is often rushed by the water has been protected by masonry of concrete and bricks. The color of the materials for the bund wall and the pictures of the masonry can be blended with the ancient elements of the Yellow River. Besides, changing the straight line of the bund and highlighting the water fall with a certain fall and the waterfront platform or the fountain plaza can not only meet people's needs but also avoid the Yellow River shoreline becoming too garish. Thirdly, the material of the material for the road surface hardening processing should mainly be beautiful, permeable and durable. It is not only convenient for the surface water circulation but also can prevent the road surface slip. The color, the size and the texture of the mat formation should be designed integrated with the bund to construct a comfortable and beautiful footpath environment.

2) the fountain plaza

Under the condition of flood prevention, if there is a comparatively large square stretching towards the Yellow River, for the people it undoubtedly step closer to the mother river. Build a large area of hard fountain plaza integrating the bund. Set up chairs, parterre and lamps on the fountain plaza. Integrating the ancient culture of the Yellow River embellish human landscape into it. This will enable the fountain plaza mixed with the environment closely. Thus, when people are standing on the plaza facing the Yellow River, they can feel the embrace of the mother river and experience the ancient and modern beauty enhance each other.

3) The Beach

The Beach is some beach below the bund. The water of the Yellow River seems to stop to have a rest here. People stand in the water with their trousers legs rolling up. They are taking photos and amusing themselves fervently. Combining with the bund construct the beamy steps level by level downwards to the lower water shallow and usher in a quality hydrophilic beach in the safe zone for people to paddle and amuse themselves. The hydrophilic beach belongs to the soft piece surface hydrophilic landscape, its distance away from the water is nearly zero and can greatly satisfy people's need to be close to the water and nature.

4) the cultural landscape

The cultural landscape along the Yellow River is the critical point of the whole scenic spot. The ancient origin of our country's civilization, the early capital aggregation and the branch of the Chinese civilization are also the content of the Yellow River。 The theme stature and stature gallery can be modeled on the fountain platform. And on the road of the bund, the totem-poles showing the ancient civilization of the Yellow River or the sculpture carved with Chinese family names seals can be designed. On the beach interesting feature sketch such as Yellow River carps can be designed.

IV. MODEL THE LANDSCAPE WITH REGIONAL FEATURES AND FOCUS ON THE AUDIENCES' AESTHETIC SENSE

The landscape is an art integrated all kinds of elements, and it is the carrier of the local history、 culture and customs. As a visual image, its design should stand on the side of the audience, thus when designing the landscape first thing is to start out from the audiences' mentality and plan the adopted elements from the micro to macro and from the part to the whole.

The sense of aesthetics is a kind of mental activity of aesthetic awareness and emotion aroused on the base of the environmental consciousness when the tourists are traveling. "Consciousness is the entire reflection on each part and attribute of the objective substances that directly act on people's brain"[2]The history、 culture and customs in the structure of the landscape is of regional tendency and its image attraction and identifiablity combing with the feelings of the tourists to form the sense of beauty. And this kind of beauty conforms to people's aesthetic interest. Besides the visual aesthetic feeling people hope for understanding of the connotation behind the material. Thus the visual and the mental aspects will resonate and thus the landscape is not only a form but of features of humanity.

V. CONCLUSION

Waterfront landscape optimization design should coordinate the relationship between man and nature first, and deal with the utilization issues of different plots. And then, planning and creating the landscape optimization design by the rhythm, balance, contrast and unification of aesthetics based on the aesthetic psychology of the public. The good visual not only can satisfy the basic visual aesthetic, but also can meet people's spirit desire for the water, as well as the understanding and love for the cultural folk. Thus, it can create the waterfront space which features with individuality and multi-levels and enhance the scenic beauty experience of beauty spots. The innovation of this paper is: For the Zhengzhou City waterfront landscape research and practice provide a realistic theoretical basis, which will also cover the areas of Henan Province, has long been weak, and seek for similar subjects in-depth play a positive theory of reference for.

REFERENCES

[1] Edited by Japan Society of Civil Engineers, Waterfront landscape optimization design, Translated by Sun Yizen, DaLian, Dalian University of Technology Press, 2002-11 (1) : 79

[2] Chiefly edited by Guo Liyan, Psychology, Nanjing: Nanjing University Press, 2002-08 (1) : 99

[3] Liu Binyi, "Modern Landscape Planning and Design," Nanjing: Southeast University Press,1999.

[4] Deng Tao, "Principles of Landscape Design Tour," Beijing: China Building Industry Press,2007.

[5] Steven • C. Brazzaville, "Landscape Aesthetics," Translated by Feng Peng, Beijing: Beijing University Press,2008.

Study on Spatial Planning Strategy of the River-Crossing Cities in the Middle and Lower Reaches of Yangtze River Based on Low Carbon

Chuancheng Li
School of Architecture and Urban Planning
Huazhong University of Science and Technology
Wuhan, China
Email: chuanchengli@163.com

Jin Yu
Civil Engineering and Architecture School
Wuhan University of Technology
Wuhan, China
Email: yujin613@126.com

Abstract:The sustainable growth of CO_2 emission has enormously threatened the globe ecological environment, and the reducing of CO_2 emission consequently became a focus question concerned by the whole world. Studies indicate that the urban spatial structures has partly restricted the urban carbon emission and the sustainable development. This paper analyses the spatial planning strategy of the river-crossing cities in the middle and lower reaches of Yangtze river from three aspects of urban traffic, urban form and industrial pattern.

Keywords:The Middle and Lower Reaches of Yangtze River, River-Crossing Cities, Low Carbon, Spatial Planning

Rapid economic development and accelerating process of urbanization inevitably lead to the increase of the total amount of urban energy consumption and CO_2 emission. China has been the country with the largest amount of CO_2 emission. For the need of being responsible for world morality, energy security and environmental protection to our own, China needs to pursue an ideal model for urban planning to resolve the hazards to society brought by high-carbon cities [1]. Domestic scholars give the idea of low-carbon city concept icluding both economic and society parts which advocates goverment to transform the economic mode into a low-carbon one and the citizen to adopt the low-carbon lifestyle mode.

I Low -Carbon City Spatial Planning Strategy

A. *Study on Low-Carbon City*

Internationally, many countries begins to study the low-carbon urban planning theory and implements low-carbon strategy at all levels in their own cities. At present, the experience of developed country to build low-carbon city include developing new clean technology and clean energy, spreading the use of sustainable development design and construction, carrying out highly efficient transportation planning, promoting resource recovery and utilization and green consumption behavior[2]. In January 2008 the World Wide Fund for Nature (WWF) officially launched China's Low-Carbon City Development Project in Beijing. Shanghai and Baoding(in Hebei province)are selected as the first batch of experimental city. In the same month Sino-British Chongming Dongtan eco-city project had been put into practice[3].Meanwhile, the United Kingdom was going to Strengthen cooperation with China in environmental protection. In January 2009 International Climate Organizations promulgated China Low Carbon Leadership: City report in Beijing. It firstly proposed an urban leadership system of low-carbon economy according to the research on development cases of twelve cities represented by Beijing city. In November 2009, the new version of "Wuhan Urban Master Plan (2009-2020)" obtained Global Outstanding Contribution Award issued by the International Institute of Urban and Regional Planners (ISOCARP) because of its low-carbon city concept.

B. *The Content and Strategy of Low Carbon Urban Planning*

The implementation of low-carbon urban planning embodied in two levels, one is management level composed of government and planning department; the other is pubic level composed of resident and enterprises. The management controls urban spatial form through rational planning of construction, transportation and industry. Through changing residents' lifestyle and trip mode as well as enterprise's mode of production which work together in the following four areas, that is, the use of renewable energy, improvement of energy efficiency, control of energy demand and improvement of the carbon capture and sink, in order to achieve the ultimate goal of reducing urban carbon emissions. The study focuses on the management level, that is, the research of urban spatial planning.

Urban spatial planning mainly reveals the rules of the formation, evolution and development of urban space. The research results are reflected in urban planning, which controls and guides the urban space development through urban planning mechanism. So far, there isn't any planning type named by urban spatial planning until the appreance of strategically planning centers on urban spatial development strategic research in recent years[4]. It mainly consists of the control of urban density, construction of urban traffic system, assurance of rational use and multi-function of urban land as well as industry distribution. Studies indicate that the urban statically structures has partly restricted the urban carbon emission and the sustainable development. To develop an ideal low-carbon city, low-carbon urban spatial planning theory and methods are necessary key technologies.

According to China's Sustainable Development Strategy Report, the basic support system of low-carbon city is divided into the following four aspects:

- Industrial structure system: to accomplish the transformation of industry to service industry and heavy chemical industrialization to high-processing degree is beneficial for China to reduce energy consumption and develop low-carbon economy.

- Infrastructure system: to make the overall planning of urban infrastructure in advance and guarantee the design of urban infrastructure being low-carbon.
- Consumption support system: in order to achieve city's low-carbon development, people should change their previous lifestyle of high consumption and high-waste.
- Policy and institutions system: to establish reasonable and proper institutions and policies, prop and integrate the existing policy systems and means, identify long-term goals of the low-carbon city development, then show mass society the government's resolution of realizing carbon emission reduction or zero emission with the whole society .

C. *Low-Carbon Urban Planning Strategy*

Studying the current cases of low-carbon city planning at home and abroad, the conventional method of low-carbon city planning is divided into the following two:

1) Low-Carbon Space Strategy

Implement low-carbon building and low-carbon communities gradually. Low-carbon buildings means that throughout the life cycle of manufacturing building materials and equipment, construction, utilization of architecture, we reduce the use of fossil energy, improve energy efficiency and reduce carbon dioxide emissions.

Develop the mass transit, reduce the proportion of private car journey. Take Hangzhou for example, subway, buses, taxies, aquatic buses, and free bicycles will construct Hangzhou's futuristic big public transportation system. What match it are the construction of bicycle special-purpose road and especially the construction of slow traffic system of urban river course. At the time of expanding the restricted driving area of high- pollution motor vehicle, we need to expand new energy automobile application test and so on[5].

Make use of land intensively. The limiting degree of city land utilization is inversely correlated with the amount of carbon emissions of city resident remarkably. The stricter of the constraint for land utilization in city planning is, the lower lever of the amount of the carbon emission from residents' livelihood is.

Increase the forestation to absorb carbon dioxide. Constructing ecological city and forest city is an important way for carbon sink. Systematized forest resource is possible to meliorate the climatic condition of entire city, reduce the tropical island effect, improve air quality and enhance water purification, also it can reduce the light pollution, promote human settlement.

2) Low-Carbon Energy Strategy

Adopt renewable green energy. Replacing the traditional energy with solar energy and wind energy is the most direct and effective method to change city's energy structure[6]. The research and development of new energy also concerns the use of new energy vehicles, the construction of energy conservation architecture and the optimization of industrial structure.

Save energy, recycle water resources, reclaim and reuse garbage. Denmark's Copenhagen is the first country to realize it. They make a precise classification of 65% of the garbage for reclaiming and recycling. 27% of the garbage is used in refuse-burning power; the left 8%, the incombustible metal and virulent plastic and so on, will be buried[7]. This garbage disposal is one part of Copenhagen's low-carbon life style.

Advocate thrifty life style, reduce commodity waste. It Includes: use the air conditioning sparingly, recycle garbage and make garbage classification to the best of our ability, never use the disposable tableware, select the public transport means as far as possible and walk more, use energy conservation light bulb to substitute tungsten lamp, plant more trees and flowers and so on, which help to establish low-carbon mode of life for city resident.

II. The Characteristic Analysis on the River-crossing City in the Middle and Lower Reaches of Yangtze River from the Perspective of Low Carbon

Because of the differences in cities' location, morphology, scale and economic development level, different customized methods should be adopted when carrying on the low-carbon plan. The mechanically application of general low-carbon urban planning method is unable to achieve the goal of reducing the urban carbon emission. The middle and lower reaches of Yangtze River can also be called the middle and lower plain of Yangtze River. The plain of middle reaches of Yangtze River include：Jianghan plain in Hubei province, Dongting lake plain in Hunan province (the two be gathered as two lake plain), Poyang lake plain in Jiangxi province; the plain of lower reaches include: Yangtze River coastal plain in Anhui province and Lake Chaohu plain in the middle of Anhui province as well as Yangtze Delta between Jiangsu, Zhejiang and Shanghai.

A. *The Scale and Morphology of Cities in the Middle and Lower Reaches of Yangtze River*

The middle and lower reaches of Yangtze River is china's one of the most densely populated regions and one of the most economically developed regions. It involves six provinces and a municipality directly under the central government. Six provinces refer to Hubei Province, Hunan Province, Anhui Province, Jiangxi Province, Zhejiang Province and Jiangsu Province; the municipality refers to Shanghai municipality. The central cities in the middle and lower reaches of Yangtze River include Shanghai, Nanjing, Wuhan, Nanchang, Suzhou, Wuxi, Changzhou, Nantong, Wuhu and Changsha.

（Table 1，Fig. 1，Fig. 2）

The study of the spatial development strategy for the low carbon dioxide cities is aiming at the river-crossing cities in the middle and lower reaches of yangtze river such as Shanghai, Nanjing, Wuhan, Changsha and Nanchang. There are many cities of this kind, with intensive population, large amount of carbon emissions, distinct features and prosperous economy, which can provide sufficient economic foundation to develop themselves into low carbon cities.

B. *Large Energy Consumption and Obvious Heat Island Effect*

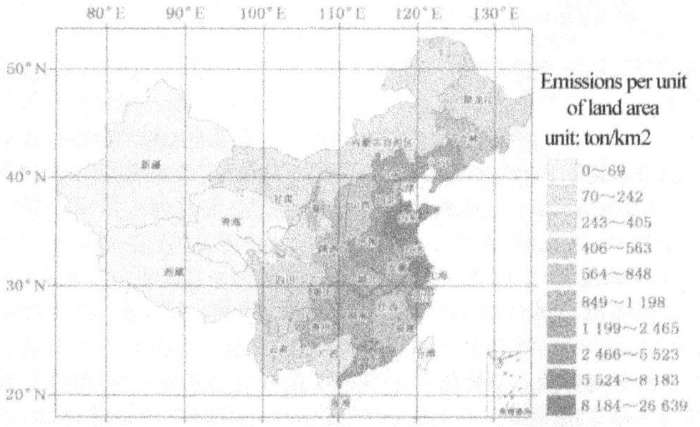

position while Nanjing Wuhan Changsha and Nanchang are all important provincial cities in china. These cities have large population and intensive motor vehicles, owning lots of energy-consuming equipment such as boilers and heaters. Both these machines and human activities consume massive energies and generate massive CO_2 . The consumed energies are transferred into the atmosphere in the form of heat and enormous artificial structures such as concrete, asphalt changed the thermal properties of the underlying surface, which lead to the increasingly obvious heat island effect.

C. *Rich Water Resources and Distinct Climate Features*

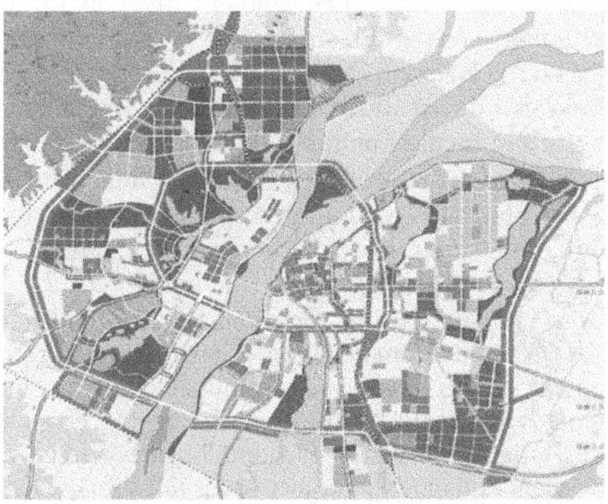

Fig. 4: The Master Planning of the Central District of NanChang

In the middle and lower reaches of yangtze river, the cities and the rivers crisscross each other while lakes smatter here and there. Most of the river networks in shanghai are part of the Huangpu river system, mainly with the Huangpu river itself and its tributary rivers such as Suzhou river, Chuanyang river and Dianpu river. Not only the Yangtze river, Hanjiang river, Daoshui lake, Nieshui lake and Jushui lake, there are also more or less 166 lakes in the city of Wuhan. Besides the Xiangjiang river in the city of Changsha, the capital of Hunnan province, there are other 15 tributary rivers, mainly of which are the Lliuyang river, Laodao river, Lejiang river and Wei river. The main rivers in Nanjing are the Yangtze river and the Qinhuai river, with the Chuhe river in the north, the Chunsu canal and the Tiansheng bridge river in the south. The city of Nanchang is with intensive river networks, having Ganjiang river, Fuhe river, jade-belt river and Jing river crisscrossing through the city and hundreds of lakes. The lakes in the urban areas mainly consist of four lakes outside the city and other four lakes within the city. The city surrounded by lakes and the lakes in the city coexists harmoniously.

Most of these cities lie in the northern subtropical monsoon climate zone, whose climate is characterized by "hot in summer and cold in winter." Compared with other districts at the same latitude, cities in the middle and lower reaches of yangtze river are much hotter in summer and colder in winter, therefore, it is hard to meet the comfortable living standards. Still, some cities such as Wuhan and Changsha have the typical geographical features of Lacustrine, so the ecological systems of these cities are vulnerable. In order to develop these cities, they should on one hand deal with the peak energy-consumption in both summer and winter and the constantly deteriorating climate conditions; on the other hand, they have to protect the vulnerable urban wetland.

D. Massive River-crossing Traffic Volumes Due to the River-crossing Development of Cities

These cities are naturally divided into two parts since rivers or lakes run through them. Shanghai is divided into Puxi which includes 9 districts and Pudong which includes Pudong new area, primary Nanhui area, part of Minghang area and the eastern part of Fengxian area. Wuhan, by the separation of the Yangtze river, is to be two parts with Wuchang district on one side of the Yangtze river and Hankow and Hanyang on the other side. Nanjing is divided into the northern part and the southern part by Ganjiang river. And Changsha turns into the eastern and western districts since the winding through of the Xiangjiang river while Nanchang is divided into the southern city and the northern city by Ganjiang river. The construction of bridges and river-crossing channels make the barriers of river to be nothing and change the development mode from the former river-along development to the new river-crossing development.

Different reasons such as the history, various functions, condition of the place or the sequence of the development cause the imbalance development on two sides of rivers. The need of the space on the other side and the potential advantages of the land can be realized through the construction of bridges and river-crossing tunnels. The river-crossing transportation can narrow the gap between the two sides to form a new pattern of river-crossing development, with the integration of both sides, the functional complementation, the mutual advancement and prosperity. However, it is these advantages that lead to the enormous increase in the volume of river-crossing traffic. Therefore, these newly-built river-crossing bridges and tunnels have become an important part in the new round general plan of these cities.

III. The Spatial Development Strategy for Low-Carbon Cities

A. *The Strategy for Transportation Planning*

The main method of calculation of bp carbon-dioxide in carbon network is to multiply the distance by the carbon-dioxide emission index M of the corresponding mode of transportation. M is the amount of carbon-dioxide emission of per person per kilometer under the given mode of transportation. And each mode of transportation has its own index.

Table 2: The Value Table of Carbon Dioxide

Value Range of M	Transportation Means and Value
0<M<0.05	electric vehicle:0.043;subway:0.042; minibus:0.055
0.05<M<0.08	Deisel Train:0.06; Auto-Rickshaw: 0.061; Vintage Trains: 0.065; Heeper: 0.069; Mopeds: 0.073
0.08<M<0.1	Snow mobile: 0.094; autocycle:0.094
0.1<M<0.2	Compact Car:0.11;streetcar or trolley-bus::0.115; Medium car:0.133; Full-size car:0.183; Hydrogen Fuel Cell car:0.17
0.5<M<0.6	Ferry:0.53;yacht:0.53

For big cities, they should develop TOD mode for the economic use of land and the energy-saving and emission-reduction. Center on the central area, it takes public traffic as its growing axis with the axis center structure which links the main city center and secondary city centre[8]. The nearer the axis center, the higher the density, and vice versa. Among axises, gardens and forest are integrated. Through this way, the spatial distance is increased , however, with the rapid transit on the axis, time distance is shorter on the contrary, cities become compact and energy use get integrated. In terms of planning theory and approach, with the mode of public traffic-oriented development, it practically integrated the urban land use planning and urban road and transportation planning. And this model follows the internal demands of low-carbon city planning.

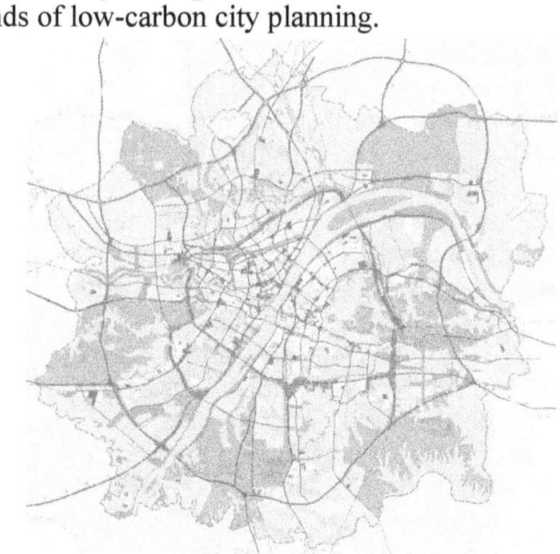

Fig. 5: The Master Planning of Wuhan

Take Wuhan city for example, the main urban area of it is divided by Yangtze river and Hanjiang river.The planning of river-crossing traffic is the significant and difficult point of the urban traffic planning.Yangtze River Bridge and the Second Yangtze River Bridge accounts for more than 90% of the total demand. It predicts that the vehicle's daily demand for river-crossing will come up to 300,000 to 350,000 and 450,000 to 500,000[9].In order to solve the problem of river-crossing traffic, enlarge the scale of center district and realize the urban intensive development, it must accelerate the constriction of river-crossing tunnel. Through the improvement of the river-crossing transportation condition, it will also promote to increase the river-crossing traffic capacity of private cars at the same time, and thus bring about the explosion of CO_2 emission. Therefore, low carbon subway and light rail for river crossing must take precedence when carrying on the planning of river-crossing transportation.

B. *Strategy for Urban Morphology*

It will cause the air circulation lower if too many high-rise residential housings or intensive structures locate in the city proper of the metropolis. In the city planning, city channel should be considered for ventilation to effectively relieve the heat island effect. If the trend of urban roads is the same as the dominate wind direction of the city, wind from rivers and lakes can go through the main urban district serving as an wind machine. Thus,the regional microclimate is improved, while the usage frequency of air conditioner is reduced at the same time, and then the goal of carbon emission reduction can be achieved.

The urban rivers not only possess the function of water draining, but also constitute the significant wind regimes. In the city planning, it should give full play to the functions of these waters to keep the balance of city ecosystem, regulate the regional microclimate, form the beautiful cityscape, build pleasant waterfront space, and transmit the history culture.

In 2005, cooperating with Huazhong University of Science and Technology, Wuhan Planning Bureau carried out a Research on the Relation Between Urban Structure Planning which was already involved in the new general planning of Wuhan city and focused on the six wind regimes of more than ten lakes of the six ecological water system .The strategy has obtained recognition from international planning community by its function of carbon emission reduction.

C. *Strategy for Industry Layout*

Industry layout theory researching on the industry's spatial layout is a national economic destruction and strategic arrangement which involved in multi-level, multi-industry and effected by variety factor. From the energy consumption structure, industry, civilian use and business have always been the main consuming subjects over the years. Therefore, It should reduce carbon emission through the adjustment of urban industry structure, and accerrate the transformation from high energy-consuming industry to low energy-consuming one into the low-carbon spatial planning. River-crossing cities, relying on its bank lines, can vigorously develop waterfront industry with urban character and tertiary industry

With the successive exploitation of riversides located in Hankow district, Hanjiang district, Qingshan district and Wuchang district, Jiangtan parks of the two rivers and four banks become new scenic, attracting hordes of visitors both at home and abroad and promoting the development of tourism in Wuhan city.Moreover,Wuhan municipality implements the renovation project of Yanjiang Avenue which requires the government agencies to collectivity remove in order to vacate the golden place for attracting investment and build a business district along the river which integrates high-end enterprise headquarters and commercial enterprises as a whole. Besides, Wuhan government introduces foreign capital for Wuhan Newport expansion project, expecting to build the largest inner harbor of China.It is predicted that, by 2020, the cargo handling capacity of Wuhan's Newport will reach 210,000,000 tons and container throughput will reach 3,800,000TEU.The series of these industry distribution strategies not only optimize the industry structure, accelerate the economic growth of Wuhan city, but also reduce the total amount of carbon emission of the city.

[1] Gu Chaolin, Tan Zongbo,Han Chunqiang, *Climate Change and Urban Low-Carbon Planning,* Southeast University Press, 2009.

[2] Wang Zhuanglin. *International Strategy and Experience on Low-Carbon City.* Resources and Human Environment, 2009,(15),pp.49-51

[3] Liu Zhilin, Dai Yixin,Dong Changgui. *Low-Carbon City Concept and International Experience.* Research on the City Development . 2009,(06),pp.2-7.

[4] Ding Chengri. *Urban Spatial Planning:Theoretic Approach and Practice.* Higher Education Press (HEP).2007.

[5] Pan Haixiao, Tang Yang, Wu Jinyu,Lu Yuan, Zhang Yangfei. *The Spatial Distribution Strategy of Low-Carbon Cities of China,* Journal of City Planning.2008,(06),pp.57-63

[6] Crawford J, Susmita ,et al ,A low-carbon future: spatial planning's role in enhancing technological innovation in the built environment,. Energy Policy,(36),pp.4575-4579.

[7] Ministry of the Environment.2007.Building a low Carbon Society.Japan:December.

[8] Chen Guowei. *Basic Theory and Practice of Low-Carbon City,* City Planning of Jiangsu Province, 2009,(07),pp.41-44.

[9] Sun Xiaoli. *Research on River-crossing Traffic Problem of Wuhan City.*Traffic and Transportation,2002,(05),pp.22-24.

	Shanghai	Nanjing	Wuhan	Nanchang	*Suzhou*
Designated function of the city	Municipality	Provincial capital	Provincial capital	Provincial capital	*Prefecture-lever ci ty*
Ranking of GDP	1	16	12	50	*5*
Dimension of urba n areas	2648	5844	1171	617	*1650*
Population of urba n areas	1453	541	560	359	*238*
Development mode	River-crossing dev elopment	River-crossing development	River-crossing development	River-crossing development	Single side development
	Wuxi	Changzhou	Nantong	Wuhu	*Changsha*
Designated function of the city	Prefecture-lever city	Prefecture-lever city	Prefecture-lever city	Prefecture-lever city	*Provincial capital*
Ranking of GDP	8	32	27	114	*23*
Dimension	1659	1669	1692	720	*969*
Population	267	226	216	105	*289*
Development mode	*Single side development*	*River-crossing development*	*River-crossing development*	*River-crossing development*	*River-crossing development*

Dimension: square kilometers Population: Ten thousand people

Table 1: The Development of Central Cities in the Middle and Lower Reaches of Yangtze River in 2008

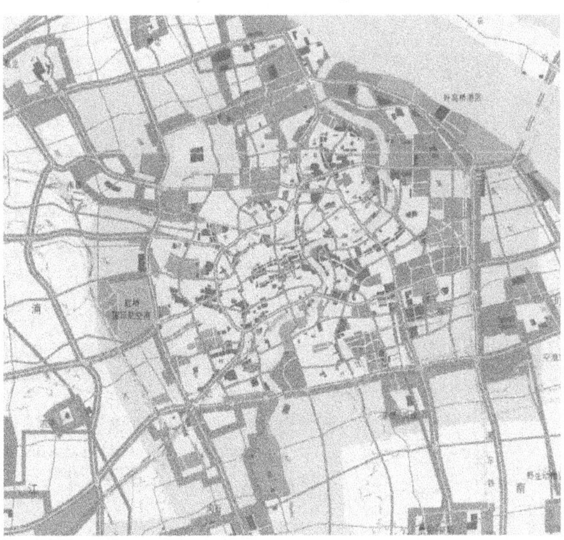

Fig. 1: The Land Capability Map of Main Urban Areas of Shanghai

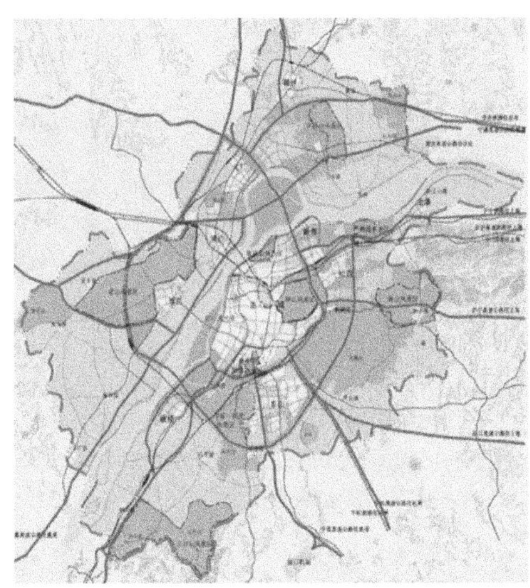

Fig.2: The Planning Map of Urban Development Areas of Nanjing

Study on the Relationship between High Technology Inclination of Fashion Design & Humanity Origin

Liu Zhouhai
Art & Design Department
Zhongshan Polytechnic
Zhongshan, China
Email: lzhouhai@163.com

Fan Fei
Art & Design Department
Guangdong Baiyun Institute
Guangzhou,China
Email: fanfeicute@126.com

***Abstract*—Oneness and reciprocity between high-technology trend and humanity origin is discussed with analysis of the relationship between high-technology trend of fashion design & humanity origin, viewpoint of connection of high-technology inclination and humanity origin in future fashion design is gained.**

Keywords-high technology; humanity origin; interaction

I. INTRODUCTION

Fashion design is a creative behavior to construct human and culture environment of garments through garments products orientation, visual scheme, marketing, design of style, design of color and flower pattern, structure and technique design, propaganda design etc. High-technology trend of fashion design gets more and more evident with advancement and development of technology, from high-technology fashion material, high-technology fashion marketing, intelligent design development system to high technique garments technology productive system, all of which affect design action of fashion. Art design serves for people, which is the functional purpose established from the produce and beginning of art, There is no exceptions for fashion design, and because fashion design action is a creative action that is inseparable with people, during design period, it's very essential to let humanized design ideas melt into fashion design, especially for fashion design develops towards the trend of high technique in nowadays, it has practical significance to study on the relationship between high technology trend of fashion design & humanity origin.

II. APPLICATION OF HIGH TECHNIQUE AND HUMANITY ORIGIN IN FASHION DESIGN

A. Application of high technique in fashion design

Application of high technique in fashion design embodies mainly in three facets including costume material, costume technology and garments marketing, the most direct facets are costume material and costume technology. In a way, fashion design starts from costume material firstly, the development of costume material is result of science technology development, making a general observation of costume material development history, we could notice that it's science technology which promotes the development of costume material, as well as science technology is changing people's thinking [1],including design thinking. Appear of New thermal fabric, temperature-sensitive fabric, carbon fiber fabric, health care fabric changed fashion design traditional ideas, which makes design create infinite possibility. Besides, science and technology promotes costume technology equipments as well as methods, high technology costume technology equipments could improve costume technology quality and efficiency, making costume technology methods varied, to promote diversified choice of technology methods of design, for example computerized embroidery, computerized carve, computerized printing, which provide various choice for decorative technique, it's undoubtedly promote diversity choice of design elements. Besides, high technology affects fashion marketing methods directly and indirectly, fashion marketing methods changed from selling goods by bawling to internet buying and selling, high technology has changed marketing methods of garments, therefore, application of high technology in fashion is the most close and the most main method, and it's the trend in future fashion design. High technology here is not high technology style design that is pure meaning, but techniques which have a sense of high technology or design thinking inclination of applying high technology, otherwise, it's rather biased to comprehend high technology inclination of fashion design from pure high technology style.

B. Application of humanity origin in fashion design

- Origin means initial point, here it means fashion design creative initial point, humanized origin indicates that "taking humans as the emphasis", initial point for serving the people, it might be comprehended as "taking humans as the emphasis" or "humanized design" in fashion design.
- Art design serves for people, the final target was fixed at the beginning of art design produce, whose connotation is rich change, from beautify people's life to "taking humans as the emphasis", from man-machine relationship to fitting people, all of which embodies changes of design thinking.
- Fashion design is design that has direct relation with people, application of humanity origin in design in fashion is necessary, it mainly embodies in garments' functional design, garments' function includes two facets that are substance function and

spirit function, substance function mainly means garments' using purpose and utility, as well as all of the factors that matches substance function, for example garment structure, using method, Man-Machine Engineering etc. Spirit function mainly indicates design need of psychology need in garments using period. Here it mainly indicates tradition and culture fusion in fashion design. Humanity origin is important content of design creativity, no matter on garments' substance function and spirit function, design is optimum method that will save people in people's future living especially for present society which is in a state of city expansion, environmental pollution, resource exhausting, insufficient ideological faith. Taking humans as essentials, green design, humanized design have to be main inclination in future design. Application of humanized design in fashion will be general day by day.

III. Unity of High Technology and Humanity Origin in Fashion Design

A. *Unity between high technology and humanity*

- Origin Modern design advanced quickly with high technology, the close relationship of design and new material, new technology, new environment, new cultural life and new social structure is the inclination of design development, humanized design will be the inclination of future design development, therefore, there are something in common between high technology and humanity origin under inclination of future design development, which is especially prominent in fashion design action, it shows the development unity between high technology and humanity origin from application of high technology in fashion design as well as application of humanity origin in fashion design discussed preceding, that is both high technology and humanity origin are inclination of nowadays and future, and there isn't any contradict and opposition between them.
- Firstly, watching points of agreement from science angle. Science takes humans as essentials, which is the spirit science development advocate, making a general observation of development history of science and technology, it could be said that all of the scientific technology develop toward humanized direction, it can be proved amply. Present application of high technology nanotechnology materials in garments make garments have multi-function, for example, antibacterial, waterproof, antifouling, which is necessary to satisfy humanity, these garments could be prophylactic, you don't need to bring umbrella when going out, clothes needn't to be washed etc. Therefore, there is inherent unity between high technology and humanity origin [2].
- Secondly, watching points of agreement from art design angle. Art design takes humans as essentials, which is final target of art design, for the final target of art design is serving people, so does humanized purpose of science and technology. Some essential elements of humanity origin are included in high technology by analyzing the application of high technology and humanity orient in fashion design, however, high technology application doesn't equal to humanity origin characteristic, for example, if one high technology costume material couldn't combine with human engineering related essential elements, high technology function may not be able to work out well, interaction will be gained no matter from scientific angle or art design angle, if combine high technology with humanity.

IV. Interaction of High Technology and Humanity Orient in Fashion Design

Creative conception is the most process in fashion, during this period, designers need to process various information, and combine own experience and accomplishment to gain design inspiration, this period is rather difficult, it's also a period facing various choices, different choices will affect succeed and failure of products directly. Design is oriented to future, there isn't any exception for costume, and fashion design is a design that is very strict with time, fashion design products must have time prospect, or it'll reduce to be out of date, it's fatal for fashion business, therefore, design concept must be prospective, high science and technology represents the most forward products and mentality as preceding paper discussed, if make high technology elements as conception elements in fashion design, then prospect orients of fashion products will be grasped firmly in fact, aggressive possibilities of fashion products will be gained meanwhile, it could only say possibilities here [3],because it refers to high technology design elements integration problem, fashion design products could be advanced in respects of costume material and techniques by applying various elements of high technology, for example high technology material, high technology techniques and high technology finishing etc while idea space of design is widened, which is undoubtedly the most wisdom choice. External problem of design is settled by combination with high technology, then consider from interior design request that is person's nature request, humanity is final target of design is discussed in preceding paper, as design, to satisfy people's need is the most critical, humanized design initial point is the most critical if you want to make design products have wide market, and be favored by people. Why do we discuss this problem in respect of initial point? Because several design is narrow sense of design, for example consider problems only in respect of market when design costumes, it's ok when they 're popular, and price is good, they don't mind whether costumes do damage to health and environment, part of children's wear in some areas do damage to health because of exceed azo published by the State Bureau of Quality, examples like these are too numerous to list, therefore, these design are not successful design in respect of serving people, the real successful design has macroscopic design concept, firstly, design starts from humanity orient, which is much more advanced than narrow sense of design no matter in respects of pure concept angle and design prospect. Humanity orient is not a complete function, it is a concept and feeling, it could lead designers to combine each requests of people themselves into design products spontaneously, then the consciousness of serving for people essentially will be enforced, having this design consciousness, design products must be prospective products, and they'll also gain favor of people, design products also will be prospective if combine humanity orient into fashion design action. Since both high technology and humanity orient are inclination of future design, both of them have prospect, then how does the interaction of them in fashion design? It should be said that design rationalization is guided by humanity orient in macroscopic, design rationalization is promoted by high technology in respect of practice, in fashion design, design rationalization could be better obtained by applying high technology material or techniques as well as macro humanized design concept, fashion design development is propelled by both of them. Development and serving direction of high technology is indicated by humanity orient essentially, while target is established for fashion design; high technology not only enrich connotation of fashion design, but also promote fashion design developing towards more humanized, both of which interacts, to promote fashion design development.

V. CONCLUSION: COMBINATION OF HIGH TECHNOLOGY AND HUMANITY ORIENT IS INEVITABLE INCLINATION OF FUTURE FASHION DESIGN

High technology inclination of fashion design is essential for society development, humanity orient of fashion design is also the inclination of future design, final function of design-function of serving for people will be weakened if considering design in respects of pure high technology and humanity orient. Here introduce the points of view in <Discussion on Function of Art in Scientific Culture>written by Shelton. Reichnad, America. He thought that irrational concept is provided for science by art imagination, these concepts are cleared up and come true by science, they'll dent no matter science leave art or art leave science. Although he talked about the relation between science and art, as important category of art design, fashion design could take over this point to illustrate the inter relationship between high technology and humanization more or less.

Applying the combination of high technology and humanity orient in fashion design will not only widen design space but also will grasp design mainstream direction by analyzing the relationship between high technology and humanity orient in fashion design, final target that is "taking humans as essentials" will come true better, which is development inclination of future design.

REFERENCES

[1] LV Feng, LIAN Yi, YAN Yinglin. "history of Art Design.Liaoning Fine Arts Publishing House.June 1996,pp. 312-342.

[2] (America) Susan Lange. Problems of Aart. Nanjing Publishing House. Jan 2006, pp. 112-142.

[3] TENG Shourao. Sociology Description of art. Nanjing Publishing House.Dec 2006, pp. 425-453.

Strategic Research on Optimization of Ecological Waterfront Landscape with Sustainability

A Study of Wuhan - The Hot Summer and Cold Winter Area in China

Jing Wu
Huazhong University of Science & Technology
School of Architecture and Urban Planning
Wuhan, China
Email: wuerna_777@hotmail.com

Baofeng Li
Huazhong University of Science & Technology
School of Architecture and Urban Planning
Wuhan, China
Email: libaofeng@gmail.com

Hao Xie
Huazhong University of Science & Technology
School of Architecture and Urban Planning
Wuhan, China
Email: xiehao1978@hotmail.com

***Abstract*—the paper takes Wuhan, a historic city rich in resources in China, as an example to study the issue on how to effectively cope with the special climate problems associated with hot summer and cold winter regions. It is a study based on the global water shortage and people's growing concern about protecting and reviving the safe patters of waterfront landscape based on EI construction and making good use of the water resources in Wuhan to improve the local climate.**

The paper deals with the special geographical features of Wuhan as a water abundant area and the fact that it is hot in summer and cold in winter. The writer proposes a scientific method applying landscape ecology to improve such a climate situation in Wuhan by ways such as hardening the wetland, building an integrated eco-system between water and wetland, using cooling and heating by soil and sub terrain soil, building a wind guiding system through plant clustering, studying the differences between different landscape elements and the study of reviving the water ecological system. It is hoped that the present climate condition in Wuhan can be tremendously improved through these measures which can change the micro-climate circle and revive a comparatively integrated multi eco-system.

Keywords*- *hot summer and cold winter region; Wuhan China; water ecological landscape; climatic improvement sustainability

I. DIVSION AND DEFINITION OF REGIONS HOT IN SUMMER AND COLD IN WINTER OF CHINA

A. Scope and climatic characteristics of regions hot in summer and cold in winter

The majority of the regions locating along and around Yangtze River Basin belong to the hot in summer and cold in winter (Picture1-2). The general scope of this area is south from Longhai Line, north from Nanling, and east from Sichuan basin, including the two municipalities directly under the Central government, Chongqing and Shanghai; Hubei, Hunan, Anhui, Zhejiang and Jiangxi province; the eastern part of Sichuan and Guizhou province; the southern half of Jiangsu and Henan province, the northern Guangdong province and Guangxi province, which involves 16 provinces, counties and municipalities directly under the Central government This area embraces highly developed economy and culture as well as important position with almost 1/3 population of the country covering about 1,800,000km2 and taking up 48% of DGP.

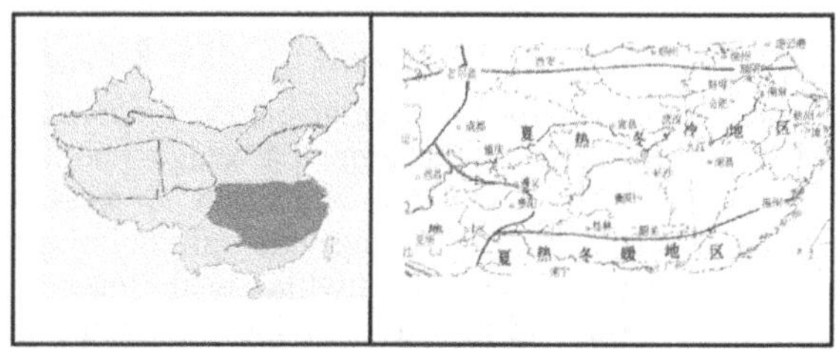

Figure 1. The summer-hot and winter-cold area in china

B. The typical region which is hot in summer and cold in winter—Wuhan

The geographical position of Wuhan city is 113'411'-115'05'of east longitude and 29'58'-31'22' of north latitude which belongs to the north subtropical monsoon (humid) climate, having plenty of rain and rich in heat throughout the year and is the typical region which is hot in summer and cold in winter. （The figures in Table 1 and Table 2 shows） the average temperature in July and August of Wuhan is 29.8 ℃ and the highest extreme temperature reaches 39.4℃; while the average temperature in January and February is 3℃ and the lowest extreme temperature reduces to -18.1℃. However, during these two periods, both of the average air humidity is near to 80%. The coexistence of humidity and heat as well as that of humidity and coldness make the outside environment in Wuhan quite tough in summer and winter.

II. City hot in summer and cold in winter present situation of water environment in Wuhan

A. Present situation of water environment in Wuhan

Due to the constant expansion of cities and rapid development of economy, the environment transformation changes with each passing day and the relationship between human beings and living environment becomes increasingly tense. Thus the sustainable ecological plan must be put forward. As for the present satiation of waters in Wuhan, the hardened and channelized phenomena are quite obvious and there is not the possibility for sustainable development, which is a big obstacle for the formation of ecological landscape.

As the city with hundreds of lakes, Wuhan is located in the center of China, close to rivers and lakes. Among 189 of them, 43 are distributed in the central area of the city. Currently, the water resource is enough. However, the water quality problem is rather striking. 56% rivers and 89% lakes are polluted to some extent throughout the city. The division of water system, the withering of ecological function of water and the increase of water pollution greatly prevents the improvement of living environment. Many lake net systems exist in name only.

B. The present situation of city heat island in Wuhan

The city climate effect in Wuhan is very notable, esp. after 80s of last century, the annual average temperature increases with more rain and less wind speed. At present, the city scale is constantly expanding and tall buildings are standing in great numbers. Meanwhile, with the radiated energy from the sun and the city activities, the gathering of large amount of makes part sections of the city show the hot island effect. Now, the "heat island" of Wuhan is mainly appearing in the old district and Qingshan industrial district with highly dense population. Because of the lack of ecological vegetation as well as the holding back and absorbing of water surface, the city cyclone turns "scorching" into "stifling" and at the same time brings out city "rain island effect", "fog island effect ", etc.

III. The strategy to improve living environment by using water environment

A. The establishment of wetland system to improve living environment

Wetland system not only has the highest production of living beings in the ecological system, which can produce a great amount of microorganisms rich on vegetation community or plants, but good to further purify water environment and then improve microclimate to optimize the living environment.

In Wuhan, the ecological wetland should be recovered for the over-developed and unduly big water environment. The hard revetment should be changed into soft one with the blend of natural water and land. Suitable development around the undeveloped wetland can also be done, keeping the integral of its ecology to the utmost. Some water plants such as canna, reed, etc., which not only provides flourishing living environment, forms blending landscape of human and nature, but also has active effect to the improvement of microclimate. The wetland environment can really optimize and improve living environment.

B. Use underground water as the resource of cold-producing and warm-keeping

By using the cold-producing and heating potential of soil environment 2M underground, the energy can de reduced by 30%-100%. The soil temperature 2M underground is almost the same to the annual average temperature in this area. In Wuhan, the highest extreme temperature is 39.4 ℃ in summer, the coldest extreme temperature is -18.1 ℃ in winter and the annual average temperature is 16.4℃. By the way of energy storing and burying, the living environment can be evidently improved. Using the temperature change at the bottom of waters between the daytime and night, the temperature difference on the land surface can be ameliorated.

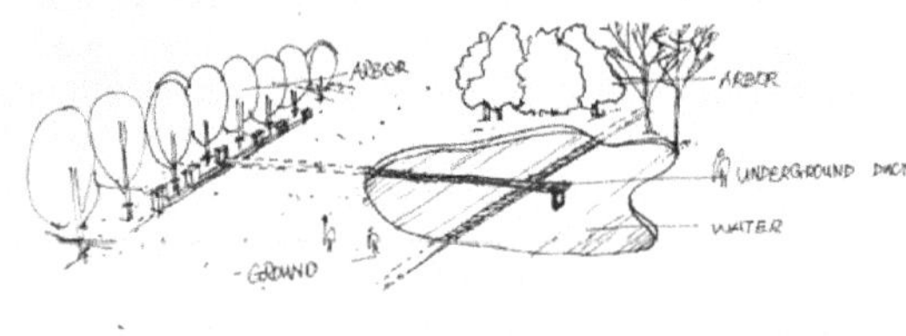

Figure 2.

In winter, the hot pump absorbs heat from the underground cycling water and releases it indoor and outdoor, increasing the temperature inland surface, creating the comfort of human body and improving living environment.

C. Summer ventilation and winter wind breaking of waters

In the area hot in summer and cold in winter, the water surface is better arranged windward of the leading wind, which is benefit to reduce the air humidity around the water and the scope of low temperature, can be expanded. The summer ventilation can be led to the surface by tall deciduous arbor. (Figure 3)

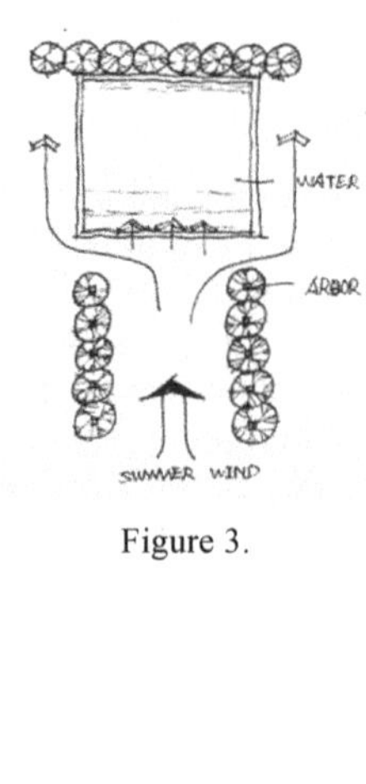

Figure 3.

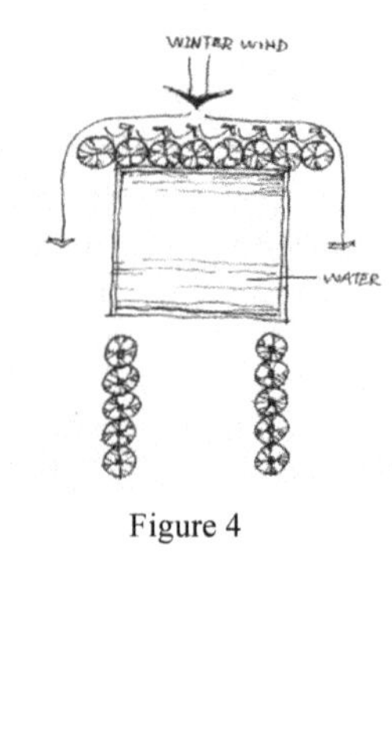

Figure 4

Planting evergreen arbor in the windward of the leading wind in summer can prevent negative wind and prevent the loss of heat from the transpiration. (Figure 4)

D. *Water design of defending sunshine in summer and receiving it in winter*

Defending-sunshine water design outside buildings: it is better to plant aquatic plants. Thus it can not only prevent the water surface receiving excessive radioactive heat from the sun, but reduce the dazzling light from water surface in summer and can create richer landscape effect. Species should be selected, according to the ecological feature of aquatic plants and the demand of landscape. The rootstock of the water-floating plants, such as lotus, water lily, etc, grows from the water soil. In order to guarantee the effect of the uneven density of aquatic plants and not influence the appreciation of the inverted image besides water, it is better not to plant those plants on the whole bank, esp. the emergent plants (such as reed, water bamboo, etc.), with more clumps and smaller scope.

E. *The importance of living environment from water transpiration and irradiating angle of the sun*

Waters arranged windward in the summer leading wind are good to reduce the air humidity and expand the scope. In summer, the tall deciduous arbor can lead air to the water surface. Planting evergreen arbor windward in winter leading wind can prevent negative wind and reduce the loss of heat from water steaming. Water design can combine the defending-sunshine design by itself and that by surroundings, avoiding the water surface receiving too much radiation and can reduce the dazzling light in summer. When the pond's bottom is paved, the changeable design strategy can be adopted to solve the heat contradict in winter and summer.

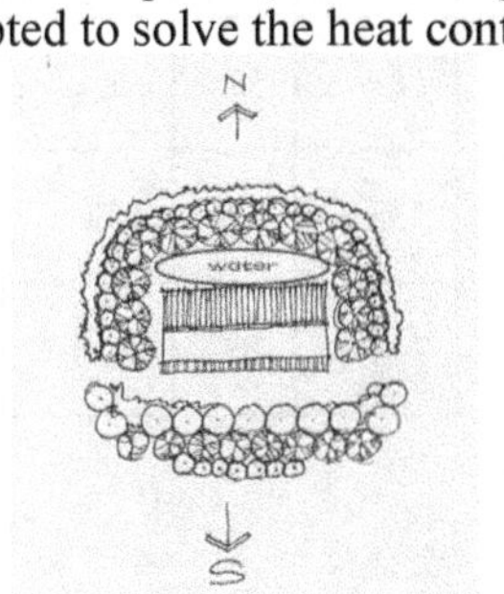

Figure 5

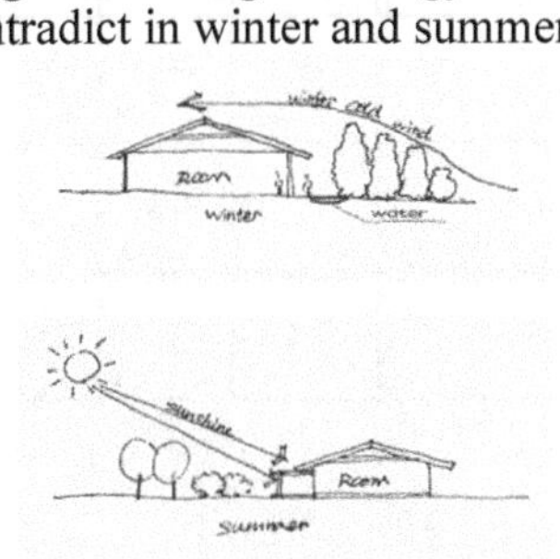

Figure 6,7

In summer, the southern part of the house can prevent sunshine or radiation by the wave structure; in winter, the obtaining of sun energy should be increased. The best way is to plant deciduous arbor. In winter, only dead branches are left, so the sun can directly radiate on the outside of the building. The heat storing work can be transformed to the indoor through the walls and keep warm. In summer, the arbor has thriving branches and the sun cannot directly radiate on the outside of the building. The plants function to prevent heat. (Figure 5,6,7)

F. *The influence of constituting for our space of waterfront buildings to the wind environment*

The constituting way for out space of waterfront buildings has something to do with natural ventilation, which can be considered from plane and stereoscopic aspects. From the angel good to the ventilation of single building, the long side of the building is better vertical to the summer's leading wind; but from the angle good to the ventilation of building complex, it will seriously affect the summer ventilation of the bank building. Therefore, the angel should be better controlled from 30 to 60 between the planned orientation and summer leading wind. (Figure 8)

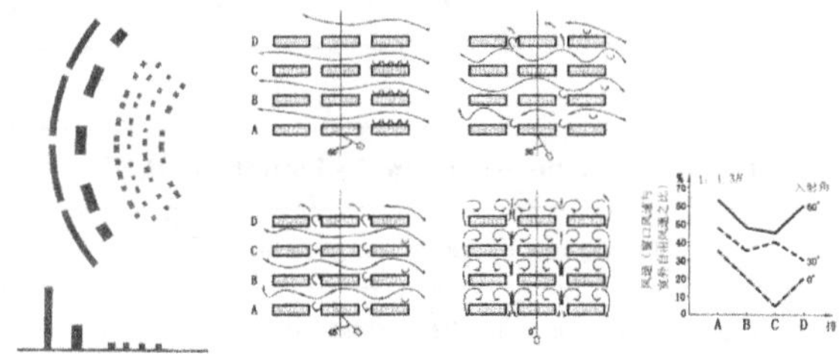

Figure 8

IV. Conclusion

Evaluating whether a region or city is comfortable or not, climate is a very crucial factor. Then that is the water environment and terrain condition, i.e. forming integral ecological environment. As a passive design element, water can be designed to reduce the temperature in summer and increase in winter as well as change climate cycling. In the region which is hot in summer and cold in winter, water environment plays an active role in optimizing and improving the living surrounding within fairly big size. According to different size, different section and different physiognomy, the possibility that water, as the main body, improves the cycling of microclimate will also change. Wuhan, as a typical region where is hot in summer and cold in winter, has big scale, rich waters and special climate. The key point illustrated in this paper is how to scientifically use water environment to create comfortable and suitable environment for human beings themselves as well as establish sustainable close water landscape.

REFERENCES

[1] Thomas, Herzog. Solar Energy In Architecture And City Planning Prestel, 1998.

[2] IAN L McHARG. Design With Nature, Doubleday / Natural History Press Doubleday & Company. Inc. Garden City , New York 1994

[3] David A hardy. Visions of space—Artists Journey through The Cosmos, Dragon's World Ltd. 1989

[4] John Julius Norwich. The World Atlas of Architecture, Edited by Mitchell Beazley International Ltd. 1984

[5] Francisco Asensio Cerver. The World of Contemporary Architecture, Konemann verlagsgesellschaft mbH, copyright 2000

[6] Mcihael, J, Crosbie. Green Architecture A guide to sustainable design, Washington Dc: AIA Press, 1994.

[7] James, Shay, Christopher, Irion. New architecture San Francisco, San Francisco: Chronicle Books, 1989.

[8] Heinz W. Krewinkel. Glass Buildings, Birkhauser – Publishers for Architecture, Switzerland, 1998

[9] Tropical Architecture in the Dry and Humid Zones.Robert E. Kriger Pblishing Company, Malabar, Florida 1982

TABLE 1. 9 CITYS' SUMMER MTEOROLOGICAL PARAMETER IN SUMMER-HOT AND WINTER-CPLD AREA

	Shanghai	Hangzhou	Nanjing	He fei	Nan chang	Wu han	Chang sha	Chong qing
The average temperature in the hottest month/℃	27.8	2.6	28.0	28.3	29.6	29.8	29.3	28.1
The highest extreme temperature /℃	38.9	39.9	40.7	41.0	40.6	39.4	40.6	40.2
The average temperature in the hottest month at 14'o clock /℃	32	33	32	32	33	33	33	33
The days the highest temperature is below 35℃ in a year	8.7	22	15.8	16.3	27.7	21	30	
The average relatively humidity in the hottest month%	83	80	81	81	75	79	75	71

TABLE 2. 9 CITYS' WINTER METEOROLOGICAL PARAMETER IN SUMMER-HOT AND WINTER-COLD AREA

	Shanghai	Hang zhou	Nan jing	He fei	Nan chang	Wu han	Chang sha	Chong qing
The average temperature in the coldest month/℃	3.5	3.8	2.0	2.1	5.0	3.0	4.7	7.2
The days the average temperature is below 5℃in a year	56	51	77	72	18	59	32	
The lowest temperature is lower than 5℃ which the number of days in the year	801	714	1155	1080	239	856	432	
The average relatively Humidity in the coldest month%	75	77	73	75	74	76	81	83

Sustainable Urban and Space Planning Strategy

Zhang Zhong Hua
Architecture College
Xian's University of Architecture and Technology
Xian, China
Email: zzhdeai@163.com

Abstract: **This paper concludes and analyses the main contents of sustainable urban space researches, including various viewpoints and spatial conceptions. Key issues and trends of sustainable urban space planning are analyzed from the aspects of space, material, society and culture. This study will further clarify the characteristic development elements involved in urban planning guidelines, promote planning of an ecological and socially sustainable new urban, and enhance the completive edges of the urban in global urban hierarchies.**

Key Word: sustainability, urban space, planning strategy, spatial relations

I. INTRODUCTION

Ever since sustainable development was brought forward for the first time in Brundtland's report in 1987, sustainable development has been an important topic drawing the attentions of the world. The early stage researches were mainly focused on levels of environment, resource and economy. And the core issue is selection of urban sustainable indicator and establishment of indicator system (Portney and Kent, 2003). In recent years, the research has been further deepened aspects related to social equalities and cultures of the abovementioned aspects. And the focuses have been shifted to how to measure and ensure the sustainability of urban space and space strategies for social developments. The research methods have been shifted from macroscopical discussions to microcosmic fields (Witten, 2003; Ley, 1983). Presently, studies on how to establish sustainable urban space and the relations between sustainable urban and relation affecting factors are being attached importance by governments and academic circles.

II. CONNOTATION OF SUSTAINABLE URBAN

As defined by the World Wildlife Fund (WWF), sustainable urban is "to enhance living qualities within the ranges affordable by the social-ecological systems". And according to the World Bank, sustainable urban "must leave no burdens for developments of offspring's; care for sustainability of environment, equality and economy; care for current health and safety; balance economic developments and protect environments." Girardet (Girardet,1999) holds that, a sustainable urban "should meet the requirements of the citizens and cause no harms no matter on natural environments where they live in or living conditions for others, no matter for present and in future". Smith claims that "to take measures environmental management and protection, economic coordination and cooperation, equal social participations and space reconstruction, etc., to make urban a more attractive residential area, while the ecological footprints must be reduced." (Smith,1996).

Foley and Webber hold that, sustainable urban space is a kind of introspection of human beings on traditional urban space development mode; and also an effective manner to change the status of traditional urban space structure development, which is material and economy oriented, and alter the original development modes of extensive, overspreading and in disorders. Moreover, it is an exploration of ideal urban space modes and ideal urban spaces (Foley, 1998, Webber, 1998). Bourne is the pioneer to bring forward opinions from ecosystem theories that urban spaces are formed under the mutual effects of urban morphology and urban space. Urban morphology refers to spatial distribution modes of various urban factors (including physical infrastructures, social groups, economic activities and public organizations). The mutual effects of urban refer to the correlations of urban elements, which integrate different types and behaviors of land utilization, group activities into various entities of different functions, also known as subsystem (Bourne, 1982).

III. GOAL AND PRINCIPLE FOR SUSTAINABLE URBAN

A. Spatial Goal System for Sustainable Urban

Sustainability researches on urban space mainly involve the following contents: equalities between same generations, equalities between different generations (including social equalities, regional equalities and management equalities); protections on natural environments (and lives within the range of environmental capabilities); applications on nonrenewable resources to the lowest extent; feasibilities and diversities of economies; changes in urban citizen lives and consumption patterns; self-reliance of communities; sustainable development or urban land uses; sustainable urban traffics; happiness of community members; satisfactions of basic urban citizen demands, etc. International Institute for the Urban Environment in Netherland (Wilkins, 2003) has established an evaluation model consisting of several indices such as environment, Greenland, resource, availability factor, openness and feasibility, economic, social and cultural activeness, community participation, social justness, citizen lives and welfares, etc., to evaluate the sustainability of urban space (as shown in Table Ⅰ).

TABLE I. INDICATOR SYSTEM FOR SUSTAINABLE URBAN SPACE EVALUATIONS

Respect on local historical and cultural resources	People-oriented neighborhood	Better urban traffic system
Make sound conservations on local landmarks, buildings, streets, archeological materials, recreation areas, cultures and traditions to offer unique space landscapes.	Design new residential areas and invest on old street blocks to promote mixed commercial and residential functions. Gathered pedestrian and beautified environment; mutual promotion of environment and residences.	Improve public traffic system, bicycle- and pedestrian-friendly streets, improve the roads (landscape, lighting, pedestrian path, bus stop), prompt replacements of automobiles.
Positive response to local communities and governments	Sufficient urban green space for recreation and entertainment	Make organic urban renewal, restrain urban overspreading.
Participate in social volunteer services of citizens, community constructions, response to the calls of governments; organize relations between citizen investments and governments.	Urban in recreation facilities and green spaces, including streets, regional parks, public spaces, community gardens, bicycle and pedestrian paths, linearity, riverside parks, trees and urban landscapes.	Development of urban cores under elaborate design and management instead of sharp extension and diffusion of urban fringes.
Reduce poverty and offer fairer chances.	Protected natural environments	Safe community spaces
Make fair distributions on spatial resources, create chances, and fight against poverty and inequality in society and economy.	Protect urban ecological system, recreation and entertainment areas, and wildlife resources.	Means to provide people with a sense of security in community and make them with risk conceptions.
Outstand public educations	Urban on-the-job training qualities	Harmonious and healthy families
Optimize education qualities on various stages from youth to adult (including occupation, life abilities, culture and citizen trainings)	Education, training and skill development, leading to high life quality and wages.	Provide chances, services and conditions to support families and youths.
Powerful local enterprises	Better job chances with higher wages	Clean air and pure water
Local economies, especially small-sized enterprises	Wages, job qualities, job viabilities, improvements on life standards	Reduce urban pollutions and provide clean drinking water.
Effective utilizations on natural resources	Success urban center	……
Protect resources and use sustainable resources	Cultural and commercial urban center	

Data source: (Wilkins, 2003)

B. *General Principle for Sustainable Urban*

- Draw long-term urban goals according to inter-generational sustainability principle and equality principles of society, economy and politics.
- Achieve long-term economic growths and social security for urban.
- Know the internal values of biodiversity and natural ecological system, and take protection and recovery measures.
- Equip the community with capability to reduce ecological footprints.
- Develop and cultivate healthy sustainable urban according to the features of ecological system.
- Know and rely on the special features of humanistic and cultural values, histories, natural systems, etc. to make developments.
- Strengthen the power of citizens, have the citizens to participate in urban development.
- Establish and extent cooperation networks and join hands towards the trends of sustainable development.
- Prompt sustainable productions and consumptions according to environmental friendly technologies and effective demand managements.
- Achieve the goal of continuous improvements based on the principles of responsible, open and favourable.

IV. RESEARCHES ON SUSTAINABLE URBAN ELEMENTS

A. *Sustainable Urban Land Use*

Land use planning refers to systematic evaluations on potentials or land and water, selection methods on land use, economic and social statuses, to select and accommodate optimal land uses. At least three aspects of factors should be considered in the planning: (1) Agriculture: Long-term considerations should be made on durability, exploitability and continuity of natural resources; (2) Mutual dependence: It is mainly focused on breaking and splitting of lands and relations between different types of lands; (3) Morality: It mainly refers to responsibilities assumed for offspring. Values such as franchise values, heritage values, existing values, etc. may all be interpreted by morality. The main goal of urban land use is to advocate comprehensive utilization of lands; limit residential, industrial and traffic land uses; give priorities to develop and reutilize the brown fields in the urban before developing the green fields; save building spaces, optimize urban density; attach importance to urban planning; balance land management measures of local government; protect and enlist open spaces favorable for ecology; enhance cooperation of urban and surrounding areas.

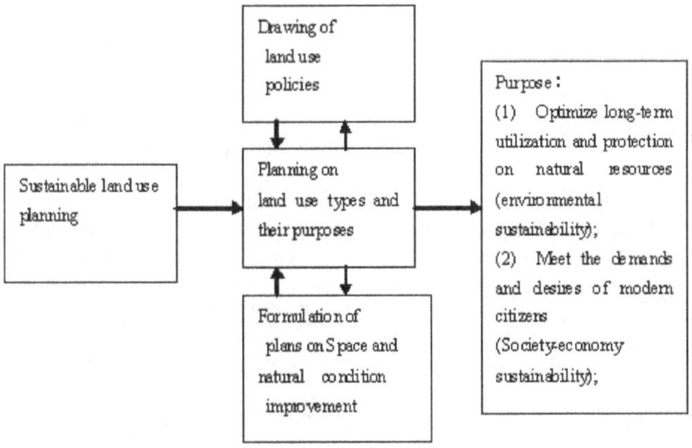

Figure 1. Sustainable Land Use Planning

B. Sustainable Urban Traffic

Under the trends of globalization, sustainable urban traffic has been one of the focuses on urban space. Rapid urbanization and vehicle-based development bear direct influences on sustainable urban development. With the sharp increment of traffics and vehicles, the development mode is required to be shifted into sustainable traffic system. Various factors should be regarded for a sustainable traffic system, such as air and noise pollutions, accidents, jam, social exclusion, etc., to meet current and future spatial mobility requirements without causing excessive negative influences for traffic space. Moreover, sustainable traffic should establish dynamic balances between environmental protection, social equality and economic efficiency and not damage the development benefits for offspring (Nijkamp, 1997).

Presently, the global attentions on space strategy for sustainable development are mainly focused on the following aspects:

- Change urban morphologies and land use modes. Provide various optional traffic modes based on “smart” urban development to balance the correlations between work and residence and connections between work and urban center (Katz, 1994).
- Integrate traffic and land use planning. Traffic planning must be a part of land use planning and land development. Comprehensive and clear viewpoints are required to analyze the urban we anticipate to reside in. Different urban structures require different traffic planning systems. Urban space structures and urban traffic systems should furthest improve social welfare and sustainable development (Banister, 1994).
- Implement transit-oriented development (TOD) development strategies; formulate, consummate and implement urban traffic laws and policies; restrict the amount of cars by economic means; as for urban, safety and guarantee, maintain pedestrian path and arrange some green separated space for bicycle; provide related facilities for non-motorized traffic, such as pedestrian path, pedestrian crossing, proper crossing signal, street lighting, special bike-way, to care for the interests of underprivileged urban groups and achieve the goal of social equality in space.

C. Sustainable Urban Morphology

As the conception of sustainable developments is being widely accepted, many urban researchers and urban planners are taking urban morphology as the core and focus of the researches. For a long time, they have been trying to find a sustainable modern urban space morphology model. Some researchers hold that the main advantages of compact urban form are low car dependency, low exhaust emission, low energy consumption, consummated public transit facilities, developed urban traffics, reutilization on original land resources, regenerate old urban areas, high quality of life, green space protection, more consummated social environments on commercial and business activities, etc. (Jacobs, 1961; Owens, 1992). Newman and Kenworth (1989) claim more compact land use mode, more centralized activity methods and developments in higher density. The compact development should be made with the boundaries of the existing urban area and cannot exceed the limits. With effective and comprehensive utilizations and developments of spaces in different densities, a balance points can be found within high quality life styles, energy saving design and open space planning. Peter Newton, scholar from Australia, has discussed relations between urban morphology and some key environmental indicators with integrated land-traffic-environment model, including, traffic energy consumption, pollutant discharging, greenhouse gasses emissions, and air quality (fresh air and particle). The findings tend for a more centralized urban morphology (i.e. a broad sense of compact type urban, such as compact urban, boundary urban, corridor urban, etc.). No matter what the definition is, the entire benefits of urban environments will be improved (Newton, 2000).

D. Sustainable Urban Community

Ever since the 1970s, the developed western countries had entered post-industrial and high consumption stages. The viewpoints of urban sustainability were shifted from urban land use to people-oriented urban life space. From the aspect of spatial cognition, the subjective opinions of citizens on urban life space quality are an important method to evaluate urban sustainability. Researching on sustainable urban community is called ecology for evil (LEY, 1983).

Researches on sustainable urban must take into account the internal mechanism of urban space structures and various urban communities and the regularity of such internal relations. Only by this means scientific basis may be provided for urban sustainable development. Present researches on such subject of the researcher represent the following features:

- Economy-centered. Research the internal relations between urban constructions and developments of urban life space arrangement and seek for regularities, to prompt sustainable development of urban communities.

- People-oriented. Research the harmoniousness of internal structural elements of the urban life space integration - between human and nature, individual and individual, individual and group, group and group, time and space — to prompt sustainable development of urban communities.
- Take improving urban life space quality as goal; research urban community space planning; provide scientific basis for urban community reconstructions.
- Take equality as guideline, research the equalities of residences to make urban community citizens proper living spaces; research equalities between different generations to have the developments and demands of urban community citizens not damaging the life space quality of offspring, and to ensure sustainable developments of urban.

TABLE II. SUSTAINABLE COMMUNITY PLANNING

Community Traffic Roads	Planning Strategy
(Build for Everyone)	Urban environment lacks humanism; a healthy space should reflect the amount of users and the diversities of groups passing by, staying; an urban center should be formed by many children, youths, elders, and disables.
(Scale For People)	Towns or a part thereof as neighborhood must be provided with proper scale for pedestrian; a town is formed by various complete neighborhoods; as proved by history, a radius of 0.25 mile is the optimal distance for people to establish relations.
(Create Many Linkages)	Modify available corridors, waterways and other public spaces into pedestrian paths and bicycle paths according to the land use forms; increase the accessibilities to shops, parks and schools on foot or by bike.
(Streets have Multiple Uses)	More than basic functions of carrying mobility and transportation facilities, a street should be provided with many other functions; as for main streets, the functions should involve "outdoor living room" function, health function, sense of community enhancement function, education functions, communication area, security function, etc. The design of streets should conform to the values and attitudes that people hold on the streets and communities they reside in.
(Comfortable Sidewalks)	The sidewalk should not be too narrow and should be provided with buffer areas and edges. Proper maintenance should be provided.
(Build Green Streets)	Road landscapes are presently regarded as one of the important parts for urban traffic quality. Pedestrians are quite in need of greens, shades and street atmospheres. With the same widths, streets with trees will reduce the driving speeds by 10 to 15 miles. Street tree is a main manner to transfer and reduce hazard wastes emitted by vehicles.
(Build Bike Lanes)	A effective and safe road should be provided with side aisles and bicycle paths on both sides. The side aisles on both sides can be used for bicycling and walking of people and as temporary car parks for drivers. Increase turning radius and improve sight triangle.
(Build Public Space)	Most of public streets should be available as public spaces. As required by urban constructions, all families should be no greater than 1/8 mile from a national park.
(Build With Proper Size and Scale)	Design community service facilities with proper sizes and scales to facilitate the daily lives of residents and increase accessibilities of community resources for residents. The sizes and layouts of community schools and community centers are the most important in design of community service facilities.

V. STRATEGY FOR SUSTAINABLE URBAN PLANNING

A. Philosophies of Sustainable Urban Planning

More than reflecting the environmental levels and space diffusion statuses, space is also a geological platform exhibiting conflicting problems between urban land use managements and space planning. According to papers on urban planning theories, clear evidences can be drawn on some sustainable planning conceptions. Such philosophies aim to achieve the most ideal decision-making effects in public domains:

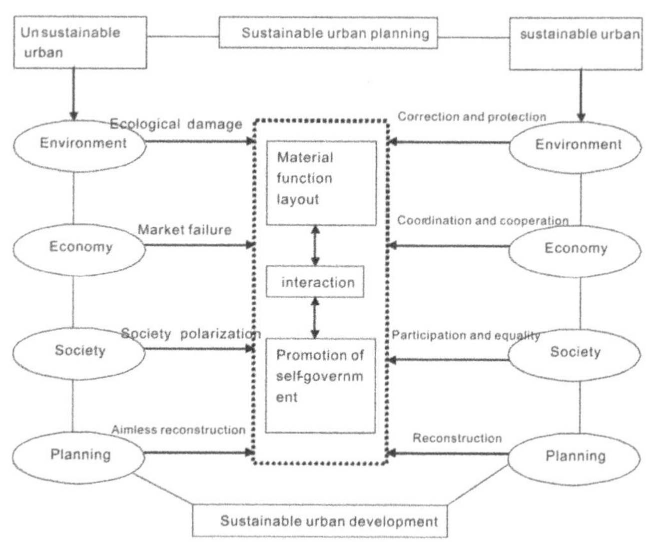

Figure 2. Sustainable Urban Planning

- Optimization philosophy: The philosophy assumes a definite target function, which is provided with given limiting factors and M L information.
- Satisfying behavior: Seek for sub-optimal solutions based on expensive enforcement costs (such as results caused by conflict of interest and insufficiency of information) for assumed optimal results.
- Multi Dimensional Decision-Making Philosophy: Such philosophy claims that participants from different fields may make comprehensive considerations for different alternatives according to

different goals (as reflected in diversity standard analyses). An optimal trade-off decision will be achieved in the end.

- Accountability Philosophy: The major task of planning is to ensure the rationalities of decision and policy results according to general rules, procedures or practices, instead of being subjected to any optimal standards.

B. International Sustainable Urban Planning Mode

Presently, sustainable urban planning has been dynamically rising in some developed occident countries and many planning assumptions and principles drawn from practices have been generated (as shown in Table III).

TABLE III. INTERNATIONAL SUSTAINABLE URBAN PLANNING MODE

Sustainable urban planning mode	Country	Contents
Green urban	U.K.	a. Harmonious correlation between biological materials and cultural resources. b. Be provided with complete viability in nature c. Protect natural resources, eliminate or reduce wastes under least requirement principle. Make recycling of unavoidable wastes d. Be provided with vast open spaces and other species coexisting with human being e. Stress human health, encourage green foods and advocate rational eating.
Ideal ecological urban planning mode	Australia	a. Sufficient amalgamation of technologies and natures b. Maximum exerting the creativities and productivities of human beings. c. Offer maximum protections on physical and mental health of citizens and environmental qualities. d. Establish coordinated developments based on ecological principles; make highly effective utilizations on materials, energies and information; create a habitation for human beings with favorable ecological circulations.
Ecological planning principles of sustainable planning	U.S.	a. Protection, preservation and recovery of natural environment. b. Establish net value system as the base of economic vitality, which is to say, more than reflecting the current availability status, price should be established from long-term, recyclable and systematic viewpoints. c. Support local agriculture and local industry and commerce, service industry. d. Develop agglomeration, multifunctional and pedestrian-based ecological communities. e. Apply advanced traffic, communication and production systems. f. Protect and develop renewable resources as possible g. Establish recycling programs and recyclable material industry. h. Support universal education of planning participation.
Methodology principle of ecological sustainable environment evaluations and protection strategy plans	Russia	a. Proportions of planning layout and processing technologies in urban natural protection issues. b. Layout relations, functional relations and population scales of urban area and surrounding area. c. Urban ecological subareas, to restrict pollution influences and loads of people of each subarea, and reduce the extent of influence. d. Land use functions and layouts that reduce environmental hazard e. Infrastructures such as urban traffic, energy, etc. conforming to sustainable ecology requirements. f. Rational proportions between buildings and green spaces with "green" as framework. g. Sustainable urban building spaces.

C. Methodology of Sustainable Urban Planning

1) Compact Urban Planning

Compact urban planning was originated in the report of Measures for Sustainable Environmental Planning issued by Sustainable Development Study Group of Town and Country Planning Association. As indicated by the report, the most sustainable urban morphology should be applied to boost urban developments, while density is a key element for sustainable urban morphology. The conception of compact urban is therefore adopted and taken as core element of national land use planning policies. Compact urban planning supports enhancing urban densities, claims to apply new land use planning to reduce mileages required for motor vehicles, and advocate sustainable transportations facilities such as public traffic, bicycle, on foot, etc. A sustainable urban "must be provided with space morphology and scale easy for pedestrians, non-motorized vehicles and establishments of public transit facilities; and provided with tightness to some extent for social interactions of people".

2) Metropolis Area Planning

Ever since the 1990s, most planning and practices of metropolis areas in Europe have turned to the goal of sustainable development. For example in the European Metropolitan Area Planning hold by German Transportation, Construction and Housing Department in 1999, the planning was basically developed concentrating on the five core goals for the sustainable development of the metropolis areas: *a) balancing spatial structures of metropolis; b) improving urban life qualities; c) preserving regional characteristics, boosting renaissances of cultures and traditions; d)implementing integration management and cooperating within the regional infrastructure network, establishing new cooperation relations in the course of planning and implementing (Xie Shouhong,2006).* Balance in metropolitan space strategies mainly involves two aspects: decentralized concentration layout and green belt constructing. Decentralized concentration layout is a method to establish a balanced urban system. Cities and towns in the metropolis area are growth poles of regional economic base and centers for social activities, and provided with favorable infrastructures, public facilities and cultural lives. By exerting such functions of the cities and towns, the attractive forces of urban system center areas can be enhanced and the excessive expansions of center urban will be avoided.

VI. CONCLUSION

The developed western countries have realized the importance of sustainable development and are positively exploring space strategies that can reduce energy consumption, improve environment, represent urban life force, and boost social equality and economic development. The current achievements of the researches are of great promotions for sustainable urban space studies; and are of importance references for global urban constructions. This paper lays stresses on the analyses and conclusions of international urban sustainable development philosophies and related elements, to support further studies combining urban characteristics in China and help in correct space decision-making.

REFERENCES

[1] Banister,D.:Transport planning. E&FN Spon,London,1994.

[2] Bourne, L .S. :Internal structure of the City: Reading on Urban Form,Growth and Policy. 2nd ed. Oxford University Press,New York ,1982.

[3] Finco,A., Nijkamp,P.:Sustainable Land Use: Methodology and Application. In Future Land Use in Europe. ed J. Stilwell and H. Scholten, s Kluwer, Dordrecht,2000.

[4] Foley,D.L.: An Approach to Metropolitan Spatial Structure.Explorations into The Sustainable Urban Structure. University of Pennsy Press, Philadelphia,1998.

[5] Girardet, H.:Sustainable Cities: A Contradiction in Terms?. In Sustainable Cities, ed. D Satterthwaite, Earthscan, London,1999.

[6] Jacobs,J.: The Death and Life of Great American Cities.Vintage Books/Random House, New York,1961.

[7] Katz,P.:The New Urbanism: toward an architecture of community. McGraw-Hill,New York,1994.

[8] Ley,D.A.:Social geography of the city. Harper and Row publisher,New York,1983

[9] Newman,P.G.,Kenworthy,J.R.:Cities and Automobile Dependency. An international Soourcebook.Gower Technical ,Aldersho,1989.

[10] Newton,P.:Urban Form and Environmental Performance.Achieving Sustainable Urban Form. E&FN Spon, London and New York ,2000.

[11] Nijkamp,P.:Social change and sustainable transport. J. Journal of Transport Geography. 1997, Biol. pp215-218.

[12] Owens,S.:Land Use Planning Ploicy and Climate Change, HMSO.London,1991.

[13] Portney,A.,Kent,E.: Take Sustainable Seriously, Economic Development the Environment and Quality of Life in American Cities. MA: MIT Press,Cambridge,2003.

[14] Smith, N.:The New Urban Frontier: Gentrification and the Revanchist City.Routledge,New York,1996.

[15] Sohail,M.:Current situation (Section 2).Urban public transport and sustainable livelihoods for the poor, a case study: Karachi, Pakistan. WEDC, Loughborough University, UK,2000.

[16] Webber,M.M.:The Urban Place and the Nonplace Urban Realm. WebberMM. Explorations into Sustainable Urban. University of Pennsylvania Press, Philadelphia,1998.

[17] Wilkins,H.:The need for subjectivity in EIA:discourse as a tool for sustainable development.J.Environmental Impact Assessment Review,Biol. 2003, pp401-414.

[18] Witten,K.:Eneter,D.:Field A. The quality of urban environments:maping variation in access to community resources.J.Urban Studies, 2003, Bio3.161-177.

[19] Xie Shouhong.,Wangzhen.:Study On Overseas Urban Sustainable Development. J.Journal of Shaoyang University. 2006, Bio4 . pp69.

[20] Zhang Zhonghua.,Zhang Pei.:Development of Overseas Sustainable Urban Space Studies. J.Urban Planning Forum. 2009, Bio3, pp99-107.

The analysis & Design of Urban Wetland

With the Water Garden in Portland & Living Water Park in Chengdu as Case Studies

Zhongzhong Zeng
School of Architecture and Urban Planning
Huazhong University of Science & Technology
Wuhan, China
Email: zengzhongzhong@gmail.com

Baofeng Li
School of Architecture and Urban Planning
Huazhong University of Science & Technology
Wuhan, China
Email: libaofeng@gmail.com

Abstract—Problems caused by urbanization such as inadequate water content in urban foundation, ecological imbalance in the soil and heat island effect, have become issues that people pay great amount of attention to. Two projects, cited as being successful in integrating urban landscape planning and wetland ecology, are the Water Garden in Portland, USA, and the Living Water Garden in Chengdu, China. Artificial wetland is effective to establish urban ecology. It may not only solve the problem of urban water crisis, but also bring a series of good eco-environmental effects, such as to conserve groundwater, to regulate climate, to extend green areas, to purify the air, to beautify a city, and even effectively control flood damages, etc.

Keywords- rainwater infiltration, artificial wetland, city ecology

I. INTRODUCTION

With the over exploitation of soil resources in urban development, length and breadth of land which was utilized for agricultural or forestry purposes or as natural habitats of animals or plants, have been unearthed by architectural changes in many landscape areas. Industrial land, traffic site, large numbers of concrete buildings as well as other sorts of land impervious to water have taken the place of the original land to become the main part of urban foundation.
This phenomenon seriously affects the natural infiltration of precipitation and causes surface runoff to collect relatively together which may result in urban floods, serious water shortage in urban foundations and disturbs the balance of the soil ; thus cutting off the natural cycle in an urban environment. Meanwhile, a fog is formed by exhaust gases that covers the city and traps radiation heat in the air which cause higher temperatures to the city and leads to the formation of an urban heat island. Moreover, the hardening of the urban land leads to ecological imbalance and pollution. All of the above factors focuses on what occurs daily.

II. WETLAND & WATER GARDEN

Lately, in the process of urbanization, wetland resources have become more and more relevant to the city not only because of its site but also in view of social, economic, and ecological reasons. As a result, the existence of wetland faces great challenges. It warns us to re-explore and re-consider the relationship between humans and nature in the urban developing process.

Wetland as a special ecotype first came into research in the 1970s. To be brief, wetland is an ecotype between land and water. The conception of Wetland Park is similar to that of a small protection zone, yet somewhat different from the general sense of natural protection areas and parks. According to the recent trend of domestic and foreign wetland protection and management, wetlands involve species and habitats protection, ecotourism, as well as environmental education under the entitlement of Wetland Park.

Wetland Park can prosper a local economy and offer a site for citizens to recreate and enjoy wetland sceneries. What's more, Wetland Park may become a special area for local residence to protect their culture formed in the long-term process of living around wetland; it might even become an important element and component of a city's features, distributing benefits to various communities.

Figure 1: Water garden of Portland

III. WATERGARDEN, PORTLAND,OREGON, USA

About ten years ago, a water infiltration experiment was conducted in Portland, Oregon concerning the annual months of almost continuous rain that brought people to conceptualize their original solution that's spreading and being imitated throughout Portland. In this experiment, Murase, a Portland landscape designer, with his cooperators designed the rain garden, pervious to most or all of the runoff from the parking lot.

This extraordinary rain park, designed by Reed Meyer Architect, can infiltrate water from the roof of the Oregon Convention Center, the largest of the tested buildings, East Bank building. In this project, rainwater is collected in the 33.38-acre roof, delivered to the garden through tubes on the conference south center's side and then seeps into the ground. In many parts of Portland , prevention of rainwater problems caused by drainage overload shall be listed as a principal issue. However, this is not the case here. With Huila River flowing ten meters away, rainwater can be discharged into the river smoothly. The designers main goal is to improve water quality by absorbing rainwater and filtrating micro-organisms inhabiting plants, roots, and soil (the majority of the micro-organisms are motes in vehicle exhaust gases).

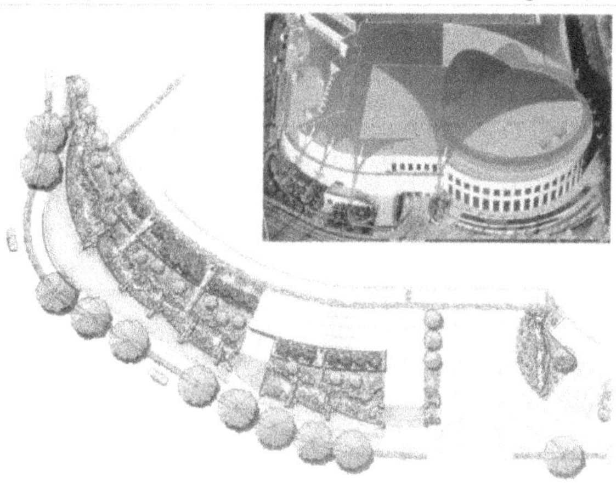

Figure 2: Plan of Water garden

The Portland Garden contains the following systems:

A. Pool System

Rain garden, through a series of shallow, small waterfalls as well as series-wound pools separated by basalt weirs, slows down the speed of the water flow. When each pool is filled to

Figure 3: Pool System capacity, the rain water then

overflows falling 46 centimeters to the next pool and leaving sediments stably in the primary one. The whole artificial ditch is 969 meters long and 1.8 meter wide on the average. In this way, the pools slow down the speed of water flow and not only save water but also provide sufficient time for water to infiltrate into the underground.

B. Stone System

The bottom of the ditch in the rain water park is mainly made up of pale green slate , allowing water to flow smoothly to the top. Meanwhile cobblestones on the edge of the ditch make way for water to infiltrate into the underground while rough basalts walling the ditch create an artless and natural atmosphere. All these vivid and dramatic details combining both the dynamic and the static embody the designer's profound understanding on the law of nature, penetrating discernment and high standard of design. Meyer Reed's design of the main ditch can be viewed as perfect in the composition of the Northwest Pacific stones, embodied in the following five aspects:

The bottom of the ditch made of hard slates produced in a quarry on the border of Washington and Idaho.

Figure4: Stone System Figure 5: Plant System

River stones utilized to fix the slope on the side of the whole system comes from Montana's colorful quarry, with diameters varying between 2 centimeters and 10 centimeters.

In order to direct the spillway to spread along the length, the origin of the ditch uses a retaining wall, which is built up with solid stones from Montana.

Vertical elements along the ditch's length is composed of cylindrical bronze basalts from the district of Moses Lake in Washington.

The wide and dry wall as well as the eroding pool of the deposition pond on the building's south apply basalts outside of Camas, Washington.

C. Plant System

Lots of aquatic plants are planted on both sides of the drain to create natural ecological environment of artificial wetlands. These plants, growing up from gaps between pebbles and gravels, add a green color to the rain garden and make it more vivid, lively and natural, but also play the role of a depurative agent, absorbing a variety of harmful pollutants, such as grease washed down from the adjacent roads. In addition, the plants and its root system can firmly fix gravel and sand to prevent against a long-time water erosion.

Theoretically speaking, the systems mentioned above are proved to have the capability to allow plenty of water to infiltrate into the underground, especially during short continuous flurries. What is the rate of rain infiltrating into the soil during the heavy downpour? It is difficult to cite convincing data to answer this question, for in practice no one is there to inspect the performance of the system. However, Bill Cop, the project engineer, has introduced plenty of signs such as the phenomena that water does not stay in the pool for a long time, showing that lots of water are being absorbed. In a heavy rain, water overflowing the cascading pools finally influx into the lowest pool. Water runs over the lower pool through a siphon and flows into a 76 centimeter public drainage pipe.

Generally speaking, the rain garden in the extension of the Oregon Convention Center deals with the storm water issue very successfully. It won the annual award for the best water conservation in Portland. As a project combining water, supervision, and celebration; the rain garden deserves wide dissemination as a model.

IV. LIVING WATER GARDEN, CHENGDU, SICHUAN PROVINCE, CHINA

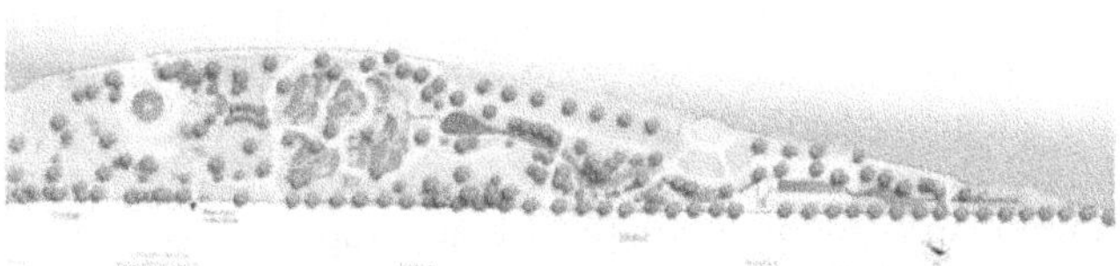

Figure 6: Plan of living water garden

The garden, an appealing urban park that grew out of a larger project to improve the water quality of the Fu-Nan River, covering six acres along the banks of the Fu-Nan River near the center of the city of Chengdu in western China, is an elegant example of environmental education. [2]

A. Overview

Living Water Park is China's first water theme park of the city's ecological landscape. Water is collected from the polluted Fu-Nan River, and flows through the park's wetland purification system, and finally returns into the river. Battersea Damon, an American artist and water protector, first proposed the conception of the living water garden and won Chengdu's government support. She organized artists as well as water conservancy, city planning and landscape engineers from China, America, and South Korea to design and construct the garden, which covers an area of 2.4 hectares and has great cultural, artistic, and ecological significance.

Figure 7/8: Anaerobic settling tank

B. The Water-Treatment System

Wetland Park covers an area of 0.2 hectares, and the capacity of sewage purification is 200 cubic meters each day. It removes bacterial pollutants and heavy metals in the water and returns it to the river. Contaminated water (usually equal to or lower than the water quality GB3838V) is pumped into anaerobic sedimentation tanks from the Fu-Nan River where anaerobic micro-organisms decomposition occurs in physical sedimentation.

The process allows most of the solid matter in the water to accumulate at the bottom or float on the surface, in which part of the solvable organic matter will be broken down into CO_2,CH_4,H_2O and N_2 while some of the simple organic matters will be degraded. Water coming out from the anaerobic pond flows through sculptures and into the oxygen tanks, and gets degraded once more by aerobic micro-organisms and then flows into two sets of plant-ponds and plant-bed treatment system. This system is composed of five lively pond plants and 12 plant-beds. Water pollutants, when flowing through the whole system, experience adsorption, filtration, oxidation, revivification, decomposition and gradually degrade into nutrients available for flora and fauna. Water coming from the treatment system flows through several water sculptures and goes into three fish ponds successfully. To ensure that dissolved oxygen in the water meets the needs of the growth of fish, two circumfluence equipment are used to return part of the water into the plant pond to increase dissolved oxygen and up-grade the water quality to meet GB3838--88III. Part of the water from the fish pond can be pumped back into the park for landscape use while the rest inflows Paddling Pool and back to Fu-Nan River. Every day up to 200 cubic meters of water flows through the water park. Although this is not enough to change the quality of water in Fu-Nan River, it does allow visitors to see the process in which "dead water" is gradually activated, purified, and finally flows as clean water.

Figure 9: living water Figure10: sculpture of living water

C. The Art of the Garden

The living water park's core park-----the plant river as well as the plant-bed group are located in the fish stomach. This shape imitates Huanglong travertine pool group on Sichuan. Wood trestle set up between ponds are in rural style. Three fish ponds and fountains as well as paddling ponds show at the tail of the fish. Plant blocks, plant-bed group , planted duckweed, water hyacinth, pond lilies, etc. which are tested out to have strong water purification capacity, adaptability and visual value, park slopes imitating Emeishan Mountain's micro-climatic and natural vegetation, plant pond, raised beds and restocking of fish ponds with ornamental fish and frog, etc. all form a symbiotic system of the fauna and flora in the water environment. Visitors can see the water treating process and the phenomena that the quality of water upstream is poor which accommodates sparsely any life, for example, no fish but only few duckweed algae and very few plants exist in the anaerobic pool and oxygen tank. With step by step purification, the higher the grade of the flora and fauna, the better the growth of plants, indicating that life and water are interdependent with each other. As the garden provides a living environmental education classroom, it attracts groups of teachers and students. Some parents take their children to gather around the pool for an "eco-environmental class" experience. Some middle school students even apply to participate in water analysis and protection.

The living water garden is not the one and only public facility for education on water purification. Water laboring park designed by Lorna Jordan (referring to "the art of cleaning", Landscape

Architecture, Jan. 1997) as well as College Padillas Elementary in L.A. designed by Campbell Okuma Perkins associates, play a similar role to that of the living water garden in Washington and New Mexico. There are also some public facilities on water purification education, such as Wuhan hydrobiology institute.

However, the living water garden wins special achievements since it is located in the center of the city and has strong affinities. Other projects locating remotely, for example, the municipal waste water factory, cannot provide as sites for everyday entertainment and relaxation while the living water garden can. Water purification details inspire people's instinctive curiosity towards science and natural environment. Sculptures, platforms as well as all other sorts of design elements can be appreciated just for their beauty. Paths in the woods, verdant plants and lawn offer a perfect place for recreation in the city. This park is designed to upgrade citizens' organization to protect the environment. As a result, designers did not try to reforge a natural environment. Visitors clearly know the fact that they are in the artificial water park environment. The conception of the park is close to public art rather than natural protection in most artists' eyes.

Both the rain water park and the living water park are achievements with efforts of landscape designers, biologists, as well as architects. Designers integrate water bodies, introduce favorable elements to the parks, and pay adequate attention to the park's merits and demerits, both allowing plants, soil, as well as water to function properly in urban space. Both parks are upgraded rather than harm the living environment of the districts they're located in. Soil, water, and plants stay in harmony with the people in the city space.

V. THE ECOLOGICAL SIGNIFICANCE OF URBAN WETLAND

A. wetland increases the capacity of water storage and purification and water content of city foundation.

Many problems exist in city foundations, such as too much hard ground and serious water loss . As a result, it is difficult for rain to return to earth, causing a serious damage to the capacity of natural water retention and resulting in shallow underground water that quickly dries out. Rain garden and living water garden can solve this problem effectively: on rainy days, living water garden and rain garden can store water. They slow down the speed of water with shoals, small waterfalls, and series-wound pools. Moreover, they can accommodate plenty of water that infiltrates into the underground. Water superfluously flows through open ditches and into public drainage pipes, releasing much pressure on the public drainage systems and other municipal facilities.

B. Vegetation's role with regards to regulating climate leads to reducing pollution and controlling urban heat island effect.

Because of the impelling force produced by photosynthesis, plant pump can transport most of the water from the roots to the leaves, as water evaporates into the air and takes the heat away. There are sharp differences of temperature between each side of the leaves and the plant pump consumes a lot of heat in the process of transporting water from roots to leaves, enabling effective absorption of large amounts of radiation and heat in cities, as well as regulation of air, temperature, and humidity.

The photosynthesis of plants and its features such as the capacity to store and filtrate water, ability to regulate temperature, radiation, air humidity, as well as its cleansing ability makes an impact on urban management and digestion of city noises which all perform to improve eco-city micro-environment.

As far as the green house effect caused by compaction of urban space, vegetation photosynthesis is effective to regulate concentration of carbon gases and produce oxygen in the center of cities. Furthermore, vegetation can absorb urban radiation, regulate air temperature and humidity, and improve urban climate.

What's more, vegetation in wetland park has adsorption and absorption capacities and can degrade air pollutants and partly absorb the air and rainwater contaminants containing nitrates or other harmful substances. Vegetation growth which consumes a lot of carbon and releases oxygen, improve urban air quality. And vegetation can effectively muffle noise and reduce the impact of noises on citizens' life. Contaminants adsorbed and clotted can be utilized and absorbed as nourishment.

C. The Role of Biodiversity Conservation

Vegetation corridor and urban water system can be combined as the "urban ecological corridor" of the urban structure. Urban ecological corridor and urban cold bridge coexist and provide fresh air for cities, reducing the impact imposed on cities by urban heat island effect and restoring urban ecological chain which has been destroyed.

Plants and waters in the living water park and the rain garden provide a site for birds and insects unusual in cities to habitat, and extend as ecological corridors. The scene that children kneel, sit, or lie beside water, observing insects and fish living and existing in every wetland park, promoting children's instinct to be curious, though nowadays not many cities offer a water park that satisfies the longing to observe nature directly.

D. The Educational Value Of Ecological Environment

The water purification system of the living water garden is not aimed at great improvement in water quality. Take the Fu-Nan River for example, the water quality can be greatly improved only by reducing the amount of discharged pollutants. Actually, the living water garden is designed to help Chengdu citizens to realize that the progress of their environmental awareness along with a sense of pride for their city can inspire them to protect the environment and improve it.

In many aspects both of the gardens located in China and the U.S interpret the environmental theme. They are successful and can be viewed as integrated and vivid examples of urban ecological wetland design and have profound influence on people's environmental sense.

VI. CONCLUSION

Urbanization is progressing with accelerated rate in China. In this process, outskirt wetlands such as lakes, rivers, pools and swamps are resources irreproducible once they are destroyed. The protection and utilization of wetland resources, such as to reconstruct them into urban wetland parks or urban ecological parks or other sorts of public green spaces, are favorable for the formation of a city's characteristics.

The living water garden as well as the rain garden in the city can effectively protect, clean-up, and beautify nature and offer citizens a chance to get close to it. They create a good ecological environment for various urban species with comfortable, healthy, safe and nice living conditions for residents. Of course, the anfractuous and ever-changing urban environmental issues are impossible to be accomplished in an action. Final settlement of the problems of cities depends on humanity's reflection and the development of science and technology to the comprehension management of urban environment, recovery and regeneration of the urban eco-virtuous circle. Yet in any case, living water park and rain water garden provide as important examples in the field of creating urban ecological spaces, proving that people can effectively regulate and improve urban development through design as well as by utilizing landscape conception and urban ecological engineering principals. In this way, people can restore satisfying urban environment and living space. Living water park and rain water garden are fruits of the combination of biological science and landscape architecture. They start a revolution in the process of urbanism in "landscape, biological science, and architecture", and provide an effective way in creating urban ecology.

ACKNOWLEDGMENT

I would like to thank Mr. Xingliang Zhang for his company when I visited the living water garden in Chengdu.

REFERENCES

[1] J, William., Thompson. 2004, 9. Landscape architecture.

[2] Mari, Padova. 2004, 9. Landscape architecture.

[3] H. Takabayashi, M. Moriyama, Urban heat island phenomena influence by sea breeze, AIJ. Technol. Des.No.21, 2005.6

The Application of Virtual Reality Technology in the Preservation of Mining and Metallurgy Culture in Huangshi Region

Zhou Jinlin
Arts and Design College
Wuhan University of Technology
Wuhan, China
Email: zjiny_778@163.com

Liu Manzhong
Arts and Design College
Huangshi Institute of Technology
Huangshi, china
Email: zjiny_778@163.com

Abstract—**The application of virtual reality technology in the preservation of mining and metallurgy culture in Huangshi region will greatly promote the sharing of traditional culture resources in the whole society. It will provide spiritual impetus and intellectual support in promoting advanced culture, establishing a learning society and improving the national quality comprehensively. This paper focuses on the study of the operation mode of virtual reality technology in the preservation of mining and metallurgy cultural heritage in the Huangshi region concerning on the development of mining and metallurgy culture in Huangshi region as well as the application of virtual reality technology.**

Keywords- virtual reality; mining and metallurgy culture; preservation

China has a long history of more than 3500 years of smelting and it is one of the earliest countries smelting the ore. China enjoys an abundant mining and metallurgy culture, and the mining and metallurgy sites are everywhere. Huangshi City, which is located in the southeast of Hubei Province, PRC, was once a national heavy industry base, full of mineral resources and mining and metallurgy ancient ruins. The excavation of ancient copper mine site in the Bronze Green Mountain (Tong Lv Shan), Daye of this area shocked the world, attracting great attention of scholars and experts both home and abroad. Mr. Li Xianjun, an archaeologist, once said: "Daye bronze is the carrier of Chinese bronze culture. The relationship between the Daye bronze and Chinese bronze culture is inseparable. The bronze ware of the Chu State in the Spring and Autumn and the Warring States Periods was the best of all in the world in that period whether in quantity or in quality. Chu State created a second peak in the Chinese bronze culture since the Shang dynasty. Daye bronze made irreplaceable contributions to the bronze culture of the Chu State." Today, Huangshi City has been listed as a resource-exhausted city by China. In order to accelerate the enterprise transformation and to improve service to the economic development in Huangshi region, there must be a systematic excavation and preservation of the historical mining and metallurgy cultural achievements. Meanwhile, the culture and arts brand in this region should be built to serve the development strategy of the whole economy and industry.

In the 21st century, the advent of the digital era has undergone tremendous changes in the way of information dissemination. The Internet technology, multimedia technology and virtual reality technology have become the three most representative key technologies of the digital era [1]. Huangshi City (region) will put the mining and metallurgy culture as the key task in the development of distinctive culture and make it as the forefront in the study and preservation of mining and metallurgy culture in the province as well as in the whole nation. We must learn to use new technologies and become good at exploring new ideas. Specifically, we should adopt the digital methods with the Internet as a vehicle and use the virtual reality technology to help spread the bronze culture with Huangshi characteristics. The virtual simulation should be adopted to represent the fragmented ancient mining and metallurgy relics and smelting scene with a long history. The necessary interactive operation should be added to the virtual mining and metallurgy scene by the virtual reality technology. With the help of the multimedia, the mining and metallurgy scene through digital virtual will be truly displayed in front of visitors. There are three major forms of performance:

I. THE DIGITAL THREE-DIMENSIONAL RECOVERY OF THE ANCIENT RUINS AND THE ROAMING IN THE VIRTUAL SCENE

In recent years, the development of hardware and network technology and the improvement of the efficient 3D computing ability of the computer have gradually made it possible to apply the three-dimensional virtual scene by virtual reality technology. As the virtual reality technology display has broken through the restrictions of the plane surface, which is not only limited to the text, pictures and images of two-dimensional space, the problem of the lack of the third dimension in the display of the image and text has been solved. [2]. Some representative bronze smelting sites in the Huangshi region can be chosen to be recovered by the digital three-dimensional technology, including the digital three-dimensional recovery of the ancient ruins and the roaming in the virtual scene. For example, Daye Tonglvshan Museum of Ancient Bronze Mining Sites can adopt the virtual reality to three-dimensional model building. It can also adopt the approach of panoramic images to build a virtual three-dimensional simulation environment. The users can view the virtual objects through three-dimensional effects by the interactions of a variety of sensors and multi-dimensional information environment from any angle to achieve the understanding both in sensory and rational way. The digital three-dimensional recovery and virtual scene roaming of the mining and metallurgical sites is a completely new concept of exhibition design. It allows the audience to get out from the computed results of simple observation and indulge into an environment which is created by the computers. It provides a new platform for exchanging information which the other traditional methods can not be compared with.

II. Digital Virtual Simulation of the Ancient Bronze Smelting Process

The ancient smelting technology of Daye Tonglvshan has reached a fairly high level. The discovery and excavation of the ancient bronze mining sites preliminarily answered the important historical subject of how to excavate and smelt in the Bronze Age in China, providing a number of valuable materials for the study of the development history of China's mining and metallurgy technology, which convincingly shows that the ancient China adopted a set of independent development system of mining and smelting technology. The adoption of shallow well prospecting and panning plate to measure heavy concentrate is an important mining prospecting method in the development of ancient Tonglvshan mine. The fact of the local ancient mining and metallurgy sites distribution in bonanza proved its scientificity. Another feature of Tonglvshan ancient bronze mine is the integration of excavation and smelting. The blast furnace was adopted to smelt the bronze as early as in the early stage of the Spring and Autumn Period. In addition, there were a lot of innovation in the mining methods, mine hoist, ventilation, drainage and tool making, etc,. These advanced technologies are precious heritage of China's mining and metallurgy culture. We can use virtual reality technology to faithfully represent the technical process of these bronze smelting, such as smelting process and smelting methods. It is a combination of Chinese traditional culture and modern digital information technology as well as the concrete manifestation of advanced productivity. It is a powerful tool to spread advanced culture and to meet people's growing cultural needs more effectively, which reflects the fundamental interests of the majority of the people in the field of cultural building.

III. The Establishment of Huangshi Virtual Digital Museum

Virtual Digital Museum refers to a comprehensive information system using the digital technology to collect, store and process the information of heritage (both movable and immovable) in an all-round and multi-format way, applying network connection and a series of relevant provisions of agreements to realize the sharing, efficient use and scientific management of heritage information, so as to provide services for users in digital-aided decision-making, scientific research, exhibition and demonstration, cultural exchanges, education and training as well as games and entertainment, etc,. Mining and metallurgical sites and heritages in Huangshi region are very rich. Since 1981, more than 2000 pieces of bronze and iron device heritage have been collected by archaeological excavations, site acquisition, dispersed collection and waste selection. 33 heritages have been identified by National Commission for the Identification of Cultural Heritage as Class A heritage property, including the full set of 13 "bronze weights" of The Warring States period, "Liang Jin Yi Zhu", "Liang Jin Si Zhu " the copper coins of Chu State during the Warring States period, "Ti Liang" of Shang Dynasty and "Copper Axe" the mining tool of the Warring States period. In addition, a large number of implements of production and living utensils have been unearthed in the ancient mines, such as copper axes, copper adze, the shovel, boat-shaped wooden Chinese peck, wooden ladle, bamboo baskets, ropes, pottery fragments, bamboo baskets, etc, [3]. All these cultural relics can be preserved and displayed by the digital three-dimensional scanning to establish the Huangshi virtual digital museum. The existing museum can also be virtual digitalized duplicating the cultural relics in the collection of cultural institution by the virtual reality technology. Due to the Internet technology, the virtual museums can be displayed on the Internet by the way of collections and virtual display, etc. so that more people can know and understand the mining and metallurgy culture more timely so as to further promote the economic and cultural development in Huangshi region.

IV. Conclusion

With the advent of the digital era, using digital technology to integrate national cultural resources to develop "information content industry," has been one of the important strategies of all the countries in the world in the 21st century to seize the political, economic, and cultural interests. Although Daye region of Huangshi City has an abundant resources of mining and metallurgy culture, the limitations of heritage preservation and display conditions hasn't make these cultural resources of mining and metallurgy in to fully utilization, and some are gradually fell into decline or extinction. For a long time, as the limitations of display conditions, a number of cultural relics preservation unit in Huangshi region are difficult to give full play to the functions of social education and cultural services. The application of virtual reality technology use of preservation of mining and metallurgy culture in Huangshi region will greatly promote the sharing of traditional culture resources in the whole society. It will provide spiritual impetus and intellectual support which Huangshi can in promoting advanced culture, establishing a learning society and improving the national quality comprehensively. At the same time, Huangshi City will further enhance cultural awareness and play an important role in creating an ecological and cultural tourism of bronze culture and landscape garden city.

References

[1] Kang Fengju. Modern Simulation Technology and Its Application [M]. National Defense Industry Press, 2001:9

[2] Hu Xiaoqiang. Virtual Reality Technology [M]. Beijing University of Posts and Telecommunications Press, 2005:7

[3] Li Xunxiang. Construction Concept and Its Significance of the Virtual Museum of Construction of Mining and Metallurgy Culture [J]. Journal of Huangshi Institute of Technology, 2007, 9(2): 12-14

The Brand Creation of Chinese Underwear in Terms of the View of the Cultural Pedigree

Sun Enle

School of Art and Design
Guangdong University of Technology
Guangzhou , China
Email: enlesun@sina.com

***Abstract*—This paper analyzes the development status of Chinese underwear and the importance of brand creation and proposes establishing the cultural pedigree of Chinese underwear and its brand, which is combined with the spirits of Chinese culture, diversified open thinking and innovative attitude. These three elements become the foundation of the three-dimensional coordinates and build up the development space for the underwear brand. The characters of such a three-dimensional space are contradictory and unified, antagonistic and harmonic at the same time, which are generally compatible with the cultural attitudes needed by underwear design and contained with tradition and fashion, inheritance and innovation.The author holds that a new thinking will be brought into the design and brand creation of underwear.**

Keywords-Chinese underwear ;design;brand;cultural Pedigree

I. THE PROBLEMS—THE DEVELOPMENT OF CHINA UNDERWEAR DESIGN NEEDS TO TAKE THE ROAD OF BRAND URGENTLY.

In the early 20th century, with the introduction of western Bra and Bikini, Chinese gradually came to accept the varieties western styles of underwear. In the mean time, traditional Chinese underwear stomachers are out of market slowly as the lack of the body shaping functions. The real modernization and commercialization of Chinese underwear start with the reform and opening up policies, which begun from 80's. Numbers of famous international underwear productions were available at China market, and became the objects pursued by Chinese customers because of the strong attractions of the styles, the fabrics and the quality, which meets the psychological and spiritual demands of consumers. The real development of Chinese underwear brand is at 90's; it gradually has been developed to a comprehensive industry covering with design, production, processing and distribution, through the imitation and absorption of western business. It has obvious status of industrial clustering and brand development trend and become an integral part of the garment industry.

The development of Chinese underwear has been leaded to the pathway of original design and brand creation, but it's still in the initial stage. The cup, plastic waist, trousers shape have continuously been improved and the advanced fabrics have been used to make underwear products more reasonable and more comfortable according to shape features of the Chinese people during the underwear design of domestic brand. But most of them are also sharing a same fault: Homogenization phenomenon became common, lack and weakness of personality can not effectively satisfy the diverse psychological needs of consumers, thus which will lose the advantage of the design differentiation. Although some brands have consciously done some individuation work on standing out the vividness and personality, but it is not systematic and professional enoughly.In particular, it is still relatively weak on the creation of unique traits which could be effectively identified by customers. There are only few designs that could do so.

The development of Chinese underwear industry must be committed to raising design level and building ourselves brand to increase market influence and grasp the speaking right for Chinese underwear industry at home and abroad. Design is known as purposeful creative activity; Brand is the pass to the market and the culture is an important symbol and soul of the design & the brand. Modern underwear as a typical product with foreign culture characteristics, must be characterized with some point views of national culture, integrated with it and obtain support from it, which is the right way to guide the Chinese underwear to the world, increase the competitiveness in the international market. The top priority to establish the Chinese own national brand underwear is to integrate and contrast Eastern and Western cultures, explore cultural demands of underwear design, establish the cultural pedigree of China's underwear design.

II. ESTABLISH THE UNDERWEAR DESIGN SYSTEM OF CHINA WHICH CONTAINS UNIQUE CULTURAL PEDIGREE

The author believes that the Chinese underwear cultural pedigree should be clear core, open and diversity. Pedigree is a term of socio-cultural studies. The inheriting and innovation of culture always interlink each other and is mutually reinforcing. In a number of successful product design,we can always find the inner culture pedigree for successful commercial brands. The existence of cultural pedigree and temperament enable them to gain long-term vitality and competitiveness and become the true flagship.

A. *Spirit——the Core elements of cultural pedigree of china underwear design*

To establish the cultural pedigree of the Chinese underwear brand must treat the spirit of national culture as the core elements. An indisputable fact, the vivid, charm focus and popular elements are the unique national cultural spirit. Through the development history of the underwear at home and abroad, it could be easy to find out that cultural heritage is the living source of survival for underwear brand and nationality cultural are the soil of brand survival. without the unique national culture elements, the design can not be to the world. The 21st century is a culture, science and technology-driven era. The civilization of mankind transfers from the industry civilization to culture civilization. "The country's cultural development strategies" has been proposed in many countries. Regional culture, as the indivisible part of the international culture, has powerful life-force with pressure under the background of globalization and modernization. Contemporary famous Chinese philosopher Zhang Dainian pointed out that Chinese nation has their own national spirit, that's why it is able to continuously develop for thousands of years[1]. This is the aggressive spirit which makes us stronger and the tolerant spirit which makes us virtuous, and these spirits are decent with subtle, elegant with magnificent, harmonious with intensive, mellow with passional, such as Chinese traditional cloth, cheongsam, stomachers Bra. They show the cultural phenomenon at a certain extent. With progress of the times, the spirit is bound to be developed and updated. Overall, it has no doubt that the creation of the underwear brands can reflect and demonstrate the values of a culture. For one individual brand, it seems to be unintentional action. However, the life style, attitude, spirit and culture of the nation or region are actually exerted in a subtle but strong way. Two strong domestic underwear brand which is named as "Aimer"and"Ordifen", showed up on the international fashion week in China, both of them chose the "China Culture" to deductive their attitude coincidentally. This is not only contain many "first" ever and also exhibits the coming of new era of China's underwear.

B. *Open and pluralistic cogitation and tolerance——the basic element of cultural pedigree of china underwear design*

The cultural pedigree of Chinese underwear brand must reflect open and pluralistic ideas and tolerance. Chinese contemporary well-known sociologist Fei Xiaotong – a famous Chinese economist believes that one important feature of Chinese culture is that the China culture can mix up the different element and become an union of opposites. We reflect the reasons that why Chinese culture is still sustainable and increasingly flourishing after the long years, so we know that magnanimous temperament and character are the vitality of the Chinese culture. An open circumstance should be constructed for the underwear brand creation to reflect the idea of "union of diversity", which contains the attitude of "appreciation of own beauty". Only we find out the beauty from the difference between ourselves and others, then "appreciation of others' beauty" could be really done from deep heart and true feelings and which is the only way to contain each other, generate the united power and create the combination of different nations and countries in peaceful and prosperous way[2]. We are used to distinguish the "East" and "West" on region and folkways, and on the development of underwear culture as well. The underwear of east style is emphasized on artistic conception, style and class, as well as the underwear of west is emphasized on shaping, composition and crazy. In the process of the human civilization, people's aesthetic philosophy will be constrained by period, ethnic, nationality and person and so on. The standard of beauty is being changed in accordance with the changing of the social trends and popular ideas. Now the fashion of the underwear could also be the clear statement of that. In the background of globalization. We need to review the position and direction of development for culture and design art of each nationality and area. This is a multiple development direction with localization and globalization. Globalization is a kind of development with double faces, which needs to develop the national culture reasonably, and develop concordantly and healthy with other area of the world as well. "Diversity" of culture and design is always the major model of development, but it has the advantages and disadvantages. Some scholars pointed out that we don't have a better way out except involved In the process of globalization deeper, but we can explore how to hasten benefit and avoid loss and how to protect the safety of china effectively and extend the economics and culture benefit[3].The beneficial experiences and principles should be found out from the advanced pattern of the international underwear brand. To pursue and reflect the value and influence of brand impression and the accumulation of reputation caused by the deduction of national style from the point of personality, to explore the touch of mind and it's effect of creative from the point of humanization, to pursue aesthetic taste of design and the shocking of visual from the point of performance. Excellent designs have been recognized by all over the world because of following these principles.

C. *Innovation continuously——the crucial elements of cultural pedigree of china underwear design*

The cultural pedigree of Chinese underwear design and brand also needs to produce innovative element continuously. Zhang Dainian points out what is innovation. "Innovation means there are different with China traditional culture and modern western culture. Because it is a new culture with Chinese characteristics, also it is a highly democratic and highly scientific culture in the human history." " Innovation is definitely not "fracture", but it is continuance and development of excellent tradition. The real innovative should sum up the strengths of Chinese and Western culture and meld them to one piece[4]."The history and reality repeatedly proved that Chinese people have the a rich diversity of regional cultural and historic humanism resources and China also have art and cultural cohesion and innovative ability of the "union of diversity", which is the reason why the vitality of Chinese national culture and art can be maintained and never decadent, why it can occupies a special place in the art treasure-house of the international culture and why it has the irreplaceable value and charm in human history of art culture. The development of Chinese underwear industry has entered into a critical stage with the thinking of "how to be itself" and "how to make an effort for building up international reputation of Chinese underwear brand", it is also moving into a stage which emphasizes personality and responsibility. There are many ways for create the brand personality, one of them is to make the design personality meets the personality pursuit of consumer more possibly. The building of personality needs shaping, the realization is a growing process which follows established policy and perseverance is the principle of "creation of brand personality". With constant enriching of "own story" , the brand also began to make itself. If there were only strategy but without shaping work, the consumers would not appreciate the kindness of the boasting strategy because it relies on the mutual recognition from both sides. So the brand without unique personality would be difficult to leave memories and impressions in the heart for the customers, and it would never become recognized brand for the target consumers without characters. Brand culture is not changeless; we need to mix with new elements according to the changes of market. The consolidation and upgrade actually is the re-examining and energy refilling for the brand. The designs and brands will be rigidly stagnant without above action, or even exist only in short period[5].

III. CONCLUSION——THE CULTURAL PEDIGREE OF THREE-DIMENSIONAL INTEGRATED IS THE NEW WAY OF COGITATION OF CHINA UNDERWEAR DESIGN DEVELOPMENT.

In conclusion, we can discover that the basic pattern of cultural pedigree in the process of creation of Chinese underwear brand is combined with the spirits of Chinese culture, diversified open thinking and innovative attitude. These three elements become the foundation of the three-dimensional coordinates and build up the development space for the underwear brand. The characters of such a three-dimensional space are contradictory and unified, antagonistic and harmonic at the same time, which are generally compatible with the cultural attitudes needed by underwear design and contained with tradition and fashion, inheritance and innovation. The author holds that a new thinking will be brought into the China design and brand creation of underwear.

REFERENCES

[1] Zhang,Dainian . Corpus of Zhang,Dainian Vol. [6] Tsinghua University Press 1995. pp.492.

[2] Fei.Xiaotong About Consciousness of Culture[J], Qunyang publish Press 2003.02. pp.7.

[3] Li,Qing Overview of Globelation & Sino Fine Art[J] Art Discovery Press 2008.10. pp.30–35.

[4] Zhang,Dainian . Corpus of Zhang,Dainian Vol. [6] Tsinghua University Press 1995. pp.490.

[5] Hu,Chuanni. China Branding[J] Art Journal Press. 2005.02. pp.43–44.

The Consideration about Setting of Topics in Animation Creation

Zhang Wei
Arts and Design College
Wuhan University of Technology
Wuhan, China
Email: power900@sina.com

Zhu Mingjian
Arts and Design College
Wuhan University of Technology
Wuhan, China
Email: power900@sina.com

Abstract—**In teaching of making short Animation film, a fundamental problem is to choose the topics and how to cutting in the topics of the consideration in not valid, it would influence the whole work and effect. The setting off of topics is so important that it is the crucial work about the creativity and whole work. In teaching process, to induce the student understand this importance of setting the topics, is vital to his creation and future development. Here we present some of the dilemma we face and the solution we probe into the selection and topics and we set the fundamentals in order to supply some of experience we could utilize.**

Keywords-Animation Creation; Teaching; Topics; Topics Setting Principle

I. Preliminary Remarks

In the process of Animation Creation teaching, we face different problem in different processing. But the crucial problem is the initial step the Drama. Our choice of topics and cutting in effect, if we failed to illuminate this process we will bring in negative effect in our maneuver and final effect. The failure of the drama will make our effort futility. We say Drama is the root of a story.

The three essential of a drama is topics, story and characters, the topic is very one. When we set the topics, all our details will serve this topic, all roles serve too. We will transfer this as our purpose, so as conclusion topics in an intrinsic important.

II. Same problems and solution in facing the main topics as for Animation creation

Wherever Times is specified, Times Roman or Times New Roman may be used. If neither is available on your word processor, please use the font closest in appearance to Times. Avoid using bit-mapped fonts if possible. True-Type 1 or Open Type fonts are preferred. Please embed symbol fonts, as well, for math, etc.

A. several problems when we select adapted topics

The problem we faced in adapted one there are three kinds of adaptation: Intrinsic to the original topic, Recreated and Reformed. The setting off of the first kind is normal. The second and the Third have more probability.

- We edging in differently whence we select differently. In third kind we maintain the frame of story and the designation of personality, to make sure maintaining the originality. But as to same topics, different personality selects different proceeding so the creation makes utterly different aspect making quite very different expression. This is because different people understand to different standard and the time-space makes it too. Basic style is different so we make decision how to cut into the topics. We set an example in Animation "Enduring Bitterness" , the same story and personality, we may set differently objectively or we make the unfortunate king as the vision, or his rival counterpart, or his concubine his loyal subject, that make big difference.
- Recreation, The use of recreating makes essentially transform to the original. We only keep some aspect as to the style of his topics, the personality, to frame the topics of the relationship of guys. But as for the intrinsic of story, casting the role, principle of ideas, they make their own sense and made recreation maybe forging some other story. But the style and the title unified, in this kind it approach a standard of original. But whether successful will relay on their choice. In the student's woke "Pajia and fire god", the author want combine. Two ethnic minority legends, but the two stories have different widely. One relate a unorthodox girl get educated and know reverence for deity, the other story related a fire-bird exchange the fire with human wing in a bargain. But they (my student) could not find a proper point to cutting and edging, the just keep some time space personality of minority people and fire itself. Finally their story topics fire is beneficiate and both dangerous. The whole story is in this stage.
- The story we innovate. The story innovate has two kind but not to refrain its time, incidence, personality the second kind is to discard the old story and taken its inner sense and satirize the circumstance, for instance the student work "Push the plant into his growth", he absolutely abolish the old details to reveal a normal nightmare a fundamental school child join all kind of Extra Curriculum in every holiday again other will. But there is nothing to do

with growing plant, but the implications in the same. The student woke “Wu Song and the tiger”. They change the environment to nowadays; they just imply the thing we saw in our daily views, for instance forgery and pollution. The student woke “Stealing the Bell”. He keep the original story frame, but change the original time, space personality, and in the way they use very, fashionable actions, make into comics.

B. Several problems when we select the original topics

The original topics must be selected from the aspect of panorama, selecting the edging point define the topics, we induce them but not make for them the decision of the students. The selected topics, if not basically wrong, we usually intend not to deny or overthrow. We do our best to help them fulfill or justify so that their passion could be sustained. In teaching and practicing the following is the student liability.

- The slurred Animation framework but the purpose is not definite. In the student work “Story in a Farm”, the author try to imitated “Crazy Farm” and “Chicken Run” to create a story about the farm. They set a colt the main character, the farm is transferred to a patrimonial merchant, the animals and in crisis, the colt spray the farm and make the resurrection redeem. In such a story is too difficulty to be express in a 20 minutes Animation film, and the author don’t understand the style and details of a farm and the story can confines, the logics is difficulty to insure the audience. There is only an ambiguous outskirt could not make the topic which is essential of creativity. In maintaining the story structure must we help him a definite understandable and controllable. Main topics if we set in this way. The main character is a farmer. He just inherited the ranch from his dead father. He is passionate about the ranch and the animals, but he is in confident considering he could not match his father's expectation. In this way the rancher confide the farm to a manager. But soon he fined the manager failed to match his expectation. So he cancels the bargain and faces his own responsibility and does a wonderful job. The main route is a in confident rancher evasive to responsibility, in this case when the routine is set, all good story trends and their character definite, the story go up straight, all irrelevant details are deleted the relevant is rescued, the insufficient is fortified. The whole Animation becomes notable.
- When notification is certain they fail to select the details. in the creation of “Pajia and fire god”, when he decides the implication, he need to set the contrast personnel and details, what intrinsic show. The fire is beneficial and dangerous. After resetting, the main character is a aboriginal boy naughty with everything, whose name is Pajia he is mischievous with the fire trick and almost end in disaster. In order to teach Pajia, the fire god changes himself into a fire ball and induce Pajia into an ancient Rock Cave, in the surface of the rock there is painting, Relating man's relationship with fire. This in sequence: The mischievous with fire show we normally dealing with fire. Almost come into disaster show the dangerous effect of the fire. And rock painting show the rough and tough way we emphasis fire is important to men. The Animation emphasis the topic though Pajia is making his progress.
- There is definite character but sense and story are obscure. Student woke “Fruit and Vegetable Story”, He want to exasperate the war between fruit and vegetable, the story and relationship is funny and creative. But his fail to relate the story and topics, so he delay his creation until the Troy story come into his mind, and the solution is here ,the story is O.K. the topics is O.K. furious war, Love or fame, the end is a delicious food in human table.

III. Selecting Stuff and Topic Setting

A few experience principle in teaching Animation, whence selecting and choice the main topics, there is ordinary some problems we suggest.

A. We say the clothes go with the body

A normal deficient, we lack this clothe body conception. The student’s creation Animation require two men for a 1-3 minute Animation, the graduation design requires 2 men an 8-10 minute production.

Minute and medium constructions are different, their expression and topics to resign are different, the expression and topics to resign are different, the Medium could completely recite a story, and by the story show the topics, but the mini one is difficulty only for a little implication and fable when we use this mini Animation as our object since story has a intricate personnel we could not with hold and resolve the problem. This we minimize the story.

Another solution we set the scope into the figures vision, and do this with his sensation. Such as the student woke “ Enduring Bitterness”, select the expelled Kings sensation as focus this avoid many unnecessary personnel, sequence and scenery, with minor scenery and Camera Set in the deep feeling and inner feeling with short scope and big material. This is advisable. The student woke “Exodus” is a enormous subject, originally unfit for Mini-Animation but my student insist this trial. This we do with sensations, he complete this we do with sensations, he complete this with “Water”, “Fire”, “lightning”, “Rock”, etc. The abstract way to reveal moods and sensation, mixed with real personnel and the personnel utilize light and shadow to express emotion, the make it a success.

B. selecting familiar topics to express comprehensible emotion

A very formidable point, we must select our familiar topics to express our comprehensible emotion in the mini-Animation creation. Because only familiar things we could relay on details, the comprehensible emotion we know to seize, if we do what we don’t understand, the audience could not take the sense and don't expect them to have any sympathy.

Nowadays in Magazine Flamboyant story is always a hit, relating live and death, seemingly sensational, but after a decent consideration we know it is not logical and profaned, work craft is apparent sometimes student favors to select this topics say it is sensational. But as for the emotions you hardly understand without profoundness of sympathy, with only imagination we hardly expect others will respond, only make your work infantile and obscure.

The motion picture Animation is agreeable a tremendous piece, if we do an unjustified story, what a waste.

C. Animation Businessman should understand their social Responsibility

The first lesson of architectural student is to inform. Architects have social Responsibility because they design the real environmental of human life, so the motion Animation designer designs the spiritual environmental, this is even far more beyond, so Animation professionals must sense their social responsibility.

In the ancient, not everyone could read and write, but everyone know "Liu-Pei", "Guan Gong", "Zhang Fei", "Yue Fei", "General Yong" and their story, and what implied Loyalty ,Justice, Faith etc. This is essential for "drama", "depicting story" art form concern so a illiterate man knows the reason, The art piece could influence social value inclination, so a art work have social responsibility.

Nowadays Motion picture producer have further impact, because every family have television (TV) People watch TV everyday the TV product have very influence on everyone. The indulgent students has their vision of history mainly from the TV they watch. Once I have my student argue the costume and living habit of ancient Greece people and their insistence relays on the material mot historical chronically but a movie he insist his vision, say it is the history. As about for the classics of China, the circumstances are about the same. In the 80's many Comics and Animation creation are serious and responsible as for the ancient costume, talks, living style, there is a justification of every detail. But Nowadays, they are in Ancient Costume to Behave Modern Show, not only in "comics" but the Normal ones, if we persist in this way, the history we know will be changes, where will the next generation be guided.

The traditional has defects, but mostly positive, never obsolete, Kindness, Justice, Curtsey, Wisdom, Trust; Humble, Goodness, Reverence, Frugality, Retreating; Loyalty, Obedience, Benevolence, affection, Courteous, Justice, Cleanness, Sense of honor. There is many adorable stories, which process positive effect to human or society, the ancient man's wisdom experience and teaching is not to be discarded.

So when we selected tradition material we should understand and justify ancient men's action and mentality and don't set it in a modern way. Or else we probably slander it and mislead our audience. Nowadays it is crucial in motion Animation picture creation.

IV. CONCLUSION

Good drama and an appropriate topics is only the first step of Animation, this can't make sure the short motion picture a success, many other factor decide their success, the painting skill, the Animation skill, maneuver Motion, Color, Video, Compliance. A good drama don't attain a good Animation is so much a shade.

But the more shade is due to an not so proper drama, some product in the expense of large time and energy, the fundamentals are good picture style scenery, motion color is so good. But limited by the drama composition, don't make an up-right product, even doesn't in makes. So a good drama composition makes sure the fundamental. Even the other parts did not attain the purpose as for the Animation it make sense. Vice-Versa we could not make and a superb Animation when we don't make a good drama composition. However nicely we make it so the topics of the drama are half success a Animation Motion.

In teaching to induce our student concedes with the drama and topics are essential the future creation and development. This is practical.

REFERENCES

[1] Ge Jing, "Animation film Script writing", Beijing, Ocean Press, 2005.4.

[2] [France] Pierre. Let, "Techniques in film Script writing", Beijing, Chinese film Press, 2005.12.

The Construction of Chinese Tea Culture Tourism Landscape

Li Zhiyong
School of Civil Engineering and Architecture
Wuhan University of Technology
Wuhan, China
Email: dyxlzy@163.com

Chen Ming
School of Civil Engineering and Architecture
Wuhan University of Technology
Wuhan, China
Email: cm-1966@163.com

***Abstract*—The rich tea resource of China is the important carrier of tea culture landscape which is also a feature of Chinese landscape.In order to promote development of tea culture tourism, it optimizes and improves tea natural landscape resources through the research on natural environment and culture characteristics of tea, then constructs a set of complete system. Thus, the quality of tea culture landscape and the diversity of tourist attractions get improved.**

Keywords-Tea;Tea Culture; Tourism;Landscape;

I. Introduction

The history of tea collecting and drinking can be traced back to 1100 BC, the Zhou Dynasty. Over the long course of development, tea has not only boomed in physical form but also dept into the cultural and spiritual dimensions, and then formed unique tea culture which dominates and promotes the formation and development of global tea culture.

Tea culture tourism is a tourism way that combines both tea resources and tourism in an organic way. It integrates the ecological environment and natural resources, production and cultural connotation of tea to develop a tea culture tourism. Taking advantage of beautiful and quiet environment and based on the tea production, tea culture tourism landscape takes tea as its carrier to make rich cultural connotation and colorful folk customs activities which relies on the natural and specific historical and cultural landscape. Most importantly, the tea ceremony centers on "harmony" which advocates us to choose a sincere attitude of life, to socialize with courtesy, to build a new type interpersonal relationships with peaceful coexistence, mutual respect and interest. Tea culture promotion is conducive to develop socialist culture and ethics and construct a harmonious society.

From the angle of tourist economics, attracting the tourist to carry on the travel activity is a region 's tourist resources which must have the distinctive and ornamental quality regardless of being visible or invisible. The differences of tourist resources and other resources lies in that the tourist resources can give tourists aesthetic enjoyment conforming to their physical and psychological demands, making people's spirit, disposition, the quality and so on find the objected performance in tourist resources with the most beauty quality.

II. Optimization of Natural and Ecological Landscape of Tea

At present, tea landscapes of our country are mainly formed by large scale of natural scape of tea mountains which only provide for the tea growing but cannot adapt to people's demands for high-quality of living and touring. Thus, it should carry a sound optimization and upgrading of tea landscapes of our country. In the process of optimization of tea ecological landscape, tea mountains should be re-planning to , based on the preservation of original natural beauty of them, be a eco-tourism landscape of tea which is not only for tea growing but also for people's travelling by the design approach of modern landscape.

A. Optimization of The Natural Landscape of Tea

Combining the natural landscape of tea with tea hill to plan tridimensional tea plantation, we may consider: intercrop tea and trees, integrate tea and medicine, companion plant flower and tea, undersow tea and fruits. Widen thoughts and bring in improved variety of seed to make tea plantation become the natural green botanical garden, the demonstration garden of famous tea variety and the novel and unique tridimensional garden. Moreover, under the foundation of natural beauty, set the artificial scenic spot skillfully, which is situated at the foot of a hill and near a stream, also takes the advantages of environment. With the thatched shed, the bamboo house, the small log cabin, the simple tea shed strewn over the spot consciously, the tourist could taste tea, have a rest and take photograph as souvenir. In the tea garden, the variety of plant could increase our experience, make us feel the fresh and pleasant of green. When placing oneself to such “scenery of tea garden” which constitutes the fairyland in the bosom of green mountains and clean water, you may feel the utmost rustic charm of countryside.

B. Optimization of The Eco-Agriculture Landscape

The eco-agriculture landscape consists of natural environment, ecological resources, production means, production activities, rural culture and lifestyle and so on. It posts enormous attraction to modern urbanite through keeping the relatively original production activities. Moreover, establish eco-agriculture tourism area——the tea amusement garden, and under the foundation of the improved variety demonstration base of tea, design a series of activities of seeing and enjoying tea garden scenery,explaining tea culture knowledge, tea plucking, making tea, tea making ritual , tasting the tea meal, sales of famous tea and tea sets and so on. The culture value of tea is the most important one among all farm products

The exploitation of tea culture eco-tourism includes a good natural environment, tea garden and tea hill; production goods, tea; production activities, tea-picking, weeding and tea-fried;life style, tea chant, tea dance, tea sipping and taste of cate made of tea; agricultural culture, the knowledge, literary quotation and appreciation of tea.

III. OPTIMIZATION OF THE HISTORICAL-DEVELOPMENT LANDSCAPE OF TEA

Tea of our country has a long history with amount of legend stories which are only promoted by books and media without any landscape value. At present, by the study of historical and cultural values of tea, it integrates landscape elements and the historical development of tea culture, then builds a intersting tourist attraction of tea.

A. Optimization of The Historical-Origin Landscape of Tea

The origin of tea has a long history with rich tea history , anecdotes and legends.The legend of Yan Di(Emperor Yan),the Shen Nong(the Immortal Farmer) told that Shen Nong tasted many kinds of herbs, and often tasted several poisonous herbs daily, one day he ate some herb by accident,then detoxified, this herb namely tea. Tea Classics wrote by Lu Yu , intensively study on the history of tea culture, narrates that tea men were studying hard and undaunted by repeated setbacks.All these stories were combined with the the natural environment of tea and many of them became the materials of frescos and inscriptions on precipices.To some extent, these historic origins are also the reflections of some people's values and esthetic sentiment of tea culture.Searching for them can find the joy in our own heart.

B. Optimization of The Historical-Heritage Landscape of Tea

The formation of tea culture has a long historic standing. The self-development of tea and the communication and transmission of tea man's anecdotes, tea sets and tea culture with archaeological significance have developed without the restriction of the tea producing area.Tea in every period, always combined with one or several artistic forms, presents some beauty characteristics， such as the Grand View Garden of Tea Culture and Tea-Horse Road.These resources are almost spiritual, intangible and with comparatively high cultural taste .Compared with sigh-seeing tourism and ecological tourism, they can better satisfied with people's demands for opening their mind,seeking culture sources and tasting different cultures.

IV. OPTIMIZATION OF THE CULTURE LANDSCAPE OF TEA

Tea culture of our country are very rich, but seldom reflected in the tea landscape. In the present construction of tea landscape system, it optimizes tea culture, and integrates culture and landscape by using techniques of landscape elements, landscape architecture or landscape performance, finally forms a particular tea culture landscape.

A. Optimization of Tea Ceremony Landscape

In the process of development of Chinese culture, a unique tea culture has formed by constantly absorbing the philosophy spirit of Confucianism, Taoism and Buddhism,which plays an important role in carrying forward the traditional virtues, cultivating the moral character and nourishing the nature,promoting social progress and developing economic and trade. Chinese tea ceremony originated from Tang Dynasty, flourished in Song Dynasty and Ming Dynasty, declined in the Qing Dynasty. It stresses on the beauty of tea leaves,tea water, duration and degree of heating,tea sets and natural environment.

Cultivation methods of tea men refers to elaboration, nobility,thrift and benevolence which are the basic demands raised by Lu Yu. These are also standards of judging tea men 's thinking, moral character, behaviour, beliefs, etc.Lu Yu had sublimated tea morality from simple physical level to spirit level.

Tea shows its rich connotation whether you make or drink it that tea-making reflects the "The Doctrine of the Mean" while tea-tasting reflects modest propriety. To combine tea fragrance with elegance and grace of refined aristocrats and scholars to form Scribes Tea which can best embody the essence of tea culture and tea ceremony.Taosit tea ceremony is unique for its respect of God and Heaven, which pursues a state of quietist mind and a return to basics.

B. Optimization of The Culture Landscape of Tea Custom

The majority of tea areas have a long tea planting history. Songs and dances about tea plucking with obvious local characteristics carry down from generation to generation. The traditions of tea drinking, tea legends and tea wares become the spiritual wealth which gradually takes form over a long course of development.Different ethnic groups have variety of using methods of tea and remain a lot of historical heritages. All these aspects we discussed are obvious representations of folk customs of a region and also become the characteristic forms of ethnic tourism.

The cultural life of tea custom is abundant in content with variable characteristics such as tea and wedding ceremony, tea and sacrifice, and tea-drinking customs (tea-serving,tea-grinding,"three-course tea","guest-serving tea", "guest-detaining tea","blessing tea" and "the tea monopoly language" and etc).The tea monopoly language,which has many modes of expression such as tea proverbs, tea ballads, tea puzzles,tea songs and tea dances,often passes on by oral transmission .All these cultural heritages are precious possessions with a lot of categories and far and wide spreading and show human landscape of tea.

V. OPTIMIZATION OF THE TEA CULTURE STREET LANDSCAPE.

Street landscapes in China are numberless, but most of them lack of particular tea culture features. Through re-planning of these streets that have some relation with tea and integration of the whole street and tea culture, it builds a street landscape with particular tea culture features.

Enjoying tea ceremony and attaining the truth of tea in the teahouse, the tourist could express their interests and soul under the quiet and graceful artistic conception. Set up all kinds of special famous tea shop along the street, where the tea of QiHong, TunLv, MuDan, MoJu ,DeFang, MaoJian, the flat tea and scented tea vie with each other for glamour. People may invariantly marvel at it at the same time of purchasing eagerly. Moreover, various kinds of tea sets of red porcelain, ancient pottery and ceramic white ware and so on, together with unique Xinan tea calligraphy, the tea poem and painting can initiate the exciting interests of tourists with high grade. The tea song is delightful, the tea advertisement is striking, the tea ceremony hall, the tea cure center, the tea exposition hall and so on, which makes this tea culture street, permeated with local character and style, manifests the tea culture street landscape by its fresh tea fragrance .

VI. Optimization of The Faming-Family Landscape of Tea

Nowadays, farmhouse landscape can be seen everywhere in villages of our country. Farmhouses near tea gardens generally only possesse the function for peasants' s living, but not for sight-seeing. By the design optimization, these farmhouses not only satisfy people' s living ,but also provide a particular sight-seeing for tourists.

Tourists are singing tea songs with villagers in the melody of tea plucking, experiencing joy of being a tea man, field travelling to watch technicians making tea by hand, and trying to do maintain some kinds of tea-making skills. The primitive instrument are simple and the original craft are exquisite. In the process of being rich in dynamic appeal and pleasant of creation, we may feel the profoundness of the tea industry, a primitive ancient industry, and realize the abstruseness of tea industry which considered as the best natural health care drink long lasting millennium.Tourists should be taking as the guests in this small village, learn tea history and the knowledge of purchase and storage from tea farmers, eat from the same Big pot,drink stall tea, experience rural life,or teach children by permitting them to be close to the nature,thus they will know more about the ecological environment.

VII. Conclusion

At present, our tea culture tourism has not entered the period of maturation, the construction of tea culture tourism landscape is helpful to enrich the distinctive and ornamental quality of tea culture tourism, and also it highlights the tea's characteristic and guarantees the richness and perfection of tourism landscape. Only in this way that our country's tea culture tourism is possible to develop and expand healthily.

References

[1] Zhuang Xuce and Huang Hongfa , "Status Quo Problems and Countermeasures in the Development of Special Towns with Tourist Characteristics:With Special Towns in the Development of Urban Circle of Wuhan Municipality as Examples,"Development of Small Cities & Towns.

[2] Lin Limin, "Consumption Trend and Development Strategy for Leisure and Holiday-Making Tourism Real Estate Market," China High Technology Enterprises,2009.04.

[3] Lu Yu(original) and Zi Tu(compilation), "Illustration of Tea Classics:To Realize the Chinese Tea Ceremony,"Nanhai Press ,2007.

[4] Wang Zhongcheng and Li Jinlian, "Research on the Development Mode of Small Cities and Towns,"City Planning of Jiangsu Province,2009.03.

[5] Lu Wenrong, "Carry out Tea Cultural Eco-tourism Activities to Serve for the Construction of Well Developed Eco-Economic City,"Journal of Tea,2002,8(1) ,pp.2-5.

[6] Lin Fuming, "Develop the Tea-Culture Tourism to promote the Development Tea Industry ,Tea Science and Technology," 2007.3.

[7] Zhang Linjie, "The Research of Tea Culture Tourism,Journal of Tea,"2007,pp.183-186.

[8] Han Shuyuan and Bi Caihong, "Eco-Tourism of Tea village Bring about the Regional Economic Development,"Science and Technology of Sichuan Agriculture,2007.2.

The Contextual Characteristics of the Special Landscape of City Intersections

An Qi
School of Arts & Design
Hubei University of Technology
Wuhan, China
Email: anqi2777@163.com

***Abstract*—As junction of stereoscopic new style road, city intersections not only have the function of systematizing traffic and keep it smooth, but also work as a landscape node for the whole city's image. This paper focuses on the study of the contextual characteristics of landscape design on the base on the scientific and artistic analysis of the special space form --- intersections. By doing so, we try to find a solution for those questions we may face when doing related landscaping design and then to help us to analyze and research the contextual direction of intersection design.**

Keywords-Special Landscape of City; Characteristics;, bridge construction; overpass

With the quick development of city economy and the population explosion, there is a sharp increase in the number of car ownership and inevitably the traffic jam has become more and more a severe problem in most cities. Consequently numerous intersections are built at the vital communication lines to release the huge traffic pressure on the road. So the intersection has become one of the most important symbols of modernized cities. As junction of a stereoscopic new-style road system, intersections not only have the function of organizing traffic and keep it smooth, but also work as a landscape node which is of great significance for the conformity of city landscape. City intersections are the communication tools between city and the world not only for traffic but also for culture. But is it possible that they are also the tools for city to communicate with the nature? Actually, the space beneath intersections, due to its special function and Special pattern, are arousing increasing concern on the issue of city environment.

Ⅰ.THE SPECIAL CHARACTERISTICS OF CITY INTERSECTION

Intersection is one of the bridge construction, it is the aerial construction built on different levels on the cross of two or more than two roads to avoid the mutual interference of vehicles.(Encyclopedia of Chinese Civil Construction---Bridge Construction). The main function of city intersections is to systematize traffic and keep it smooth. Because of their bulky outline and unique formation, intersections are actually regarded as both important transportation means and significant landscape parts. Meanwhile, due to their different design purposes, they also have great radiant effects on the environment around. Different kinds of intersections have various radiant effects and requirements on their environment around because of their own different form and location.

From the junction form, there are mainly the following intersections at junction of three or four roads:

Junction of three roads: horn-shaped, half-clover leaf-shaped,T-shaped, double Y-shaped;

Junction of four roads: clover leaf-shaped, half clover leaf- shaped,
lozenge-shaped, half directional and totally directional (picture a, b, c, d).

Taking cross shape into consideration, we also have flyover and tunnel bridge. What's more, the cross interflow mode is another main factor which has final effect on junction dimensions. According to the mode of the stereoscopic interflow of traffic direction, all intersections are the following: entire interflow, half interflow, simple interflow, non-interflow and discrete interflow. A complete intersection system may have the following components: the circumjacent buildings, the main body of the bridge, the greenbelt surrounding the bridge and along the ramp.

Therefore, based on the above points, we see that the Special pattern of intersection area has the characteristics of convolution and germination which produce unique landscape angel of landscape and bring people splendid and fancy experience. It is the first kind of construction which offers us both faction and appreciation.

II. The Special Landscape of New-style City Intersections

When taking intersection as one effective way to solve the current traffic problems, we have to take the following questions into consideration: we have to make it serve as much as possible both in the function side and as unique sight resources; we must reduce the environmental pressure it may lead to and try best to keep the original features of environment so as to achieve harmony between the bridge and its surroundings in both sight and context. Most city intersections are built in reinforced concrete and in most cases in boring dark colors. And the space beneath bridge is usually designed molofuctionally as a passway or simply closed as a dead corner.

But in the city of Chengdu, Sichuan Province, China, most intersections are delectable to see. They almost have formed unique regional intersection culture and have become the most brilliant artistic view line of the whole city. Intersections in Chengdu are usually designed more like a park or even a museum than a simply bridge. They have naturally become the name card of this beautiful city now. Taking Yangxi Intersection as example, we may have a chance to taste the Special landscape of new-style city intersections(picture e,f).

The landscape space of Yangxi Intersection in Chengdu is designed in the form of science gallery. Down the Yangxi Intersection, you see huge pillars standing in front and three doors in different colors open one after another. Four groups of sculpture representing the theme of ancient, history, present and future are presenting in front of visitors. All visitors here are guided to enjoy an easy travel across time and space. The huge model of numerator structure in stainless steel, Dolly Sheep on the ground, gene chains, satellites and rockets...all these art pieces are evidence of the sweatiest fruit of modern science and technology. Standing in front of the blue-purple door of future and looking back at the other three doors, you feel that you have just completed a mysterious trip through the tunnel of time and have just come back to the modern world from the fartherest uncivilization. This impresses on you how fast the development of our cities is.

Yushuang Intersection, which is located among several provincial buildings, is another example. The intensive cultural landscape under the bridge vividly reflects the changes of information culture in the city of Chengdu. In those models, you see the oldest television aerial people used in the last century, you see the mineral radio aerial on civilian's roofs, you see the latest satellite receiver through which you eye-witness the developing and the prosperity of information culture.

Mozi Intersection on Yihuan Road is located in the central part of the biggest electronic products mall in western China. Conceptions like future, science, globalization and so on are embodied in the stereoscopic pictures on the bridge body. There is circuit board in alloy material in here, there is transition of Chinese counting method (e.g. rope knots, oracle bone inscriptions, compass and computers). All these pictures create strong visual impacts on their viewers with their experienced line work and lifelike shape.

III. The Contextual Characteristics of Special Landscape of City Intersection

At almost every transportation junction of modern cities, there are intersections. If we take advantages of the space under the huge bridges at those pivots to represent the geographic and contextual features of different city regions, we will bring fresh vitality to this cold construction and thus create a lively name card for cities. The problem is how shall we inject rich contextual information into the intersection space which is located in different parts of the city?

A. The Characteristics of the Space of Intersections at the Entrance of a City

The development of modern cities has endowed the entrance of cities new contents and new features. Modern city entrance not only has the symbolizing function, but also plays a role of transition region in the city's exchange with outside world in materials, energy and information. So the entrance view is naturally the first impression of the city's landscape perception system. The most striking feature of the city entrance landscape is a kind of non-experienced visually sensed view. The process of being sensed is dynamic, and the aesthetic appreciation and memory are the main aesthetic effect. The aesthetic effect of city entrance landscape is mainly embodied in the landscape itself and the expending, simplicity, broadness and unaffectedness of its space.

As the important transportation space which plays the role of the first impression of a city, city entrance should be designed firstly from the angel of its dynamic perception when designers are developing its landscape construction. Comprehensive consideration should be taken into from the views' dynamic visual features. Only by doing so can the designers achieve the strong visual impact on the aesthetic subject to the first impression of the city. Special landscape takes city as the main body, it explores the history of the city, manifests the features of the city, and carries forward culture of the city.

B. The Space Characteristics of Intersections in Commercial Zone

The development of city commercial zone and the development of transportation are always complementary and inter-constrained to each other. The improvement of the latter is always accompanied by the rapid development of the former. Early city commercial zones were usually located on the pivots of main roads. And just because of this superiority of transportation in its location, city intersections as the new-style transportation means have finally replaced the traditional complicate traffic pattern. This has exerted a profound influence on regional transportation.

City commercial zone is a multi-functional area where streams of people and vehicles are intensively interwoven. Both motor vehicles and walkers exist here. So the intersections in commercial zones should meet the requirements of both vehicle means of transportation and walking means of transportation. The corresponding Special characteristics are being crowded, narrow and of high headroom. For the drivers, they may enjoy the skyscrapers in the commercial zone, the excellent landscape sky line, active commercial atmosphere, pending city pulse and so on. All of these are the intensive embodiment of city vitality. Fundamentally speaking, intersections in commercial zones have strong commercial atmosphere, but they don't have any commercial advantages themselves. Anyhow they represent a kind of landscape atmosphere of variety. The rich landscape structure in region, more than one view highlights with strong sense of wonder. Moreover, each city has its own specific traditional commercial features, some of which may have lost in modern commercial activities. Then, the intersection space in commercial zones may act as the role of the recorder, the advocator and the educator of these commercial traditions.

C. The Space Characteristics of Intersections in City Residential Area

Residential zone is the most direct human settlement and the most human area. It is an resting harbor independent of city noise where we can relax both our body and soul. It must be people oriented so that we may feel the belongingness, sense of security and comfort here. It is a really nice space for residency. So intersections in residential area must be integrated into the environmental atmosphere of the community and take the inhabitants' visual perception into consideration when being designed. The inhabitants' requirements are the consideration priorities on the base of humanism.

When the intersections are facing the problems like heavy traffic and then affecting residents' vision and influencing people's outdoor activities, they become a sticky issue for designers. Their enormous shape becomes a huge "wall" which not only visually but also mentally. In other words, we can take intersections in residential areas as solid construction. From another point of view, the height of intersections in residential areas is the key issue. Here spaghetti junction is suggested because it allows more space utilization, more open space and less land occupancy. The intersection space can also become the extended part of the internal environment of the residential areas, and the connecting point between the residential areas and city context.

D. The Space Characteristics of Intersections in City Industrial Area

The construction in city industrial areas is mainly workshop, ware house and few office buildings. Due to the restrictions in produce technologies and manufacture method, industrial buildings are mainly arranged according to the product lines. So the primary impression they give people is massiveness. In industrial areas, green land is comparatively lacking. Intersections in industrial areas then have to face the rigorous reality and pay attention to the effects they may create on the regions. The design of intersections in industrial areas then, must try best to reduce the side effect they may cause to the environment and at the same time exploit their own landscape advantages. Luckily, the graceful lines, multi-layered structure and the strong dynamism of intersections are in sharp contrast to the comparatively simple form of industrial buildings and this may inject more vitality to regional landscape.

The space of the intersections in industrial areas and the surrounding buildings has the same relation as that between aspect and entirety. In industrial areas where people mainly see continuous and orderly view, the dynamic modeling and beautiful lines of intersections are a kind of visual buffer and conciliation. The design of intersections in industrial areas should take landscaping into its main consideration. Designers should firstly think about the matching of industrial pollution and the cleaning function of plants. The development of industry, the application of hi-technology can also be used as landscape factors in design. This kind of design can help the intersection space to fulfill its function both ecologically and contextually.

In nowadays, the economy and science are keeping developing all the time, more and more space form emerge in endlessly. This means we need to reform, develop and use more and more humanized space. With the continuous perfection of city function, human settlement mode has also being changed. It will take the designers quite a lot of time thinking about the question of how to utilize different space to serve people to achieve better and better life in the future.

REFERENCES

[1] J. Sun mingchun, zhou weina"Urban-Landscape Design". Xi'an Jiaotong University Press,2007.

[2] M. John. O. Simmonds, "Landscape Architecture - Site Planning and Design Manual". China Building Industry Press ,2000.

[3] G. Liang jin, "City Interchange Landscape Design and Reform of the principles of the lot " Huazhong University of Science and Technology press, 2004.

[4] J. Dai zhizhong,zheng shengfeng, "City Bridge Space" Southeast University Press, 2003.

[5] G. Liu jun, liu chen, "City overpass to the subsidiary principle of space utilization of the landscape" Journa of Chongqing Architecture University , 2007.

(a)

(b)

(c)

(d)

(e)

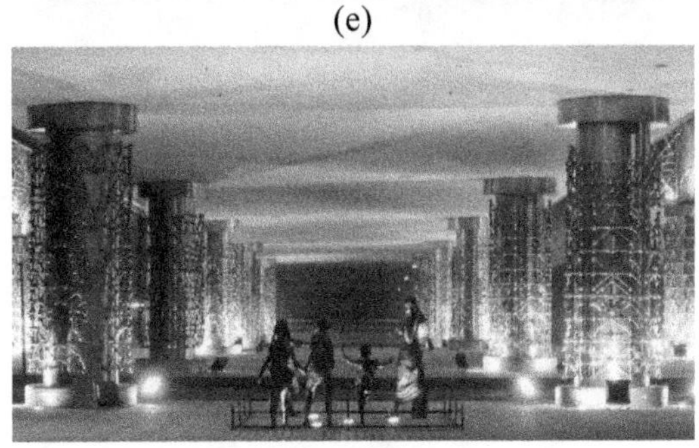

(f)

The Design of Urban Green Space Based on the Landscape Pattern Optimization

Lu Shizhu
Academy of Arts
Jiangxi University of Finance and Economics
Nanchang, China
Email: nbbox@163.com

***Abstract*—Urban greenbelt is a variety of green space in urban planning area, is a green space which plays the ecological balance of features in a living environment and closely relates to our human life. The design of urban greenbelt based on the landscape pattern optimization is the intersected science research of landscape ecology and landscape design, this paper analyzes the urban planning ideas and landscape ecology ideas of the urban greenbelt design, illustrates the pattern of landscape ecological optimization must be studied in methods, makes sure about the process and mechanism of the landscape patterns ' evolution, proposes the methods that make use of ecological planning and design, along with the way of designing the green space scaled from the spatial pattern and ecological factors.**

Keywords-green space; landscape ecology; landscape patterns; optimal design

I. INTRODUCTION

The system of urban greenbelt are powerful, it is an integral part of the city and it reflects the city's natural attribute, what's more, it plays the important role to incarnate the promoting of urban natural features. Human beings make good use of green space to improve the urban environment and shape the city's character; urban green spaces have the function to define the urban space, ecology, using, aesthetic value and so on. These function of the greenbelt related closely to its spatial pattern. Urban green space pattern reflects the urban form directly, which can organize urban space effectively and create a city's character.

II. THE STUDY OF URBAN GREEN SPACE DESIGN IDEAS

A. Urban planning ideas

At present, China has not generally used the urban planning system to programme the urban greenbelt and make the appropriate changes. Firstly, a typical urban green space planning analysis the planning area based on the basic data, the next step is targeting, and then plan its functional layout in accordance with its function, distribute the various functional areas and industrial belts rationally. Generally, it is divided into core areas, demonstration areas and radiation zone. After the functional division, plan the land according to factors such as land needs, and the road for transportation, then organize and associate all functional areas, to make sure about the size and scope of various land. Finally, plan and design each type of urban green space with in the district according to land suitability and so on, and express it on a clear drawing. This urban playing idea which based on the way of using lands is mainly about spatial layout planning, it mostly emphasis social and economic benefits, however, the key elements of the urban greenbelt planning is to find a urban green space system that make the needs of urban landscape, make a comprehensive analysis of planning with the organizational structure, operation mechanism, economic, social and ecological benefits of urban green space, then put it into layouts. Therefore, this line of thought has not been planned and reflected the eco-efficiency that the planning of urban green space emphasized.

B. Landscape ecological planning ideas

landscape ecological planning refers to the use of landscape ecological principles, a regional landscape ecosystem overall optimization is the basic goal, based on the landscape ecological analysis, synthesis and evaluation, establish the spatial structure and pattern which can make optimal use of regional landscape ecosystems. It emphasizes the landscape spatial pattern can control and influence the process, and tried to change the pattern of flows to maintain the landscape function healthy and safe. [3] The essence of landscape ecological planning is to arrange the landscape ecology spatial pattern reasonability, to achieve sustainable use of landscape as a whole. According to principles and methods of landscape ecology, plan the landscape pattern rationally at the macro and design the appropriate ecological conditions at the micro, to promote the rational use of resources and sustainable development. Landscape ecological planning and design includes two parts which are the level of the process and vertical process design. The level of a process refers to the optimization build of landscape pattern, vertical process known as the unit landscape land-use patterns design. According to principles of landscape ecological pattern optimization, the most important one is to classify the landscape, which is to classify the use patterns of lands in the planning area, and then plan the landscape functions through the analysis of landscape pattern, the last one is a reasonable allocation of landscape pattern. To adjust and re-build the functional unit in the planning of regional landscape, including the adjustment of the original landscape pattern, and the introduction of new landscape elements(Ecological Buffer and so on), establish a sustainable pattern of landscape ecology.

C. A comparative study of two kinds of planning ideas

Through the comparative study of two kinds of planning ideas, finding that both in planning methods and planning ideas, they are very different from each other, both of the planning ideas have focused on different aspect. Therefore, it is proposed to mix the two kinds of planning ideas, planning the urban greenbelt through the landscape pattern optimization to achieve the coordination of the overall functionality and results, this function primarily achieved through the city's green corridors and green patches of the spatial layout. Green corridor(street trees, green belt) its overall structure of the whole shape is essential to the creation of a city, natural corridor(rivers, vegetation zones, parks and green spaces, farmland and so on) it can limit the uncontrolled urban development of the share of the pie. Greater London's green belt and rural ring defines the central city of London and the surrounding satellite metro, ring green belt is wedge-type distributed and make the city green space at all levels excel networked through the green wedge, green corridors and river, etc. to form the pattern of Greater London. [1]

III. THE MECHANISM OF THE EVOLUTION OF URBAN LANDSCAPE PATTERN

A. Data acquisition

Obtaining information of green space is the first step for the study of the urban green space pattern. The scope of small-scale can be surveyed with the field trips, for large-scale example, this workload is too heavy to only rely on the field trips, it is very difficult to complete. However, recent years urban construction developed rapidly in all countries, the change of green space pattern became bigger, it also took so long to census data, so it is too difficult to reflect the real situation accurately. Airborne remote sensing data and satellite remote sensing data can acquire information on green space more accurately. For large-scale (large cities), these remote-sensing data can be largely reflected the overall structure of urban green space, however, the major part of urban green space system- green corridor and river roads are more difficult to extract. In September 1999, the IK0N0S satellite which was launched by the U.S. Space company and which was launched by the United States such as the Quick Bird satellite shot a very high spatial resolution images (lm or less). It can reflect the situation of urban green space very accurately, on this basis, it is a high reliability to study for the spatial pattern. In home, Wuhan Iron and Steel Industrial Park, Liu, and Zhuzhou City have used this kind of high-resolution satellite images to analyze the structure of green space, the results were very good.

It is essential to classify the accessed information about green areas for the analysis of urban green space, the classification is different, so the results of analysis will be different on this basis. Currently, the green category which was done before structural analysis of urban green space are divided into two main categories, a class classification according to the size of green area, it is divided into four classes or seven classes. The second category largely classified by the green space function, this classification based primarily on the national green space classification standards, generally include public green spaces, green residential area, sub green space, roads, protection green land. These two types of classification have their own characteristics. Analysis of the results is significantly different, but no matter which classification method is used, the analysis can be largely reflecting the spatial pattern of urban green space. Urban public green areas, urban green coverage rate of urban green space and urban green space targets can be used for green analysis. However, in the evaluation of different plant species and the green function of spatial structure, especially when analysising the ecological benefits in systems, it is very difficult to estimate it accurately. The amount of urban green indicator refers to the stem and leaves of plants growing in the space occupied volume (pm), have overcame the lack of two-dimensional green values and reflected the ecological benefits of urban green space more effectively, to make the evaluation of green space step forward the 3D technology, but it can not reflect the overall spatial distribution of urban green space. Landscape pattern analysis indicators can analysis the urban green space pattern commendably, in the analysis of urban green space at present, the diversity index was almostly used. These indicators can analysis the spatial structure of urban green space quantitatively.

B. Data analysis

For spatial data sources, using ArcGIS software in the ArcMap module, for different time periods of the Green Map, using spatial overlay analysis, have acquired different periods of land used changing maps, then extract data with ArcCatalog software, according to the type of distribution of green space and its changing environmental effects, merge the original green using the classified data, and have acquired types of green space in the cities region, as well, the natural landscape was sub-typed to obtain the dynamic evolution of urban green space data. Using green space dynamic model and classification index of changed model to explore the temporal evolution of the natural landscape within the study area, and then identify the main driving force factor, simulated the evolution trend of the natural landscape dynamically. [2]

In the late 80s of the 20th century, landscape ecology began to flourish in the world, its development provides some basic principles and quantitative analysis of index for the landscape pattern analysis, such as landscape heterogeneity index, landscape diversity index, landscape dominance index, landscape connectivity index, corridor density index, plaque index and fractal analysis. Researchers begin to integrate these indicators applied to the quantitative analysis of the spatial pattern of urban green space system. Benxi City and Jinan City analyzed the structure of green space with the heterogeneity of landscape ecology index and homogeneity index, to analyze that the two cities were out of the uneven distribution of green space, suburban living green is more than the center area of green space. Landscape ecology indices can better reflect the spatial structure of urban green space and analyze the problems of urban green space.

IV. THE DESIGN APPROACH TO URBAN GREEN SPACE LANDSCAPE PATTERN OPTIMIZATION

Make sure about the goal of building green system green building as an eco-city construction system is an important aspect, its building goal should be the urban ecological green space system, urban ecological green space system planning should be carried out. Through analysis the natural landscape dynamics of model, to identify spatial patterns of landscape change and the driving force of the mechanism, t to analyze the relationship between changes in landscape pattern and ecological environmental impact, the establishment is the basis for ecological pattern design optimization. Landscape pattern optimization is divided into two research-scales, one is the optimization of scale spatial pattern, and the other one is an optimal allocation of ecological scale factor. The study of spatial pattern optimization start from ecological function areas (ecologically sensitive unit), niche and ecological zone, ecological planning and design strategic point; optimal allocation of ecological factors starts from the biological diversity, ecological and cultural landscapes.

A. The approach to the scale spatial pattern optimization research

The ways of ecological function area planning is committed to distinguish between ecological function zoning and regional service function of human activities. Ecological service function areas are carried out by geo-spatial partition by the importance of regional ecological service functions, similarities and differences. According to the potential capacity of the ecosystem and human's choosing values to plan the study of ecological function. It is generally used the importance of ecosystem services, ecological suitability analysis, combined with ecological sensitivity analysis.

The important ecosystem services of the ecosystem services include environmental features, biological functions and human features. These include: air conditioning, climate mediation, interfere with regulation, hydrological regulation, water regulation, erosion control and sediment conservation, soil formation, nutrient recycling, waste management; biological features include: pollination, biological control, shelter, food production, raw materials, gene pool, recreation and culture. [3]

Ecological sensitivity analysis of ecologically sensitive areas is to maintain which was conducive to safeguarding the ecological system of humankind and its own ecological value. Environmentally sensitive areas can also serve as a buffer and be linked to those who as a higher level of protected areas, eco-systems have a variety of climatic variations in the region, the sensitivity of human activities. That is when ecosystems are subjected to interference, the probability of the size of the ecological balance. It refers to the specific ecological processes in the natural conditions, the size of the potential. In general, the more complex ecosystem structure and function is, the stronger nti-interference ability and self-healing function is. The more single ecosystem structure and function is, the more vulnerable to human activities, the more sensitive to disturbance.

Ecological location is a unit concept which was proposed to establish a research indicator next to the ecological function areas, a basic settlement unit belongs to the biological community, including the various landscape elements that were connected with the strong interaction, a single ecological neighborhood can meet the need of biological life. Each ecological unit has its own relatively independent ecological functions and the basic function can be achieved, however, complementarities between them are enhanced to achieve functions optimized of the ecological function zones.

B. The approach to the ecological factors optimal research

Ecological spatial patterns can provide the necessary ecological basis for the ecological green space system planning, at the same time, it is also a necessary requirement for eco-city construction. Ecological factors include biological factors and non-living elements, namely elements of the physical environment. In previous studies, non-biological factors, including soil, air, water, easy access to optimal allocation of the standards and methods, optimization method is also relatively clear, the salient point are sources of pollution control and environmental remediation applications, etc. The optimal distribution of biological elements such as the establishment of plant and animal diversity, mostly remain in the analysis phase, lacking of specific planning and design method.

1) Optimization of plant diversity

The allocation of plant diversity optimized from two aspects which are plant composition and spatial patterns. Plant composition indicators include the proportion of native plants and alien plants, the proportion of evergreen plants and deciduous plants, the proportion of forestry, irrigation and grass; and the application of functional plant such as the tree species which absorb sulfur dioxide, sterilization tree species, dust species, heavy metal-polluted ecological protection green tree species. Spatial patterns of plants, including the maintenance of biological diversity, flora patterns, forest health of plant community patterns of plant community model of Scenic tours, camping and picnic leisure patterns of plant community, plant community model of science education, the environment model functional community, cultural display plant community model, wet health and community patterns of woody plants, aquatic herb community patterns. In the ecological green space system planning, we should focus on biodiversity conservation to achieve green tree species diversity and create a rich habitat for plant communities to create conditions for various animals habituating and breeding. To construct the city, it often requires the introduction of species from the field, the introduction of non-invasive alien species appropriately can enrich local biodiversity. Refer to tree species selection, it is better to select landscape trees and the native tree species with the regional characteristics, both spectator and the economy should be gived consideration. At the same time, carry out some researches about the basic work on the protection of old trees and the technology of conservation and management, formulate some corresponding technical regulations and norms, establish a scientific and systematic management system to protect old trees. [4]

2) Optimal distribution of animal diversity

Animals diversity mainly plan from two aspects, one is birds diversity, the other one is insect diversity. The methods are providing food sources, such as the cultivation of berry plants; creating a natural ecological environment, such as remain the state of vegetation and wildlife habitat as far as possible; increase the number of hidden places where wild animals hide in., wild animals can survive in one place, in addition to the food, they must have some place to hide, if stacking some stones, poplar and straw in the belt, it will attract more animals to settle them.; increase plant diversity and so on. Insect diversity and ecosystem stability are closely related to insects, mainly are plants Ming and butterflies.

3) The planning for Eco-Cultural Landscape Construction

Plant landscaping in the urban greening, although constrained by the common factors of regional climate and urban micro-climate, but due to the cities' underlying surface are different, there are greater differences existed between different sections. Using site condition namely spatial heterogeneity of the landscape elements of environmental resources can create diverse landscapes and gardens in the city. Landscape diversity also includes the complexity of the landscape mosaic formed by the difference of the vertical space. This diversity is often achieved through the plant configuration of the different biological characteristics. This is not only true in a large public garden, in other areas, such as garden landscaping, residential quarters greening, green walls and even a smaller-scale indoor plants decorative areas, also can take advantage of the creation of landscape diversity such as environmental diversity and plant diversity. Meanwhile, it should not be confined to the protection of urban natural ecology, but the collaborative of natural ecological system and the artificial ecosystem. Addition to considering the relationship between urban physical space and natural ecology, what's more, the relationship between the natural elements and spatial perception should be emphasized and the ecological and cultural landscape construction should be focused on.

V. CONCLUSION

Based on the landscape pattern optimization of the design of urban green space landscape is the intersection science research of landscape ecology and landscape design, from the perspective of ecological processes, urban green space is the transition region which from semi-natural ecosystems to urban man-made fragile ecosystems. In the ecosystem, a large number of uncertain landscape elements interwoven, the relationship of ecological function is confounding, the normal ecological processes have been cut, broken and missed, which has led to a sharp conflict between the environment, the unpredictable nature of urban development has led to the disordered environment and fragmented landscape patterns. Landscape ecological pattern optimization must be studied in methods, understand the mechanism and process of landscape patterns' evolution, to use the ecological planning and design methods design urban green space scaled from the spatial pattern and ecological factors, it is the research methods to explore the urban landscape pattern optimization, which is the best approach to establish a spatial dynamic of urban green space pattern optimization design.

REFERENCES

[1] He Ping, Bian Xin-min, "Agricultural park planning and design based on the landscape pattern optimization," Jiangsu Agricultural Sciences, 2006, No. 5.

[2] Zhao Hong Xia, Tang Geng Guo, "Pattern of urban green space and the progress of its function research," Journal of Shandong Agricultural University (Natural Science) pp.155 — 158,2007, 38(1).

[3] Che Sheng Quan, Ma Xiao Xing, "The way to Shanghai semi-urban regional natural landscape ecological pattern optimized design, "Shanghai Jiao tong University (Agricultural Science Edition).

[4] Li Chang Hao, "The green space system planning research for the eco-city, Urban and Ecological," 1006- 3862(2007)02- 0039- 05.

The Development and Integration of Historical Culture and Geographical Characteristics

On the Outdoor Environmental Planning Program Design of Hubei Provincial Museum

Wang Gang
School of Art & Design
Wuhan University of Technology
Wuhan, China
Email:wang_gang919@163.com

***Abstract*: The historical and cultural landscape architecture planning is like architectural design, shouldering the mission of historical culture inheritance and reflecting the regional characteristics. In this paper, the outdoor environmental planning of Hubei Provincial Museum is taken as the example, taking consideration of the natural, geographical, historical and cultural features, then to develop the traditional art by such means as abstract symbols, the evolution of deformation , and integrating into the new material, to combine the new modeling with history, culture and geographical characteristics. Therefore, the author of this paper aims to explore and study the planning and design ideas of historical and cultural construction of landscape.**

Keywords*: *historical culture; regional characteristics; development and integration

I. THE INTRODUCTION OF THE PROJECT

Architecture is a freezing art. Both constructions and landscape can be retained for quite a long time for people to appreciate and judge. If the construction is integrated into a number of ethnic and local characteristics, it will be richer in personality and will be retained for longer time. It is true for museum-type building in particular.

Museum constructions shoulder the arduous mission of the history and culture inheritance. In style and form, it is more necessary to pay attention to the unity of function and form. Taking Hubei Provincial Museum as an example, the author of this paper attempts to explore and study how historical and cultural features are applied in the planning and design ideas in construction of landscape.

With its deep historical and cultural accumulation in the central plains area, Hubei province has fostered a lot of literature talents. The Chu culture also plays an important role in Chinese history. So the design of Hubei Provincial Museum is not an ordinary landscape design, it needs to reflect more local features to complete the dissemination of Chu culture.

Hubei Provincial Museum expansion project is the national and provincial "The Tenth Five-Year Plan" major cultural infrastructure project and it is also the symbol of the provincial government works and Wuhan city landmark expansion project. Outdoor environment planning is one of the key parts of the expansion plan. The purpose is to give visitors a leisure, peaceful, and comfortable place for outdoor recreation activities, to create a harmonious and complementary effect with the main building of the Museum. Therefore, it can achieve the purpose of making the main building and the environment an overall effectiveness.

II. GENERAL IDEA OF THE DESIGN

The outdoor environmental planning program of the Hubei Provincial Museum is a combination of the outdoor expansion project environmental planning and design requirements, conducting a scientific, rational planning and design to meet the needs of urban planning, public security, fire, water, electricity, transportation, gardening and the special requirements of different departments. Considering the local conditions and reality, the designers take advantage of the natural topography and vegetation, and give full consideration to the special geographical location environment, the rich historical and cultural connotation of the East Lake Scenic Area for a reasonable layout and to the co-ordination arrangements.

Following the people-oriented spirit, rationally using the museum venues, the planning and the design fully satisfy the audience's physical, spiritual, psychological, behavioral norms needs. Putting the convenience and the service for the visitors in first place, the project tries to create a diverse and comfortable outdoor environment for visitors from different ages, different classes and different nationalities. "Fig. 1"

Figure 1. The Museum Panorama.

Designed by 3ds MAX (Note: A computer-aided design software)

III. DETAILS TREATMENT OF THE DESIGN

In the Provincial Museum outdoor environment planning and design, designers take into account the layout of axial symmetry of the environment to maintain the basic symmetry and balanced unified style among the green tree species, crown, shape, size, hedges, shrubs, flower beds etc.. The formation of the main building meets the requirement of "one body with two wings", ("one body" refers to the Comprehensive gallery; "two wings" includes the Bell House and Chu Cultural Center), and axial symmetry of the architectural layout.

The outdoor environmental planning program fully reflects the local characteristics of Chu culture in order to make the deep cultural connotation consistent with the Provincial Museum. In the central part of the main building, there is an attraction of feature--- Fish Marsh Fly Beams. Covered bridges and water exist side by side, a strong momentum century bronze Ding (a big bronze pot) stands in the central area, on whose body inscribed the ancient Chu culture patterns and words, symbolizing a long and profound history in Chu culture. Around there are four gray marble relieves of celebrities, respectively of China's four famous figures and talented women. Next to the celebrity relieves are four slightly smaller ancient seal characters embossed side of stones, fully reflect the history of the museum's implied accumulation. There are also stone relief flowers of four seasons around, ensuring that flowers are blooming in every season , so as to enrich the color level of central landscape area. "Fig. *2*"

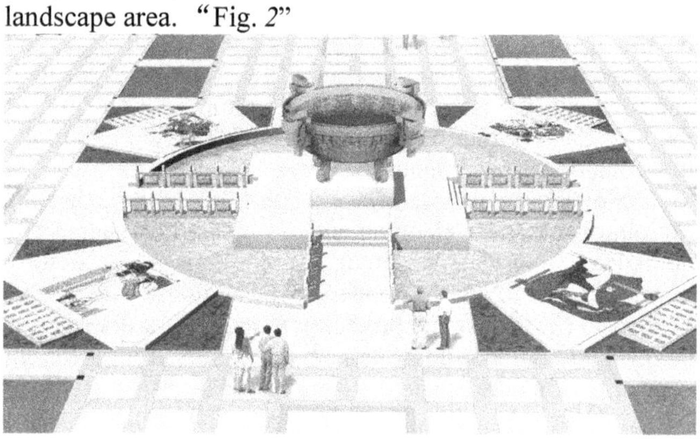

Figure 2. Fish Marsh Fly Beams. (Designed by 3ds MAX)

Between the Fish Marsh Fly Beams and the main entrance, another feature of the landscape is displayed: eight tall totem poles take a seat on the eight symbols of traditional arts patterns, creating a sacred lofty and solemn atmosphere of the entrance. The totem poles are carved with Chu cultural patterns and the ancient tripod shapes which are simplified, abstract, exaggerated, distorted, and finally to achieve a mystery, ancient, solemn shape effect. Ding shapes are also added with lighting components. So Ding not only becomes a decorative element tripod, but also a lighting element. .The floral patterns are consisted of two groups: Shades of two colors are used to create four kinds of artistic symbols---marble Dragon, White Tiger, Zhuque (a red bird or fire phoenix) and Basaltic. Thus a strong arts scene is formed. "Fig 3, Fig 4, Fig 5"

Figure 3. Chu Culture Patterns.

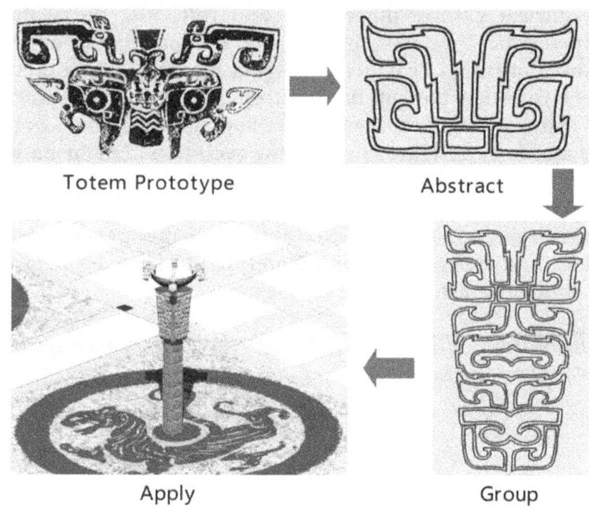

Figure 4. Deductive Procedure. (Designed by 3ds MAX)

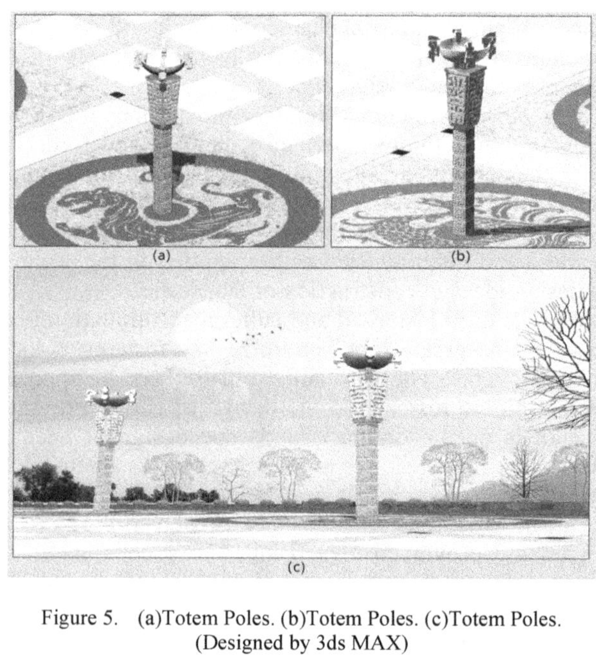

Figure 5. (a)Totem Poles. (b)Totem Poles. (c)Totem Poles. (Designed by 3ds MAX)

In the museum planning and design in the outdoor environment, an architectural sketch of the design is an important element. The architectural sketch is a function of simple, physical volume small, chic style, and with the mood, rich in unique building components. It is the art of handling, processing and formal beauty, and with environmental groups, the ingenious architectural configuration, and it can constitute a certain appreciation of the value of the screen. Therefore, the design of an appropriate layout should be added with some small building pieces, such as public tables, chairs, stools, green lights, litter bins, environmental signs, signage, promotional billboards, fences bollards , fences, pavilion, flower racks, flower beds, planting containers, background music, speakers, fire hydrants, green spray tap, building lighting spotlights, garden lights, street lamps, floor paving, steps and so on. To be integrated with pieces of construction with the elements of Hubei regional cultural history, the traditional form and folk art, the design can not only satisfy the psychological and behavioral characteristics of public use, but also reflect the historical and cultural connotation in the leisure space environment.

The Provincial Museum outdoor environment arranges its layout by using the characteristics of landscape type and garden-style, and tries to reflect cultural characteristics of Hubei province and Hubei Chu artistic styles in the gardens and landscapes. In the program, the lawn hedges on the right part of the entrance are made into the map of Hubei provincial shape. It can not only enrich the grass facade effect, but also demonstrate the concept of geography of Hubei Province.

IV. COORDINATION AND CONTROL OF OVERALL SPACE

Full attention is paid to the landscape design with the coordination and harmonization of the main building. So the landscape can play a beautified, contrast, decorative role in the main building landscaping. By controlling the height of the trees around the main building to avoid blocking the main building facade effect, and by making efforts to make the proportion of the landscape-scale, the structure, process, plane and elevation of the design to meet the requirements of the main building, the planning of the Museum finally achieved an effect of being ancient, stately, beautiful and magnificent.

V. CONCLUSION

To sum up, in historical and cultural construction of landscape planning, the essence of the miraculous traditional artistic elements should be fully absorbed. Meanwhile, the design should reflect the regional characteristics, and should be integrated with new materials, be combined with the new shape of history cultural and geographical characteristics, so that the ordinary can be transformed into a legend. Considering the above mentioned, the designers have made a good attempt in the outdoor environmental planning and design of the Museum in this area.

REFERENCES

[1] Wu Liangyong, Basic Concepts. Regional Culture. Paradigm of Our Times—Reflection of Future Development of Architecture in China, Architectural Journal, 2002(2), pp.6-8.

[2] Wen Xutao, Lu Wei, Conception of protection and recycle for modern architecture heritage, vol. 23 (2), Journal of Dalian University of Technology (Social Science Edition) 2002, pp.68-72.

[3] Li Xiaomin, Zhong Xunzheng, Green Landscape and Architecture Environment——Examination on the Development of Modern Design Theory, New Architecture, 2004 (3), pp.78.

[4] Jiang Xiaoping, Traditional Chinese Architectural Art, Southwest China Normal University Press, 1998, pp.1-10.

[5] Dai Bixiang, Li Jikai, Art Introduction, Culture and Arts Publishing House, 1983, pp.1-10.

[6] Xu Jin, On Architecture Culture and Its Social and Psychology, vol. 22 (2), Sichuan Architectural, 2002, pp.38-39.

[7] Zhu Guo-xing, A Rational Thought for Regional Traditional Culture and Tourism Development——Taking Huizhou Culture for Instance, vol. 17 (3), Human Geography, 2002, 79-81.

The Effects of Transformational Leadership and Team Climate on Employee Creativity

A Cross-Level Model

Guo Wei
Economics and Management School
Research Center for China Industry-University-Research Institute
Wuhan University
Wuhan, China
Email: guowei0827@126.com

Li Yanping
Economics and Management School
Research Center for China Industry-University-Research Institute
Wuhan University
Wuhan, China
Email: lypwhu@yahoo.com.cn

***Abstract*—Nowadays, team working is considered as a popular model to enhance the organization's flexibility and innovation. The roles of employee's are changes frequently, and the responsibilities have become blurred. So the creativity of employee is very important. In this paper we introduce the concept of transformational leadership, team climate for innovation and employee creativity into analytical framework. And then we explore assumptions about the levels of analysis embedded in the literature on employee creativity in organization. We propose a multi-level model of creativity, as its defining feature, examines the role team climate for innovation plays in the transformational leadership and employee creativity. Finally, it points out the insufficiency and the future research direction.**

***Keywords*—Transformational leadership; Team climate for innovation; Employee creativity**

I. INTRODUCTION

With the advent of the knowledge economy, companies tend to choose the team as a working model.Previous studies on innovative team-building mostly confined to the individual level or team level, less meanwhile from multiple perspectives of team-level, individual level and cross-level research. This paper provides a multi-level analysis of the innovative team, organizational level of the team climate for innovation and individual dimensions of transformational leadership and employee creativity relationship.

In recent years, domestic and foreign scholars began to focus on transformational leadership and employee creativity, (Geng Xin, 2009; Shin &ZHOU, 2003), the relationship between the transformational leadership and team climate for innovation (Zhu Shao-Ying, 2008). Team climate for innovation belongs to the team level, while the transformational leadership and employee creativity have a multi-level nature, especially the leadership personality is even more variable exists in the individual level, which the team level and organizational level.

II. LITERATURE REVIEW

A. Transformational leadership

Since the 1980s, transformational-charismatic leadership has received increasing attention of Western scholars. Some scholars proposed the concept of transformational leadership involved leaders gaining the respect, trust, and confidence from others and transmitting a strong sense of mission to them. It includes leaders communicating a vision with fluency and confidence, increasing optimism and enthusiasm, and setting high expectations for themselves and their followers (Bass, 1985, francis J. Yammarino, 1997). High levels of transformational leadership can make subordinates exceed their expectations, inspire their creativity and enhance their productivity.

Transformational leadership is comprised of four dimensions: idealized influence, individualized care, intellectual stimulation and vision inspired. Idealized influence refers to the leadership which demonstrated a strong personal charisma, a clear vision of the future representation, so employees would be proud of what they do, the leadership receive respects from employees in this way can trust to employees; personalized attention refers to give individuals more attention for those who have been neglected according to the specific problems of each subordinate, and stressing the future development of subordinates. Intellectual stimulation refers to encouraging subordinates to think about a new point of view, to be creative.

B. Team climate for innovation

The concept of team climate has received attention from applied psychologists and organizational scholars over the last four decades; it means the cognition of team members to the working environment which affects their ability of innovation. Some academics proposed a four-factor model of work group innovation, hypothesized that four major factors of climate are predictive of innovativeness (West, 1990).

There are three key assumptions: first, market, globalization, internationalization and environment brought innovations in the organization and in its structure. Particular,

This research was support by the National Social Science Foundation of China (07AJY023)

team climate for innovation is to stimulate innovation in the organization. Second, new forms of organization have led the organization to the team-building and adopt application of the rapid growth in demand. Third current academic research lags behind the practitioners of the team and organization of effective measurement of the phenomena actual demand. Subsequently, some scholars have pointed out that the team climate of the measure of organizational innovation should be based on the individual closest to the work of the team as a unit. This team is defined as the closest permanent or semi-permanent work team (Anderson & West, 1998).

In this study, we have adopted the team climate of the four factors innovative structural model (West, 1990): Vision, Participative Safety, Task Orientation, and Support for Innovation.

C. *Employee creativity*

Creative ideas increase the likelihood that other employees will apply the ideas into theirs own work, further develop the ideas, until these ideas into staff creativity, and ultimately for business use. Considerable evidence now suggests that employee creativity can substantially contribute to organizational innovation, effectiveness, and survival (Amabile, 1996; Nonaka, 1991).Over the past two decades, most theorists have defined creativity as the development of ideas about products, practices, services or procedures that are novel and potentially useful to the organization(Zhou&Shelly,2003). In recent years, scholars were interested in employee creativity from the initial under an organizing framework, personal characteristics, contextual characteristics, interactions among personal and contextual characteristics, by turning intrinsic motivation as a mediator, mood states, self-efficacy and creative role identity, creative role models, creative process, creativity in international contexts, team creativity.(Shally C.E.,2004).

Chinese scholars study creativity of employees significantly less than abroad. Some scholars within the motivation based on the theory, analysis how transformational leadership behavior impact on the on subordinate creativity (Ding Lin, 2008). Some scholars from the location, system, evaluation perspective, upgrading the set of high-tech creative talent to build a system orientation, and put forward corresponding countermeasures system design (LI Shu-ye, 2008).Another studies were conducted by individual level of creativity to the team of creative mechanism of transforming (ZHOU Yao-lie, 2007). In addition, knowledge-sharing, for example, team norms impact of R & D personnel on the team creativity as an intermediary mechanism of empirical studies (Wang Duan-xu, 2009).

However, there it has not yet to emerge into the creativity of staff organizations, individuals and organizations to individuals to consider three dimensions of an integrated system.

III. MODEL

Creative team emphasizes internal management innovation and flexibility in organizations as well as responds quickly to the changing environment; which is widely used mode business. Members in innovative teams show the vitality of the team, and play vital for the survival and development of enterprises. Creative works are of great risks, employees can use the previous work in different ways or different ways to solve the problem, there will still exist some difficult such as whether new ways are effective or work etc. At this point, transformational-charismatic leaders support on innovation and understanding is vital.

Transformational-charismatic leaders' abilities to get subordinates to transcend their own interests and to encourage their own innovation, so that subordinates could recognize the importance and urgency of innovation, and to make the team and create a better vision, inspire subordinates to propose new ideas, to achieve common purpose. Transformational leaders encourage subordinates to actively innovate, and can tolerate the failure in the process of innovation, prompting the team to create a good climate for innovation. When the team members perceived themselves in a harmony climate for innovation, they will agree with this climate, and are going to be inspired, the better to stimulate employee creativity. Then, all members continue to challenge themselves, and to be creative. While the team creativity of each employee are fully motivated, team creativity was been strengthened.

- H1: Transformational leadership will be positively related to team climate for innovation.
- H2: Team climate for innovation will moderate the relationship between transformational leadership and employee creativity.
- H3: Transformational leadership will be positively related to employee creativity.

Through the above analysis, we formed the next graph theoretical model (Figure 1).

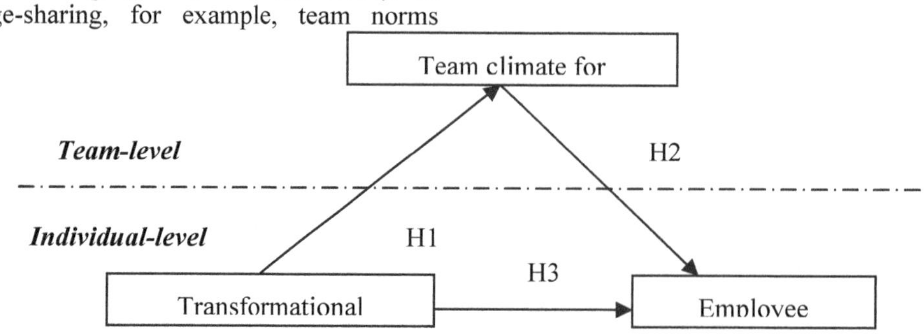

Figure 1 Transformational Leadership and Team Climate on Employee Creativity Models

IV. CONCLUSIONS AND OUTLOOK

This article integrates the concept of transformational leadership, team climate for innovation, and employee creativity. According to research needs, we select the corresponding variable levels of study, then establishment of a cross between the three levels of research model.

Further more, we point out the future research direction.

First of all, we can use the data to analyze the relationship of three empirical studies. Foreign studies have the maturity scale of transformational leadership, team climate for innovation, and employee creativity. For example, the measurement of transformational leadership of the MLQ(Multifactor Leadership Questionnaire), employee questionnaires, as well as creativity. Anderson & West (1994) made by Team Climate Inventory (TCI) scale. In order to fit the context of Chinese culture, cross-cultural research, Chinese scholars had a lot of revision on transformational leadership, and empirical analysis shows good reliability and validity. Taking into account the differences between Chinese and Western cultural, this paper proposed adaptation of Chinese scholars have been widely used off the scale, such as Li Chao-ping and SHI Kan transformational leadership questionnaire (TLQ), the questionnaire including a total of 26 projects. In the individual-level research, using SPSS software, the regression analysis test transformational leadership and employee creativity relationship between the cross-level researches. Multi-linear model using HLM to examine the team climate for innovation in the transformational leadership of the workers affected in the process of innovation as intermediary role.

Secondly, this paper gave a preliminary exploration on transformational leadership, team climate for innovation, and employee creativity. The individual level of transformational leadership, employee creativity could be aggregated into the team level of transformational leadership and team-level creativity, allowing the team-level study on transformational leadership and team climate.

In this paper, transformational leadership, team climate for innovation, employee creativity will not only help to understand triadic relationship of three, but also contribute to innovative team-building. Furthermore contribute to the further study on analyzing, comparing the three variables and other variables' relationship.

REFERENCES

[1] GENG Xin, SHI Jintao, ZHANG Wenqin. : The Effects of Transformational Leadership and Team Climate on Organizational Citizenship Behavior: A Cross-Level Model. Science of science and management of S.&T., Vol.9, 2009,pp184-187.

[2] Shin & ZHOU, J. Transformational leadership, conservation, and creativity: Evidence form Korea. Academy of Management Journal, Vol.46, 2003, pp703-714.

[3] ZHU Shao-Ying, Qi Er-shi, Xu Yu. An Empirical Study of the Relationship among Transformational Leadership Style, Team Climate, Knowledge Sharing and Performance of Team Innovation.Soft science, Vol.11, 2008 ,1-4,9.

[4] Bass B M. Leadership and Performance Beyond Expectations. M. New York: Free Press,1985.

Francis L Yammarino, Alan J. Dubinsky, Lucette B. Comer, Marvin A.Joloson, Women and transformational and contingent reward leadership: A multiple-levels-of –analysis perspective. Academy of management Journal, 1997, vol14,No.1,pp205-222.

[5] West, M. A., & Farr, J.L. Innovation at work: pp3-13. Chichester, England: wiley,1990.

[6] Anderson, N. R., & west, M. A. Measuring climate for work group innovation: Development and validation of the team climate inventory.Journal of Organizational Behavior ,Vol.19, 1998,pp235-258.

[7] Amabile, T. M. Creativity in context. Boulder, CO: West view(1996)

[8] Nonaka, I.. The creative process in virtual teams. Creativity research journal, Vol.14, pp69-83 (1991).

[9] Shalley, C.E., &Gilson, L. L. What leaders need to know: A review of social and contextual factors that can foster or hinder creativity. Leadership Quarterly, Vol.15, 2004,pp33-53.

[10] DING Lin, XI You-min. The Mechanism of Transformational Leadership's Effect on Employee Creativity . Journal of management sciences, Vol. 12, pp40-46 (2008)

[11] LI Shu-ye, ZHOU Bin. High-tech System for Creative Talent, Upgrading the Building Orientation and Counter Measures.Forum science and technology in china, Vol.6, 2008,pp121-124,129.

[12] ZHOU Yao-lie, YANG Teng-jiao. Research on the "lens" model of transformation mechanism about personal creativity to team creativity, Studies in science of science, Vol.12, 2007,pp409-413.

[13] WANG Duan-xu, ZHU Xiao-jing, WANG Zi-wei. Study on the Mechanism of Group Norms and R&D Professional Creativity ---The Mediating Role of Knowledge Sharing. Tongji University Journal Social Science Section, Vol.8, 2009,pp119-124.

[14] Anderson, N. R. and M.A. West. Team climate inventory: manual and user's guide. M. Windsor, Berkshire: NFER NELSON,1994.

The Evaluation System of Industrial Product Design Based on Comprehensive Fuzzy Evaluation Method

Changzhong Ke
College of Art and Design
Wuhan University of Technology
Wuhan, China
Email: 63242450@qq.com

Wenyu Wang
College of Art and Design
Wuhan University of Technology
Wuhan, China
Email: wwy_002@sina.com

Abstract—**Evaluation system of industrial product design is vague, nondeterministic and difficult to quantify, which makes it very difficult to construct. In order to overcome this problem, we take the theory of fuzzy mathematics into our research. In this research we discuss the possibility that whether the comprehensive fuzzy evaluation method can be used in the construction of evaluation system of industrial product design though building and apply the model which based on the fuzzy mathematics, and finally we find it is feasible. This is a kind of attempt using mathematical methods to optimize the industry design.**

Keywords - evaluation system of industrial product design; comprehensive fuzzy evaluation; fuzzy mathematics

I. INTRODUCTION

It is well-known that the evaluation of industrial product design plays a very important role in the entire design process. It directly affects the final choice of the design. In the other words, to choose the best design is the ultimate goal of product evaluation. So an appropriate, efficacious and comprehensive evaluation system is essential, which can help us to choose the best one.

However, there are a lot of difficulties to construct such kind of evaluation system. First, there are millions kinds of industrial products, each of them has its own characteristic, which directly leads to the result that each evaluation system needs different details and focus. Second, the design process of one product always involves kinds of fields. All of these fields have their own standards and methods of evaluation, which are not unified. Third, part of the evaluation indicators based on the emotional judgments. They belong to qualitative evaluations, which is difficult to be put into quantitative calculation. Our task of research is trying to find a method which can solve all of these problems. And comprehensive fuzzy evaluation method meets our requirement quite good.

Comprehensive fuzzy evaluation method is a kind of comprehensive evaluation method based on membership functions in the theory of fuzzy mathematics. Because of its advantages, such as it can transform qualitative evaluations into quantitative evaluations, give an overall assessment to the case even affected by a variety of factors, and can also get a very clear, unequivocal and systematic result, it can perfectly be used to solve the problems which people would encounter when they use the traditional evaluation method, such as the evaluations are vague, nondeterministic or the conclusions are difficult to quantify..

II. MODEL BUILDING

Fuzzy comprehensive evaluation method is mainly based on four elements. They are factor set U, appraisal set V, the weight set A, and single-factor evaluation matrix R.

A. Establish the factor set

$$U = \{u_1, u_2, u_3, \ldots, u_n\}$$

Before we do the evaluation, we should find and selected all the factors needed to form an assessment factor set. If the evaluation factors are too complicated, we can set the multi-evaluation factors. For example, if we set $U_1 = \{u_{11}, u_{12}, u_{13}, \ldots, u_{1k}\}$ as our evaluation factors set, then $u_i(i = 1,2,..n)$ is the first-level evaluation factor set, $u_{ij}(i = 1,2,...n; j = 1,2,...k)$ is the subset of the first-level evaluation factor set. It can be also called the second-level evaluation factor.

B. Establish the appraisal set

$$V = \{v_1, v_2, v_3, \ldots, v_n\}$$

Appraisal set provides different options to be selected by people when they do the evaluation. It contains some appraisals which are divided into several levels by a certain way. Because the appraisal set have two kinds of factors, hard factors and soft factors, so we use different ways to divide them. For hard factors, we can transform the index data directly into the value of evaluation factors. Such as $V = \{0.2, 0.4, 0.6, 0.8, 1\}$. The value of soft targets can be given by experts, and it can be divided into several levels by degree. Such as $V = \{bad,\ poor,\ averige,\ good,\ excellent\}$. Direct ratio is the first chose in principle.

C. Determine the weight vector

$$A = \{a_1, a_2, a_3, \ldots, a_n\}\ (0 \le a \le 1, \sum_{i=1}^{n} a_i = 1)$$

Weight vector set shows the degree of importance of the different factors in the factor set. Each level of the factor set has its own weight vector set. Weight vector set can be made by the experienced experts for specific projects.

D. Single-factor evaluation matrix

$$R = \begin{Bmatrix} r_{11} & r_{12} & r_{13} & \cdots & r_{1m} \\ r_{21} & r_{22} & r_{23} & \cdots & r_{2m} \\ \cdots & \cdots & \cdots & \cdots & \cdots \\ r_{n1} & r_{n2} & r_{n3} & \cdots & r_{nm} \end{Bmatrix}$$

We can get the single-factor evaluation matrix through the fuzzy mapping from set U to set V. The $r_{ij}\left(0 \le r_{ij} \le 1, j = 1,2,\ldots,m\right)$ shows membership from the performance of factor $u_i\left(i = 1,2,\ldots,n\right)$ to the appraisal set, $v_j(j = 1,2,\ldots,m)$. In other words, it shows how much factor u_i belongs to appraisal set v_j.

III. APPLICATION

In order to test and verify the model, we take a mobile telephone to do the evaluation. The steps are as follows.

A. Determine the factors and establish the set

The first thing we should do is to select the appropriate factors for the mobile telephone's evaluation from all the factors. Before this test, we have already made a table listed almost contain all the factors which an industrial products evaluation would need. See Table II. We make this table mainly based on the advices from experienced experts and professional publications. And this table can also be expanded to march with the specific situations and be complemented with the development of industrial design.

Table II shows how many and what kind of factors we have chosen. Altogether we get three levels of evaluation factors to do the comprehensive fuzzy evaluation.

B. Determine the appraisal set

Since we have already known that the evaluation factors are diverse and flexible at the introduction part, so now, we must find a uniform standard for the evaluation, which can be used to appraise both hard factors and soft factors, to overcome that difficulty. And the method of founding a uniform standard is to define a uniform standard first, and then transform the different standards into this one. At this mobile phone case, we set the standard appraisal set as: $V = \{1,2,3,4,5\}$. To appraise the soft factors such as "Modeling", "color", "Decorative" and so on, we can use set $\{bad,\ poor, general,\ good, excellent\}$. And the set $\{none, 1,2,3,\ above\ 4\}$ can be used to appraise the hard factors, such as "Motivation of Purchase". Both of them can be transformed to the standard set $V = \{1,2,3,4,5\}$, so a uniform standard is found.

C. determine the weighs

In order to get the more reasonable weights, we could invite several experienced experts to give the weights together. Through recording and dealing with the values which are given by those experts, we can get more reasonable weights. Here are the specific steps. Suppose we invited 20 experts to give the weights of "Economy", we will get 20 values of weights. Among them, the maximum value is 0.9 and the minimum is 0.5. Divided these data into 5 groups, and the interval is 0.08. Table I shows the distribution clearly.

TABLE I. STATISTICS OF WEIGHTS

Group	Number of experts	Frequency
0.50-0.57	1	5%
0.58-0.65	3	15%
0.66-0.73	5	25%
0.74-0.81	10	50%
0.82-1.00	1	5%

From the Table I, we can see that group 0.74-0.81 has the most frequency. Therefore we take 0.8 as the weight of the factor "Economy". Under the first level "Social", there are two subsets "Economy" and "Aesthetic". Because of the restraint: $0 \le a \le 1, \sum_{i=1}^{n} a_i = 1$, so we take the weight of the factor "Aesthetic" as 0.2. Other weights could be determined by the same method. And the weights value of factors in the third level can be distributed into equal parts. Finally we get weights of all the selected factors like Table II shows.

D. Determine the single-factor evaluation matrix

We have already determined a five levels' appraisal set before. If the person who makes the evaluation considers that one factor belong to the third degree in the appraisal set, we count 1 under this level. The form is like this:

$$\begin{Bmatrix} 0 & 0 & 0 & 1 & 0 \\ 0 & 0 & 0 & 0 & 1 \\ 0 & 0 & 0 & 0 & 1 \\ 0 & 0 & 1 & 0 & 0 \\ 0 & 0 & 0 & 1 & 0 \end{Bmatrix}$$

Suppose there are 30 people to make the evaluation of this mobile phone, the evaluation matrix is:

$$\begin{Bmatrix} 0 & 0 & 3 & 15 & 12 \\ 0 & 0 & 3 & 9 & 18 \\ 0 & 0 & 6 & 15 & 9 \\ 0 & 0 & 9 & 15 & 6 \\ 0 & 0 & 12 & 9 & 9 \end{Bmatrix}$$

This single-factor is the result of factor "Function". Then normalize this matrix to get a new matrix:

$$R_1 = \begin{Bmatrix} 0 & 0 & 0.1 & 0.5 & 0.4 \\ 0 & 0 & 0.1 & 0.3 & 0.6 \\ 0 & 0 & 0.2 & 0.5 & 0.3 \\ 0 & 0 & 0.3 & 0.5 & 0.2 \\ 0 & 0 & 0.4 & 0.3 & 0.3 \end{Bmatrix}$$

So that the result of evaluation to the factor "function" is:

$$C_1 = A_1 R_1 = (0.2,0.2,0.2,0.2,0.2) \begin{Bmatrix} 0 & 0 & 0.1 & 0.5 & 0.4 \\ 0 & 0 & 0.1 & 0.3 & 0.6 \\ 0 & 0 & 0.2 & 0.5 & 0.3 \\ 0 & 0 & 0.3 & 0.5 & 0.2 \\ 0 & 0 & 0.4 & 0.3 & 0.3 \end{Bmatrix}$$

$$= (0,0,0.22,0.42,0.36)$$

Use the same method to find the other evaluation results of factions in level 3, they are:

$$C_2 = (0,0.06,0.54,0.20,0.20)$$

$$T_1 = (0,0,0.40,0.56,0.04)$$

$$T_2 = (0,0,0.12,0.2,0.68)$$

$$S_1 = (0,0,0.22,0.30,0.48)$$

$$S_2 = (0,0.06,0.10,0.62,0.22)$$

The results of factions in the second level are:

$$C = (0,0.036,0.412,0.288,0.264)$$

$$T = (0,0,0.344,0.488,0.168)$$

$$S = (0,0.012,0.196,0.364,0.428)$$

Then we get the final result of this mobile phone's evaluation. It is:

$$B = AR = (0.3,0.3,0.4) \begin{Bmatrix} 0 & 0.036 & 0.412 & 0.288 & 0.264 \\ 0 & 0 & 0.344 & 0.488 & 0.168 \\ 0 & 0.012 & 0.196 & 0.364 & 0.428 \end{Bmatrix}$$

$$= (0,0.1128,0.3052,0.3784,0.3008)$$

From this result, we can see that the evaluation of this mobile phone belongs to the forth degree more, which take the 37.84%. But this result is less intuitive than a specific value. So we can use the appraisal set $V = \{1,2,3,4,5\}$ to count the final result. It is:

$$D = (0,0.1128,0.3052,0.3784,0.3008) \begin{pmatrix} 1 \\ 2 \\ 3 \\ 4 \\ 5 \end{pmatrix} = 4.1588$$

Also we can get the evaluation value of another mobile phone by the same way. Through comparing the two values, we can easily make the choice, which design is better.

IV. Conclusion

The mobile phone design evaluation system before is very complicated and can not give a clear, comprehensive and objective result to us for doing the choosing. But now, from the example we can find that the comprehensive fuzzy evaluation is applied successfully in the mobile phone design evaluation system. It indeed gives us a clear data that can help us do the horizontal comparison. In this case, the result is 4.1588. Our practice shows all the advantages of comprehensive fuzzy evaluation method we have introduced above, and proves that comprehensive fuzzy evaluation method can be well applied in the evaluation system of industrial product design. In this practice, we take good use of the flexibility of the factor sets and the weight sets. This also tells us that through choosing the appropriate factors and weights, we can construct the different evaluation system to fit different industrial products. The advantages of the flexibility of the factor sets and the weight sets are not only reflected in the area of constructing the different evaluation system of industrial products, but also in the system of one product. Some of the data in the process of evaluation we find can be picked out alone to do the comparing. For example, in this mobile telephone case we can only just compare factor "innovative" or "scientific and technical" as we need. Otherwise, based on this algorithm we can also use the computer to develop a new tool to optimize the industrial design evaluation system. This will reduce the work of evaluation greatly.

References

[1] Chen Tung-sheng, Huang Shou-kai, and Tan Rui-mei, Linear algebra and analytic geometry of space. Beijing: China science and culture Press, 2003.

[2] Li Hanbang, and Xu Jiwei, "Evaluation model of college learning and teaching characteristics based on Fuzzy measurement", China Higher Education Research, pp. 15-17, Mar. 2004.

[3] Dong Yan, and Huang Ronghuai, "Network software programs recorded in the certification of quality standards", Audio-Visual Education, pp. 65-70, June, 2003.

[4] Hu Yonghong, He Sihui. Comprehensive evaluation methods[M]. Beijing: Science press. 2000.

[5] Qin Shoukang. Comprehensive evaluation principle and application[M]. Beijing: Publishing House of Electronics Industry, 2003.

[6] Andre Jone. Fuzzy sets theory and applications[M]. Dordrecht: D.Reidel Pub, 1986.

TABLE II. THE SYSTEM OF INDUSTRIAL PRODUCTS DESIGN EVALUATION

Factor Level 1	weight	Factor No.	Factor Level 2	Factor No.	weight	Factor Level 3					
Innovative ■	0.3	C	Function ■	C1	0.4	use ■	Construction ■	Interface ■	Materials ■	Technology ■	Else □
						0.2	0.2	0.2	0.2	0.2	
			Appearance ■	C2	0.6	Modeling ■	Decorative ■	Color ■	Texture ■	Else	
						0.25	0.25	0.25	0.25		
			intellectual property rights ■	C3							
Scientific and Technical ■	0.3	T	Function ■	T1	0.8	Functional Design■	use □	Agreeableness ■	Security □		
						0.33		0.33			
						Versatility ■	Sustainability □	Modular □	Else		
						0.33					
			Manufacture ■	T2	0.2	change of Technology■	Product cycle ■	Life Cycle □	product family ■		
						0.25	0.25		0.25		
						Technology Risk - Design□	Market Positioning ■	Technology Risk - Product□	Security □		
							0.25				
						Production stage □	Production efficiency □	Else			
Social ■	0.4	S	Economy ■	S1	0.8	Cost estimates ■	The definition of market■	Market relevance □	Product / market suitability■		
						0.2	0.2		0.2		
						existed product's market share□	Urgency of demand □	Consumer Contacting □	motivation of Purchase ■		
									0.2		
						Number of consumers■	understanding of Purchase effects □	Pricing plane □	Else		
						0.2					
			Aesthetic ■	S2	0.2	Traditional Culture■	Pop Style ■	Brand Style ■	Special style■		
						0.2	0.2	0.2	0.2		
						Emotional □	Psychology■	Else			
							0.2				

The Mathematical Beauty of Traffic Circles

An Exploration of the Role of Traffic Circles in Controlling Traffic Flow

Hao Zhang
College of Landscape Architecture
Nanjing Forestry University
Nanjing, China
Email: zhzh88222@yahoo.com.cn

Hengjia Song
College of Civil Engineering
Nanjing Forestry University
Nanjing, China
Email: songhengjia2008@163.com

Fei Qiu
College of Science
Nanjing Forestry University
Nanjing, China
Email: qiufei1989@126.com

***Abstract*—Determine how best to control traffic flow in, around, out of a circle is currently a hard problem. We provide 3 models here to solve it. Combined with Chinese actual situation of traffic circle, its radius usually varies from 20 to 50 meters and the channelization level is low. We would choose model I: the single annular lane without signal control if the traffic volume varies from 0 to 1000 pcu/h. And we choose model II and model III: the dual annular lanes without signal control if the traffic volume varies from 1000 pcu/h to 3500 pcu/h.**

Keywords-traffic circle; traffic flow; model

I. INTRODUCTION

The radius of a traffic circle is between 20~50 m and the channelization level is low. The vehicles will line up to enter the circle when the volume of traffic is great. Traffic jams mainly occur in the entrance of a traffic circle.

II. MODELS

A. Model I

1) Terms, Definitions and Symbols

The part of traffic flow on loop driveways will tie the team transit by minimum driving time-distance in t_m s when the traffic flow on loop driveways is great. Suppose α represents the proportion of unlimited flow when time-headway is greater than t_m and the traffic flow on loop driveways is q pcu/s. The probability that the time-headway of the traffic flow on loop driveways is greater than or equal to t_m respectively are α and $1-\alpha$.

2) Assumptions

Vehicles entering the circle can enter when the flow of vehicles on loop driveways appears greater than some critical gap t_c s. Otherwise they must wait. However, vehicles in the circle can pass through its conflict spots directly without delay.

3) The Foundation of Model

The time-headway of the loop driveway flow of vehicles has the following probability density function [1]

$$f(t)=\alpha\lambda e^{-\lambda(t-t_m)}\ (t>t_m) \tag{1}$$

$$\lambda=\alpha q/(1-t_m q).$$

Suppose t_f s is the following up time of flow of vehicles entering the circle. Namely the time-headway of two adjacent vehicles lining up to enter the traffic circle when time-headway of traffic flow in the circle lane is large and allows over two vehicles to enter.

One vehicle is allowed to enter the traffic circle when $t_c<h<t_c+t_f$. Vehicles the in amount of k area are allowed to enter the traffic circle when $t_c+(k-1)t_f<h<t_c+kt_f$. Suppose the probability is p_k if the flow of vehicles on loop driveways satisfies $t_c+(k-1)t_f<h<t_c+kt_f$. Then,

$$p_k=p\left[h\ge t_c+(k-1)t_f\right]-p\left(h\ge t_c+kt_f\right)$$
$$=\alpha e^{-\lambda\left[t_c+(k-1)t_f-t_m\right]}-\alpha e^{-\lambda\left(t_c+kt_f-t_m\right)} \tag{2}$$

Suppose the number of vehicles being able to enter the traffic circle lane each hour is c_n .Then we figure out the transit capability equation as follows

$$c_n=3600\sum_{k=1}^{\infty}p_k kq=3600\frac{\alpha q e^{-\lambda(t_c-t_m)}}{1-e^{-\lambda t_f}} \tag{3}$$

B. Model II

1) Extra Symbols

Suppose the amount of vehicles that are able to enter the traffic circle from both the left and right side of the entrance lane are respectively c_{n1} and c_{n2} .Then,

$$c_n=c_{n1}+c_{n2} \tag{4}$$

The calculation method of c_{n2} is as same as (3). The only difference is the parameters in (4) express the parameters of outside traffic flow on loop driveways.

2) Additional Assumptions

The relations obey shifted negative exponential distribution when the time-headway of the equivalent flow of vehicles is greater than t_m. They obey uniform distribution when the equivalent time-headway is less than t_m.

The traffic volume of the equivalent flow is equal to the sum of the traffic volume on two traffic lanes. Namely, $q=q_1+q_2$. In this equation, q_1, q_2 respectively represent the inside and outside traffic flow on loop driveways;

Suppose the time-headway of the equivalent flow of vehicles is less than the probability t_m. Then,

$p(t \le t_m) = t_m/h$. In the equation, h is supposed to be the average of time-headway of two patterns of traffic flow on loop driveways. Namely,

$$h = \frac{1}{2}\left(\frac{1}{q_1} + \frac{1}{q_2}\right)$$

$$p(t \le t_m) + p(t > t_m) = 1 \cdot$$

3) The Foundation of Model

Based on the above supposition, we can figure out the time-headway of the equivalent flow of vehicles having the following distribution form

$$f(t) = \begin{cases} \frac{2q_1q_2}{q_1+q_2} & 0 \le t \le t_m \\ \lambda\left(1 - \frac{2q_1q_2}{q_1+q_2}t_m\right)e^{-\lambda(t-t_m)} & t_m \le t \end{cases} \quad (5)$$

$$\lambda = \lambda_1 + \lambda_2 = \frac{\alpha_1 q_1}{1 - t_m q_1} + \frac{\alpha_2 q_2}{1 - t_m q_2}$$

According to the deduction of (3), we figured out the left side traffic lane's transit capability equation is as follows:

$$c_{n1} = 3600\frac{qe^{-\lambda(t_c - t_m)}}{1 - e^{-\lambda t_f}}\left(1 - \frac{2q_1q_2}{q_1+q_2}\right) \quad (6)$$

The total transit capability equation is as follows:

$$c_n = 3600\frac{qe^{-\lambda(t_c-t_m)}}{1-e^{-\lambda t_f}}\left(1 - \frac{2q_1q_2}{q_1+q_2}t_m\right) + 3600\frac{\alpha_2 q_2 e^{-\lambda_2(t_c-t_m)}}{1-e^{-\lambda_2 t_f}} \quad (7)$$

C. Model III

1) Terms, Definitions and Symbols

We take one-fourth of a traffic circle as a test subject. The subject includes two lanes in the entrance, an intersection including an inside circle and an outside circle connected with the entrance and two lanes linked to the intersection in the exit .Based on the results above, we simplify all the traffic flows into four parts. (Fig. 1)

First, we consider the two patterns of flow coming out of the circle as one outgoing flow A. Then we consider the flow which is always around the circle as a surrounding flow D. We consider the two patterns of flow turning right as one flow C which turns right in general. At last, we take the two patterns of flow entering the circle as flow B.

We present the parameters in the model as follows:

q_{sA}, q_{sB}: The saturated rate of flow A and flow B, q_A, q_B: The arriving rate of flow A and flow B, X_A: The probability that the time-headway of flow A is greater than or equal to t_{cB}, X_B: The probability that the time-headway of flow B is greater than or equal to t_{cA}, T_u: The time spent on the random section, T_s: The time spent on the saturated section .

2) The Foundation of Model

a) The theory of method of platoon analysis

We suppose the outgoing flow as flow A and the incoming flow as flow B. When flow A is running and there is an acceptable critical time-headway t_c, the vehicles of flow B will go through it so that flow A has to stop until there's a time-headway which allows flow A to pass.

As discussed above, flow A and flow B have the same priority to go ahead. According to the statistic of actual traffic data, we can get a probable conclusion that if flow A and flow B are not less than 150 pcu/h, then there must be a vehicle from flow A waiting to go at least when flow B is going out. According to the characteristics of traffic circles, we find that the method of platoon analysis is accurate.

Now, we divide each motorcade that go through the conflict spot into two parts: one part is first-pass and being delayed, it is called the saturated section; the other part which is not being delayed is called the random section. Assume the expectation of the two sections are N_s and N_u, the corresponding times of transit are T_s and T_u .We can get:

$$N_A = N_{sA} + N_{uA}, N_B = N_{sB} + N_{uB} \quad (8)$$

$$T_A = T_{sA} + T_{uA}, T_B = T_{sB} + T_{uB} \quad (9)$$

(N_A, N_B : the expectation of one motorcade's amount of flow A and flow B).

Obviously, if flow A and flow B pass through the conflict spot by form of motorcade, according to the length of the motorcade and the amount of motorcades that go through it per hour, we can calculate the maximum amount of vehicles that pass the conflict spot per hour of flow A and flow B. The equation is given by:

$$Q = Q_A + Q_B = 3600(N_A + N_B)/(T_A + T_B) \quad (10)$$

In the random section, the time-headway normally refers to that of a vehicle usually arrives without being delayed. From this model: we know that the time-headway of the random section of flow A varies from 0 to the acceptable gap t_{cB} of flow B. The random section of flow B is the same as flow A, but its time-headway varies from the minimal time-headway t_m to the acceptable gap t_{cA}. By integrating the time-headway of flow A and B, we will obtain the average time-headway of the random section. The result is given by:

$$N_A = T_{sA} * q_{sA} + \frac{1}{X_A} - 1 \quad (11)$$

$$N_B = T_{sB} * q_{sB} + \frac{1}{X_B} \quad (12)$$

$$T_A = \frac{(q_{sB} - q_B) * [q_A * (T_{uB} - T_{uA}) + q_{sA} * T_{uA}]}{q_{sA} * q_{sB} - q_A * q_{sB} - q_B * q_{sA}} \quad (13)$$

$$T_B = \frac{(q_{sA} - q_A) * [q_B * (T_{uA} - T_{uB}) + q_{sB} * T_{uB}]}{q_{sA} * q_{sB} - q_A * q_{sB} - q_B * q_{sA}} \quad (14)$$

b) The basic transit capability of the roundabout

We define the basic transit capability of the roundabout as the reasonable maximum amount of vehicles under ideal traffic and road conditions with a fixed time. Ideal traffic conditions refer to all the vehicles in traffic are supposed as small vehicles without any bikes, pedestrians and non-motorized vehicles. The ideal road condition refers to a traffic circle with two annular lanes, a road with four directions and two in-and-out lanes.

If the incoming vehicles continually increase, they have to queue up at the conflict spot when the conflict spot reaches its transit capability. Because the incoming vehicles include not only the vehicles of the left roadway in the entrance, but also the vehicles of the right roadway in the entrance, the incoming vehicles will lead to a traffic jam.

The ultimate purpose of the traffic circle is to unblock the flow of traffic, thus, the transit capacity of the whole circle is equal to the maximum incoming or outgoing traffic volume, that is:

$$X_1 + X_2 + X_3 + X_4 \text{ or } Y_1 + Y_2 + Y_3 + Y_4$$

Because the intention of the traffic circle is to help unblock the outgoing volume of traffic, we use $Y_1 + Y_2 + Y_3 + Y_4$ to describe the transit capacity of the whole circle.

If the traffic circle is under normal circumstances, the incoming traffic volume $X_1 + X_2 + X_3 + X_4$ is equal to the outgoing traffic volume $Y_1 + Y_2 + Y_3 + Y_4$.

Suppose the flow of the roundabout between entrance 1 and exit 2 is W_1. Suppose the flow of the roundabout between entrance 2 and exit 3 is W_2. We will get:

$$\begin{cases} W_1 = x_{12} + x_{13} + x_{14} + x_{32} + x_{42} + x_{43} \le C_w \\ W_2 = x_{21} + x_{23} + x_{24} + x_{13} + x_{43} + x_{14} \le C_w \\ W_3 = x_{34} + x_{31} + x_{32} + x_{14} + x_{24} + x_{21} \le C_w \\ W_4 = x_{41} + x_{42} + x_{43} + x_{31} + x_{21} + x_{32} \le C_w \end{cases} \quad (15)$$

x_{ij} : the traffic volume from entrance i to exit j.

$$Y_j = \sum_i x_{ij} \qquad i \neq j$$

(i=j is a situation that the vehicles turn back, and we will not take into consideration.)

From the analyses above, we can get:

$$Y_1+Y_2+Y_3+Y_4=x_{21}+x_{31}+x_{41}+x_{12}+x_{32}+x_{42}+x_{13}+x_{23}+x_{43}+x_{14}+x_{24}+x_{34} \quad (16)$$

After the transformation, we can get:

$$Y_1+Y_2+Y_3+Y_4 \le 2C_w+\frac{1}{2}(x_{12}+x_{23}+x_{34}+x_{41}-x_{14}-x_{21}-x_{32}-x_{43}) \quad (17)$$

From Fig. 2, we find that $x_{12}+x_{23}+x_{34}+x_{41}$ is the traffic volume of the vehicles turning right and $x_{14}+x_{21}+x_{32}+x_{43}$ is the traffic volume of the vehicles turning left. Suppose the proportion of turning right is r and the proportion of turning left is l. We will get:

$$Y_1 + Y_2 + Y_3 + Y_4 \le \frac{4C_w}{2 - r + l} \quad (18)$$

The transit capacity of the traffic circle is given by :

$$C = \max(Y_1 + Y_2 + Y_3 + Y_4) = \frac{4C_w}{2 - r + l} \quad (19)$$

If r =0.2 and l =0.3, then C=3200 pcu/h.

III. CONCLUSIONS

The conclusions of the problem and the applications of our models are as follows:

Model I has only one circle. When applying it , the transit capacity should vary from 0 pcu/h to about 1000 pcu/h;

Model II has two circles. The transit capacity should vary from about 1000 pcu/h to 3500 pcu/h when applying it and will reduce if the proportion of unlimited moving traffic increases. The transit capability will also decrease if the proportion of the right-turning direction of traffic flow that enters the circle increases.

Model III has two circles and describes the weaving section of the roundabout. The capacity is affected by the acceptable gap and the proportion of turning right and turning left.

REFERENCES

[1] P. Hughesb, “So You Think You Understand Gap Acceptance,” Australian Road Research, Victorary, 119(3), pp. 195-204, Australia, 1989.

[2] R M. Kimber, “Gap-acceptance and Empiricism in Capacity Prediction,” Transportation Science, 127(2), pp. 100-111, Baltimore, U.S.A., 1989.

[3] W. Wang, “Platoon Method of Analyzing Capacity and Delay of Non-Signalized Intersection,” Journal of Chongqing Jiaotong University, pp. 36-44, China, 1995.

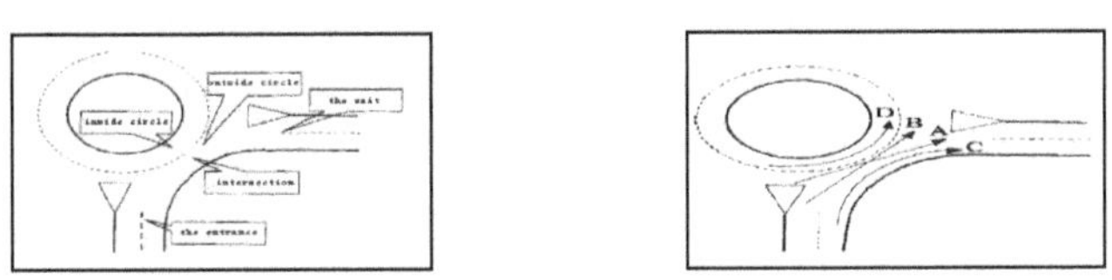

Figure 1. The simplified model hint picture and the one-fourth of the traffic circle

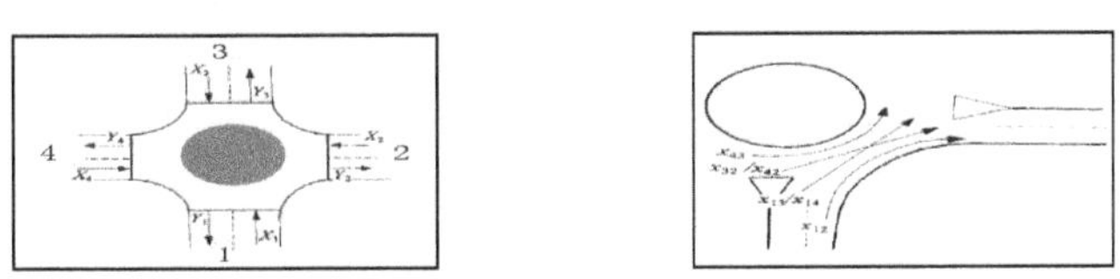

Figure 2. The hint picture of leading the traffic and the deduction picture of leading the traffic

The Preliminary Exploration of the Sustainable Research on the Urban and Rural Construction in the Western Mountainous Area of China

The Case Study of the Planning of Traditional Chinese Medicine Industrial Park in Southern Shaanxi Mountainous Area

Degang Duan
Xi'an University of Architecture and Technology
Xi'an, China
Email: haliduan@163.com

Jin Wang
Xi'an University of Architecture and Technology
Xi'an, China
Email: wangjin815@163.com

Chengchen Wang
Xi'an International Studies University
Xi'an, China
Email: ern_1989@163.com

***Abstract*—Based on the resources condition of southern Shaanxi mountainous area, this paper employs the choice of industry, presenting the theory and method of regional cooperation, to form agglomeration benefits and utilize the resources in a scientific and effective way. Meanwhile, this paper carries out a research on the planning layout of traditional Chinese medicine (TCM) industrial park, in combination with the natural features and the developmental level of socio-economy, attempting to build up a set of layout model and construction index which is appropriate for the geographical features in southern Shaanxi mountainous area, in expectation to seek a rational, well-organized layout method on the premise of protecting eco-environment and promoting the comprehensive economic development in local society, as references to the construction of western mountainous cities.**

Keywords-sustainable development; southern Shaanxi mountainous area; traditional Chinese industrial park; layout model; construction index

Since China's reform and opening up in the 1970s, the development of the western and eastern part of China has been out of balance and the gap between them has been widening, which also happens in the urban and rural construction because of the restrictions of locational factors and transportation condition. Resulting from the slow development of western part of China, for about 30 years, the urban and rural construction hasn't led tremendous destruction to the eco-environment on which we rely for existence. Meanwhile, that's the reason why people from the eastern part of China prefer traveling to the west and enjoying the primitive scenery after the industrialization. However, local governments in western part of China, urgent to solve the issue of poverty, are making the following common preparations: during the process of adjusting industrial structure in the eastern part of China (suppress the second industry and develop the third industry), western urban and rural areas take over the eliminated industries with low land price and manpower cost, in the master plan of China's western cities, such kind of thing can be revealed from the blind and larger-area-seeking attitude toward the formulation of urban land scale. As the author of this paper, I think this tendency is gravely mistaken. Nowadays, global warming led by industrialization and world financial crisis make us begin to suspect the supporting function of service-producing industry, so the developmental strategy of the western urban and rural areas of China lies in how to really take backwardness advantages, treasure and make full use of local resources, and take a western characteristic urban and rural development road. Nevertheless, we still have a better preserved eco-environment, but, at the same time, it's fragile. The "experience" -----"pollution first, treatment later" to eastern part of China is what a total catastrophe to western part.

The selected southern Shaanxi mountainous area[a] in this paper is rather representative in the western mountainous area of China. Southern Shaanxi mountainous area is situated in the Qinling Mountains, belonging to Yangtze River basin. Qinling Mountains, known as the Central National Park of China, is rich in animal and plant resources and its vegetation cover rate is exceedingly high, even more than 90% in some conservation zones. But as a matter of fact, the soil horizon is thin there, once the vegetation is destroyed, the whole eco-environment would be deteriorate rapidly. Simultaneously, the southern Shaanxi mountainous area is one of the slowest in the development of urban and rural construction in China, which is faced with not only the frequent natural disasters and the heavy burden of eco-charges, but also the poor condition of information and transportation, the backwardness in economy, the scarcity of land resources, and the high cost of the urban and rural construction.

How to seek out a sustainable way to exploit the southern Shaanxi mountainous area lies in: firstly, selecting the appropriate industries; secondly, formulating the rational layout model, centered on the previously chosen industries. This paper researches on the regional development strategy of TCM industrial parks in southern Shaanxi mountainous area and the layout model, with the aim of functioning as references to the urban and rural construction in this area.

I. CHOICE OF INDUSTRY

The southern Shaanxi mountainous area, known as "China's natural drug storage", possesses incomparable resources endowments. As the transfer quantity increases in the international TCM market in recent years, the central government of China shows increasing concerns and attention to the TCM industry in this area, and Shaanxi province also publicizes a series of relevant policies to support the development, which means shaking off poverty and shifting eco-dominance into advantage of backwardness to this area. Therefore, the TCM industry there will become the dominant direction of development in urban and rural areas.

Nowadays, the TCM industry in southern Shaanxi mountainous area has already begun, some TCM industrial parks have been building, while many counties and cities are engaged in preparing for it. However, there're some issues: the industrial collaborative partnership hasn't been formulated; the industry type is convergent and the concentration degree low; the construction of parks destroys the fragile eco-environment in southern Shaanxi mountainous area.

II. REGIONAL COOPERATION

The transportation between the southern Shaanxi mountainous area and other regions is inconvenient all the time, while the construction of TCM industrial parks in each region is expected to cover all sides of a matter. It's difficult to formulate agglomeration benefits because of the restriction of land utilization, the small scale of industrial park and the indistinctive features.

A. The Current Situation of TCM Industry in the Southern Shaanxi Mountainous Cities

TABLE I. THE CURRENT SITUATION OF TCM INDUSTRY IN HANZHONG, ANKANG, SHANGLUO

District		Hanzhong
Characteristic biological resources		gastrodia, eucommia-bark, agaric polyporus, gen-seng
Industry basis	*Drug Source Base*	The state/provincial key planting base of gastrodia, agaric polyporus, eucommia bark, rhizomacorydalis radixaconiti, panax quinguefolium, radix codonopsitis, white paeony root, coptis root,etc.
	Enterprise Overview	12 enterprises (7 of which certified by GMP); AAOV : about 810,000,000RMB; Product: gastrodia, eucommia bark, lovage rhizome, cornus officinalis, agaric polyporus, gen-seng, gingko
	Park Construction	Hanzhong Traditional Chinese Medicine Industry Park (the structure of it is monotonous, lacks of scientific research and related marketing enterprise; the scale small)
	Core product	Gastrodin Tablets, "Shudan" Tablets
Technology Strength		Hanzhong Institute of Medical Sciences

District		Ankang
Characteristic biological resources		yellow ginger, gynostemma, radix puerariae, radix isatidis
Industry basis	*Drug Source Base*	The state/GAP/provincial planting base of gynostemma, radix isatidis, yellow ginger, radix puerariae,etc.
	Enterprise Overview	10 enterprises (2 of which certified by GMP); AAOV: about 750,000,000 RMB; Product: gynostemma, yellow ginger, radix puerariae, radix isatidis, rhizoma polygonati
	Park Construction	Ankang Bio-science and Technology Industrial Park (the industrial real estate is engaged in speculation and profiteering, a lot of land remains idle; the infrastructure needs amplifying; the visual effect is bad because of the scarcity of greening)
	Core product	Gypenosides (certified as the first second-class medicine)
Technology Strength		Ankang Beijing Medical University Holding Co. Ltd; Zheng-da Pharmaceutical Research Institute, The Yellow Ginger Institution

District		Shangluo
Characteristic biological resources		salvia miltiorrhiza, cornus officinalis, bupleurum chinense, honeysuckle
Industry basis	*Drug Source Base*	The state/provincial planting base of salvia miltiorrhiza, dioscorea opposita, taxus chinensis,etc. The planting base of the ten key species (TCM)in this area: bupleurum chinense, yellow ginger,gastrodia, honeysuckle, platycodon radix,etc.
	Enterprise Overview	9 enterprises (5 of which certified by GMP); AAOV: about 510,000,000 RMB; Product: salvia miltiorrhiza, bupleurum chinense, yellow ginger, gastrodia, honeysuckle, platycodon radix, hizandra berry
	Park Construction	Shangluo Bio-medical Science and Technology Park[b], Panlong Ecological Medicine Industrial Park in Zhashui County (the land price is high [c]; scientific manpower is weak; the environment needs improvement)
	Core product	-
Technology Strength		Shangluo University, institute of agricultural sciences in Shangluo, Panlong Bio-medical Industrial Park

B. Regional Cooperation——Establishment of Industrial Cluster Relationship

The gradual improvement of regional transportation makes the regional industrial cooperation possible. (figure 1) Through the on-the-spot investigation, evaluation and analysis, this paper employs the factors-evaluation of the counties and cities in southern Shaanxi mountainous area[d]. The evaluation is based on the biological resources, environmental condition, transportation condition, industrial basis, technical strength, policy advantage, infrastructure, and industrial back-up land, etc. The configuration results are showed in the table II and figure 2.

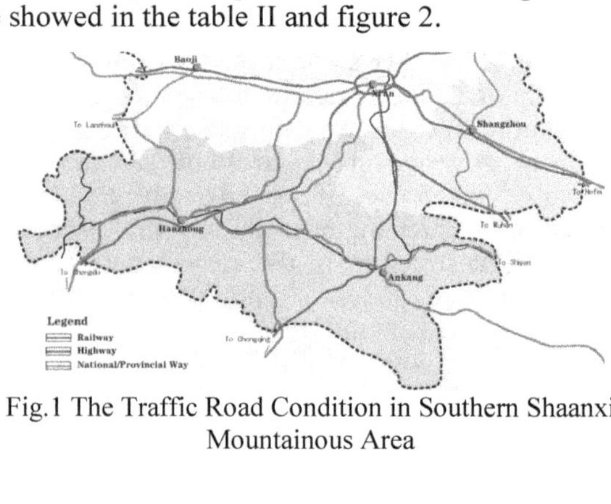

Fig.1 The Traffic Road Condition in Southern Shaanxi Mountainous Area

TABLE II. THE REGIONAL ALLOCATION OF TCM INDUSTRY OF HANZHONG, ANKANG, SHANGLUO IN SOUTHERN SHAANXI MOUNTAINOUS AREA

Project	Hanzhong
The Development Strategy	On the basis of Hanzhong Traditional Chinese Medicine, enlarge the scale and add up the preferential policies, to attract more enterprises
Supporting Industry	Hanzhong Institute of Medical Sciences + Hanwang Chinese Medicine Co.Ltd
The Position of the 3 cities in the Regional Industrial Chain	In the position of processing, extraction, and pharmaceutics. Processing : The current industries have solid foundation Extraction : It's the key part of development, providing the basis of diversified products Pharmaceutics: The comprehensive strength of Hanwang Chinese Medicine Co.Ltd is more powerful as it adjoins to Sichuan, one of the most biggest pharmaceutical cities; the market competition stimulates the development

Project	Ankang
The Development Strategy	Develop the Ankang Bio-science and Technology Industrial Park; set up the criterions for enterprise access; improve the facilities and the environment
Supporting Industry	Ankang Beijing Medical University Holding Co.Ltd + Zhengda Pharmaceutical Research Institute,
The Position of the 3 cities in the Regional Industrial Chain	In the position of collection, processing and research. Collection : 4 formulated drug source bases Processing: 4 formulated processing bases Research : The strength from Ankang Beijing Medical University Holding Co.Ltd

Project	Shangluo
The Development Strategy	On the basis of existed industrial parks, the strategy should combine with travel demands to improve the facilities and service environment
Supporting Industry	Tasly Group + Panlong Ecological Medicine Industrial Park
The Position of the 3 cities in the Regional Industrial Chain	In the position of pharmaceutics and marketing[e] Pharmaceutics: The industry begins later, two formulated pharmaceutical industrial parks Marketing: The transportation advantage; less back-up land; fit for developing related sale part; government of Shangluo put forward policies to seek co-development, combined with tourist industry

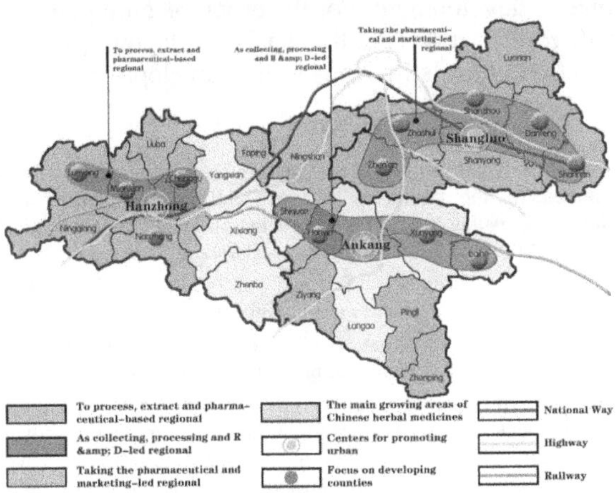

Fig.2 The Regional Layout Diagram of TCM Industry in Southern Shaanxi Mountainous Area

In Michael E. Porter's *Clusters and the new economics of competition,* he explores that the competitive advantages formed by the industrial chain, which is constituted of the related comprehensive upstream and downstream enterprises, are stable and reliable. Therefore, it's good for utilizing resource advantages and complementing each area's advantages by strengthening the regional cooperation and improving the industrial chain relationship from the aspects of planting, collection, processing, extraction, researching, pharmaceutics, and marketing. At the same time, it's vital to the improvement of whole competitiveness of southern Shaanxi mountainous area.

III. ADAPTATION TO THE LOCAL CONDITION: THE SUITABLE CONSTRUCTION MODEL OF TCM INDUSTRY FOR SOUTHERN SHAANXI AREA

More than 90% of Southern Shaanxi is mountainous, as the saying goes "The geographical feature is ninety percent in mountains, while, each five percent in rivers and fields". The shortage of land is the major issue faced with this area. Meanwhile, southern Shaanxi mountainous area, situated in the Qinling Mountains and Bashan Mountain, is the national eco-conservation zone, the water sources protection zone of the middle route project in south-to-north water transfer project of China, and the area of concentrated reflection of bio-diversity. The natural eco-environment is excellent but fragile, so the premise and basis of the construction of TCM industrial park in this area are to protect the eco-environment.

A. The Internal Structure of the Industrial Park

The existed TCM industrial parks in southern Shaanxi mountainous area haven't achieved industry cluster[f], so the primary issue is to formulate the internal structure. Firstly, the multi-industrial chain should be built up. For instance, product A, made up of both the upstream and downstream enterprises, A1, A2 ... ... An, etc, is designed to complete the production of A. Every enterprise has the products that can be marketed directly, which can promote the formulation of

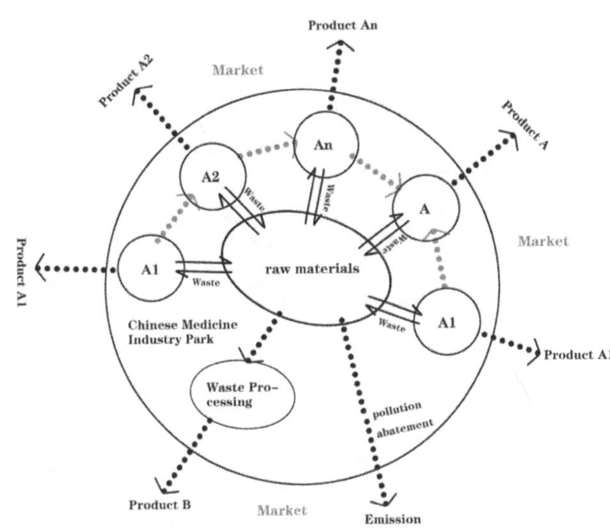

Fig.3 The Organization Model of TCM Industrial Park

the enterprises' symbiosis network (figure 3). Secondly,

through the recycling process of materials, the reduction of resource consumption and the pollution abatement (figure 4), the industrial parks are geared to protect the regional eco-environment and achieve the balanced development

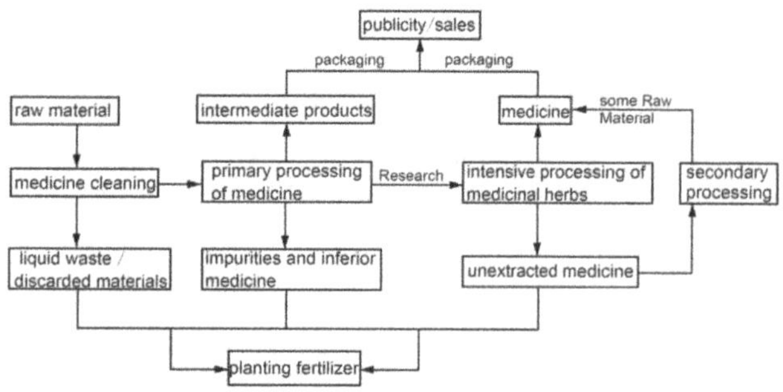

Fig.4 The Schematic Diagram of Recycling Process of TCM

among economy, environment, resources, and society.

B. Planning Layout Model

Most cities in southern Shaanxi mountainous area are of band-shaped pattern, the geographic feature is described as "'two' mountains with 'one' river in between". The natural factors strongly affect the urban spatial arrangement, making the layout of land a reflection of the regional features.

The layout features of southern Shaanxi mountainous cities: (1).The primary development trend of the cities always extends along the rivers; (2).The cities are related to the natural environment tightly; (3).The urban transportation extends in the band-shaped pattern, the road network has a great density. (4). The clustering feature is distinct.

Based on the layout features of the southern Shaanxi mountainous cities and the parks, the pattern "eco-core + eco-network + functional groups + peripheral environment" is more suitable for the parks to establish a friendly eco-relationship with the peripheral mountains and rivers. (figure 5)

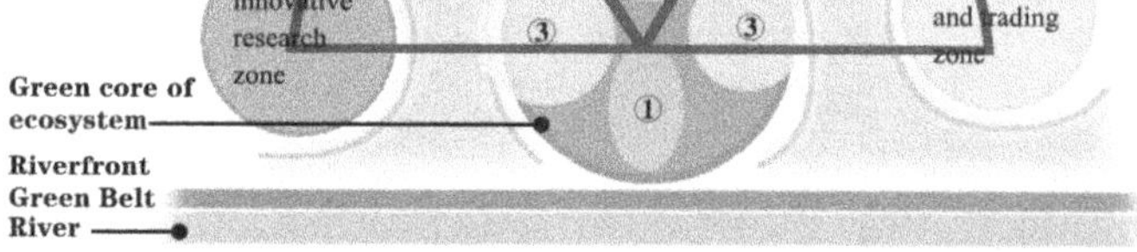

Notes: ①Core space ②administrative zone ③supporting service zone

Fig.5 The Spatial Layout Pattern of TCM Industrial Park in Southern Shaanxi Area

Among them, eco-network includes green plot, mountain, waterbody, open space, leisure and communication area, etc. The linked groups have formulated concentrated eco-space and provided high-quality environment for medicine production, which is also the control of the scale of the parks' land. Relatively, each group is independent, separated by eco-space, through the perfect transportation system to communicate with the others. (On the basis of the division of labor in regional cooperation, different regions emphasize on different functional groups. [g])

C. Indexes Research

Through the analysis of related medicine industrial parks in China, integrated *The Controlling Indexes of Construction Land Use for Industrial Projects*, the related regulations in Shaanxi province, the natural features and the socio-economic developmental level of southern Shaanxi mountainous area, this paper presents the guiding indexes of the construction of TCM park in southern Shaanxi mountainous area in the aspects of eco-environment, industry type, economic benefits, etc. (Table III is put back at the end of the paper)

Besides standardizing the construction model of TCM industrial park, the key to a successful industrial park is to strengthen the management intensity, as the following examples: standardizing the land using guide for TCM industry to allocate the resources appropriately throughout the entire area and strengthen the construction of industrial chain; establishing the enterprise access system to enhance the land use capability; working out the preferential policies to attract investment; establishing the eco-compensation to protect the environment in southern Shaanxi mountainous area.

IV. CONCLUSION

Since the UNCED in 1992, China has implemented the strategy of sustainable development, which includes an important act: Westward Expansion. Therefore, the development of western part needs to study and weigh the situation, erect the scientific development concept, make full use of its own resources and transform the extensive economy, and in combination with each city's condition, to form the policy environment which is propitious for sustainable development. For the urban planning researchers, their research work, on the basis of layout pattern and construction index, still needs future development.

NOTES

a. Southern Shaanxi mountainous area: In the southern part of Shaanxi, situated in the southern foot of Qinling Mountains, including Hanzhong, Ankang, Shangluo, totalling up to 28 counties (cities) with a population over 9,000,000 and a total area of 70,000 km^2.

b. The Shangluo Bio-medical Science and Technology Park hasn't been built up completely because of the highway.

c. About 1,400,000 RMB/mou (a Chinese unit of measure, 1 mou equals to 666.67m^2); the preferential price in the park is 700,000 RMB/mou.

d. The score of factors-evaluation is 2, 1, 1, 1.5, 1.5, 1, 1, 1 in turn, the environment condition, transportation condition, policy advantages factors are almost equivalent in the evaluation (except Shangluo, which is more favorable in transportation condition and emphasizes on the combination of tourism and the development of TCM park)

e. The posterior position in industrial chain makes higher profit, the industrial development emphasizes on the aspects of pharmaceutics, selling and also the co-ordination. (Shangluo is the most underdeveloped cities in southern Shaanxi)

f. Industrial cluster doesn't mean spatial agglomeration, it's rational division of labor of upstream and downstream, as well as, supporting enterprises, centered on the related enterprises.

g. The suggested functional layout: Hanzhong, eco-core + sight-seeing zone + scientific research planting zone + high-quality goods processing zone + supporting service zone; Ankang, eco-core + innovative research zone + primary finished goods processing zone + supporting service zone; Shangluo, eco-core + sight-seeing zone + marketing and selling zone + supporting service zone.

h. The volume fraction of Shangluo Bio-medical Science and Technology Park is 0.42, while the volume fraction of Panlong Ecological Medicine Industrial Park is 0.85, which means the latter is superior.

i. The ratios of the area of all the buildings and structures use for production and served for it and the total area, within the land use boundary.

References

[1] *The Controlling Indexes of Construction Land Use for Industrial Projects* [S], authorized by The ministry of land and resources P.R.C. [2008] NO.24.

[2] *Guide for Established of Eco-industrial Parks Planning* [S], authorized by The Ministry of Environmental Protection of P.R.C. (HJ/T409-2007)

[3] *Discussion on Effects of Industrial Land Standard* [J], FAN Jun-yong, *City Planning Review*, 2008. (32).3:50-55.

[4] *Exploration of the Ecological Planning for Partial Section: Case Study of Shangluo Bio-pharmaceutical Industry Park Planning* [J], Gen-wei HONG, De-gang DUAN, *Architectural Journal*, 2006. (2) : 30-32.

[5] *Functional Group: the Effective Choice of China's Sustainable Development,* Yan SHU, *China Pharmaceuticals*, 2007. (16).6:4-5.

[6] *Clusters and New Economics of Competition* [J]. Michael E Porter. *Harvard Business Review*, 1998(11):77-90.

[7] *On the Industrial Ecology Potential in Asian Developing Countries* [J], Anthony S.F.Chiu, Yong GENG, *Journal of Cleaner Production*, 2004.(12):1037-1045.

TABLE III. RELATED INDEX

Project	General Index	The Suitable Index for Shaanxi	Basis
The Scale of Land	3～10 km^2	2～8 km^2	limited construction land and comprehensive strength of the city
Volume Fraction	≥ 0.6	0.7～1.0	lower limit: limited land resources and ensure the development intensity [h] higher limit: avoiding the destruction of eco-environment
Building Coverage	≥ 30%	30%～50%	lower limit: intensified development of land higher limit: emphasis of eco-environment protection
Network Density	2～3km/km^2	3～4km/km^2	restricted terrain
Greening Rate	≥30%	35%～40%	lower limit: the emphasis of eco-environment construction higher limit: restricted by the intensity of development
The Per Capita Scale of Land	100～160 m^2/person	80～120 m^2/person	limited construction land; limited industrialization level(industry:30-45m^2/person; resident:20-30m^2/person; public service facility: 15-25m^2/person; others: 15-20m^2/person)
Green GDP	≥18000RMB / person	≥15000RMB / person	average rate of growth of green GDP: 9%;the comprehensive strength is lower than national level
Investment Intensity	7,800,000～10,35 0,000 RMB/hm^2		Hanzhong, Ankang, Shangluo each belongs to the city of 9th , 11th , 13th class [i]
Land-output Ratio	≥9,000,000 RMB / hm^2		based on the related income of the industrial park and the current development of southern Shaanxi mountainous area
The scale of industry population	19000～72000 RMB		land area 2～8km^2 , volume fraction: 0.8 industrial land: 35～40% per capita construction area: 30－35m^2(including service population)
Service Radius of Auxiliary Facilities	within 20 minutes' walking		within 2km^2 the resources are shared, beyond the range the cost will be higher

The Relationship between the Function and the Form In Product Shape Optimization Design

Tang Jun
School of Art and Design
Wuhan University of Technology
Wuhan, China
Email: tiao16@163.com

Abstract—**The optimization design process of product shape is a complex system engineering determined by a variety of factors such as the function factor, the form factor and so on. The features of product shape are determined by the intersection effect of two factors. For optimization design of the different products, we should analyze the characteristics of the product shape trends, using an integrated multi-disciplinary cross-evaluation method which focuses on the application environment, the use factors, production factors and so on.**

Keywords-shape of product; optimization design; function; form

I. SCHEMATISM IN THE OPTIMIZATION DESIGN PROCESS OF PRODUCT SHAPE

A product not only has the practical functions, but also has the mental function. The mental function means the art taste which satisfies our aesthetic needs and brings us the enjoyment of the beauty. The good products have the beautiful shape, structure and color, spread the beautiful information, stimulate and satisfy our art interesting. The aesthetic needs become consumer motivation easily. Society needs more and more innovative products which satisfies not only our aesthetic needs but also functional needs. Every product may have many kinds of shape, so designer focuses on choosing the most rational one, considering all factors. Because the shape design deals with many integrated factors, such as model, color, material and manufacturing process, the optimization design process has the complex schematism. The eye-catching appearance of one product can attract consumers to buy.

Fig.1 shows us that the function location of product design has the useful function and the aesthetic function. From which we can sum up, good products should be considered with physics, manufacturing, social, use, beauty, affinity, national feelings, and many other aspects of a combination of factors.

These factors involved in a wide range of discipline and technology. At the same time, these factors are interdependent, correlative complex, discrepant, such as easy installing and firmness, convenient making and no polluting. Therefore, the optimization design process of product shape is a systematic work of interdisciplinary. With the progress of human society, the solution turned to multi-disciplinary cross-system integration method from single-disciplinary research in the past. The optimization design study is such a new independent engineering discipline and has been widely applied in practice.

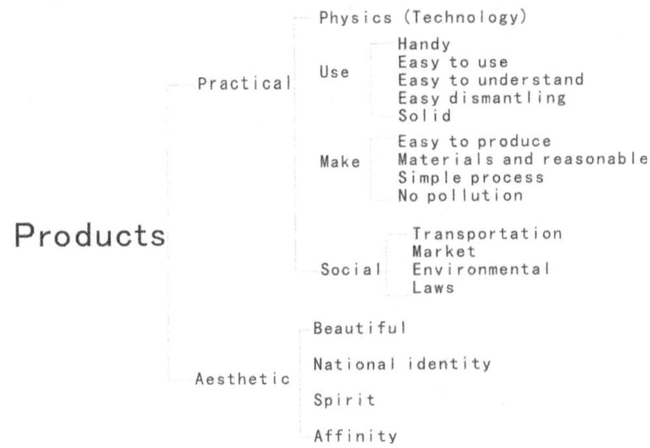

Figure 1.

II. THE LATENT FUNCTION FACTORS AND FORM FACTORS OF PRODUCT SHAPE

Product appearance is an actual carrier meeting the functional requirements (whether psychological or physical), is also a basic form completing the design purpose. It is both the bearer of the external structure and the transmitter of the internal functions. All of this needs a certain shapes and materials to express, the different shapes have different sense of beauty.

Therefore, under a specified condition, the environment or context, the entity is created to complete the intended function. And it shows that a product ultimately achieve its design goal. From the results of natural evolutionary, those animals and plants that be able to survive and multiply to today should have rational characterized. Their shapes well demonstrate their survival function; their rich natural bodies give us a lot of inspiration. By studying their shapes and nature, we found that their shapes almost are related to functions, most of which are closely related to their survival instinct (Fig. 2-1 "C Fig. 2-5).

Fig 2-1

Fig 2-2

Fig 2-3

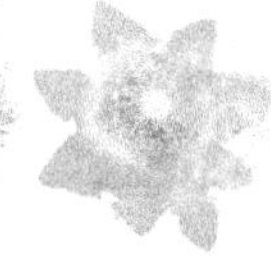

Fig 2-4

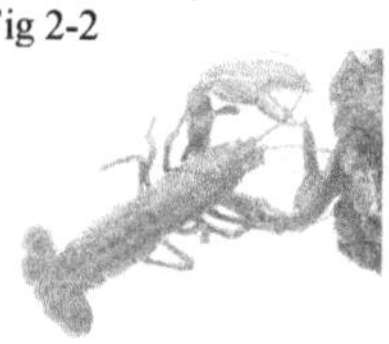

Fig 2-5

For example, kangaroo has a long and powerful tail. Triangle support can make it stand tall and stand a long time. It helps kangaroo to get the leaves of small shrubs, to detect the more distant potential hazards timely. When kangaroo is running, it is also a balancer and a rudder. Crocodile and shrimp also have this characteristic appearance, which are closely related to their survival function. In plants, the dandelion flower likes a parachute. When wind is blowing, they can make their own seeds migrate far away. This is the evolution of morphological characteristics is to produce offspring. Product is a kind of physical human-made body. It embodies creators¡¯ ideas to meet people's physical and psychological function, physical body plays an important role in meeting people's mental function and physical function. The composition of the natural body is mainly to meet the needs of physiological function. In bionics, we make maximum use the enlightenment of the natural shape. Using bionics principles, we combine model and innovation to create a tool for transforming the objective world (Fig. 2-6 ¨C Fig. 2-7).

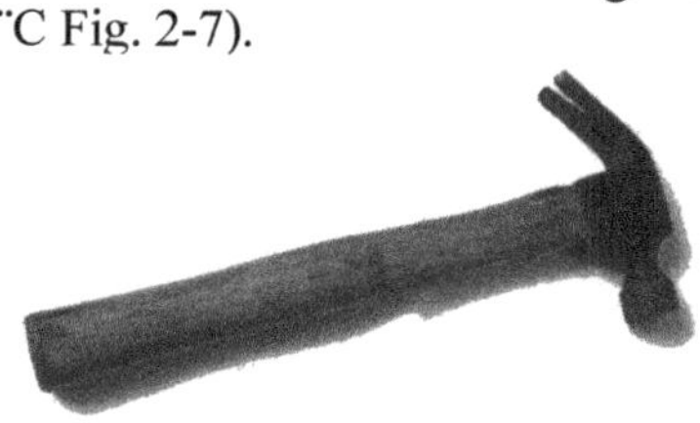

Fig. 2-6

Fig. 2-7

III. THE CASE ANALYSIS FOR THE RELATIONSHIP BETWEEN THE FUNCTION AND THE FORM

For the optimization design of different using products, how do we make our choice between the function factor and the form factor? Below, we give two design cases for analysis. One is a hand drill's optimization design (Figure 3-1 ~ Figure 3-3), another is an eyeglass's optimization design (Figure 3-4 ~ Figure 3-9). First of all, these two products should be consistent with their basic functions: hand drill to be able to punch, glasses to be able to care people's eyes; Secondly, in the shape optimization process, these two products should strive to shape beautiful, but the difference is that a hand drill as a professional and operational tool, functional requirements of component should be more than shape or color requirements. At present, a hand drill and its force direction is not a same line, thus easy makes a problem that is the wandering hole. So, the shape structure of a hand drill was improved and Figure 3-3 shows us the improved shape. As opposed to hand drill, glasses only have a basic functional requirement to fix on the glasses lens. It is very easy to implement, and now its basic structure is very mature, very reasonable, and does not require many changes. Therefore, the optimized design of glasses focuses on its appearance and fashion. In the optimization design process, designers will extract and absorb the fashion elements, then improve its form and color, etc.

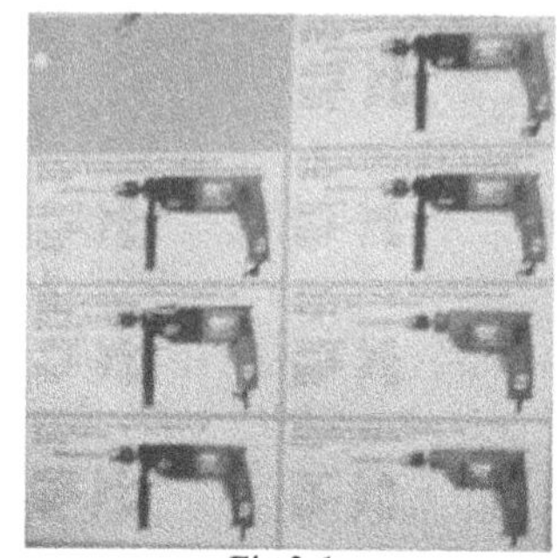

Fig 3-1

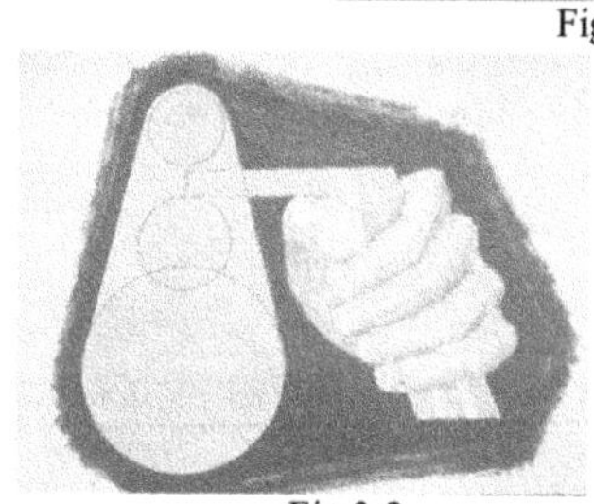

Fig 3-2

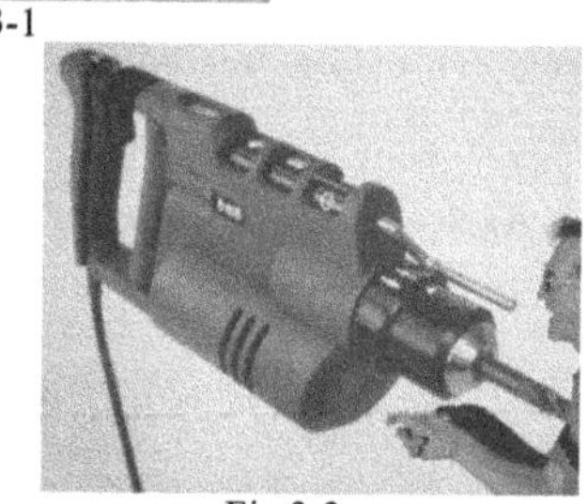

Fig 3-3

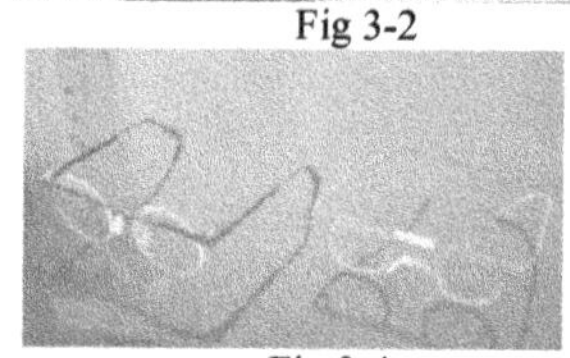

Fig 3-4

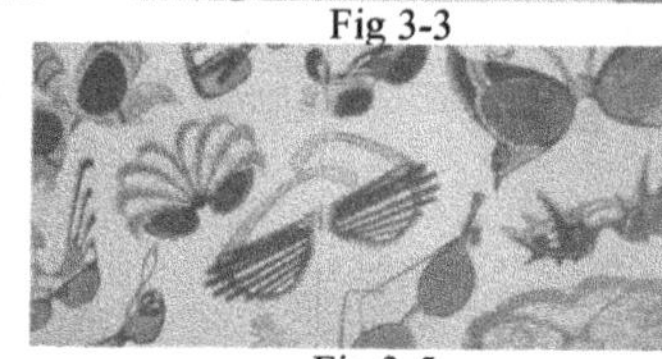

Fig 3-5

Fig 3-6

Fig 3-7

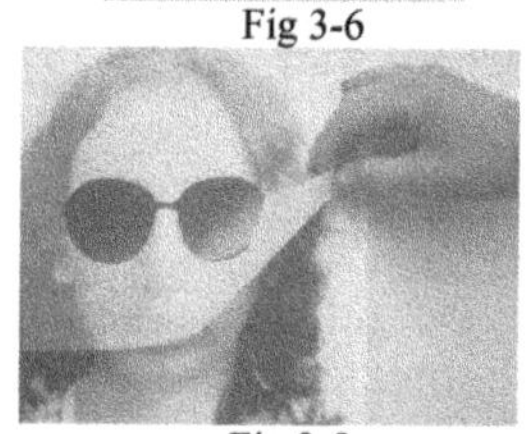

Fig 3-8

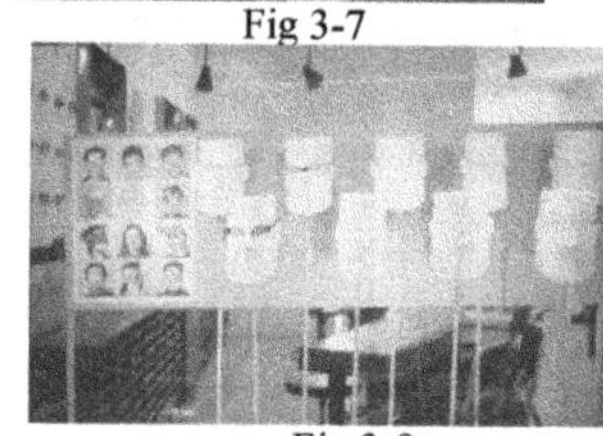

Fig 3-9

From the above cases, we can conclude that different products have different shapes, use patterns, environment, and positioning and evaluation systems should have a corresponding change in optimizing their design process. Some design should focus on shape, but some design should focus on the reasonable function and structure. This is the reason that the different products with different function shave different shapes. The functional design of any product must be achieved by a shape. No matter how simple or complex its body is, its appearance has play an important role.

IV. Conclusion

Consumers have dual requests for products' shape, and one is spiritual, another is material. Shape design should be similar to the products' natural bodies, its function can meet the material needs; At the same time, it also should include the mental functions, such as social, age, culture, folklore, religion, and aesthetics and so on. Mental function is an important part of product shape and it is constituted by many factors. They do not like the functional as obvious form, but it contains more and more complex meaning and content. Therefore, in the optimization design process of product shape, the relationship between function factors and form factors is contradictory, antagonistic, unified, the reliant each other. Starting from two factors of the optimization design, we choose the best option from a variety of programs. Basing on the systematic optimization theory, the computer as a means, design performance as an objective, we seek an optimal design scheme by using objective functions with the various specified constraint conditions.

References

[1] Liu Guanzhong, "Design Basic of the Compositive Shape" , Higher Education Press, 2009

[2] Tang Jun, "The Basic of the Industry Design Shape", Tsinghua University Press, 200607

The Research of the Optimized Planning Behavior on Industrial Design Procedure

Yueqin Wu & Hu Sun
Department of Industrial Design
Xihua University
Chengdu, China
Email: 12596929@qq.com & Email: 14344380@qq.com

***Abstract*—The view of design comes from system theory and introduces planning into the Industrial Design. It proposes some planning action in design procedure by explaining the means of planning and analyzing the relationship between planning and design. It re-considers the choice and control of materials, technology and color, and further elaborates the optimized role of the planning behavior in industrial design procedure, and try to make the product development more effective and more adaptive to the times.**

Keywords- planning; design procedure; optimizatio

Industrial design is an all-dimensional and multi-level systematic project. To introduce planning into industrial design can effectively promote the design and deal with the whole design process with a systematic view. The design artistry shall be in the inseparable integration with application demand, material processing capacity, structural rationality and color suitability.

I. CONCEPT OF PLANNING

For GUIHUA, the Chinese meaning of planning, GUI means laws & regulations and, HUA as spearing, means dividing. As per the nature of content, it can be divided as general planning and specific planning. Planning is similar with plan, but with the difference that planning is long-term, overall, directional and general. Planning is to plan for the development of a thing, event or system in a complete and long-term way. It is the consideration on the future integration, long-term and fundamental problems. It considers and designs the complete action process for the future.

II. RELATIONSHIP BETWEEN PLANNING AND DESIGN

Design means to manifest plan into signals and summarize it under certain intentions, with the meaning of plan, operation and planning. It is an orderly and methodic technical modeling activity according to specific aims. It seeks for better coordination between things and human beings and creates environment in conformance with the physical and psychological goals of human society and, can reach specific process in a viewable manifestation.

Planning is a method for further optimization of industrial design under the guidelines of system theory. Architects' planning contains the understanding of living and working of people in the architecture, and designs the space to support and ease them. For industrial designers, planning contains the understanding of the living and working of customers of the products and, designs the appearance and conducts of products to support and ease the living style. And, in the whole design process, it also includes the selection of materials, process and technology, control of cost, pairing of color, etc. Various planning in industrial design process finally serve the lives and living environment of human beings, which is a dynamic process under the joint effect of government, enterprises and society.

III. PLANNING IN DESIGN

Model, color, material and function are the most fundamental elements in the composition of a product, in which material and technology are the material base for the basic pattern of a product and color is an important element for the appearance of a product. Thus, selection and control of material, technology and color necessarily becomes a planning conduct.

A. Material Planning

Material, in short, is the combination of material and tactile impression. Industrial design once was explained as bestowed material shape, color and function for certain aim. It can be observed that the correlation degree between material and design is far more than all modeling elements. Application of material must meet the proposed aim. Material selection in design process reflects the purposefulness of industrial design. Material planning is a most important section closely related with product modeling in the whole industrial design. As per the property of material, material planning is shown in Fig.2.

The project supported by the Sichuan Provincial Education Department Natural Science Foundation, NO.09ZC018

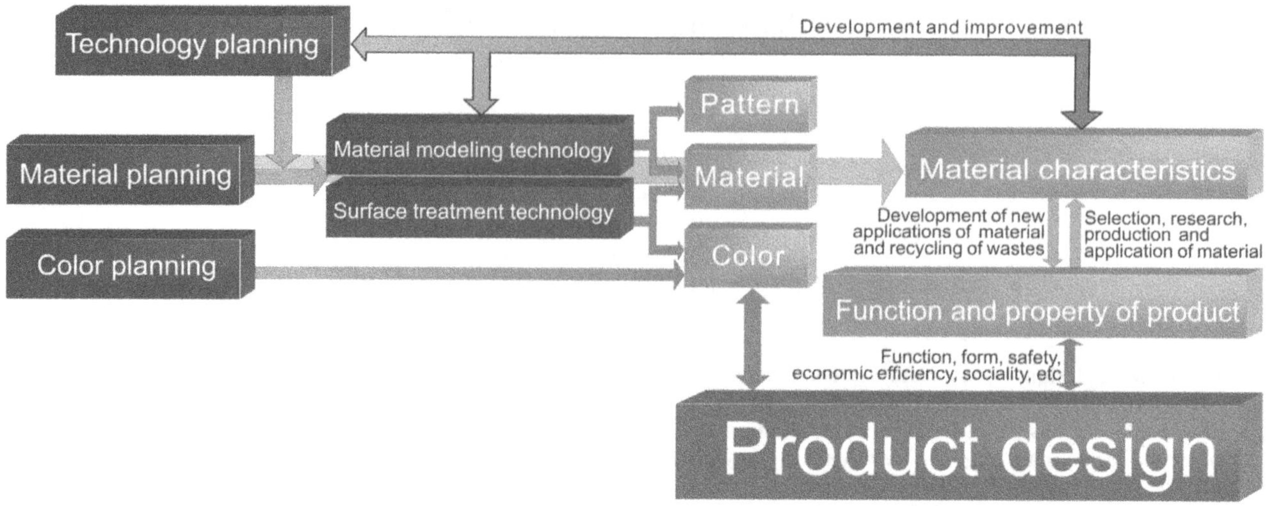

Figure 1. The relationship between material, technology, color planning and the pre-product design

Material planning in design includes two aspects: one is to start from the design and select proper material as per the function of product; in case no proper material, improve and redesign material and structure and create new material meeting the requirement of product; in the commercialization process, research and develop application technologies, such as material modeling and surface treatment. Another aspect is to start from the property of material, makc effort to develop new applications besides traditional ways, such as recycling of wastes, changing waste materials into resources. The base for material planning in design is the comparison and weighing between the performance of material and functions & performance required by the product so as for best matching relationships. Refer to Fig.1.

B. Technology Planning

In this paper, it mainly refers to material technology, that is, engineering technology closely related with material processing, mainly including production technology, engineering technology and processing technology, etc. In industrial design, the support of technology enables integration among practicability, artistry and economical efficiency. Production technology refers to technology on the production of certain industry or enterprise; engineering technology refers to the technology related with certain engineering; processing technology refers to technology for one or several materials. Processing technology, closely related with industrial design, can be divided into material modeling technology and surface treatment technology. One is the way for the physical modeling of product and another is the condition for perfect product modeling. They are well combined, penetrated and promoted with each other. Technology planning in industrial design is shown in Fig.2. Technology planning is usually simultaneous and parallel with material planning instead of being isolated or simplex. They jointly impact all phases of product design. Technology planning shall be connected with material, cost, production cycle, surface effect of the product, etc. Refer to Fig. 3.

C. Color Planning

In the era of scientific progress and developed commercialization, majority products are popularizing and the consumption market is striding into the maturity phase. Perfect color planning can create unique image of product and satisfy the 'individuation, differentiation and diversification' requirements. The trend of color planning shall not only conform to the future color trend and artistic requirement, but also integrate marketing tactics, fully absorb relative information, consider the company's overall

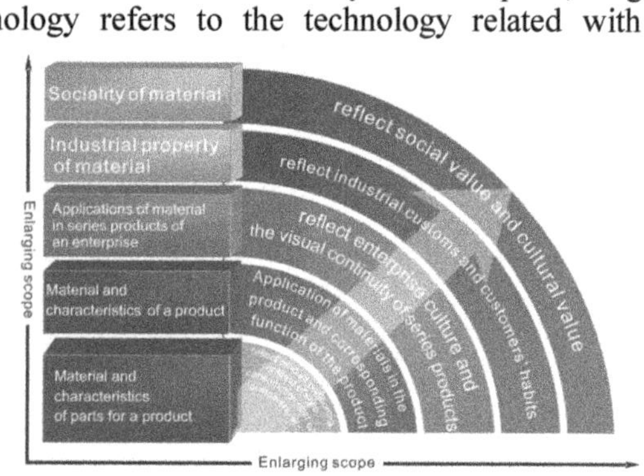

Figure 2. Material Planning

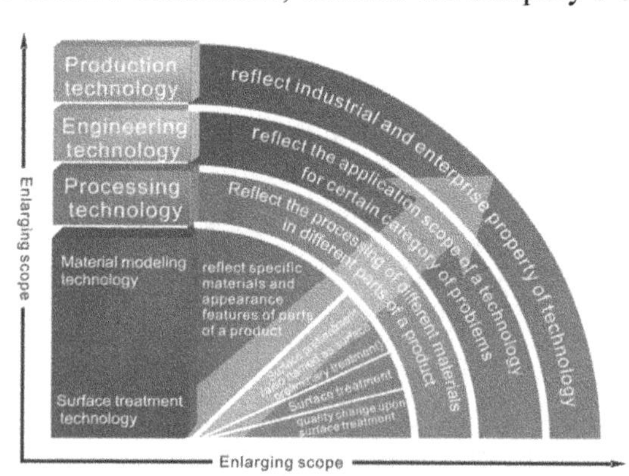

Figure 3. Technology planning

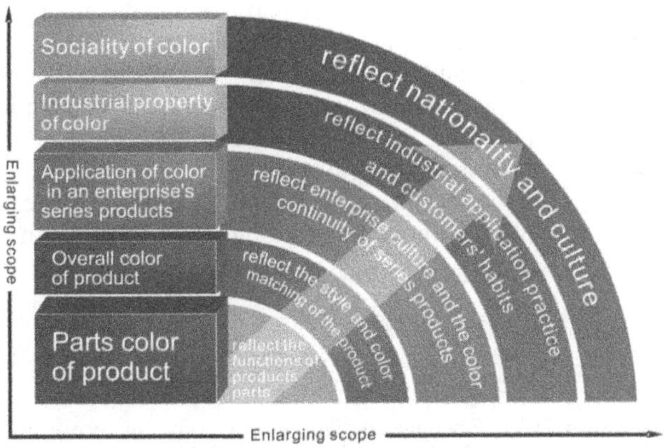

Figure 4. Color Planning

image and finally bestow the most proper color for the product. According to the property of color, color planning of product is shown in Fig. 4.

Color selection for a new product by enterprises usually follows the following steps: ① to sort out and analyze color information collected from different areas; ② to understand and summarize the popularization regularity of color in different years, areas and industries; ③ to make strategic decision and, based on visual image, extract visual observation drawing of single color and visual drawing of color matches; ④ to use single color palette and matching color palette to make basic color and popularizing color; ⑤ to develop a color via simulated experiment; to make model via computer software or to make physical product model to test and observe the sensation to human beings from the observation and then, decide the final color via meetings.

IV. Material Planning, Technology Planning and Color Planning Have Good Optimizing Effect to Industrial Design Process

Under the continuously developed historical background, modern design shall not only consider new material, new technology and new process, but also integrate relative resources and information and, constantly optimize corresponding design process so as to advance with the times.

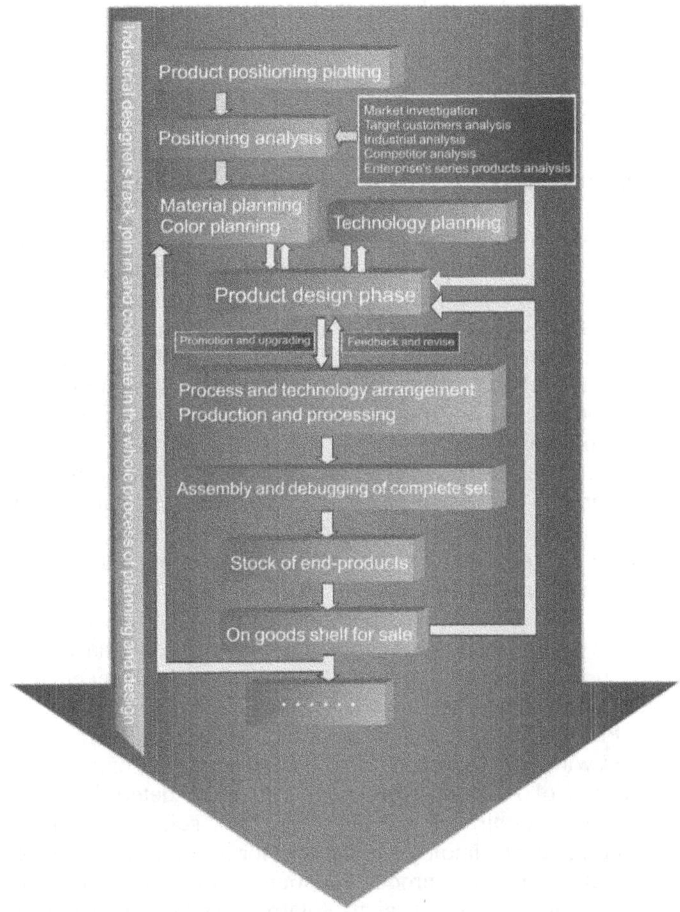

Figure 6. Optimized industrial design process

A. *Traditional industrial design process*

Majority of traditional industrial design processes are in one-way linear structures. The enterprise owner joins in the planning of product positioning while the sales staff finally put the product on the market, industrial designers shall not only be responsible for the work in product design phase, but also track, join in and cooperate in the whole process of design so as to enable the design conforming to the

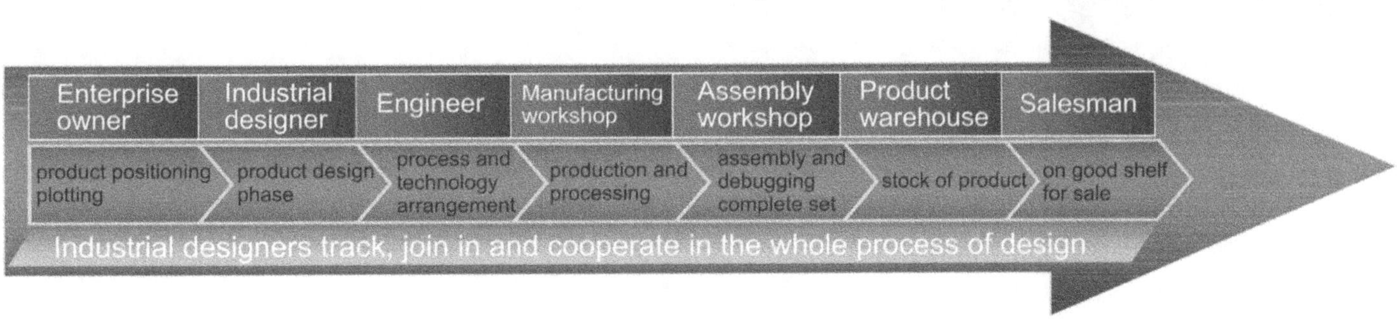

Figure 5. Traditional industrial design process

conditions of different phases and be suitable for industrial production and meet the demand of target customers, refer to Fig.5.

B. Industrial design process with introduction of planning

Introduce the concept of planning into industrial design. The planning contained in the design process mainly includes product positioning planning, material planning, color planning and technology planning, etc. The design process is not idealized and in one-way linear structure. Each phase of the design will touch upon various contents, including selection of new material, new technology and new process. It will cause series changes of humane, artistry, consumption and traditional technology. Therefore, based on new material and new technology, part of the design process shall be in circular structure and move forward one by one, refer to Fig.6.

Industrial designers shall not only be responsible for the work in product design phase, but also track, take part in and assist the design in the whole process. They should play their role in the early stage of design and take part in the planning of all elements in the design process. Thus, the industrial design process can not only be optimized, but also ensure the design conforming to the conditions of different phases, fit the requirement of industrial production and meet the demand of target customers in the maximum degree.

Planning in industrial design process has good optimizing effect with effective promotion to the design. To some extent, selection of material, technology and color determines the success or failure of a design. Sightless selection of new material, new technology and new color is not necessarily for the coming of a new product. Resources, such as production, living, manufacturing, society, enterprise and culture shall be well integrated.

V. Conclusion

The design process is related with the smoothness of design activities, as well as decides the production efficiency and quality of product development in certain degrees. As for the complexity and fuzziness in the early stage of design, complex activities and numerous participants, the concept of planning is introduced into industrial planning. Define the selection of material, technology, color and control in the design process as material planning, technology planning, color planning, etc. so as to relieve or eliminate the complexity and fuzziness in the early stage of design. Planning has the optimization effect to industrial design process, as well as certain verification to the design result.

References

[1] Norman.D.A, "Emotional design—why we love or hate everyday things," New York: Basic Books, 2004,pp.20-34.

[2] Gerwin.D.and Tarondeau.J.C, " Case studies of computer integrated manufacturing systems: a view of uncertainty and innovation processes," Journal of Operations Management, 1982, 2(2), pp. 87-99.

[3] Kruger.C, "Solution driven versus problem driven design: strategies and outcomes,".Design Studies,2006, 27(5),pp. 527-548.

[4] Cross.N, "Science and design methodology: a review ," Research in Engineering Design, 1993, 5(2),pp. 63-69.

[5] Dorst K and Cross N, "Creativity in the Design Process:Co- evolution of Problem Solution," DesignStudies,2001,122(5),pp.425-437.

[6] AKEN VAN J E, "Valid knowledge for the professional designof large and complex design processes," Design Studies,2005, 26(4),pp.379-404.

[7] Luo Shijian, Sun Shouqian and Pan Yunhe, "Research on Product Information Design and Cognition. 3rd International Conference on CAIN&CD, " Hong Kong, 2000(in Chinese)

[8] Liu Zheng, Sun Shouqian, Wu Jianfeng, Wuqun. Model of product form innovation design knowledge based on users cognition. Computer Integrated Manufacturing Systems. 2009,15(2),pp.265-270

[9] MENG Kaining, SUN Hu. The industry design procedure under new material and new technology background[J]. Packaging Engineering. 2009.10 （6） :P121-123

[10] JIANG Ping, ZHANG Huangao, CHEN Zishun, TAN Runhua. Product Platform Design Method Based on Axiomatic Design. JOURNAL OF MECHANICAL ENGINEERING. 2009.45(10) P216-221

The Role of Public Participation on Environmental Evaluation in Residential Areas

Bin Mao
School of Urban Design
Wuhan University
Wuhan, China
Email: maobin200888@126.com

Abstract—**The residential environment is taken as a direct carrier for the public life. The method of public participation has now been gradually regarded as an indispensable measure that adapts to urban planning practice in China. In this paper, the new and old residential areas in Wuhan are surveyed and analysed through the method of public participation to create the factors of environmental evaluation in residential areas and predict the changes that the residents need for residential environments.The optimized design and development model of residential environment that conform to the reality and the demands of the residents were established in order to enhance the operability and accuracy of planning and implementation in residential areas.**

Keywords-residential area; environmental evaluation; public participation

I. INTRODUCTION

With the improvement of housing development and human residential environments, The housing demands for people in town and country are transforming from quantity-type consumes pursuing simple existence to quality-type consumes. At present, when we make plans and construction of residential areas, it is very important to know what is needed for modern housing? What environmental atmosphere would they like? What kind of integrated service should be provided? And to what extent can we do ? What are the current residential value orientation of the Chinese people in the 21st century? How will the environment make impact on the people's physiological behavior and mantal sense? How will we evaluate residential environment today? All these problems will cause extensive concern.

Today, the method of public participation has now been gradually regarded as an indispensable measure adaptable to urban planning practice in China. To the public, the method of public participation is not a kind of passive understanding and acceptance of urban planning result, but a kind of positive participation in urban planning process. This is a kind of exchange and integration process of the conception and idea. In urban environmental evaluation, the methods of public participation include public attitudes to life, relationships with the Government, the interests of enterprise groups, the roles of experts and so on.

It is well known that the environment in an residential area is taken as a direct carrier for the public life. Some constructive suggestions through the method of pubic participation are put forward in planning and construction in residential areas, which greatly improve the accuracy of policy decision on the development of urban environment. The method of pubic participation is not only consistent with the wishes of the public, but also enhance the operability and comprehensive coordination of urban planning.

In recent years, in some planning and design projects that I have undertaken, the writer has preliminarily made attempt to urban planning integrated with public participation, thus also arouses some thoughts. We applied the method of public participation in the reconstruction of an old residential area and in the development of a new residential area.The questionnaire surveys were conducted by means of random sampling. We sent out altogether 250 questionnaires to the residents and got back 234. The ratio of callback of valid questionnaire was 93.6% and the results of the surveys were basically valid.

II. THE ESTABLISHMENT OF ENVIRONMENTAL EVALUATION FACTORS IN RESIDENTIAL AREAS

The residential area is not only a place of residence, but also an optimum environment that is created for the life of local resident to the fullest extent possible.The environmental evaluation factor of the residential area is intended to descript and evaluate whether it is safe, comfortable, harmonious, or convenient in residential areas, and meet the need for suitable residence and sustainable development. Therefore, the factor of evaluation never simply go in for quantification, but mostly establish suitable human inhabitance environment from public visual properties and viewpoints of "people-oriented" and "Sustainable development". Under the guidances of these principles, some factors such as location, accessibility, building groups, individual building, green landscapes, road traffic, public facilities, cultural environment, management, environmental characteristics and images, and environmental qulity are selected as main evaluation factors. Every factor also includes one or more single elements, which can be seen in Table Ⅰ.

TABLE I. THE BASIC FRAMEWORK OF ENVIRONMENTAL EVALUATION FACTORS IN RESIDENTIAL AREAS

Main Evaluation Factors	One or More Single Elements
Location	the site of the city, house price
Accessibility	service facilities, road traffic around residential areas
Building groups	building density, height, orientation, style, space between buildings
Individual building	the types of house orientation, areas, structures and floors
Green landscapes	central flower garden, green lawn, public plaza, natural vegetation, water landscape
Road traffic	the convenience of road, the quantity of parking lots, firefighting access, walking street
Public facilities	the integrity of supporting infrastructure (water and electricity system, telecommunications service, etc.) the perfection and convenience of service facilities (business, hospital, market, cultural recreation, etc.)
Cultural environment	neighborhood harmony, the sense of community order, the sense of psychological belongingness, cultural value identity
Management	The quality of property management
Environmental characteristics and images	building style and characteristic, the sense of historical and traditional places, the protection of historic site
Environmental quality	outdoor noise, air quality, dispose of rubbish

III. THE ANALYSIS FOR THE STATISTICAL CHARACTERISTICS OF QUESTIONAIRE SURVEYS

The sample statistics through questionaire surveys are seen in Table II.

TABLE II. THE RESULTS OF SAMPLE STATISTICS

	New Residential Areas (%)	Old Residential Areas (%)
Location	4.75	7.25
Accessibility	5.93	7.30
Building groups	12.13	10.06
Individual building	13.36	10.10
Green landscapes	15.11	12.08
Road traffic	11.46	9.37
Public facilities	14.85	12.25
Cultural environment	6.31	11.39
Management	5.24	3.12
Environmental characteristics and images	4.81	12.25
Environmental quality	6.05	4.83

(1) The statistical results of questionnaire surveys show that the residents pay more attention to green landscapes in an new residential area, which reflect that the residents pursue a kind of emotional living grade and attach more importance to the perfect integration of technology, art and environment in housing development on the basis of more advanced technology of housing development in China. All these aspects also show the affinity, adapatability and the sense of homeland to household environment.

As far as environmental design is concerned, "green" is not a kind of simply "existence", not only remain green space ratio increased, but also emphasize how to raise artistic quality of building and environment and enhance comprehensive quality of the residential area. The objective of the residents' house purchasing mainly compares with the soft environment of each property management. Moreover, they further consider public leisure space (public square, green space and water landscape, etc.) and visual quality, which increasingly make ecological consciousness strengthen.

The sureys also found that the developers have improved their cognitive ablilities and laid more emphasis on the construction of the qulity of residential environment, which change the shortsighted behavior of pursuing excessively high economic benefit.

(2) From another aspect, the residents pay more attention to the design for the types of house, which make a great impact on purchasing power. After interviewing with some residents, we discovered that three-room types are received the admiration from them. the customers would like to purchase highly types of house with a three-bedroom and two living room whose area ranges from 90 to 110 square meter, as well as multifloor. Some relevent data can be seen in table III, tableIV and table V.

TABLE III. THE AREAS OF HOUSES PRAISED BY THE CONSUMERS

The Areas of Houses	Proportions (%)
Below $80m^2$	12
$80\sim90m^2$	20
$90\sim110m^2$	32
$110\sim130m^2$	21
Above $130m^2$	15

From the table above, we can see that the maximum provision quantity of housing market is presently three-room apartments with multi floor whose area rangs from 90 to 110 square meter in all types of house, which can be afforded by the ordinary working man or high income inhabitant, Therefore, it will be the most popular type of house in the consumers who have purchaseing powers. In addition, much larger types of house are restricted by demand and they can not be develop greatly. Thus it can be seen that three-room types whose area rangs from 90 to 110 square meter will become dominant in housing market. However, the types of house with larger and the smaller areas will be developed as auxiliary types. It is a realistic reflection that the real estate market will increasingly close to the consumers and grow into maturity.

TABLE IV. THE TYPES OF HOUSES PRAISED BY THE CONSUMERS

The Types of Houses	Proportions (%)
One-bedroom and one-living room	9
Two-bedroom and two-living room	30
Three-bedroom and two-living room	38
Four-bedroom and two-living room	12
Compound apartment	7
Others	4

TABLE V. THE FLOORS PRAISED HIGHLY BY THE CONSUMERS

The Floors	Proportions (%)
Multifloor (below 6 floors)	35
Middle and high floor (6~9 floors)	21
Small high floor (9~12 floors)	28
High floor (above 12 floors)	16

With regards to the floor of house, the houses with multifloor are still praised highly by the majority of the population and the houses with small high floor (the floor ranges from 9 to12) are popularly accepted by some middle and high income groups. With the development of social economy and the improvement of living standards for the residents, the houses with small high floor will become extensive types of house in urban housing market in Wuhan.

(3) The surveys demonstrate that an traditional old residential area has better community culture environment. Neighborhood is harmonious and has a strong sense of belongingness. These residents pay more attention to spatial environments with modern facilities and high quality when they attach great importance to historical and cultural identity. As far as a new residential area is concerned, with the accelerated population growth rate of non-native population, the extensive sources of the local have decided the characteristics of an advanced degree of residential member dissimilation. When non-native population enter into a new and strange community, they need injection with new ideas and methods to promote sociability among people who live in a new community, and develop the senses of identity and belongingness to the community.

(4) For the needs for public service facilities, there are certain differences due to their different ages. Generally speaking, The elderly demand high levels of living conditions such as medical health care, business nets and market fairs and the distributions of these service facilities are accessible to the community. But the middle-aged and young residents have greater needs for other public facilities like school, culture and recreation. These residents deliberate on their children needs for going to school and entertainment and can fully enjoy their lives. Accordingly, we should lay stress on the construction of living service networks combined with regional features, historical evolution features and inhabitant structure features.

(5) There is a high density of population in an old residential area where the infrastructure is backward and the structures of some buildings are dilapidated and the outdoor environment is poor. In the surveys, the residents have strong demands for old infrastructure reconstruction and young residents yearn more for moving out of the old residential area. Of course, these residents first concern about their own living space. Consequently, It is significant to build necessary municipal infrastructure and public service facilities and establish a set of intact and effective organization and management institution.

IV. PUBLIC PARTICIPATION AND PARTICIPATION METHODS FOR DIFFERENT LEVEL

The levels of public participation on planning practice in two residential areas above involve in planning agency, developers, urban management bureau, experts, planning units and the public like ordinary residents, enterprise workers, retired employees, educators, leaders of sub-district offices and community residents' committees. The methods of public participation consist of exhibition, questionnaire, discussion, random interview, developer appraisal, etc.

In environmental evaluation of an old residential area, Public views and comments on residential environment are directly collected by community residents' committees. We visit some family to investigate and understand residential conditions and housing situations, integrate the information with policies and principles of urban administration departments and urban planning departments and work out the development plans for residential areas on these bases.

Urban planning agency and design units develop very useful work on public participation. They think that it is necessary for the public to understand vividly the content of the plan for residential areas through the publication of plan processes and design scheme. The preliminary achievements of the plans for residential areas are openly exhibited together with videotape and model. They solicit comments and suggestions from the public on the spot and answer questions from them. The feedback from the public is revised, complemented and perfected after careful analysis and selection. The government and the public get a common understanding on many problems through the method of public participation, so relevent plans for residential areas are approved and put into effect by the government.

The problems that is reflected by public participation will give main references for planning and design. The design for the types of houses and the construction of green landscapes and public facilities are further emphasized, which not only strengthen the practice character of planning project, but also provide convictive basis for promoting the implementation of plan projects.

The methods of participating in the exhibition, filling in the questionnaires and having an informal discussion show that the public are enthusiastic about the functions, standards and images of future residential areas. Because the public has an deep acquaintance with public participation, furthermore, the plans for residential areas also directly concern with the interests of the public, thus the designed plans through public participation are commonly approved and accepted.

V. Conclusion

Through residential planning practice combined with the method of public participation above, some main conclusions are obtained as follows:

(1) the methods of public participation through different levels and organizations are proven to find some effective measures. The approaches to the method of interactive and open public participation contribute to optimum project design and implementation and governmental development decision, which will achieve new concepts and communication modes of solving problems.

(2) It is necessary that we need sufficient financial support and close coordination in planning and management, and also need to strengthen spontaneous behavior of the public like discussion meeting in the course of public participation.

(3) The setting of survey content and participation approach, as well as the methods for analysis and evaluation will make survey results have some limitations and one-sidedness. But the results of the surveys of public attitudes are not equal to that of planning and design. Nevertheless, these opinions and suggestions still reflect the wishes and desires of the public, which should be reasonably adopted in planning and will become a necessary and useful supplement to the compilation and achievement of the plan for the city.

(4) Most residents pay attention to their immediate interests. However, their understandings of the overall situation are still superficial, which will impact on the depth and the effect of public participation.

(5) It is necessary that public participation combine with the use of new technology. For example, virtual reality technology is widely used in the new residential area planning and the old residential area reconstruction with new visual characteristics and become the best means of planning projects optimization. The use of virtual reality technology achieve authentic simulation from outdoor environment in the new residential area to diversified types of residential houses and interior design like rooms with different styles, which provide the residents with a real and intuitive experience of whole spatial environment in future residential areas.

Moreover, the shape and height distribution of old residential building, along with relationship between the image of an traditional or modern area are thought over according to current spatial form and profile in old residential areas by means of virtual reality technology, which can compare landscape images before transformation with after and conduce to decision-making support for optimizing design for residential areas.

Virtual reality technology provides aided design, the analysis of query, design scheme comparison and reflect the feature of digital design of multi-angle and dynamic, which not only contribute to improve the quality of the project design and speed up the process of project implementation, but also provide the investment, evaluation, and consultation with project program.

The design and implementation of an new and old residential area project are a urban planning decision-making process that governments, developers and residents participate together by means of a dynamic interactive mode of virtual system.

In a word, two residential planning cases through the method of public participation in Wuhan once again demonstrate that urban construction will achieve great success once the plans for residential areas integrate with the method of public participation and take feasible decision.

Acknowledgment

The author would like to express her appreciation to some graduate students for valuable work on the survey for the data of this research.

References

[1] GE Jian, TEKNOMO Kardi, LU Jiang, HOKAO Kazunori, "Computer support system for residential environment evaluation for citizen participation," Journal of Zhejiang University SCIENCE, 2005 6A (8), pp. 894-899.

[2] LI Wei-guo, YANG Jing, WANG Juan, "Analysis of residents' evaluation of residential area environment in a small city," Journal of Zhejiang University (Engineering Science), Vol.40 No.1, Jan. 2006, pp. 92-95.

The Smoke Movement Prediction and Ventilation Optimization for Fire in Long Tunnel

Li Tingting
Highway School
Chang'an University
Xi'an, China
Email: lt198244@163.com

Tang Xian
Department of Highway Engineering
Shaanxi College of Communication Technology
Xi'an, China
Email: helian_xicheng@163.com

***Abstract*—Fire accident broken in the long highway tunnels always leaded to heartbreaking tragedy. It was essential to understand the characteristics of the burning process and smoke movement in the tunnel, which were critical for the optimization of the design of the ventilation system and the arrangement of cross evacuation tunnels. A three-dimensional simplified model of long highway tunnel was built up based on the Qinling Zhongnashan King-size Highway Tunnel. Numerical studies on the characteristics of the burning process and smoke movement were performed using CFD software. The effects of positions of fire, blowing velocity and direction, and burning intensity were considered, that were helpful for the optimum of the ventilation system and evacuation tunnels. Computational results showed that high temperature appeared with concentrated smoke spread with it. Ventilation had important influence on the smoke movement which would prevent reverse flow and force the smoke move to downstream area. The upstream region was easier to be damaged than downstream area in the present model once no ventilation was provided.**

Keywords- Fire in tunnel; Ventilation system; Evacuation tunnels; Smoke movement; Reverse flow; Ventilation optimum

I. INTRODUCTION

The tunnel transportation is developing quickly with the increment of the highway and railroad mileage. Fires in tunnels have obtained special attention during recent years due to several catastrophic fires in tunnels[1]. Because of the limited area of the cavity, it is found that fire brigade is difficult to get near to the ignition sources and the fire is hard to control. Thus two main issues should be resolved when designing a safe tunnel. One is the safety of the people inside the tunnel, where the hot smoke in the tunnel is of great concern. The second is the endurance of the structure capability under the effect of a fire. Understanding the detailed information of how the fire develops, how it propagates and how the smoke spreads in an enclosed space is critical for tunnel designers. Relevant information about the fire environment can be obtained by full-scale experiments, which are extraordinary expensive to carry out. Computational fluid dynamics simulations are increasingly used to estimate the effect of fires in tunnels and corridors, which advantage is that it can predict the air flow pattern, the temperature and the smoke contours. Studies on tunnel fires had been reported in the literature since the mid-1980s. Experimental studies on scaled models have been performed by Vantelon et al[2]. Full-scale experiment research and theoretical study for fires in tunnels with roof openings were conducted by Wang Yanfu et al[3]. They found early fire-detection systems and adequate warning signs were required to reduce the risk of being trapped in the smoke plume. Numerical simulations with computational fluid dynamics also have been reported. The sensitivity of simulations to input parameters were tested by Brandeis and Bergman[4], who carried out two dimensional simulations of a forced ventilated tunnel under varies ventilation schemes. A multi-zone model had been proposed by Charter et al[5-6], but both its experimental verification and the probable area of application had to be examined carefully. Simulation of temperature and smoke distribution of a tunnel fire based on modifications of multi-layer zone model was performed by Chen Xiaojun[7], he took the radiation heat loss of fire source into account and introduced a four-surface radiation heat transfer model. Fire zone models were developed for appropriate simulation of tunnel fires. W.k.Chow used a zone model to simulate tunnel fires[8]. He found the model was much simpler in nature and included almost all physical principles of a compartment fire, once a stable stratified hot layer was found in the tunnel. Fire properties in near field of square fire source with longitudinal ventilation in tunnels were studied by Hitoshi Kurioka et al[9]. They confirmed empirical formulae are sufficiently applicable for predicting fire phenomena in the near field of square fire source. Shorab Jain et al compared the prediction results of CFAST and CFX, which proved that CFX could captured more detailed features than CFAST[10].

In view of above, numerical studies were performed extensively to simulate tunnel fires. This paper presented the Fluent simulation of a fire inside a forced ventilated tunnel. Varies ventilation velocity and fire position were taken into consideration. The process of the development of far field smoke flow and impact of ventilation flow was analyzed. It would be helpful to optimize the ventilation system and evacuation tunnels in long tunnels.

II. METHOD OF SIMULATION

As shown in Fig.1, a three-dimensional simplified model was built up based on the Qinling Zhongnashan King-size Highway Tunnel. The distance between the main and parallel tunnel was 25m. The interval of evacuation tunnels which connect the main and parallel tunnel was 500m. The diameter of the tunnel vault was 5.5m, while the length, width and height of the main tunnel were 1500m, 11m and 7.6m respectively. The evacuation tunnel had a 45°angle to the main tunnel. No gradient of the tunnel was concerned in present study. The fire was assumed to break out in the main tunnel, which position was 400m to the tunnel inlet and 100m to the evacuation tunnel. The ventilation direction was from left to right.

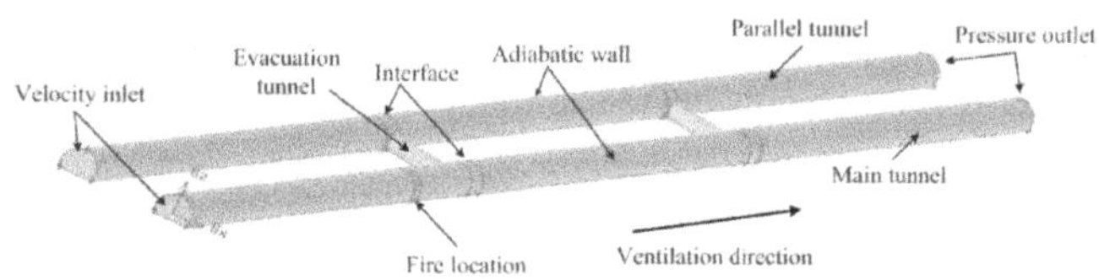

Figure 1. The three-dimensional simpled model of a fire in the tunnel and the boundary conditions

A computational dynamics software Fluent was used to simulate the three-dimensional smoke movement and heat transfer. The high temperature smoke could be treated as a unsteady three-dimensional flow and heat transfer problem, in which the Naves-Stoke equation were involved. The burning process was simulated by species reaction, and the buoyancy effect was taken into account which would have great influences on the flow field structure and species transportation.

The establishing of three-dimensional model of the fire in long tunnel was accomplished by Gambit, so as the grid generation. The boundary conditions of the entrance of the main tunnel and exit of the parallel tunnel were set as velocity inlet, while those of the opposite end were set as pressure outlet. The vaults, side walls and bottom walls of the tunnels were treated as adiabatic wall, while the connected surface of main tunnels and evacuation tunnels were set as interface. Fire was introduced in the tunnel as a source zone, in which 0.1 kg gasoil per second was released into the air. The gasoil was considered that reacted fully with the oxygen in the air, thus only two produced species, water vapor and carbon dioxide, were generated. And the air was supposed to be consisted by 0.23 percent of oxygen and 0.77 percent of nitrogen. The computational mesh and calculated zones were shown in Fig.2, in which upstream region and downstream region of the fire location were defined. Cooper mesh was applied in the model and the total mesh number was about 1,400,000.

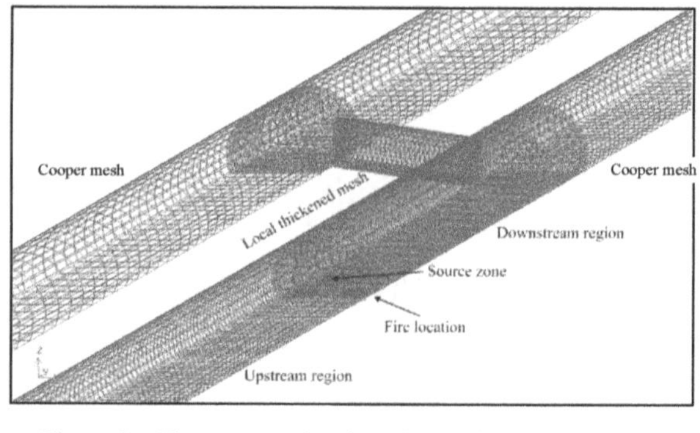

Figure 2. The computational mesh and simulation area

Three ventilation velocities of 0, 1, and 2 m/s were studied in the present work. The evacuation tunnels could be opened and closed with altering relevant boundary conditions. In this paper, only the results of opened evacuation tunnels were presented.

III. SIMULATION RESULTS AND DISCUSSIONS

The computational results of the movement of the smoke, and the temperature distribution of the vault without ventilation were given in Fig.3, while the upper picture indicated the smoke movement and the lower picture denoted temperature distribution. The air pressure near the fire was increased where with the air was heating, which caused the air volume to expand with thermal effect and new combustion productions, then accelerated the smoke to spread in the tunnel. Since the heated air obtained stronger buoyancy force, firstly the smoke rose up to the vault of the tunnel above the fire and leaded to an obvious local high temperature area, then it was concentrated to move to both upstream region and downstream region pulling by pressure gradient. It seemed that more smoke spread to upstream region, which was nearer to the open area, that leaded to severe reverse flow. Constrained by the vault of the tunnel, clear high temperature smoke layer could be observed in the top space of the tunnel, which covered a considerable range in the tunnel. The velocity varied violently in the burning area, while the highest temperature had not appeared in this region but drifted with the generated smoke. The vault of the main tunnel was heated to a considerable high temperature above where fire broke out, which could caused severe damage to the tunnel structure. The temperature of the vault increased with the smoke spreading. In this case, there was almost no smoke entered the parallel tunnel that its vault was little affected. The influence of the smoke in evacuation tunnel was not severe, that was convenient for the rescue action.

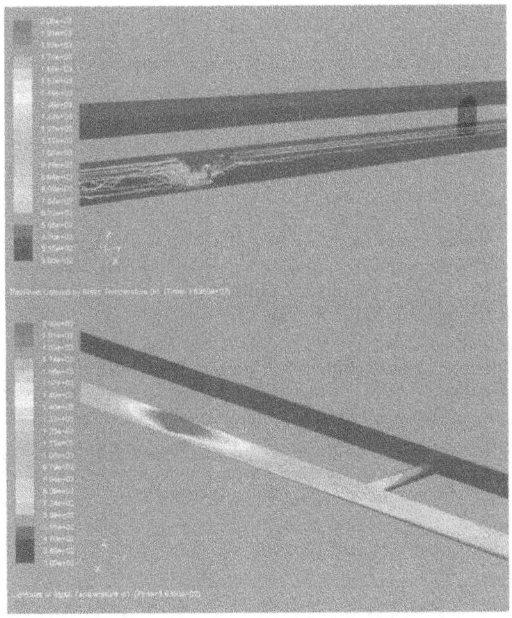

Figure 3. The contours of the smoke movement and temperature distribution of the vault without ventilation

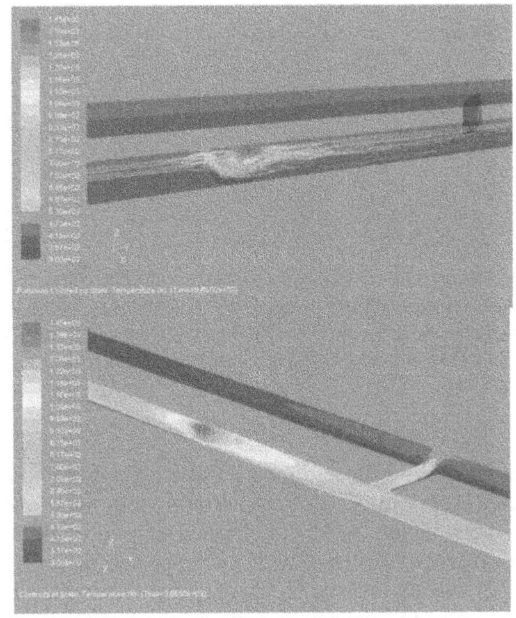

Figure 4. The contours of the smoke movement and temperature distribution of the vault with ventilation velocity of 1 m/s

The movement and temperature distribution of the smoke with the ventilation velocity of 1 m/s was given in Fig.4. Compared with the computational results without ventilation, it could be found clearly that the maximum temperature of the smoke was decreased since more cool air was mixed into the smoke. The concentration region of the high temperature mixtures was drifted to downstream area. It became harder for the smoke to diffuse to upstream region because of the flow resistance brought by the ventilation, while the flow in downstream region was enhanced. Since flowing in a limit space in the tunnel, the smoke impact the vault with the buoyancy effect and was rebounded down, then was lifted up again. That caused the flow in the downstream region became more turbulent. The smoke almost fully occupied the tunnel near the downstream area of the fire, its temperature decreased with the expanding process. It would take long distance for the smoke to form distinctly even temperature layer in the tunnel. The parallel tunnel was soon affected by the fire when the smoke overspread the joint evacuation tunnel. It was clear in the right picture that smoke had poured into the parallel tunnel which vault was heated. The incidence of the smoke extended with ventilation. The ventilation in the parallel tunnel still could not stop the upstream smoke although spreading in a long distance already made it exhausted. The temperature in the evacuation tunnel rose up with more smoke entered.

As shown in Fig.5, the hottest smoke was driven to farther downstream region and the temperature was decreased when the ventilation velocity increased to 2 m/s. The reverse flow of the smoke shrank to a narrow area above the fire location and attended to be eliminated. Coolant air held the upstream area of the tunnel and forced the smoke to be concentrated to downstream region. As the result, the downstream flow became even faster and more turbulent. Thus the smoke would be transport to larger extent and affected distant area in the parallel tunnel. As shown in the right picture, more heat was taken away by the injected air thus the temperature of the vault was declined. But the high temperature core was extended to affect further downstream region. The smoke expanded into the parallel tunnel could not form reverse flow any more. It turned to downstream direction as soon as it entered the tunnel.

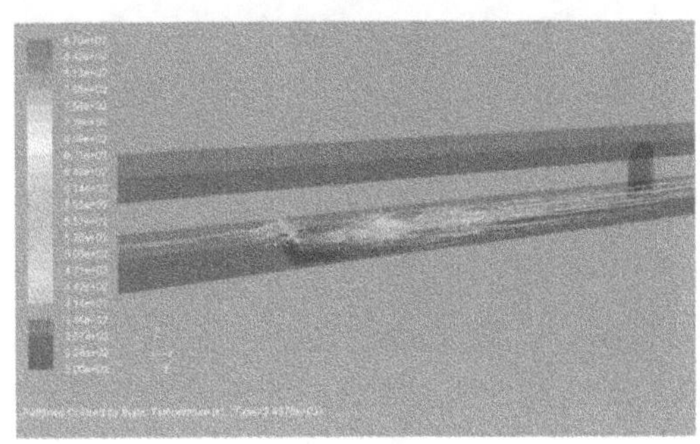

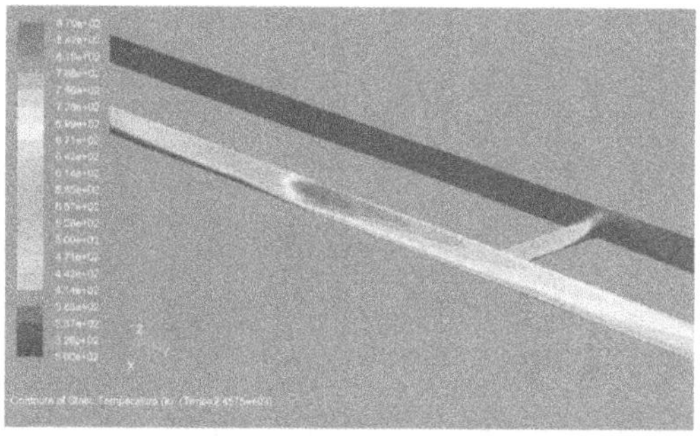

Figure 5. The contours of the smoke movement and temperature distribution of the vault with ventilation velocity of 2 m/s

IV. Conclusion

The simplified three-dimensional tunnel fire model was conducted to optimize the ventilation system and evacuation tunnels arrangement in long tunnels. Detailed computational simulation was performed with different ventilation velocity when the evacuation tunnels were open. The results were summarized as follows:

(1) High temperature appeared in the concentration zone of the smoke and spread with the smoke, which would cause severe damage to the vault and brought critical problem to the rescue action.

(2) Ventilation was helpful to weaken the reverse flow of the smoke and decrease the smoke temperature, as well as to accelerate the smoke to spread in downstream area.

(3) Smoke would bring great troubles in downstream regions with ventilation while upstream regions suffered more without ventilation.

(4) Evacuation tunnels should be closed in the condition of low velocity ventilation, while with high ventilation velocity, it was hard for the smoke to enter the parallel tunnel and the evacuation tunnel could still operated for rescue action.

Acknowledgment

The authors wish to acknowledge support of this study by Shaanxi College of Communication Technology.

References

[1] Elisabeth Krausmann, Fesil Mushtaq, "Learning lessons from tunnel accidents-Recommendations in support of the implementation of Article 15 on Reporting of the EU Directive 2004/54/EC," Safety Science, vol. 48, 2010, pp. 230-237.

[2] Vantelon, J.P., Guelzion, A., Quach, D., Son, D.K., "Investigation of fire-induced smoke movement in tunnels and stations: an application to the Paris Metro," Fire Safety Science-Proceeding of the Third International Symposium, Elsevier, London, 1991, pp. .907-918.

[3] YanfuWang, Juncheng Jiang , DezhiZhu, "Full-scale experiment research and theoretical study for fires in tunnels with roof openings," Fire Safety Journal, vol. 44, 2009, pp. 339-348.

[4] Brandeis, J., Bergman, D.J., "A numerical study of tunnel fires," Combustion of Sci. and Technol., vol. 35, 1983, pp. 133-155.

[5] Charters, D.A., Gray, W.A., McIntosh, A.C., "A computer model to assess fire hazards on tunnels(FASIT)," NFPA-Fire Technology, First Quarter, 1994, pp. 134-154.

[6] Charters, D.A., McIntosh, A.C., "Tunnel fire safety assessment, Proceeding of Asiaflam'95 Conference," March 15-16, Hong Kong , 1995, pp. 77-86.

[7] Chen Xiaojun, "Simulation of temperature and smoke distribution of a tunnel fire based on modifications of multi-layer zone model," Tunnelling and Underground Space Technology, vol. 23, 2009, pp. 75-79.

[8] Chow, W.K., "Simulation of tunnel fires using a zone model," Tunnelling and Underground Space Technology, vol. 11, 1996, pp. 221-236.

[9] Hitoshi Kurioka, Yasushi Okab, Hiroomi Satoha, Osami Sugawa, "Fire properties in near field of square fire source with longitudinal ventilation in tunnels," Fire Safty Journal, vol. 38, 2003, pp. 319-340.

[10] Shorab Jain, Shashi Kumar, Surendra Kumar, Sharma T.P., "Numerical simulation of fire in a tunnel: Comparative study of CFAST and CFX predictions,", Tunnelling and Underground Space Technology, vol. 23, pp. 2008, 160-170.

The Stormwater Use Technology of Caofeidian International Eco-city

Wang Gang
Urban and Rural Planning Bureau of Tangshan
Tangshan, China
Email: luckywg@yahoo.cn

Abstract—**According to the ecological theories, the Caofeidian International eco-city establishes a set of the green technique systems which sustainably manages fresh water, energy, discards etc. and increases recycling and recovery efficiency. The system is an effective way to promote the city to be sustainable. In this text, the author introduces the city of stormwater collection systems in details and the functions of RMC（Resource Manage Center）briefly.**

Keywords-Caofeidian; Eco-City; Stormwater; RMC

I. INTRODUCTION

The Caofeidian is an island of nearby the south inshore area of Tang Shan city. Because of its natural land closest to the international deep water flight; it becomes the new strategy area for national competition and external intercommunion. Along with moving of the Capital Steel to it, the Caofeidian industry area and the building of deep water harbor uncover the new era of the northern heavy industry development of China. At the same time, as its service base, the developments of the Caofeidian International eco-city startups powerfully.

According to the sustainable principles, Caofeidian International Eco-city will explore to solve the questions of resource destroying, energy over-consuming and environment polluting during the urbanization of China. In this text, the author will introduce the Storm water collection technique system of the Caofeidian International Eco-city in detains with the planning principles of Sweden SWECO for a reference.

II. STORM WATER COLLECTION AND GREEN LANDSCAPE

The existing green structure of the Caofeidian area is very poor. There is a lack of vegetation of all kind because of recent human activities, rough climate etc. For the new development of the plan, all new green structures have to be recreated. That means all landscaping have to be manmade in a positive way with an ecological knowledge.

The ecological systems will in best sense be more varied and richer than the existing one. New manmade green areas and parks will increase the biodiversity and the experience in the city.

It is lack of the freshwater and the limited rainfall to produce a restraint with green park and the green land system. The rain fall is very small and has to be stored during the rain seasons and used for irrigation during the dry seasons. To find out a new green structure that manages to adapt to this situation and in the same time be a nice place to live in ought to be carefully studied to fit the estimated amount of rainfall and stored storm water.

Lots of efforts have to be done in ambition to minimize the use of fresh water for irrigation taken from the water supply. Basically the rainfall should be collected and used together with purified grey water for all kind of outdoor needs such as irrigation of parks and trees, flushing the streets, carwash etc. That causes certain demands on coordination of the storm water system and set up a limitation for possible new green areas. Due to the conditions different measures have to be taken.

III. STORM WATER COLLECTION SYSTEM

The overall blue and green structure of the 30 sqkm of Caofeidian International Eco-City is mainly based on new manmade parks and green areas. Only minor green areas exist within the area or in the vicinity. Therefore, the shape of the land surface must be carefully planned. Topological variations in the landscape will not only create exciting city spaces, it will also have an ecological function with regards to water collection.

To design the storm water system it is important to consider the three circles of eco-cycle hydrological-cycle and Waterscape-cycle, and also include the issue of quantity and quality of storm water and the responding recipients: rivers, groundwater, lagoons, sea and soil.

In order to as much as possible to collect and store rain water in public places, Caofeidian International Eco-city will combine wetlands of infiltration, storage and irrigation function with rainwater underground storage and waterscape design. First of all, it is to re-design the ground plane, change the flat state, make the formation of low-lying wetlands in order to collect and store rainwater *"Fig. 1"*. This does create an overall ecological structure together with the Caofeidian landscape systems.

IV. STORM WATER TREATMENT SYSTEM

A. Conceptual Design

To be able to store as much rain as possible from public areas the suggested system combines wetlands for infiltration/storage/irrigation, with storage underground and waterscape. The system is also divided in clean and dirty stormwater that is treated differently.

A flooding system is designed to secure the buildings and important public places. At more extreme rain events the streets will be a part of the public places. At more extreme rain events the streets will be a part of the flooding system.

B. Clean Stormwater (Roof Runoff)

- Stormwater from roofs can be stored at block and district level to use for irrigation of local green areas and other purposes.
- By choosing the right building materials (as for example avoiding unpainted metal surfaces) and chemicals (pesticides, high degree of fertiliser) in the outdoor environment, stormwater from roofs, green areas and less dense housing can be infiltrated to the groundwater in green infiltration areas/wetlands without further treatment.
- The clean stormwater can also go directly to underground storage for percolation to the groundwater. These storages can be made of for example gravel, infiltrations cassettes and concrete basins.

C. Dirty Stormwater

Stormwater from dense city areas and high trafficed roads contain pollutions. Before storage or infiltration a stormwater treatment is needed.

There are two major different ways of proceeding this treatment:

1) *Sedimentation basins underground*
2) *Wetlands/filterstrips with vegetation*

To be able to take out the sediments these basins should be made of concrete. It is possible to combine basins with other treatments for example natural vegetaion filters. See *"Fig. 2"* for example.

D. Wetlands/Filterstrips

Filterstrips can be used for the less dirty stormwater. The filterstrips are built up with material that works as cleaning filters, for example different kind of gravels, natural absorbents like tree product and minerals like chalk to buffer if the rain is acid. See *"Fig. 3"* for examples.

V. RESOURCE MANAGEMENT CENTER OF CAOFEIDIAN ECO-CITY

The city's various eco-cycle sub-systems do not exist in isolation. Only reliance on engineering technology way to solve the problem mechanically, it will greatly enhance the eco-cycle cost of the entire city and also led to inefficiencies. Therefore, from the level of city, this paper proposes a solution of integrated ecological circulatory system.

Reference on Caofeidian International Eco-city experience, the author believes that the city plans to build a unified resource management center will be the rational model of urban ecological cycle.

A. Introduction of RMC

To provide public utilities, such as fresh water, waste water treatment, waste treatment and energy supply and utilization, adequate space has to be allocated. The proposed Resources Management Center includes all the main necessities for a City. The co-location gives important advantages:

- The understanding of the system synergies and objectives is enabled
- The different systems can be optimized to enable operational and staff synergies
- The co-location visualizes and communicates the synergies also fornon-specialists

B. Location of the RMC

The following criteria are essential when choosing the location of the Resources Management Center:

- Short distance to agricultural areas – for reuse of liquid and solid plant nutrients, irrigation.The treated food waste, black water and grey water contain too much plant nutrients to use in the city parks. A pipe-line is planned to pump the treated nutrient water from the RMC to storages upstream to rice fields.
- RMC can cause risk of dust, odour, noise, micro organisms disturbances, hence the prevalent wind direction is important.A green zone-safety zone around the RMC area is necessary. It is beneficiary not to enclose the RMC with housing areas.

C. Concept of RMC

Resource Management Center is to integrate the city's energy use, resource use, pollution, processing into an overall which handles resource management systematically. In particular, the resource management center is to establish a unified system, the system includes eco-cities recycling facility of waste water treatment, biogas use, waste incineration, combined heat and power supply, water reuse, waste recycling etc at the same location. The city's ecological cycle is holistically managed by the resource management center; Resource Management Center is to provide fresh water energy, pollutant treatment and other routine public services for the entire city.

Resource Management Center describes the contact and interaction between reduction of energy, water ,waste etc in demand with providing high-quality of them .It is not only conducive to coordination between the various facilities to achieve efficient use but also will play the role of visual demonstration.

VI. SUMMARIES

In order to create a city of "low energy consumption, low pollution, low-carbon emission", Caofeidian International Eco-city adopts thoroughly domestic and foreign advanced technology and management concepts of clean energy, water resources, comprehensive utilization, and waste separation. Particularly the use of rainwater harvesting achieves more eye-catching achievements and puts forward creatively concept of resource management center, so it has a certain amount of promotional value.

References

[1] SWECO, Conceptual Plan for 30 sqkm of Caofeidian International Eco-City, 2008, unpublished.

[2] SWECO, Sustainability Guidelines of Caofeidian International Eco-City, 2008, unpublished.

[3] SEWCO, Preliminary Design of Caofeidian Sustainability Center,2008,unpublished.

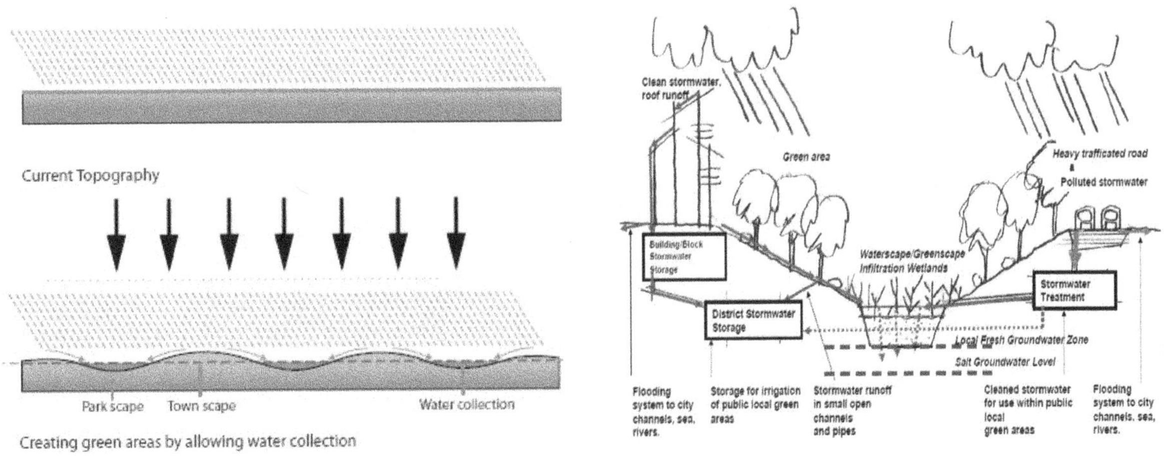

Figure 1. Shaping the surface of Caofeidian International Eco-city

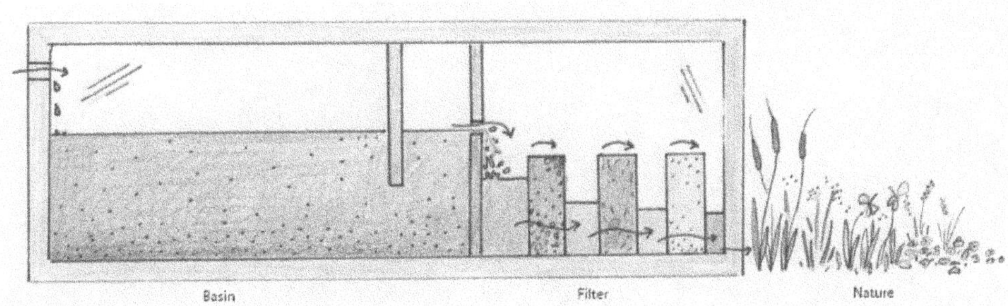

Figure 2. An example of a sedimentation basin.

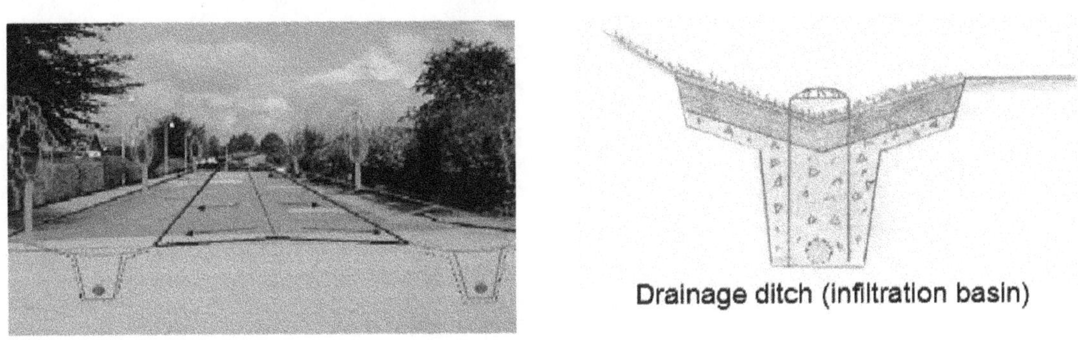

Figure 3. An example of wetland and drainage ditch.

Theoretical Analysis and Application of Pile Shaft Resistance Strengthening Effect

Hu Dehua
School of Environmental and Civil Engineering
Wuhan Institute of Technology
Wuhan, China
Email: piglet1@163.com

Hu Guoxiang
School of Environmental and Civil Engineering
Wuhan Institute of Technology
Wuhan, China
Email: hgxcug@126.com

Abstract*—Based on the view of distribution laws of internal forces of pile and internal forces keep in a balance between rock-socketed pile and pile rock interface, pile shaft resistance strengthening effect — rock-socketed pile shaft resistance increased with the increasing of pile tip rock uniaxial compressive strength or elastic modulus and load at pile top was analyzed; mathematical expression of quantitative indicators of pile shaft resistance strengthening effect was established, including pile shaft resistance strengthening coefficient to pile shaft rock internal friction angel, uniaxial compressive strength or elastic modulus and load at pile top; The concept of the expression was clear, easy to use and consistent with the relevant experimental results. With engineering examples, we proposed that rock-socketed pile shaft resistance strengthening effect should be considered in calculating ultimate bearing capacity of single pile to reduce the depth of rock-socketed and save construction investment during the period of designing.

Keywords-pile shaft resistance; the strengthening effect; theoretical analysis; application research

I. INTRODUCTION

Pile shaft resistance strengthening effect is that the pile shaft friction resistance increases with the increasing of depth, especially at pile tip of rock-socketed piles, as the tip resistance play, pile shaft resistance will increase both with the increasing of uniaxial compressive strength of rock and the load at pile top[1, 2]. In recent years, has many experts and scholars have obtained a number of valuable conclusions on the phenomenon of pile shaft resistance strengthen by model tests and engineering analysis of measured data in China. For example, the phenomenon of pile shaft resistance strengthen increases with the increasing of rock strength at pile tip, and decreases with the declining of rock strength at pile tip, especially when the length of pile is small and thick weak layer is exited at the tip of pile (such as unclean hole with thick sediment), the bearing capacity test value will be a sharp decrease, less than 1/3 of normal value[3]; moreover, the pile shaft resistance is also a very uneven distribution of strengthening segment, the frictional resistance at the place to enter rock is much larger than at the bottom[4]. However, the conclusions above are mostly works from the experimental results or engineering measured data, the mathematical relationship between pile shaft resistance strengthening effect and the rock strength around the pile the load at pile top is not very clear, therefore, the pile shaft resistance strengthening effect theory is still imperfect essentially.

II. PILE FORCE ANALYSIS

As shown in Fig.1, under the action of the load P_0 at pile top, the axial force N at any depth z of pile is that the load at pile top minus the total friction above point **z**, that is:

$$N = P_0 - \pi d \sum h_i q_{sik} \tag{1}$$

And cross-section stress as follows:

$$\sigma_{(z)} = N / A_p \tag{2}$$

Where: P_0 —load at pile top, h_i —thickness of the i-layer soil, q_{sik} — pile shaft friction resistance of the i-layer soil, d —diameter of the pile, A_p —cross-section area of the pile.

We can obtain the conclusion from (1) and (2) that the axial force and cross-section stress of pile decrease with the increasing of length. The axial force and stress at top surface of pile are the largest, the smallest at the tip of pile.

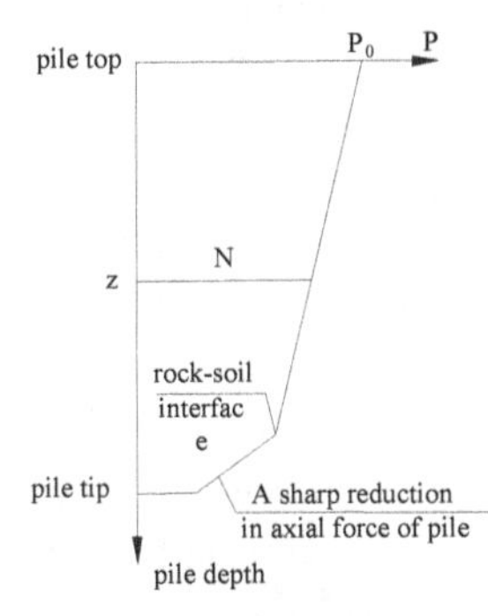

Figure 1 Pile axial force diagram

III. ANALYSIS OF PILE SHAFT RESISTANCE STRENGTHENING

In order to facilitate the visual analysis of the phenomenon of pile shaft resistance strengthening, and to simplify the secondary factors, it is assumed that:

- The concrete materials of both pile and rock are homogeneous, isotropic elastomers, elastic modulus are E_r, E_p respectively;
- Poisson's ratio is the same, namely $\mu = 0.2$, lateral pressure coefficients: $\xi = \frac{\mu}{1-\mu} = 0.25$.

As shown in Fig.2, when the pile axial force at the rock interface $N=0$, the pile shaft confining pressure is subjected to $\sigma_{03}=\xi_1 q=\xi_1\gamma_0 h$, where $\xi_1=\frac{\mu_1}{1-\mu_1}$ is soil pressure coefficient, μ_1 is Poisson's ratio of soil above the rock stratum, γ_0 is weighted-gravity, h is the depth of rock.

When $N>0$, the pile will produce vertical compression strain ε_v and axial displacement s. Now we will illustrate the impact on the pile shaft resistance of compression strain and axial displacement.

A. The Impact of Vertical Compression Strain ε_v on Pile Shaft Resistance

When vertical compression strain ε_v of pile occurs, expansion strain ε_p will occur in the radial direction, and it will generate lateral pressure around the pile, and then make compressive strain ε_r emerge.

Assumed that the radial strain didn't occur, the initial lateral expansion stress is σ_h, once expansion occurs, σ_h begins to decrease, and the rock stress around the pile σ_r begins to increase.

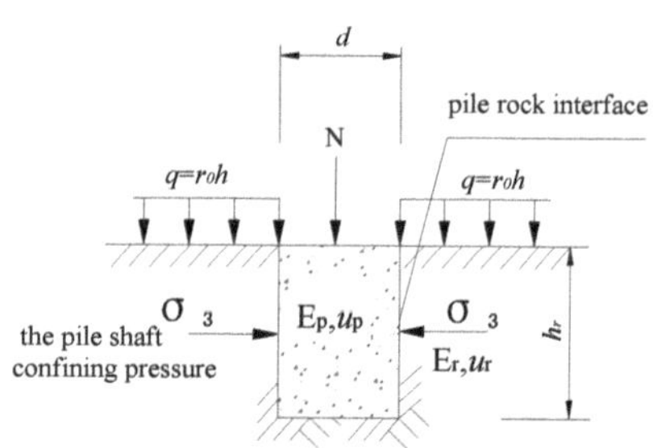

Figure 2 The pile shaft stress of rock-socketed part

However, if the radial expansion strain ε_p occurred, the decreased value of σ_h was $\Delta\sigma_h$, then $\Delta\sigma_h=E_p\varepsilon_p$, when the pile keeps in balance, rock pile the actual interface normal stress as follows:

$$\Delta\sigma=\sigma_h-E_p\varepsilon_p=E_r\varepsilon_r \tag{3}$$

When the radial expansion deformation is not large, we can suppose that the expansion strain is equal to compressive strain, that is $\varepsilon_p=\varepsilon_r=\varepsilon$, we can get (4) from (3):

$$\varepsilon=\frac{\sigma_h}{E_r+E_p} \tag{4}$$

Let $E_r=E_p$, we can obtain the strain at the homogeneous situation: $\varepsilon=\xi\varepsilon_v=0.25\varepsilon_v=\frac{0.25\sigma_{(z)}}{E_p}$, then substitutes in (4):

$$\sigma_h=0.5N/A_p \tag{5}$$

So we can get the pile shaft additional confining pressure the rock generates is:

$$\Delta\sigma=\frac{0.5E_r}{E_p+E_r}\times\frac{N}{A_p} \tag{6a}$$

We suppose that ε_r is the strain when the rock reaches to uniaxial compressive strength σ_c, so $E_r=\frac{\sigma_c}{\varepsilon_r}$, then E_r is substituted in (6a):

$$\Delta\sigma=\frac{0.5\sigma_c}{E_p\varepsilon_r+\sigma_c}\times\frac{N}{A_p} \tag{6b}$$

The formulas above indicate that:

- At the same loads, pile shaft additional confining pressure increases with the increasing of elastic modulus and uniaxial compressive strength; and vice versa, especially when thick mud sediment exits or cavity at the bottom of pile, $E_r\to 0$, $\sigma_c\to 0$, then $\Delta\sigma\to 0$, pile shaft resistance decreases together with the sharp declining of pile tip resistance. Some experiments have proved that.
- Pile shaft additional confining pressure $\Delta\sigma$ and the pile axial force are proportional, it means that the larger the axial force is, the larger the pile additional confining pressure is and more adequate the lateral resistance plays;
- As the pile shaft resistance to strengthen, rock-socketed pile section stress $\sigma(z)$ decreases sharply and reached the end of pile, the impact of pile stress resistance on the pile shaft strengthening effect can be basically ignored, so the shaft friction resistance strengthening coefficient at the part of entering into the rock should be considered average.

B. The Impact of Axial Displacement s on Pile Shaft Resistance

Pile hole and pile are analyzed separately, as shown in Fig.3, if the pile produces axial displacement s, "bending" deformation will occur in the rocks around pile, the rock at the edge of the pile hole will squeeze into the hole, which can be seen from Fig.3:

- The deformation trend is related to the depth of pile into the bearing stratum, the larger the depth is, and the greater the deformation trend is. As the pile constraints exist, the depth into the bearing stratum is greater, pile lateral confining pressure is larger, and the increase range of pile shaft resistance is larger. The distanc

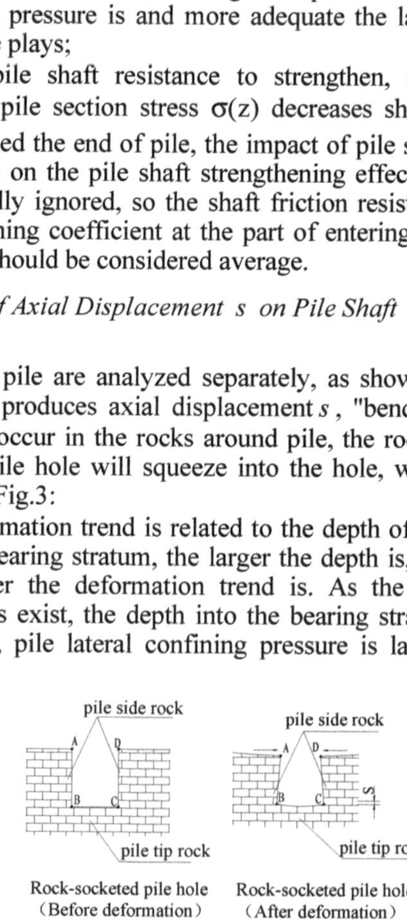

Figure 3 Diagram of rock-socketed pile hole deformation

e increases with the increasing of the depth h_r of the pile tip into rock at the place pile shaft resistance significantly enlarged (as points A and D in Fig.3), pile shaft resistance is almost no increase at points B and C near the pile tip.

- The larger the load at pile top is, the larger the axial displacement resulted is, the greater the pile deformation tend is and therefore the greater the pile shaft resistance is.
- When a thick pile of mud sediment exits, due to the low intensity of mud sediment, large deformation (more than pile settlement limit 40 mm), so the above-mentioned deflection trend will not occur before the mud sediment is not pressed compact, and objectively it demonstrates the pile shaft resistance greatly reduces with the decreasing of pile tip resistance.

IV. Theoretical Analysis of Pile Shaft Resistance Strengthening Effect

In order to facilitate quantitative analysis, the quantitative indicators of pile shaft resistance strengthening effect is defined as the pile shaft resistance strengthening coefficient ξ_k: the ratio of the actual value of rock-socketed pile shaft resistance and pile shaft resistance calculated value no considering the rock characteristics and pile load. When we design, the pile shaft resistance is Q_{rs1} according to the limit standard value supported by engineering geological investigation, measured pile shaft resistance is Q_{rs2}, then $\xi_k = Q_{rs2} / Q_{rs1}$. Now ξ_k would be analyzed from two aspects of pile compression strain and axial displacement respectively.

A. *The Impact of pile compression strain on pile shaft resistance strengthening coefficient* ξ_k

At the interface of rock and pile, pile shaft additional confining pressure is generated by pile vertical compression strain ε_v at the part of rock-socketed, the shaft pressure generated by casing layer is σ_{03}, so the total confining pressure at the place of entering the rock is $\sigma_3 = \sigma_{03} + \Delta\sigma$, due to the strengthening of pile shaft resistance, pile axial force shows the rapid decay at the place of entering the rock, until to the bottom of pile, the impact of pile axial force on pile shaft resistance strengthening effect can be ignored, so the average of pile shaft additional confining pressure at the place of entering the rock is that:

$$\sigma_3 = \sigma_{03} + 0.5\Delta\sigma \qquad (7)$$

We make $\Delta\sigma / \sigma_{03} = \beta$, so the average growth coefficient of pile shaft additional confining pressure is $\lambda_1 = \sigma_3 / \sigma_{03}$, then (6) is substituted in it, we can obtain the average growth coefficient:

$$\lambda_1 = 1 + 0.5\beta \qquad (8)$$

B. *The impact of pile axial displacement* s *on pile shaft resistance strengthening coefficient* ξ_k

As can be seen from Fig.3, due to the axial displacement, passive lateral pressure is applied on the rock around pile, when it reaches equilibrium, according to earth pressure theory, pile shaft resistance strengthening coefficient is passive earth pressure coefficient $\kappa = \tan(45 + \phi/2)$, where ϕ is related with internal friction angle of rock, the value of medium and micro weathering rock can be 35~50°[5].

At the tip of pile, pile shaft resistance is nearly no increase, so the average the average growth coefficient of lateral pressure is as follows:

$$\lambda_2 = (1 + \kappa)/2 \qquad (9a)$$

Then κ is substituted in (9a), we can obtain the impact of axial displacement on pile shaft resistance growth by simplified:

$$\lambda_2 = \frac{1}{1 - \sin\phi} \qquad (9b)$$

Therefore, pile shaft resistance strengthening coefficient ξ_k is that:

$$\xi_k = \lambda_1 \lambda_2 \quad \text{or} \quad \xi_k = \frac{1 + 0.5\beta}{1 - \sin\phi} \qquad (10)$$

It is easy to see that with the increase of internal friction angle of rock, pile top load, elastic modulus of rock, pile shaft resistance strengthening effect is also growing.

Although the Formulas are based on the rock, but they have been largely applicable for sand, clay and other soil, just change the "rock-socketed part" to "into the bearing stratum part".

For example, if the bearing stratum of pile tip is sand, silt layer and so on, internal friction angle $\phi = 30°$, $\lambda_1 = 2.0$, $\lambda_2 = 1.0$, $\xi_k = 2.0$. The result is very similar to test pile of a power station in Changshu, Jiangsu Province.

V. Example of Engineering Practice

A. *Comprehensive Analysis of Ultimate Bearing Capacity of Single Pile*

Ultimate bearing capacity of single pile is controlled by two factors: the pile material and strata bearing capacity. Once the pile quality can be ensured, the pile ultimate bearing capacity of the material can be determined; the other hand, the bearing capacity of subsoil on the pile depends on the pile shaft resistance and tip resistance. According to the analysis conclusion of pile shaft resistance strengthening effect, ultimate bearing capacity of rock-socketed single pile can be determined as follows:

- When rock strength of tip pile is greater than the concrete strength of the pile, if pile cross-section stress at the tip of pile reaches f_c, we can know by (1) and pile shaft resistance strengthening phenomenon that cross-section stress must be greater than f_c in the interface of rock-socketed pile, and therefore some conclusions can be determined, ultimate bearing capacity of single pile is controlled by pile material strength when the depth of pile into the rock is not less than 0.5 d, that is to say, ultimate bearing capacity of single pile is the ultimate bearing capacity of pile material.

- When the rock strength at the tips is less than concrete strength of the pile itself: ultimate bearing capacity of single pile also is calculated based upon the specifications, but pile shaft resistance strengthening effect should be considered at the part of entering into the rock, finally bearing capacity of single pile should be reviewed by the pile strength.
- The sediment at the bottom of pile should be cleared to ensure stable and reliable bearing capacity of rock-socketed pile during the construction period, if difficult exits indeed; post-grouting procedure should be taken at the bottom of pile to cure the sediment.

B. Example of Engineering Practice

An inpatient building is a 16-story high-rise building; the maximum column axial force is 12MN. Land surface is miscellaneous filling, the entire site is distributed, and the thickness is 2.2~4.0 meters. The other layer and pile foundation design parameters follow as Table Ⅰ: saturated uniaxial compressive strength of rock $f_{rk} = 6.58\text{MPa}$, elastic modulus $E_r = 0.48\text{GPa}$, $\phi = 39°$, using bored piles, the concrete strength grade was C30, $E_p = 30\text{GPa}$, $f_c = 14.3\text{MPa}$, the diameter of pile is 800 millimeters, the length of pile is 14 meters, the depth of medium and micro weathering rock embedded is 3 meters, the standard value of ultimate bearing capacity is 5400 kN. During the period of construction, it was extremely difficult to drill into the medium and micro weathering rock, construction party presented the views of reducing the depth of rock-socketed to the pile foundation. Originally they planned to expand the diameter of pile to reduce the embedding depth, but it related bearing platform adjustment and other related issues, therefore further analysis was done by pile shaft resistance strengthening effect:

TABLE Ⅰ PILE FOUNDATION DESIGNING PARAMETERS

No. of Layer	Name of Soil Layer	Depth of Soil Layer	q_{sik}	f_{rk}
(2)	Silty clay	1.4~3.5	60	
(3)	Silty clay	1.9~4.6	49	
(4)	Round gravel	3.0~4.4	130	
(5-1)	Strong weathering Silty mudstone	0.5~3.7	140	
(5-2)	Medium and micro weatheringsilty mudstone	>3.8	240	6580

1) The Impact of pile compression strain on pile shaft resistance strengthening coefficient ξ_k: the natural depth of rock was 16 m, soil weighted-gravity, $\gamma_0 = 19\text{kN/m}^3$, and rock self-weight stress $\sigma_z = 16\times19 = 304\text{KPa}$.

The medium and micro weathering argillaceous rock interface soil: $\mu_1 = 0.30$, $\xi_1 = \dfrac{\mu_1}{1-\mu_1} = 0.4286$,

$\sigma_{03} = 0.4286\times304 \approx 0.130\text{MPa}$, the load at pile top $P_0 = 5400\text{KPa}$, the axial force of pile and cross-section stress in the interface of rock and soil:

$$N = 5400 - 0.8\times3.14\times(38\times3.0 + 60\times3.0 + 49\times3.0 + 130\times3.0 + 140\times3.0) \approx 2961\text{KPa};$$

$$\sigma_{(z)} = N / A_p \approx 5.88MPa;$$

$$\Delta\sigma = \frac{0.5E_r\sigma_{(z)}}{E_r + E_p} = \frac{0.5\times0.48\times5.88}{30+0.48} \approx 0.046\text{MPa};$$

$$\beta = \Delta\sigma / \sigma_{03} = 0.046/0.130 \approx 0.35;$$

$$\lambda_1 = 1 + 0.5\beta \approx 1.18.$$

2) The impact of pile axial displacement s on pile shaft resistance strengthening coefficient ξ_k: the friction angle of the medium and micro weathering rock, according to (7b):

$$\lambda_2 = \frac{1}{1-\sin\phi} = \frac{1}{1-\sin 39°} \approx 2.70$$

Pile shaft resistance strengthening coefficient: $\xi_k = \lambda_1\lambda_2 = 1.18\times2.70 \approx 3.19$

3) Ultimate bearing capacity of single rock-socketed pile:

We changed the depth of rock-socketed into 0.8 m, $\xi_p = 0.73$ [6], the total limit resistance was:

$$Q_{uk} = Q_{sk} + Q_{rk}$$

$$Q_{sk} = \mu\sum q_{sik} l_i = 0.8\times3.14\times(3\times49 + 3\times130 + 1\times140) = 1700(\text{kN})$$

$$Q_{rk} = Q_{rs} + Q_{rp} = \xi_k q_{sik}\pi dh + \xi_p f_{rk}\frac{\pi}{4}d^2 = 3.19\times240\times3.14\times0.8\times0.8 + 0.73\times6580\times3.14\times0.8^2/4 \approx 3952(\text{kN})$$

$$Q_{uk} = Q_{sk} + Q_{rk} = 3952 + 1700 = 5652(kN) > 5400(kN).$$

Bearing force of pile: Encryption spiral stirrups at the top of pile $\phi8@100$, steel bar $HRB335$, $8\Phi18(A_s = 2036mm^2)$, $\psi_c = 0.7$,

$$N = \psi_c \times f_c A_p + 0.9 f_y' A_s' = 0.7\times14.3\times502655 + 0.9\times300\times2036 \approx 5581(kN) > 5400(kN)$$

The results of test pile was shown in Table Ⅱ, (the actual depth of rock-socketed was 1.2m, due to Construction Party). Because destructive tests hadn't been done, we could judge according to Table Ⅱ, the standard value of ultimate bearing capacity of single pile was larger than 5400 kN, which could meet the requirements of designing.

From the results above and (6) we have known when the elastic modulus of pile is much larger than the rock around the pile, pile shaft resistance strengthening coefficient mainly depends on axial displacement of pile. When $E_r \geq E_p$, that is to say, pile shaft resistance strengthening coefficient is mainly controlled by compression strain of the pile.

TABLE II STATIC LOAD TEST RESULTS OF PROJECT PILE

No. of Test Pile	The Diameter of Pile (mm)	Depth of Rock-socketed (m)	The Maximum Test Load(kN)	
			Load	*Settlement (mm)*
1	800	1.5	5400	8.441
2	800	2.2	5400	20.04
3	800	3.2	5400	10.38
4	800	1.2	5400	10.53

VI. CONCLUSION

The phenomenon of pile shaft resistance strengthening is the inherent characteristic of pile tip resistance, whether it is concrete grouting pile, the preset static pile, or DJM[7, 8], field experiments and model tests have proven this point. But today, the literature referring to the theoretical analysis of pile shaft resistance strengthening effect is still rare. We carried out preliminary exploration based on certain assumptions, and described in detail combing engineering examples, although there may be some imperfections, but various factors to generate pile shaft resistance strengthening effect have obtained an objective and comprehensive quantitative reflection, physical and mechanical properties of the rock around pile (including elastic modulus, unconfined compressive strength, Poisson's ratio, internal friction angle, etc.), the material properties of pile (including pile elastic modulus, Poisson's ratio, interface dimensions, etc.), the thickness of covered soil at the rock layer of pile tip, gravity and load at the top of pile. Therefore, we still hope a broader and deeper concern among the colleagues and scholars, so as to try our best to promote the level of existing pile foundation design.

REFERENCES

[1] Dong Jinrong. Enhanced and Weakened Effect of Skin Friction of Cast-in-situ piles. Chinese Journal of Geotechnical Engineering, vol.31, no.5, 2009, pp, 658-662.

[2] Lei Xiaozang, He Siming. Study on Ultimate Resistance of Rock-socket Pile. Journal of SiChuan University (Engineering science edition), vol.37, no.4, 2005, pp, 7-10.

[3] Liu Liming. Theory Progress and Engineering Practice of Pile Foundation Engineering. China Building Material Industry Publishing House. 2003, pp, 62-64, 67-72, 211-212.

[4] Wang Yaohui， Tham L G, etc. Test and Numerical Analysis of Model Rock-socketed Pile. Chinese Journal of Rock Mechanics and Engineering, vol.8, no.26, 2007, pp, 1691-1697.

[5] Editor: Huang Xingchun, Associate Editor: Tao Jinlian. Rock Mechanics. Higher Education Press, 2005, pp, 32-39.

[6] Technical Code for Building Pile Foundations JGJ94–2008. China Architecture & Building Press, 2008.

[7] Du Haijin, Ye Hongdong, etc. Analysis on Strengthening Effect of Skin Friction of Deep Jetting Mixing Pile. Chinese Journal of Rock Mechanics and Engineering, vol.23, no.32, 2004, pp, 3775-3779.

[8] Zhou Hongbo, Chen Zhuchang. Abnormal Effect of Skin Friction of Bored Pile in Shanghai. Chinese Journal of Geotechnical Engineering, vol.29, no.2, 2007, pp, 174-179.

University Construction of Sustainable Development

Tu Yuansheng
Academy of Fine Arts
Hubei Normal University
Huangshi, China
Email: toysun777@yahoo.com.cn

***Abstract*—Since the late twentieth century, university education has been vigorously developed in China, with large-scale constructions in succession on both old and new campuses. In order to ensure the sustainable development of the university construction, we are committed to adopting regenerated or environment-friendly materials of energy conservation, to constructing on the basis of green buildings and ecological environment, to rationally using water resources and energy, and to conducting scientific operations and management. Among them, campus planning and architectural design, which are vital factors in university construction of sustainable development, play particular important roles.**

Keywords-sustainable development; campus planning; college and university construction; design; association space

I. INTRODUCTION

Since China's acceleration of the development of higher education at the end of the twentieth century, universities and colleges have expanded enrollment in a large scale, causing the total size of higher education more than doubled in three years. In 2009, the number of the students enrolled has reached 27 million. Following this is a large-scale construction of old and new campuses in universities and colleges, which is not only a huge construction project but also a space protection to accommodate to the new model of modern teaching, and which also has a direct impact on the city's landscape and economic and spiritual life.

Large-scale construction, consuming large quantities of resources and energy, will inevitably produce an effect on the natural environment. Among China's various industries, the influence of the construction industry on the climate occupies more than 12%. Therefore, the construction industry bears a very great responsibility for the sustainable development of human society. The sustainable development of colleges and universities building, which is an important part of the whole construction industry, is reflected on practicability, durability, environmental protection, beautifulness, and harmony. In addition, it may also influence potentially teachers and students' views on scientific development. Here, let's first learn about the concept of sustainable construction.

II. THE CONCEPT OF SUSTAINABLE CONTSTRUCTION

Sustainable Construction, also known as the Green Construction, derived from the concept of Sustainable Development, which was formally proposed in the research report *Our Common Future* in 1987 by the World Commission on Environment and Development. In the First International Conference on Sustainable Building held in Florida, the United State in 1994, Dr. Charlse Kibert put forward the concept of sustainable construction: "follow the principle of effective using resources on the base of ecology so as to establish a healthy building environment."

The concept of sustainable architecture can be interpreted as the following: in the activities of the construction industry (including the site selection, planning, designing, construction, operation, alteration and demolition of various types of buildings and the environmental field, as well as building materials, components and the production, transportation, installation of equipments, etc.), the current needs as well as the future needs should be considered; contemporary people's interests should not be met at the expense of future generations'; nor should the local interests be met at the expense of others'; not only should the purposes of economic development and improving people's lives be achieved, but also its negative impact on the environment should be kept smallest.

III. UNIVERSITY PLANNING AND ARCHITECTURAL DESIGN AS AN IMPORTANT ELEMENT OF SUSTAINABLE DEVELOPMENT

Universities, places of transmitting and creating knowledge, lead the development of human civilization and social technology, and meanwhile are important places and institutions to cultivate talents and serve the society. Therefore, university planning, architectural design and environmental landscape design should be adapted to the needs of national and regional development, and on the principle of promoting the effective conduction of teaching, academic research and exchanges with foreign countries, combined with teachers and students' psychological characteristics and the basic law of their behavior, they also should always reflect humanistic feelings on campus, the green concept of caring and educational idea, thus truly serving cities and communities to achieve the harmonious co-construction of universities and cities.

A. *Campus Planning*

With the "green campus" philosophy and "people-oriented" purposes, make rational use of land resources, and optimize the building density and cubage rate; divide the functional areas of universities according to local conditions; develop reasonable and effective campus transportation system, and create a comfortable and beautiful campus environment, so as to reflect the features of independently-based research and open communication. We should fully respect the existing topography, and conduct sustainable development and utilization to achieve a harmonious blend of man and nature, of architecture and nature.

To respond to the idea of colleges and universities serving the city and the community, we need to do a good job in planning and designing public resources-based architecture and open space. For buildings like libraries, gymnasiums, stadiums, and applied research venues, characteristics of sharing with the city could be considered. Hence, develop the appropriate construction scale and establish effective links of urban transport system, thus reducing the sense of distance between societies and universities, and between the public and universities so as to create an intensive system of urban colleges and universities.

Campus planning and construction should not simply be several plans offered by designers, nor should it be determined by administrative leaders' shaking or nodding heads. It should take into consideration the rights of the whole faculty and students, for it is only teachers and students as the terminals who know very clearly about the campus teaching and living function.

"Planning", a rigorous and meticulous work for the benefits of generations and generations, is the base of sustainable construction.

B. *Architecture Design*

It's necessary to design on the primary principle of use function. Under the premise of fully meeting the use function, arrange reasonably the building layout, space height, structure selection, landscape layout, architectural decoration, equipment installation and so on..

Generally speaking, teaching buildings higher than seven layers are not advisable, because high-rise buildings increase the building cost and resource consumption, hence, prone to waste. Construction material should be economical and practical. A rich and harmonious architectural landscape can be created through good space design, appropriate building form and changeable monomer mix, such as (1) changes of different unit combinations; (2) changes of different unit forms; (3) increasing the scale of humanization and enriching the visual image; (4) increasing small components and creating a cordial and pleasant teaching place.

As regard to architectural decorations, there is no need for the large area use of expensive advanced materials on the walls and the ground. Rational and changeable use of common building materials can create good results as well. Of course, everything should be based on durability, practicability and beautifulness. Laboratories generally do not need drop ceiling (unless there are special requirements). All kinds of pipelines can be laid under the floor or along the corners in different colors for easy identification and maintenance. For the building construction, actively adopt environment-friendly wall materials; use less or no clay to protect arable land; reasonably select wall thermal insulation materials to reduce heat conduction losses. Windows and doors should be suitably arranged to ensure natural ventilation and lighting, thus saving energy consumption, and closing the distance between man and nature.

Humanized association space within the teaching building brings the building breath of life, making it serve people in the building in a long term, which is helpful to teachers and students' physical and mental health, and which may have the subtle influence on teachers and students to promote their concept of scientific development in a healthy way. Creating a good association space inside the overall teaching has become one of the central links in the design of teaching architecture. Association space, which can be said to be the new functional requirements on teaching buildings proposed by modern higher education, indicates sustainable relations between the construction and human.

The association space of colleges and universities refers to the appropriate space environment for free exchanges, academic discussions and group activities between teachers and students, between students and between teachers.

The social associations within teaching buildings are mainly between teachers and students, between students, between teachers, and interdisciplinary exchanges. Apart from some highly specialized interior space, such as classrooms, study rooms, laboratories, and offices, other public spaces are all areas for mutual exchanges between teachers and students (especially between students).

Due to the diversity of Students' learning and activities, their association space also has the characteristics of multi-function and diversity. In accordance with the number of people for services, functional requirements and privacy features, relatively independent spaces of different size can be arranged, reflecting relative territory and privacy but not being fully closed. Walls and columns are made the transition and division of space through changes in layout; relatively independent divisions are made through a variety of partition (such as decorative molding, panels, plants, railings or chairs enclosure), which can keep comfortable contacts with channels and halls. Another is the completely open space, which can trigger more interaction and communication. The scale of the association space can be reasonably arranged on the base of the psychological demands of student groups.

The distribution of association space should be reasonably arranged with the size and layout of the teaching building fully integrated. Its interspersed use in the construction plane can also enrich the spatial modeling, creating flexible building facade and interior space.

1) *Corridors and stairwells*

Corridor is the main level communication space of the construction, as well as the place where students stays most often in between classes. With people to and fro, it is possible for large amounts of communication to occur in the corridor. Therefore, the corridor already bears composite functions. First of all, it is necessary to ensure adequate area for transportation in the corridor. At the same time, totally or partially extend its space to provide relatively independent sites for interaction, thus creating more platforms for learning and exchange.

As the vertical channel, stairwells, which is the most crowded area, also provide more opportunities for exchanges, so expanding the rest platform, arranging additional semi-open space next to the stairs, and adjusting the shape of the rest platform, and so on can create a relatively independent association space

2) Independent platform for communication

Independent association space can be arranged between classrooms or at the end of the corridor, with one or more than one on each floor, creating an association space which is more intimate with nature and more relatively independent.

Additional association space can also be arranged at the interchange between disciplines to promote interdisciplinary exchanges.

3) The lobby, atrium and roof terraces as available ideal places for association space

Lobby is the transportation portal of the inside and outside space of a building. With large flows of people, it is the most noticeable place for open communication. Therefore, besides ensuring necessary area for transportation, arranging certain lecture space, display space, interactive space, etc. can greatly increase the possible factors activating the academic atmosphere and academic thinking.

The service area or the display area can be set in atrium with some casual tables and chairs, creating more communication convenience.

The roof platform with safe protection facilities can bring students more association space to contact with nature by the additional equipment of corresponding out-door desks and chairs and adequate landscape elements

The association space offers limited place, but brings us an open space and flying ideas. If more interactive facilities (such as an automatic book borrow machine — equivalent to an additional small library to the teaching building, the human-computer interactive display wall, the network information room, etc.) are equipped, the limited association space will be extended to a greater degree.

IV. Other Elements of Sustainable Construction

The scientific layout of pipe network, the rational use of water resources and energy, the conduction of scientific operation and management, the utilization of regenerated or environment-friendly materials of energy conservation, the construction based on ecological construction and ecological environment, and the rational use of underground space are all effective insurance for construction of sustainable development.

Optimize water supply and drainage system. According to different functions of water using, adopt the technology and equipments of frequency constant pressure water supply and water treatment. Meanwhile, using water-saving devices can not only ensure the effectiveness, but also save operating costs. Natural lakes or man-made lakes and pools can be used to collect rainwater to meet the daily needs of non-potable water for greening, landscape, etc.. It can also affect the microclimate within the campus as well

For the integrated services and the management needs of the campus, a complete digital communication network system can be set up. Via the intelligent central management system, the remote monitoring on terminals and information exchanges can be enhanced to effectively strengthen interrelation, coordination and unity between executive departments and teaching and research units in universities and colleges.

In campus construction, we should choose green building materials, promote the utilization of new recycling building materials, and select renewable materials and products according to local conditions, so as to reduce material and energy consumption and environmental pollution as well.

Rational use of underground space for the construction of car parks, canteens, activity centers and other public venues can save valuable land resources, and can increase the space levels and the landscape modeling of the campus.

V. Summary

Although the sustainable development of university buildings is reflected in many aspects like the site selection, planning, architectural design, construction process, material using, equipments, water and energy conservation, operation and management, etc., the space layout and modeling results stemming from the planning and architectural design bring permanent physical and psychological effects. Good planning and design will create vitality in architecture and space, and can inspire creative awareness in every user; human-oriented design provides the construction with breath as well as temperature sense, thus infecting every user, making research more rigorous and creative thinking more active. Therefore, colleges and universities construction on a combination of sense and sensibility is bound to shine the permanent vitality.

References

[1] Bruno Zevi, "Architecture as Space," Beijing: China Building Industry Press, 1985, pp.9-19.

[2] Jan Gehl, "Life Between Buildings," Beijing: China Building Industry Press,1992, pp.13-53.

[3] Tu Huijun, "The Overall Design of University Campus—Planning, Landscape, Construction," Beijing: China Building Industry Press, 2007.

[4] Ma Guang, "An Introduction on Environment and Sustainable Development," Beijing: Science Press, 2000.

Article in a journal:

[5] Liu Huajie, Liu Xian, and Zheng Zhi, "People-oriented Intermediate Space in University Teaching Buildings," Chinese and Overseas Architecture, April, 2005.

[6] Hu Xiaoming and Wu Weifeng, "Aggregation and Dispersion — New Trends in the Development of Modern Comprehensive University Campuses," Architectural Journal, April, 2002, pp.18-19.

[7] He JingTang and Huang Peining, "Retrospect and Prospect of University Campus Planning and Design in the New Era," Architecture and Culture, May, 2007, pp.52-55.

Urban Fire Safety Evaluation and Optimization Based on GIS

Zhu Wenlong
Department of Architecture
Xi'an Univ. of Architecture and Technology
Xi'an, China
Department of Architecture
China Univ. of Mining and Technology
Xuzhou, China
Email: zhuk9@163.com

Chen Dong
Technical Service Center of Planning Bureau
Planning Bureau of Xuzhou City
Xuzhou, China
Email: c369d@qq.com

Abstract—**Fire has been one of the most often and common threat to public safety in cities. Considering the disadvantages of combining urban fire risk and fire protection capacity in current Urban Evaluation System, we propose to evaluate and compare urban fire risk and regional fire protection capacity by using GIS technology, analyzing main factors which affect urban safety and offering optimization strategy in areas with inadequate fire safety.**

Key words-**GIS; Security Evaluation; Optimization**

I. PREFACE

Cities are the concentration of population, production and wealth. Due to the high density of population, and buildings in urban areas, and the rapid development of economic construction, new energy, materials and equipment, cites are exposed to multiple risk factors. Fire has gradually become one of the most frequent and common threat to public safety, economic development and social stability. Correct fire risk assessment is the first step to improve the urban fire safety. The method of combining of fire risks and fire protection capacity for fire assessment exists large defects, so there is a important practical significance to evaluate separately and compare with the city's fire risk and fire protection capacity.

II. GIS INTRODUCTION

GIS is short for Geographic Information Systems. It is the high-tech which integrates the computer graphics and databases, storage and due with spatial information. It is the combination of location and associated attributes. According to actual needs, output to the user accurately to meet the urban construction, business management and residents' requirements of spatial information, making the auxiliary decisions through its unique function of spatial analysis and expression of visualization. Above-mentioned characteristics of GIS make very different from traditional methods to solve the problem. As an essential infrastructure of modern society, it permeates into every detail of production and living.

III. CITY FIRE SAFETY ASSESSMENTS

City Fire Safety Assessment (also known as the fire risk assessment) is an important part of fire science. Fire Safety Assessment in urban areas also has its own characteristics which have various environment and complex structures, so we use reasonable methods of fire safety analysis of urban areas, correct evaluation of city fire safety conditions, to conduct scientific and effective management of urban fire, to construct and improve the urban fire safety conditions and to promote the coordinated development of urban socio-economic.

Fire safety assessment of urban areas is divided into several methods, a single target, multi-target micro-and macro – evaluation. Their characters are the integration of city fire capability and fire risk and evaluation. Separate studying and comparing city fire capability and fire risk will improve the study of fire safety conditions, as well as reduce the disadvantages of existing methods. There are two states for fire safety in urban units: lack of capacity in the protection and over-capacity in protection. This is a qualitative analysis of the urban unit. The city unit which lacks the city's fire protection can be optimized in two ways, including seeking small risk in urban fire and large capacity in urban fire protection.

A Fire risk assessment in urban areas

Determining the evaluation index system or risk factors is the basic work and core tasks of urban fire risk assessment. Influencing factors and the degree of risk of the probability of fire risk in urban areas depend primarily on two aspects: one is characteristics of urban areas; the other is the status of the fire hazard.

1) Characteristics of the influencing factors in urban areas

The factors of fire, fire fighting and fire effects in urban areas are mainly reflected in the following aspects: 1) The population density; 2) Security level of building; 3) The building density.

2) Fire hazard conditions

According to related theory about the source of fire danger, we can see: Urban regional fire system form a fire by fire hazard source and the affected object. One of the hazards refers to energy or hazardous substance, the other hazard refers to insecurity factors leading to the lapse or damage of binding energy or hazardous substances.

3) Urban Area Fire Risk Assessment System

Urban unit fire risk evaluation index system can be divided into three layers. The first layer is the overall goal layer, containing a target object: City fire risk evaluation index system, with A expressed. The Second layer is the criterion layer, composed by characteristics of urban areas and fire hazard conditions, with B1, B2 expressed. The third layer is the basic index level, composed by density of population, building safety rating, density of building, class I and II types of fire hazards, with C1, C2, C3, C4, C5 expressed. We establish fire risk evaluation of urban areas index system chart (Shown in Fig. 1).

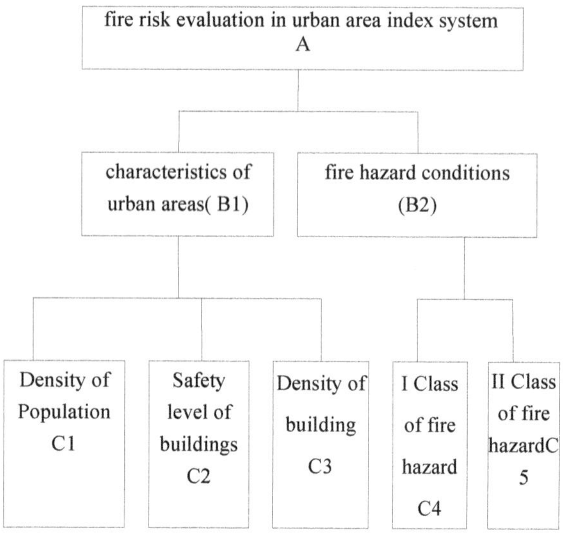

Figure 1. Fire risk evaluation of urban area index system chart

4) Determination of characteristic values evaluation index

During the process of assessment, in order to reasonably reflect the result and impact of overall evaluation, we need to adjust reasonably with each indicator to determine the results of overall evaluation. The paper adopts AHP to determine each indictor. We invite 5 experts and divide the task into 7 steps: 1)Provide basic indicators for experts to estimate the evaluate sheet; 2)A number of experts convened by;3)Experts filled in form anonymous; 4)Set out a summary, returning to the experts; 5)Have a meeting to discuss the large differences; 6)Expert review summary and submit another anonymous sheet; 7)Repeat 4-6; 8)Until it reaches the consistent of minimum and maximum, identify the basic indicators of this article(Table 1).

TABLE I. BASIC INDICATORS FOR WEIGHT LAYER FACTORS

Indicators of basic level layer	Weight
Density of population C1	0.21
Safety level of buildings C2	0.28
Density of buildingsC3	0.22
I class fire hazard C4	0.10
II class fire hazard C5	0.19

5) Urban unit fire risk assessment model

Based on fire risk assessment of city unit, firstly treat the urban area as grid, and then assign each factor in each city unit, according to fire risk evaluation index system in urban area. According to the different factors, evaluate fire risk of urban unit. Equation (1)

$$FXZ_i = \sum Y_j Q_j (j = 1,2,\cdots n) \quad (1)$$

FXZ_i, refers to the comprehensive score of urban unit fire risk

Y_j, refers to City unit No. j items dimensionless score

Q_j, refers to Index system of the weight of the value of No. j

B *The assessment of fire-fighting capabilities in urban areas*

The factor of fire capacity in urban area can be divided into two layers. The first layer is the overall objective layer, including one objective: fire capability in urban area, with a expressed. The second layer is the indicator layer, composed by the five factors which impact fire capacity in urban area, including urban road traffic conditions, fire station layout, communications equipment building, firefighters, construction of fire vehicles and protective equipment and construction of fire water supply facilities, with B1, B2, B3, B4, B5 expressed. As shown in Fig. 2.

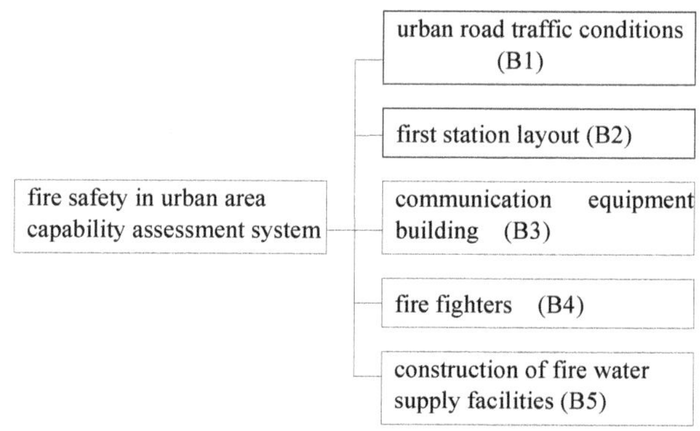

Figure 2. Urban area fire capability assessment system chart

1) Determine the weight of index-layer of factors

Adopt AHP to determine each indictor.

TABLE II. THE WEIGHT OF BASIC INDICATOR LAYER

Indicator of basic indicator layer	***weight***
Urban road traffic conditions B1	0.14
fire station layout B2	0.40
Construction of City Fire Services Communication Equipment B3	0.08
Firefighters, vehicles and protective equipment B4	0.28
Construction of fire water supply facilities B5	0.10

2) The building of model based on fire capacity assessment in urban area

Based on fire risk assessment of city unit, firstly treat the urban area as grid, then assign each factor in each city unit, determining fire capacity condition in urban area. Seen in equation (2)

$$FHZ_i = \sum Y_j Q_j (j = 1,2,\cdots n) \quad (2)$$

FHZ_i, refers to the comprehensive score of urban unit fire capacity

Y_j, refers to City unit No. j items dimensionless score

Q_j, refers to Index system of the weight of the value of No. j

C City Capacity Evaluation of Fire Safety

1) Comprehensive Evaluation Model for Urban Fire Safety

Urban Safety optimization is guided by principles in the system, compared with fire risk in urban unit assessment and the result of fire capacity assessment, finding fire capacity inadequate area; Analyze the reasons of insufficient fire protection and overcapacity fire protection, finding the main reasons and related optimization strategy(Fig. 3):

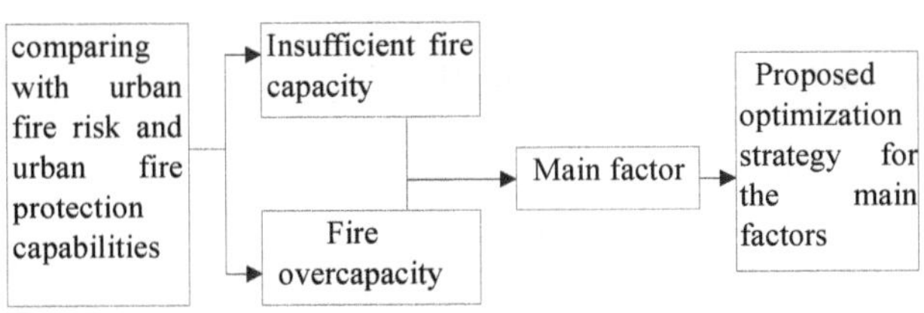

Figure 3. Optimization Process of Fire

Through the comparison of fire protection and city fire, get the evaluation of city security. See Equation (3).

The result of fire security evaluation=

$$\begin{cases} \textit{Insufficient fire capacity,} & XFAQ_i < 0 \\ \textit{Risk and balance of fire capacity,} & 0 \le XFAQ_i < 10\% \times \frac{XFAQ_i}{FHZ_i} \\ \textit{Fire capacity slightly stronger,} & 10\% \times \frac{XFAQ_i}{FHZ_i} \le XFAQ_i < 30\% \times \frac{XFAQ_i}{FHZ_i} \\ \textit{Fire capacity somewhat stronger,} & 30\% \times \frac{XFAQ_i}{FHZ_i} \le XFAQ_i < 50\% \times \frac{XFAQ_i}{FHZ_i} \\ \textit{Fire capacity very,} & 50\% \times \frac{XFAQ_i}{FHZ_i} \le XFAQ_i \end{cases}$$

(3)

$XFAQ_i = FHZ_i - FXZ$ refers to Fire resistance and fire risk comprehensive evaluation of the difference scores

FXZ_i refers to the comprehensive score of urban unit

FXZ_i refers to comprehensive evaluation of fire risk

2) The dominant factor analysis

There are two steps to analyze the dominant factor of each urban security unit, firstly to determine the status of the city unit, (it is lack of capacity of fire protection or overcapacity, which is a qualitative analysis of the urban unit); then, if that city is lack of capacity of fire protection, to enhance or reduce the risk of fire protection ability should be considered. If the city has overcapacity of fire protection, it is not necessary to weaken. If it is lack of capacity of fire protection in neighboring region, the re-allocation of protection resources should be considered.

a) The mathematical model for urban fire risk, there are two ways to solve the problem of inadequate capacity of fire protection in urban area, seek and compare with a small urban fire risk with a large capacity of urban fire protection. See formula (4)

$$F_i = \min(Z_{i1}, Z_{i2}, \ldots, Z_{im})$$

$$= \min\left\{\left[\left(\frac{\sum_{i=1}^{n} F_{i1}}{n} - F_{i1}\right) \times Q_1\right], \left[\left(\frac{\sum_{i=1}^{n} F_{i2}}{n} - F_{i2}\right) \times Q_2\right], \cdots, \left[\left(\frac{\sum_{i=1}^{n} F_{im}}{n} - F_{im}\right) \times Q_m\right]\right\}$$

(4)

b) The mathematical model for fire protection ability:See formula (5)

$$H_i = \max(Z_{i1}, Z_{i2}, \ldots, Z_{im})$$

$$= \max\left\{\left[\left(H_{i1} - \frac{\sum_{i=1}^{n} H_{i1}}{n}\right) \times Q_1\right], \left[\left(H_{i2} - \frac{\sum_{i=1}^{n} H_{i2}}{n}\right) \times Q_2\right], \cdots, \left[\left(H_{im} - \frac{\sum_{i=1}^{n} H_{im}}{n}\right) \times Q_m\right]\right\}$$

(5)

After solving F_i and H_i, compare absolute value F_i with H_i. If the absolute value of F_i larger is than H_i, it means the reason why the fire safety in this city unit is weak is mainly fire risk. On the other hand, it is lack of fire protection ability.

The same principle can be used to get dominant factor of overprotection capability in city areas.

IV. Examples of Fire Safety in Xuzhou City

A Fire Risk Assessment in Xuzhou City

We adopt assessment of city fire risk to carry out the research. Through the collection of relative data of city fire risk in Xuzhou City, GIS spatial analyzation with city units, and calculation, we get standard integrated scores of city fire risk.

According to comprehensive evaluation model, we calculate all the city units, get the comprehensive evaluation diagram of city fire risk, (Fig. 4). From the figure we can see the relationship between the distribution of fire risk and the utilization of urban planning and construction land. The low fire risk area is mainly water, green and suburban.

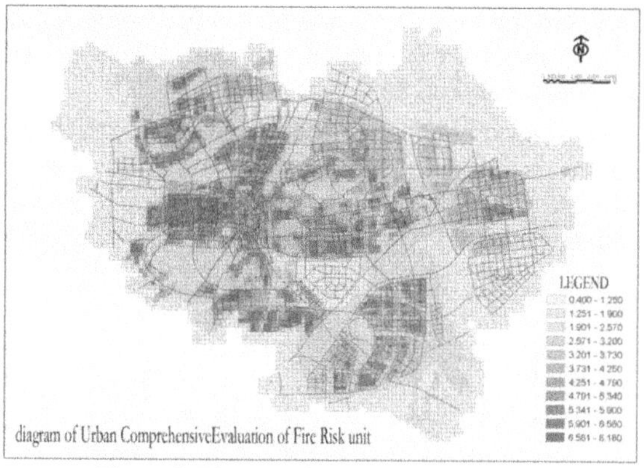

Figure 4. Diagram of Urban Comprehensive Evaluation of Fire Risk unit

B The assessment of City fire-fighting capabilities

According to comprehensive evaluation model, we calculate all the city units get the comprehensive evaluation diagram of city fire risk (Fig. 5).

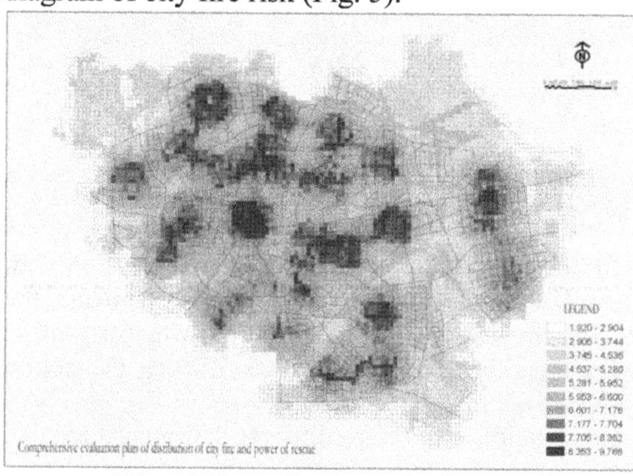

Figure 5.Comprehensive evaluation plan of distibution of city fire and power of rescue

Through the comparison of fire protection ability and the level of city fire risk, we can get starting point of improvement of the fire safety in each city. In other words, we begin with the reduction of urban fire risk or increase of the city's fire protection capability. From the figure 6 we can see the importance to sovle the problem of the fire safety in Xuzhou City is to reduce the urban fire risk. Only some urban fringe areas need to increase the city's fire protection capability.

Then we can see the problem of city fire safety, is not only to be sovled by increasing the fire input. Beginning with urban planning, reducing the risk, is the importance point to improve the fire risk situation in Xuzhou City (Fig. 6).

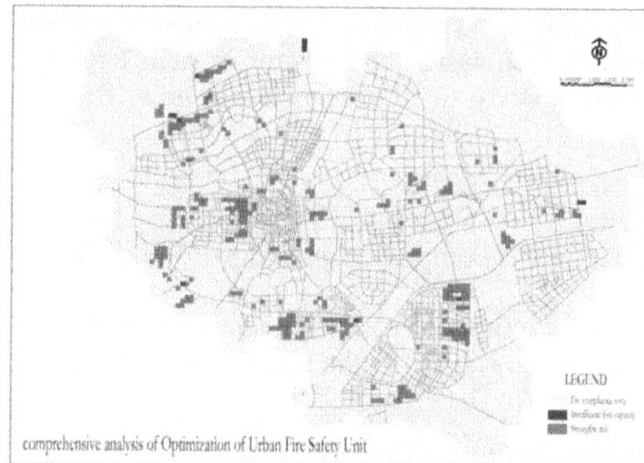

Figure 6. Comprehensive analysis of optimization of Urban Fire Safety Unit

After the analysis of comprehensive city safety optimization, considering the dominant factors, the requirement of fire safety in different areas in Xuzhou should be different. Leading solution in old city areas is to evacuate the population. It is needed to enhance the fire safety in new city and the configuration of interior fire-fighting facilities. In urban fringe areas, it should connect with the outskirts and surrounding areas of fire protection, inhacing the input of Fire-fighting facilities, espacially fire station building(Fig. 7).

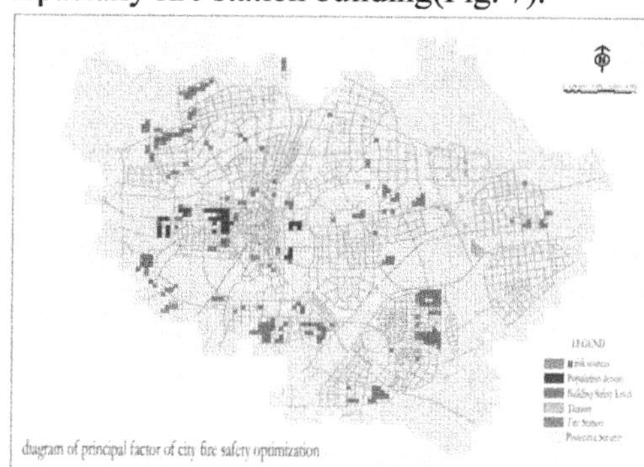

Figure 7. Diagram of principal factor of city fire safety optimization

V. CONCLUSION

After the anlysis of dominate facotors, we can make the different coping strategies with different areas of the city by spatial clustering method.

According to different dominant factors, we can find and make optimization the main factor of lackness of fire safety in Xuzhou City. According to the calculation of the different GIS results, we can optimize the areas with snsufficient fire capacity from the following aspects: 1) layout of fire station; 2) layout of urban space; 3) Optimization of urban architecture; 4) Optimization of fire water supply; 5) Optimization of urban fire exits; 6) Optimization of urban fire-fighting equipment; 7) Education of public fire safety; 8) Construction of Enterprises and institutions, full-time fire brigade.

REFERENCES

[1] NORSTRROM J G.Use of Precautionary Loss Functions In Risk analysis [J] , IEEE Transactions on Reliability , 1996 , 45(3):400-403.

[2] Entec. Review of High Occupancy Risk Assessment [M] Toolkit 2000

[3] Larsson D. Developing the Structure of a Fire Risk Index Method for Timber -frame Multistory Apartment Buildings[R]. Lund: Department of Fire Safety Engineering, Lund University, Report 5062, 2000.

[4] USFA. US Fire Administration Offers New Risk, Hazard and Value Evaluation Program for Local Officials [EB/OL]. (2001-11-19). http: //www.usfa.fema.gov.

[5] Michael S W. Development and Trial of a Risk Assessment Toolkit for the UK Fire Service [M]. London: Home Office Fire Research and Development Group, 1998, pp.129-137.

[6] Annex. Social Exclusion and the Risk of Fire [EB/OL]. (2004 - 08 - 31). http: //www. odpm. gov.uk.

[7] Duncanson M, and Woodward A,"Socioeconomic Deprivation and Fatal Unintentional Domestic Fire Incidents in New Zealand 1993-1998 "[J]. Fire Safety Journal,Vol.37,Feb. 2002, pp. 165-179.

[8] Yi Lixin. "the design of indexes system for assessing the urban fire risk"[J].Journal of Catastrophology, Vol.15,Dec.2000,pp.90-93

[9] Yang Rui, and Hou Zunze."Researches for the Method on the City Zone Fire-Safety Evalutaion"[J].Journal of the Chinese People's Armed Police Force Academy,Vol.19,Oct.2003,pp.17-20

Variables and Invariants in Product Image Design Management

Case Study on the Development of "Swatch"

Congying Guan
The School of Arts and Communication
Southwest Jiaotong University
Chengdu, China
Email:gcy796@126.com

Wenjing Bai
Economy and Management Department
Hunan Electrical Vocational and Technical College
Xiangtan, China
Email: baizhaneal@163.com

***Abstract*—From the aspect of long-term development of enterprises, constructing and implementing the product identity system should be the core content of design management. The paper analyzed the level relationship between the elements of product identity system, deployed the elements and researched them in detail. Based on that, studied the development process of "Swatch" brand, analyzed the variable and invariants in product image design management, and discussed the interrelation between them.**

Keywords-Corporate Image; Product Image; Image Identity; Design Strategy; Design Policy

I. INTRODUCTION

As market competition intensifies, most businesses have realized that product should be the core of commercial competition, whether the product can catch the hearts of consumers or not is the most critical. Nowadays, some companies put too much emphasis on the enterprise's name, logo, packaging, promotion, advertising and other brand extension in the process of establishing corporate image, but ignore the product itself as the the content of brand, which is the key media of the exchanges between business and customers. In my opinion, it's wise to import the product identification system in the enterprise, and to inject the philosophy and connotation of the brand into the products. The brand building process of Swatch, a watch company in Swiss, proves that point of view correctly, this paper analyses the variables and invariants in product image design management via Swatch to provide some references for native enterprise brand image establishing.

II. THE CLASS RELATIONSHIP OF PRODUCT IDENTIFICATION SYSTEM

Fig.1 shows the general class relationship of product identification system, the three levels in the figure are not self-governed, but the relationship of the top-down guidance and bottom-up communication, adhere to the corporate image as a guide and the brand image as the core in the process, and through product image to achieve the goals.

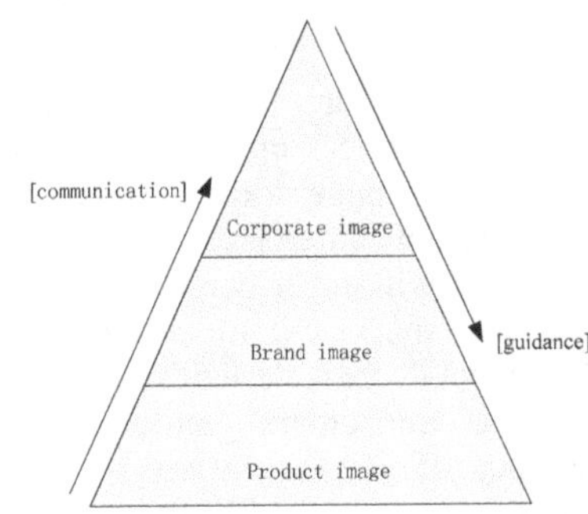

Figure 1. Class relationship graph of product identification

As the intuitionistic performance of enterprise operation philosophy, CI's relation with brand image represents in two forms: on the one hand, brand image obeys the company's overall image, on the other, brand image influences corporate image. Brand image is created first, followed by the maintenance of brand image[1]. When carrying on brand orientation, it necessary to take into account the right communication of corporate image and product image, to catch the characteristics of product, and to consider the ways to convey product value.

Product is the carrier of brand image communication, and represents the overall goals of an enterprise. Product image design is a complex system, Fig.2 is the deployment mapping of product identification system. The superstratum of product image design belongs to design strategy including the deployment of brand strategy under the guide of corporate strategy, brand orientation and the extending connotation of product image, which are all reduced to exterior image identification in this paper. The substrate of product image design belongs to design policy including the components of product identification, the core philosophy of product image and the extending philosophy of product image and the embodiment of product image, which are reduced to interior image identification in this paper.

In the respect of exterior image identification, product image design represents the "differences" reflected the difference of a company's products from other companies, to highlight the company's personality. On the contrary, product image design represents the "uniformity" in the respect of interior image identification. This difference and uniformity reveal the variables and invariants in product image design management.

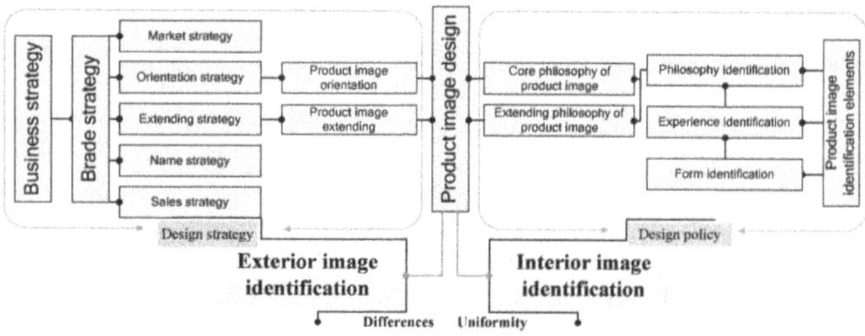

Figure 2. The deployment mapping of product identification system

III. VARIABLES IN PRODUCT IMAGE DESIGN MANAGEMENT

The variables of product image design management mainly reflects the differences in product image, that is the differences of exterior image identification in the design strategy level. Different enterprises are bound to show different images, IBM will never become APPLE, this is the inevitable result of market competition. This requires emphasis on the special characteristics distinguishing from competitors, so that consumers can recognize the very products in the face of a wide variety of products in the market. This variable is embodied in two aspects, one is the stress of difference distinguishing from similar products, the other is the extending of product image along with the development of the company itself.

Firstly, establish the special product image orientation to enhance consumers' awareness. Swatch father -- Nicholas G • Hayek established Swatch Group in 1985. In order to distinguish itself from other swatch companies in Swiss, Swatch broke the conventional comprehension of watch, pointed that Swatch not only presented a tool used for timekeeping, it also interpreted the concept of "time" in a unique way. The image positioning of Swatch is innovation, novelty, weird, fun, fashion, humour, avant-garde, etc., which formed the special image belonging to Swatch in the minds of consumers. Fig.3 is the product structure diagram of Swatch, from general watch series to jewelry series and then to the special section series, Swatch always makes time so cool via different styles and creation, attracts everyone using the style of "permanent change"[5].

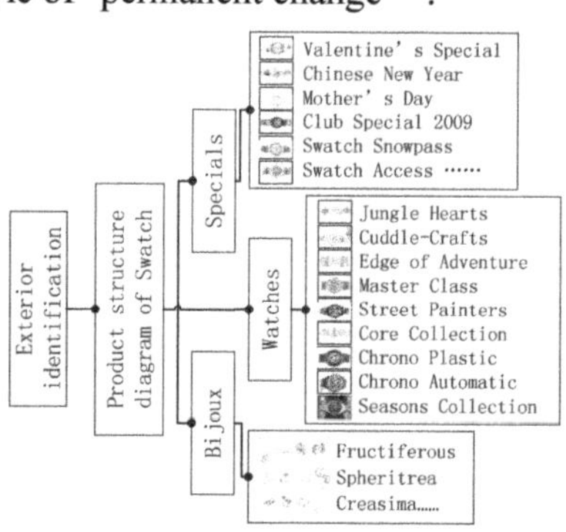

Figure 3. Variable product series of Swatch

Secondly, product image should keep developing forwardly. Swatch leads the fashion trend, known as "vanguard of the trend"[6].In the development process of Swatch, from Irony to Swatch SKIN Chrono, Swatch Snowpass, Swatch Infinity and Swatch Beat[3], Swatch constantly updated the product image, and never stopped. We can see Swatch even in other fields, a typical representative is smart -- the creative collaboration of Swatch and Mercedes-Benz.

Swatch extends product image to the perspective of corporate culture. Swatch is good at spreading corporate culture with product as a powerful tool, which representing the leading role of design. For instance, Swatch holds regular auctions. Swath only established 12 years ago, but it already get the status of "modern antiques"[5].Swatch has some special display units in Lisbon Museum, and owns an exhibition valued millions dollar,named "Swatch emotional experience"[5].

In a word, whether the product image positioning or the extending management are a process of constantly creation, which means "variables" in this paper. Whether the style, function or the technology all follow consumption needs, making consumer's interest reach its peak, "last year's Swatch can not substitute for this year's Swatch"[4], that's Swatch's goal. Swatch continues to launch new style watches every year, product has been completed 9 months earlier, and will be cut-off 5 months after they were put into the market, then Swatch began to develop new products. This product strategy make many customers look forward to the new products.

IV. INVARIANTS IN PRODUCT IMAGE DESIGN MANAGEMENT

The invariant in product image design mainly represents the uniformity, which belongs to design strategy level. The development of enterprise usually shows the law of life in the long term, with a fair amount of continuity. This paper studied the invariants from vertical and horizontal perspectives.

The components which constitute product image identification system includes philosophy identification, form identification (shape, color, grain, etc.), experience identification (function, operation interface, etc.), and every element changes along with time and environment. How to keep the uniformity of the corporate image in the contimusous changes? It's necessary to establish a clear CI design policy under the guide of design strategy, including product identification treats and a set of criterion for judgement. Then the senior managers can make reasonable judgement with the criterions.

Swatch has continued a unified visual image, even though each product demands a different focus, but consumer awareness of information are the same, making consumer recognize the consistent personality of Swatch when they are in face of various Swatch products.

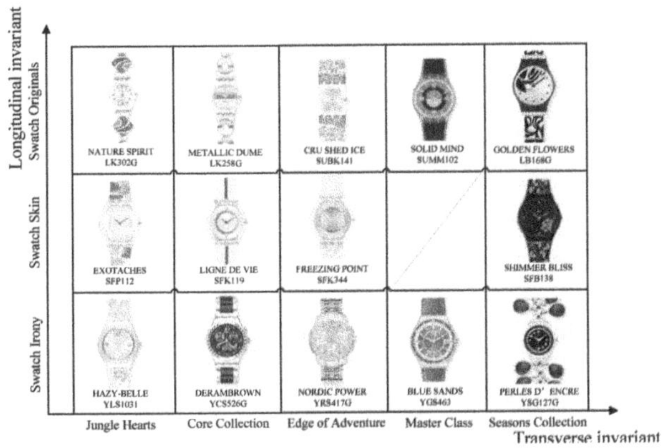

Figure 4. Swatch product image list

A. *Horizontal invariants*

From the horizontal point of view, the concept of product line is introduced first of all. Product line refers to a group of closely related products, because they play a role in a similar manner, sold to the same customer cluster, sale through the same type of channels, or their price changes within a certain range[2]. The spread of a single product line's span should retain the uniformity of product image. There are two key points here: the unification of the product elements themselves, which means the uniformity of product form, experience and philosophy; and the continuity of the uniformity among different products.

For instance, the Swatch Ultra-thin Series[3] product line in Fig.4, which is reputed the thinnest watch in the world, 0.39cm thickness, 3.4cm diameter, synthetic resin watchcase, waterproof 30m and shock absorbers. Each product of this series shows the identification treat, conveying Swatch's product image of alternatives, fashion and high-quality.

B. *Vertical invariants*

From the vertical point of view, different product lines should maintain the uniformity of image, so that all the products can own the identification personality which is common, unique and special. Meanwhile, the identification personality should coincide with the connotation of brand image and culture, operation philosophy and design strategy. While the Swatch slogan is fashion, free,divers and eccentric, but that dose mot make the product image unsystematic, which is a key point why Swatch is superior.

Fig.4 shows Swatch grouped products in different series with different styles such as Jungle HeartsCuddle-Crafts, Edge of Adventure, Master Class, Street Painters, all products were put into metal series, ultra-thin series and original Series orderly, which is a key point to image identification personality. Fig.5 is the latest Swatch Christmas 2009 Series[3] products.The identification features of Christmas were respectively carry out to metal series, original series, ultra-thin series product lines and even to jewelry series, showing the uniform continuity of 09 black golden series from shape, color, texture and other identifying elements.

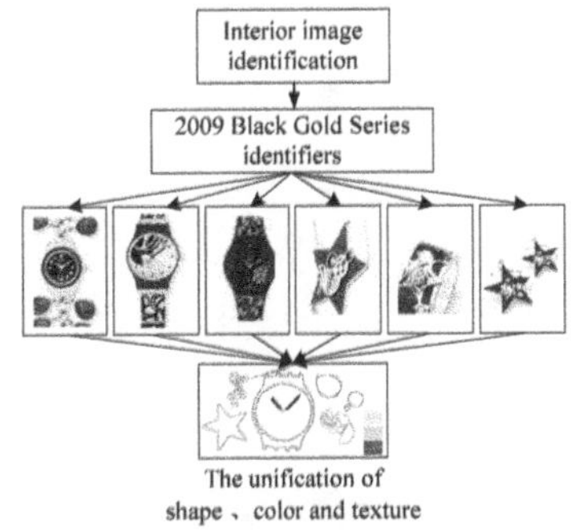

Figure 5. The continuous unification of Swatch

V. THE RELATIONSHIP BETWEEN VARIABLES AND INVARIANTS

According to the affiliation of business strategy, design strategy and design policy, it can be concluded that business strategy determines the variables of product image design management, the content indicates the changes under the overall strategy, and the measure taken to achieve the purpose is to maintain the invariants in product image design management.

Variables are the purpose of invariants, invariants are part of the elements to achieve variables. The key issue of product image design management is the coordination of the variables and invariants.

VI. CLOCLUSION

"birds of a feather flock together", products never exist alone, so companies should introduce product image identification system, and develop the characteristics of product identification, determine the design style from the long term perspective. Only in this way, can products maintain the unification and enhance consumers' awareness of brand and corporate image.

This paper analyzed the class relationship of product identification system, researched the establishing process of Swatch brand system in detail, and analysed the variables and invariants in product image design management taking Swatch for instance.

REFERENCES

[1] WANG Xiao-jie, JIN Hai. "Design Management". Beijing: China Light Industry Press,2008,3: 88-102

[2] HUA Jing-yong. "Design Management: Corporation's Products Identity Design".Beijing:Beijing Institute of Technology Press, 2007,1:147-168

[3] http://www.swatch.com

[4] JIN Shi. "Swatch Alternative Marketing". China Industry and Commerce News,2006-07-18

[5] ZHANG Qing, WANG An-xiao. "The significance of design management in moder corporate". Policy Research & Exploration,2008, 32 (5) : 58-59

Parse the Visual Performance of the Real Estate Ads

Xiao Wei
Huazhong University of Science and Technology Wuchang Brand
Wuhan, China
Email: xiaowei105@126.com

Abstract—**This paper focus on the visual manifestation of the real estate ads conducted a systematic,holistic, dynamic, comprehensive study, and by extension to the form and vision, the relationship between image and content analyzed. Mainly from the Virtual advertising, ink advertising, education theme, green, cartoons start discussion, thinking open, at the present stage of our ongoing systematic theory of the real estate advertising and marketing launch a direct guiding significance.**

Keywords- real estate advertising*; *visual expression

I. INTRODUCTION

At present, the real estate advertising real estate industry has become a major propaganda tool and an important pillar of the real estate industry in many cities has become the largest advertisers. Real estate advertising is designed to quickly and accurately to the timely transmission of real estate information to consumers, to persuade consumers to purchase, thus promoting sales; while the art is based on sales polish. First, from advertising the need to allow consumers to see what products to bring them the benefits of the target consumer advertising should be preoccupied with is a product of interest, rather than simply advertising art. If the design does not adhere to this point, leading to look at ads are fun to find fun and enjoy the artistry, it would greatly contrary to advertise in mind.

II. REAL ESTATE ADVERTISING FORMS

The so-called "form is neither a product of human will, it is not only a purely objective response." Outstanding works of the content and form of advertising is an organic unity, to better express the theme advertising to attract the attention of consumers to achieve good aberration effect to create a strong visual impact, the form of advertising works must focus on the use of effects and grasp. Manifestations constitutes an integral part of the image, but it is inherent in the image of being, and with the appearance of the image is essentially different from the composition, forms are divided into two categories: "One is the form of effects acting on the visual perception, one is and image and content of the performance of combining methods, the effect of turning feelings, fun and significance. The former is a regular pattern in the form to explore, which is in the graphics on the creative thinking. " [1] Therefore, the visual form of advertising for real estate analysis from these two aspects separately.

III. THE VISUAL PERFORMANCE OF REAL ESTATE ADVERTISING

Real estate marketing, advertising is the main tool in today's real estate industry in China has become the main pillars of media advertising in many cities, real estate are the largest advertisers. Numerous advertising agencies rely on the real estate fortune, and many of advertising through a real estate fame. A number of successful advertising firm in the real estate has been completely stage, they only carry real estate ads, for other sectors dismissive. Some well-known advertising itself stand out from the real estate industry, they are more proficient than some developers, real estate, in such a backdrop, China's contemporary real estate advertising is heading towards crowning glory.

A. Virtual Class

In previous years, without exception, all real estate ads to use the Virtual Community, as the center of focus images can be said to rage. The scenes with plot as a design element to advertising as a marketing selling points, highlight the district environmental advantages and delight of life, can resonate with consumers through advertising so that consumers understand that advertising will be able to convey intent. Shenyang Rhine River, is a residential area, it is rich in amenities to make the screen.

The ad's selling points, pictures of cute little children to play in order to highlight their cute, also reflects the district has provided us with a good educational environment and entertainment. At the same time, another picture is a beautiful residential old people playing Taijiquan, highlighting the concern of the elderly, reflecting the face of a harmonious community.

The same time, Yangchun March, Wuhan Jiangnan home two predicate City introduces new houses, the main screen while not accompanied by a reasonable figure in the environment, but it is unique, as its facade and Virtual Community.

Time to buy is also able to attract many consumers. Building also can be said that we have a common real estate advertising. Technique of expression. Architecture is an important part of human culture, and many buildings have a historical background, many of Architectural symbol of the forces have a long tradition, of which implies a huge artistic charm. Unfortunately, many advertising on the building without a deeper understanding and awareness building itself did not make an effort to understand the cultural meaning and heritage. If you're in the construction of the balcony on the performance characteristics, we can come to reflect the big picture windows, the characteristics of real estate, it should be based on the performance of a close-up lens window so that we can highlight the real estate selling point.

Virtual advertising must in real estate and construction performance of organic together, so that the two closely together to achieve effective advertising for propaganda purposes to highlight characteristics of the project.

B. Ink Class

"Chinese ink painting is one of the forms of Chinese painting. Generally refers to water and ink painting, made by. From the ink of the coke, thick, heavy, light, clear generate a wealth of changes in the performance of object-image, with unique artistic effect. As the ink emphasizes natural, days of fun, informal, pay attention to freehand. " In order to better highlight the regional characteristics and cultural background, real estate ads," ink "as a visual expression of the practices, real estate can be a perfect expression of the human environment.

With ink as a design concept of the real estate real estate advertising is also being increased, such as landscapes Dragon Baoan, Baoan South Village, through the performance of Chinese ink painting techniques to achieve the advertising, and thus to increase the visual personalization and differentiation, to achieve marketing purposes.

Ink's performance is always combined with the humanities and ink on paper is a form of cultural expression. Real estate is located in a well-known streets, this street has its own history, stories, legends, today's business climate has both cultural characteristics. Including an arch, a screen wall, one ancient, one tablets pavilions, all implies that endless ages. If the real estate surrounded by the ancient city wall defects, sparse in the old city gate, after Binghuo ancient pagoda, ancient temples, all reflect the cultural characteristics of a good subject.

For the understanding of cultural characteristics can have a broader understanding. Guangzhou Pearl River real estate Roman homes, this group of advertising is to ancient Roman statue of Julius Caesar and similar well-known Greek sculptures in Europe as the main lens, fastening the name of the Roman home, highlighting the cultural pages of color in the dark refractive status of real estate's not bad. A city should have a city contextual, in-depth to explore the local contextual, in the ad to be appropriate in the performance of the interpretation of the ad designer's responsibility. This would be the first lies in the very in-depth understanding of local contextual analysis. For example, Xi'an, 13 on behalf of the imperial capital, the famous ancient town, is not the performance of Xi'an's geographical Context-bound to highlight the Han and Tang Dynasties era of ancient buildings, and even the Terracotta Warriors? Not necessarily so. Our geographical location should be based on a project to tap the Context of its history, to seize the project and accurate positioning.

C .Green

With people's living standards increase, for the pursuit of the living environment is also becoming stronger. This determines that in recent years in the high-end luxury real estate, especially in stressed importance of the environment. The environment is a composite indicator, where the green environment is the most attractive factors. Suddenly, due to urban life far away from nature, people are accustomed to seeing in the city of gray cement bald bald forests, noisy roads, congested traffic, so people are yearning for the green on the extra and pursuit.

A green-based real estate hit a selling point if, in fact, can be issued in conjunction Pina sound creativity. For example, Guangzhou large garden of this group to the "big tree health of others allow the theme" ads:

Series 1: In the large garden, more than 1,000 square meters natural oxygen bar, free of charge every day Renxiang, open all night in order to make cups of green leaf to promote the image of the selling points.

Series 2: In the Everbright Garden, so that the natural anion release device, the less there are 8 billion, fresh feeling of a natural stream to leaves as the main visual image.

Series 3: The green leaves make headphones prominent selling point, at Everbright garden, so that the noise absorption of natural anti-shielding more than 1000 every day just for you drop audio de-noising, cut off from downtown hustle and bustle for you.

This group ad highlighting a strong garden owned by Everbright of more than forty or fifty years of age more than 1,000 large banyan trees and 15 million square meters of green area, indeed in the visual effect of the thick enough to create a breathtaking green mood. The green real estate, there are two, one is natural, one is artificial. Naturally more simple, more character and beauty heaven, but most of all the properties for sale around, you can borrow but not exclusive; artificial green environment more beautiful, mostly flats in the house, but a little more on the taste of decorative carvings, in particular, community environment in recent years, the quest for like a park, so a lot of real estate are "a home in the park" do the demands. In fact, such a purely man-made gardens and nature of the natural night of fun, and how could the same breath, only to meet the people as a psychological comfort. In some real estate, the house itself is an excellent building, but if the green environment more features, it should be decisively relegated to a secondary position to the construction, with the main space to highlight the green environment.

D. The theme of education classes

China's one-child policy, the direct consequences, for every household is its own "little emperor", to enable parents to feel hopeful more urgent, so that parents invest in their children's education, intelligence is more willing to spend money, so called "universal al sub-taught "and not excessive. The use of such a nation, society as a whole the situation there

Knot to do the article, "Education is the eternal theme of the property" will be inevitable. In the major cities can be seen: the vicinity of any of those in the elite real estate sales are good. The South China plates are broader market also inject huge sums of money into the Indian elite, to add color for their own district. In the market research can be clearly felt that there is indeed a considerable buyers, the choice of real estate when the next generation of educational environment on the status of paramount importance. Shenzhen banks of the Rhine of Peking University's affiliated high school of the Tsinghua Experimental School in Taoyuan Ju, Guangzhou Star River's Zhixin Middle School, the introduction of these prestigious schools are in the real estate sales has a decisive significance. Especially the rural areas the broader market of consumers to abandon the city's prosperity and transport facilities, must be compensated to gain something, in which the convenience of their children to school must be a priority focus. Scenic names of colleges and universities located in the door, Wuhan University of Technology, Wuhan University of Technology, Central China Normal University, for the sale of real estate has brought greater opportunities.

The performance of educational resources, advertising is also very rich and colorful past two years, some of the success of the case out. China Everbright Garden in Guangzhou to introduce a large middle school north Experimental College, as did a group of education, the theme of a series of ads. The performance of the theme of education, pay attention to the wording on the subtle and can not be the words of too straightforward, can not be home buyers do not meet the actual commitment. The theme of education in general to make a series of ads a group of real estate, is the subject of stage, especially before and after June 1 of each year, the exam, after the use of such themes are more usually it is difficult to see the developer put the theme of education through the use of .

E. Class cartoons

With cartoon characters as a means of real-estate advertising, creativity clear, accurate positioning can be the first time capture the target consumer group. Among the many, many real estate advertising real estate is located in the city center, usually cartoons as the primary means of visual expression, highlighting the taste and grade the entire project, to sell a house Purpose. Cartoons of the features are: vivid, vivid, easy to understand, interesting and strong. SOHO Small Apartment is also in the form of cartoons to achieve good visual effects. In 2006, the Optical Valley Walking Street launched the "still have" small-family, In order to attract more consumers to buy homes, it would take young people are like cartoons in the form of visual propaganda in the commercial center and places the flow of people and more able to highlight the "still have" small-family characteristics. Thus, at the open day of a sold out house, to a very good sales.

IV. Summary:

The performance of the visual method of real estate ads to tell us that in the advertising market, you must go through a long-term market research, market as a whole to do a thorough analysis, comprehensive grasp the market context, identify the entry point, can reflect the personality characteristics of the project. Research aim is to clear the visual manifestations, the performance of specific types of vision, so that there is to put loss.

References

[1]"100 the details of the real estate marketing,"Zhangxiaoguangauang Guangdong rovince, a group Guangdong Economic Press, 2005.3

[2] "real estate strategies for implementation of the workshop were heads of enterprises of the" expert groups with, Guangzhou Sound Publishing, 2005.7 p. 210

[3] "China's commercial real estate operations," Feng Hui, Liu a, Jinan University Press, 2004.11

[4] "Real Estate Planning - the classic case of all recorded real" Chiu Cheuk Man Forward, Guangdong Economic

Press, 2006.7 p. 349

[5] "real estate strategies for implementation of the workshop were heads of enterprises of the" expert groups with, Guangzhou Sound Publishing,2005.7 .20

[6] "visual representation" South Kumoharu Jia The Huang Lei and other translation China Youth Press, 2004.1

[7] "real estate track" Vincent Mo Yu Huang The Economic Management Publishing House, 2007.3

[8]"Integrated Marketing Communication" [America] Terence A · Simplon The Lim Xiao et al translated, Beijing University Press, 2005.1

[9] "Ad new thinking" [South Africa] Erik du Plessis Li Ying Liu soil with translation Renmin University of China out of , 2007.3

10]"Principles of Advertising" has been Zhenhua Hu Hua Huang Qinghua, a, Jinan University Press, 2006.10

[11] "real estate strategies for implementation of the workshop were heads of enterprises," Zhang Xiaoguang forward, Guangdong Province, a group Guangdong Economic Press, 2005.3

[12] "Consumer psychology" Wen Xiao-Qing, Shi had spring editor, Tianjin University Press, 2004.7

The Application of Environmental Psychology to the Environment Art Design under the Greenhouse Effect

Xiaozhou Tang
College of Art and Design
Wuhan University of Technology
Wuhan, China
Email: 16790653@qq.com

Zhi Ding
College of Art and Design
Wuhan University of Technology
Wuhan, China
Email: 695779195@qq.com

Abstract—**With the deepening of the greenhouse effect, more and more people begin to pay attention to the environment, as well as feel anxious over the environment in which they exist. In order to achieve harmonious development between the men and the environment and safeguard sustainable development of the environment at the same time, the use of environmental psychology approach appropriately to guide people in arranging and coordinating the various elements rationally. These are not only the urgent tasks for the time being on the one hand, but required by the environmental development in the future on the other.**

Keywords-greenhouse effect; environmental changes; low carb)

I. INTRODUCTION

In 2009, from December 7 to 18, the World Climate Conference ("the United Nations Framework Convention on Climate Change," the 15th meeting of contracting parties) held in the Danish capital Copenhagen has been paid much attention by the Governments, non-governmental organizations, academics, the media and the public. In recent years, there are endless reports on the various types of natural disasters resulted from environmental degradation, while the greenhouse effect has become the most important problem that will cause the natural disasters and affect the future existing environment.

Current environment has undergone significant changes over the past a few decades. It not only reveals in the survival of the physical environment, but also extends in the psychological environment of a deeper level. Changes in the environment require us to rethink the interdependent relationship between the human, environment, society and the economic development. Thus, if you reassess the future direction of the environment art design development from the perspective of environmental psychology, you may obtain something valuable unexpectedly.

II. THE FORMATION AND DEVELOPMENT OF ENVIRONMENTAL PSYCHOLOGY

In the 1960s, accompanied by the emergence of baby boom, the rapid process of industrialization, environmental pollution and degradation, and many other issues, some scholars began to reflect on the problems between population and environment, and their focus shifted from the "the influence of environment on human behavior" to "the correlation of human behavior and the natural environment". Environmental psychology is formed on such basis. It is an interdisciplinary subject that combines anthropology, demography, architecture, natural sciences, geography, sociology, psychology and many other subjects.

Environmental psychology studies the mutual relations of individual behavior and their environment, especially the relationship between environment and psychology, that is to use the psychological methods to analyze the interrelationship and mutual influence between human experience, activities and their social environment (especially the physical environment), and reveal the occurrence of human psychology and the law of its development under the various environmental conditions. According to the definition of environmental psychology, we can understand how individual and environment affect each other, then use and transform the environment, as well as solve the human behavior problems caused by the environment.

The contents of environmental psychology research are not static; they are changing over time, and becoming more specific and rich with the deepening of people's awareness, with the progress of psychology, and with constant changes of the natural and social environment. During the 30 years after the formation of environmental psychology, its study mainly focused on the following aspects: 1. The influence of environment on human psychology and behavior, involving specific studies on crowding, noise, temperature, air pollution and so on; 2. The impact of environmental factors on people's work and the quality of life, related to the studies on architectural design and urban planning; 3. The interaction of the environment and human behavior, involving the studies on environmental stress, stress response, and environmental load and so on; 4. The effect of people's behavior on the surroundings and ecosystems, related to research on the behavior and psychology of environmental protection; 5. The sustainable development issues of environmental psychology with nature and society.

III. PURPOSES AND METHODS OF ENVIRONMENTAL PSYCHOLOGY RESEARCH E

As environmental psychology is a subject used for studying the correlation of individuals and their environment, then its purpose is definitely included in its research content, that is, by understanding how individual and environment interact, we can use and transform the environment, and also use psychological means to adjust human behavior in order to achieve the aim of protecting the environment. Here, the concept of environment is not the natural environment in a narrow sense, but includes the natural environment, social environment, psychological environment and some general environment of various occasions. The concept of natural and social environment, unlike the psychological principle of environment and people's psychological phenomenon and consciousness of environment under different environmental conditions, is prominently shown in the noise, temperature, wind, climate, air pollution and other natural changes; social environment is mainly embodied in the personal space, geographical concepts, social trends, social culture, interpersonal relationships, and other relationship between human and society, human and culture, and human themselves; psychological environment is primarily through the association to convey the influence of environment on the people's awareness and the psychology, and the impact of psychological changes in environmental pollution on human information transfer, the working efficiency, even on the level of people's psychological self-adjustment and feedback under different environmental conditions. Thus, the concept of environment, in environmental psychology, has the dual characteristics of objectivity and subjectivity.

As a new discipline, environmental psychology can conduct related research by observation, experiment, investigation and archival methods. Observation method is based on behaviorism, observing and recording the facts objectively without any intervention. Researchers need to know clearly about the purposes and requirements, and make observation procedures clear. Notes, recordings, videos, paintings, sketches can be used in form. While experimental method needs a certain degree of comparison, calculation, reasoning and other means to verify whether a factor of some variable can affect the other factor (dependent variable) in a certain environment. According to the selected environmental sites and control methods in different degrees, experimental method can be divided into different categories, such as laboratory experiment, natural experiment and field experiment. Survey method is the most commonly used in different types of studies. Its basic approach is to develop a range of issues as required, then make a questionnaire, and put forward to the informant, ask them to make answers and finally draw results after arranging the materials. Archival research method can be described as a particular research method, which differs from other methods in that it doesn't draw conclusions by analyzing the existing archival materials, but analyze newspaper reports, government documents or groups records or non-professional publications, such as personal letters, speeches, logs. That's why these materials need to be refined, classified and arranged.

IV. Environmental Changes under the Greenhouse Effect

Environmental changes, not only refer to the changes of the objective natural environment, but also encompass the changes of social environment and psychological environment. The changes in the natural environment promote the changes in social environment accordingly, which will exert a subtle influence on people's psychological environment at the same time. It is such a way how greenhouse effect affects people's environmental changes.

Since the industrial revolution, the whole world especially some developed countries in Europe and in the northern hemisphere all benefit from it. However, there are potential threats among the rising civilization, which have been profoundly proved today, after more than a century. At present, the Fourth Assessment Report of the Intergovernmental Panel on Climate Change showed that concentrations of carbon dioxide in the atmosphere has risen from 280PPM (parts per million) before the industrial revolution to 379PPM in 2005, which has passed the range of natural variation in the recent 65 years. In addition, the average temperature of global land surface has risen by 0.74 degrees in about 100 years. The report said that since 1750, the global carbon dioxide emission has accumulated more than 1 trillion tons, in which developed countries accounted for 80%.

Global warming has become one of the most significant issues related to human existences that the people are care about most. Scientists predict that if people do not take immediate action to reduce greenhouse gas emission, the earth's surface temperature will raise 4℃ whether in 2050 for a short time, or until the end of this century for a long time. By that time, many coastal countries and regions in the earth will be swallowed by sea, such as the Maldives, Tuvalu, 1/3 land of Shanghai, most coastal parts of New Zealand, and there are only about 1 billion people can survive.

In order to protect human beings from the threat of global warming, the 3rd Conference of "United Nations Framework Convention on Climate Change" held in Kyoto, Japan in December, 1997. Representatives of 149 countries and regions passed the "Kyoto Protocol" for the purpose of limiting greenhouse gas emission of the developed countries to curb global warming.

On February 16, 2005, the "Kyoto Protocol" came into effect. This is the first time to limit greenhouse gas emission in the form of laws and regulations in our history. The establishment and function of the "Kyoto Protocol" mark the people's concern about the greenhouse effect has reached an unprecedented height, in which numbers are set for limitation of carbon emission. People will think about before their move due to the confines of the system, and their acts, to a certain extent, have been limited. Therefore, the big environment of reducing greenhouse gas emission and conserving energy will be formed. People begin to consider what kind of approach is the safest, the most environmental-friendly, the most harmonious and the most healthy in our life and production.

Greenhouse effect has changed our contemporary way of life, and such change will exert profound effect on our future life. Here, the changes of greenhouse effect make many people concern about their future existing environment. The obvious representation of greenhouse effect is that the land surface temperature becomes higher gradually, and more and more warm in winter and more and more roasting in summer, then people are increasingly inseparable from the accompany of air-conditioners. At the same time, fluoride emissions of air-conditioners and refrigerators have further damaged the ozone layer, forming a vicious cycle of environmental degradation. In addition, with the high demand for cars in developing countries, carbon dioxide and exhaust emissions are also intensified. To maintain their life, people showed great resistance and psychological conflict in the vehicles, electrical appliances, and energy projects that emit carbon dioxide, sewage and other harmful substances. People also began to consider a cleaner, more sustainable and environmental-friendly way of life. For example, a great number of people will use electric cars, bicycles, and even walking instead of cars. And in the environment art design, particularly in the field of architectural design, the concept of low-carbon building achieves universally appreciation and consensus.

V. Low-carbon Building ---- The New Direction of the Environment Art Design in the Future

Low carbon means less of the greenhouse gases (mainly carbon dioxide) emission. Such a concept, not put forward from the field of environment art design, but a sustainable development model proposed due to the aggravation of the greenhouse effect. The concept of low-carbon economy covers all industrial sectors, also known as the "the fifth wave of the global industry." It is an economic model based on low energy consumption, low pollution and low emission, which is another great progress following the agricultural civilization and industrial civilization in human society. Besides, there are some other concepts when expanding its meaning, such as low-carbon society, low-carbon economy, low-carbon production, low-carbon consumption, low-carbon life, low-carbon city, low-carbon communities, low-carbon home, low-carbon travel, low-carbon culture, low-carbon philosophy, low carbon art, low-carbon music, low-carbon life, low-carbon existing doctrine and so on.

Low-carbon building refers to during the whole cycle of manufacturing of building materials and equipment, construction and building use, we try to reduce fossil energy use, increase energy efficiency and lower carbon dioxide emission. Nowadays, the low-carbon building has gradually become the main trend of the international construction industry. However, an often overlooked fact is that in the total carbon dioxide emission, the construction industry almost accounts for 50%, which is far higher than the proportion of transportation and industry occupied. On the way of developing low-carbon economy, "energy conservation" and "low carbon" of construction are destined to be the closely related topics.

Low-carbon building advocates interior design with the principle of natural ventilation and natural lighting, minimize the use of fans, air-conditioners and lights. Generally speaking, among the energy loss of the whole building, about half is in the windows and doors. Insulating glass not only blocks heat waves and cold snap outside, but also isolates the noise and reduces energy consumption. Low-carbon building promotes small-family homes, because small-family homes have advantages over large-family homes in terms of saving building materials, energy, construction and cost, and so does carbon emission. Meanwhile, except for using the environmental-friendly and recyclable building materials, low-carbon building will take the factors of reducing carbon emission into consideration in its design form, including color, texture and space structures, so as to reduce people's fear and panic about the greenhouse effect from the environmental and psychological point of view.

Climate conference in Copenhagen in 2009, is a chance for popularizing the knowledge of the necessity for environmental improvement rather than the formulation of low-carbon emission reduction plan over the next 10 or 20 years. Through this conference, people express their fear and anxiety over the future development of the environment, and these feelings we also called Environmental Stress Reaction. The emergence of such feelings indicates that people have realized the urgency and necessity of environmental protection. But the serious long-term anxiety can easily lead to emotional instability and aggressive behavior, too. Studies show that the temperature will also affect people's psychology and behavior, and hot weather can cause people to have aggressive tendency. So, the existence of anxiety is a potential threat for personal health and social stability. At the moment, if some countries and regions only aim at their own development, and deliberately ignore the urgency of emission reduction and energy conservation, it will inevitably lead to a wide range of distrust of the authorities and of the governments. Then, as soon as the outbreak of environmental disasters, such as storms, typhoons, floods and heat waves, people's discontent with the society will be easily caused.

The effective way of removing people's anxiety and worries under the greenhouse effect is to advocate and correctly guide the concept of low-carbon building and promote environmental-friendly design and development. The greenhouse effect will exist for a long period of time, and will exert widespread and long-term effect on people's psychological scope and the temporal duration. The effective strategy is that try to persuade people to reduce carbon emission, and promote more environmental-friendly and more appropriate life style based on sustainable development.

The development of science and technology can promote the conduction of low carbon and emission reduction as well. With the consumption of energy, scientists have gradually come to realize that the oil, coal, natural gas and other non-renewable resources will no longer support the continuation of human civilization, and maybe, to a certain extent, they will be the stumbling blocks on the way of the development of human civilization. Bio-energy and new energy sources such as solar, wind and nuclear energy will be widely used and will become the main form of future human life. We are looking forward to the arrival of the fifth revolution of science and technology. It will be an unprecedented leap, which is bound to affect the development of environment art design in the future.

Reference:

[1] R. Baker. Ecological Psychology [M] . Palo Alto: Standford University Press, 1968.

[2] Kenneth Craik. Environmental Psychology [J] . Annual Reviews of Psychology, 1973.

[3] Daniel Stokols. Environmental Psychology [J] . Annual Reviews of Psychology, 1978.

[4] Howard G S. Adapting human lifestyles for the 21st century [J] . American Psychologist, 2000 (55) : 509- 514.

[5] Stern P C. Psychology and the science of human- environment interactions [J] . American Psychologist, 2000 (55) :523- 530.

[6] Stuart Oskamp. A Sustainable future for humanity: how can psychology help? [J] . American Psychologist, 2000 (55) :496- 508.

[7] Daniel Stokols. The Paradox of Environmental Psychology [J] . American Psychologist, 1995 (50) : 821- 837.

[8] Eric Sundstrom. Environmental Psychology [J] . Annual Reviews of Psychology, 1994.

[9] Yu Guoliang& Wang Qinglan& Yang Zhiliang. *Environmental Psychology*. [J]. Beijing: People's Educational Press. 2000.

[10] http://www.chinanews.com.cn/gn/news/2009/08-25/1833857.shtml, "Chinese action of global warming: face the challenge, control the discharge", Chinanews.com.cn, 25th, Aug. 2009.

[11] Foresee of scientist: There are only 1 billion people in the world. http://tech.qq.com/a/20090303/000048.htm.

[12] *Kyoto Protocol*, Xinhua.net, http://news.xinhuanet.com/ziliao/2002-09/03/content_548525.htm.

[13] Liu Jianxin& Gao Lan. *Brief Introduction to the Formation and Development of Psychology*. Academic Study, 2005(11).

[14] Ma Xunfeng& Huang Dongmei& Huang Yu& He Ziguang. *Some Discussion on Environmental Psychology*. Journal of JiLin Normal University(Natural Science), 2004,2(1).

[15] The Development of Chinese Low-carbon Building Needs Support of Certain Policy. Chinese Business Times, 2009.

[16] http://www.kechixu.com/web/Family/like_term7.htm. Today, Are You Low-carbon?. Chinese Sustainable Development Website.

www.ingramcontent.com/pod-product-compliance
Lightning Source LLC
LaVergne TN
LVHW062111300926
849161LV00010B/2433